T0281249

Dynamics of Particles and Rigid Bodies: A Systematic Approach

Dynamics of Particles and Rigid Bodies: A Systematic Approach is intended for undergraduate courses in dynamics. This work is a unique blend of conceptual, theoretical, and practical aspects of dynamics generally not found in dynamics books at the undergraduate level. In particular, in this book the concepts are developed in a highly rigorous manner and are applied to examples using a step-by-step approach that is completely consistent with the theory. In addition, for clarity, the notation used to develop the theory is identical to that used to solve example problems. The result of this approach is that a student is able to see clearly the connection between the theory and the application of theory to example problems. While the material is not new, instructors and their students will appreciate the highly pedagogical approach that aids in the mastery and retention of concepts. The approach used in this book teaches a student to develop a systematic approach to problem-solving. The work is supported by a great range of examples and reinforced by numerous problems for student solution. An Instructor's Solutions Manual is available.

Anil V. Rao earned his B.S. in mechanical engineering and A.B. in mathematics from Cornell University, his M.S.E. in aerospace engineering from the University of Michigan, and his M.A. and Ph.D. in mechanical and aerospace engineering from Princeton University. After earning his Ph.D., Dr. Rao joined the Flight Mechanics Department at The Aerospace Corporation in Los Angeles, where he was involved in mission support for U.S. Air Force launch vehicle programs and trajectory optimization software development. Subsequently, Dr. Rao joined The Charles Stark Draper Laboratory, Inc., in Cambridge, Massachusetts. During his six years at Draper, Dr. Rao led numerous projects related to trajectory optimization, guidance, and navigation of aerospace vehicles. Concurrent with his tenure at Draper, Dr. Rao was a Lecturer and Senior Lecturer of Aerospace and Mechanical Engineering at Boston University where he taught the core undergraduate engineering dynamics course continuously for five years. During his time at Boston University Dr. Rao was voted Department Faculty Member of the Year (during the 2001–2002 and 2005–2006 academic years) and was voted College of Engineering Professor of the Year for outstanding teaching (during the 2003–2004 academic year). In July 2006 Dr. Rao joined the faculty in the Department of Mechanical and Aerospace Engineering at the University of Florida where he currently holds the rank of Associate Professor. Dr. Rao's research interests are in the areas of computational optimal control, nonlinear optimization, and vehicle guidance.

DYNAMICS
OF PARTICLES AND RIGID BODIES

A SYSTEMATIC APPROACH

ANIL V. RAO

University of Florida
Gainesville, FL

CAMBRIDGE
UNIVERSITY PRESS

CAMBRIDGE UNIVERSITY PRESS
Cambridge, New York, Melbourne, Madrid, Cape Town,
Singapore, São Paulo, Delhi, Tokyo, Mexico City

Cambridge University Press
32 Avenue of the Americas, New York NY 10013-2473, USA

Published in the United States of America by Cambridge University Press, New York

www.cambridge.org
Information on this title: www.cambridge.org/9780521187909

First published 2006
First paperback edition 2011

A catalogue record for this publication is available from the British Library

ISBN 978-0-521-85811-3 Hardback
ISBN 978-0-521-18790-9 Paperback

Cover illustration: image courtesy of NASA/JPL-CalTech.

Cover design by Alice Soloway

Fig. 1–4 used by permission of DiracDelta Consultants, UK.

Fig. 3–3 adapted from Figure 4.6 on page 49 of O. O'Reilly, *Engineering Dynamics: A Primer*,
2001 by permission of Oliver O'Reilly and Springer-Verlag, Heidelberg, Germany.

Drawing of bulldozer appearing in Fig. P5-3 used by permission of the artist, Richard Neuman
(URL: http://www.richard-neuman-artist.com).

Questions 3–9, 5–1, 5–7, and 5–14 and Examples 5–8, 5–10, 5–14, and 5–17 adapted from Greenwood,
D. T., *Principles of Dynamics*, 2nd Edition, 1988, by permission of Donald T. Greenwood
and Pearson Education, Inc., Upper Saddle River, NJ (Pearson Education Reference Number: 103682).

The illustrations in this manuscript were created using CorelDRAW® Version 12. CorelDRAW®
is a registered trademark of Corel Corporation, 1600 Carling Avenue, Ottawa, Ontario K1Z 8R7, Canada.

This manuscript was typeset using the teTeX version of LaTeX 2ε using the Lucida Bright Math®,
Lucida New Math®, and Lucida Bright Math Expert® fonts.

Vakratunda Mahaakaaya Soorya Koti Samaprabha
Nirvighnam Kuru Mein Deva Sarva Kaaryashu Sarvadaa

I dedicate this book with love to Anita, Vikram, and Divya,
and to my parents, Saroj and Rajeswara

Contents

Preface

The subject of dynamics has been taught in engineering curricula for decades, traditionally as a second-semester course as part of a year-long sequence in engineering mechanics. This approach to teaching dynamics has led to a wide array of currently available engineering mechanics books, including Beer and Johnston (1997), Bedford and Fowler (2005), Hibbeler (2001), and Merriam and Kraige (1997). From my experience, the reasons these books are adopted for undergraduate courses in engineering mechanics are threefold. First, they include a wide variety of worked examples and have more than 1000 problems for the students to solve at the end of each chapter. The variety of problems provides instructors with the flexibility to assign different problems every semester for several years. Second, these books are generic enough that they can be used to teach undergraduates in virtually any branch of engineering. Third, they cover both statics and dynamics, thereby making it is possible for a student to purchase a single book for a year-long engineering mechanics course. Using these empirical measures, it is hard to dispute that these books cover a tremendous amount of material and enable an instructor to tailor the material to the needs of a particular course. Given the vast array of undergraduate dynamics books already available, an obvious question that arises is, why write yet another book on the subject of undergraduate engineering dynamics? While it is clear that the availability of another book on the subject would clearly add to the *number* of choices available to instructors, it may be difficult at first glance to see how the addition of another book would add *value* to the existing literature. However, after my experience over the past several years of teaching dynamics, not only do I now believe that there is room for another book, but I feel strongly that the paradigm used to teach the subject of dynamics needs to be completely overhauled.

Before I ever taught undergraduate dynamics, I, too, believed that the existing books on engineering mechanics were more than adequate and that an additional book would add little to no value to the existing literature. Consequently, without giving it much thought, the first time I taught engineering dynamics (course EK302 at Boston University) I randomly chose one of the standard undergraduate textbooks. Given my notions at the time, it never occurred to me that the book I chose for my class would pose so many difficulties for my students. However, not more than a few weeks into my first semester of teaching, I was met by vehement complaints from my students regarding the textbook. Given their frustration and my sincere desire to keep them motivated, I began investigating more thoroughly why my students found the textbook so difficult to follow and what I could do to help them overcome their frustration.

My investigation began by carefully reading each of the aforementioned engineering mechanics textbooks. My conclusion from reading these books was that the frustration my students were experiencing emanated from two sources. First, I found an enormous inconsistency between the presentation of the theory and the application of the theory to examples. Second, I found the approach to problem solving was highly formulaic and did not place an emphasis on understanding. Essentially, I concluded that these books lacked the pedagogy required for a student to master the key concepts and, instead, promoted an ad hoc approach to problem solving. More importantly, because of the inconsistency between the presentation and the application of the theory, I found that these books make it difficult for a student's understanding of the material to grow as the course progressed. Consequently, rather than solving problems systematically from first principles, my students were trying to solve homework problems either by emulating a problem solved in the book, by reverse engineering a solution using the answers at the back of the book (by analogy to a boundary-value problem, I call this approach the "shooting" method for finding a solution to a problem), or by searching for formulas from which they could "plug in" the information that they are given. The worst part was that, given a new problem (however similar it may appear to be to previous problems), they were at a loss as to how to proceed because they had not truly understood the key concepts.

My desire to write this book has grown out of my experience that undergraduate engineering dynamics needs to be taught in an extremely systematic and highly explicit manner. My approach has been put to the test over the past several years while teaching the core undergraduate engineering dynamics course at both Boston University and the University of Florida. I consider my approach to dynamics to be a significant departure from any of the existing books on undergraduate engineering dynamics. First, different from the aforementioned books, I have developed a highly rigorous presentation of the concepts. Second, the level of rigor in solving problems is *identical* to that used in developing the theory. Using my approach, it is possible for a student to see clearly the connection between the theory and the application of the theory. To this end, I have adopted a more advanced (but what I believe is a significantly more descriptive) notation than is commonly found in other undergraduate engineering dynamics books. Third, I have kept the material at the undergraduate level, i.e., the types of problems that are included share similarities with those found in many other engineering dynamics books. With regard to notation, with the exception of second-order tensors, the only mathematical prerequisite for this book is vector calculus (with regard to tensors, I believe that, given a few simple explanations and without losing a step, the basics of tensor algebra can be handled by a fourth-semester undergraduate student in mechanical or aerospace engineering). Fourth, in absolutely every topic covered in this book, I use a step-by-step vector mechanics approach to solving problems. I have found through experience that the approach I have chosen works extremely well in practice. In particular, I am able to see substantial growth in the thought process of my students from the first week of class to the final exam.

This book is intended for undergraduate students who want a systematic and rigorous approach to the subject of particle and rigid body dynamics. Because of the intended audience, certain topics in this book have been intentionally omitted. In particular, I do not cover the topics of systems where mass is gained or lost. Furthermore, I cover three-dimensional kinetics of rigid bodies in a relatively limited manner. In the case of systems that gain or lose mass, to teach this topic correctly requires a basic course on fluid mechanics, which many students do not have upon entering an undergraduate engineering dynamics course. With regard to three-dimensional kinetics of a rigid body, it is simply not possible to cover this entire topic in a one-semester undergraduate engineering dynamics course.

The material presented in this book is not new. However, I believe strongly that my approach is highly pedagogical, truly aids in mastering the key concepts, and promotes retention of the material well beyond the duration of the course. As I have already said, my approach is a significant departure from approaches used in other books. To motivate my approach, I have attempted throughout the book to include a sufficient number of worked examples and have included a wide range of problems at the end of each chapter for a student to solve. Most of the problems are ones that I have constructed myself while others are based on problems from the beautifully written book by Greenwood (1988). Finally, the notation I have adopted for kinematics is based on the notation developed by Kane and Levinson (1985).

Finally, I would like to re-emphasize that this book has been written with the student in mind. To this end, everywhere possible I have attempted to provide explicit guidance so that the student is able to follow clearly both the theory and the examples. It is my sincere hope that students everywhere will benefit from this book.

Anil V. Rao
Gainesville, Florida
April 2011

Acknowledgments

Writing a textbook is an arduous task and I have many people to acknowledge for their inspiration and support. First, I am indebted to all of the teachers I have had in my life, but particularly to my high school calculus teacher, Mr. David Bock, for giving me the inspiration to want to be a teacher, and to my Ph.D. thesis advisor, Dr. Kenneth D. Mease, for encouraging me to develop a rigorous approach to research and for teaching me by example the true value of expressing my thoughts in as clear a manner as possible.

With regard to the evolution of this book, I acknowledge my friend and colleague, Dr. Scott Ploen, for helping me greatly to improve both my perspective on the subject of dynamics and to develop pedagogical approaches to motivate students to learn the subject. I also acknowledge my former students, Theresia Becker and Kimberley Clarke, for taking enormous amounts of time and effort to carefully examine the manuscript for typographical errors and for providing helpful suggestions for improving the discussions in the text. Next, I would like to thank my friend, Mr. David Woffinden, and my teaching assistants, Christophe Lecomte and Josh Burnett, for providing me with valuable feedback about the content, style, and clarity of the manuscript. In addition, I would like to acknowledge Dr. John G. Papastivridis for his help in obtaining an accurate historical reference to the parallel-axis theorem. Finally, I gratefully acknowledge Dr. Donald T. Greenwood, Dr. Oliver M. O'Reilly, and Dr. David Geller for taking the time to carefully read and provide constructive criticism of the manuscript. I particularly thank Dr. O'Reilly for helping me gain insight into the Euler basis and the dual Euler basis and for helping me arrive at an accurate description of a conservative torque.

With regard to making this book a possibility, I owe a special acknowledgment to Dr. John Baillieul for giving me the opportunity to teach at Boston University. Without Dr. Baillieul's help, I would never have been able to do something that has turned out to be so fulfilling and would never have had the opportunity, let alone the inspiration, to write this book.

I also thank my dear parents, Saroj and Rajeswara, whose lifelong efforts made it possible for me to obtain a high quality education and have made a work such as this a reality. Finally, I thank my beloved wife, Anita, for her encouragement and support during the time when I was working on this manuscript. I realize only now just how much time my writing this book took from other things in our lives, and I am grateful for her patience throughout this long endeavor.

Anil V. Rao
Gainesville, Florida

Nomenclature

Symbols

$\otimes$	=	Tensor product between two vectors
$\odot$	=	Vector direction out of page
$\otimes$	=	Vector direction into page
∇	=	Gradient operator
$\mathcal{A}$	=	General reference frame
$\mathcal{B}$	=	General reference frame
$\mathcal{C}$	=	General reference frame
$\mathbb{E}^3$	=	Three-dimensional Euclidean space
$\mathcal{F}$	=	Fixed inertial reference frame
$\mathcal{N}$	=	General inertial reference frame
$\mathcal{R}$	=	Rigid body
$\mathcal{R}_1, \mathcal{R}_2$	=	Rigid body
$\mathbb{R}$	=	Set of real numbers
$\mathbb{R}^n$	=	n-dimensional real space
$\mathbb{R}^{n\times n}$	=	$n \times n$-dimensional real space
$\frac{{}^{\mathcal{A}}d\mathbf{b}}{dt}$	=	Rate of change of **b** as viewed by an observer in reference frame $\mathcal{A}$

Scalars

E	=	Total energy
g	=	Magnitude of acceleration due to gravity
G	=	Universal constant of gravitation
H	=	Hamiltonian
ℓ	=	Length of linear spring
ℓ_0	=	Unstretched length of linear spring
L	=	Lagrangian
m	=	Mass
M	=	Mass
q	=	Generalized momentum
Q	=	Generalized force
q	=	Generalized coordinate
$\dot{q}$	=	Rate of change of generalized coordinate
r	=	Magnitude of position or radius
R	=	Magnitude of reaction force or radius
s	=	Arc-length

t = Time
T = Kinetic energy
u = Dummy variable of integration
U = Potential energy or potential function
v = Speed
x = First component of Cartesian position
y = Second component of Cartesian position
z = Third component of Cartesian position
β = Angle
δ = Virtual displacement operator
κ = Curvature of trajectory
λ = Lagrange multiplier
μ = Coefficient of friction
μ_d = Coefficient of dynamic friction
μ_s = Coefficient of static friction
θ = Angle
$\dot{\theta}$ = Angular rate
ω_1 = First component of angular velocity
ω_2 = Second component of angular velocity
ω_3 = Third component of angular velocity
ϕ = Angle
$\dot{\phi}$ = Angular rate
τ = Torsion of trajectory

Vectors and Tensors

${}^{\mathcal{A}}\mathbf{a}$ = Acceleration as viewed by an observer in reference frame $\mathcal{A}$
${}^{\mathcal{A}}\bar{\mathbf{a}}$ = Acceleration of center of mass as viewed by an observer in reference frame $\mathcal{A}$
$\mathbf{a}$ = General vector
$\mathbf{b}$ = General vector
$\mathbf{c}$ = Constant vector
$\mathbf{e}_1$ = First basis vector
$\mathbf{e}_2$ = Second basis vector
$\mathbf{e}_3$ = Third basis vector
$\mathbf{e}_x$ = First Cartesian basis vector
$\mathbf{e}_y$ = Second Cartesian basis vector
$\mathbf{e}_z$ = Third Cartesian basis vector
$\mathbf{e}_t$ = Unit tangent vector
$\mathbf{e}_n$ = Principle unit normal vector
$\mathbf{e}_b$ = Principle unit bi-normal vector
$\mathbf{f}_{ij}$ = Force exerted by particle j on particle i
$\mathbf{g}$ = Local acceleration due to gravity
$\mathbf{n}$ = Unit normal to surface
$\mathbf{r}$ = Position
$\bar{\mathbf{r}}$ = Position of center of mass
$\mathbf{u}$ = Unit tangent vector
${}^{\mathcal{A}}\mathbf{v}$ = Velocity as viewed by an observer in reference frame $\mathcal{A}$
${}^{\mathcal{A}}\bar{\mathbf{v}}$ = Velocity of center of mass

		as viewed by an observer in reference frame $\mathcal{A}$
${}^{\mathcal{A}}\bar{\mathbf{v}}_C^{\mathcal{R}}$	=	Velocity of point C on rigid body $\mathcal{R}$ as viewed by an observer in reference frame $\mathcal{A}$
$\mathbf{w}$	=	Unit tangent vector
$\mathbf{E}_1$	=	First basis vector
$\mathbf{E}_2$	=	Second basis vector
$\mathbf{E}_3$	=	Third basis vector
$\mathbf{E}_x$	=	First Cartesian basis vector
$\mathbf{E}_y$	=	Second Cartesian basis vector
$\mathbf{E}_z$	=	Third Cartesian basis vector
${}^{\mathcal{N}}\mathbf{G}$	=	Linear momentum in inertial reference frame $\mathcal{N}$
$\frac{{}^{\mathcal{N}}d}{dt}\left({}^{\mathcal{N}}\mathbf{G}\right)$	=	Rate of change of linear momentum inertial reference frame $\mathcal{N}$
${}^{\mathcal{N}}\mathbf{H}_Q$	=	Angular momentum in inertial reference frame $\mathcal{N}$ relative to point Q
${}^{\mathcal{N}}\mathbf{H}_O$	=	Angular momentum in inertial reference frame $\mathcal{N}$ relative to point O fixed in $\mathcal{N}$
${}^{\mathcal{N}}\bar{\mathbf{H}}$	=	Angular momentum in inertial reference frame $\mathcal{N}$ relative to center of mass
$\frac{{}^{\mathcal{N}}d}{dt}\left({}^{\mathcal{N}}\mathbf{H}_Q\right)$	=	Rate of change of ${}^{\mathcal{N}}\mathbf{H}_Q$ in inertial reference frame $\mathcal{N}$ relative to point Q
$\frac{{}^{\mathcal{N}}d}{dt}\left({}^{\mathcal{N}}\mathbf{H}_O\right)$	=	Rate of change of angular momentum in inertial reference frame $\mathcal{N}$ relative to point O fixed in $\mathcal{N}$
$\frac{{}^{\mathcal{N}}d}{dt}\left({}^{\mathcal{N}}\bar{\mathbf{H}}\right)$	=	Rate of change of angular momentum in inertial reference frame $\mathcal{N}$ relative to center of mass
$\mathbf{I}^{\mathcal{R}}$	=	Moment of inertia tensor of a rigid body $\mathcal{R}$
$\mathbf{I}_Q^{\mathcal{R}}$	=	Moment of inertia tensor of a rigid body $\mathcal{R}$
	=	relative to point Q
$\bar{\mathbf{I}}^{\mathcal{R}}$	=	Moment of inertia tensor of a rigid body $\mathcal{R}$ relative to center of mass of $\mathcal{R}$
$\mathbf{M}$	=	Moment
$\mathbf{M}_O$	=	Moment relative to point O
$\mathbf{M}_Q$	=	Moment relative to point Q
$\bar{\mathbf{M}}$	=	Moment relative to center of mass
$\mathbf{N}$	=	Reaction force
$\mathbf{S}$	=	Symmetric tensor
$\mathbf{T}$	=	General tensor
$\mathbf{U}$	=	Identity tensor
$\mathbf{R}$	=	Reaction force
${}^{\mathcal{A}}\boldsymbol{\alpha}^{\mathcal{B}}$	=	Angular acceleration of reference frame $\mathcal{B}$ as viewed by an observer in reference frame $\mathcal{A}$
${}^{\mathcal{A}}\boldsymbol{\omega}^{\mathcal{B}}$	=	Angular velocity of reference frame $\mathcal{B}$ as viewed by an observer in reference frame $\mathcal{A}$
$\boldsymbol{\rho}$	=	Relative position
$\boldsymbol{\tau}$	=	Pure torque

Chapter 1

Introductory Concepts

The scientist does not study nature because it is useful; he studies it because he delights in it, and he delights in it because it is beautiful. If nature were not beautiful, it would not be worth knowing, and if nature were not worth knowing, life would not be worth living.

- Jules Henri Poincaré (1854-1912)
French Mathematician and Physicist

Mechanics is the study of the effect that physical forces have on objects. *Dynamics* is the particular branch of mechanics that deals with the study of the effect that forces have on the motion of objects. Dynamics is itself divided into two branches called Newtonian dynamics and relativistic dynamics. *Newtonian* dynamics is the study of the motion of objects that travel with speeds significantly less than the speed of light while *relativistic* dynamics is the study of the motion of objects that travel with speeds at or near the speed of light. This division in the subject of dynamics arises because the physics associated with the motion of objects that travel with speeds much less than the speed of light can be modeled much more simply than the physics associated with the motion of objects that travel with speeds at or near the speed of light. Moreover, nonrelativistic dynamics deals primarily with the motion of objects on a macroscopic scale while relativistic dynamics deals with the study of the motion of objects on a microscopic or submicroscopic scale. The objective of this book is to present the underlying concepts of Newtonian dynamics in a clear and concise manner and to develop a systematic framework for solving problems in classical Newtonian dynamics.

As with any subject that is based on the laws of physics, Newtonian dynamics needs to be described using mathematics. More specifically, it must be possible to describe the physical laws in a way that is independent of the particular coordinate system in which one chooses to formulate a particular problem. The mathematical approach that gives us the freedom to develop a coordinate-free approach to Newtonian mechanics is that of *vector and tensor algebra.*

Once the physical laws have been described in a coordinate-free manner, the next step is to formulate the particular problem of interest. While the basic laws themselves are coordinate-free, to solve a particular problem it is necessary to specify all relevant quantities using a coordinate system of choice. While in principle it is possible to use any coordinate system to describe the motion of a material body, choosing a particular coordinate system could vastly simplify the particular problem under consideration. The remainder of this chapter is devoted to providing a review of the vector and tensor algebra required to formulate and analyze problems in nonrelativistic mechanics. While this chapter provides a mathematical overview, it is not intended as a substitute for a book on engineering mathematics. For a more in-depth presentation of engineering mathematics, the reader is referred to a standard text in undergraduate engineering mathematics such as that found in Kreyszig (1988).

1.1 Scalars

A *scalar* is any quantity that is expressible as a real number. We denote a scalar by a non-boldface character and denote the set of real numbers by $\mathbb{R}$, i.e., we say that the (non-boldface) quantity a is a scalar if

$$a \in \mathbb{R}$$

Scalars satisfy the following properties with respect to addition and multiplication:

1. Commutativity: For all $a \in \mathbb{R}$ and $b \in \mathbb{R}$,

$$\begin{aligned} a+b &= b+a \\ ab &= ba \end{aligned}$$

2. Associativity: For all $a \in \mathbb{R}$, $b \in \mathbb{R}$, and $c \in \mathbb{R}$,

$$\begin{aligned} (a+b)+c &= a+(b+c) \\ a(bc) &= (ab)c \end{aligned}$$

3. Zero Scalar: There exists a scalar 0 such that for all $a \in \mathbb{R}$,

$$\begin{aligned} a+0 &= 0+a=a \\ 0(a) &= (a)0=0 \end{aligned}$$

4. Unit Scalar: there exists a scalar 1 such that for all $a \in \mathbb{R}$,

$$1(a)=(a)1=a$$

5. Inverse scalar: For all $a \neq 0 \in \mathbb{R}$, there exists a scalar $1/a$ such that

$$\frac{1}{a}(a)=a\frac{1}{a}=1$$

6. Negativity: There exists a scalar -1 such that for all $a \in \mathbb{R}$

$$\begin{aligned} -1(a) &= a(-1)=-a \\ a+(-a) &= (-a)+a=0 \end{aligned}$$

1.2 Vectors

A *vector* is any quantity that has both *magnitude and direction.* A vector is denoted by a *boldface* character, i.e., a quantity $\mathbf{a}$ is a vector. Because the study of Newtonian mechanics focuses on the motion of objects in *three-dimensional Euclidean space,* throughout this book we will be interested in three-dimensional vectors. Three-dimensional Euclidean space is denoted $\mathbb{E}^3$. Consequently, the notation

$$\mathbf{a} \in \mathbb{E}^3$$

means that the vector $\mathbf{a}$ lies in three-dimensional Euclidean space.

The *length* of a vector $\mathbf{a} \in \mathbb{E}^3$ is called the *magnitude* of $\mathbf{a}$. The magnitude or *Euclidean norm* of a vector $\mathbf{a}$ is denoted $\|\mathbf{a}\|$ and is a scalar, i.e., $\|\mathbf{a}\| \in \mathbb{R}$. A vector whose magnitude is zero is called the *zero vector.* We denote the zero vector by a boldface zero, i.e., the zero vector is denoted by $\mathbf{0}$. The *direction* of a nonzero vector $\mathbf{a}$ is the vector divided by its magnitude, i.e., the direction of the vector $\mathbf{a}$, denoted $\mathbf{u}_a$, is given as

$$\mathbf{u}_a = \frac{\mathbf{a}}{\|\mathbf{a}\|}$$

Furthermore, the direction of a nonzero vector is called a *unit vector* because its magnitude is unity, i.e., $\|\mathbf{u}_a\| = 1$. Two vectors are said to be equal if they have the same magnitude and direction.

1.2.1 Types of Vectors

While geometrically a vector is any quantity with magnitude and direction, the physical effect of a vector $\mathbf{a}$ on a mechanical system may depend in addition on a particular line of action in $\mathbb{E}^3$ or a particular point in $\mathbb{E}^3$. In particular, vectors arising in mechanics fall into one of three categories[1]: (a) free vectors; (b) sliding vectors; and (c) bound vectors. Each type of vector is now described in more detail.

A *free vector* is any vector $\mathbf{b}$ with no specified line of action or point of application in $\mathbb{E}^3$. Figure 1-1 shows an example of two identical free vectors $\mathbf{b}$ and $\mathbf{b}'$. While $\mathbf{b}$ and $\mathbf{b}'$ have the same direction and magnitude, they do not share the same start or end point. In particular, $\mathbf{b}$ starts at point Q and ends at point P while $\mathbf{b}'$ starts at point $Q' \neq Q$ and ends at point $P' \neq P$. However, because $\mathbf{b}$ and $\mathbf{b}'$ have the same direction and magnitude, they are identical free vectors. Examples of free vectors are the angular velocity of a reference frame or a rigid body, a pure torque applied to a rigid body, and a basis vector.

A *sliding vector* is any vector $\mathbf{b}$ that has a specified *line of action* or *axis* in $\mathbb{E}^3$, but has no specified point of application in $\mathbb{E}^3$. Figure 1-2 shows two identical sliding vectors $\mathbf{b}$ and $\mathbf{b}'$. As with free vectors, $\mathbf{b}$ and $\mathbf{b}'$ have the same magnitude and direction. However, while the vector $\mathbf{b}$ starts at point Q and ends at point P, the vector $\mathbf{b}'$ starts at point $Q' \neq Q$ and ends at point $P' \neq P$ (where the points P, Q, P', and Q' are colinear). Consequently, $\mathbf{b}$ and $\mathbf{b}'$ are identical sliding vectors, but are different free vectors. An example of a sliding vector is the force applied to a rigid body.

A *bound vector* is any vector that has both a specified line of action in $\mathbb{E}^3$ and a specified point of application in $\mathbb{E}^3$. From its definition, it can be seen that a bound

[1] An excellent description of free, sliding, and bound vectors can be found in either Synge and Griffith (1959) or Greenwood (1988).

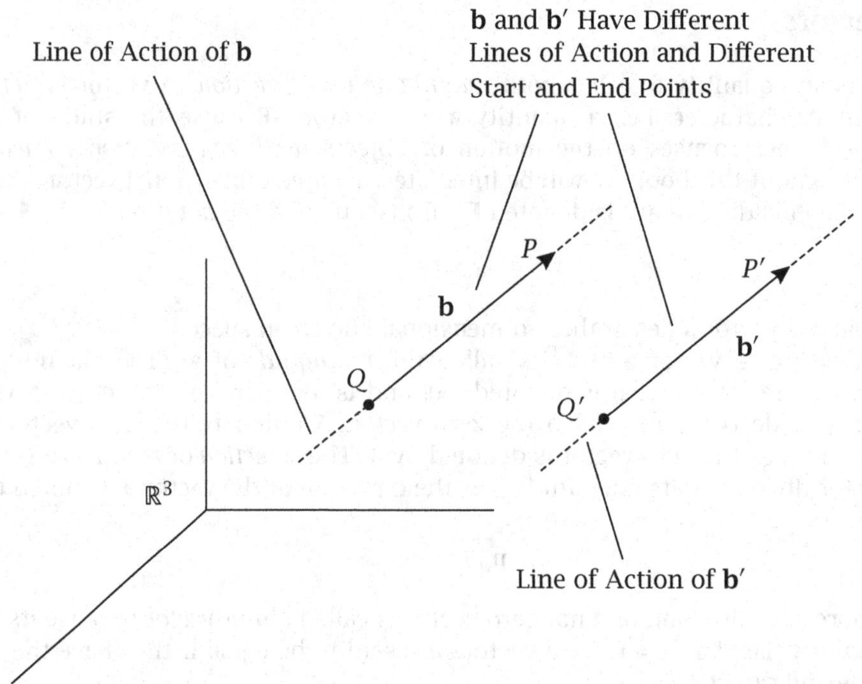

Figure 1–1 Two equal free vectors $\mathbf{b}$ and $\mathbf{b}'$ that have the same direction and magnitude, but different lines of action and different start and end points.

vector is unique, i.e., only *one* vector can have a specified direction, magnitude, line of action, and origin. An example of a bound vector is the force acting on or exerted by an elastic body (e.g., the force exerted by a linear spring); in the case of an elastic body, the deformation of the body depends on the changing point of application of the force.

It should be noted that vector algebra is valid only for free vectors. However, because all vectors are defined by their direction and magnitude, vector algebra can be performed on sliding and bound vectors by treating them *as though* they are free vectors. Consequently, the result of any algebraic operation on vectors, regardless of the type of vector, results in a free vector. From this point forth, unless otherwise stated or additional clarification is necessary, all vectors will be assumed to be free vectors.

1.2.2 Addition of Vectors

Let $\mathbf{a}$ and $\mathbf{b}$ be vectors in $\mathbb{E}^3$. Then the *sum* of $\mathbf{a}$ and $\mathbf{b}$, denoted $\mathbf{c}$, is given as

$$\mathbf{c} = \mathbf{a} + \mathbf{b} \tag{1–1}$$

Vector addition has the following properties:

1. Commutativity: For all $\mathbf{a} \in \mathbb{E}^3$ and $\mathbf{b} \in \mathbb{E}^3$,

$$\mathbf{a} + \mathbf{b} = \mathbf{b} + \mathbf{a}$$

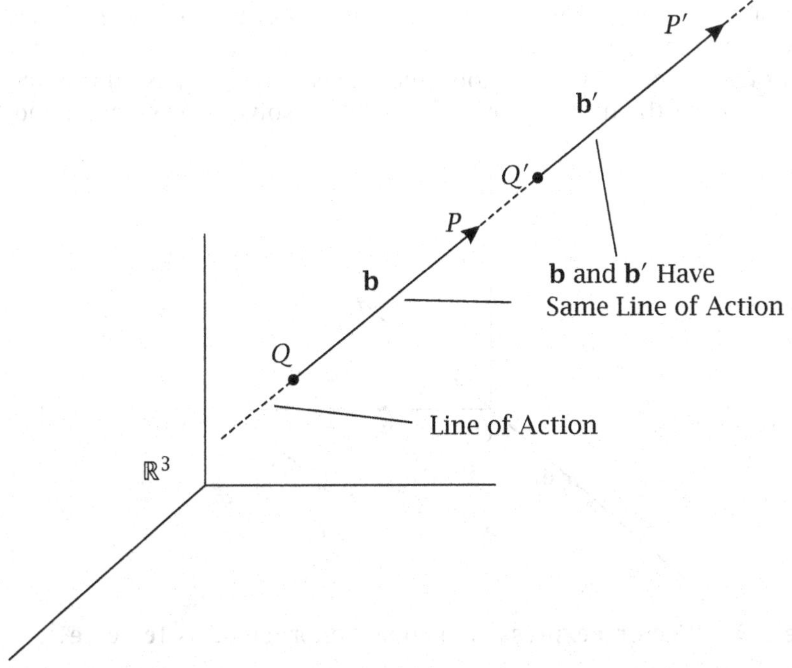

Figure 1–2 Two equal sliding vectors **b** and **b**′ that have the same direction, magnitude, and line of action, but different start and end points.

2. Associativity: For all $\mathbf{a} \in \mathbb{E}^3$, $\mathbf{b} \in \mathbb{E}^3$, and $\mathbf{c} \in \mathbb{E}^3$,

$$(\mathbf{a} + \mathbf{b}) + \mathbf{c} = \mathbf{a} + (\mathbf{b} + \mathbf{c})$$

3. Zero vector: There exists a vector $\mathbf{0}$ such that for all $\mathbf{a} \in \mathbb{E}^3$,

$$\mathbf{a} + \mathbf{0} = \mathbf{a}$$

4. For all $\mathbf{a} \in \mathbb{E}^3$, there exists $-\mathbf{a} \in \mathbb{E}^3$ such that

$$\mathbf{a} + (-\mathbf{a}) = \mathbf{0}$$

1.2.3 Components of a Vector

Any vector $\mathbf{a} \in \mathbb{E}^3$ can be expressed in terms of three noncoplanar vectors $\mathbf{e}_1$, $\mathbf{e}_2$, and $\mathbf{e}_3$ called *basis vectors*. Correspondingly, any noncoplanar set of vectors $\{\mathbf{e}_1, \mathbf{e}_2, \mathbf{e}_3\}$ is called a *basis* for $\mathbb{E}^3$. In terms of the basis $\{\mathbf{e}_1, \mathbf{e}_2, \mathbf{e}_3\}$, the vector $\mathbf{a}$ can be written as

$$\mathbf{a} = a_1\mathbf{e}_1 + a_2\mathbf{e}_2 + a_3\mathbf{e}_3 \tag{1-2}$$

where a_1, a_2, and a_3 are the *components* of $\mathbf{a}$ in the basis $\{\mathbf{e}_1, \mathbf{e}_2, \mathbf{e}_3\}$. Generally speaking, it is preferable to use a basis of mutually orthogonal vectors. Any basis consisting of mutually orthogonal vectors is called an *orthogonal basis.* Even more specifically, it is most preferable to use a basis consisting of mutually orthogonal *unit* vectors. A basis consisting of mutually orthogonal unit vectors is called an *orthonormal basis.* In the remainder of this book, we will restrict our attention to orthonormal bases. To this end, we will use the term "basis" to mean specifically an orthonormal basis. The representation of a vector **a** in an orthonormal basis $\{\mathbf{e}_1, \mathbf{e}_2, \mathbf{e}_3\}$ is shown schematically in Fig. 1-3. Using the basis $\{\mathbf{e}_1, \mathbf{e}_2, \mathbf{e}_3\}$, we can resolve two vectors **a** and **b** into

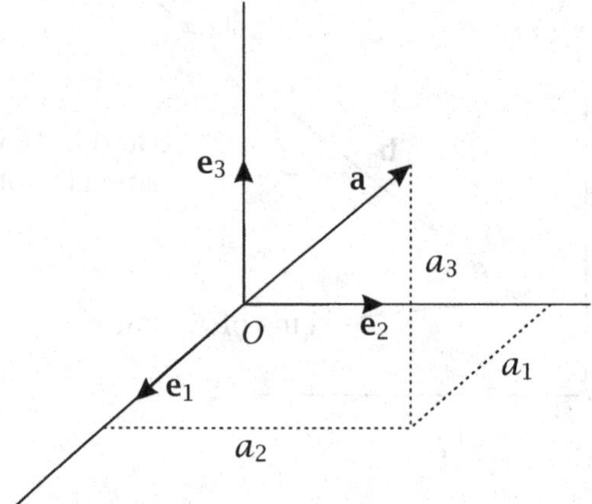

Figure 1-3 Vector **a** expressed in an orthonormal basis $\{\mathbf{e}_1, \mathbf{e}_2, \mathbf{e}_3\}$.

$\{\mathbf{e}_1, \mathbf{e}_2, \mathbf{e}_3\}$ as follows:

$$\begin{aligned} \mathbf{a} &= a_1\mathbf{e}_1 + a_2\mathbf{e}_2 + a_3\mathbf{e}_3 \\ \mathbf{b} &= b_1\mathbf{e}_1 + b_2\mathbf{e}_2 + b_3\mathbf{e}_3 \end{aligned} \tag{1-3}$$

Then the sum of **a** and **b** is given in terms of $\{\mathbf{e}_1, \mathbf{e}_2, \mathbf{e}_3\}$ as

$$\mathbf{c} = (a_1 + b_1)\mathbf{e}_1 + (a_2 + b_2)\mathbf{e}_2 + (a_3 + b_3)\mathbf{e}_3 \tag{1-4}$$

1.2.4 Multiplication of a Vector by a Scalar

Let **a** be a vector in $\mathbb{E}^3$ and let $k \in \mathbb{R}$ be a scalar. Then the product of **a** with the scalar k, denoted $k\mathbf{a}$, has the following properties:

1. $\|k\mathbf{a}\| = |k|\|\mathbf{a}\|$
2. $\dfrac{k\mathbf{a}}{\|k\mathbf{a}\|} = \dfrac{\mathbf{a}}{\|\mathbf{a}\|}$ if $k > 0$ and $\mathbf{a} \neq \mathbf{0}$
3. $\dfrac{k\mathbf{a}}{\|k\mathbf{a}\|} = -\dfrac{\mathbf{a}}{\|\mathbf{a}\|}$ if $k < 0$ and $\mathbf{a} \neq \mathbf{0}$
4. $k\mathbf{a} = \mathbf{0}$ if either $\mathbf{a} = \mathbf{0}$ or $k = 0$
5. $k(\mathbf{a} + \mathbf{b}) = k\mathbf{a} + k\mathbf{b}$
6. $(k_1 + k_2)\mathbf{a} = k_1\mathbf{a} + k_2\mathbf{a}$
7. $k_2(k_1\mathbf{a}) = k_2k_1\mathbf{a}$
8. $(1)\mathbf{a} = \mathbf{a}(1) = \mathbf{a}$
9. $(0)\mathbf{a} = \mathbf{a}(0) = \mathbf{0}$

10. $(-1)\mathbf{a} = \mathbf{a}(-1) = -\mathbf{a}$

Finally, if $\mathbf{a}$ is expressed in the basis $\{\mathbf{e}_1, \mathbf{e}_2, \mathbf{e}_3\}$, then $k\mathbf{a}$ is given as

$$k\mathbf{a} = ka_1\mathbf{e}_1 + ka_2\mathbf{e}_2 + ka_3\mathbf{e}_3 \tag{1-5}$$

1.2.5 Scalar Product

Let $\mathbf{a}$ and $\mathbf{b}$ be vectors in $\mathbb{E}^3$. Then the *scalar product* or *dot product* between $\mathbf{a}$ and $\mathbf{b}$ is defined as

$$\mathbf{a} \cdot \mathbf{b} = \|\mathbf{a}\| \|\mathbf{b}\| \cos\theta = ab\cos\theta \tag{1-6}$$

where θ is the angle between $\mathbf{a}$ and $\mathbf{b}$. The scalar product has the following properties:

1. $\mathbf{a} \cdot \mathbf{b} = \mathbf{b} \cdot \mathbf{a}$

2. $\mathbf{a} \cdot (k\mathbf{b}) = k\mathbf{a} \cdot \mathbf{b}$ where $k \in \mathbb{R}$

3. $(\mathbf{a} + \mathbf{b}) \cdot \mathbf{c} = \mathbf{a} \cdot \mathbf{c} + \mathbf{b} \cdot \mathbf{c}$

Two nonzero vectors are said to be *orthogonal* if their scalar product is *zero*, i.e., $\mathbf{a}$ and $\mathbf{b}$ are orthogonal if

$$\mathbf{a} \cdot \mathbf{b} = 0 \quad (\mathbf{a}, \mathbf{b} \neq \mathbf{0}) \tag{1-7}$$

A set of vectors $\{\mathbf{a}_1, \ldots, \mathbf{a}_n\}$ is said to be *mutually orthogonal* if

$$\mathbf{a}_i \cdot \mathbf{a}_j = 0 \quad (i \neq j,\ i, j = 1, \ldots, n) \tag{1-8}$$

Finally, the magnitude of a vector $\mathbf{a}$ is equal to the square root of the dot product of the vector with itself, i.e.,

$$\|\mathbf{a}\| = \sqrt{\mathbf{a} \cdot \mathbf{a}} \tag{1-9}$$

Suppose now that $\mathbf{a}$ and $\mathbf{b}$ are expressed in a particular basis $\{\mathbf{e}_1, \mathbf{e}_2, \mathbf{e}_3\}$ as

$$\begin{aligned} \mathbf{a} &= a_1\mathbf{e}_1 + a_2\mathbf{e}_2 + a_3\mathbf{e}_3 \\ \mathbf{b} &= b_1\mathbf{e}_1 + b_2\mathbf{e}_2 + b_3\mathbf{e}_3 \end{aligned} \tag{1-10}$$

Then the scalar product of $\mathbf{a}$ with $\mathbf{b}$ is given as

$$\mathbf{a} \cdot \mathbf{b} = (a_1\mathbf{e}_1 + a_2\mathbf{e}_2 + a_3\mathbf{e}_3) \cdot (b_1\mathbf{e}_1 + b_2\mathbf{e}_2 + b_3\mathbf{e}_3) \tag{1-11}$$

Because we are restricting attention to orthonormal bases, the basis $\{\mathbf{e}_1, \mathbf{e}_2, \mathbf{e}_3\}$ satisfies the properties that

$$\mathbf{e}_i \cdot \mathbf{e}_j = \begin{cases} 1\ (i = j) \\ 0\ (i \neq j) \end{cases} \quad (i, j = 1, 2, 3) \tag{1-12}$$

Consequently, we have

$$\mathbf{a} \cdot \mathbf{b} = a_1b_1 + a_2b_2 + a_3b_3 \tag{1-13}$$

Using Eq. (1-13) and the definition of the magnitude of a vector as given in Eq. (1-9), the magnitude of a vector $\mathbf{a}$ can be written in terms of the components of $\mathbf{a}$ in the basis $\{\mathbf{e}_1, \mathbf{e}_2, \mathbf{e}_3\}$ as

$$\|\mathbf{a}\| = \sqrt{a_1^2 + a_2^2 + a_3^2} \tag{1-14}$$

1.2.6 Vector Product

Let $\mathbf{a}$ and $\mathbf{b}$ be vectors in $\mathbb{E}^3$. Then the *vector product* or *cross product* between two vectors $\mathbf{a}$ and $\mathbf{b}$ is defined as

$$\mathbf{c} = \mathbf{a} \times \mathbf{b} = \|\mathbf{a}\| \, \|\mathbf{b}\| \sin\theta \, \mathbf{n} \tag{1-15}$$

where $\mathbf{n}$ is the unit vector in the direction orthogonal to both $\mathbf{a}$ and $\mathbf{b}$ in a right-handed sense and θ is the angle between $\mathbf{a}$ and $\mathbf{b}$. The term "right-handed sense" arises from the fact that the vectors $\mathbf{a}$, $\mathbf{b}$, and $\mathbf{c}$ assume an orientation that corresponds to the index finger, middle finger, and thumb of the right hand when these three fingers are held as shown in Fig. 1–4.

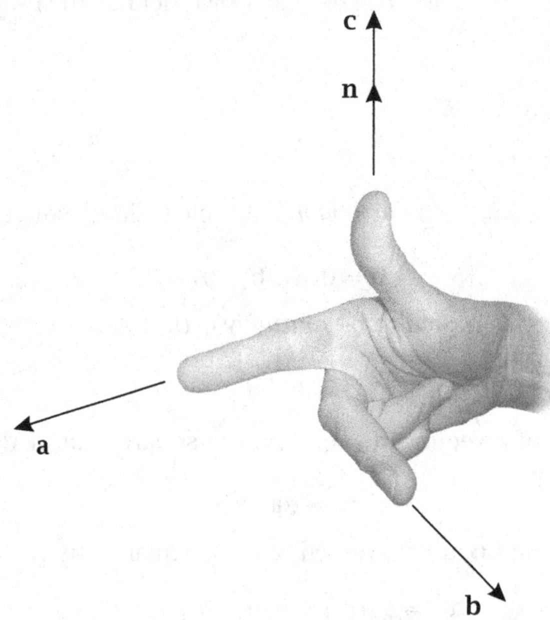

Figure 1–4 Schematic of right-hand rule corresponding to the vector product of two vectors using the index finger, middle finger, and thumb of the right hand.

The magnitude of the vector product of two vectors is given as

$$\|\mathbf{c}\| = \|\mathbf{a}\| \, \|\mathbf{b}\| \sin\theta \tag{1-16}$$

The vector product has the following properties:

1. $\mathbf{a} \times \mathbf{a} = \mathbf{0}$
2. $\mathbf{a} \times \mathbf{b} = -\mathbf{b} \times \mathbf{a}$
3. $(k\mathbf{a}) \times \mathbf{b} = k(\mathbf{a} \times \mathbf{b}) = \mathbf{a} \times (k\mathbf{b})$ where $k \in \mathbb{R}$
4. $(\mathbf{a} + \mathbf{b}) \times \mathbf{c} = (\mathbf{a} \times \mathbf{c}) + (\mathbf{b} \times \mathbf{c})$

Now suppose that $\mathbf{a}$ and $\mathbf{b}$ are expressed in terms of the basis $\{\mathbf{e}_1, \mathbf{e}_2, \mathbf{e}_3\}$, i.e.,

$$\begin{aligned} \mathbf{a} &= a_1\mathbf{e}_1 + a_2\mathbf{e}_2 + a_3\mathbf{e}_3 \\ \mathbf{b} &= b_1\mathbf{e}_1 + b_2\mathbf{e}_2 + b_3\mathbf{e}_3 \end{aligned} \tag{1-17}$$

Then the cross product of **a** and **b** is given as

$$\mathbf{a}\times\mathbf{b} = (a_1\mathbf{e}_1 + a_2\mathbf{e}_2 + a_3\mathbf{e}_3) \times (b_1\mathbf{e}_1 + b_2\mathbf{e}_2 + b_3\mathbf{e}_3) \tag{1-18}$$

Expanding Eq. (1–18), we obtain

$$\begin{aligned}\mathbf{a}\times\mathbf{b} &= a_1b_2\mathbf{e}_1\times\mathbf{e}_2 + a_1b_3\mathbf{e}_1\times\mathbf{e}_3 + a_2b_1\mathbf{e}_2\times\mathbf{e}_1 \\ &+ a_2b_3\mathbf{e}_2\times\mathbf{e}_3 + a_3b_1\mathbf{e}_3\times\mathbf{e}_1 + a_3b_2\mathbf{e}_3\times\mathbf{e}_2\end{aligned} \tag{1-19}$$

Again we remind the reader that we are restricting our attention to orthonormal bases. Furthermore, suppose that the basis $\{\mathbf{e}_1, \mathbf{e}_2, \mathbf{e}_3\}$ forms a right-handed set, i.e., $\{\mathbf{e}_1, \mathbf{e}_2, \mathbf{e}_3\}$ satisfies the following properties:

$$\begin{aligned}\mathbf{e}_1\times\mathbf{e}_2 &= \mathbf{e}_3 \\ \mathbf{e}_2\times\mathbf{e}_3 &= \mathbf{e}_1 \\ \mathbf{e}_3\times\mathbf{e}_1 &= \mathbf{e}_2\end{aligned} \tag{1-20}$$

Then $\mathbf{a}\times\mathbf{b}$ is given as

$$\mathbf{a}\times\mathbf{b} = (a_2b_3 - a_3b_2)\mathbf{e}_1 + (a_3b_1 - a_1b_3)\mathbf{e}_2 + (a_1b_2 - a_2b_1)\mathbf{e}_3 \tag{1-21}$$

In terms of a right-handed basis, Eq. (1–21) can also be written as the following determinant (Kreyszig, 1988):

$$\mathbf{a}\times\mathbf{b} = \begin{vmatrix}\mathbf{e}_1 & \mathbf{e}_2 & \mathbf{e}_3 \\ a_1 & a_2 & a_3 \\ b_1 & b_2 & b_3\end{vmatrix} \tag{1-22}$$

1.2.7 Scalar Triple Product

Given three vectors **a**, **b**, and **c**, the *scalar triple product* is defined as

$$\mathbf{a}\cdot(\mathbf{b}\times\mathbf{c}) \tag{1-23}$$

The scalar triple product has the following properties:

1. $\mathbf{a}\cdot(\mathbf{b}\times\mathbf{c}) = (\mathbf{a}\times\mathbf{b})\cdot\mathbf{c} = \mathbf{b}\cdot(\mathbf{c}\times\mathbf{a}) = \mathbf{c}\cdot(\mathbf{a}\times\mathbf{b})$
2. $\mathbf{a}\cdot(k\mathbf{b}\times\mathbf{c}) = k\mathbf{a}\cdot(\mathbf{b}\times\mathbf{c})$

Suppose that the vectors **a**, **b**, and **c** are each expressed in an orthonormal basis $\{\mathbf{e}_1, \mathbf{e}_2, \mathbf{e}_3\}$ as

$$\begin{aligned}\mathbf{a} &= a_1\mathbf{e}_1 + a_2\mathbf{e}_2 + a_3\mathbf{e}_3 \\ \mathbf{b} &= b_1\mathbf{e}_1 + b_2\mathbf{e}_2 + b_3\mathbf{e}_3 \\ \mathbf{c} &= c_1\mathbf{e}_1 + c_2\mathbf{e}_2 + c_3\mathbf{e}_3\end{aligned} \tag{1-24}$$

Then the scalar triple product can be written as

$$\mathbf{a}\cdot(\mathbf{b}\times\mathbf{c}) = a_1(b_2c_3 - b_3c_2) + a_2(b_3c_1 - b_1c_3) + a_3(b_1c_2 - b_2c_1) \tag{1-25}$$

The scalar triple product can also be written as the following determinant:

$$\mathbf{a}\cdot(\mathbf{b}\times\mathbf{c}) = \begin{vmatrix}a_1 & a_2 & a_3 \\ b_1 & b_2 & b_3 \\ c_1 & c_2 & c_3\end{vmatrix} \tag{1-26}$$

Finally, the scalar triple product can be written as

$$\mathbf{a}\cdot(\mathbf{b}\times\mathbf{c}) = \|\mathbf{a}\|\,\|\mathbf{b}\times\mathbf{c}\|\cos\theta \tag{1-27}$$

where θ is the angle between the vector **a** and the vector $\mathbf{b}\times\mathbf{c}$.

1.2.8 Vector Triple Product

Given three vectors **a**, **b**, and **c**, the *vector triple product* is given as

$$\mathbf{a}\times(\mathbf{b}\times\mathbf{c}) \tag{1-28}$$

The vector triple product can be written as

$$\mathbf{a}\times(\mathbf{b}\times\mathbf{c}) = (\mathbf{a}\cdot\mathbf{c})\mathbf{b} - (\mathbf{a}\cdot\mathbf{b})\mathbf{c} \tag{1-29}$$

Suppose that the vectors **a**, **b**, and **c** are each expressed in an orthonormal basis $\{\mathbf{e}_1, \mathbf{e}_2, \mathbf{e}_3\}$ as

$$\begin{aligned}\mathbf{a} &= a_1\mathbf{e}_1 + a_2\mathbf{e}_2 + a_3\mathbf{e}_3 \\ \mathbf{b} &= b_1\mathbf{e}_1 + b_2\mathbf{e}_2 + b_3\mathbf{e}_3 \\ \mathbf{c} &= c_1\mathbf{e}_1 + c_2\mathbf{e}_2 + c_3\mathbf{e}_3\end{aligned} \tag{1-30}$$

The vector triple product can then be written as

$$\begin{aligned}\mathbf{a}\times(\mathbf{b}\times\mathbf{c}) = {} & (a_1c_1 + a_2c_2 + a_3c_3)(b_1\mathbf{e}_1 + b_2\mathbf{e}_2 + b_3\mathbf{e}_3) \\ & - (a_1b_1 + a_2b_2 + a_3b_3)(c_1\mathbf{e}_1 + c_2\mathbf{e}_2 + c_3\mathbf{e}_3)\end{aligned} \tag{1-31}$$

1.3 Tensors

A *tensor* (or second-order tensor[2]), denoted **T**, is a linear operator that associates a vector $\mathbf{a} \in \mathbb{E}^3$ to another vector $\mathbf{b} \in \mathbb{E}^3$, i.e., if **T** is a tensor and $\mathbf{a} \in \mathbb{E}^3$ is a vector, then there exists a vector **b** such that

$$\mathbf{b} = \mathbf{T}\cdot\mathbf{a} \tag{1-32}$$

It is noted that the binary operator "·" in Eq. (1-32) is different from the scalar product between two vectors in that the "·" denotes the operation of the tensor **T** on the vector **a**. Now, because tensors are linear operators, they satisfy the following properties:

1. For all $\mathbf{a}, \mathbf{b} \in \mathbb{E}^3$,

$$\mathbf{T}\cdot(\mathbf{a}+\mathbf{b}) = \mathbf{T}\cdot\mathbf{a} + \mathbf{T}\cdot\mathbf{b}$$

2. For all $\mathbf{a} \in \mathbb{E}^3$ and $k \in \mathbb{R}$,

$$\mathbf{T}\cdot(k\mathbf{a}) = k\mathbf{T}\cdot\mathbf{a}$$

3. There exists a *zero tensor*, denoted **O**, such that for every $\mathbf{a} \in \mathbb{E}^3$,

$$\mathbf{O}\cdot\mathbf{a} = \mathbf{0} \tag{1-33}$$

   where **0** is the zero vector.

4. There exists an *identity tensor* or *unit tensor*, denoted **U**, such that for every $\mathbf{a} \in \mathbb{E}^3$,

$$\mathbf{U}\cdot\mathbf{a} = \mathbf{a} \tag{1-34}$$

[2]Strictly speaking, the tensor defined in Eq. (1-32) is a *second-order* tensor. While tensor algebra generalizes well beyond second-order tensors, in this book we will only be concerned with second-order tensors. Consequently, throughout this book we will use the term "tensor" to mean "second-order tensor".

1.3.1 Important Classes of Tensors

A tensor T is said to be *invertible* if there exists a tensor T^{-1} such that

$$\mathrm{T} \cdot \mathrm{T}^{-1} = \mathrm{T}^{-1} \cdot \mathrm{T} = \mathrm{U} \tag{1-35}$$

If T is invertible, then T^{-1} is called the *inverse* of T. The *transpose* of a tensor T, denoted T^T, satisfies the property that for all $\mathbf{a} \in \mathbb{E}^3$ and $\mathbf{b} \in \mathbb{E}^3$,

$$(\mathrm{T}^T \cdot \mathbf{a}) \cdot \mathbf{b} = \mathbf{a} \cdot (\mathrm{T} \cdot \mathbf{b}) \tag{1-36}$$

A tensor T is said to be *symmetric* if it is equal to its transpose, i.e., T is symmetric if

$$\mathrm{T} = \mathrm{T}^T \tag{1-37}$$

A tensor T is said to be *skew-symmetric* if it is equal to the negative of its transpose, i.e., T is skew-symmetric if

$$\mathrm{T} = -\mathrm{T}^T \tag{1-38}$$

Finally, an invertible tensor T is said to be *orthogonal* if its transpose is equal to its inverse, i.e., T is orthogonal if

$$\mathrm{T}^T = \mathrm{T}^{-1} \tag{1-39}$$

1.3.2 Tensor Product Between Vectors

Let **a** and **b** be vectors in $\mathbb{E}^3$. Then the *tensor product* of **a** and **b**, denoted $\mathbf{a} \otimes \mathbf{b}$, is defined as

$$\mathrm{T} = \mathbf{a} \otimes \mathbf{b} \tag{1-40}$$

where T is the tensor that results from the tensor product between **a** and **b**. It can be seen that a tensor is obtained from an operation on a *pair* of vectors. Then, using Eq. (1-32), the scalar product between the tensor $\mathbf{a} \otimes \mathbf{b}$ and the vector **c** is defined as

$$(\mathbf{a} \otimes \mathbf{b}) \cdot \mathbf{c} \equiv (\mathbf{b} \cdot \mathbf{c})\mathbf{a} \tag{1-41}$$

$$\mathbf{c} \cdot (\mathbf{a} \otimes \mathbf{b}) \equiv (\mathbf{a} \cdot \mathbf{c})\mathbf{b} \tag{1-42}$$

It can be seen that, unlike the scalar product between two vectors, the scalar product between a tensor $\mathbf{a} \otimes \mathbf{b}$ and a vector **c** depends on the order of the operation, i.e., in general we have that

$$(\mathbf{a} \otimes \mathbf{b}) \cdot \mathbf{c} \neq \mathbf{c} \cdot (\mathbf{a} \otimes \mathbf{b}) \tag{1-43}$$

Suppose now that we let $\{\mathbf{e}_1, \mathbf{e}_2, \mathbf{e}_3\}$ be an orthonormal basis for $\mathbb{E}^3$. Then, using Eq. (1-41), we have

$$(\mathbf{e}_i \otimes \mathbf{e}_j) \cdot \mathbf{e}_k = (\mathbf{e}_j \cdot \mathbf{e}_k)\mathbf{e}_i \quad (i, j, k = 1, 2, 3) \tag{1-44}$$

Substituting the expression for $\mathbf{e}_i \otimes \mathbf{e}_j$ from Eq. (1-12) into Eq. (1-44), we obtain

$$\mathbf{e}_i \otimes \mathbf{e}_j \cdot \mathbf{e}_k = \begin{cases} \mathbf{e}_i \ (j = k) \\ 0 \ (j \neq k) \end{cases} \quad (i, j, k = 1, 2, 3) \tag{1-45}$$

1.3.3 Basis Representations of Tensors

Suppose now that we let $\mathbf{a} \in \mathbb{E}^3$ and $\mathbf{b} \in \mathbb{E}^3$. Then $\mathbf{a}$ and $\mathbf{b}$ can be expressed in the orthonormal basis $\{\mathbf{e}_1, \mathbf{e}_2, \mathbf{e}_3\}$ as

$$\begin{aligned} \mathbf{a} &= a_1\mathbf{e}_1 + a_2\mathbf{e}_2 + a_3\mathbf{e}_3 \\ \mathbf{b} &= b_1\mathbf{e}_1 + b_2\mathbf{e}_2 + b_3\mathbf{e}_3 \end{aligned} \tag{1-46}$$

The tensor product between $\mathbf{a}$ and $\mathbf{b}$ is then given as

$$\mathbf{a} \otimes \mathbf{b} = (a_1\mathbf{e}_1 + a_2\mathbf{e}_2 + a_3\mathbf{e}_3) \otimes (b_1\mathbf{e}_1 + b_2\mathbf{e}_2 + b_3\mathbf{e}_3) \tag{1-47}$$

Expanding Eq. (1–47) term-by-term, we have

$$\mathbf{a} \otimes \mathbf{b} = \sum_{i=1}^{3}\sum_{j=1}^{3} a_i b_j \mathbf{e}_i \otimes \mathbf{e}_j \tag{1-48}$$

Now, because $\mathbf{a}\otimes\mathbf{b}$ is a tensor, it is seen from Eq. (1–48) that a tensor $\mathbf{T}$ can be expressed in an orthonormal basis $\{\mathbf{e}_1, \mathbf{e}_2, \mathbf{e}_3\}$ as

$$\mathbf{T} = \sum_{i=1}^{3}\sum_{j=1}^{3} T_{ij} \mathbf{e}_i \otimes \mathbf{e}_j \tag{1-49}$$

Eq. (1–49) is called the *representation* of the tensor $\mathbf{T}$ *in the basis* $\{\mathbf{e}_1, \mathbf{e}_2, \mathbf{e}_3\}$. Using Eq. (1–49), the transpose of $\mathbf{T}$ is obtained from $\mathbf{T}$ as

$$\mathbf{T}^T = \sum_{i=1}^{3}\sum_{j=1}^{3} T_{ij} \mathbf{e}_j \otimes \mathbf{e}_i \tag{1-50}$$

In particular, it is seen that the representation of $\mathbf{T}^T$ is obtained from the representation of $\mathbf{T}$ by interchanging the order of the tensor products $\mathbf{e}_i \otimes \mathbf{e}_j$ $(i, j = 1, 2, 3)$. Now, in terms of the basis $\{\mathbf{e}_1, \mathbf{e}_2, \mathbf{e}_3\}$, the identity tensor $\mathbf{U}$ is given as

$$\mathbf{U} = \sum_{i=1}^{3}\sum_{j=1}^{3} U_{ij} \mathbf{e}_i \otimes \mathbf{e}_j = \sum_{i=1}^{3} \mathbf{e}_i \otimes \mathbf{e}_i \tag{1-51}$$

In other words, for the identity tensor, we have

$$U_{ij} = \begin{cases} 1 & \text{if} \quad i = j \\ 0 & \text{if} \quad i \neq j \end{cases} \tag{1-52}$$

Example 1–1

Show that the vector triple product $\mathbf{a} \times (\mathbf{b} \times \mathbf{c})$ can be written in the tensor form

$$\mathbf{a} \times (\mathbf{b} \times \mathbf{c}) = (\mathbf{b} \otimes \mathbf{a}) \cdot \mathbf{c} - (\mathbf{c} \otimes \mathbf{a}) \cdot \mathbf{b}$$

Solution to Example 1–1

Recall from Eq. (1–29) that the vector triple product is given as

$$\mathbf{a} \times (\mathbf{b} \times \mathbf{c}) = (\mathbf{a} \cdot \mathbf{c})\mathbf{b} - (\mathbf{a} \cdot \mathbf{b})\mathbf{c} \tag{1-53}$$

Then, using the property of the tensor product as given in Eq. (1–41), we have

$$(\mathbf{a} \cdot \mathbf{c})\mathbf{b} = (\mathbf{b} \otimes \mathbf{a}) \cdot \mathbf{c} \tag{1-54}$$

$$(\mathbf{a} \cdot \mathbf{b})\mathbf{c} = (\mathbf{c} \otimes \mathbf{a}) \cdot \mathbf{b} \tag{1-55}$$

Finally, substituting the results from Eqs. (1–54) and (1–55) into Eq. (1–53), we obtain

$$\mathbf{a} \times (\mathbf{b} \times \mathbf{c}) = (\mathbf{b} \otimes \mathbf{a}) \cdot \mathbf{c} - (\mathbf{c} \otimes \mathbf{a}) \cdot \mathbf{b} \tag{1-56}$$

■

Example 1–2

Given two vectors **a** and **b** expressed in the orthonormal basis $\{\mathbf{e}_1, \mathbf{e}_2, \mathbf{e}_3\}$, show that the vector product $\mathbf{a} \times \mathbf{b}$ can be written as $\mathbf{T} \cdot \mathbf{b}$, where **T** is the tensor

$$\begin{aligned}\mathbf{T} = &-a_3\mathbf{e}_1 \otimes \mathbf{e}_2 + a_2\mathbf{e}_1 \otimes \mathbf{e}_3 \\ &+ a_3\mathbf{e}_2 \otimes \mathbf{e}_1 - a_1\mathbf{e}_2 \otimes \mathbf{e}_3 \\ &- a_2\mathbf{e}_3 \otimes \mathbf{e}_1 + a_1\mathbf{e}_3 \otimes \mathbf{e}_2\end{aligned}$$

Solution to Example 1–2

In terms of the basis, the product of the tensor **T** with the vector **b** is given as

$$\begin{aligned}\mathbf{T} \cdot \mathbf{b} = (&-a_3\mathbf{e}_1 \otimes \mathbf{e}_2 + a_2\mathbf{e}_1 \otimes \mathbf{e}_3 + a_3\mathbf{e}_2 \otimes \mathbf{e}_1 - a_1\mathbf{e}_2 \otimes \mathbf{e}_3 \\ &-a_2\mathbf{e}_3 \otimes \mathbf{e}_1 + a_1\mathbf{e}_3 \otimes \mathbf{e}_2) \cdot (b_1\mathbf{e}_1 + b_2\mathbf{e}_2 + b_3\mathbf{e}_3)\end{aligned} \tag{1-57}$$

Then, expanding Eq. (1–57) using Eq. (1–45) on each term, we have

$$\mathbf{T} \cdot \mathbf{b} = a_3b_1\mathbf{e}_2 - a_2b_1\mathbf{e}_3 - a_3b_2\mathbf{e}_1 + a_1b_2\mathbf{e}_3 + a_2b_3\mathbf{e}_1 - a_1b_3\mathbf{e}_2 \tag{1-58}$$

Grouping terms in Eq. (1–58), we obtain

$$\mathbf{T} \cdot \mathbf{b} = (a_2b_3 - a_3b_2)\mathbf{e}_1 + (a_3b_1 - a_1b_3)\mathbf{e}_2 + (a_1b_2 - a_2b_1)\mathbf{e}_3 \tag{1-59}$$

It can be seen that Eq. (1–59) is identical to Eq. (1–21), i.e., we have

$$\mathbf{T} \cdot \mathbf{b} \equiv \mathbf{a} \times \mathbf{b} \tag{1-60}$$

which implies that the tensor **T** is equivalent to the operator "$\mathbf{a}\times$," i.e.,

$$\mathbf{T} \equiv \mathbf{a}\times \tag{1-61}$$

■

1.4 Matrices

1.4.1 Systems of Linear Equations

Consider a system of equations of the form

$$\begin{aligned} a_{11}x_1 + a_{12}x_2 + \cdots + a_{1n}x_n &= b_1 \\ a_{21}x_1 + a_{22}x_2 + \cdots + a_{2n}x_n &= b_2 \\ &\vdots \\ a_{n1}x_1 + a_{n2}x_2 + \cdots + a_{nn}x_n &= b_n \end{aligned} \tag{1-62}$$

where a_{ij} $(i,j = 1,\ldots,n)$ and b_i $(i = 1,\ldots,n)$ are *known* and x_i $(i = 1,\ldots,n)$ are *unknown*. Equation (1-62) is called a system of n *linear* equations in the n unknowns x_i $(i = 1,\ldots,n)$. This system can be written in *matrix form* as

$$\left\{\begin{array}{ccc} a_{11} & \cdots & a_{1n} \\ a_{21} & \cdots & a_{2n} \\ & \vdots & \\ a_{n1} & \cdots & a_{nn} \end{array}\right\} \left\{\begin{array}{c} x_1 \\ x_2 \\ \vdots \\ x_n \end{array}\right\} = \left\{\begin{array}{c} b_1 \\ b_2 \\ \vdots \\ b_n \end{array}\right\} \tag{1-63}$$

Suppose now that we make the following substitutions:

$$\mathbf{A} = \left\{\begin{array}{ccc} a_{11} & \cdots & a_{1n} \\ a_{21} & \cdots & a_{2n} \\ & \vdots & \\ a_{n1} & \cdots & a_{nn} \end{array}\right\} \tag{1-64}$$

$$\mathbf{x} = \left\{\begin{array}{c} x_1 \\ x_2 \\ \vdots \\ x_n \end{array}\right\} \tag{1-65}$$

$$\mathbf{b} = \left\{\begin{array}{c} b_1 \\ b_2 \\ \vdots \\ b_n \end{array}\right\} \tag{1-66}$$

Then Eq. (1-63) can be written compactly as

$$\mathbf{A}\mathbf{x} = \mathbf{b} \tag{1-67}$$

The quantity $\mathbf{A} \in \mathbb{R}^{n\times n}$ is called a *matrix* while the quantities $\mathbf{x} \in \mathbb{R}^n$ and $\mathbf{b} \in \mathbb{R}^n$ are called *column-vectors*.

1.4.2 Classes of Matrices

Suppose we are given a matrix $\mathbf{A} \in \mathbb{R}^{n\times n}$ defined as

$$\mathbf{A} = \left\{\begin{array}{ccc} a_{11} & \cdots & a_{1n} \\ a_{21} & \cdots & a_{2n} \\ & \vdots & \\ a_{n1} & \cdots & a_{nn} \end{array}\right\} \tag{1-68}$$

Then $\mathbf{A}$ is said to be *invertible* if there exists a matrix $\mathbf{A}^{-1}$ such that

$$\mathbf{A}\mathbf{A}^{-1} = \mathbf{A}^{-1}\mathbf{A} = \mathbf{I} \tag{1-69}$$

where

$$\mathbf{I} = \begin{Bmatrix} 1 & 0 & 0 & \cdots & 0 \\ 0 & 1 & 0 & \cdots & 0 \\ \vdots & \vdots & \vdots & \vdots & \vdots \\ 0 & 0 & 0 & \cdots & 1 \end{Bmatrix} \tag{1-70}$$

is the $n \times n$ *identity matrix.* If $\mathbf{A}$ is invertible, the $\mathbf{A}^{-1}$ is called the *inverse* of $\mathbf{A}$. The *transpose* of a matrix $\mathbf{A}$, denoted $\mathbf{A}^T$, is defined as

$$\mathbf{A}^T = \begin{Bmatrix} a_{11} & \cdots & a_{n1} \\ a_{12} & \cdots & a_{n2} \\ & \vdots & \\ a_{1n} & \cdots & a_{nn} \end{Bmatrix} \tag{1-71}$$

It can be seen that the matrix transpose is obtained by interchanging the rows and the columns in the matrix $\mathbf{A}$. Using the definition of the transpose of a matrix, a *row-vector* is defined as the transpose of a column-vector, i.e., if $\mathbf{x}$ is the column-vector

$$\mathbf{x} = \begin{Bmatrix} x_1 \\ x_2 \\ \vdots \\ x_n \end{Bmatrix} \tag{1-72}$$

then

$$\mathbf{x}^T = \begin{Bmatrix} x_1 & x_2 & \cdots & x_n \end{Bmatrix} \tag{1-73}$$

is a row-vector. A matrix is said to be *symmetric* if

$$\mathbf{A} = \mathbf{A}^T \tag{1-74}$$

A matrix $\mathbf{A}$ is said to be *skew-symmetric* if

$$\mathbf{A} = -\mathbf{A}^T \tag{1-75}$$

Finally, an invertible matrix is said to be *orthogonal* if

$$\mathbf{A}^T = \mathbf{A}^{-1} \tag{1-76}$$

1.4.3 Relationship Between Tensors and Matrices

Consider the equation

$$\mathbf{T} \cdot \mathbf{x} = \mathbf{b} \tag{1-77}$$

where $\mathbf{T}$ is a tensor and $\mathbf{x} \in \mathbb{E}^3$ and $\mathbf{b} \in \mathbb{E}^3$ are vectors. Suppose now that we choose to express each of the quantities in Eq. (1-77) in terms of an arbitrary orthonormal basis

$\{\mathbf{e}_1, \mathbf{e}_2, \mathbf{e}_3\}$ as

$$\begin{aligned} \mathbf{T} &= \sum_{i=1}^{3}\sum_{j=1}^{3} T_{ij}\mathbf{e}_i \otimes \mathbf{e}_j \\ \mathbf{x} &= \textstyle\sum_{i=1}^{3} x_i\mathbf{e}_i \\ \mathbf{b} &= \textstyle\sum_{i=1}^{3} b_i\mathbf{e}_i \end{aligned} \tag{1-78}$$

Substituting the expressions from Eq. (1-78) into Eq. (1-77), we have

$$\left[\sum_{i=1}^{3}\sum_{j=1}^{3} T_{ij}\mathbf{e}_i \otimes \mathbf{e}_j\right] \cdot \left[\sum_{i=1}^{3} x_i\mathbf{e}_i\right] = \sum_{i=1}^{3} b_i\mathbf{e}_i \tag{1-79}$$

Then, using the property of Eq. (1-45) in Eq. (1-79), we obtain

$$\sum_{i=1}^{3}\sum_{j=1}^{3} T_{ij}x_j\mathbf{e}_i = \sum_{i=1}^{3} b_i\mathbf{e}_i \tag{1-80}$$

Equation (1-80) can be written in matrix form as

$$\left\{\begin{array}{ccc} T_{11} & T_{12} & T_{13} \\ T_{21} & T_{22} & T_{23} \\ T_{31} & T_{32} & T_{33} \end{array}\right\} \left\{\begin{array}{c} x_1 \\ x_2 \\ x_3 \end{array}\right\} = \left\{\begin{array}{c} b_1 \\ b_2 \\ b_3 \end{array}\right\} \tag{1-81}$$

More compactly, Eq. (1-81) can be written as

$$\{\mathbf{T}\}_{\mathrm{E}}\{\mathbf{x}\}_{\mathrm{E}} = \{\mathbf{b}\}_{\mathrm{E}} \tag{1-82}$$

where

$$\{\mathbf{T}\}_{\mathrm{E}} = \left\{\begin{array}{ccc} T_{11} & T_{12} & T_{13} \\ T_{21} & T_{22} & T_{23} \\ T_{31} & T_{32} & T_{33} \end{array}\right\} \tag{1-83}$$

is called the *matrix representation* of the tensor in the basis $\mathbf{E} = \{\mathbf{e}_1, \mathbf{e}_2, \mathbf{e}_3\}$ while the quantities

$$\{\mathbf{x}\}_{\mathrm{E}} = \left\{\begin{array}{c} x_1 \\ x_2 \\ x_3 \end{array}\right\} \tag{1-84}$$

and

$$\{\mathbf{b}\}_{\mathrm{E}} = \left\{\begin{array}{c} b_1 \\ b_2 \\ b_3 \end{array}\right\} \tag{1-85}$$

are called the *column-vector representations* of the vectors $\mathbf{x}$ and $\mathbf{b}$ in the basis $\mathbf{E}$.

The previous discussion shows the relationship between tensors and matrices. In particular, Eq. (1-81) shows that a column-vector is *not* a vector. Furthermore, Eq. (1-81) shows that a matrix is *not* a tensor. Instead, a column-vector and a matrix are *representations* of a vector and a tensor, respectively, *in a particular basis.* Consequently, both a column-vector and a matrix are specific to a particular basis, whereas vectors and tensors are *independent* of any particular basis.

Another way of viewing the relationships between vector and column-vectors and tensors and matrices is as follows. Suppose that a basis $\mathbf{U} = \{\mathbf{u}_1, \mathbf{u}_2, \mathbf{u}_3\}$ is chosen that is *different* from the basis $\{\mathbf{e}_1, \mathbf{e}_2, \mathbf{e}_3\}$. Then the column-vector representations of the vectors $\mathbf{x}$ and $\mathbf{b}$ and the matrix representation of the tensor $\mathbf{T}$ in the basis $\{\mathbf{u}_1, \mathbf{u}_2, \mathbf{u}_3\}$ will be different from their respective representations in the basis $\{\mathbf{e}_1, \mathbf{e}_2, \mathbf{e}_3\}$. Mathematically, if $\mathbf{E} = \{\mathbf{e}_1, \mathbf{e}_2, \mathbf{e}_3\}$ and $\mathbf{U} = \{\mathbf{u}_1, \mathbf{u}_2, \mathbf{u}_3\}$ are *distinct* bases, then

$$\begin{aligned} \{\mathbf{T}\}_{\mathrm{E}} &\neq \{\mathbf{T}\}_{\mathrm{U}} \\ \{\mathbf{x}\}_{\mathrm{E}} &\neq \{\mathbf{x}\}_{\mathrm{U}} \\ \{\mathbf{b}\}_{\mathrm{E}} &\neq \{\mathbf{b}\}_{\mathrm{U}} \end{aligned} \tag{1-86}$$

Now, while the matrix representation of a tensor or the column-vector representation of a vector is different in different bases, the underlying tensor or vector itself is the same *regardless* of the basis. In other words, a tensor and a vector are *coordinate-free* quantities while the matrix representation of a tensor or the column-vector representation of a vector depends on the particular coordinate system (i.e., basis) in which the tensor or vector is expressed.

Example 1–3

Given two vectors $\mathbf{a}$ and $\mathbf{b}$ and an orthonormal basis $\mathbf{E}$, show that the scalar product of $\mathbf{a}$ with $\mathbf{b}$ is equal to the product of the transpose of the column-vector representation of $\mathbf{a}$ with the column-vector representation of $\mathbf{b}$, i.e., show that

$$\mathbf{a} \cdot \mathbf{b} = \{\mathbf{a}\}_{\mathrm{E}}^T \{\mathbf{b}\}_{\mathrm{E}}$$

Solution to Example 1–3

The vectors $\mathbf{a}$ and $\mathbf{b}$ can be decomposed in the basis $\mathbf{E} = \{\mathbf{e}_1, \mathbf{e}_2, \mathbf{e}_3\}$ as

$$\mathbf{a} = a_1\mathbf{e}_1 + a_2\mathbf{e}_2 + a_3\mathbf{e}_3 \tag{1-87}$$

$$\mathbf{b} = b_1\mathbf{e}_1 + b_2\mathbf{e}_2 + b_3\mathbf{e}_3 \tag{1-88}$$

Then the scalar of $\mathbf{a}$ with $\mathbf{b}$ is given as

$$\mathbf{a} \cdot \mathbf{b} = a_1b_1 + a_2b_2 + a_3b_3 \tag{1-89}$$

Furthermore, the column-vector representations of $\mathbf{a}$ and $\mathbf{b}$ in the basis $\mathbf{E}$ are given, respectively, as

$$\{\mathbf{a}\}_{\mathrm{E}} = \left\{ \begin{array}{c} a_1 \\ a_2 \\ a_3 \end{array} \right\} \tag{1-90}$$

$$\{\mathbf{b}\}_{\mathrm{E}} = \left\{ \begin{array}{c} b_1 \\ b_2 \\ b_3 \end{array} \right\} \tag{1-91}$$

Consequently, we have

$$a_1b_1 + a_2b_2 + a_3b_3 = \left\{ \begin{array}{ccc} a_1 & a_2 & a_3 \end{array} \right\} \left\{ \begin{array}{c} b_1 \\ b_2 \\ b_3 \end{array} \right\} \tag{1-92}$$

Furthermore, the transpose of $\{\mathbf{a}\}_{\mathrm{E}}$ is given as

$$\{\mathbf{a}\}_{\mathrm{E}}^T = \left\{ \begin{array}{ccc} a_1 & a_2 & a_3 \end{array} \right\} \tag{1-93}$$

Therefore, the scalar product of **a** with **b** in the basis **E** is given as

$$\mathbf{a} \cdot \mathbf{b} = \{\mathbf{a}\}_{\mathrm{E}}^T \{\mathbf{b}\}_{\mathrm{E}} \tag{1-94}$$

■

Example 1–4

Given two vectors **a** and **b** expressed in the orthonormal basis $\mathbf{E} = \{\mathbf{e}_1, \mathbf{e}_2, \mathbf{e}_3\}$, determine the matrix representation of the tensor product $\mathbf{a}\otimes\mathbf{b}$ in the basis E, i.e., determine

$$\{\mathbf{a} \otimes \mathbf{b}\}_{\mathrm{E}}$$

Solution to Example 1–4

The vectors **a** and **b** can be decomposed in the basis $\mathbf{E} = \{\mathbf{e}_1, \mathbf{e}_2, \mathbf{e}_3\}$ as

$$\mathbf{a} = a_1\mathbf{e}_1 + a_2\mathbf{e}_2 + a_3\mathbf{e}_3 \tag{1-95}$$

$$\mathbf{b} = b_1\mathbf{e}_1 + b_2\mathbf{e}_2 + b_3\mathbf{e}_3 \tag{1-96}$$

Then, the tensor product between **a** and **b** is given as

$$\mathbf{a} \otimes \mathbf{b} = (a_1\mathbf{e}_1 + a_2\mathbf{e}_2 + a_3\mathbf{e}_3) \otimes (b_1\mathbf{e}_1 + b_2\mathbf{e}_2 + b_3\mathbf{e}_3) = \sum_{i=1}^{3}\sum_{j=1}^{3} a_i b_j \mathbf{e}_i \otimes \mathbf{e}_j \tag{1-97}$$

Finally, extracting the coefficients of the tensor $\mathbf{a}\otimes\mathbf{b}$ in Eq. (1-97), the matrix representation of $\mathbf{a}\otimes\mathbf{b}$ in the basis **E** is given as

$$\{\mathbf{a} \otimes \mathbf{b}\}_{\mathrm{E}} = \left\{ \begin{array}{ccc} a_1 b_1 & a_1 b_2 & a_1 b_3 \\ a_2 b_1 & a_2 b_2 & a_2 b_3 \\ a_3 b_1 & a_3 b_2 & a_3 b_3 \end{array} \right\} \tag{1-98}$$

■

Example 1–5

Determine the matrix representation of the tensor **T** from Example 1-2.

Solution to Example 1–5

Recall from Example 1-2 that the tensor **T** is given as

$$\begin{aligned}\mathbf{T} = &-a_3\mathbf{e}_1 \otimes \mathbf{e}_2 + a_2\mathbf{e}_1 \otimes \mathbf{e}_3 \\ &+ a_3\mathbf{e}_2 \otimes \mathbf{e}_1 - a_1\mathbf{e}_2 \otimes \mathbf{e}_3 \\ &- a_2\mathbf{e}_3 \otimes \mathbf{e}_1 + a_1\mathbf{e}_3 \otimes \mathbf{e}_2\end{aligned} \tag{1-99}$$

Then, using the result of Eq. (1–83), the matrix representation of the tensor **T** as given in Eq. (1–99) in the basis $\mathbf{E} = \{\mathbf{e}_1, \mathbf{e}_2, \mathbf{e}_3\}$ is

$$\{\mathbf{T}\}_{\mathbf{E}} = \begin{Bmatrix} 0 & -a_3 & a_2 \\ a_3 & 0 & -a_1 \\ -a_2 & a_1 & 0 \end{Bmatrix} \tag{1-100}$$

It is seen that $\{\mathbf{T}\}_{\mathbf{E}}^T = -\{\mathbf{T}\}_{\mathbf{E}}$, i.e., $\{\mathbf{T}\}_{\mathbf{E}}$ is a skew-symmetric matrix.

■

1.4.4 Transformation of Column-Vectors and Row-Vectors

Suppose now that we choose to express a vector $\mathbf{a} \in \mathbb{E}^3$ in terms of two orthonormal bases $\mathbf{E} = \{\mathbf{e}_1, \mathbf{e}_2, \mathbf{e}_3\}$ and $\mathbf{E}' = \{\mathbf{e}_1', \mathbf{e}_2', \mathbf{e}_3'\}$. Then **a** can be written as

$$\mathbf{a} = a_1\mathbf{e}_1 + a_2\mathbf{e}_2 + a_3\mathbf{e}_3 \tag{1-101}$$

$$\mathbf{a} = a_1'\mathbf{e}_1' + a_2'\mathbf{e}_2' + a_3'\mathbf{e}_3' \tag{1-102}$$

Next, we can express the basis $\{\mathbf{e}_1, \mathbf{e}_2, \mathbf{e}_3\}$ in terms of the basis $\{\mathbf{e}_1', \mathbf{e}_2', \mathbf{e}_3'\}$ as

$$\begin{aligned}\mathbf{e}_1 &= c_{11}\mathbf{e}_1' + c_{21}\mathbf{e}_2' + c_{31}\mathbf{e}_3' \\ \mathbf{e}_2 &= c_{12}\mathbf{e}_1' + c_{22}\mathbf{e}_2' + c_{32}\mathbf{e}_3' \\ \mathbf{e}_3 &= c_{13}\mathbf{e}_1' + c_{23}\mathbf{e}_2' + c_{33}\mathbf{e}_3'\end{aligned} \tag{1-103}$$

where

$$c_{ij} = \mathbf{e}_i' \cdot \mathbf{e}_j \qquad (i, j = 1, 2, 3) \tag{1-104}$$

Using the property of the scalar product as given in Eq. (1–6), c_{ij} can be written as

$$c_{ij} = \|\mathbf{e}_i'\| \|\mathbf{e}_j\| \cos\gamma_{ij} \qquad (i, j = 1, 2, 3) \tag{1-105}$$

where γ_{ij} is the angle between $\mathbf{e}_i'$ and $\mathbf{e}_j$. Now, because $\mathbf{e}_i'$ $(i = 1, 2, 3)$ and $\mathbf{e}_j$ $(j = 1, 2, 3)$ are each unit vectors, Eq. (1–105) reduces to

$$c_{ij} = \cos\gamma_{ij} \qquad (i, j = 1, 2, 3) \tag{1-106}$$

Because c_{ij} $(i, j = 1, 2, 3)$ can be written exclusively as the cosine of the angle between $\mathbf{e}_i$ and $\mathbf{e}_j'$, the coefficients c_{ij} $(i, j = 1, 2, 3)$ are called the *direction cosines* between the basis vectors in $\mathbf{E}$ and the basis vectors in $\mathbf{E}'$. Substituting the results of Eq. (1–103) into Eq. (1–101), we obtain

$$\begin{aligned}\mathbf{a} = &\, a_1(c_{11}\mathbf{e}'_1 + c_{21}\mathbf{e}'_2 + c_{31}\mathbf{e}'_3)\\ &+ a_2(c_{12}\mathbf{e}'_1 + c_{22}\mathbf{e}'_2 + c_{32}\mathbf{e}'_3)\\ &+ a_3(c_{13}\mathbf{e}'_1 + c_{23}\mathbf{e}'_2 + c_{33}\mathbf{e}'_3)\end{aligned} \tag{1–107}$$

Rearranging Eq. (1–107), we obtain

$$\begin{aligned}\mathbf{a} = &\, (c_{11}a_1 + c_{12}a_2 + c_{13}a_3)\mathbf{e}'_1\\ &+ (c_{21}a_1 + c_{22}a_2 + c_{23}a_3)\mathbf{e}'_2\\ &+ (c_{31}a_1 + c_{32}a_2 + c_{33}a_3)\mathbf{e}'_3\end{aligned} \tag{1–108}$$

Then, setting the expression for $\mathbf{a}$ from Eq. (1–108) equal to the expression for $\mathbf{a}$ from Eq. (1–102), we obtain

$$\begin{aligned}a'_1 &= c_{11}a_1 + c_{12}a_2 + c_{13}a_3\\ a'_2 &= c_{21}a_1 + c_{22}a_2 + c_{23}a_3\\ a'_3 &= c_{31}a_1 + c_{32}a_2 + c_{33}a_3\end{aligned} \tag{1–109}$$

Using matrix notation, we can write Eq. (1–109) as

$$\left\{\begin{array}{c} a'_1\\ a'_2\\ a'_3\end{array}\right\} = \left\{\begin{array}{ccc} c_{11} & c_{12} & c_{13}\\ c_{21} & c_{22} & c_{23}\\ c_{31} & c_{32} & c_{33}\end{array}\right\}\left\{\begin{array}{c} a_1\\ a_2\\ a_3\end{array}\right\} \tag{1–110}$$

Now we see that the coefficient matrix transforms column-vectors from the basis $\mathbf{E}$ to the basis $\mathbf{E}'$. In order to specify the direction of the transformation unambiguously, we use the following notation:

$$\{\mathbf{C}\}_{\mathbf{E}}^{\mathbf{E}'} = \left\{\begin{array}{ccc} c_{11} & c_{12} & c_{13}\\ c_{21} & c_{22} & c_{23}\\ c_{31} & c_{32} & c_{33}\end{array}\right\} \tag{1–111}$$

where the right subscript denotes the basis *from* which the transformation is performed while the right superscript denotes the basis *to* which the transformation is performed. In other words, the quantity $\{\mathbf{C}\}_{\mathbf{E}}^{\mathbf{E}'}$ is the matrix that transforms column-vectors from the basis $\mathbf{E}$ to the basis $\mathbf{E}'$. The matrix $\{\mathbf{C}\}_{\mathbf{E}}^{\mathbf{E}'}$ is called the *direction cosine matrix* from the basis $\mathbf{E}$ to the basis $\mathbf{E}'$. In terms of the direction cosine matrix, we can write Eq. (1–110) compactly as

$$\{\mathbf{a}\}_{\mathbf{E}'} = \{\mathbf{C}\}_{\mathbf{E}}^{\mathbf{E}'}\,\{\mathbf{a}\}_{\mathbf{E}} \tag{1–112}$$

where $\{\mathbf{a}\}_{\mathbf{E}}$ and $\{\mathbf{a}\}_{\mathbf{E}'}$ are the column-vector representations of the vector $\mathbf{a}$ in the bases $\mathbf{E}$ and $\mathbf{E}'$, respectively. Equation (1–112) provides a way to transform the components of a vector from the basis $\mathbf{E} = \{\mathbf{e}_1, \mathbf{e}_2, \mathbf{e}_3\}$ to the basis $\mathbf{E}' = \{\mathbf{e}'_1, \mathbf{e}'_2, \mathbf{e}'_3\}$. Equivalently, the direction cosine tensor $\mathbf{C}$ is given as

$$\mathbf{C} = \sum_{i=1}^{3}\sum_{j=1}^{3} c_{ij}\mathbf{e}'_i \otimes \mathbf{e}_j \tag{1–113}$$

It should be noted that the tensor $\mathbf{C}$ of Eq. (1–113) is expressed as the sum of the tensor products formed by basis vectors in *both* the basis $\mathbf{E}$ and the basis $\mathbf{E}'$.

Next, examining Eq. (1–112), we see that the transformation of column-vectors from the basis $\mathbf{E}'$ to the basis $\mathbf{E}$ is given as

$$\{\mathbf{a}\}_{\mathbf{E}} = \{\mathbf{C}\}_{\mathbf{E}'}^{\mathbf{E}} \{\mathbf{a}\}_{\mathbf{E}'} \tag{1–114}$$

where $\{\mathbf{C}\}_{\mathbf{E}'}^{\mathbf{E}}$ is the direction cosine matrix from the basis $\mathbf{E}'$ to the basis $\mathbf{E}$. Equation (1–114) provides a way to transform the components of a vector $\mathbf{a}$ from the basis $\mathbf{E}'$ to the basis $\mathbf{E}$. Finally, because the transformation from $\mathbf{E}$ to $\mathbf{E}'$ is the inverse of the transformation from $\mathbf{E}'$ to $\mathbf{E}$, we know that $\{\mathbf{C}\}_{\mathbf{E}'}^{\mathbf{E}}$ must be the inverse of $\{\mathbf{C}\}_{\mathbf{E}}^{\mathbf{E}'}$, i.e.,

$$\{\mathbf{C}\}_{\mathbf{E}'}^{\mathbf{E}} = \left[\{\mathbf{C}\}_{\mathbf{E}}^{\mathbf{E}'}\right]^{-1} \tag{1–115}$$

Furthermore, using an approach similar to that used to determine $\{\mathbf{C}\}_{\mathbf{E}}^{\mathbf{E}'}$, we know that

$$\left\{\begin{array}{c} a_1 \\ a_2 \\ a_3 \end{array}\right\} = \left\{\begin{array}{ccc} c_{11} & c_{21} & c_{31} \\ c_{12} & c_{22} & c_{32} \\ c_{13} & c_{23} & c_{33} \end{array}\right\} \left\{\begin{array}{c} a_1' \\ a_2' \\ a_3' \end{array}\right\} \tag{1–116}$$

From Eq. (1–116) it can be seen that the direction cosine matrix that transforms column-vectors from the basis $\mathbf{E}'$ to the basis $\mathbf{E}$ is the *transpose* of the direction cosine matrix that transforms column-vectors from the basis $\mathbf{E}$ to the basis $\mathbf{E}'$, i.e.,

$$\{\mathbf{C}\}_{\mathbf{E}'}^{\mathbf{E}} = \left[\{\mathbf{C}\}_{\mathbf{E}}^{\mathbf{E}'}\right]^{T} \tag{1–117}$$

Combining the result of Eq. (1–117) with the result of Eq. (1–115), we have

$$\left[\{\mathbf{C}\}_{\mathbf{E}}^{\mathbf{E}'}\right]^{-1} = \left[\{\mathbf{C}\}_{\mathbf{E}}^{\mathbf{E}'}\right]^{T} \tag{1–118}$$

Observing that Eq. (1–118) satisfies the condition of Eq. (1–76), we see that $\{\mathbf{C}\}_{\mathbf{E}}^{\mathbf{E}'}$ is an *orthogonal matrix*. Consequently, the components of a vector $\mathbf{a}$ in the basis $\mathbf{E}'$ are transformed to the basis $\mathbf{E}$ as

$$\{\mathbf{a}\}_{\mathbf{E}} = \left[\{\mathbf{C}\}_{\mathbf{E}}^{\mathbf{E}'}\right]^{T} \{\mathbf{a}\}_{\mathbf{E}'} = \{\mathbf{C}\}_{\mathbf{E}'}^{\mathbf{E}} \{\mathbf{a}\}_{\mathbf{E}'} \tag{1–119}$$

1.4.5 Transformation of Matrices

Now suppose that we are given two bases $\mathbf{E} = \{\mathbf{e}_1, \mathbf{e}_2, \mathbf{e}_3\}$ and $\mathbf{E}' = \{\mathbf{e}_1', \mathbf{e}_2', \mathbf{e}_3'\}$. Furthermore, suppose we are given the system of linear equations expressed in matrix form in terms of the basis $\mathbf{E}$ as

$$\{\mathbf{T}\}_{\mathbf{E}} \{\mathbf{x}\}_{\mathbf{E}} = \{\mathbf{b}\}_{\mathbf{E}} \tag{1–120}$$

Then we know from Eq. (1–112) that the column-vector representations of the vectors $\mathbf{x}$ and $\mathbf{b}$ in the basis $\mathbf{E}'$ are given as

$$\{\mathbf{x}\}_{\mathbf{E}'} = \{\mathbf{C}\}_{\mathbf{E}}^{\mathbf{E}'} \{\mathbf{x}\}_{\mathbf{E}} \tag{1–121}$$

$$\{\mathbf{b}\}_{\mathbf{E}'} = \{\mathbf{C}\}_{\mathbf{E}}^{\mathbf{E}'} \{\mathbf{b}\}_{\mathbf{E}} \tag{1–122}$$

where $\{\mathbf{C}\}_{\mathbf{E}}^{\mathbf{E}'}$ is the direction cosine matrix from the basis $\mathbf{E}$ to the basis $\mathbf{E}'$. Then, recalling that $\{\mathbf{C}\}_{\mathbf{E}}^{\mathbf{E}'}$ is an orthogonal matrix, we have from Eqs. (1–121) and (1–122) that

$$\{\mathbf{x}\}_{\mathrm{E}} = \left[\{\mathbf{C}\}_{\mathrm{E}}^{\mathrm{E}'}\right]^T \{\mathbf{x}\}_{\mathrm{E}'} \tag{1-123}$$

$$\{\mathbf{b}\}_{\mathrm{E}} = \left[\{\mathbf{C}\}_{\mathrm{E}}^{\mathrm{E}'}\right]^T \{\mathbf{b}\}_{\mathrm{E}'} \tag{1-124}$$

Substituting the results of Eqs. (1–123) and (1–124) into Eq. (1–120), we obtain

$$\{\mathbf{T}\}_{\mathrm{E}}\left[\{\mathbf{C}\}_{\mathrm{E}}^{\mathrm{E}'}\right]^T \{\mathbf{x}\}_{\mathrm{E}'} = \left[\{\mathbf{C}\}_{\mathrm{E}}^{\mathrm{E}'}\right]^T \{\mathbf{b}\}_{\mathrm{E}'} \tag{1-125}$$

Multiplying both sides of Eq. (1–125) by $\{\mathbf{C}\}_{\mathrm{E}}^{\mathrm{E}'}$ gives

$$\{\mathbf{C}\}_{\mathrm{E}}^{\mathrm{E}'}\{\mathbf{T}\}_{\mathrm{E}}\left[\{\mathbf{C}\}_{\mathrm{E}}^{\mathrm{E}'}\right]^T \{\mathbf{x}\}_{\mathrm{E}'} = \{\mathbf{C}\}_{\mathrm{E}}^{\mathrm{E}'}\left[\{\mathbf{C}\}_{\mathrm{E}}^{\mathrm{E}'}\right]^T \{\mathbf{b}\}_{\mathrm{E}'} \tag{1-126}$$

Then, because $\{\mathbf{C}\}_{\mathrm{E}}^{\mathrm{E}'}$ is an orthogonal matrix, we have

$$\{\mathbf{C}\}_{\mathrm{E}}^{\mathrm{E}'}\left[\{\mathbf{C}\}_{\mathrm{E}}^{\mathrm{E}'}\right]^T = \mathbf{I} \tag{1-127}$$

where **I** is the identity matrix. Consequently, Eq. (1–126) simplifies to

$$\{\mathbf{C}\}_{\mathrm{E}}^{\mathrm{E}'}\{\mathbf{T}\}_{\mathrm{E}}\left[\{\mathbf{C}\}_{\mathrm{E}}^{\mathrm{E}'}\right]^T \{\mathbf{x}\}_{\mathrm{E}'} = \{\mathbf{b}\}_{\mathrm{E}'} \tag{1-128}$$

Then, applying the result of Eq. (1–117), we have

$$\{\mathbf{C}\}_{\mathrm{E}}^{\mathrm{E}'}\{\mathbf{T}\}_{\mathrm{E}}\{\mathbf{C}\}_{\mathrm{E}'}^{\mathrm{E}}\{\mathbf{x}\}_{\mathrm{E}'} = \{\mathbf{b}\}_{\mathrm{E}'} \tag{1-129}$$

Examining Eq. (1–129), it is seen that $\{\mathbf{C}\}_{\mathrm{E}}^{\mathrm{E}'}\{\mathbf{T}\}_{\mathrm{E}}\{\mathbf{C}\}_{\mathrm{E}'}^{\mathrm{E}}$ is the matrix representation of the tensor **T** in terms of the basis **E**′. Consequently, we can write

$$\{\mathbf{T}\}_{\mathrm{E}'} = \{\mathbf{C}\}_{\mathrm{E}}^{\mathrm{E}'}\{\mathbf{T}\}_{\mathrm{E}}\{\mathbf{C}\}_{\mathrm{E}'}^{\mathrm{E}} \tag{1-130}$$

Equation (1–130) provides a way to transform the matrix representation of an arbitrary tensor **T** between two coordinate systems whose orientations are different.

1.4.6 Eigenvalues and Eigenvectors

Let **A** be an $n \times n$ matrix. Then there exists a column-vector **u** such that

$$\mathbf{A}\mathbf{u} = \lambda\mathbf{u} \tag{1-131}$$

The vector **u** and scalar λ that satisfy Eq. (1–131) are called an *eigenvector* and *eigenvalue* of the matrix **A**, respectively. We can rewrite Eq. (1–131) as

$$\mathbf{A}\mathbf{u} = \lambda\mathbf{I}\mathbf{u} \tag{1-132}$$

where **I** is the $n \times n$ identity matrix. Rearranging Eq. (1–132) gives

$$(\mathbf{A} - \lambda\mathbf{I})\mathbf{u} = \mathbf{0} \tag{1-133}$$

Then, using the general expression for an $n \times n$ matrix and an n-dimensional column-vector as given in Eq. (1–63), we can write Eq. (1–133) in expanded form as

$$\begin{bmatrix} a_{11} - \lambda & a_{12} & \cdots & a_{1n} \\ a_{21} & a_{22} - \lambda & \cdots & a_{2n} \\ \vdots & \vdots & \ddots & \vdots \\ a_{n1} & a_{n2} & \cdots & a_{nn} - \lambda \end{bmatrix} \begin{Bmatrix} u_1 \\ u_2 \\ \vdots \\ u_n \end{Bmatrix} = \begin{Bmatrix} 0 \\ 0 \\ \vdots \\ 0 \end{Bmatrix} \tag{1–134}$$

Now, it is well known that the solution of Eq. (1–134) is obtained by solving for those values of λ that are the roots of the so-called *characteristic equation*:

$$\det(\mathbf{A} - \lambda\mathbf{I}) = \det \begin{Bmatrix} a_{11} - \lambda & a_{12} & \cdots & a_{1n} \\ a_{21} & a_{22} - \lambda & \cdots & a_{2n} \\ \vdots & \vdots & \ddots & \vdots \\ a_{n1} & a_{n2} & \cdots & a_{nn} - \lambda \end{Bmatrix} = 0 \tag{1–135}$$

where $\det(\cdot)$ is the determinant function (Kreyszig, 1988). Because of the properties of the determinant, Eq. (1–135) is a polynomial in the scalar λ with real coefficients a_{ij}, $(i, j = 1, 2, \ldots, n)$. Then, from the fundamental theorem of algebra, it is known that the roots of Eq. (1–135) are either real or occur in complex conjugate pairs. Once the eigenvalues of the matrix $\mathbf{A}$ have been found, the eigenvectors can be obtained. In general, no analytic methods exist for computing eigenvalues and eigenvectors. Consequently, for most matrices the eigenvalues and eigenvectors are computed numerically.

1.4.7 Eigenvalues and Eigenvectors of a Real-Symmetric Matrix

Now consider the special case where $\mathbf{A}$ is a *real-symmetric* matrix, i.e., the coefficients of $\mathbf{A}$ are real and

$$\mathbf{A} = \mathbf{A}^T \tag{1–136}$$

Suppose further that we let λ_i and λ_j be two distinct eigenvalues of $\mathbf{A}$ (i.e., $\lambda_i \neq \lambda_j$) with corresponding eigenvectors $\mathbf{w}_i$ and $\mathbf{w}_j$. Then from Eq. (1–131) we have

$$\mathbf{A}\mathbf{w}_i = \lambda_i \mathbf{w}_i \tag{1–137}$$

$$\mathbf{A}\mathbf{w}_j = \lambda_j \mathbf{w}_j \tag{1–138}$$

Multiplying Eqs. (1–137) and (1–138) by $\mathbf{w}_j^T$ and $\mathbf{w}_i^T$, respectively, we have

$$\mathbf{w}_j^T \mathbf{A}\mathbf{w}_i = \mathbf{w}_j^T \lambda_i \mathbf{w}_i \tag{1–139}$$

$$\mathbf{w}_i^T \mathbf{A}\mathbf{w}_j = \mathbf{w}_i^T \lambda_j \mathbf{w}_j \tag{1–140}$$

Now, because λ_i and λ_j are scalars, we have

$$\mathbf{w}_j^T \lambda_i \mathbf{w}_i = \lambda_i \mathbf{w}_j^T \mathbf{w}_i \tag{1–141}$$

$$\mathbf{w}_i^T \lambda_j \mathbf{w}_j = \lambda_j \mathbf{w}_i^T \mathbf{w}_j \tag{1–142}$$

Furthermore, since $\mathbf{w}_i$ and $\mathbf{w}_j$ are column-vectors, we know that $\mathbf{w}_i^T$ and $\mathbf{w}_j^T$ are row-vectors. Consequently, the quantities $\mathbf{w}_j^T \mathbf{w}_i$ and $\mathbf{w}_i^T \mathbf{w}_j$ are scalars and we have

$$\mathbf{w}_i^T \mathbf{w}_j = \mathbf{w}_j^T \mathbf{w}_i \tag{1–143}$$

Substituting the result of Eq. (1–143) into (1–139) and (1–140), we obtain

$$\mathbf{w}_j^T \mathbf{A} \mathbf{w}_i = \lambda_i \mathbf{w}_j^T \mathbf{w}_i \tag{1–144}$$

$$\mathbf{w}_i^T \mathbf{A} \mathbf{w}_j = \lambda_j \mathbf{w}_j^T \mathbf{w}_i \tag{1–145}$$

Next, because $\mathbf{A}$ is symmetric, we have

$$\left[\mathbf{w}_j^T \mathbf{A} \mathbf{w}_i\right]^T = \mathbf{w}_i^T \mathbf{A}^T \mathbf{w}_j = \mathbf{w}_i^T \mathbf{A} \mathbf{w}_j \tag{1–146}$$

Substituting the result of Eq. (1–146) into (1–144) and (1–145) gives

$$\mathbf{w}_i^T \mathbf{A} \mathbf{w}_j = \lambda_i \mathbf{w}_j^T \mathbf{w}_i \tag{1–147}$$

$$\mathbf{w}_i^T \mathbf{A} \mathbf{w}_j = \lambda_j \mathbf{w}_j^T \mathbf{w}_i \tag{1–148}$$

Subtracting Eq. (1–148) from (1–147), we obtain

$$(\lambda_i - \lambda_j)\mathbf{w}_j^T \mathbf{w}_i = 0 \tag{1–149}$$

Now, since we assumed that $\lambda_i \neq \lambda_j$, we have from Eq. (1–149) that

$$\mathbf{w}_j^T \mathbf{w}_i = 0 \tag{1–150}$$

Equation (1–150) implies that $\mathbf{w}_i$ and $\mathbf{w}_j$ are orthogonal. Now, as it turns out that the eigenvectors associated with two eigenvalues whose values are the same are orthogonal, i.e., if two eigenvalues of $\mathbf{A}$ are such that $\lambda_i = \lambda_j = \lambda$, then $\mathbf{w}_i$ and $\mathbf{w}_j$ are orthogonal. Consequently, a so-called *complete* set of mutually orthogonal eigenvectors can be obtained for a real-symmetric matrix. Suppose now that we denote the matrix of eigenvectors as

$$\mathbf{W} = \{\mathbf{w}_1 \cdots \mathbf{w}_n\} \tag{1–151}$$

Then it is known that the matrix $\mathbf{W}$ *diagonalizes* the matrix $\mathbf{A}$, i.e.,

$$\mathbf{W}^{-1}\mathbf{A}\mathbf{W} = \left\{\begin{array}{ccccc} \lambda_1 & 0 & \cdots & \cdots & 0 \\ 0 & \lambda_2 & 0 & \cdots & 0 \\ 0 & 0 & \ddots & & \vdots \\ \vdots & \vdots & & \ddots & 0 \\ 0 & 0 & \cdots & 0 & \lambda_n \end{array}\right\} \tag{1–152}$$

Now, since the eigenvector matrix $\mathbf{W}$ consists of mutually orthogonal vectors, each column of $\mathbf{W}$ can be normalized such that each eigenvector is of unit magnitude. Consequently, the matrix $\mathbf{W}$ can be made to be an *orthogonal matrix*. Equation (1–152) can then be written as

$$\mathbf{W}^{T}\mathbf{A}\mathbf{W} = \left\{\begin{array}{ccccc} \lambda_1 & 0 & \cdots & \cdots & 0 \\ 0 & \lambda_2 & 0 & \cdots & 0 \\ 0 & 0 & \ddots & & \vdots \\ \vdots & \vdots & & \ddots & 0 \\ 0 & 0 & \cdots & 0 & \lambda_n \end{array}\right\} = \mathbf{\Lambda} \tag{1–153}$$

where we have used the fact that $\mathbf{W}^{-1} = \mathbf{W}^T$ for an orthogonal matrix.

1.5 Ordinary Differential Equations

Suppose that $(x_1(t),\ldots,x_n(t))$ is a set of scalar functions of the scalar parameter t. Furthermore, suppose that $(f_1(t),\ldots,f_m(t))$ is a set of *known* functions of time. Finally, let $(\alpha_1,\ldots,\alpha_p)$ be a set of known parameters. Suppose now that we define the following three column-vectors:

$$\mathbf{x} = \begin{bmatrix} x_1(t) \\ \vdots \\ x_n(t) \end{bmatrix} \tag{1–154}$$

$$\mathbf{f} = \begin{bmatrix} f_1(t) \\ \vdots \\ f_m(t) \end{bmatrix} \tag{1–155}$$

$$\boldsymbol{\alpha} = \begin{bmatrix} \alpha_1 \\ \vdots \\ \alpha_p \end{bmatrix} \tag{1–156}$$

Then the functions $(x_1(t),\ldots,x_n(t))$ form a system of differential equations if they can be written as

$$\mathbf{G}(\mathbf{x},\dot{\mathbf{x}},\ldots,\mathbf{x}^{(n)},t;\boldsymbol{\alpha},\mathbf{f}) = \mathbf{0} \tag{1–157}$$

where $\mathbf{x}^{(i)}$ is the ith derivative of the function $\mathbf{x}$ with respect to the parameter t and

$$\mathbf{G} = \begin{bmatrix} g_1 \\ \vdots \\ g_n \end{bmatrix} \tag{1–158}$$

is a vector function of $\mathbf{x}$ and any derivatives of $\mathbf{x}$ with respect to t, the scalar t itself, and the vector $\mathbf{f}$. It is noted that we can write Eq. (1–157) in terms of the scalars $g_1,\ldots,g_n$ as

$$\begin{aligned} g_1(\mathbf{x},\dot{\mathbf{x}},\ldots,\mathbf{x}^{(n)},t;\boldsymbol{\alpha},\mathbf{f}) &= 0 \\ g_2(\mathbf{x},\dot{\mathbf{x}},\ldots,\mathbf{x}^{(n)},t;\boldsymbol{\alpha},\mathbf{f}) &= 0 \\ &\vdots \\ g_n(\mathbf{x},\dot{\mathbf{x}},\ldots,\mathbf{x}^{(n)},t;\boldsymbol{\alpha},\mathbf{f}) &= 0 \end{aligned} \tag{1–159}$$

It is important to note that the system of Eq. (1–157) (or, equivalently, Eq. (1–159)) can be considered a system of differential equations only if all functions and parameters (other than $\mathbf{x}$ and its derivatives with respect to t) are *known*. Throughout the remainder of this book, particularly from Chapter 3 onward, it will be important to derive differential equations of motion for various systems. In these problems it will be necessary to identify those quantities that are known and those that are unknown so that a system of equations can be obtained that contains purely *known* information.

Chapter 2

Kinematics

Geometry existed before the Creation. It is co-eternal with the mind of God. Geometry provided God with a model for the Creation. Geometry is God Himself.

- Johannes Kepler (1571–1630)
German Astronomer

The first topic in the study of dynamics is *kinematics*. Kinematics is the study of the geometry of motion *without* regard to the forces that cause the motion. For any system (which may consist of a particle, a rigid body, or a system of particles and/or rigid bodies) the objectives of kinematics are fourfold: to determine (1) a set of reference frames in which to observe the motion of a system; (2) a set of coordinate systems fixed in the chosen reference frames; (3) the angular velocity and angular acceleration of each reference frame (and/or rigid body) resolved in the chosen coordinate systems; and (4) the position, velocity, and acceleration of each particle in the system. In order to develop a comprehensive and systematic approach, the study of kinematics given in this Chapter is divided into two parts: (1) the study of kinematics of particles and (2) the study of kinematics of rigid bodies.

This Chapter is organized as follows. First, both a qualitative and precise definition of a reference frame is given. In particular, it is discussed that a reference frame provides a perspective from which to observe the motion of a system. Next, a coordinate system is defined and provides a way to measure the motion of a system within a particular reference frame. Then the rate of change of a vector function in a particular reference frame is defined. Using the definitions of a reference frame and a coordinate system, a key result, called the *rate of change transport theorem*, is derived that relates the rate of change of a vector in one reference frame to the rate of change of that same vector in a second reference frame where the second reference frame rotates relative to the first reference frame. Using the rate of change transport theorem, expressions are derived that relate the velocity and acceleration of a particle between two reference frames that rotate relative to one another. In addition, expressions are derived for the velocity and acceleration of a particle in terms of several commonly used coordinate systems. Next, expressions are derived that relate the velocity and acceleration of a particle between two reference frames that simultaneously rotate *and* translate relative to one another. Finally, the kinematics of a rigid body are discussed. In particular, it is described that a rigid body has six degrees of freedom, and, therefore, both its translational and rotational motion must be described in order to fully describe the kinematics. In particular, the rotational kinematics of a rigid body is discussed and a set of parameters, called *Eulerian angles*, are derived that describe the orientation of a rigid body. Throughout the Chapter examples are given to illustrate the key concepts.

2.1 Reference Frames

2.1.1 Definition of a Reference Frame and an Observer

The first step in kinematics is to choose a set of *reference frames*. Qualitatively, a reference frame is a *perspective* from which observations are made regarding the motion of a system. Using this qualitative notion, a reference frame is defined as follows. Let C be a collection of *at least three* noncolinear points that move in three-dimensional Euclidean space, $\mathbb{E}^3$. Next, let P and Q be two arbitrary points in C. Then the points P and Q are said to be *rigidly connected* or *rigidly attached* if the distance between P and Q, denoted d_{PQ}, is constant regardless of how P and Q move in $\mathbb{E}^3$. The collection C is then said to be a reference frame if the distance between every pair of points in C is rigidly connected, i.e., a reference frame is a collection of at least three noncolinear points in $\mathbb{E}^3$ such that the distance between any two points in the collection does not change with time (Tenenbaum, 2004).

In order to visualize a reference frame, let C be an arbitrary collection of points that move in $\mathbb{E}^3$ as shown in Fig. 2-1. Furthermore, suppose we choose two points P and Q in C and let $d_{PQ}(t_1)$ denote the distance between P and Q at some instant of time $t = t_1$. In general, at some later time $t_2 \neq t_1$, every point in C will have moved to a new location in $\mathbb{E}^3$ and the distance between P and Q at $t = t_2$ will be $d_{PQ}(t_2)$. Then, if $d_{PQ}(t_1) = d_{PQ}(t_2)$ for *all* values of t_1 and t_2, the collection C will be a reference frame.

Using the aforementioned definition, many common objects can be used as reference frames. For example, any three-dimensional rigid body (e.g., a cube, a sphere, or a cylinder) can be chosen as a reference frame. In addition, any planar rigid object (e.g., a square, a circle, or a triangle) can be chosen as a reference frame. However, given the definition, it is seen that an isolated point in $\mathbb{E}^3$ does not qualify as a reference frame. Furthermore, strictly speaking, a line in $\mathbb{E}^3$ also does not qualify as a reference frame because the points on a line are, by definition, colinear.[1] Finally, for most of the applications considered in this book, a reference frame will typically be chosen based on physically or geometrically meaningful objects.

In order to use reference frames systematically both in the theoretical development and the application of the theory to problems, we will use a calligraphic letter (e.g., $\mathcal{A}$, $\mathcal{B}$) to denote a reference frame.[2] Consistent with this notation for a reference frame, we will use the terminology *as viewed by an observer in reference frame* $\mathcal{A}$ or, more simply, *in reference frame* $\mathcal{A}$ to describe observations made about the motion of a vector relative to reference frame $\mathcal{A}$. Furthermore, we define an *observer in reference*

[1] While it is true that a line does not satisfy the definition of a reference frame, in certain applications, by abuse of the definition of a reference frame, we will sometimes take a line to be a reference frame. In such instances, the reference frame will be clear by context.

[2] It is noted that Kane and Levinson (1985) use a Roman italic letter to denote a reference frame. However, in order to provide more clarity, we use a calligraphic letter to denote a reference frame.

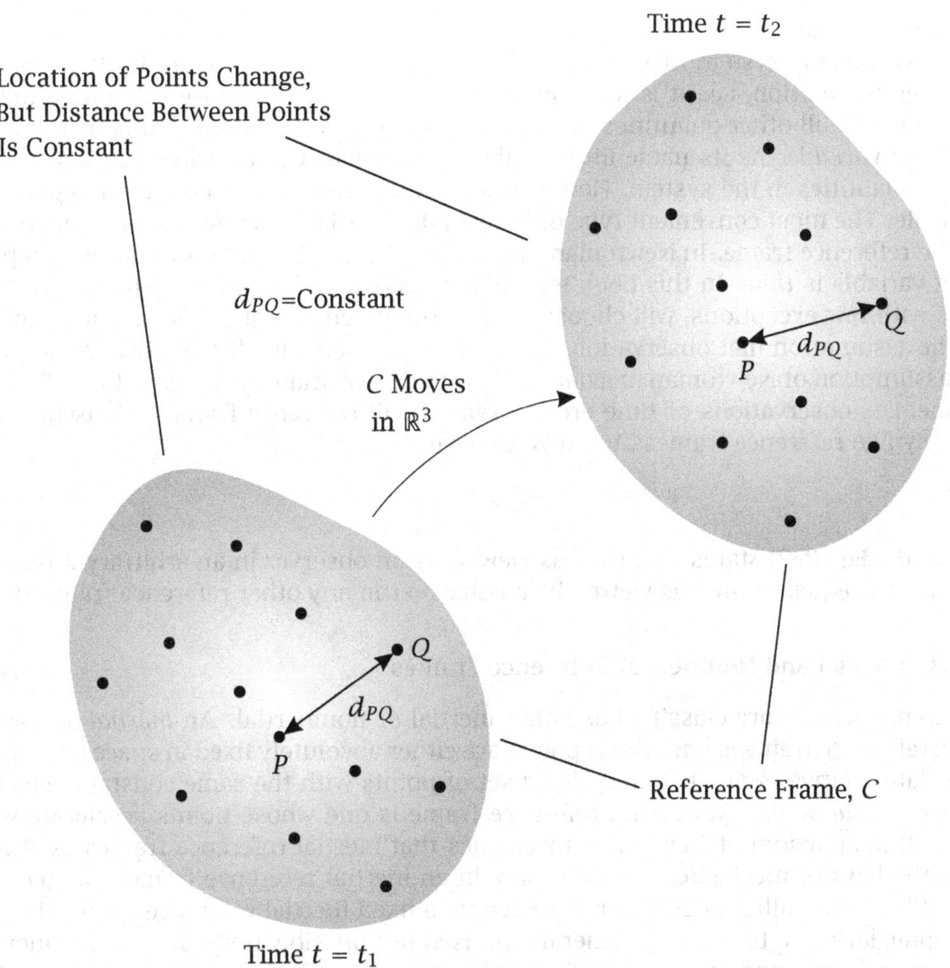

Figure 2-1 Collection of points C that define a reference frame. The distance between any two points P and Q in the collection is constant regardless of how the points in C move in $\mathbb{E}^3$.

frame $\mathcal{A}$ as any device that is rigidly attached to reference frame $\mathcal{A}$ and, thus, observes the motion of a system in reference frame $\mathcal{A}$.

2.1.2 Invariance of Space and Time in Different Reference Frames

An assumption of Newtonian mechanics is that space is invariant with respect to changes in reference frames, i.e., observations made of space are the same in all reference frames. As a result of the assumption of the invariance of space, observations of an arbitrary vector **b** are the same regardless of the reference frame. Consequently, for any two reference frames $\mathcal{A}$ and $\mathcal{B}$ we have

$$^{\mathcal{A}}\mathbf{b} = {}^{\mathcal{B}}\mathbf{b} \equiv \mathbf{b} \tag{2-1}$$

In words, Eq. (2-1) states that a vector **b** as viewed by an observer in an arbitrary reference frame $\mathcal{A}$ is equal to that same vector as viewed by an observer in any other reference frame $\mathcal{B}$.

Next, for any system of interest, it is desirable to specify a quantity that parameterizes the motion, i.e., it is desirable to specify a quantity whose value determines the values of all other quantities of interest. Such a quantity is referred to as the *independent variable.* As its name implies, the independent variable does not depend on other quantities in the system. However, other quantities depend on the independent variable. The most convenient type of independent variable is one that is independent of the reference frame. In Newtonian mechanics, the most commonly chosen independent variable is *time.* In this book we will generally use the variable t to denote time and, with few exceptions, will choose time as the independent variable. Then, similar to the assumption that observations of space are the same in all reference frames, it is an assumption of Newtonian mechanics that time is invariant with respect to reference frame, i.e., observations of time are the same in all reference frames. Consequently, for any two reference frames $\mathcal{A}$ and $\mathcal{B}$ we have

$$^{\mathcal{A}}t = {}^{\mathcal{B}}t \equiv t \tag{2-2}$$

In words, Eq. (2-2) states that time as viewed by an observer in an arbitrary reference frame $\mathcal{A}$ is equal to time as viewed by an observer in any other reference frame $\mathcal{B}$.

2.1.3 Inertial and Noninertial Reference Frames

Reference frames are classified as either inertial or noninertial. An *inertial* or *Newtonian* reference frame is one whose points are either absolutely fixed in space or at most translate relative to an absolutely fixed set of points with the same constant velocity. A *noninertial* or *non-Newtonian* reference frame is one whose points accelerate with time. It is an axiom of Newtonian mechanics that inertial reference frames exist and that the laws of mechanics are valid only in an inertial reference frame.[3] In general, we will use the calligraphic letter $\mathcal{F}$ to denote a fixed inertial reference frame, the calligraphic letter $\mathcal{N}$ to denote a general (nonfixed but possibly uniform velocity) inertial reference frame, and any other calligraphic letter (e.g., $\mathcal{A}$, $\mathcal{B}$) to denote a noninertial reference frame.

2.2 Coordinate Systems

Once a set of reference frames has been chosen to make observations regarding the motion of a system, the next step is to quantify or realize the motion of the system by choosing a set of *coordinate systems* that are fixed in each of the chosen reference frames. It is important to understand that, while they may seem to be the same, a reference frame and a coordinate system are distinct entities. We recall that a reference frame is a collection of points such that the distance between any two points is constant. However, it is seen that the definition of a reference frame does not directly provide any means of measuring the observations that may be made of the motion by an observer fixed in the reference frame. A coordinate system provides such a measurement system. In particular, the following information is required in order to specify a coordinate system that is fixed in a particular reference frame $\mathcal{A}$: (a) a point

[3]The principle of *Galilean invariance* or *Newtonian relativity* states that the fundamental laws of mechanics are the same in all inertial reference frames.

fixed in $\mathcal{A}$ called the *origin*, the point from which all distances are measured in reference frame $\mathcal{A}$, and (b) a *basis*, which is a set of three linearly independent directions that are fixed in reference frame $\mathcal{A}$; the basis provides a way to resolve vectors in $\mathbb{E}^3$.

Given the definition of a coordinate system, it is seen that, not only must a coordinate system be fixed in a *unique* reference frame, but it is also the case that a coordinate system *cannot* be fixed simultaneously in two distinct reference frames. Furthermore, while in principle any set of three linearly independent vectors can be chosen as a basis for a coordinate system, recall from Chapter 1 that it is most convenient to use a so-called *orthonormal basis* (i.e., a set of mutually orthogonal unit vectors) in which to resolve vectors. Because of its convenience, from this point forward we will restrict our attention to orthonormal bases. Consequently, the basis must satisfy the following properties:

Property 1: The basis vectors $\mathbf{e}_1$, $\mathbf{e}_2$, and $\mathbf{e}_3$ are unit vectors, i.e.,

$$\begin{aligned} \mathbf{e}_1 \cdot \mathbf{e}_1 &= 1 \\ \mathbf{e}_2 \cdot \mathbf{e}_2 &= 1 \\ \mathbf{e}_3 \cdot \mathbf{e}_3 &= 1 \end{aligned} \tag{2-3}$$

Property 2: The basis vectors $\mathbf{e}_1$, $\mathbf{e}_2$, and $\mathbf{e}_3$ are mutually orthogonal, i.e.,

$$\begin{aligned} \mathbf{e}_1 \cdot \mathbf{e}_2 &= \mathbf{e}_2 \cdot \mathbf{e}_1 = 0 \\ \mathbf{e}_1 \cdot \mathbf{e}_3 &= \mathbf{e}_3 \cdot \mathbf{e}_1 = 0 \\ \mathbf{e}_2 \cdot \mathbf{e}_3 &= \mathbf{e}_3 \cdot \mathbf{e}_2 = 0 \end{aligned} \tag{2-4}$$

Property 3: The basis vectors $\mathbf{e}_1$, $\mathbf{e}_2$, and $\mathbf{e}_3$ form a **right-handed system**, i.e.,

$$\begin{aligned} \mathbf{e}_1 \times \mathbf{e}_2 &= \mathbf{e}_3 \\ \mathbf{e}_2 \times \mathbf{e}_3 &= \mathbf{e}_1 \\ \mathbf{e}_3 \times \mathbf{e}_1 &= \mathbf{e}_2 \end{aligned} \tag{2-5}$$

We note that a basis that satisfies Properties 1 – 3 is called a *right-handed orthonormal basis*. Figure 2-2 shows a schematic that provides a way to remember the right-hand rule of Eq. (2-5). The arrows in Fig. 2-2 indicate the direction of positive orientation of the vector products given in Eq. (2-5); any vector product taken in the direction of the arrows yields a positive result while any vector product taken in the direction opposite the arrows yields a negative result. It is noted that a right-handed system satisfies the property

$$\mathbf{e}_3 \cdot (\mathbf{e}_1 \times \mathbf{e}_2) = 1 \tag{2-6}$$

It is important to understand that the set $\{\mathbf{e}_1, \mathbf{e}_2, \mathbf{e}_3\}$ must always be ordered such that Eq. (2-6) is satisfied (we note that the right-handed system as described in Eq. (2-5) is consistent with the orientation shown in Fig. 1-4 on page 8). Then, given a coordinate system fixed in a particular reference frame $\mathcal{A}$, any vector $\mathbf{b}$ in $\mathbb{E}^3$ can be expressed in the basis $\{\mathbf{e}_1, \mathbf{e}_2, \mathbf{e}_3\}$ corresponding to the chosen coordinate system as

$$\mathbf{b} = b_1\mathbf{e}_1 + b_2\mathbf{e}_2 + b_3\mathbf{e}_3 \tag{2-7}$$

In other words, given a reference frame $\mathcal{A}$ and a basis $\{\mathbf{e}_1, \mathbf{e}_2, \mathbf{e}_3\}$ corresponding to a coordinate system fixed in $\mathcal{A}$, for every vector $\mathbf{b}$ there exist scalar coefficients b_1, b_2, and b_3 (which, in general, change with time), such that $\mathbf{b}$ can be represented in the basis $\{\mathbf{e}_1, \mathbf{e}_2, \mathbf{e}_3\}$.

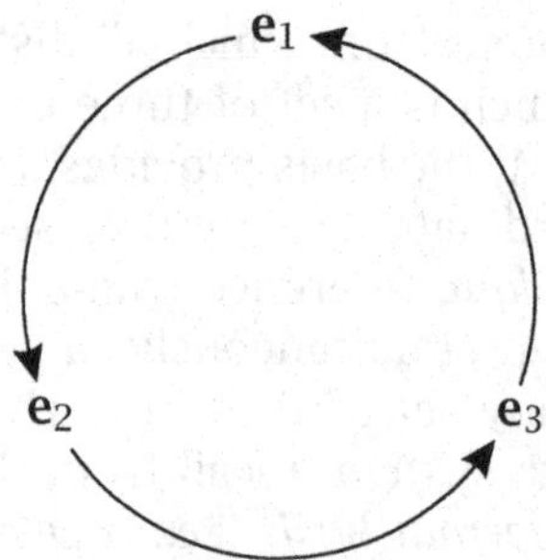

Figure 2–2 Schematic of right-hand rule corresponding to Eq. (2–5).

2.3 Rate of Change of Scalar and Vector Functions

2.3.1 Rate of Change of a Scalar Function

Suppose we are given a scalar function $x(t)$ where $x(t) \in \mathbb{R}$ is a function of the parameter $t \in \mathbb{R}$. Then the rate of change of $x(t)$ is defined as

$$\frac{dx}{dt} \equiv \lim_{\Delta t \to 0} \frac{x(t + \Delta t) - x(t)}{\Delta t} \tag{2–8}$$

It is seen from the limit in Eq. (2–8) that, because $x(t)$ does not depend on any particular direction in $\mathbb{E}^3$, its rate of change dx/dt is *independent of the reference frame.* As a result, observers in all reference frames see exactly the same rate of change of a scalar function, and the notation dx/dt has only one meaning (namely, that of the well known calculus derivative).

2.3.2 Rate of Change of a Vector Function

It is seen from the previous discussion that observers in different reference frames all see the same rate of change of a scalar function of time. Likewise, given an arbitrary vector function of time, at every instant of time observers in all reference frames see the same vector. In other words, a vector is independent of the reference frame in which it is observed. However, unlike a scalar function of time, observers in different reference frames do *not* see the same rate of change of a vector function of time. Contrariwise, observers fixed in different reference frames will, in general, make different observations about the rate of change of a vector function of time. As a result, the phrase "rate of change of a vector" must always be qualified by stating explicitly the reference frame in which the observation is being made.

A precise notion of the rate of change of a vector function of time is given as follows. Given an arbitrary reference frame $\mathcal{A}$, the rate of change of a vector in reference frame $\mathcal{A}$ is defined as

$$\frac{{}^{\mathcal{A}}d\mathbf{b}}{dt} = \overset{\mathcal{A}}{\lim_{\Delta t \to 0}} \frac{\mathbf{b}(t + \Delta t) - \mathbf{b}(t)}{\Delta t} \tag{2–9}$$

where the notation

$$\overset{\mathcal{A}}{\lim_{\Delta t \to 0}} \tag{2–10}$$

means that the limit is taken by an observer fixed in reference frame $\mathcal{A}$ and ${}^{\mathcal{A}}d/dt$ is the rate of change operator as viewed by an observer in reference frame $\mathcal{A}$. In other words, the limit in Eq. (2–9) is being computed based on how the vector $\mathbf{b}$ appears to be changing to an observer in reference frame $\mathcal{A}$. Using the definition of Eq. (2–9), a vector $\mathbf{b}$ is said to be *fixed* in a reference frame $\mathcal{A}$ if the following condition holds for all time:

$$\frac{{}^{\mathcal{A}}d\mathbf{b}}{dt} = \mathbf{0} \tag{2-11}$$

In other words, if a vector $\mathbf{b}$ always appears to always have the same magnitude and direction to an observer in reference frame $\mathcal{A}$, its rate of change in reference frame $\mathcal{A}$ is zero.

Suppose now that we choose to express an arbitrary vector $\mathbf{b}$ in terms of an orthonormal basis $\{\mathbf{e}_1, \mathbf{e}_2, \mathbf{e}_3\}$ (see Section 2.2 for the definition of an orthonormal basis). Then $\mathbf{b}$ can be written as

$$\mathbf{b} = b_1\mathbf{e}_1 + b_2\mathbf{e}_2 + b_3\mathbf{e}_3 \tag{2-12}$$

Furthermore, suppose that the basis $\{\mathbf{e}_1, \mathbf{e}_2, \mathbf{e}_3\}$ is fixed in reference frame $\mathcal{A}$. Then the rate of change of $\mathbf{b}$ as viewed by an observer in reference frame $\mathcal{A}$ is defined by the following limit:

$$\frac{{}^{\mathcal{A}}d\mathbf{b}}{dt} = \lim_{\Delta t \to 0} \sum_{i=1}^{3} \frac{b_i(t + \Delta t) - b_i(t)}{\Delta t}\mathbf{e}_i \tag{2-13}$$

Observing that the limit and the summation are interchangeable, Eq. (2–13) can be written as

$$\frac{{}^{\mathcal{A}}d\mathbf{b}}{dt} = \sum_{i=1}^{3} \lim_{\Delta t \to 0} \frac{b_i(t + \Delta t) - b_i(t)}{\Delta t}\mathbf{e}_i \tag{2-14}$$

Now, because b_1, b_2, and b_3 are scalar functions of t, we have

$$\frac{db_i}{dt} = \lim_{\Delta t \to 0} \frac{b_i(t + \Delta t) - b_i(t)}{\Delta t}, \quad (i = 1, 2, 3) \tag{2-15}$$

where we note that b_i, $(i = 1, 2, 3)$ are scalar functions of t and hence their rates of change are independent of the choice of the reference frame. Consequently, ${}^{\mathcal{A}}d\mathbf{b}/dt$ can be written as

$$\frac{{}^{\mathcal{A}}d\mathbf{b}}{dt} = \frac{db_1}{dt}\mathbf{e}_1 + \frac{db_2}{dt}\mathbf{e}_2 + \frac{db_3}{dt}\mathbf{e}_3 = \dot{b}_1\mathbf{e}_1 + \dot{b}_2\mathbf{e}_2 + \dot{b}_3\mathbf{e}_3 \tag{2-16}$$

Now, consistent with the definition given in Eq. (2–11) of a vector fixed in a reference frame, it is observed in Eq. (2–16) that, because $\mathbf{b}$ is expressed in a basis $\{\mathbf{e}_1, \mathbf{e}_2, \mathbf{e}_3\}$ that is fixed in reference frame $\mathcal{A}$, the rate of change $\mathbf{b}$ in reference frame $\mathcal{A}$ is obtained by differentiating only the components of $\mathbf{b}$ (because, by Eq. (2–11), the rates of change of the basis vectors in reference frame $\mathcal{A}$ are all *zero*). Consequently, when a vector is expressed in a basis fixed in a reference frame $\mathcal{A}$, it appears to an observer in reference frame $\mathcal{A}$ that only the components of the vector are changing with time.

From the definition of the rate of change of an arbitrary vector in an arbitrary reference frame, it is seen that observers in different reference frames will see different rates of change of the vector $\mathbf{b}$. In other words, for two distinct reference frames $\mathcal{A}$ and $\mathcal{B}$, it will generally be the case that

$$\frac{{}^{\mathcal{A}}d\mathbf{b}}{dt} \neq \frac{{}^{\mathcal{B}}d\mathbf{b}}{dt} \tag{2-17}$$

Now, because the fundamental postulates of classical Newtonian mechanics, namely Newton's and Euler's laws, are valid only when the motion is viewed by an observer in an *inertial* reference frame, we will ultimately be interested in obtaining the rate of change of a vector as viewed by an observer in an inertial reference frame. Thus, to avoid confusion in presenting the theory, the symbol $\mathcal{N}$ will always be used to denote an observation in an inertial reference frame.

2.3.3 Rate of Change of a Scalar Function of a Vector

A particular class of scalar functions whose rate of change is of interest in dynamics is the class of scalar functions of vectors where the vectors are themselves functions of time. Consider a scalar function F of a single vector $\mathbf{b}$, i.e., $F = F(\mathbf{b})$. Furthermore, let $\mathbf{b}$ be a function of time, i.e., $\mathbf{b} = \mathbf{b}(t)$. As seen above, because F is a scalar function, its rate of change is independent of the reference frame and is given as dF/dt. Suppose now $\mathbf{b}$ is resolved in a basis $\{\mathbf{e}_1, \mathbf{e}_2, \mathbf{e}_3\}$, where $\{\mathbf{e}_1, \mathbf{e}_2, \mathbf{e}_3\}$ is fixed in a reference frame $\mathcal{A}$. Then $\mathbf{b}$ can be expressed as

$$\mathbf{b} = b_1\mathbf{e}_1 + b_2\mathbf{e}_2 + b_3\mathbf{e}_3 \tag{2-18}$$

where b_1, b_2, and b_3 are the components of $\mathbf{b}$ in the basis $\{\mathbf{e}_1, \mathbf{e}_2, \mathbf{e}_3\}$ that is fixed in $\mathcal{A}$. Moreover, because the output of the function F is a scalar, we have

$$F(\mathbf{b}) = F(b_1, b_2, b_3) \tag{2-19}$$

Equation (2-19) merely states that, because F is a scalar function, it depends only on the components of $\mathbf{b}$ and not on the basis vectors. Then, computing the rate of change of F in Eq. (2-19) and applying the chain rule, we obtain

$$\frac{dF}{dt} = \frac{\partial F}{\partial b_1}\frac{db_1}{dt} + \frac{\partial F}{\partial b_2}\frac{db_2}{dt} + \frac{\partial F}{\partial b_3}\frac{db_3}{dt} \tag{2-20}$$

Using the fact that the basis $\{\mathbf{e}_1, \mathbf{e}_2, \mathbf{e}_3\}$ forms a mutually orthogonal set, Eq. (2-20) can be rewritten as

$$\frac{dF}{dt} = \left(\frac{\partial F}{\partial b_1}\mathbf{e}_1 + \frac{\partial F}{\partial b_2}\mathbf{e}_2 + \frac{\partial F}{\partial b_3}\mathbf{e}_3\right) \cdot \left(\frac{db_1}{dt}\mathbf{e}_1 + \frac{db_2}{dt}\mathbf{e}_2 + \frac{db_3}{dt}\mathbf{e}_3\right) \tag{2-21}$$

Using the definition of rate of change of a vector in a reference frame $\mathcal{A}$ as given in Eq. (2-16), Eq. (2-21) can be written more compactly as

$$\frac{dF}{dt} = \nabla_{\mathbf{b}} F \cdot \frac{{}^{\mathcal{A}}d\mathbf{b}}{dt} \equiv \frac{\partial F}{\partial \mathbf{b}} \cdot \frac{{}^{\mathcal{A}}d\mathbf{b}}{dt} \tag{2-22}$$

where

$$\nabla_{\mathbf{b}} F = \frac{\partial F}{\partial \mathbf{b}} \tag{2-23}$$

is the *gradient* of F with respect to $\mathbf{b}$. Now, because both the gradient of F and the vector $\mathbf{b}$ can be expressed in a basis fixed in any reference frame, the result of Eq. (2-22) is valid in any reference frame.

2.3.4 Rates of Change of Common Functions of Vectors

Let **b** and **c** be vectors in $\mathbb{E}^3$ and let k be a scalar. Then the following is a list of basic rules for computing the rates of change of common functions of vectors in an arbitrary reference frame $\mathcal{A}$:

$$\frac{^{\mathcal{A}}d}{dt}(\mathbf{b}+\mathbf{c}) = \frac{^{\mathcal{A}}d\mathbf{b}}{dt} + \frac{^{\mathcal{A}}d\mathbf{c}}{dt} \tag{2-24}$$

$$\frac{^{\mathcal{A}}d}{dt}(k\mathbf{b}) = k\frac{^{\mathcal{A}}d\mathbf{b}}{dt} \tag{2-25}$$

$$\frac{^{\mathcal{A}}d}{dt}(\mathbf{b}\times\mathbf{c}) = \frac{^{\mathcal{A}}d\mathbf{b}}{dt}\times\mathbf{c} + \mathbf{b}\times\frac{^{\mathcal{A}}d\mathbf{c}}{dt} \tag{2-26}$$

$$\frac{^{\mathcal{A}}d}{dt}(\mathbf{b}\cdot\mathbf{c}) = \frac{^{\mathcal{A}}d\mathbf{b}}{dt}\cdot\mathbf{c} + \mathbf{b}\cdot\frac{^{\mathcal{A}}d\mathbf{c}}{dt} \equiv \frac{d}{dt}(\mathbf{b}\cdot\mathbf{c}) \tag{2-27}$$

$$\frac{^{\mathcal{A}}d}{dt}(\mathbf{b}\otimes\mathbf{c}) = \frac{^{\mathcal{A}}d\mathbf{b}}{dt}\otimes\mathbf{c} + \mathbf{b}\otimes\frac{^{\mathcal{A}}d\mathbf{c}}{dt} \tag{2-28}$$

We remind the reader that, because the rate of change of a scalar function is independent of the reference frame, we have stated explicitly that the rate of change of the function $\mathbf{b}\cdot\mathbf{c}$ in Eq. (2-27) does *not* depend on the choice of the reference frame $\mathcal{A}$.

2.4 Position, Velocity, and Acceleration

A *particle* (or *point*) is defined as an object that lies in three-dimensional Euclidean space $\mathbb{E}^3$ and has no dimension, i.e., a particle occupies no space. Consider a particle moving in $\mathbb{E}^3$ relative to a reference frame $\mathcal{A}$ as shown in Fig. 2-3.

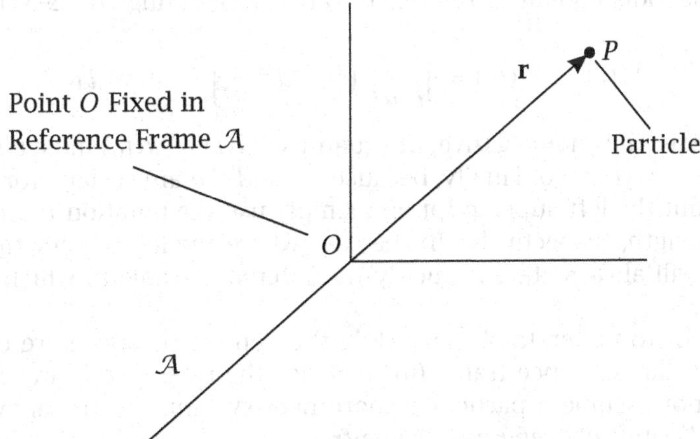

Figure 2-3 Particle moving relative to a reference frame $\mathcal{A}$ in three-dimensional Euclidean space $\mathbb{E}^3$.

The *position* of the particle describes the location of the particle relative to a point O fixed in $\mathcal{A}$ and is denoted $\mathbf{r}$. The position of a particle is a function of a scalar parameter t (generally speaking, time), i.e., $\mathbf{r} = \mathbf{r}(t)$. By varying t, the *path* or *trajectory* of the particle is defined. In particular, when the motion of the particle is observed in reference frame $\mathcal{A}$, we say that the particle subtends a trajectory relative to reference frame $\mathcal{A}$. Given $\mathbf{r}(t)$, the *velocity of the particle as viewed by an observer in reference frame* $\mathcal{A}$, denoted ${}^{\mathcal{A}}\mathbf{v} = {}^{\mathcal{A}}\mathbf{v}(t)$, is the rate of change of the position relative to $\mathcal{A}$ and is defined as

$$ {}^{\mathcal{A}}\mathbf{v} = \frac{{}^{\mathcal{A}}d\mathbf{r}}{dt} = {}^{\mathcal{A}}\lim_{\Delta t \to 0} \frac{\mathbf{r}(t+\Delta t) - \mathbf{r}(t)}{\Delta t} \tag{2-29} $$

where we have used the definition of the limit as given in Eq. (2-10).[4] Given ${}^{\mathcal{A}}\mathbf{v}(t)$, the *speed of the particle in reference frame* $\mathcal{A}$, denoted ${}^{\mathcal{A}}v$, is defined as

$$ {}^{\mathcal{A}}v = \|{}^{\mathcal{A}}\mathbf{v}(t)\| = \sqrt{{}^{\mathcal{A}}\mathbf{v}(t) \cdot {}^{\mathcal{A}}\mathbf{v}(t)} \tag{2-30} $$

Given ${}^{\mathcal{A}}\mathbf{v}(t)$, the *acceleration of the particle in reference frame* $\mathcal{A}$, denoted ${}^{\mathcal{A}}\mathbf{a} = {}^{\mathcal{A}}\mathbf{a}(t)$, is the rate of change of ${}^{\mathcal{A}}\mathbf{v}$ and is defined as

$$ {}^{\mathcal{A}}\mathbf{a} = \frac{{}^{\mathcal{A}}d}{dt}\left({}^{\mathcal{A}}\mathbf{v}\right) = \frac{{}^{\mathcal{A}}d}{dt}\left(\frac{{}^{\mathcal{A}}d\mathbf{r}}{dt}\right) = {}^{\mathcal{A}}\lim_{\Delta t \to 0} \frac{{}^{\mathcal{A}}\mathbf{v}(t+\Delta t) - {}^{\mathcal{A}}\mathbf{v}(t)}{\Delta t} \tag{2-31} $$

A point P in $\mathbb{E}^3$ is said to be *fixed* relative to reference frame $\mathcal{A}$ if its velocity and acceleration in reference frame $\mathcal{A}$ are both zero, i.e., P is fixed in $\mathcal{A}$ if ${}^{\mathcal{A}}\mathbf{v} \equiv \mathbf{0}$ and ${}^{\mathcal{A}}\mathbf{a} \equiv \mathbf{0}$. The distance traveled by a particle in a reference frame $\mathcal{A}$ during a time interval $t \in \left[t_0, t_f\right]$ is called the *arc-length in reference frame* $\mathcal{A}$. The arc-length is denoted by the scalar variable ${}^{\mathcal{A}}s$, where ${}^{\mathcal{A}}s$ satisfies the differential equation

$$ \frac{d}{dt}\left({}^{\mathcal{A}}s\right) = {}^{\mathcal{A}}\dot{s} \equiv {}^{\mathcal{A}}v \tag{2-32} $$

and ${}^{\mathcal{A}}v$ is obtained from Eq. (2-30); we note once again that, because ${}^{\mathcal{A}}s$ is a scalar, its rate of change is independent of reference frame. Integrating Eq. (2-32) from t_0 to t, we obtain

$$ {}^{\mathcal{A}}s(t) - {}^{\mathcal{A}}s(t_0) = \int_{t_0}^{t} \frac{d}{dt}\left({}^{\mathcal{A}}s\right) d\tau = \int_{t_0}^{t} {}^{\mathcal{A}}v(\tau)d\tau \tag{2-33} $$

Note that, because ${}^{\mathcal{A}}v$ is nonnegative, the quantity ${}^{\mathcal{A}}s(t) - {}^{\mathcal{A}}s(t_0)$ must also be nonnegative, i.e., ${}^{\mathcal{A}}s(t) - {}^{\mathcal{A}}s(t_0) \geq 0$. Finally, because ${}^{\mathcal{A}}s$ and ${}^{\mathcal{A}}v$ are scalars, for compactness we will often omit the left superscript and simply use the notation v and s to denote speed and arc-length, respectively. In the case where the left superscript is omitted, the discussion will always state explicitly the reference frame in which observations are being made.

It is important to understand that, while the definitions above are computed relative to a particular reference frame (in this case the reference frame $\mathcal{A}$), the above definitions do not assume a particular coordinate system, i.e., the above definitions are *coordinate-free* but *not reference frame-free.*

[4]It is important to understand that the notation ${}^{\mathcal{A}}\mathbf{v}$ is merely a *shorthand* for ${}^{\mathcal{A}}d\mathbf{r}/dt$. In other words, when we write ${}^{\mathcal{A}}\mathbf{v}$, we mean that the vector ${}^{\mathcal{A}}\mathbf{v}$ *arises* from taking the rate of change of another vector (in this case position) in reference frame $\mathcal{A}$. Furthermore, it is important to understand that, although ${}^{\mathcal{A}}\mathbf{v}$ arises from taking the rate of change of $\mathbf{r}$ in reference frame $\mathcal{A}$, the quantity ${}^{\mathcal{A}}\mathbf{v}$ is itself a vector. Consequently, by assumption (see Section 2.1.2) observations of the vector ${}^{\mathcal{A}}\mathbf{v}$ are the *same* in all reference frames.

2.5 Degrees of Freedom of a Particle

Because the position of a particle is a vector in $\mathbb{E}^3$, it is seen in general that *three* independent quantities are required in order to fully describe the *configuration* of a particle. More specifically, regardless of the coordinate system chosen to express the motion of the particle, three scalar quantities are required in order to specify the position of the particle (i.e., one scalar quantity for each of the three components of position). As such, the position of a particle in $\mathbb{E}^3$ can be expressed as

$$\mathbf{r} = r_1\mathbf{e}_1 + r_2\mathbf{e}_2 + r_3\mathbf{e}_3 \tag{2-34}$$

where $\{\mathbf{e}_1, \mathbf{e}_2, \mathbf{e}_3\}$ is a basis fixed in an arbitrary reference frame $\mathcal{A}$. The scalar quantities r_1, r_2, and r_3 are called the *coordinates* of the position in the basis $\{\mathbf{e}_1, \mathbf{e}_2, \mathbf{e}_3\}$.

Now, when a particle is free to move in $\mathbb{E}^3$ without any constraints, we say that the particle has *three degrees of freedom* (one degree of freedom for each coordinate used to describe the position of the particle). In the case where the motion of the particle is constrained, the number of degrees of freedom is reduced by the number of constraints. For example, in the case of a particle constrained to move in a fixed plane, one of the components of position does not change. Consequently, for case of motion in a fixed plane, only two variables are required to specify the position of the particle and the particle has only *two* degrees of freedom. In general, the number of degrees of freedom for a particle is equal to the difference between the number of coordinates required to describe the position and the number of independent constraints imposed on the motion. Denoting the number of coordinates by N, the number of constraints by P, and the number of degrees of freedom by M, we have

$$M = N - P \tag{2-35}$$

Example 2-1

A circular disk of radius R rotates in the plane relative to the ground (where the ground constitutes a fixed inertial reference frame) with a constant angular rate ω about the fixed point O. Determine the rate of change of a point P fixed to the edge of the disk as viewed by observers fixed to the ground and fixed to the disk.

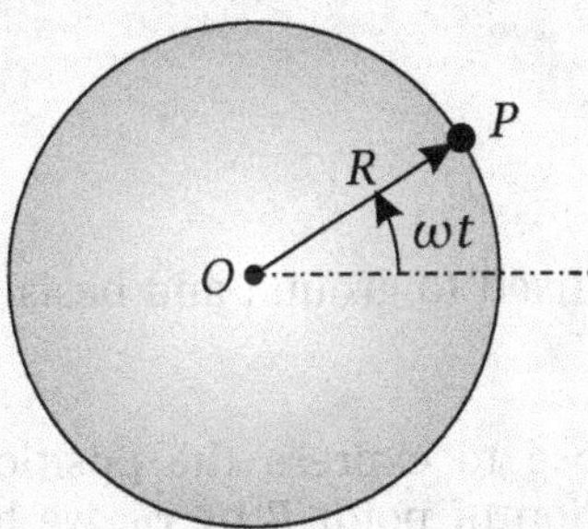

Figure 2-4 Disk of radius R rotating with angular rate ω.

Solution to Example 2–1

This problem is a simple but excellent example of how observers in different reference frames see different rates of change of the same vector. To solve this problem, let $\mathcal{F}$ be a reference frame that is fixed to the ground and let $\mathcal{A}$ be a reference frame that is fixed to the disk. Corresponding to reference frame $\mathcal{F}$, we choose the following coordinate system:

	Origin at O	
$\mathbf{E}_1$	=	Along OP when $t = 0$
$\mathbf{E}_3$	=	Out of page
$\mathbf{E}_2$	=	$\mathbf{E}_3 \times \mathbf{E}_1$

Furthermore, corresponding to reference frame $\mathcal{A}$, we choose the following coordinate system:

	Origin at O	
$\mathbf{e}_1$	=	Along OP fixed in disk
$\mathbf{e}_3$	=	Out of page (= $\mathbf{E}_3$)
$\mathbf{e}_2$	=	$\mathbf{e}_3 \times \mathbf{e}_1$

The bases $\{\mathbf{E}_1, \mathbf{E}_2, \mathbf{E}_3\}$ and $\{\mathbf{e}_1, \mathbf{e}_2, \mathbf{e}_3\}$ are shown in Fig. 2–5, where it is seen that reference frame $\mathcal{A}$ rotates relative to $\mathcal{F}$.

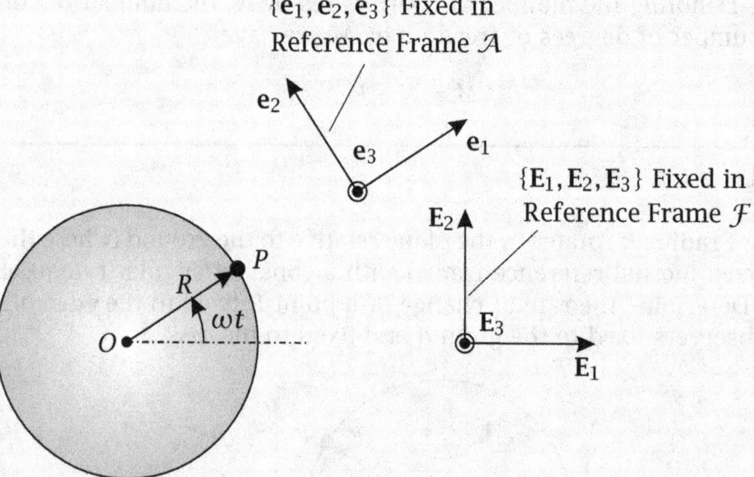

Figure 2–5 Basis $\{\mathbf{E}_1, \mathbf{E}_2, \mathbf{E}_3\}$ fixed to ground and basis $\{\mathbf{e}_1, \mathbf{e}_2, \mathbf{e}_3\}$ fixed to disk for Example 2-1.

Now suppose that we choose to express the position of point P in the basis $\{\mathbf{E}_1, \mathbf{E}_2, \mathbf{E}_3\}$. Denoting the position of point P by $\mathbf{r}_P$, we have in terms of $\{\mathbf{E}_1, \mathbf{E}_2, \mathbf{E}_3\}$ that

$$\mathbf{r}_P = R\cos\omega t\mathbf{E}_1 + R\sin\omega t\mathbf{E}_2 \tag{2-36}$$

where we see that $\mathbf{r}_P$ is obtained by taking components in the directions $\mathbf{E}_1$ and $\mathbf{E}_2$. Then, because $\{\mathbf{E}_1, \mathbf{E}_2, \mathbf{E}_3\}$ is fixed to the ground (i.e., reference frame $\mathcal{F}$), the rate of change of $\mathbf{r}_P$ as viewed by an observer in reference frame $\mathcal{F}$ is obtained using Eq. (2–16) as

$$\frac{{}^{\mathcal{F}}d}{dt}(\mathbf{r}_P) = -R\omega\sin\omega t\mathbf{E}_1 + R\omega\cos\omega t\mathbf{E}_2 \tag{2–37}$$

where it is emphasized in Eq. (2–37) that the basis $\{\mathbf{E}_1, \mathbf{E}_2, \mathbf{E}_3\}$ is fixed in reference frame $\mathcal{F}$.

Next, suppose that we express the position of point P in terms of the basis $\{\mathbf{e}_1, \mathbf{e}_2, \mathbf{e}_3\}$. We then have

$$\mathbf{r}_P = R\mathbf{e}_1 \tag{2–38}$$

where we see from Eq. (2–38) that, at every instant of the motion, $\mathbf{r}_P$ lies along the direction $\mathbf{e}_1$. Then, because $\mathbf{e}_1$ is fixed in reference frame $\mathcal{A}$, the rate of change of $\mathbf{r}_P$ in reference frame $\mathcal{A}$ is obtained using Eq. (2–16) as

$$\frac{{}^{\mathcal{A}}d}{dt}(\mathbf{r}_P) = \frac{dR}{dt}\mathbf{e}_1 = 0 \tag{2–39}$$

where we note that R is constant and, hence, $dR/dt = 0$.

■

2.6 Relative Position, Velocity, and Acceleration

Consider two points Q and P with positions $\mathbf{r}_Q$ and $\mathbf{r}_P$, respectively. Furthermore, let $\mathcal{A}$ be a reference frame in which observations are made about the motion of points Q and P. Then the position of P relative to Q, denoted $\mathbf{r}_{P/Q}$, is defined as

$$\mathbf{r}_{P/Q} = \mathbf{r}_P - \mathbf{r}_Q \tag{2–40}$$

Given $\mathbf{r}_{P/Q}$, the velocity of P relative to Q as viewed by an observer in reference frame $\mathcal{A}$, denoted ${}^{\mathcal{A}}\mathbf{v}_{P/Q}$, is defined as the rate of change of ${}^{\mathcal{A}}\mathbf{r}_{P/Q}$ as viewed by an observer in reference frame $\mathcal{A}$, i.e.,

$${}^{\mathcal{A}}\mathbf{v}_{P/Q} = \frac{{}^{\mathcal{A}}d}{dt}(\mathbf{r}_{P/Q}) = {}^{\mathcal{A}}\mathbf{v}_P - {}^{\mathcal{A}}\mathbf{v}_Q \tag{2–41}$$

Finally, the acceleration of P relative to Q as viewed by an observer in reference frame $\mathcal{A}$, denoted ${}^{\mathcal{A}}\mathbf{a}_{P/Q}$, is defined as the rate of change of ${}^{\mathcal{A}}\mathbf{v}_{P/Q}$ in the reference frame $\mathcal{A}$, i.e., is defined as

$${}^{\mathcal{A}}\mathbf{a}_{P/Q} = \frac{{}^{\mathcal{A}}d}{dt}\left({}^{\mathcal{A}}\mathbf{v}_{P/Q}\right) = \frac{{}^{\mathcal{A}}d^2}{dt^2}(\mathbf{r}_{P/Q}) = {}^{\mathcal{A}}\mathbf{a}_P - {}^{\mathcal{A}}\mathbf{a}_Q \tag{2–42}$$

2.7 Rectilinear Motion

The simplest motion is that of a particle P moving in a straight line. Such motion is called *rectilinear motion*. Suppose that we choose to describe the rectilinear motion of a particle as viewed by an observer in a fixed reference frame $\mathcal{F}$. Because the motion is one-dimensional, it is sufficient to define a coordinate system consisting of an origin and a *single* basis vector. In particular, suppose we choose a fixed point O as the origin of a fixed coordinate system and define a basis vector $\mathbf{E}_x$ to be in the direction of positive motion as shown in Fig. 2-6. In terms of the coordinate system defined by

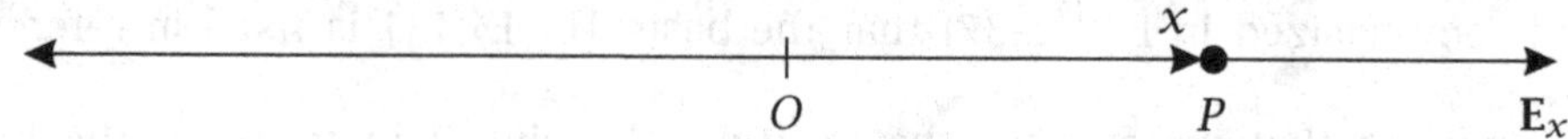

Figure 2-6 Rectilinear motion of a particle P.

O and $\mathbf{E}_x$, the position of the particle can be written as

$$\mathbf{r} = x\mathbf{E}_x \tag{2-43}$$

where $x = x(t)$ is the displacement of the particle from the point O. Given $\mathbf{r}$ from Eq. (2-43), the velocity and acceleration of the particle are given, respectively, as

$$\begin{aligned} {}^{\mathcal{F}}\mathbf{v}(t) &= \frac{{}^{\mathcal{F}}d\mathbf{r}}{dt} = \dot{x}(t)\mathbf{E}_x = v(t)\mathbf{E}_x && \text{(2-44)} \\ {}^{\mathcal{F}}\mathbf{a}(t) &= \frac{{}^{\mathcal{F}}d^2\mathbf{r}}{dt^2} \equiv {}^{\mathcal{F}}\ddot{\mathbf{r}}(t) = {}^{\mathcal{F}}\dot{\mathbf{v}}(t) = \ddot{x}(t)\mathbf{E}_x = a(t)\mathbf{E}_x && \text{(2-45)} \end{aligned}$$

where we note once again that, because $x(t)$ is a scalar function of time, its rate of change is independent of the reference frame. Furthermore, for the case of rectilinear motion, the arc-length ${}^{\mathcal{F}}s(t)$ satisfies the equation

$$\frac{d}{dt}\left({}^{\mathcal{F}}s\right) = \|{}^{\mathcal{F}}\mathbf{v}\| = \|v(t)\mathbf{E}_x\| = |v(t)| = |\dot{x}| \tag{2-46}$$

Finally, it can be seen for rectilinear motion that the motion is completely specified by the scalar functions $x(t)$, $v(t)$, and $a(t)$. In particular, the acceleration $a(t)$ of a particle in rectilinear motion is specified in one of the following three ways:

(1) Acceleration as a function of time

(2) Acceleration as a function of speed

(3) Acceleration as a function of position

We now analyze each of these cases.

Case 1: Acceleration as a Function of Time

Suppose we are given acceleration as a function of time, i.e., $a = a(t)$. Then the velocity can be computed as

$$v(t) = v(t_0) + \int_{t_0}^{t} a(\eta)d\eta \tag{2-47}$$

where η is a dummy variable of integration. Using Eq. (2-47), the position is computed as

$$x(t) = x(t_0) + \int_{t_0}^{t}\left[v(t_0) + \int_{t_0}^{\nu} a(\eta)d\eta\right]d\nu \tag{2-48}$$

where the variables η and ν are dummy variables of integration.

Case 2: Acceleration as a Function of Velocity

Suppose we are given acceleration as a function of velocity, i.e., $a = a(v)$. Then, from the chain rule (Kreyszig, 1988) we have

$$a = a(v) = \frac{dv}{dt} = \frac{dv}{dx}\frac{dx}{dt} = v\frac{dv}{dx} \tag{2-49}$$

Furthermore, from Eq. (2–49) we have

$$v\frac{dv}{dx} = a(v) \tag{2-50}$$

Separating variables in Eq. (2–50) gives

$$dx = \frac{vdv}{a(v)} \tag{2-51}$$

Then, integrating both sides of Eq. (2–51) we obtain

$$x - x_0 = \int_{v_0}^{v} \frac{\eta d\eta}{a(\eta)} \tag{2-52}$$

where η is a dummy variable of integration. Finally, solving Eq. (2–52) for x, we obtain

$$x = x_0 + \int_{v_0}^{v} \frac{\eta d\eta}{a(\eta)} \tag{2-53}$$

Also, via separation of variables we have

$$dt = \frac{dv}{a} = \frac{dv}{a(v)} \tag{2-54}$$

Integrating Eq. (2–54) we obtain

$$t - t_0 = \int_{v_0}^{v} \frac{d\eta}{a(\eta)} \tag{2-55}$$

Rewriting Eq. (2–55), the time can be solved for as

$$t = t_0 + \int_{v_0}^{v} \frac{d\eta}{a(\eta)} \tag{2-56}$$

Case 3: Acceleration as a Function of Position

Suppose we are given acceleration as a function of position, i.e., we are given $a = a(x)$. Then we can write

$$a(x) = \frac{dv}{dt} = \frac{dv}{dx}\frac{dx}{dt} = v\frac{dv}{dx} \tag{2-57}$$

Separating variables in Eq. (2–57) we have

$$vdv = a(x)dx \tag{2-58}$$

Integrating both sides of Eq. (2–58) gives

$$\int_{v_0}^{v} \eta d\eta = \int_{x_0}^{x} a(\eta)d\eta \tag{2-59}$$

where η is a dummy variable of integration. Therefore,

$$\frac{v^2}{2} - \frac{v_0^2}{2} = \int_{x_0}^{x} a(\eta)d\eta \tag{2-60}$$

which implies that

$$v^2 = v_0^2 + 2\int_{x_0}^{x} a(\eta)d\eta \tag{2-61}$$

Solving for v gives

$$v(x) = \left[v_0^2 + 2\int_{x_0}^{x} a(\eta)d\eta\right]^{1/2} \tag{2-62}$$

Moreover, because $v = dx/dt$ we have

$$dt = \frac{dx}{v} \tag{2-63}$$

Integrating both sides of Eq. (2-63) gives

$$t - t_0 = \int_{x_0}^{x} \frac{d\eta}{v(\eta)} \tag{2-64}$$

which implies that

$$t = t_0 + \int_{x_0}^{x} \frac{d\eta}{v(\eta)} \tag{2-65}$$

where the expression for $v(x)$ from Eq. (2-62) is used in Eqs. (2-64) and (2-65).

2.8 Using Noninertial Reference Frames to Describe Motion

While using an inertial coordinate system is appealing because of its simplicity, it is often the case that observations regarding the motion of a system are made in a noninertial reference frame. The two types of noninertial reference frames that are commonly encountered in dynamics are *accelerating* and *rotating* reference frames. Using the definitions of a reference frame and a coordinate system from Section 2.2, in this section we derive an expression for the time derivative of a vector when observations of the vector are made in a rotating reference frame.

2.9 Rate of Change of a Vector in a Rotating Reference Frame

Let $\mathcal{A}$ and $\mathcal{B}$ be two reference frames such that $\mathcal{B}$ rotates relative to $\mathcal{A}$. Furthermore, let $\{\mathbf{E}_1, \mathbf{E}_2, \mathbf{E}_3\}$ and $\{\mathbf{e}_1, \mathbf{e}_2, \mathbf{e}_3\}$ be orthonormal bases fixed in reference frames $\mathcal{A}$ and $\mathcal{B}$, respectively. In addition, let O be a fixed point in $\mathbb{E}^3$ common to both reference frames $\mathcal{A}$ and $\mathcal{B}$. Next, let $\mathbf{b}$ be a vector in $\mathbb{E}^3$. A schematic of the reference frames $\mathcal{A}$ and $\mathcal{B}$, the orthonormal bases $\{\mathbf{E}_1, \mathbf{E}_2, \mathbf{E}_3\}$ and $\{\mathbf{e}_1, \mathbf{e}_2, \mathbf{e}_3\}$, and the vector $\mathbf{b}$ is shown in Fig. 2-7. Then $\mathbf{b}$ can then be expressed in the basis $\{\mathbf{e}_1, \mathbf{e}_2, \mathbf{e}_3\}$ as

$$\mathbf{b} = b_1\mathbf{e}_1 + b_2\mathbf{e}_2 + b_3\mathbf{e}_3 \tag{2-66}$$

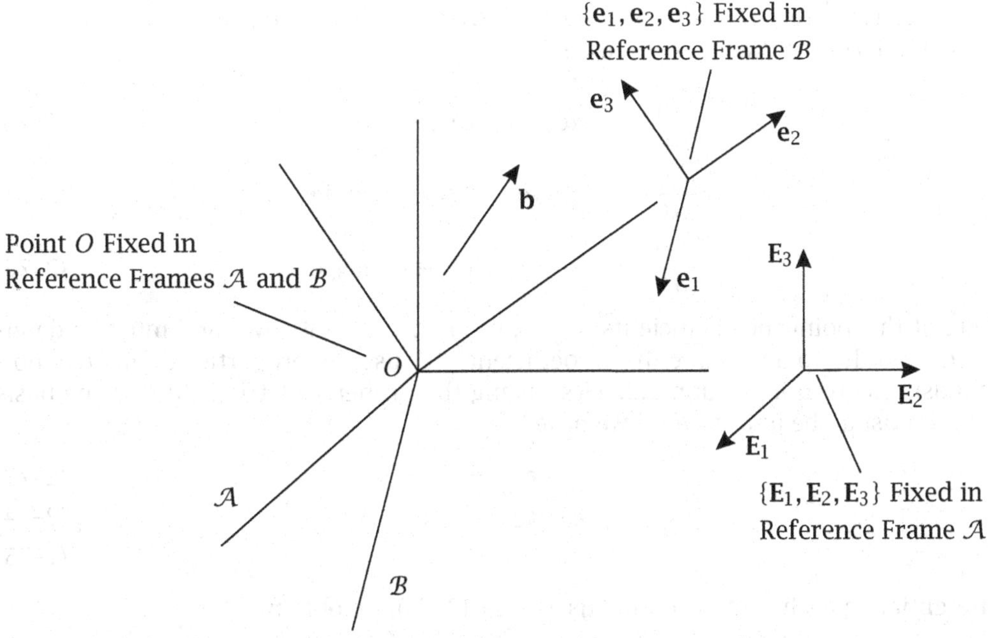

Figure 2-7 Reference frame $\mathcal{B}$ that rotates relative to a reference frame $\mathcal{A}$ and a general vector $\mathbf{b}$ in $\mathbb{E}^3$.

The rate of change of $\mathbf{b}$ with respect to an observer in reference frame $\mathcal{A}$ is then given as

$$\frac{{}^{\mathcal{A}}d\mathbf{b}}{dt} = \frac{db_1}{dt}\mathbf{e}_1 + \frac{db_2}{dt}\mathbf{e}_2 + \frac{db_3}{dt}\mathbf{e}_3 + b_1\frac{{}^{\mathcal{A}}d\mathbf{e}_1}{dt} + b_2\frac{{}^{\mathcal{A}}d\mathbf{e}_2}{dt} + b_3\frac{{}^{\mathcal{A}}d\mathbf{e}_3}{dt} \tag{2-67}$$

We note again that, because b_1, b_2, and b_3 are scalars, their rates of change, db_1/dt, db_2/dt, and db_3/dt, are independent of the reference frame. On the other hand, because $\mathbf{e}_1$, $\mathbf{e}_2$, and $\mathbf{e}_3$ are vectors, it is necessary to state explicitly the reference frame relative to which the rate of change is being computed. Because in this case the rate of change is being taken with respect to an observer in reference frame $\mathcal{A}$, we write ${}^{\mathcal{A}}d\mathbf{e}_1/dt$, ${}^{\mathcal{A}}d\mathbf{e}_2/dt$, and ${}^{\mathcal{A}}d\mathbf{e}_3/dt$. Now, because $\{\mathbf{e}_1, \mathbf{e}_2, \mathbf{e}_3\}$ is fixed in reference frame $\mathcal{B}$, from Eq. (2-16), we have

$$\frac{{}^{\mathcal{B}}d\mathbf{b}}{dt} = \frac{db_1}{dt}\mathbf{e}_1 + \frac{db_2}{dt}\mathbf{e}_2 + \frac{db_3}{dt}\mathbf{e}_3 \tag{2-68}$$

The rate of change of $\mathbf{b}$ in reference frame $\mathcal{A}$ is then given as

$$\frac{{}^{\mathcal{A}}d\mathbf{b}}{dt} = \frac{{}^{\mathcal{B}}d\mathbf{b}}{dt} + b_1\frac{{}^{\mathcal{A}}d\mathbf{e}_1}{dt} + b_2\frac{{}^{\mathcal{A}}d\mathbf{e}_2}{dt} + b_3\frac{{}^{\mathcal{A}}d\mathbf{e}_3}{dt} \tag{2-69}$$

Next, because ${}^{\mathcal{A}}d\mathbf{e}_1/dt$, ${}^{\mathcal{A}}d\mathbf{e}_2/dt$, and ${}^{\mathcal{A}}d\mathbf{e}_3/dt$ are all vectors, they can each be expressed in terms of the basis $\{\mathbf{e}_1, \mathbf{e}_2, \mathbf{e}_3\}$ as

$$\frac{{}^{\mathcal{A}}d\mathbf{e}_1}{dt} = \omega_{11}\mathbf{e}_1 + \omega_{21}\mathbf{e}_2 + \omega_{31}\mathbf{e}_3 \quad (2\text{–}70)$$
$$\frac{{}^{\mathcal{A}}d\mathbf{e}_2}{dt} = \omega_{12}\mathbf{e}_1 + \omega_{22}\mathbf{e}_2 + \omega_{32}\mathbf{e}_3 \quad (2\text{–}71)$$
$$\frac{{}^{\mathcal{A}}d\mathbf{e}_3}{dt} = \omega_{13}\mathbf{e}_1 + \omega_{23}\mathbf{e}_2 + \omega_{33}\mathbf{e}_3 \quad (2\text{–}72)$$

where at this point the coefficients ω_{ij}, $i, j = 1, 2, 3$ are unknown and must be determined. In order to determine these coefficients, we use the properties of an orthonormal basis as given in Section 2.2. First, using the property of Eq. (2–3) that the basis vectors must all be unit vectors, we have

$$\mathbf{e}_1 \cdot \mathbf{e}_1 = 1 \quad (2\text{–}73)$$
$$\mathbf{e}_2 \cdot \mathbf{e}_2 = 1 \quad (2\text{–}74)$$
$$\mathbf{e}_3 \cdot \mathbf{e}_3 = 1 \quad (2\text{–}75)$$

Differentiating each expression in Eqs. (2–73)–(2–75), we obtain

$$\frac{{}^{\mathcal{A}}d\mathbf{e}_1}{dt} \cdot \mathbf{e}_1 + \mathbf{e}_1 \cdot \frac{{}^{\mathcal{A}}d\mathbf{e}_1}{dt} = 0 \quad (2\text{–}76)$$
$$\frac{{}^{\mathcal{A}}d\mathbf{e}_2}{dt} \cdot \mathbf{e}_2 + \mathbf{e}_2 \cdot \frac{{}^{\mathcal{A}}d\mathbf{e}_2}{dt} = 0 \quad (2\text{–}77)$$
$$\frac{{}^{\mathcal{A}}d\mathbf{e}_3}{dt} \cdot \mathbf{e}_3 + \mathbf{e}_3 \cdot \frac{{}^{\mathcal{A}}d\mathbf{e}_3}{dt} = 0 \quad (2\text{–}78)$$

Furthermore, because the scalar product is commutative, Eqs. (2–76)–(2–78) can be simplified to

$$\frac{{}^{\mathcal{A}}d\mathbf{e}_1}{dt} \cdot \mathbf{e}_1 = 0 \quad (2\text{–}79)$$
$$\frac{{}^{\mathcal{A}}d\mathbf{e}_2}{dt} \cdot \mathbf{e}_2 = 0 \quad (2\text{–}80)$$
$$\frac{{}^{\mathcal{A}}d\mathbf{e}_3}{dt} \cdot \mathbf{e}_3 = 0 \quad (2\text{–}81)$$

Then, substituting the expressions for ${}^{\mathcal{A}}d\mathbf{e}_1/dt$, ${}^{\mathcal{A}}d\mathbf{e}_2/dt$, and ${}^{\mathcal{A}}d\mathbf{e}_3/dt$ from Eqs. (2–70)–(2–72) into (2–79)–(2–81), respectively, we obtain

$$(\omega_{11}\mathbf{e}_1 + \omega_{21}\mathbf{e}_2 + \omega_{31}\mathbf{e}_3) \cdot \mathbf{e}_1 = 0 \quad (2\text{–}82)$$
$$(\omega_{12}\mathbf{e}_1 + \omega_{22}\mathbf{e}_2 + \omega_{32}\mathbf{e}_3) \cdot \mathbf{e}_2 = 0 \quad (2\text{–}83)$$
$$(\omega_{13}\mathbf{e}_1 + \omega_{23}\mathbf{e}_2 + \omega_{33}\mathbf{e}_3) \cdot \mathbf{e}_3 = 0 \quad (2\text{–}84)$$

From Eqs. (2–82)–(2–84) we obtain

$$\omega_{11}\mathbf{e}_1 \cdot \mathbf{e}_1 + \omega_{21}\mathbf{e}_2 \cdot \mathbf{e}_1 + \omega_{31}\mathbf{e}_3 \cdot \mathbf{e}_1 = 0 \quad (2\text{–}85)$$
$$\omega_{12}\mathbf{e}_1 \cdot \mathbf{e}_2 + \omega_{22}\mathbf{e}_2 \cdot \mathbf{e}_2 + \omega_{32}\mathbf{e}_3 \cdot \mathbf{e}_2 = 0 \quad (2\text{–}86)$$
$$\omega_{13}\mathbf{e}_1 \cdot \mathbf{e}_3 + \omega_{23}\mathbf{e}_2 \cdot \mathbf{e}_3 + \omega_{33}\mathbf{e}_3 \cdot \mathbf{e}_3 = 0 \quad (2\text{–}87)$$

Then, using the orthogonality property of Eqs. (2–3) and (2–4), Eqs. (2–85)–(2–87) simplify to

$$\omega_{11}\mathbf{e}_1 \cdot \mathbf{e}_1 = 0 \quad (2\text{–}88)$$
$$\omega_{22}\mathbf{e}_2 \cdot \mathbf{e}_2 = 0 \quad (2\text{–}89)$$
$$\omega_{33}\mathbf{e}_3 \cdot \mathbf{e}_3 = 0 \quad (2\text{–}90)$$

From Eqs. (2–88)–(2–90) we have

$$\omega_{11} = 0 \quad (2\text{–}91)$$
$$\omega_{22} = 0 \quad (2\text{–}92)$$
$$\omega_{33} = 0 \quad (2\text{–}93)$$

Also, from the orthogonality property of Eq. (2–4), we have

$$\mathbf{e}_1 \cdot \mathbf{e}_2 = \mathbf{e}_2 \cdot \mathbf{e}_1 = 0 \quad (2\text{–}94)$$
$$\mathbf{e}_1 \cdot \mathbf{e}_3 = \mathbf{e}_3 \cdot \mathbf{e}_1 = 0 \quad (2\text{–}95)$$
$$\mathbf{e}_2 \cdot \mathbf{e}_3 = \mathbf{e}_3 \cdot \mathbf{e}_2 = 0 \quad (2\text{–}96)$$

Computing the rates of change of Eqs. (2–94)–(2–96), we obtain

$$\frac{^{\mathcal{A}}d\mathbf{e}_1}{dt} \cdot \mathbf{e}_2 + \mathbf{e}_1 \cdot \frac{^{\mathcal{A}}d\mathbf{e}_2}{dt} = 0 \quad (2\text{–}97)$$
$$\frac{^{\mathcal{A}}d\mathbf{e}_1}{dt} \cdot \mathbf{e}_3 + \mathbf{e}_1 \cdot \frac{^{\mathcal{A}}d\mathbf{e}_3}{dt} = 0 \quad (2\text{–}98)$$
$$\frac{^{\mathcal{A}}d\mathbf{e}_2}{dt} \cdot \mathbf{e}_3 + \mathbf{e}_2 \cdot \frac{^{\mathcal{A}}d\mathbf{e}_3}{dt} = 0 \quad (2\text{–}99)$$

Once again, substituting the expressions for $^{\mathcal{A}}d\mathbf{e}_1/dt$, $^{\mathcal{A}}d\mathbf{e}_2/dt$, and $^{\mathcal{A}}d\mathbf{e}_3/dt$ from Eqs. (2–70)–(2–72), respectively, and using the results of Eqs. (2–91)–(2–93), we obtain

$$(\omega_{21}\mathbf{e}_2 + \omega_{31}\mathbf{e}_3) \cdot \mathbf{e}_2 + \mathbf{e}_1 \cdot (\omega_{12}\mathbf{e}_1 + \omega_{32}\mathbf{e}_3) = 0 \quad (2\text{–}100)$$
$$(\omega_{21}\mathbf{e}_2 + \omega_{31}\mathbf{e}_3) \cdot \mathbf{e}_3 + \mathbf{e}_1 \cdot (\omega_{13}\mathbf{e}_1 + \omega_{23}\mathbf{e}_2) = 0 \quad (2\text{–}101)$$
$$(\omega_{12}\mathbf{e}_1 + \omega_{32}\mathbf{e}_3) \cdot \mathbf{e}_3 + \mathbf{e}_2 \cdot (\omega_{13}\mathbf{e}_1 + \omega_{23}\mathbf{e}_2) = 0 \quad (2\text{–}102)$$

Computing the scalar products in Eqs. (2–100)–(2–102) gives

$$\omega_{21}\mathbf{e}_2 \cdot \mathbf{e}_2 + \omega_{31}\mathbf{e}_3 \cdot \mathbf{e}_2 + \omega_{12}\mathbf{e}_1 \cdot \mathbf{e}_1 + \omega_{32}\mathbf{e}_1 \cdot \mathbf{e}_3 = 0 \quad (2\text{–}103)$$
$$\omega_{21}\mathbf{e}_2 \cdot \mathbf{e}_3 + \omega_{31}\mathbf{e}_3 \cdot \mathbf{e}_3 + \omega_{13}\mathbf{e}_1 \cdot \mathbf{e}_1 + \omega_{23}\mathbf{e}_1 \cdot \mathbf{e}_2 = 0 \quad (2\text{–}104)$$
$$\omega_{12}\mathbf{e}_1 \cdot \mathbf{e}_3 + \omega_{32}\mathbf{e}_3 \cdot \mathbf{e}_3 + \omega_{13}\mathbf{e}_2 \cdot \mathbf{e}_1 + \omega_{23}\mathbf{e}_2 \cdot \mathbf{e}_2 = 0 \quad (2\text{–}105)$$

Then, using the orthogonality property of Eq. (2–4), Eqs. (2–103)–(2–105) simplify to

$$\omega_{21} + \omega_{12} = 0 \quad (2\text{–}106)$$
$$\omega_{31} + \omega_{13} = 0 \quad (2\text{–}107)$$
$$\omega_{32} + \omega_{23} = 0 \quad (2\text{–}108)$$

Consequently,

$$\omega_{21} = -\omega_{12} \quad (2\text{–}109)$$
$$\omega_{31} = -\omega_{13} \quad (2\text{–}110)$$
$$\omega_{32} = -\omega_{23} \quad (2\text{–}111)$$

The expressions for ${}^{\mathcal{A}}d\mathbf{e}_1/dt$, ${}^{\mathcal{A}}d\mathbf{e}_2/dt$, and ${}^{\mathcal{A}}d\mathbf{e}_3/dt$ can then be simplified to

$$\frac{{}^{\mathcal{A}}d\mathbf{e}_1}{dt} = \omega_{21}\mathbf{e}_2 + \omega_{31}\mathbf{e}_3 \tag{2-112}$$

$$\frac{{}^{\mathcal{A}}d\mathbf{e}_2}{dt} = -\omega_{21}\mathbf{e}_1 + \omega_{32}\mathbf{e}_3 \tag{2-113}$$

$$\frac{{}^{\mathcal{A}}d\mathbf{e}_3}{dt} = -\omega_{31}\mathbf{e}_1 - \omega_{32}\mathbf{e}_2 \tag{2-114}$$

It can be seen from Eqs. (2-112)-(2-114) that only three of the original nine coefficients remain, namely, ω_{21}, ω_{31}, and ω_{32}. For simplicity, the following substitutions are now made:

$$\begin{aligned} \omega_1 &\equiv \omega_{32} \\ \omega_2 &\equiv -\omega_{31} \\ \omega_3 &\equiv \omega_{21} \end{aligned} \tag{2-115}$$

Equations (2-112)-(2-114) can then be written as

$$\frac{{}^{\mathcal{A}}d\mathbf{e}_1}{dt} = \omega_3\mathbf{e}_2 - \omega_2\mathbf{e}_3 \tag{2-116}$$

$$\frac{{}^{\mathcal{A}}d\mathbf{e}_2}{dt} = -\omega_3\mathbf{e}_1 + \omega_1\mathbf{e}_3 \tag{2-117}$$

$$\frac{{}^{\mathcal{A}}d\mathbf{e}_3}{dt} = \omega_2\mathbf{e}_1 - \omega_1\mathbf{e}_2 \tag{2-118}$$

Next, taking the scalar products of ${}^{\mathcal{A}}d\mathbf{e}_2/dt$, ${}^{\mathcal{A}}d\mathbf{e}_3/dt$, and ${}^{\mathcal{A}}d\mathbf{e}_1/dt$ in Eq. (2-117), (2-118), and (2-116), respectively, with $\mathbf{e}_3$, $\mathbf{e}_1$, and $\mathbf{e}_2$, respectively, and applying the results of Eqs. (2-97)-(2-99), we obtain ω_1, ω_2, and ω_3 as

$$\omega_1 = \frac{{}^{\mathcal{A}}d\mathbf{e}_2}{dt} \cdot \mathbf{e}_3 \equiv -\mathbf{e}_2 \cdot \frac{{}^{\mathcal{A}}d\mathbf{e}_3}{dt} \tag{2-119}$$

$$\omega_2 = -\frac{{}^{\mathcal{A}}d\mathbf{e}_1}{dt} \cdot \mathbf{e}_3 \equiv \mathbf{e}_1 \cdot \frac{{}^{\mathcal{A}}d\mathbf{e}_3}{dt} \tag{2-120}$$

$$\omega_3 = \frac{{}^{\mathcal{A}}d\mathbf{e}_1}{dt} \cdot \mathbf{e}_2 \equiv -\mathbf{e}_1 \cdot \frac{{}^{\mathcal{A}}d\mathbf{e}_2}{dt} \tag{2-121}$$

Suppose now that we define the vector

$${}^{\mathcal{A}}\boldsymbol{\omega}^{\mathcal{B}} = \omega_1\mathbf{e}_1 + \omega_2\mathbf{e}_2 + \omega_3\mathbf{e}_3 \tag{2-122}$$

The quantity ${}^{\mathcal{A}}\boldsymbol{\omega}^{\mathcal{B}}$ is called the *angular velocity* of reference frame $\mathcal{B}$ as viewed by an observer in reference frame $\mathcal{A}$ or, more simply, the angular velocity of $\mathcal{B}$ in $\mathcal{A}$. In terms of the angular velocity, the vectors ${}^{\mathcal{A}}d\mathbf{e}_1/dt$, ${}^{\mathcal{A}}d\mathbf{e}_2/dt$, and ${}^{\mathcal{A}}d\mathbf{e}_3/dt$ can be written as

$$\frac{{}^{\mathcal{A}}d\mathbf{e}_1}{dt} = \omega_3\mathbf{e}_2 - \omega_2\mathbf{e}_3 = {}^{\mathcal{A}}\boldsymbol{\omega}^{\mathcal{B}} \times \mathbf{e}_1 \tag{2-123}$$

$$\frac{{}^{\mathcal{A}}d\mathbf{e}_2}{dt} = -\omega_3\mathbf{e}_1 + \omega_1\mathbf{e}_3 = {}^{\mathcal{A}}\boldsymbol{\omega}^{\mathcal{B}} \times \mathbf{e}_2 \tag{2-124}$$

$$\frac{{}^{\mathcal{A}}d\mathbf{e}_3}{dt} = \omega_2\mathbf{e}_1 - \omega_1\mathbf{e}_2 = {}^{\mathcal{A}}\boldsymbol{\omega}^{\mathcal{B}} \times \mathbf{e}_3 \tag{2-125}$$

Substituting the results from Eqs. (2–123)–(2–125) into (2–69), the quantity ${}^{\mathcal{A}}d\mathbf{b}/dt$ can be written as

$$\frac{{}^{\mathcal{A}}d\mathbf{b}}{dt} = \frac{{}^{\mathcal{B}}d\mathbf{b}}{dt} + b_1({}^{\mathcal{A}}\boldsymbol{\omega}^{\mathcal{B}} \times \mathbf{e}_1) + b_2({}^{\mathcal{A}}\boldsymbol{\omega}^{\mathcal{B}} \times \mathbf{e}_2) + b_3({}^{\mathcal{A}}\boldsymbol{\omega}^{\mathcal{B}} \times \mathbf{e}_3) \tag{2–126}$$

Then, using the distributive property of the vector product, Eq. (2–126) can be written as

$$\frac{{}^{\mathcal{A}}d\mathbf{b}}{dt} = \frac{{}^{\mathcal{B}}d\mathbf{b}}{dt} + {}^{\mathcal{A}}\boldsymbol{\omega}^{\mathcal{B}} \times (b_1\mathbf{e}_1 + b_2\mathbf{e}_2 + b_3\mathbf{e}_3) \tag{2–127}$$

Observing that $b_1\mathbf{e}_1 + b_2\mathbf{e}_2 + b_3\mathbf{e}_3 \equiv \mathbf{b}$, Eq. (2–127) simplifies to

$$\boxed{\frac{{}^{\mathcal{A}}d\mathbf{b}}{dt} = \frac{{}^{\mathcal{B}}d\mathbf{b}}{dt} + {}^{\mathcal{A}}\boldsymbol{\omega}^{\mathcal{B}} \times \mathbf{b}} \tag{2–128}$$

Equation (2–128) is referred to as the *rate of change transport theorem* or, more simply, the *transport theorem*. As can be seen, the transport theorem is used to compute the rate of change of a vector in a reference frame $\mathcal{A}$ when observations are made about the rate of change of the vector in a reference frame $\mathcal{B}$, where the angular velocity of reference frame $\mathcal{B}$ as viewed by an observer in reference frame $\mathcal{A}$ is ${}^{\mathcal{A}}\boldsymbol{\omega}^{\mathcal{B}}$. The transport theorem can be stated in words as follows: Given two reference frames $\mathcal{A}$ and $\mathcal{B}$ such that the angular velocity of reference frame $\mathcal{B}$ as viewed by an observer in reference frame $\mathcal{A}$ is ${}^{\mathcal{A}}\boldsymbol{\omega}^{\mathcal{B}}$, the rate of change of a vector $\mathbf{b}$ as viewed by an observer in reference frame $\mathcal{A}$ is equal to the sum of the rate of change of $\mathbf{b}$ as viewed by an observer in reference frame $\mathcal{B}$ and the cross product of ${}^{\mathcal{A}}\boldsymbol{\omega}^{\mathcal{B}}$ with the vector $\mathbf{b}$.

2.9.1 Further Explanation of the Transport Theorem

It is important to understand that, in applying the transport theorem of Eq. (2–128), $\mathbf{b}$ can be *any* vector whatsoever. The important point is that the vector $\mathbf{b}$ is known in terms of a reference frame $\mathcal{B}$ that rotates relative to another reference frame $\mathcal{A}$. Consequently, the rate of change of $\mathbf{b}$ as viewed by an observer in reference frame $\mathcal{A}$ must account for the motion of the rotation of the reference frame $\mathcal{B}$ relative to reference frame $\mathcal{A}$.

Next, it is seen that the only assumption made in the derivation of this section is that there exists a point that is common to both reference frame $\mathcal{A}$ and reference frame $\mathcal{B}$. Consequently, $\mathcal{A}$ and $\mathcal{B}$ can be any *rotating* reference frames whatsoever. Moreover, because $\mathcal{A}$ and $\mathcal{B}$ are arbitrary reference frames, the order of the reference frames can be interchanged to give

$$\frac{{}^{\mathcal{B}}d\mathbf{b}}{dt} = \frac{{}^{\mathcal{A}}d\mathbf{b}}{dt} + {}^{\mathcal{B}}\boldsymbol{\omega}^{\mathcal{A}} \times \mathbf{b} \tag{2–129}$$

where ${}^{\mathcal{B}}\boldsymbol{\omega}^{\mathcal{A}}$ is the angular velocity of reference frame $\mathcal{A}$ as viewed by an observer in reference frame $\mathcal{B}$. It is noted that the angular velocities ${}^{\mathcal{A}}\boldsymbol{\omega}^{\mathcal{B}}$ and ${}^{\mathcal{B}}\boldsymbol{\omega}^{\mathcal{A}}$ are related as

$${}^{\mathcal{A}}\boldsymbol{\omega}^{\mathcal{B}} = -{}^{\mathcal{B}}\boldsymbol{\omega}^{\mathcal{A}} \tag{2–130}$$

2.9.2 Addition of Angular Velocities

Let $\mathcal{A}_1$, $\mathcal{A}_2$, and $\mathcal{A}_3$ be reference frames. Furthermore, let ${}^{\mathcal{A}_1}\boldsymbol{\omega}^{\mathcal{A}_2}$ be the angular velocity of $\mathcal{A}_2$ as viewed by an observer in $\mathcal{A}_1$ and let ${}^{\mathcal{A}_2}\boldsymbol{\omega}^{\mathcal{A}_3}$ be the angular velocity of $\mathcal{A}_3$ as viewed by an observer in $\mathcal{A}_2$. Then from Eq. (2–128) we have

$$\frac{{}^{\mathcal{A}_1}d\mathbf{b}}{dt} = \frac{{}^{\mathcal{A}_2}d\mathbf{b}}{dt} + {}^{\mathcal{A}_1}\boldsymbol{\omega}^{\mathcal{A}_2} \times \mathbf{b} \tag{2–131}$$

$$\frac{{}^{\mathcal{A}_2}d\mathbf{b}}{dt} = \frac{{}^{\mathcal{A}_3}d\mathbf{b}}{dt} + {}^{\mathcal{A}_2}\boldsymbol{\omega}^{\mathcal{A}_3} \times \mathbf{b} \tag{2–132}$$

Now, substituting Eq. (2–132) into (2–131), we obtain

$$\frac{{}^{\mathcal{A}_1}d\mathbf{b}}{dt} = \frac{{}^{\mathcal{A}_3}d\mathbf{b}}{dt} + {}^{\mathcal{A}_2}\boldsymbol{\omega}^{\mathcal{A}_3} \times \mathbf{b} + {}^{\mathcal{A}_1}\boldsymbol{\omega}^{\mathcal{A}_2} \times \mathbf{b} \tag{2–133}$$

Equation (2–133) can be rewritten as

$$\frac{{}^{\mathcal{A}_1}d\mathbf{b}}{dt} = \frac{{}^{\mathcal{A}_3}d\mathbf{b}}{dt} + \left({}^{\mathcal{A}_1}\boldsymbol{\omega}^{\mathcal{A}_2} + {}^{\mathcal{A}_2}\boldsymbol{\omega}^{\mathcal{A}_3}\right) \times \mathbf{b} \tag{2–134}$$

Now, from the transport theorem of Eq. (2–128), we also have that

$$\frac{{}^{\mathcal{A}_1}d\mathbf{b}}{dt} = \frac{{}^{\mathcal{A}_3}d\mathbf{b}}{dt} + {}^{\mathcal{A}_1}\boldsymbol{\omega}^{\mathcal{A}_3} \times \mathbf{b} \tag{2–135}$$

Setting Eqs. (2–135) and (2–134) equal, we obtain

$$\boxed{{}^{\mathcal{A}_1}\boldsymbol{\omega}^{\mathcal{A}_3} = {}^{\mathcal{A}_1}\boldsymbol{\omega}^{\mathcal{A}_2} + {}^{\mathcal{A}_2}\boldsymbol{\omega}^{\mathcal{A}_3}} \tag{2–136}$$

Equation (2–136) is called the *angular velocity addition theorem* and states that the angular velocity of a reference frame $\mathcal{A}_3$ in reference frame $\mathcal{A}_1$ is the sum of the angular velocity of $\mathcal{A}_3$ in $\mathcal{A}_2$ and the angular velocity of $\mathcal{A}_2$ in $\mathcal{A}_1$. Moreover, the angular velocity addition theorem can be extended to more than three reference frames as follows. Suppose we are given reference frames $\mathcal{A}_1, \ldots, \mathcal{A}_n$ such that ${}^{\mathcal{A}_i}\boldsymbol{\omega}^{\mathcal{A}_{i+1}}$ $(i = 1, \ldots, n-1)$ are the angular velocities of reference frames $\mathcal{A}_{i+1}$ $(i = 1, \ldots, n-1)$ in reference frames $\mathcal{A}_i$ $(i = 1, \ldots, n-1)$, respectively. Then the angular velocity of reference frame $\mathcal{A}_n$ in reference frame $\mathcal{A}_1$, denoted ${}^{\mathcal{A}_1}\boldsymbol{\omega}^{\mathcal{A}_n}$, is given as

$$\boxed{{}^{\mathcal{A}_1}\boldsymbol{\omega}^{\mathcal{A}_n} = {}^{\mathcal{A}_1}\boldsymbol{\omega}^{\mathcal{A}_2} + {}^{\mathcal{A}_2}\boldsymbol{\omega}^{\mathcal{A}_3} + \cdots + {}^{\mathcal{A}_{n-1}}\boldsymbol{\omega}^{\mathcal{A}_n}} \tag{2–137}$$

2.9.3 Angular Acceleration

The angular velocity defined in Eq. (2–122) defines the manner in which a reference frame $\mathcal{B}$ rotates with time relative to another reference frame $\mathcal{A}$. Now suppose that we apply the rate of change transport theorem of Eq. (2–128) to the angular velocity itself, i.e., we apply Eq. (2–128) to ${}^{\mathcal{A}}\boldsymbol{\omega}^{\mathcal{B}}$. We then have that

$$\frac{{}^{\mathcal{A}}d}{dt}\left({}^{\mathcal{A}}\boldsymbol{\omega}^{\mathcal{B}}\right) = \frac{{}^{\mathcal{B}}d}{dt}\left({}^{\mathcal{A}}\boldsymbol{\omega}^{\mathcal{B}}\right) + {}^{\mathcal{A}}\boldsymbol{\omega}^{\mathcal{B}} \times {}^{\mathcal{A}}\boldsymbol{\omega}^{\mathcal{B}} \tag{2–138}$$

Now, because ${}^{\mathcal{A}}\boldsymbol{\omega}^{\mathcal{B}} \times {}^{\mathcal{A}}\boldsymbol{\omega}^{\mathcal{B}} = \mathbf{0}$, Eq. (2–138) reduces to

$$\frac{{}^{\mathcal{A}}d}{dt}\left({}^{\mathcal{A}}\boldsymbol{\omega}^{\mathcal{B}}\right) = \frac{{}^{\mathcal{B}}d}{dt}\left({}^{\mathcal{A}}\boldsymbol{\omega}^{\mathcal{B}}\right) \tag{2–139}$$

It is seen from Eq. (2–139) that the rate of change of ${}^{\mathcal{A}}\boldsymbol{\omega}^{\mathcal{B}}$ in reference frame $\mathcal{A}$ is the *same* as the rate of change of ${}^{\mathcal{A}}\boldsymbol{\omega}^{\mathcal{B}}$ in reference frame $\mathcal{B}$. The quantity

$$\frac{{}^{\mathcal{A}}d}{dt}\left({}^{\mathcal{A}}\boldsymbol{\omega}^{\mathcal{B}}\right) \tag{2–140}$$

is called the *angular acceleration* of reference frame $\mathcal{B}$ in reference frame $\mathcal{A}$. As a convention, we use the variable $\boldsymbol{\alpha}$ to denote angular acceleration, i.e., the quantity

$$\boxed{{}^{\mathcal{A}}\boldsymbol{\alpha}^{\mathcal{B}} \equiv \frac{{}^{\mathcal{A}}d}{dt}\left({}^{\mathcal{A}}\boldsymbol{\omega}^{\mathcal{B}}\right) = \frac{{}^{\mathcal{B}}d}{dt}\left({}^{\mathcal{A}}\boldsymbol{\omega}^{\mathcal{B}}\right)} \tag{2–141}$$

is the angular acceleration of reference frame $\mathcal{B}$ in reference frame $\mathcal{A}$. Now, from Eq. (2–130) we see that

$${}^{\mathcal{A}}\boldsymbol{\alpha}^{\mathcal{B}} = \frac{{}^{\mathcal{A}}d}{dt}\left({}^{\mathcal{A}}\boldsymbol{\omega}^{\mathcal{B}}\right) = -\frac{{}^{\mathcal{A}}d}{dt}\left({}^{\mathcal{B}}\boldsymbol{\omega}^{\mathcal{A}}\right) = -{}^{\mathcal{B}}\boldsymbol{\alpha}^{\mathcal{A}} \tag{2–142}$$

which is similar to the result obtained for angular velocity as shown in Eq. (2–130).

Suppose now that we compute the rate of change of ${}^{\mathcal{A}_1}\boldsymbol{\omega}^{\mathcal{A}_3}$ in reference frame $\mathcal{A}_1$ using the expression for ${}^{\mathcal{A}_1}\boldsymbol{\omega}^{\mathcal{A}_3}$ in Eq. (2–136). We then obtain

$$\frac{{}^{\mathcal{A}_1}d}{dt}\left({}^{\mathcal{A}_1}\boldsymbol{\omega}^{\mathcal{A}_3}\right) = \frac{{}^{\mathcal{A}_1}d}{dt}\left({}^{\mathcal{A}_1}\boldsymbol{\omega}^{\mathcal{A}_2} + {}^{\mathcal{A}_2}\boldsymbol{\omega}^{\mathcal{A}_3}\right) \tag{2–143}$$

Expanding Eq. (2–143) gives

$$\frac{{}^{\mathcal{A}_1}d}{dt}\left({}^{\mathcal{A}_1}\boldsymbol{\omega}^{\mathcal{A}_3}\right) = \frac{{}^{\mathcal{A}_1}d}{dt}\left({}^{\mathcal{A}_1}\boldsymbol{\omega}^{\mathcal{A}_2}\right) + \frac{{}^{\mathcal{A}_1}d}{dt}\left({}^{\mathcal{A}_2}\boldsymbol{\omega}^{\mathcal{A}_3}\right) \tag{2–144}$$

Applying the rate of change transport theorem of Eq. (2–128) to the second term in Eq. (2–144), we have

$$\frac{{}^{\mathcal{A}_1}d}{dt}\left({}^{\mathcal{A}_1}\boldsymbol{\omega}^{\mathcal{A}_3}\right) = \frac{{}^{\mathcal{A}_1}d}{dt}\left({}^{\mathcal{A}_1}\boldsymbol{\omega}^{\mathcal{A}_2}\right) + \frac{{}^{\mathcal{A}_2}d}{dt}\left({}^{\mathcal{A}_2}\boldsymbol{\omega}^{\mathcal{A}_3}\right) + {}^{\mathcal{A}_1}\boldsymbol{\omega}^{\mathcal{A}_2} \times {}^{\mathcal{A}_2}\boldsymbol{\omega}^{\mathcal{A}_3} \tag{2–145}$$

Then, using the definition of angular acceleration as given in Eq. (2–141), the first and second terms of Eq. (2–145) can be written as

$$\begin{aligned}\frac{{}^{\mathcal{A}_1}d}{dt}\left({}^{\mathcal{A}_1}\boldsymbol{\omega}^{\mathcal{A}_2}\right) &= {}^{\mathcal{A}_1}\boldsymbol{\alpha}^{\mathcal{A}_2} \\ \frac{{}^{\mathcal{A}_2}d}{dt}\left({}^{\mathcal{A}_2}\boldsymbol{\omega}^{\mathcal{A}_3}\right) &= {}^{\mathcal{A}_2}\boldsymbol{\alpha}^{\mathcal{A}_3}\end{aligned} \tag{2–146}$$

Equation (2–145) can then be written as

$$\frac{{}^{\mathcal{A}_1}d}{dt}\left({}^{\mathcal{A}_1}\boldsymbol{\omega}^{\mathcal{A}_3}\right) = {}^{\mathcal{A}_1}\boldsymbol{\alpha}^{\mathcal{A}_2} + {}^{\mathcal{A}_2}\boldsymbol{\alpha}^{\mathcal{A}_3} + {}^{\mathcal{A}_1}\boldsymbol{\omega}^{\mathcal{A}_2} \times {}^{\mathcal{A}_2}\boldsymbol{\omega}^{\mathcal{A}_3} \tag{2–147}$$

Finally, using the fact that

$$\frac{{}^{\mathcal{A}_1}d}{dt}\left({}^{\mathcal{A}_1}\boldsymbol{\omega}^{\mathcal{A}_3}\right) = {}^{\mathcal{A}_1}\boldsymbol{\alpha}^{\mathcal{A}_3} \tag{2-148}$$

Equation (2–147) simplifies to

$${}^{\mathcal{A}_1}\boldsymbol{\alpha}^{\mathcal{A}_3} = {}^{\mathcal{A}_1}\boldsymbol{\alpha}^{\mathcal{A}_2} + {}^{\mathcal{A}_2}\boldsymbol{\alpha}^{\mathcal{A}_3} + {}^{\mathcal{A}_1}\boldsymbol{\omega}^{\mathcal{A}_2} \times {}^{\mathcal{A}_2}\boldsymbol{\omega}^{\mathcal{A}_3} \tag{2-149}$$

It can be seen from Eq. (2–149) that, because the term ${}^{\mathcal{A}_1}\boldsymbol{\omega}^{\mathcal{A}_2} \times {}^{\mathcal{A}_2}\boldsymbol{\omega}^{\mathcal{A}_3}$ is generally *not* zero, the angular velocity addition theorem of Eq. (2–136) *does not* extend to angular acceleration, i.e., in general it is the case that

$$\boxed{{}^{\mathcal{A}_1}\boldsymbol{\alpha}^{\mathcal{A}_3} \neq {}^{\mathcal{A}_1}\boldsymbol{\alpha}^{\mathcal{A}_2} + {}^{\mathcal{A}_2}\boldsymbol{\alpha}^{\mathcal{A}_3}} \tag{2-150}$$

2.10 Kinematics in a Rotating Reference Frame

Suppose now that $\mathbf{r}$ is the position of a point P in $\mathbb{E}^3$ measured relative to a point O where O is fixed simultaneously in two reference frames $\mathcal{A}$ and $\mathcal{B}$ as shown in Fig. 2–8. Suppose further that reference frame $\mathcal{B}$ rotates with angular velocity ${}^{\mathcal{A}}\boldsymbol{\omega}^{\mathcal{B}}$ relative to

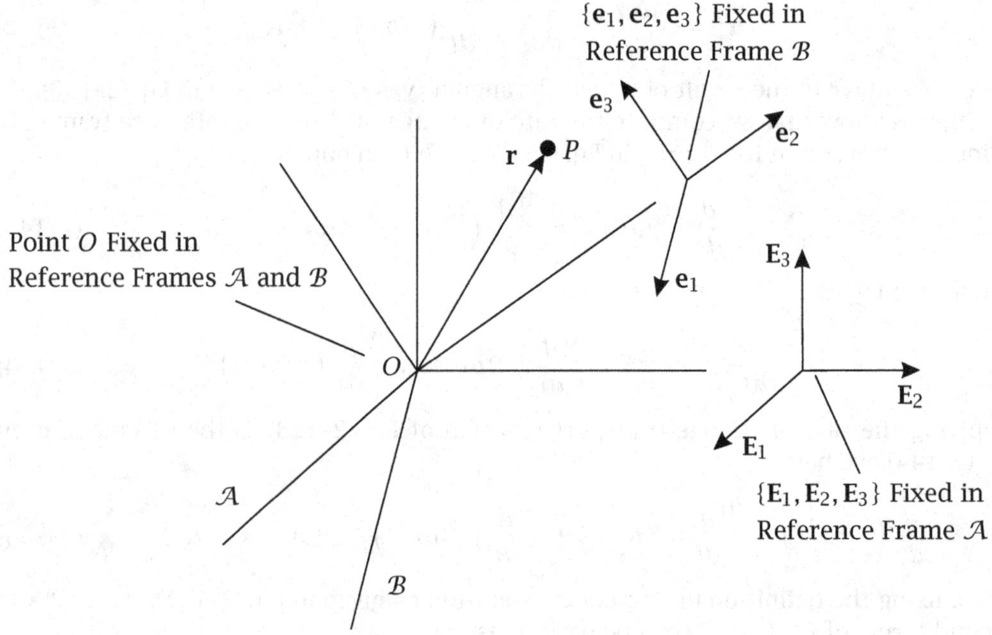

Figure 2–8 Position of point P in $\mathbb{E}^3$ measured relative to a point O that is fixed simultaneously in two reference frames $\mathcal{A}$ and $\mathcal{B}$.

reference frame $\mathcal{A}$. It is seen that, because O is fixed in both reference frames $\mathcal{A}$ and $\mathcal{B}$, the position of point P is the same vector in *both* reference frames. Then, applying the rate of change transport theorem of Eq. (2–128), the velocity of point P as viewed by an observer in reference frame $\mathcal{A}$ is given as

$${}^{\mathcal{A}}\mathbf{v} = \frac{{}^{\mathcal{A}}d\mathbf{r}}{dt} = \frac{{}^{\mathcal{B}}d\mathbf{r}}{dt} + {}^{\mathcal{A}}\boldsymbol{\omega}^{\mathcal{B}} \times \mathbf{r} \tag{2-151}$$

Equation (2–151) can be written as

$$\boxed{{}^{\mathcal{A}}\mathbf{v} = {}^{\mathcal{B}}\mathbf{v} + {}^{\mathcal{A}}\boldsymbol{\omega}^{\mathcal{B}} \times \mathbf{r}} \tag{2–152}$$

Next, applying the rate of change transport theorem to ${}^{\mathcal{A}}\mathbf{v}$, the acceleration as viewed by an observer in reference frame $\mathcal{A}$ is given as

$${}^{\mathcal{A}}\mathbf{a} = \frac{{}^{\mathcal{A}}d}{dt}\left({}^{\mathcal{A}}\mathbf{v}\right) = \frac{{}^{\mathcal{B}}d}{dt}\left({}^{\mathcal{A}}\mathbf{v}\right) + {}^{\mathcal{A}}\boldsymbol{\omega}^{\mathcal{B}} \times {}^{\mathcal{A}}\mathbf{v} \tag{2–153}$$

Substituting ${}^{\mathcal{A}}\mathbf{v}$ from Eq. (2–152) into (2–153), we obtain

$${}^{\mathcal{A}}\mathbf{a} = \frac{{}^{\mathcal{B}}d}{dt}\left({}^{\mathcal{B}}\mathbf{v} + {}^{\mathcal{A}}\boldsymbol{\omega}^{\mathcal{B}} \times \mathbf{r}\right) + {}^{\mathcal{A}}\boldsymbol{\omega}^{\mathcal{B}} \times \left({}^{\mathcal{B}}\mathbf{v} + {}^{\mathcal{A}}\boldsymbol{\omega}^{\mathcal{B}} \times \mathbf{r}\right) \tag{2–154}$$

Expanding Eq. (2–154), we obtain

$${}^{\mathcal{A}}\mathbf{a} = \frac{{}^{\mathcal{B}}d}{dt}\left({}^{\mathcal{B}}\mathbf{v}\right) + \frac{{}^{\mathcal{B}}d}{dt}\left({}^{\mathcal{A}}\boldsymbol{\omega}^{\mathcal{B}} \times \mathbf{r}\right) + {}^{\mathcal{A}}\boldsymbol{\omega}^{\mathcal{B}} \times {}^{\mathcal{B}}\mathbf{v} + {}^{\mathcal{A}}\boldsymbol{\omega}^{\mathcal{B}} \times \left({}^{\mathcal{A}}\boldsymbol{\omega}^{\mathcal{B}} \times \mathbf{r}\right) \tag{2–155}$$

Now we note that

$${}^{\mathcal{B}}\mathbf{a} = \frac{{}^{\mathcal{B}}d}{dt}\left({}^{\mathcal{B}}\mathbf{v}\right) \tag{2–156}$$

Substituting the result of Eq. (2–156) into (2–155) and expanding using the product rule of differentiation, we obtain

$${}^{\mathcal{A}}\mathbf{a} = {}^{\mathcal{B}}\mathbf{a} + \frac{{}^{\mathcal{B}}d}{dt}\left({}^{\mathcal{A}}\boldsymbol{\omega}^{\mathcal{B}}\right) \times \mathbf{r} + {}^{\mathcal{A}}\boldsymbol{\omega}^{\mathcal{B}} \times \frac{{}^{\mathcal{B}}d\mathbf{r}}{dt} + {}^{\mathcal{A}}\boldsymbol{\omega}^{\mathcal{B}} \times {}^{\mathcal{B}}\mathbf{v} + {}^{\mathcal{A}}\boldsymbol{\omega}^{\mathcal{B}} \times \left({}^{\mathcal{A}}\boldsymbol{\omega}^{\mathcal{B}} \times \mathbf{r}\right) \tag{2–157}$$

Finally, using the definition of angular acceleration as given in Eq. (2–141) on page 49, the fact that ${}^{\mathcal{B}}d\mathbf{r}/dt = {}^{\mathcal{B}}\mathbf{v}$, and combining like terms, we obtain

$$\boxed{{}^{\mathcal{A}}\mathbf{a} = {}^{\mathcal{B}}\mathbf{a} + {}^{\mathcal{A}}\boldsymbol{\alpha}^{\mathcal{B}} \times \mathbf{r} + 2\,{}^{\mathcal{A}}\boldsymbol{\omega}^{\mathcal{B}} \times {}^{\mathcal{B}}\mathbf{v} + {}^{\mathcal{A}}\boldsymbol{\omega}^{\mathcal{B}} \times \left({}^{\mathcal{A}}\boldsymbol{\omega}^{\mathcal{B}} \times \mathbf{r}\right)} \tag{2–158}$$

2.11 Common Coordinate Systems

In this section we discuss some common coordinate systems that are used as building blocks for modeling the position, velocity, and acceleration of a point in $\mathbb{E}^3$. In particular, models are developed for the kinematics of a point using Cartesian coordinates, cylindrical coordinates, and spherical coordinates.

2.11.1 Cartesian Coordinates

Let $\mathbf{r}$ be the position of a point P in $\mathbb{E}^3$ measured relative to a point O, where O is fixed in an arbitrary reference frame $\mathcal{A}$. Furthermore, let $\{\mathbf{e}_x, \mathbf{e}_y, \mathbf{e}_z\}$ be a Cartesian basis fixed in reference frame $\mathcal{A}$. Then $\mathbf{r}$ can be written in terms of the basis $\{\mathbf{e}_x, \mathbf{e}_y, \mathbf{e}_z\}$ as

$$\mathbf{r} = x\mathbf{e}_x + y\mathbf{e}_y + z\mathbf{e}_z \tag{2–159}$$

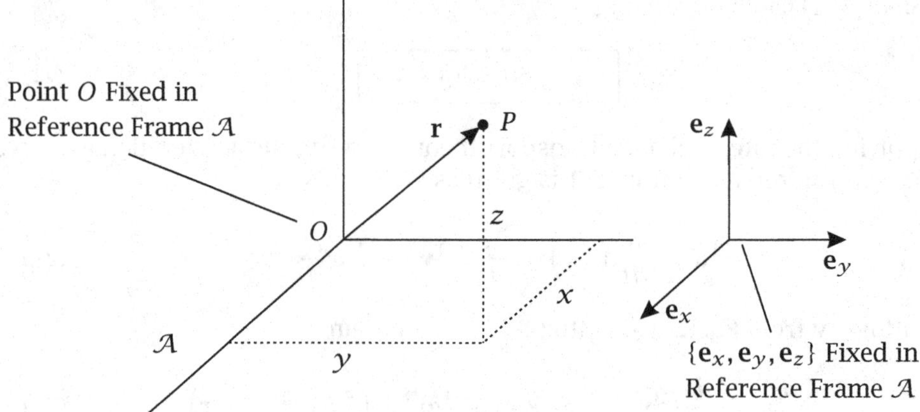

Figure 2–9 Cartesian basis $\{\mathbf{e}_x, \mathbf{e}_y, \mathbf{e}_z\}$ fixed in a reference frame $\mathcal{B}$.

where x, y, and z are the components of the position in the directions $\mathbf{e}_x$, $\mathbf{e}_y$, and $\mathbf{e}_z$, respectively, as shown in Fig. 2–9. The basis $\{\mathbf{e}_x, \mathbf{e}_y, \mathbf{e}_z\}$ is called a *Cartesian* basis because the components of the position of a point P are their distances from the planes defined by the pairs of basis vectors. More generally, when expressed in a Cartesian basis, all points in $\mathbb{E}^3$ that have a common component lie in the same plane. The velocity of point P as viewed by an observer in reference frame $\mathcal{A}$ is then given in terms of a Cartesian basis fixed in reference frame $\mathcal{A}$ as

$$^{\mathcal{A}}\mathbf{v} = \frac{^{\mathcal{A}}d\mathbf{r}}{dt} = \dot{x}\mathbf{e}_x + \dot{y}\mathbf{e}_y + \dot{z}\mathbf{e}_z \tag{2–160}$$

Finally, the acceleration of point P as viewed by an observer in reference frame $\mathcal{A}$ is given in terms of a Cartesian basis fixed in reference frame $\mathcal{A}$ as

$$^{\mathcal{A}}\mathbf{a} = \frac{^{\mathcal{A}}d}{dt}\left(^{\mathcal{A}}\mathbf{v}\right) = \ddot{x}\mathbf{e}_x + \ddot{y}\mathbf{e}_y + \ddot{z}\mathbf{e}_z \tag{2–161}$$

Example 2–2

A particle P slides on a circular disk, where the disk is oriented horizontally and rotates about the vertical direction and about its center with constant angular velocity Ω as shown in Fig. 2–10. Knowing that the position of the particle is measured relative to the orthogonal directions OA and OB, where point O is the center of the disk and points A and B are fixed to the edge of the disk, determine the velocity and acceleration of the particle as viewed by an observer fixed to the ground.

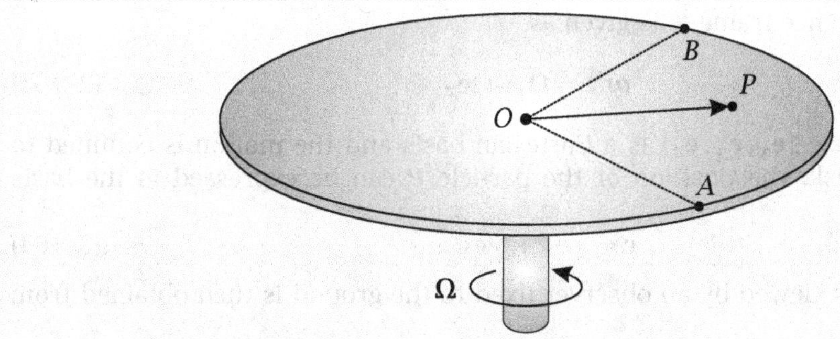

Figure 2–10 Particle sliding on a rotating disk.

Solution to Example 2–2

First, let $\mathcal{F}$ be a reference frame fixed to the ground. Then choose the following coordinate system fixed in reference frame $\mathcal{F}$:

$$\begin{array}{ccc} & \text{Origin at } O & \\ \mathbf{E}_x & = & \text{Along } OA \text{ at } t = 0 \\ \mathbf{E}_y & = & \text{Along } OB \text{ at } t = 0 \\ \mathbf{E}_z & = & \mathbf{E}_x \times \mathbf{E}_y \end{array}$$

Next, let $\mathcal{A}$ be a reference frame fixed to the disk. Then choose the following coordinate system fixed in reference frame $\mathcal{A}$:

$$\begin{array}{ccc} & \text{Origin at } O & \\ \mathbf{e}_x & = & \text{Along } OA \\ \mathbf{e}_y & = & \text{Along } OB \\ \mathbf{e}_z & = & \mathbf{e}_x \times \mathbf{e}_y \end{array}$$

The basis $\{\mathbf{e}_x, \mathbf{e}_y, \mathbf{e}_z\}$ is shown in Fig. 2-11.

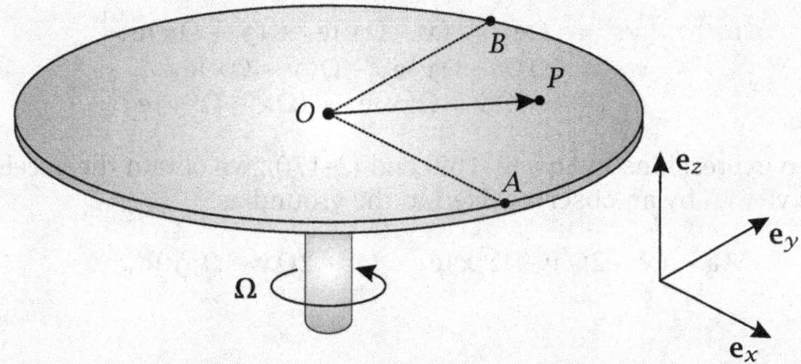

Figure 2–11 Basis $\{\mathbf{e}_x, \mathbf{e}_y, \mathbf{e}_z\}$ fixed in reference frame $\mathcal{A}$ for Example 2-2.

Now, because the disk rotates with constant angular velocity Ω relative to the ground and $\{\mathbf{e}_x, \mathbf{e}_y, \mathbf{e}_z\}$ is fixed to the disk, the angular velocity of $\mathcal{A}$ as viewed by an observer in reference frame $\mathcal{F}$ is given as

$$ {}^{\mathcal{F}}\boldsymbol{\omega}^{\mathcal{A}} = \boldsymbol{\Omega} = \Omega\mathbf{e}_z \tag{2-162}$$

Furthermore, because $\{\mathbf{e}_x, \mathbf{e}_y, \mathbf{e}_z\}$ is a Cartesian basis and the motion is confined to the plane of the disk, the position of the particle P can be expressed in the basis $\{\mathbf{e}_x, \mathbf{e}_y, \mathbf{e}_z\}$ as

$$ \mathbf{r} = x\mathbf{e}_x + y\mathbf{e}_y \tag{2-163}$$

The velocity of P as viewed by an observer fixed to the ground is then obtained from Eq. (2–128) as

$$ {}^{\mathcal{F}}\mathbf{v} \equiv \frac{{}^{\mathcal{F}}d\mathbf{r}}{dt} = \frac{{}^{\mathcal{A}}d\mathbf{r}}{dt} + {}^{\mathcal{F}}\boldsymbol{\omega}^{\mathcal{A}} \times \mathbf{r} \tag{2-164}$$

Substituting $\mathbf{r}$ from Eq. (2–163) and ${}^{\mathcal{F}}\boldsymbol{\omega}^{\mathcal{A}}$ from Eq. (2–162) into (2–164) we have

$$ \frac{{}^{\mathcal{A}}d\mathbf{r}}{dt} = \dot{x}\mathbf{e}_x + \dot{y}\mathbf{e}_y \tag{2-165}$$

$$ {}^{\mathcal{F}}\boldsymbol{\omega}^{\mathcal{A}} \times \mathbf{r} = \Omega\mathbf{e}_z \times (x\mathbf{e}_x + y\mathbf{e}_y) = \Omega x\mathbf{e}_y - \Omega y\mathbf{e}_x \tag{2-166}$$

Adding Eqs. (2–165) and (2–166), we obtain the velocity of the particle as viewed by an observer fixed to the ground as

$$ {}^{\mathcal{F}}\mathbf{v} = (\dot{x} - \Omega y)\mathbf{e}_x + (\dot{y} + \Omega x)\mathbf{e}_y \tag{2-167}$$

Next, applying Eq. (2–128) to ${}^{\mathcal{F}}\mathbf{v}$, the acceleration of P as viewed by an observer fixed to the ground is given as

$$ {}^{\mathcal{F}}\mathbf{a} \equiv \frac{{}^{\mathcal{F}}d}{dt}\left({}^{\mathcal{F}}\mathbf{v}\right) = \frac{{}^{\mathcal{A}}d}{dt}\left({}^{\mathcal{F}}\mathbf{v}\right) + {}^{\mathcal{F}}\boldsymbol{\omega}^{\mathcal{A}} \times {}^{\mathcal{F}}\mathbf{v} \tag{2-168}$$

Now, because Ω is constant in both magnitude and direction, we have

$$ \frac{{}^{\mathcal{A}}d}{dt}\left({}^{\mathcal{F}}\mathbf{v}\right) = (\ddot{x} - \Omega\dot{y})\mathbf{e}_x + (\ddot{y} + \Omega\dot{x})\mathbf{e}_y \tag{2-169}$$

$$ \begin{aligned} {}^{\mathcal{F}}\boldsymbol{\omega}^{\mathcal{A}} \times {}^{\mathcal{F}}\mathbf{v} &= \Omega\mathbf{e}_z \times \left[(\dot{x} - \Omega y)\mathbf{e}_x + (\dot{y} + \Omega x)\mathbf{e}_y\right] \\ &= \Omega(\dot{x} - \Omega y)\mathbf{e}_y - \Omega(\dot{y} + \Omega x)\mathbf{e}_x \\ &= -(\Omega\dot{y} + \Omega^2 x)\mathbf{e}_x + (\Omega\dot{x} - \Omega^2 y)\mathbf{e}_y \end{aligned} \tag{2-170}$$

Adding the two expressions in Eqs. (2–169) and (2–170), we obtain the acceleration of the particle as viewed by an observer fixed to the ground as

$$ {}^{\mathcal{F}}\mathbf{a} = (\ddot{x} - 2\Omega\dot{y} - \Omega^2 x)\mathbf{e}_x + (\ddot{y} + 2\Omega\dot{x} - \Omega^2 y)\mathbf{e}_y \tag{2-171}$$

■

2.11.2 Cylindrical Coordinates

Let $\mathbf{r}$ be the position of a point P in $\mathbb{E}^3$ measured relative to a point O, where O is fixed in an arbitrary reference frame $\mathcal{A}$. Furthermore, let $\{\mathbf{e}_x, \mathbf{e}_y, \mathbf{e}_z\}$ be a Cartesian basis for $\mathcal{A}$. Recalling from Eq. (2–159) the expression for $\mathbf{r}$ in terms of the basis $\{\mathbf{e}_x, \mathbf{e}_y, \mathbf{e}_z\}$, we have

$$\mathbf{r} = x\mathbf{e}_x + y\mathbf{e}_y + z\mathbf{e}_z \tag{2–172}$$

Suppose now that we make the following substitutions:

$$\begin{aligned} x &= r\cos\theta \\ y &= r\sin\theta \end{aligned} \tag{2–173}$$

where

$$\begin{aligned} r &= \sqrt{x^2 + y^2} \\ \theta &= \tan^{-1}(y/x) \end{aligned} \tag{2–174}$$

Then $\mathbf{r}$ can be written as

$$\begin{aligned} \mathbf{r} &= r\cos\theta\,\mathbf{e}_x + r\sin\theta\,\mathbf{e}_y + z\mathbf{e}_z \\ &= r(\cos\theta\,\mathbf{e}_x + \sin\theta\,\mathbf{e}_y) + z\mathbf{e}_z \end{aligned} \tag{2–175}$$

Now let

$$\mathbf{e}_r = \cos\theta\,\mathbf{e}_x + \sin\theta\,\mathbf{e}_y \tag{2–176}$$

Denoting Q as the projection of P into the $\{\mathbf{e}_x, \mathbf{e}_y\}$-plane, $\mathbf{e}_r$ is the direction from O to Q. The position of point P is then given in terms of $\mathbf{e}_r$ as

$$\mathbf{r} = r\mathbf{e}_r + z\mathbf{e}_z \tag{2–177}$$

Now it is seen that $\mathbf{e}_r$ is a unit vector. Furthermore, because $\mathbf{e}_r$ lies in the $\{\mathbf{e}_x, \mathbf{e}_y\}$-plane, $\mathbf{e}_r$ is orthogonal to $\mathbf{e}_z$. Consequently, the vectors $\mathbf{e}_r$ and $\mathbf{e}_z$ can be used to define an orthonormal basis. Assuming that $\mathbf{e}_r$ and $\mathbf{e}_z$ are the first and third basis vectors of this new basis, the second basis vector is obtained from the right-hand rule as

$$\mathbf{e}_\theta = \mathbf{e}_z \times \mathbf{e}_r \tag{2–178}$$

Using Eq. (2–176), the vector $\mathbf{e}_\theta$ is then obtained in terms of $\mathbf{e}_x$ and $\mathbf{e}_y$ as

$$\mathbf{e}_\theta = \mathbf{e}_z \times (\cos\theta\,\mathbf{e}_x + \sin\theta\,\mathbf{e}_y) = -\sin\theta\,\mathbf{e}_x + \cos\theta\,\mathbf{e}_y \tag{2–179}$$

Note that because $\mathbf{e}_\theta$ was obtained via the vector product of $\mathbf{e}_z$ with $\mathbf{e}_r$, it follows that $\mathbf{e}_\theta$ is orthogonal to both $\mathbf{e}_z$ and $\mathbf{e}_r$. Furthermore, because both $\mathbf{e}_z$ and $\mathbf{e}_r$ are orthogonal unit vectors, $\mathbf{e}_\theta$ is also a unit vector. The orthonormal basis $\{\mathbf{e}_r, \mathbf{e}_\theta, \mathbf{e}_z\}$ is called a *cylindrical* or *radial-transverse* basis and is shown in Fig. 2–12.

Suppose now that we define a reference frame $\mathcal{B}$ such that the basis $\{\mathbf{e}_r, \mathbf{e}_\theta, \mathbf{e}_z\}$ is fixed in $\mathcal{B}$. Then, because $\{\mathbf{e}_r, \mathbf{e}_\theta, \mathbf{e}_z\}$ rotates with angular rate $\dot{\theta}$ about the $\mathbf{e}_z$-direction, the angular velocity of $\mathcal{B}$ as viewed by an observer in $\mathcal{A}$ is given as

$${}^{\mathcal{A}}\boldsymbol{\omega}^{\mathcal{B}} = \dot{\theta}\mathbf{e}_z \tag{2–180}$$

Applying the rate of change transport theorem of Eq. (2–128) to $\mathbf{r}$ using the expressions for $\mathbf{r}$ and ${}^{\mathcal{A}}\boldsymbol{\omega}^{\mathcal{B}}$ from Eqs. (2–177) and (2–180), respectively, the velocity of the particle as viewed by an observer in reference frame $\mathcal{A}$ is given as

$${}^{\mathcal{A}}\mathbf{v} = \frac{{}^{\mathcal{A}}d\mathbf{r}}{dt} = \frac{{}^{\mathcal{B}}d\mathbf{r}}{dt} + {}^{\mathcal{A}}\boldsymbol{\omega}^{\mathcal{B}} \times \mathbf{r} \tag{2–181}$$

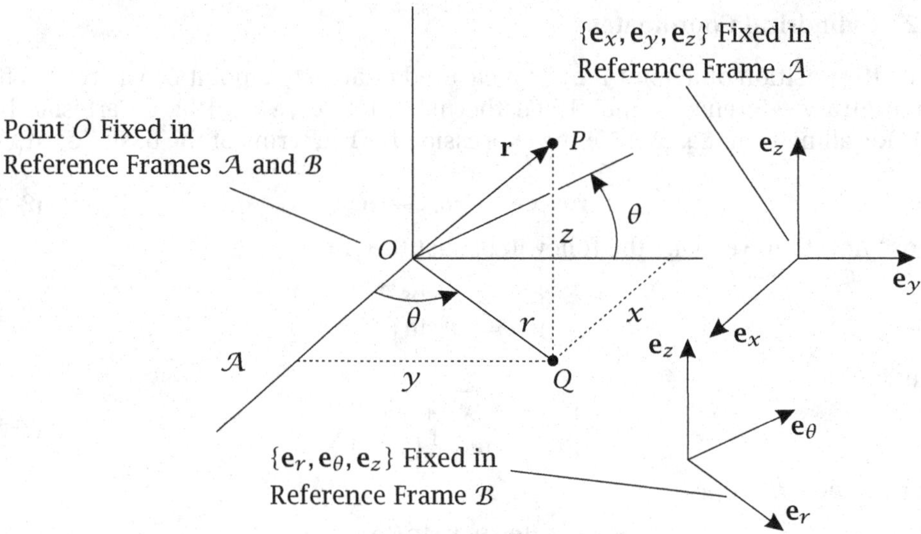

Figure 2-12 Cylindrical basis $\{\mathbf{e}_r, \mathbf{e}_\theta, \mathbf{e}_z\}$ fixed in reference frame $\mathcal{B}$ and defined relative to a Cartesian basis $\{\mathbf{e}_x, \mathbf{e}_y, \mathbf{e}_z\}$ fixed in reference frame $\mathcal{A}$.

Now we have

$$\frac{{}^{\mathcal{B}}d\mathbf{r}}{dt} = \dot{r}\mathbf{e}_r + \dot{z}\mathbf{e}_z \tag{2-182}$$

$${}^{\mathcal{A}}\boldsymbol{\omega}^{\mathcal{B}} \times \mathbf{r} = \dot{\theta}\mathbf{e}_z \times (r\mathbf{e}_r + z\mathbf{e}_z) = r\dot{\theta}\mathbf{e}_\theta \tag{2-183}$$

Adding Eqs. (2-182) and (2-183), we obtain the velocity of the particle as viewed by an observer in reference frame $\mathcal{A}$ as

$${}^{\mathcal{A}}\mathbf{v} = \dot{r}\mathbf{e}_r + r\dot{\theta}\mathbf{e}_\theta + \dot{z}\mathbf{e}_z \tag{2-184}$$

Applying the rate of change transport theorem of Eq. (2-128) to ${}^{\mathcal{A}}\mathbf{v}$, we obtain the acceleration as viewed by an observer in reference frame $\mathcal{A}$ as

$${}^{\mathcal{A}}\mathbf{a} = \frac{{}^{\mathcal{A}}d}{dt}\left({}^{\mathcal{A}}\mathbf{v}\right) = \frac{{}^{\mathcal{B}}d}{dt}\left({}^{\mathcal{A}}\mathbf{v}\right) + {}^{\mathcal{A}}\boldsymbol{\omega}^{\mathcal{B}} \times {}^{\mathcal{A}}\mathbf{v} \tag{2-185}$$

Now we have

$$\frac{{}^{\mathcal{B}}d}{dt}\left({}^{\mathcal{A}}\mathbf{v}\right) = \ddot{r}\mathbf{e}_r + (\dot{r}\dot{\theta} + r\ddot{\theta})\mathbf{e}_\theta + \ddot{z}\mathbf{e}_z \tag{2-186}$$

$${}^{\mathcal{A}}\boldsymbol{\omega}^{\mathcal{B}} \times {}^{\mathcal{A}}\mathbf{v} = \dot{\theta}\mathbf{e}_z \times (\dot{r}\mathbf{e}_r + r\dot{\theta}\mathbf{e}_\theta + \dot{z}\mathbf{e}_z) = \dot{r}\dot{\theta}\mathbf{e}_\theta - r\dot{\theta}^2\mathbf{e}_r \tag{2-187}$$

Adding Eqs. (2-186) and (2-187) and simplifying, the acceleration in reference frame $\mathcal{A}$ is obtained as

$${}^{\mathcal{A}}\mathbf{a} = (\ddot{r} - r\dot{\theta}^2)\mathbf{e}_r + (2\dot{r}\dot{\theta} + r\ddot{\theta})\mathbf{e}_\theta + \ddot{z}\mathbf{e}_z \tag{2-188}$$

Example 2–3

A collar slides freely along a rigid arm as shown in Fig. 2–13. The arm is hinged at one of its ends to a point O, where O is fixed to the ground, and rotates about a direction that is orthogonal to the plane of motion. Knowing that the position of the collar is described by the variables r and θ, where r is the displacement of the collar relative to point O and θ is the angle of the rod measured relative to a direction that is fixed to the ground, determine the velocity and acceleration of the collar as viewed by an observer fixed to the ground.

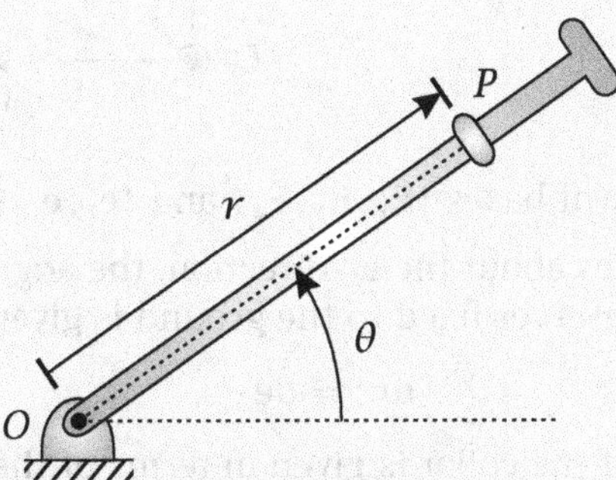

Figure 2–13 Collar sliding on a rotating arm.

Solution to Example 2–3

For this problem, it is convenient to define two reference frames. The first reference frame, denoted $\mathcal{F}$, is fixed to the ground while the second, denoted $\mathcal{A}$, is fixed to the arm. Corresponding to reference frame $\mathcal{F}$, we choose the following Cartesian coordinate system:

Origin at point O		
$\mathbf{E}_x$	=	To the right
$\mathbf{E}_z$	=	Out of page
$\mathbf{E}_y$	=	$\mathbf{E}_z \times \mathbf{E}_x$

Corresponding to reference frame $\mathcal{A}$, we choose the following cylindrical coordinate system:

Origin at point O		
$\mathbf{e}_r$	=	Along OP
$\mathbf{e}_z$	=	Out of page
$\mathbf{e}_\theta$	=	$\mathbf{e}_z \times \mathbf{e}_r$

The bases $\{\mathbf{E}_x, \mathbf{E}_y, \mathbf{E}_z\}$ and $\{\mathbf{e}_r, \mathbf{e}_\theta, \mathbf{e}_z\}$ are shown in Fig. 2–14.

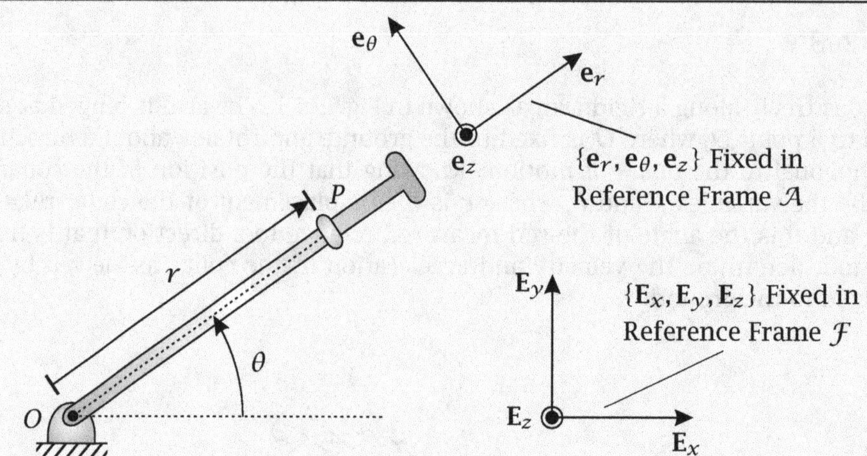

Figure 2–14 Geometry of bases $\{\mathbf{E}_x, \mathbf{E}_y, \mathbf{E}_z\}$ and $\{\mathbf{e}_r, \mathbf{e}_\theta, \mathbf{e}_z\}$ for Example 2-3.

Then, because the rod rotates about the $\mathbf{e}_z$-direction, the angular velocity of reference frame $\mathcal{A}$ as viewed by an observer fixed to the ground is given as

$$^{\mathcal{F}}\boldsymbol{\omega}^{\mathcal{A}} = \dot{\theta}\mathbf{e}_z \tag{2-189}$$

Furthermore, the position of the collar is given in terms of the basis $\{\mathbf{e}_r, \mathbf{e}_\theta, \mathbf{e}_z\}$ as

$$\mathbf{r} = r\mathbf{e}_r \tag{2-190}$$

Applying the rate of change transport theorem of Eq. (2–128) on page 47 to $\mathbf{r}$ between reference frames $\mathcal{A}$ and $\mathcal{F}$, the velocity of the collar as viewed by an observer fixed to the ground is given as

$$^{\mathcal{F}}\mathbf{v} = \frac{^{\mathcal{F}}d\mathbf{r}}{dt} = \frac{^{\mathcal{A}}d\mathbf{r}}{dt} + {^{\mathcal{F}}\boldsymbol{\omega}^{\mathcal{A}}} \times \mathbf{r} \tag{2-191}$$

Now we have

$$\begin{aligned} \frac{^{\mathcal{A}}d\mathbf{r}}{dt} &= \dot{r}\mathbf{e}_r &(2\text{-}192)\\ {^{\mathcal{F}}\boldsymbol{\omega}^{\mathcal{A}}} \times \mathbf{r} &= \dot{\theta}\mathbf{e}_z \times r\mathbf{e}_r = r\dot{\theta}\mathbf{e}_\theta &(2\text{-}193) \end{aligned}$$

Adding Eqs. (2–192) and (2–193), we obtain

$$^{\mathcal{F}}\mathbf{v} = \dot{r}\mathbf{e}_r + r\dot{\theta}\mathbf{e}_\theta \tag{2-194}$$

Applying the rate of change transport theorem of Eq. (2–128) on page 47 to $^{\mathcal{F}}\mathbf{v}$ between reference frames $\mathcal{A}$ and $\mathcal{F}$, the acceleration of the collar as viewed by an observer fixed to the ground is given as

$$^{\mathcal{F}}\mathbf{a} = \frac{^{\mathcal{F}}d}{dt}\left(^{\mathcal{F}}\mathbf{v}\right) = \frac{^{\mathcal{A}}d}{dt}\left(^{\mathcal{F}}\mathbf{v}\right) + {^{\mathcal{F}}\boldsymbol{\omega}^{\mathcal{A}}} \times {^{\mathcal{F}}\mathbf{v}} \tag{2-195}$$

Now we have

$$\begin{aligned} \frac{^{\mathcal{A}}d}{dt}\left(^{\mathcal{F}}\mathbf{v}\right) &= \ddot{r}\mathbf{e}_r + (r\ddot{\theta} + \dot{r}\dot{\theta})\mathbf{e}_\theta &(2\text{-}196)\\ {^{\mathcal{F}}\boldsymbol{\omega}^{\mathcal{A}}} \times {^{\mathcal{F}}\mathbf{v}} &= \dot{\theta}\mathbf{e}_z \times (\dot{r}\mathbf{e}_r + r\dot{\theta}\mathbf{e}_\theta) = \dot{r}\dot{\theta}\mathbf{e}_\theta - r\dot{\theta}^2\mathbf{e}_r &(2\text{-}197) \end{aligned}$$

Adding Eqs. (2–196) and (2–197), we obtain

$$^{\mathcal{F}}\mathbf{a} = \ddot{r}\mathbf{e}_r + r\ddot{\theta}\mathbf{e}_\theta + \dot{r}\dot{\theta}\mathbf{e}_\theta + \dot{r}\dot{\theta}\mathbf{e}_\theta - r\dot{\theta}^2\mathbf{e}_r \tag{2-198}$$

Equation (2–198) simplifies to

$$^{\mathcal{F}}\mathbf{a} = (\ddot{r} - r\dot{\theta}^2)\mathbf{e}_r + (2\dot{r}\dot{\theta} + r\ddot{\theta})\mathbf{e}_\theta \tag{2-199}$$

■

Example 2–4

A collar slides along a circular annulus of radius r. The annulus rotates with constant angular velocity Ω about the vertical direction as shown in Fig. 2–15. Knowing that the angle θ describes the location of the particle relative to the vertical direction, determine (a) the velocity and acceleration of the collar as viewed by an observer fixed to the annulus and (b) the velocity and acceleration of the collar as viewed by an observer fixed to the ground.

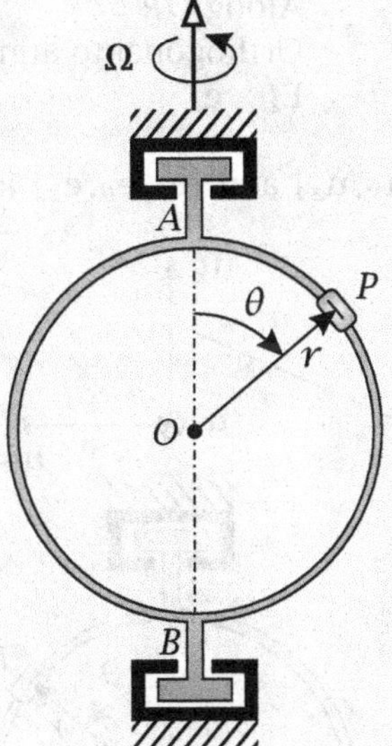

Figure 2–15 Collar sliding on a rotating annulus.

Solution to Example 2–4

Because in this problem the particle slides along the annulus and the annulus itself rotates, the motion is decomposed naturally into the motion of the annulus and the motion of the particle relative to the annulus. To this end, it is convenient to solve this problem using one fixed reference frame and two rotating reference frames. First, let $\mathcal{F}$ be a reference frame that is fixed to the ground. Then choose the following

coordinate system fixed in reference frame $\mathcal{F}$:

Origin at point O		
$\mathbf{E}_x$	=	Along OA
$\mathbf{E}_z$	=	Orthogonal to annulus (into page) at $t = 0$
$\mathbf{E}_y$	=	$\mathbf{E}_z \times \mathbf{E}_x$

Next, let $\mathcal{A}$ be a reference frame fixed to the annulus. Then, choose the following coordinate system fixed in reference frame $\mathcal{A}$:

Origin at point O		
$\mathbf{u}_x$	=	Along OA
$\mathbf{u}_z$	=	Orthogonal to annulus (into page)
$\mathbf{u}_y$	=	$\mathbf{u}_z \times \mathbf{u}_x$

Finally, let $\mathcal{B}$ be a reference frame fixed to the direction of the position of the particle. Then choose the following coordinate system fixed in reference frame $\mathcal{B}$:

Origin at point O		
$\mathbf{e}_r$	=	Along OP
$\mathbf{e}_z$	=	Orthogonal to annulus (into page)
$\mathbf{e}_\theta$	=	$\mathbf{E}_z \times \mathbf{e}_r$

The geometry of the bases $\{\mathbf{u}_x, \mathbf{u}_y, \mathbf{u}_z\}$ and $\{\mathbf{e}_r, \mathbf{e}_\theta, \mathbf{e}_z\}$ is shown in Fig. 2-16.

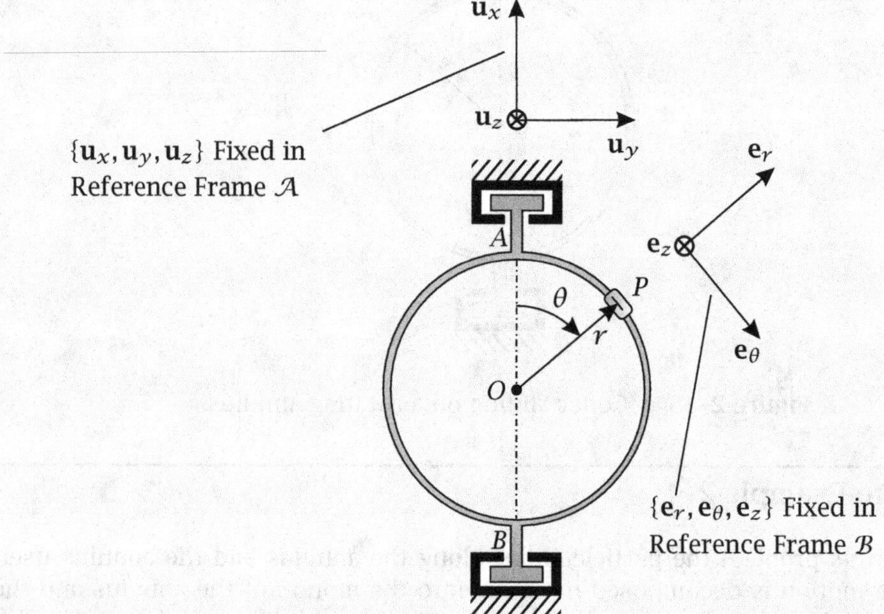

Figure 2–16 Geometry of bases $\{\mathbf{u}_x, \mathbf{u}_y, \mathbf{u}_z\}$ and $\{\mathbf{e}_r, \mathbf{e}_\theta, \mathbf{e}_z\}$ fixed in reference frames $\mathcal{A}$ and $\mathcal{B}$, respectively, for Example 2–4.

Using Fig. 2-16, we have

$$\mathbf{u}_x = \cos\theta\,\mathbf{e}_r - \sin\theta\,\mathbf{e}_\theta \quad (2\text{-}200)$$

$$\mathbf{u}_y = \sin\theta\,\mathbf{e}_r + \cos\theta\,\mathbf{e}_\theta \quad (2\text{-}201)$$

It is seen that the bases $\{\mathbf{e}_r, \mathbf{e}_\theta, \mathbf{e}_z\}$ and $\{\mathbf{u}_x, \mathbf{u}_y, \mathbf{u}_z\}$ share a common direction $\mathbf{u}_z = \mathbf{e}_z$. Also, it is seen from Eqs. (2-200) and (2-201) that the bases $\{\mathbf{e}_r, \mathbf{e}_\theta, \mathbf{e}_z\}$ and $\{\mathbf{u}_x, \mathbf{u}_y, \mathbf{u}_z\}$ are aligned when $\theta = 0$.

(a) Velocity and Acceleration of Collar as Viewed by an Observer Fixed to Annulus

The position of the particle is given in terms of the basis $\{\mathbf{e}_r, \mathbf{e}_\theta, \mathbf{e}_z\}$ as

$$\mathbf{r} = r\mathbf{e}_r \quad (2\text{-}202)$$

Now, because $\{\mathbf{e}_r, \mathbf{e}_\theta, \mathbf{e}_z\}$ is fixed in $\mathcal{B}$ and rotates with angular rate $\dot{\theta}$ about the $\mathbf{e}_z$-direction relative to the annulus, the angular velocity of reference frame $\mathcal{B}$ in reference frame $\mathcal{A}$ is given as

$$^{\mathcal{A}}\boldsymbol{\omega}^{\mathcal{B}} = \dot{\theta}\mathbf{e}_z \quad (2\text{-}203)$$

Then, applying the transport theorem of Eq. (2-128) on page 47, the velocity of the collar in reference frame $\mathcal{A}$ is given as

$$^{\mathcal{A}}\mathbf{v} = \frac{^{\mathcal{B}}d\mathbf{r}}{dt} + {^{\mathcal{A}}\boldsymbol{\omega}^{\mathcal{B}}} \times \mathbf{r} \quad (2\text{-}204)$$

Noting that r is constant, we have

$$\frac{^{\mathcal{B}}d\mathbf{r}}{dt} = \mathbf{0} \quad (2\text{-}205)$$

Also,

$$^{\mathcal{A}}\boldsymbol{\omega}^{\mathcal{B}} \times \mathbf{r} = \dot{\theta}\mathbf{e}_z \times r\mathbf{e}_r = r\dot{\theta}\mathbf{e}_\theta \quad (2\text{-}206)$$

Adding the results of Eqs. (2-205) and (2-206), we obtain the velocity of the collar as viewed by an observer fixed to the annulus, $^{\mathcal{A}}\mathbf{v}$, as

$$^{\mathcal{A}}\mathbf{v} = r\dot{\theta}\mathbf{e}_\theta \quad (2\text{-}207)$$

Applying Eq. (2-128) on page 47 to $^{\mathcal{A}}\mathbf{v}$, the acceleration of the collar as viewed by an observer to the annulus is given as

$$^{\mathcal{A}}\mathbf{a} = \frac{^{\mathcal{B}}d}{dt}\left(^{\mathcal{A}}\mathbf{v}\right) + {^{\mathcal{A}}\boldsymbol{\omega}^{\mathcal{B}}} \times {^{\mathcal{A}}\mathbf{v}} \quad (2\text{-}208)$$

Using $^{\mathcal{A}}\mathbf{v}$ from Eq. (2-207), we have

$$\frac{^{\mathcal{B}}d}{dt}\left(^{\mathcal{A}}\mathbf{v}\right) = r\ddot{\theta}\mathbf{e}_\theta \quad (2\text{-}209)$$

and

$$^{\mathcal{A}}\boldsymbol{\omega}^{\mathcal{B}} \times {^{\mathcal{A}}\mathbf{v}} = \dot{\theta}\mathbf{e}_z \times r\dot{\theta}\mathbf{e}_\theta = -r\dot{\theta}^2\mathbf{e}_r \quad (2\text{-}210)$$

Adding the results of Eqs. (2-209) and (2-210), we obtain the acceleration of the collar as viewed by an observer fixed to the annulus as

$$^{\mathcal{A}}\mathbf{a} = -r\dot{\theta}^2\mathbf{e}_r + r\ddot{\theta}\mathbf{e}_\theta \quad (2\text{-}211)$$

(b) Velocity and Acceleration of Collar as Viewed by an Observer Fixed to Ground

The velocity and acceleration of the collar as viewed by an observer fixed to the ground can be obtained using two different methods. The first method is to use a two-stage procedure in which the rate of change transport theorem is applied between reference frames $\mathcal{B}$ and $\mathcal{A}$ and again between reference frames $\mathcal{A}$ and $\mathcal{F}$. The second method is to apply the rate of change transport theorem directly between reference frames $\mathcal{B}$ and $\mathcal{F}$. Each of these methods is now applied to this example.

Method 1: Two-Stage Application of Transport Theorem

In this method, the velocity of the collar as viewed by an observer fixed to the ground is obtained by applying Eq. (2-128) on page 47 between reference frames $\mathcal{A}$ and $\mathcal{F}$ as

$$^{\mathcal{F}}\mathbf{v} = \frac{^{\mathcal{A}}d\mathbf{r}}{dt} + {^{\mathcal{F}}\boldsymbol{\omega}^{\mathcal{A}}} \times \mathbf{r} = {^{\mathcal{A}}\mathbf{v}} + {^{\mathcal{F}}\boldsymbol{\omega}^{\mathcal{A}}} \times \mathbf{r} \tag{2-212}$$

Now, we already have $^{\mathcal{A}}\mathbf{v}$ from Eq. (2-207). Then, because the collar rotates relative to the ground with angular velocity $\boldsymbol{\Omega}$ about the $\mathbf{u}_x$-direction, we have

$$^{\mathcal{F}}\boldsymbol{\omega}^{\mathcal{A}} = \boldsymbol{\Omega} = \Omega\mathbf{u}_x \tag{2-213}$$

Using the expression for $\mathbf{u}_x$ from Eq. (2-200), we have

$$^{\mathcal{F}}\boldsymbol{\omega}^{\mathcal{A}} = \boldsymbol{\Omega} = \Omega(\cos\theta\,\mathbf{e}_r - \sin\theta\,\mathbf{e}_\theta) = \Omega\cos\theta\,\mathbf{e}_r - \Omega\sin\theta\,\mathbf{e}_\theta \tag{2-214}$$

Consequently, we obtain

$$^{\mathcal{F}}\boldsymbol{\omega}^{\mathcal{A}} \times \mathbf{r} = (\Omega\cos\theta\,\mathbf{e}_r - \Omega\sin\theta\,\mathbf{e}_\theta) \times r\mathbf{e}_r = r\Omega\sin\theta\,\mathbf{e}_z \tag{2-215}$$

Adding Eq. (2-207) and (2-215), we obtain the velocity of the collar as viewed by an observer fixed to the annulus as

$$^{\mathcal{F}}\mathbf{v} = r\dot{\theta}\mathbf{e}_\theta + r\Omega\sin\theta\,\mathbf{e}_z \tag{2-216}$$

Then, applying the transport theorem of Eq. (2-128) on page 47, the acceleration of the collar as viewed by an observer fixed to the annulus is given as

$$^{\mathcal{F}}\mathbf{a} = \frac{^{\mathcal{A}}d}{dt}\left({^{\mathcal{F}}\mathbf{v}}\right) + {^{\mathcal{F}}\boldsymbol{\omega}^{\mathcal{A}}} \times {^{\mathcal{F}}\mathbf{v}} \tag{2-217}$$

Now the quantity $^{\mathcal{A}}d(^{\mathcal{F}}\mathbf{v})/dt$ can itself be obtained using the rate of change transport theorem between reference frames $\mathcal{B}$ and $\mathcal{A}$ as

$$\frac{^{\mathcal{A}}d}{dt}\left({^{\mathcal{F}}\mathbf{v}}\right) = \frac{^{\mathcal{B}}d}{dt}\left({^{\mathcal{F}}\mathbf{v}}\right) + {^{\mathcal{A}}\boldsymbol{\omega}^{\mathcal{B}}} \times {^{\mathcal{F}}\mathbf{v}} \tag{2-218}$$

Noting that r and Ω are constant and using $^{\mathcal{F}}\mathbf{v}$ from Eq. (2-216), we have

$$\frac{^{\mathcal{B}}d}{dt}\left({^{\mathcal{F}}\mathbf{v}}\right) = r\ddot{\theta}\mathbf{e}_\theta + r\Omega\dot{\theta}\cos\theta\,\mathbf{e}_z \tag{2-219}$$

Furthermore,

$$ {}^{\mathcal{A}}\boldsymbol{\omega}^{\mathcal{B}} \times {}^{\mathcal{F}}\mathbf{v} = \dot{\theta}\mathbf{e}_z \times (r\dot{\theta}\mathbf{e}_\theta + r\Omega\sin\theta\,\mathbf{e}_z) = -r\dot{\theta}^2\mathbf{e}_r \tag{2-220} $$

Adding the results of Eqs. (2–219) and (2–220), we have

$$ \frac{{}^{\mathcal{A}}d}{dt}\left({}^{\mathcal{F}}\mathbf{v}\right) = r\ddot{\theta}\mathbf{e}_\theta + r\Omega\dot{\theta}\cos\theta\,\mathbf{e}_z - r\dot{\theta}^2\mathbf{e}_r \tag{2-221} $$

Next, ${}^{\mathcal{F}}\boldsymbol{\omega}^{\mathcal{A}} \times {}^{\mathcal{F}}\mathbf{v}$ is obtained as

$$ {}^{\mathcal{F}}\boldsymbol{\omega}^{\mathcal{A}} \times {}^{\mathcal{F}}\mathbf{v} = \Omega\mathbf{u}_x \times (r\dot{\theta}\mathbf{e}_\theta + r\Omega\sin\theta\,\mathbf{e}_z) \tag{2-222} $$

Expanding Eq. (2–222), we obtain

$$ {}^{\mathcal{F}}\boldsymbol{\omega}^{\mathcal{A}} \times {}^{\mathcal{F}}\mathbf{v} = r\Omega\dot{\theta}\mathbf{u}_x \times \mathbf{e}_\theta + r\Omega^2\sin\theta\,\mathbf{u}_x \times \mathbf{e}_z \tag{2-223} $$

Then, using Eq. (2–200), we have

$$ \begin{aligned} \mathbf{u}_x \times \mathbf{e}_\theta &= (\cos\theta\,\mathbf{e}_r - \sin\theta\,\mathbf{e}_\theta) \times \mathbf{e}_\theta = \cos\theta\,\mathbf{e}_z & \text{(2-224)} \\ \mathbf{u}_x \times \mathbf{e}_z &= (\cos\theta\,\mathbf{e}_r - \sin\theta\,\mathbf{e}_\theta) \times \mathbf{e}_z = -\sin\theta\,\mathbf{e}_r - \cos\theta\,\mathbf{e}_\theta & \text{(2-225)} \end{aligned} $$

Substituting the results of Eqs. (2–224) and (2–225) into (2–223) gives

$$ {}^{\mathcal{F}}\boldsymbol{\omega}^{\mathcal{A}} \times {}^{\mathcal{F}}\mathbf{v} = r\Omega\dot{\theta}\cos\theta\,\mathbf{e}_z + r\Omega^2\sin\theta\,(-\sin\theta\,\mathbf{e}_r - \cos\theta\,\mathbf{e}_\theta) \tag{2-226} $$

Simplifying Eq. (2–226), we obtain

$$ {}^{\mathcal{F}}\boldsymbol{\omega}^{\mathcal{A}} \times {}^{\mathcal{F}}\mathbf{v} = -r\Omega^2\sin^2\theta\mathbf{e}_r - r\Omega^2\cos\theta\,\sin\theta\,\mathbf{e}_\theta + r\Omega\dot{\theta}\cos\theta\,\mathbf{e}_z \tag{2-227} $$

Adding Eqs. (2–221) and (2–227), we obtain the acceleration of the collar as viewed by an observer fixed to the annulus as

$$ \begin{aligned} {}^{\mathcal{F}}\mathbf{a} = {} & r\ddot{\theta}\mathbf{e}_\theta + r\Omega\dot{\theta}\cos\theta\,\mathbf{e}_z - r\dot{\theta}^2\mathbf{e}_r \\ & - r\Omega^2\sin^2\theta\mathbf{e}_r - r\Omega^2\cos\theta\,\sin\theta\,\mathbf{e}_\theta + r\Omega\dot{\theta}\cos\theta\,\mathbf{e}_z \end{aligned} \tag{2-228} $$

Combining terms with common components in Eq. (2–228), we obtain the acceleration of the collar as viewed by an observer fixed to the ground as

$$ {}^{\mathcal{F}}\mathbf{a} = -(r\dot{\theta}^2 + r\Omega^2\sin^2\theta)\mathbf{e}_r + (r\ddot{\theta} - r\Omega^2\cos\theta\,\sin\theta\,)\mathbf{e}_\theta + 2r\Omega\dot{\theta}\cos\theta\,\mathbf{e}_z \tag{2-229} $$

Method 2: One-Stage Application of Transport Theorem

The velocity of the collar as viewed by an observer fixed to the ground is obtained by applying the rate of change transport theorem of Eq. (2–128) on page 47 between reference frames $\mathcal{B}$ and $\mathcal{F}$ as

$$ {}^{\mathcal{F}}\mathbf{v} = \frac{{}^{\mathcal{B}}d\mathbf{r}}{dt} + {}^{\mathcal{F}}\boldsymbol{\omega}^{\mathcal{B}} \times \mathbf{r} \tag{2-230} $$

Now the angular velocity of reference frame $\mathcal{B}$ in reference frame $\mathcal{F}$ is obtained from the theorem of angular velocity addition as

$$ {}^{\mathcal{F}}\boldsymbol{\omega}^{\mathcal{B}} = {}^{\mathcal{F}}\boldsymbol{\omega}^{\mathcal{A}} + {}^{\mathcal{A}}\boldsymbol{\omega}^{\mathcal{B}} \tag{2-231} $$

Using the expressions for ${}^{\mathcal{A}}\boldsymbol{\omega}^{\mathcal{B}}$ and ${}^{\mathcal{F}}\boldsymbol{\omega}^{\mathcal{A}}$ from Eqs. (2–203) and (2–214), respectively, the angular velocity of reference frame $\mathcal{B}$ in reference frame $\mathcal{F}$ is obtained as

$$ {}^{\mathcal{F}}\boldsymbol{\omega}^{\mathcal{B}} = \Omega\cos\theta\,\mathbf{e}_r - \sin\theta\,\mathbf{e}_\theta + \dot{\theta}\mathbf{e}_z \tag{2–232} $$

Next we have

$$ \begin{aligned} \frac{{}^{\mathcal{B}}d\mathbf{r}}{dt} &= \mathbf{0} && (2\text{–}233) \\ {}^{\mathcal{F}}\boldsymbol{\omega}^{\mathcal{B}}\times\mathbf{r} &= (\Omega\cos\theta\,\mathbf{e}_r - \Omega\sin\theta\,\mathbf{e}_\theta + \dot{\theta}\mathbf{e}_z)\times(r\mathbf{e}_r) \\ &= r\dot{\theta}\mathbf{e}_\theta + r\Omega\sin\theta\,\mathbf{e}_z && (2\text{–}234) \end{aligned} $$

Adding Eqs. (2–233) and (2–234), we obtain the velocity of the collar in reference frame $\mathcal{F}$ as

$$ {}^{\mathcal{F}}\mathbf{v} = r\dot{\theta}\mathbf{e}_\theta + r\Omega\sin\theta\,\mathbf{e}_z \tag{2–235} $$

The acceleration of the collar in reference frame $\mathcal{F}$ is obtained by applying the rate of change transport theorem of Eq. (2–128) on page 47 to ${}^{\mathcal{F}}\mathbf{v}$ between reference frames $\mathcal{B}$ and $\mathcal{F}$ as

$$ {}^{\mathcal{F}}\mathbf{a} = \frac{{}^{\mathcal{F}}d}{dt}\left({}^{\mathcal{F}}\mathbf{v}\right) = \frac{{}^{\mathcal{B}}d}{dt}\left({}^{\mathcal{F}}\mathbf{v}\right) + {}^{\mathcal{F}}\boldsymbol{\omega}^{\mathcal{B}}\times{}^{\mathcal{F}}\mathbf{v} \tag{2–236} $$

Now we have

$$ \begin{aligned} \frac{{}^{\mathcal{B}}d}{dt}\left({}^{\mathcal{F}}\mathbf{v}\right) &= r\ddot{\theta}\mathbf{e}_\theta + r\Omega\dot{\theta}\cos\theta\,\mathbf{e}_z && (2\text{–}237) \\ {}^{\mathcal{F}}\boldsymbol{\omega}^{\mathcal{B}}\times{}^{\mathcal{F}}\mathbf{v} &= (\Omega\cos\theta\,\mathbf{e}_r - \Omega\sin\theta\,\mathbf{e}_\theta + \dot{\theta}\mathbf{e}_z)\times(r\dot{\theta}\mathbf{e}_\theta + r\Omega\sin\theta\,\mathbf{e}_z) \\ &= r\Omega\dot{\theta}\cos\theta\,\mathbf{e}_z - r\Omega^2\cos\theta\,\sin\theta\,\mathbf{e}_\theta \\ &\quad -r\Omega^2\sin^2\theta\mathbf{e}_r - r\dot{\theta}^2\mathbf{e}_r && (2\text{–}238) \end{aligned} $$

Adding Eqs. (2–237) and (2–238), we obtain the acceleration of the collar in reference frame $\mathcal{F}$ as

$$ {}^{\mathcal{F}}\mathbf{a} = -(r\dot{\theta}^2 + r\Omega^2\sin^2\theta)\mathbf{e}_r + (r\ddot{\theta} - r\Omega^2\cos\theta\,\sin\theta\,)\mathbf{e}_\theta + 2r\Omega\dot{\theta}\cos\theta\,\mathbf{e}_z \tag{2–239} $$

It is seen that the result of Eq. (2–239) is identical to the result obtained in Eq. (2–229) using method 1.

■

2.11.3 Spherical Coordinates

Let $\mathbf{r}$ be the position of a point P in $\mathbb{E}^3$ measured relative to a point O, where O is fixed in an arbitrary reference frame $\mathcal{A}$. Furthermore, let $\{\mathbf{e}_x, \mathbf{e}_y, \mathbf{e}_z\}$ be a Cartesian basis fixed in reference frame $\mathcal{A}$. Recalling from Eq. (2–159) the expression for $\mathbf{r}$ in terms of the basis $\{\mathbf{e}_x, \mathbf{e}_y, \mathbf{e}_z\}$, we have

$$ \mathbf{r} = x\mathbf{e}_x + y\mathbf{e}_y + z\mathbf{e}_z \tag{2–240} $$

Suppose now that we make the following substitutions:

$$\begin{aligned} x &= r\sin\phi\cos\theta \\ y &= r\sin\phi\sin\theta \\ z &= r\cos\phi \end{aligned} \tag{2-241}$$

where

$$\begin{aligned} r &= \sqrt{x^2+y^2+z^2} \\ \theta &= \tan^{-1}(y/x) \\ \phi &= \cos^{-1}(z/r) \end{aligned} \tag{2-242}$$

Then $\mathbf{r}$ can be written as

$$\mathbf{r} = r\sin\phi\cos\theta\,\mathbf{e}_x + r\sin\phi\sin\theta\,\mathbf{e}_y + r\cos\phi\mathbf{e}_z \tag{2-243}$$

Suppose now that we define an orthonormal basis $\{\mathbf{u}_r, \mathbf{u}_\phi, \mathbf{u}_\theta\}$ as follows:

$$\mathbf{u}_r = \sin\phi\cos\theta\,\mathbf{e}_x + \sin\phi\sin\theta\,\mathbf{e}_y + \cos\phi\mathbf{e}_z \tag{2-244}$$

$$\mathbf{u}_\theta = -\sin\theta\,\mathbf{e}_x + \cos\theta\,\mathbf{e}_y \tag{2-245}$$

$$\mathbf{u}_\phi = \mathbf{u}_\theta \times \mathbf{u}_r = \cos\phi\cos\theta\,\mathbf{e}_x + \cos\phi\sin\theta\,\mathbf{e}_y - \sin\phi\mathbf{e}_z \tag{2-246}$$

The basis $\{\mathbf{u}_r, \mathbf{u}_\phi, \mathbf{u}_\theta\}$ is called a *spherical basis* because the angles ϕ and θ define a locus of points on a sphere for a constant value of r. It is noted that the angle θ is commonly referred to as *longitude* or *azimuth* while the angle ϕ is commonly referred to as *co-latitude* or *polar angle.* The geometry of a spherical basis is shown in Fig. 2-17.

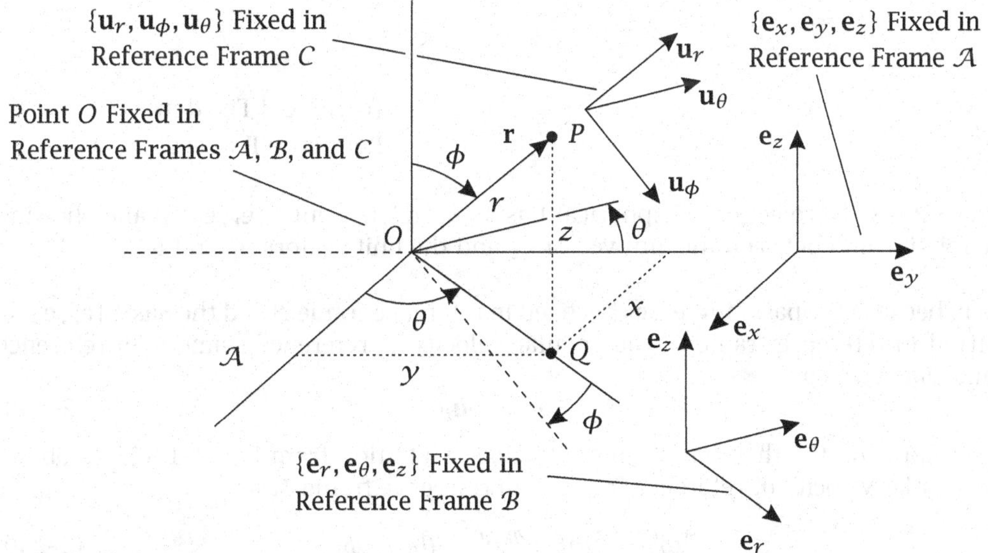

Figure 2-17 Bases $\{\mathbf{e}_x, \mathbf{e}_y, \mathbf{e}_z\}$, $\{\mathbf{e}_r, \mathbf{e}_\theta, \mathbf{e}_z\}$, and $\{\mathbf{u}_r, \mathbf{u}_\phi, \mathbf{u}_\theta\}$ fixed in reference frames $\mathcal{A}$, $\mathcal{B}$, and $\mathcal{C}$, respectively.

Now let $\mathcal{C}$ be a reference frame such that the basis $\{\mathbf{u}_r, \mathbf{u}_\phi, \mathbf{u}_\theta\}$ is fixed in $\mathcal{C}$ and suppose that we want to determine the angular velocity, ${}^{\mathcal{A}}\boldsymbol{\omega}^{\mathcal{C}}$, of reference frame $\mathcal{C}$ as viewed by an observer in reference frame C. An expression for ${}^{\mathcal{A}}\boldsymbol{\omega}^{\mathcal{B}}$ can be determined by using the theorem of angular velocity addition as follows. First, let $\{\mathbf{e}_r, \mathbf{e}_\theta, \mathbf{e}_z\}$ be the cylindrical basis as defined in Section 2.11.2 on page 55. Then, from the geometry we have

$$\mathbf{e}_\theta = \mathbf{e}_z \times \mathbf{e}_r \tag{2-247}$$

Moreover, the basis $\{\mathbf{e}_r, \mathbf{e}_\theta, \mathbf{e}_z\}$ is fixed in reference frame $\mathcal{B}$. Furthermore, it is seen from the geometry that the basis $\{\mathbf{e}_r, \mathbf{e}_\theta, \mathbf{e}_z\}$ rotates relative to the basis $\{\mathbf{e}_x, \mathbf{e}_y, \mathbf{e}_z\}$ with angular rate $\dot{\theta}$ in the $\mathbf{e}_z$-direction. Now, because $\{\mathbf{e}_x, \mathbf{e}_y, \mathbf{e}_z\}$ is fixed in reference frame $\mathcal{A}$ and $\{\mathbf{e}_r, \mathbf{e}_\theta, \mathbf{e}_z\}$ is fixed in reference frame $\mathcal{B}$, the angular velocity of reference frame $\mathcal{B}$ in reference frame $\mathcal{A}$ is given as

$${}^{\mathcal{A}}\boldsymbol{\omega}^{\mathcal{B}} = \dot{\theta}\mathbf{e}_z \tag{2-248}$$

Next, it is seen from Fig. 2-18 that the basis $\{\mathbf{u}_r, \mathbf{u}_\phi, \mathbf{u}_\theta\}$ rotates relative to the basis $\{\mathbf{e}_r, \mathbf{e}_\theta, \mathbf{e}_z\}$ with angular rate $\dot{\phi}$ in the $\mathbf{u}_\theta$-direction (that is, the $\mathbf{e}_\theta$-direction).

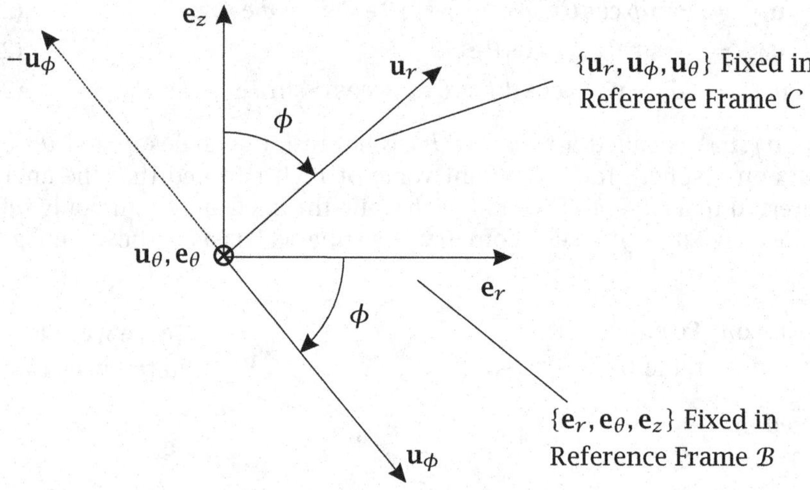

Figure 2-18 Projection of spherical basis $\{\mathbf{u}_r, \mathbf{u}_\phi, \mathbf{u}_\theta\}$ into $\{\mathbf{e}_r, \mathbf{e}_z\}$-plane showing the relationship between the unit vector $\mathbf{e}_z$ and the unit vectors $\mathbf{u}_r$ and $\mathbf{u}_\phi$.

Then, because the basis $\{\mathbf{e}_r, \mathbf{e}_\theta, \mathbf{e}_z\}$ is fixed in reference frame $\mathcal{B}$ and the basis $\{\mathbf{e}_r, \mathbf{e}_\phi, \mathbf{e}_\theta\}$ is fixed in reference frame C, the angular velocity of reference frame C in reference frame $\mathcal{B}$ is given as

$${}^{\mathcal{B}}\boldsymbol{\omega}^{C} = \dot{\phi}\mathbf{u}_\theta \tag{2-249}$$

Then, applying the theorem of angular velocity addition from Eq. (2-136), we obtain the angular velocity of reference frame $\mathcal{B}$ in reference frame $\mathcal{A}$ as

$${}^{\mathcal{A}}\boldsymbol{\omega}^{C} = {}^{\mathcal{A}}\boldsymbol{\omega}^{\mathcal{B}} + {}^{\mathcal{B}}\boldsymbol{\omega}^{C} = \dot{\theta}\mathbf{e}_z + \dot{\phi}\mathbf{u}_\theta \tag{2-250}$$

Now, in order to obtain an expression for ${}^{\mathcal{A}}\boldsymbol{\omega}^{C}$ in terms of the basis $\{\mathbf{u}_r, \mathbf{u}_\phi, \mathbf{u}_\theta\}$, we need to express $\mathbf{e}_z$ in terms of the $\{\mathbf{u}_r, \mathbf{u}_\phi, \mathbf{u}_\theta\}$. Using Fig. 2-18, it is seen that $\mathbf{e}_z$ can be written in terms of $\mathbf{u}_r$ and $\mathbf{u}_\phi$ as

$$\mathbf{e}_z = \cos\phi\mathbf{u}_r - \sin\phi\mathbf{u}_\phi \tag{2-251}$$

Substituting the result of Eq. (2–251) into (2–250), we obtain ${}^{\mathcal{A}}\boldsymbol{\omega}^{C}$ as

$$ {}^{\mathcal{A}}\boldsymbol{\omega}^{C} = \dot{\theta}\cos\phi\mathbf{u}_r - \dot{\theta}\sin\phi\mathbf{u}_\phi + \dot{\phi}\mathbf{u}_\theta \tag{2–252} $$

Using the angular velocity ${}^{\mathcal{A}}\boldsymbol{\omega}^{C}$ from Eq. (2–252), we can now compute the velocity and acceleration of point P as viewed by an observer in reference frame $\mathcal{A}$. First, the velocity as viewed by an observer in reference frame $\mathcal{A}$, denoted ${}^{\mathcal{A}}\mathbf{v}$, is obtained from the rate of change transport theorem of Eq. (2–128) as

$$ {}^{\mathcal{A}}\mathbf{v} = \frac{{}^{\mathcal{A}}d\mathbf{r}}{dt} = \frac{{}^{C}d\mathbf{r}}{dt} + {}^{\mathcal{A}}\boldsymbol{\omega}^{C} \times \mathbf{r} \tag{2–253} $$

Using the expression for $\mathbf{r}$ and the definition of $\mathbf{e}_r$ from Eqs. (2–243) and (2–244), respectively, we have

$$ \mathbf{r} = r\mathbf{u}_r \tag{2–254} $$

Then, using the expressions for $\mathbf{r}$ and ${}^{\mathcal{A}}\boldsymbol{\omega}^{C}$ from Eqs. (2–254) and (2–252), respectively, we obtain

$$ \frac{{}^{C}d\mathbf{r}}{dt} = \dot{r}\mathbf{u}_r \tag{2–255} $$

$$ \begin{aligned} {}^{\mathcal{A}}\boldsymbol{\omega}^{C} \times \mathbf{r} &= (\dot{\theta}\cos\phi\mathbf{u}_r - \dot{\theta}\sin\phi\mathbf{u}_\phi + \dot{\phi}\mathbf{u}_\theta) \times r\mathbf{u}_r \\ &= r\dot{\phi}\mathbf{u}_\phi + r\dot{\theta}\sin\phi\mathbf{u}_\theta \end{aligned} \tag{2–256} $$

Adding the expressions in Eqs. (2–255) and (2–256), we obtain the velocity as viewed by an observer in reference frame $\mathcal{A}$ as

$$ {}^{\mathcal{A}}\mathbf{v} = \dot{r}\mathbf{u}_r + r\dot{\phi}\mathbf{u}_\phi + r\dot{\theta}\sin\phi\mathbf{u}_\theta \tag{2–257} $$

Next, applying the rate of change transport theorem of Eq. (2–128) on page 47 to ${}^{\mathcal{A}}\mathbf{v}$, the acceleration of point P as viewed by an observer in reference frame $\mathcal{A}$ is obtained as

$$ {}^{\mathcal{A}}\mathbf{a} = \frac{{}^{\mathcal{A}}d}{dt}\left({}^{\mathcal{A}}\mathbf{v}\right) = \frac{{}^{C}d}{dt}\left({}^{\mathcal{A}}\mathbf{v}\right) + {}^{\mathcal{A}}\boldsymbol{\omega}^{C} \times {}^{\mathcal{A}}\mathbf{v} \tag{2–258} $$

Now we have

$$ \begin{aligned} \frac{{}^{C}d}{dt}\left({}^{\mathcal{A}}\mathbf{v}\right) &= \ddot{r}\mathbf{u}_r + (\dot{r}\dot{\phi} + r\ddot{\phi})\mathbf{e}_\phi \\ &\quad + \left[\dot{r}\dot{\theta}\sin\phi + r(\ddot{\theta}\sin\phi + \dot{\phi}\dot{\theta}\cos\phi)\right]\mathbf{u}_\theta \end{aligned} \tag{2–259} $$

$$ \begin{aligned} {}^{\mathcal{A}}\boldsymbol{\omega}^{C} \times {}^{\mathcal{A}}\mathbf{v} &= (\dot{\theta}\cos\phi\mathbf{u}_r - \dot{\theta}\sin\phi\mathbf{u}_\phi + \dot{\phi}\mathbf{u}_\theta) \\ &\quad \times(\dot{r}\mathbf{u}_r + r\dot{\phi}\mathbf{u}_\phi + r\dot{\theta}\sin\phi\mathbf{u}_\theta) \\ &= r\dot{\phi}\dot{\theta}\cos\phi\mathbf{u}_\theta - r\dot{\theta}^2\cos\phi\sin\phi\mathbf{u}_\phi + \dot{r}\dot{\theta}\sin\phi\mathbf{u}_\theta \\ &\quad - r\dot{\theta}^2\sin^2\phi\mathbf{u}_r + \dot{r}\dot{\phi}\mathbf{u}_\phi - r\dot{\phi}^2\mathbf{u}_r \\ &= -(r\dot{\phi}^2 + r\dot{\theta}^2\sin^2\phi)\mathbf{u}_r + (\dot{r}\dot{\phi} - r\dot{\theta}^2\cos\phi\sin\phi)\mathbf{u}_\phi \\ &\quad + (r\dot{\phi}\dot{\theta}\cos\phi + \dot{r}\dot{\theta}\sin\phi)\mathbf{u}_\theta \end{aligned} \tag{2–260} $$

Adding Eqs. (2–259) and (2–260), we obtain the acceleration as viewed by an observer in reference frame $\mathcal{A}$ as

$$ \begin{aligned} {}^{\mathcal{A}}\mathbf{a} &= (\ddot{r} - r\dot{\phi}^2 - r\dot{\theta}^2\sin^2\phi)\mathbf{u}_r + (2\dot{r}\dot{\phi} + r\ddot{\phi} - r\dot{\theta}^2\cos\phi\sin\phi)\mathbf{u}_\phi \\ &\quad + (r\ddot{\theta}\sin\phi + 2r\dot{\phi}\dot{\theta}\cos\phi + 2\dot{r}\dot{\theta}\sin\phi)\mathbf{u}_\theta \end{aligned} \tag{2–261} $$

Example 2–5

A particle P is suspended from one end of a rigid rod of length L as shown in Fig. 2–19. The other end of the rod is attached at a fixed point O. The position of the particle is described relative to a ground-fixed basis $\{\mathbf{E}_x, \mathbf{E}_y, \mathbf{E}_z\}$ in terms of the spherical angles θ and ϕ, where θ measured from the $\mathbf{E}_x$-direction to the direction of OQ (where Q is the projection of P onto the $\{\mathbf{E}_x, \mathbf{E}_y\}$-plane) and ϕ is measured from the $\mathbf{E}_z$-direction to the direction of OP (where OP lies along the direction of the position of the particle), determine the following quantities as viewed by an observer fixed to the ground: (a) the angular velocity of the rod; (b) the velocity of the particle; and (c) the acceleration of the particle.

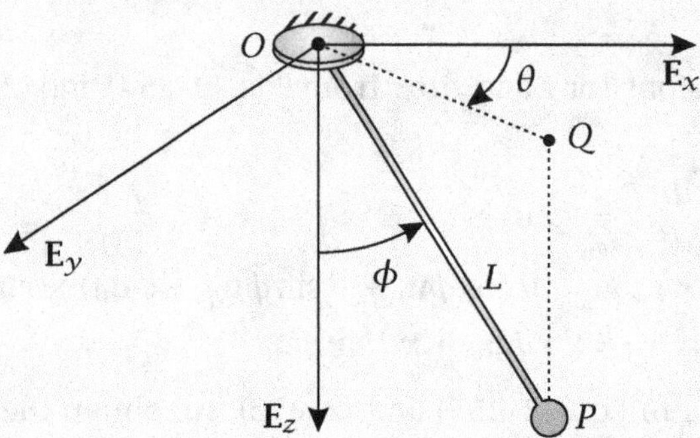

Figure 2–19 Particle suspended from end of rigid rod of length L.

Solution to Example 2–5

First, let $\mathcal{F}$ be a fixed reference frame. Then choose the following coordinate system fixed in reference frame $\mathcal{F}$:

Origin at point O

$$
\begin{array}{ccc}
\mathbf{E}_x & = & \text{As given} \\
\mathbf{E}_y & = & \text{As given} \\
\mathbf{E}_z & = & \text{As given}
\end{array}
$$

Next, let $\mathcal{A}$ be a reference frame fixed to the plane formed by the directions of OQ and $\mathbf{E}_z$. Then choose the following coordinate system fixed in reference frame $\mathcal{A}$:

Origin at point O

$$
\begin{array}{ccc}
\mathbf{e}_r & = & \text{Along } OQ \\
\mathbf{e}_z & = & \mathbf{E}_z \\
\mathbf{e}_\theta & = & \mathbf{e}_z \times \mathbf{e}_r
\end{array}
$$

Finally, let $\mathcal{B}$ be a reference frame fixed to the direction OP. Then choose the following coordinate system fixed in reference frame $\mathcal{B}$:

Origin at point O

$$
\begin{array}{ccc}
\mathbf{u}_r & = & \text{Along } OP \\
\mathbf{u}_\theta & = & \mathbf{e}_\theta \\
\mathbf{u}_\phi & = & \mathbf{u}_\theta \times \mathbf{u}_r
\end{array}
$$

The geometry of the bases $\{\mathbf{E}_x, \mathbf{E}_y, \mathbf{E}_z\}$ and $\{\mathbf{e}_r, \mathbf{e}_\theta, \mathbf{e}_z\}$ is shown in Fig. 2-20 while the geometry of the bases $\{\mathbf{e}_r, \mathbf{e}_\theta, \mathbf{e}_z\}$ and $\{\mathbf{u}_r, \mathbf{u}_\phi, \mathbf{u}_\theta\}$ is shown in Fig. 2-21. It is noted that $\{\mathbf{e}_r, \mathbf{e}_\phi, \mathbf{e}_\theta\}$ is a spherical basis.

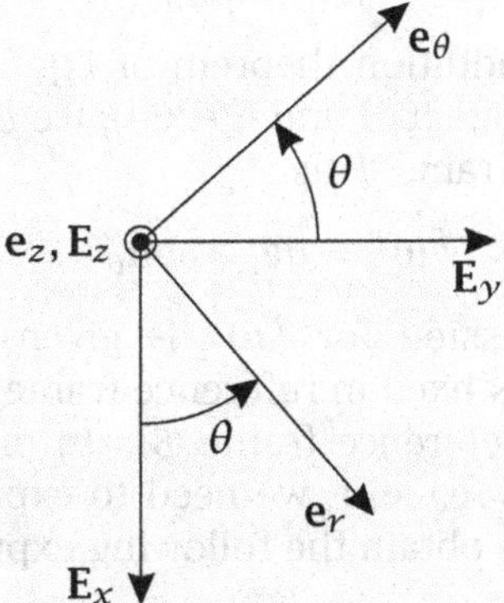

Figure 2-20 Geometry of bases $\{\mathbf{E}_x, \mathbf{E}_y, \mathbf{E}_z\}$ and $\{\mathbf{e}_r, \mathbf{e}_\theta, \mathbf{e}_z\}$ for Example 2-5.

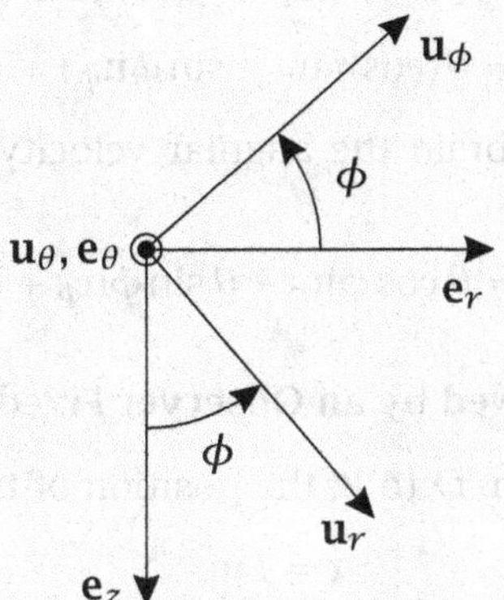

Figure 2-21 Geometry of bases $\{\mathbf{e}_r, \mathbf{e}_\theta, \mathbf{e}_z\}$ and $\{\mathbf{u}_r, \mathbf{u}_\phi, \mathbf{u}_\theta\}$ for Example 2-5.

(a) Angular Velocity of Rod as Viewed by an Observer Fixed to Ground

We note that reference frame $\mathcal{B}$ is the rod. Furthermore, it is seen that the rod rotates relative to reference frame $\mathcal{A}$ which, in turn, rotates relative to reference frame $\mathcal{F}$. Consequently, the angular velocity of the rod as viewed by an observer fixed to the ground is given as

$$^{\mathcal{F}}\boldsymbol{\omega}^{\mathcal{B}} = {}^{\mathcal{F}}\boldsymbol{\omega}^{\mathcal{A}} + {}^{\mathcal{A}}\boldsymbol{\omega}^{\mathcal{B}} \tag{2-262}$$

Now from Fig. 2-20 we see that θ is the angle between the $\mathbf{E}_x$-direction and the $\mathbf{e}_r$-direction. Furthermore, the $\mathbf{e}_r$-direction rotates about the $\mathbf{e}_z$-direction. Finally, the $\mathbf{E}_x$-direction is fixed in reference frame $\mathcal{F}$ while the $\mathbf{e}_r$-direction is fixed in reference frame $\mathcal{A}$. Consequently, the angular velocity of reference frame $\mathcal{A}$ in reference frame $\mathcal{F}$ is given as

$$^{\mathcal{F}}\boldsymbol{\omega}^{\mathcal{A}} = \dot{\theta}\mathbf{e}_z \tag{2-263}$$

Next, it is seen from Fig. 2-21 that the angle ϕ is measured from the $\mathbf{e}_z$-direction to the $\mathbf{e}_r$-direction. Furthermore, the $\mathbf{e}_r$-direction rotates about the $\mathbf{e}_\theta$-direction. Finally, the

$\mathbf{e}_z$-direction is fixed in reference frame $\mathcal{A}$ while the $\mathbf{e}_r$-direction is fixed in reference frame $\mathcal{B}$. Consequently, the angular velocity of reference frame $\mathcal{A}$ in reference frame $\mathcal{F}$ is given as

$$^{\mathcal{A}}\boldsymbol{\omega}^{\mathcal{B}} = \dot{\phi}\mathbf{u}_\theta \tag{2-264}$$

Applying the angular velocity addition theorem of Eq. (2-136) on page 48 using the angular velocities given in Eqs. (2-263) and (2-264), we obtain the angular velocity of reference frame $\mathcal{B}$ in reference frame $\mathcal{F}$ as

$$^{\mathcal{F}}\boldsymbol{\omega}^{\mathcal{B}} = \dot{\theta}\mathbf{e}_z + \dot{\phi}\mathbf{u}_\theta \tag{2-265}$$

Now it is seen that the expression for $^{\mathcal{F}}\boldsymbol{\omega}^{\mathcal{B}}$ is given in terms of both the basis $\{\mathbf{e}_Q, \mathbf{e}_\theta, \mathbf{e}_z\}$ (where $\{\mathbf{e}_Q, \mathbf{e}_\theta, \mathbf{e}_z\}$ is fixed in reference frame $\mathcal{A}$) and the basis $\{\mathbf{e}_r, \mathbf{e}_\phi, \mathbf{e}_\theta\}$ (where $\{\mathbf{e}_r, \mathbf{e}_\phi, \mathbf{e}_\theta\}$ is fixed in reference frame $\mathcal{B}$). In order to obtain an expression purely in terms of the basis $\{\mathbf{e}_r, \mathbf{e}_\phi, \mathbf{e}_\theta\}$, we need to express $\mathbf{e}_z$ in terms of the basis $\{\mathbf{e}_r, \mathbf{e}_\phi, \mathbf{e}_\theta\}$. Using Fig. 2-21), we obtain the following expression for $\mathbf{e}_z$ in terms of the basis $\{\mathbf{e}_r, \mathbf{e}_\phi, \mathbf{e}_\theta\}$:

$$\mathbf{e}_z = \cos\phi\mathbf{u}_r - \sin\phi\mathbf{u}_\phi \tag{2-266}$$

Substituting the expression for $\mathbf{e}_z$ from Eq. (2-266) into (2-265), we obtain

$$^{\mathcal{F}}\boldsymbol{\omega}^{\mathcal{B}} = \dot{\theta}(\cos\phi\mathbf{u}_r - \sin\phi\mathbf{u}_\phi) + \dot{\phi}\mathbf{u}_\theta \tag{2-267}$$

Rearranging Eq. (2-267), we obtain the angular velocity of the rod as viewed by an observer fixed to the ground as

$$^{\mathcal{F}}\boldsymbol{\omega}^{\mathcal{B}} = \dot{\theta}\cos\phi\mathbf{u}_r - \dot{\theta}\sin\phi\mathbf{u}_\phi + \dot{\phi}\mathbf{u}_\theta \tag{2-268}$$

(b) Velocity of Particle as Viewed by an Observer Fixed to Ground

Because $\mathbf{u}_r$ is the direction from O to P, the position of the particle is given as

$$\mathbf{r} = L\mathbf{u}_r \tag{2-269}$$

The velocity of the particle as viewed by an observer fixed to the ground is then obtained from the rate of change transport theorem of Eq. (2-128) on page 47 as

$$^{\mathcal{F}}\mathbf{v} = \frac{^{\mathcal{F}}d\mathbf{r}}{dt} = \frac{^{\mathcal{B}}d\mathbf{r}}{dt} + {}^{\mathcal{F}}\boldsymbol{\omega}^{\mathcal{B}} \times \mathbf{r} \tag{2-270}$$

Using the expressions for $\mathbf{r}$ and $^{\mathcal{F}}\boldsymbol{\omega}^{\mathcal{B}}$ from Eqs. (2-269) and (2-268), respectively, we have

$$\frac{^{\mathcal{B}}d\mathbf{r}}{dt} = \mathbf{0} \tag{2-271}$$

$$^{\mathcal{F}}\boldsymbol{\omega}^{\mathcal{B}} \times \mathbf{r} = (\dot{\theta}\cos\phi\mathbf{u}_r - \dot{\theta}\sin\phi\mathbf{u}_\phi + \dot{\phi}\mathbf{u}_\theta) \times L\mathbf{u}_r \tag{2-272}$$

Expanding Eq. (2-272), we obtain

$$^{\mathcal{F}}\boldsymbol{\omega}^{\mathcal{B}} \times \mathbf{r} = L\dot{\theta}\sin\phi\mathbf{u}_\theta + L\dot{\phi}\mathbf{u}_\phi = L\dot{\phi}\mathbf{u}_\phi + L\dot{\theta}\sin\phi\mathbf{u}_\theta \tag{2-273}$$

Adding Eqs. (2-271) and (2-273), we obtain the velocity of the particle as viewed by an observer fixed to the ground as

$$^{\mathcal{F}}\mathbf{v} = L\dot{\phi}\mathbf{u}_\phi + L\dot{\theta}\sin\phi\mathbf{u}_\theta \tag{2-274}$$

(c) Acceleration of Particle as Viewed by an Observer Fixed to Ground

The acceleration of the particle as viewed by an observer fixed to the ground is obtained from the transport theorem of Eq. (2-128) on page 47 as

$$ {}^{\mathcal{F}}\mathbf{a} = \frac{{}^{\mathcal{F}}d}{dt}\left({}^{\mathcal{F}}\mathbf{v}\right) = \frac{{}^{\mathcal{B}}d}{dt}\left({}^{\mathcal{F}}\mathbf{v}\right) + {}^{\mathcal{F}}\boldsymbol{\omega}^{\mathcal{B}} \times {}^{\mathcal{F}}\mathbf{v} \tag{2-275} $$

Using the expressions for ${}^{\mathcal{F}}\mathbf{v}$ and ${}^{\mathcal{F}}\boldsymbol{\omega}^{\mathcal{B}}$ Eqs. (2-274) and (2-268), respectively, we have

$$ \frac{{}^{\mathcal{B}}d}{dt}\left({}^{\mathcal{F}}\mathbf{v}\right) = L\ddot{\phi}\mathbf{u}_{\phi} + (L\ddot{\theta}\sin\phi + L\dot{\theta}\dot{\phi}\cos\phi)\mathbf{u}_{\theta} \tag{2-276} $$

$$ {}^{\mathcal{F}}\boldsymbol{\omega}^{\mathcal{B}} \times {}^{\mathcal{F}}\mathbf{v} = (\dot{\theta}\cos\phi\mathbf{u}_r - \dot{\theta}\sin\phi\mathbf{u}_{\phi} + \dot{\phi}\mathbf{u}_{\theta}) \times (L\dot{\phi}\mathbf{u}_{\phi} + L\dot{\theta}\sin\phi\mathbf{u}_{\theta}) \tag{2-277} $$

Expanding Eq. (2-277), we obtain

$$ {}^{\mathcal{F}}\boldsymbol{\omega}^{\mathcal{B}} \times {}^{\mathcal{F}}\mathbf{v} = L\dot{\theta}\dot{\phi}\cos\phi\mathbf{u}_{\theta} - L\dot{\theta}^2\cos\phi\sin\phi\mathbf{u}_{\phi} - L\dot{\theta}^2\sin^2\phi\mathbf{u}_r - L\dot{\phi}^2\mathbf{u}_r \tag{2-278} $$

Simplifying Eq. (2-278) gives

$$ {}^{\mathcal{F}}\boldsymbol{\omega}^{\mathcal{B}} \times {}^{\mathcal{F}}\mathbf{v} = -(L\dot{\phi}^2 + L\dot{\theta}^2\sin^2\phi)\mathbf{e}_r - L\dot{\theta}^2\cos\phi\sin\phi\mathbf{u}_{\phi} + L\dot{\theta}\dot{\phi}\cos\phi\mathbf{u}_{\theta} \tag{2-279} $$

Adding Eqs. (2-276) and (2-279), we obtain the acceleration of the particle as viewed by an observer fixed to the ground as

$$ \begin{aligned} {}^{\mathcal{F}}\mathbf{a} = & -(L\dot{\phi}^2 + L\dot{\theta}^2\sin^2\phi)\mathbf{u}_r + (L\ddot{\phi} - L\dot{\theta}^2\cos\phi\sin\phi)\mathbf{u}_{\phi} \\ & + (L\ddot{\theta}\sin\phi + 2L\dot{\theta}\dot{\phi}\cos\phi)\mathbf{u}_{\theta} \end{aligned} \tag{2-280} $$

■

2.11.4 Intrinsic Coordinates

Let $\mathbf{r}$ be the position of a point P in $\mathbb{E}^3$ measured relative to a point O, where O is fixed in an arbitrary reference frame $\mathcal{A}$. Furthermore, consider the motion of P relative to reference frame $\mathcal{A}$. The curve created as the point moves in $\mathbb{E}^3$ is called the *trajectory relative to reference frame* $\mathcal{A}$ and is shown in Fig. 2-22.
Suppose now that we let ${}^{\mathcal{A}}\mathbf{v}$ be the velocity of the particle as viewed by an observer in reference frame $\mathcal{A}$. Using ${}^{\mathcal{A}}\mathbf{v}$, we can define the unit vector $\mathbf{e}_t$ as

$$ \mathbf{e}_t = \frac{{}^{\mathcal{A}}\mathbf{v}}{\|{}^{\mathcal{A}}\mathbf{v}\|} \tag{2-281} $$

The unit vector $\mathbf{e}_t$ is called the *tangent vector to the trajectory relative to reference frame* $\mathcal{A}$. Using the definition of speed as given in Eq. (2-30) on page 36, we can write ${}^{\mathcal{A}}\mathbf{v}$ as

$$ {}^{\mathcal{A}}\mathbf{v} = {}^{\mathcal{A}}v\mathbf{e}_t \tag{2-282} $$

Then the acceleration of point P relative to reference frame $\mathcal{A}$ is obtained as

$$ {}^{\mathcal{A}}\mathbf{a} = \frac{{}^{\mathcal{A}}d}{dt}\left({}^{\mathcal{A}}\mathbf{v}\right) = \frac{d}{dt}\left({}^{\mathcal{A}}v\right)\mathbf{e}_t + {}^{\mathcal{A}}v\frac{{}^{\mathcal{A}}d\mathbf{e}_t}{dt} \tag{2-283} $$

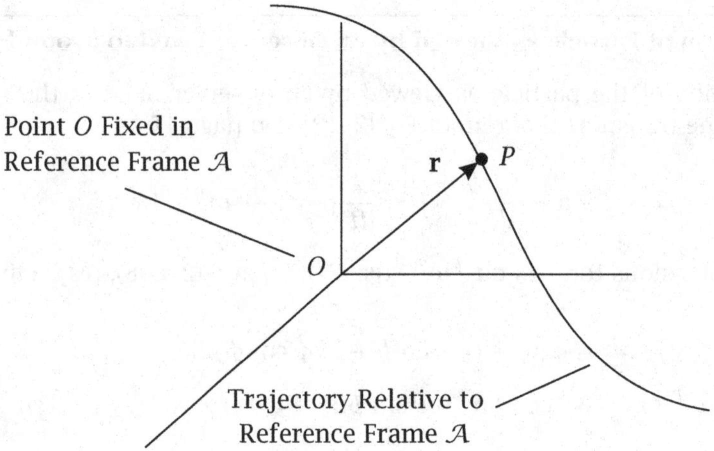

Figure 2–22 Trajectory of a particle relative to reference frame $\mathcal{A}$.

where we note once again that, because ${}^{\mathcal{A}}v$ is a scalar function, the quantity $d({}^{\mathcal{A}}v)/dt$ is independent of the reference frame. Now from Eqs. (2-79)–(2-81) on page 44 we know that ${}^{\mathcal{A}}d\mathbf{e}_t/dt$ is orthogonal to $\mathbf{e}_t$, i.e.,

$$\frac{{}^{\mathcal{A}}d\mathbf{e}_t}{dt} \cdot \mathbf{e}_t = 0 \tag{2-284}$$

Suppose we let $\mathbf{e}_n$ be the direction of ${}^{\mathcal{A}}d\mathbf{e}_t/dt$, i.e.,

$$\mathbf{e}_n = \frac{{}^{\mathcal{A}}d\mathbf{e}_t/dt}{\|{}^{\mathcal{A}}d\mathbf{e}_t/dt\|} \tag{2-285}$$

The unit vector $\mathbf{e}_n$ is called the *principal unit normal* to the trajectory of point P relative to reference frame $\mathcal{A}$. Furthermore, the plane defined by $\{\mathbf{e}_t, \mathbf{e}_n\}$ is called the *osculating plane* relative to reference frame $\mathcal{A}$. Consequently, ${}^{\mathcal{A}}d\mathbf{e}_t/dt$ must lie in the osculating plane and must point in the direction of $\mathbf{e}_n$. Finally, a third unit vector, denoted $\mathbf{e}_b$, is defined such that it completes a right-handed system with $\mathbf{e}_t$ and $\mathbf{e}_n$, i.e.,

$$\mathbf{e}_b = \mathbf{e}_t \times \mathbf{e}_n \tag{2-286}$$

The vector $\mathbf{e}_b$ is called the *principal unit bi-normal* to the trajectory of point P relative to reference frame $\mathcal{A}$. Furthermore, the plane defined by $\{\mathbf{e}_t, \mathbf{e}_b\}$ is called the *rectifying plane* relative to reference frame $\mathcal{A}$ and lies orthogonal to $\mathbf{e}_n$. The orthonormal basis $\{\mathbf{e}_t, \mathbf{e}_n, \mathbf{e}_b\}$ is called an *intrinsic basis* relative to reference frame $\mathcal{A}$.[5] The geometry of an intrinsic basis relative to an arbitrary reference frame $\mathcal{A}$ is shown in Fig. 2-23.

Suppose now that we let $\mathcal{B}$ be a reference frame such that $\{\mathbf{e}_t, \mathbf{e}_n, \mathbf{e}_b\}$ is fixed in $\mathcal{B}$. Then the angular velocity of reference frame $\mathcal{B}$ as viewed by an observer in reference frame $\mathcal{A}$ can be written as

$${}^{\mathcal{A}}\boldsymbol{\omega}^{\mathcal{B}} = \omega_t\mathbf{e}_t + \omega_n\mathbf{e}_n + \omega_b\mathbf{e}_b \tag{2-287}$$

[5]The basis $\{\mathbf{e}_t, \mathbf{e}_n, \mathbf{e}_b\}$ is often referred to as a *Serret-Frenet* basis. The Serret-Frenet basis was established by Jean-Frédéric Frenet (1816–1900) in 1847 (Frenet, 1852) and by Joseph Alfred Serret (1819–1885) in 1851 (Serret, 1851).

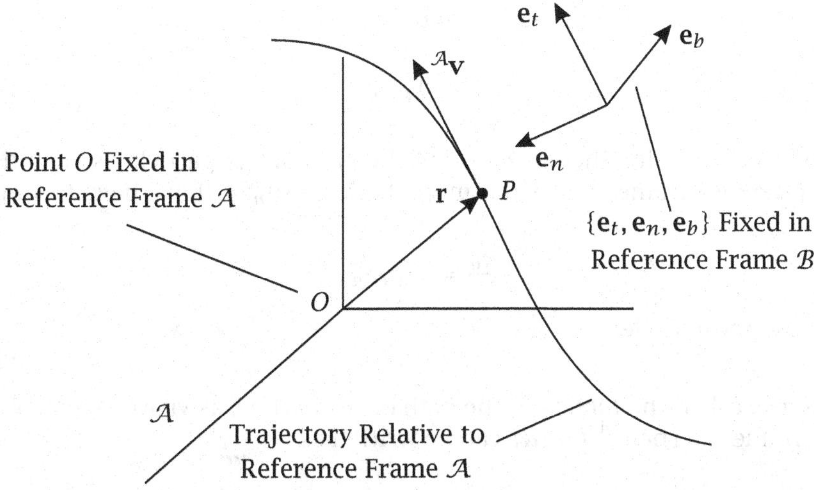

Figure 2-23 Geometry of intrinsic basis relative to an arbitrary reference frame $\mathcal{A}$.

Now, from Eqs. (2-123)-(2-125) on page 46 we have

$$\frac{{}^{\mathcal{A}}d\mathbf{e}_t}{dt} = {}^{\mathcal{A}}\boldsymbol{\omega}^{\mathcal{B}} \times \mathbf{e}_t \tag{2-288}$$

$$\frac{{}^{\mathcal{A}}d\mathbf{e}_n}{dt} = {}^{\mathcal{A}}\boldsymbol{\omega}^{\mathcal{B}} \times \mathbf{e}_n \tag{2-289}$$

$$\frac{{}^{\mathcal{A}}d\mathbf{e}_b}{dt} = {}^{\mathcal{A}}\boldsymbol{\omega}^{\mathcal{B}} \times \mathbf{e}_b \tag{2-290}$$

Substituting ${}^{\mathcal{A}}\boldsymbol{\omega}^{\mathcal{B}}$ from Eq. (2-287) into Eqs. (2-288)-(2-290), we obtain

$$\frac{{}^{\mathcal{A}}d\mathbf{e}_t}{dt} = \omega_b\mathbf{e}_n - \omega_n\mathbf{e}_b \tag{2-291}$$

$$\frac{{}^{\mathcal{A}}d\mathbf{e}_n}{dt} = -\omega_b\mathbf{e}_t + \omega_t\mathbf{e}_b \tag{2-292}$$

$$\frac{{}^{\mathcal{A}}d\mathbf{e}_b}{dt} = \omega_n\mathbf{e}_t - \omega_t\mathbf{e}_n \tag{2-293}$$

Now, because ${}^{\mathcal{A}}d\mathbf{e}_t/dt$ lies in the direction of $\mathbf{e}_n$, we see that $\omega_n = 0$. Therefore, the quantity ${}^{\mathcal{A}}d\mathbf{e}_t/dt$ can be written as

$$\frac{{}^{\mathcal{A}}d\mathbf{e}_t}{dt} = \omega_b\mathbf{e}_n \tag{2-294}$$

It is seen from Eq. (2-294) that

$$\omega_b = \left\|\frac{{}^{\mathcal{A}}d\mathbf{e}_t}{dt}\right\| \tag{2-295}$$

Suppose now that we define ω_b as

$$\omega_b = {}^{\mathcal{A}}v\kappa \tag{2-296}$$

Then, because $\omega_n = 0$, from Eq. (2-291) we have

$$\frac{{}^{\mathcal{A}}d\mathbf{e}_t}{dt} = {}^{\mathcal{A}}v\kappa\mathbf{e}_n \tag{2-297}$$

The quantity $\kappa \geq 0$ is called the *curvature* of the path of the particle as viewed by an observer in reference frame $\mathcal{A}$. Furthermore, because $\omega_n = 0$, we have from Eq. (2-293) that

$$\frac{{}^{\mathcal{A}}d\mathbf{e}_b}{dt} = -\omega_t\mathbf{e}_n \tag{2-298}$$

Furthermore, suppose we let

$$\omega_t = {}^{\mathcal{A}}v\tau \tag{2-299}$$

The quantity τ is called the *torsion* of the path of the particle as viewed by an observer in reference frame $\mathcal{A}$. Then ${}^{\mathcal{A}}d\mathbf{e}_b/dt$ can be written as

$$\frac{{}^{\mathcal{A}}d\mathbf{e}_b}{dt} = -{}^{\mathcal{A}}v\tau\mathbf{e}_n \tag{2-300}$$

Using ω_b from Eq. (2-296) and ω_t from Eq. (2-299), we obtain

$$\frac{{}^{\mathcal{A}}d\mathbf{e}_n}{dt} = -{}^{\mathcal{A}}v\kappa\mathbf{e}_t + {}^{\mathcal{A}}v\tau\mathbf{e}_b \tag{2-301}$$

We then obtain ${}^{\mathcal{A}}\boldsymbol{\omega}^{\mathcal{B}}$ as

$${}^{\mathcal{A}}\boldsymbol{\omega}^{\mathcal{B}} = {}^{\mathcal{A}}v(\tau\mathbf{e}_t + \kappa\mathbf{e}_b) \tag{2-302}$$

Consequently, the rates of change of the vectors $\mathbf{e}_t$, $\mathbf{e}_n$, and $\mathbf{e}_b$ relative to reference frame $\mathcal{A}$ are given from Eqs. (2-297), (2-301), and (2-300), respectively, as

$$\frac{{}^{\mathcal{A}}d\mathbf{e}_t}{dt} = {}^{\mathcal{A}}v\kappa\mathbf{e}_n \tag{2-303}$$

$$\frac{{}^{\mathcal{A}}d\mathbf{e}_n}{dt} = -{}^{\mathcal{A}}v\kappa\mathbf{e}_t + {}^{\mathcal{A}}v\tau\mathbf{e}_b \tag{2-304}$$

$$\frac{{}^{\mathcal{A}}d\mathbf{e}_b}{dt} = -{}^{\mathcal{A}}v\tau\mathbf{e}_n \tag{2-305}$$

We see from Eqs. (2-303)–(2-305) that the curvature and torsion are given as

$$\kappa = \frac{1}{{}^{\mathcal{A}}v}\left\|\frac{{}^{\mathcal{A}}d\mathbf{e}_t}{dt}\right\| \tag{2-306}$$

$$\tau = \frac{1}{{}^{\mathcal{A}}v}\left\|\frac{{}^{\mathcal{A}}d\mathbf{e}_b}{dt}\right\| \tag{2-307}$$

Finally, substituting the expression for ${}^{\mathcal{A}}d\mathbf{e}_t/dt$ from Eq. (2-297) into (2-283), we obtain the acceleration of the particle at point P as

$${}^{\mathcal{A}}\mathbf{a} = \frac{d}{dt}\left({}^{\mathcal{A}}v\right)\mathbf{e}_t + \kappa\left({}^{\mathcal{A}}v\right)^2\mathbf{e}_n \tag{2-308}$$

While for many applications it is most convenient to use time as the independent variable, for some problems it is more convenient to think of the trajectory relative to a reference frame $\mathcal{A}$ as spatial rather than temporal. For such problems it is convenient to use the *arc-length*, ${}^{\mathcal{A}}s$, as defined in Eq. (2-32) on page 36, as the independent variable. Suppose for convenience that we let

$$s = {}^{\mathcal{A}}s \tag{2-309}$$

Then, applying the chain rule using Eq. (2-309), the rates of change of $\mathbf{e}_t$, $\mathbf{e}_n$, and $\mathbf{e}_b$ with respect to s are given, respectively, as

$$\frac{{}^{\mathcal{A}}d\mathbf{e}_t}{ds} = \frac{{}^{\mathcal{A}}d\mathbf{e}_t}{dt}\frac{dt}{ds} \tag{2-310}$$

$$\frac{{}^{\mathcal{A}}d\mathbf{e}_n}{ds} = \frac{{}^{\mathcal{A}}d\mathbf{e}_n}{dt}\frac{dt}{ds} \tag{2-311}$$

$$\frac{{}^{\mathcal{A}}d\mathbf{e}_b}{ds} = \frac{{}^{\mathcal{A}}d\mathbf{e}_b}{dt}\frac{dt}{ds} \tag{2-312}$$

Furthermore, using ds/dt from Eq. (2-32) on page 36, we have

$$\frac{dt}{ds} = \frac{1}{{}^{\mathcal{A}}v} \tag{2-313}$$

Substituting the expression for dt/ds from Eq. (2-313) into Eqs. (2-303)-(2-305), we obtain the quantities ${}^{\mathcal{A}}d\mathbf{e}_t/ds$, ${}^{\mathcal{A}}d\mathbf{e}_n/ds$, and ${}^{\mathcal{A}}d\mathbf{e}_b/ds$, respectively, as

$$\frac{{}^{\mathcal{A}}d\mathbf{e}_t}{ds} = \kappa\mathbf{e}_n \tag{2-314}$$

$$\frac{{}^{\mathcal{A}}d\mathbf{e}_n}{ds} = -\kappa\mathbf{e}_t + \tau\mathbf{e}_b \tag{2-315}$$

$$\frac{{}^{\mathcal{A}}d\mathbf{e}_b}{ds} = -\tau\mathbf{e}_n \tag{2-316}$$

Equations (2-314)-(2-316) are called the *Serret-Frenet formulas* (Serret, 1851; Frenet, 1852) and describe the spatial rate of change of the intrinsic basis $\{\mathbf{e}_t, \mathbf{e}_n, \mathbf{e}_b\}$. From Eqs. (2-314)-(2-316), the curvature and torsion are given as

$$\kappa = \left\|\frac{{}^{\mathcal{A}}d\mathbf{e}_t}{ds}\right\| \tag{2-317}$$

$$\tau = \left\|\frac{{}^{\mathcal{A}}d\mathbf{e}_b}{ds}\right\| \tag{2-318}$$

Finally, we can define the *Darboux vector*, ${}^{\mathcal{A}}\boldsymbol{\omega}_{SF}^{\mathcal{B}}$, as

$${}^{\mathcal{A}}\boldsymbol{\omega}_{SF}^{\mathcal{B}} = \frac{{}^{\mathcal{A}}\boldsymbol{\omega}^{\mathcal{B}}}{{}^{\mathcal{A}}v} = \tau\mathbf{e}_t + \kappa\mathbf{e}_b \tag{2-319}$$

In terms of the Darboux vector, the Serret-Frenet formulas can be written as

$$\frac{{}^{\mathcal{A}}d\mathbf{e}_t}{ds} = {}^{\mathcal{A}}\boldsymbol{\omega}_{SF}^{\mathcal{B}} \times \mathbf{e}_t \tag{2-320}$$

$$\frac{{}^{\mathcal{A}}d\mathbf{e}_n}{ds} = {}^{\mathcal{A}}\boldsymbol{\omega}_{SF}^{\mathcal{B}} \times \mathbf{e}_n \tag{2-321}$$

$$\frac{{}^{\mathcal{A}}d\mathbf{e}_b}{ds} = {}^{\mathcal{A}}\boldsymbol{\omega}_{SF}^{\mathcal{B}} \times \mathbf{e}_b \tag{2-322}$$

We note that the acceleration of Eq. (2-308) is the same regardless of whether we choose a temporal or spatial description of the trajectory of point P.

It is important to understand a fundamental difference between intrinsic coordinates and other commonly used coordinate systems. In other commonly used coordinate systems (e.g., Cartesian, cylindrical, or spherical coordinates) the basis is used to describe the *position* of the particle. However, it is seen that intrinsic coordinates do not describe position, but instead describe motion along velocity (i.e., motion along $\mathbf{e}_t$) and motion orthogonal to velocity (i.e., motion along $\mathbf{e}_n$ and $\mathbf{e}_b$). Consequently, in general it is not possible to start with an expression for position in terms of an intrinsic basis and then determine velocity and acceleration using this expression for position. Instead, when using intrinsic coordinates it is necessary to define position in terms of some other coordinate system and then define the intrinsic basis in terms of this other coordinate system. As a result, it is generally not possible to use intrinsic coordinates alone; a second (auxiliary) coordinate system must be used in conjunction with intrinsic coordinates.

Example 2–6

A particle moves along a curve in the form of a fixed circular track. The equation for the track is given as

$$r = \sin\theta$$

Assuming the initial conditions $\theta(t = 0) = 0$ and $\dot{\theta}(t = 0) = \dot{\theta}_0$, determine (a) the arc-length parameter s as a function of the angle θ; (b) the intrinsic basis $\{\mathbf{e}_t, \mathbf{e}_n, \mathbf{e}_b\}$ in terms of a cylindrical basis fixed to the motion of the particle; (c) the curvature of the trajectory; and (d) the position, velocity, and acceleration of the particle in terms of the intrinsic basis $\{\mathbf{e}_t, \mathbf{e}_n, \mathbf{e}_b\}$.

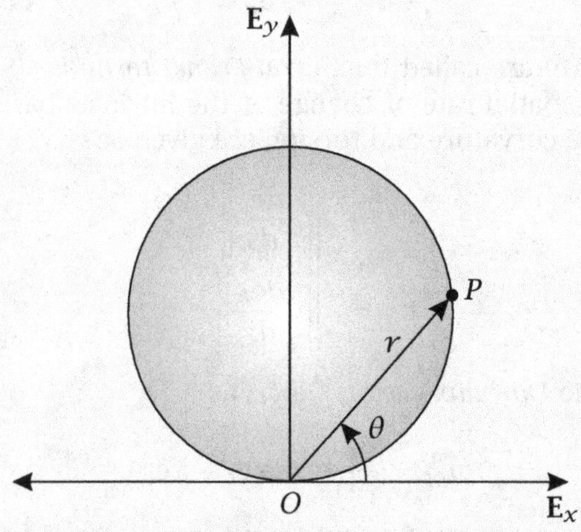

Figure 2–24 Particle moving on a circle.

Solution to Example 2–6

First, let $\mathcal{F}$ be a reference frame fixed to the circular track. Then choose the following

coordinate system fixed in reference frame $\mathcal{F}$:

Origin at point O

$\mathbf{E}_x$	=	To the right
$\mathbf{E}_z$	=	Out of page
$\mathbf{E}_y$	=	$\mathbf{E}_z \times \mathbf{E}_x$

Next, let $\mathcal{A}$ be a reference frame that is fixed to the direction of the position of the particle. Then choose the following coordinate system fixed in reference frame $\mathcal{F}$:

Origin at point O

$\mathbf{e}_r$	=	Along OP
$\mathbf{e}_z$	=	Out of page
$\mathbf{e}_\theta$	=	$\mathbf{e}_z \times \mathbf{e}_r$

The geometry of the bases $\{\mathbf{E}_x, \mathbf{E}_y, \mathbf{E}_z\}$ and $\{\mathbf{e}_r, \mathbf{e}_\theta, \mathbf{e}_z\}$ is shown in Fig. 2-25

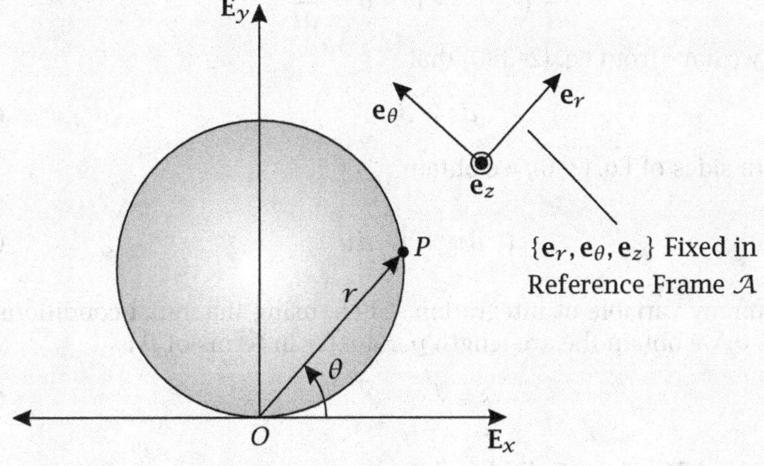

Figure 2-25 Geometry of basis $\{\mathbf{e}_r, \mathbf{e}_\theta, \mathbf{E}_z\}$ for Example 2-6.

(a) Arc-length as a Function of Angle θ

In terms of the cylindrical basis $\{\mathbf{e}_r, \mathbf{e}_\theta, \mathbf{e}_z\}$, the position of the particle is given as

$$\mathbf{r} = r\mathbf{e}_r = \sin\theta\,\mathbf{e}_r \tag{2-323}$$

Furthermore, the angular velocity of reference frame $\mathcal{A}$ relative to reference frame $\mathcal{F}$ is given as

$$^{\mathcal{F}}\boldsymbol{\omega}^{\mathcal{A}} = \dot{\theta}\mathbf{e}_z \tag{2-324}$$

The velocity of the particle as viewed by an observer in reference frame $\mathcal{F}$ is then obtained using Eq. (2-128) as

$$^{\mathcal{F}}\mathbf{v} = \frac{^{\mathcal{F}}d\mathbf{r}}{dt} = \frac{^{\mathcal{A}}d\mathbf{r}}{dt} + {}^{\mathcal{F}}\boldsymbol{\omega}^{\mathcal{A}} \times \mathbf{r} \tag{2-325}$$

Now we have

$$\begin{aligned} \frac{{}^{\mathcal{A}}d\mathbf{r}}{dt} &= \dot{\theta}\cos\theta\,\mathbf{e}_r && \text{(2-326)} \\ {}^{\mathcal{F}}\boldsymbol{\omega}^{\mathcal{A}}\times\mathbf{r} &= \dot{\theta}\mathbf{e}_z\times\sin\theta\,\mathbf{e}_r = \dot{\theta}\sin\theta\,\mathbf{e}_\theta && \text{(2-327)} \end{aligned}$$

Adding the expressions in Eqs. (2–326) and (2–327), we obtain the velocity of the particle in reference frame $\mathcal{F}$ as

$$ {}^{\mathcal{F}}\mathbf{v} = \dot{\theta}\cos\theta\,\mathbf{e}_r + \dot{\theta}\sin\theta\,\mathbf{e}_\theta = \dot{\theta}(\cos\theta\,\mathbf{e}_r + \sin\theta\,\mathbf{e}_\theta) \quad \text{(2-328)} $$

The speed of the particle as viewed by an observer in reference frame $\mathcal{F}$ is then given as

$$ {}^{\mathcal{F}}v = \|{}^{\mathcal{F}}\mathbf{v}\| = \|\dot{\theta}(\cos\theta\,\mathbf{e}_r + \sin\theta\,\mathbf{e}_\theta)\| = \dot{\theta} \quad \text{(2-329)} $$

The arc-length parameter in reference frame $\mathcal{F}$ is then obtained from Eq. (2–32) on page 36 as

$$ \frac{d}{dt}\left({}^{\mathcal{F}}s\right) = {}^{\mathcal{F}}v = \dot{\theta} = \frac{d\theta}{dt} \quad \text{(2-330)} $$

Setting ${}^{\mathcal{F}}s = s$, we have from Eq. (2–330) that

$$ ds = d\theta \quad \text{(2-331)} $$

Integrating both sides of Eq. (2–6), we obtain

$$ \int_{s_0}^{s} du = \int_{\theta_0}^{\theta} du \quad \text{(2-332)} $$

where u is a dummy variable of integration. Then, using the initial conditions $\theta(t = 0) = 0$ and $s_0 = 0$, we obtain the arc-length parameter in terms of θ as

$$ s \equiv {}^{\mathcal{F}}s = \theta \quad \text{(2-333)} $$

(b) Intrinsic Basis in Terms of Cylindrical Basis

The tangent vector $\mathbf{e}_t$ is obtained using the velocity from Eq. (2–328) and the speed from Eq. (2–329) as

$$ \mathbf{e}_t = \frac{{}^{\mathcal{F}}\mathbf{v}}{{}^{\mathcal{F}}v} = \cos\theta\,\mathbf{e}_r + \sin\theta\,\mathbf{e}_\theta \quad \text{(2-334)} $$

The rate of change of $\mathbf{e}_t$ in reference frame $\mathcal{F}$ is then obtained from the transport theorem of Eq. (2–128) on page 47 as

$$ \frac{{}^{\mathcal{F}}d\mathbf{e}_t}{dt} = \frac{{}^{\mathcal{A}}d\mathbf{e}_t}{dt} + {}^{\mathcal{F}}\boldsymbol{\omega}^{\mathcal{A}}\times\mathbf{e}_t \quad \text{(2-335)} $$

where

$$\begin{aligned} \frac{{}^{\mathcal{A}}d\mathbf{e}_t}{dt} &= -\dot{\theta}\sin\theta\,\mathbf{e}_r + \dot{\theta}\cos\theta\,\mathbf{e}_\theta && \text{(2-336)} \\ {}^{\mathcal{F}}\boldsymbol{\omega}^{\mathcal{A}}\times\mathbf{e}_t &= \dot{\theta}\mathbf{e}_z\times(\cos\theta\,\mathbf{e}_r + \sin\theta\,\mathbf{e}_\theta) \\ &= -\dot{\theta}\sin\theta\,\mathbf{e}_r + \dot{\theta}\cos\theta\,\mathbf{e}_\theta && \text{(2-337)} \end{aligned}$$

Adding the expressions in Eqs. (2–336) and (2–337), we obtain

$$\frac{^{\mathcal{F}}d\mathbf{e}_t}{dt} = -2\dot{\theta}\sin\theta\,\mathbf{e}_r + 2\dot{\theta}\cos\theta\,\mathbf{e}_\theta \tag{2–338}$$

Then, using Eq. (2–285), we have

$$\mathbf{e}_n = \frac{^{\mathcal{F}}d\mathbf{e}_t/dt}{\left\|^{\mathcal{F}}d\mathbf{e}_t/dt\right\|} \tag{2–339}$$

Substituting $^{\mathcal{F}}d\mathbf{e}_t/dt$ from Eq. (2–338) into (2–339), we obtain

$$\mathbf{e}_n = \frac{-2\dot{\theta}\sin\theta\,\mathbf{e}_r + 2\dot{\theta}\cos\theta\,\mathbf{e}_\theta}{2\dot{\theta}} = -\sin\theta\,\mathbf{e}_r + \cos\theta\,\mathbf{e}_\theta \tag{2–340}$$

Finally, the vector $\mathbf{e}_b$ is obtained from the right-hand rule as

$$\mathbf{e}_b = \mathbf{e}_t \times \mathbf{e}_n = (\cos\theta\,\mathbf{e}_r + \sin\theta\,\mathbf{e}_\theta) \times (-\sin\theta\,\mathbf{e}_r + \cos\theta\,\mathbf{e}_\theta) = \mathbf{e}_z \tag{2–341}$$

(c) Curvature of Trajectory

From Eq. (2–303) we have

$$\frac{^{\mathcal{F}}d\mathbf{e}_t}{dt} = {^{\mathcal{F}}v}\kappa\mathbf{e}_n \tag{2–342}$$

Consequently, the curvature of the trajectory is given as

$$\kappa = \frac{\|^{\mathcal{F}}d\mathbf{e}_t/dt\|}{^{\mathcal{F}}v} = \frac{2\dot{\theta}}{\dot{\theta}} = 2 \tag{2–343}$$

(d) Position, Velocity, and Acceleration of Particle in Terms of Intrinsic Basis

The position was given earlier in terms of the cylindrical basis as

$$\mathbf{r} = \sin\theta\,\mathbf{e}_r \tag{2–344}$$

Now, from Eqs. (2–334) and (2–339) we have

$$\begin{aligned} \mathbf{e}_t &= \cos\theta\,\mathbf{e}_r + \sin\theta\,\mathbf{e}_\theta \\ \mathbf{e}_n &= -\sin\theta\,\mathbf{e}_r + \cos\theta\,\mathbf{e}_\theta \end{aligned} \tag{2–345}$$

Solving Eq. (2–345) for $\mathbf{e}_r$, we obtain

$$\mathbf{e}_r = \cos\theta\,\mathbf{e}_t - \sin\theta\,\mathbf{e}_n \tag{2–346}$$

Consequently, the position of the particle is given in the intrinsic basis $\{\mathbf{e}_t, \mathbf{e}_n, \mathbf{e}_b\}$ as

$$\mathbf{r} = \sin\theta\,(\cos\theta\,\mathbf{e}_t - \sin\theta\,\mathbf{e}_n) = \sin\theta\,\cos\theta\,\mathbf{e}_t - \sin^2\theta\mathbf{e}_n \tag{2–347}$$

Then, from Eq. (2–282) we obtain $^{\mathcal{F}}\mathbf{v}$ in terms of the intrinsic basis as

$$^{\mathcal{F}}\mathbf{v} = {^{\mathcal{F}}v}\mathbf{e}_t = \dot{\theta}\mathbf{e}_t \tag{2–348}$$

Furthermore, from Eq. (2–329) we have

$$\frac{d}{dt}\left({}^{\mathcal{F}}v\right) = \ddot{\theta} \tag{2–349}$$

Substituting the expressions for ${}^{\mathcal{F}}v$, κ, and $d({}^{\mathcal{F}}v)/dt$ from Eqs. (2–329), (2–343), and (2–349), respectively, into Eq. (2–308), we obtain the acceleration as viewed by an observer in reference frame $\mathcal{F}$ expressed in the intrinsic basis $\{\mathbf{e}_t, \mathbf{e}_n, \mathbf{e}_b\}$ as

$${}^{\mathcal{F}}\mathbf{a} = \frac{d}{dt}\left({}^{\mathcal{F}}v\right)\mathbf{e}_t + \kappa\left({}^{\mathcal{F}}v\right)^2\mathbf{e}_n = \ddot{\theta}\mathbf{e}_t + 2\dot{\theta}^2\mathbf{e}_n \tag{2–350}$$

■

Example 2–7

A particle moves along a track in the shape of a cardioid as shown in Fig. 2–26. The equation for the cardioid is

$$r = a(1 - \cos\theta)$$

where θ is the angle measured from the fixed horizontal direction. Assuming that $\theta \in [0, \pi]$, determine the following quantities relative to an observer in a fixed reference frame: (a) the arc-length parameter s as a function of the angle θ; (b) the intrinsic basis $\{\mathbf{e}_t, \mathbf{e}_n, \mathbf{e}_b\}$ and the curvature of the trajectory as a function of the angle θ; and (c) the position, velocity, and acceleration of the particle in terms of the intrinsic basis $\{\mathbf{e}_t, \mathbf{e}_n, \mathbf{e}_b\}$.

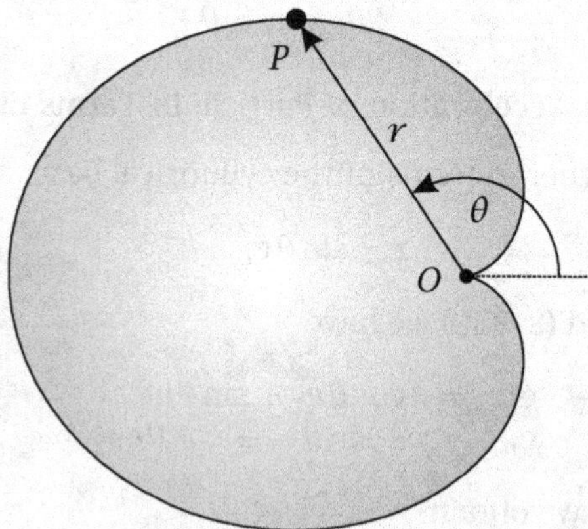

Figure 2–26 Particle moving on a track in the shape of a cardioid.

Solution to Example 2–7

First, let $\mathcal{F}$ be a reference frame fixed to the cardioid. Then choose the following

coordinate system fixed in reference frame $\mathcal{F}$:

$$
\begin{array}{lcl}
 & \text{Origin at point } O & \\
\mathbf{E}_x & = & \text{To the right} \\
\mathbf{E}_z & = & \text{Out of page} \\
\mathbf{E}_y & = & \mathbf{E}_z \times \mathbf{E}_x
\end{array}
$$

Next, let $\mathcal{A}$ be a reference frame fixed to the direction of OP. Then choose the following coordinate system fixed in reference frame $\mathcal{A}$:

$$
\begin{array}{lcl}
 & \text{Origin at point } O & \\
\mathbf{e}_r & = & \text{Along } OP \\
\mathbf{e}_z & = & \text{Out of page } (= \mathbf{E}_z) \\
\mathbf{e}_\theta & = & \mathbf{e}_z \times \mathbf{e}_r
\end{array}
$$

The bases $\{\mathbf{E}_x, \mathbf{E}_y, \mathbf{E}_z\}$ and $\{\mathbf{e}_r, \mathbf{e}_\theta, \mathbf{e}_z\}$ are shown in Fig. 2-27.

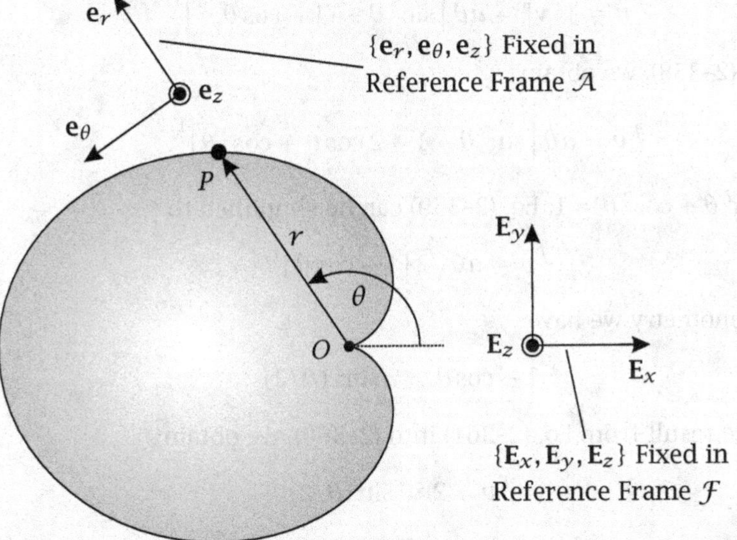

Figure 2-27 Bases $\{\mathbf{E}_x, \mathbf{E}_y, \mathbf{E}_z\}$ and $\{\mathbf{e}_r, \mathbf{e}_\theta, \mathbf{e}_z\}$ for Example 2-7.

(a) Arc-length as a Function of θ

The position of the particle in terms of the basis $\{\mathbf{e}_r, \mathbf{e}_\theta, \mathbf{E}_z\}$ is given as

$$\mathbf{r} = r\mathbf{e}_r = a(1 - \cos\theta)\mathbf{e}_r \tag{2-351}$$

Furthermore, the angular velocity of reference frame $\mathcal{A}$ in reference frame $\mathcal{F}$ is given as

$$^{\mathcal{F}}\boldsymbol{\omega}^{\mathcal{A}} = \dot{\theta}\mathbf{e}_z \tag{2-352}$$

The velocity of the particle in reference frame $\mathcal{F}$ is then obtained using the rate of change transport theorem of Eq. (2-128) on page 47 as

$$^{\mathcal{F}}\mathbf{v} = \frac{^{\mathcal{F}}d\mathbf{r}}{dt} = \frac{^{\mathcal{A}}d\mathbf{r}}{dt} + {}^{\mathcal{F}}\boldsymbol{\omega}^{\mathcal{A}} \times \mathbf{r} \tag{2-353}$$

Now we have

$$\frac{{}^{\mathcal{A}}d\mathbf{r}}{dt} = \dot{r}\mathbf{e}_r = a\dot{\theta}\sin\theta\,\mathbf{e}_r \tag{2-354}$$

$${}^{\mathcal{F}}\boldsymbol{\omega}^{\mathcal{A}} \times \mathbf{r} = \dot{\theta}\mathbf{e}_z \times a(1-\cos\theta)\mathbf{e}_r = a(1-\cos\theta)\dot{\theta}\mathbf{e}_\theta \tag{2-355}$$

Adding Eqs. (2-354) and (2-355), the velocity of the particle in reference frame $\mathcal{F}$ is obtained as

$${}^{\mathcal{F}}\mathbf{v} = a\dot{\theta}\sin\theta\,\mathbf{e}_r + a(1-\cos\theta)\dot{\theta}\mathbf{e}_\theta \tag{2-356}$$

Equation (2-356) can be rewritten as

$${}^{\mathcal{F}}\mathbf{v} = a\dot{\theta}\,[\sin\theta\,\mathbf{e}_r + (1-\cos\theta)\mathbf{e}_\theta] \tag{2-357}$$

The speed of the particle in reference frame $\mathcal{F}$ is then obtained from Eq. (2-357) as

$${}^{\mathcal{F}}v = \|{}^{\mathcal{F}}\mathbf{v}\| = a\dot{\theta}\left[\sin^2\theta + (1-\cos\theta)^2\right]^{1/2} \tag{2-358}$$

Expanding Eq. (2-358), we obtain

$${}^{\mathcal{F}}v = a\dot{\theta}\left[\sin^2\theta + 1 - 2\cos\theta + \cos^2\theta\right]^{1/2} \tag{2-359}$$

Noting that $\sin^2\theta + \cos^2\theta = 1$, Eq. (2-359) can be simplified to

$${}^{\mathcal{F}}v = a\dot{\theta}\sqrt{2}\,[1-\cos\theta]^{1/2} \tag{2-360}$$

Now from trigonometry we have

$$1-\cos\theta = 2\sin^2(\theta/2) \tag{2-361}$$

Substituting the result from Eq. (2-361) into (2-360), we obtain

$${}^{\mathcal{F}}v = 2a\dot{\theta}\sin(\theta/2) \tag{2-362}$$

Then, using the expression for arc-length from

$$\frac{d}{dt}\left({}^{\mathcal{F}}s\right) = {}^{\mathcal{F}}v = 2a\dot{\theta}\sin(\theta/2) \tag{2-363}$$

Setting $s = {}^{\mathcal{F}}s$, Eq. (2-363) can be written as

$$\frac{ds}{d\theta} = 2a\sin(\theta/2) \tag{2-364}$$

Separating the variables s and θ in Eq. (2-364) gives

$$ds = 2a\sin(\theta/2)d\theta \tag{2-365}$$

Integrating both sides of Eq. (2-365), we obtain

$$\int_{s_0}^{s} du = \int_{\theta_0}^{\theta} 2a\sin(u/2)du \tag{2-366}$$

Consequently,

$$s - s_0 = [-4a\cos(\theta/2)]_{\theta_0}^{\theta} \tag{2-367}$$

Equation (2-367) can then be written as

$$s - s_0 = -4a\,[\cos(\theta/2) - \cos(\theta_0/2)] \tag{2-368}$$

Finally, using the initial condition $\theta(t = 0) = 0$, we have $s_0 = 0$. from which Eq. (2-368) reduces to

$$s \equiv {}^{\mathcal{F}}s = -4a\,[\cos(\theta/2) - 1] \tag{2-369}$$

The arc-length in reference frame $\mathcal{F}$ is then given in terms of θ as

$${}^{\mathcal{F}}s = 4a\,[1 - \cos(\theta/2)] \tag{2-370}$$

(b) Determination of Intrinsic Basis

First, the tangent vector in reference frame $\mathcal{F}$ is given as

$$\mathbf{e}_t = \frac{{}^{\mathcal{F}}\mathbf{v}}{{}^{\mathcal{F}}v} \tag{2-371}$$

Then, using ${}^{\mathcal{F}}\mathbf{v}$ from Eq. (2-357) and ${}^{\mathcal{F}}v$ from Eq. (2-362), we obtain the tangent vector

$$\mathbf{e}_t = \frac{a\dot{\theta}\,[\sin\theta\,\mathbf{e}_r + (1 - \cos\theta)\mathbf{e}_\theta]}{2a\dot{\theta}\sin(\theta/2)} \tag{2-372}$$

Separating the $\mathbf{e}_r$ and $\mathbf{e}_\theta$ components in Eq. (2-372) gives

$$\mathbf{e}_t = \frac{\sin\theta}{2\sin(\theta/2)}\mathbf{e}_r + \frac{1 - \cos\theta}{2\sin(\theta/2)}\mathbf{e}_\theta \tag{2-373}$$

Now from trigonometry we have

$$\sin\theta = 2\sin(\theta/2)\cos(\theta/2) \tag{2-374}$$

Substituting Eq. (2-374) and the earlier trigonometric identity from Eq. (2-361) into (2-373), we obtain

$$\mathbf{e}_t = \frac{2\sin(\theta/2)\cos(\theta/2)}{2\sin(\theta/2)}\mathbf{e}_r + \frac{2\sin^2(\theta/2)}{2\sin(\theta/2)}\mathbf{e}_\theta \tag{2-375}$$

Canceling out the common factor of $\sin(\theta/2)$ from the numerator and denominator of each term in Eq. (2-375), we obtain the tangent vector

$$\mathbf{e}_t = \cos(\theta/2)\mathbf{e}_r + \sin(\theta/2)\mathbf{e}_\theta \tag{2-376}$$

Next, from the Serret-Frenet formulas we have

$$\frac{{}^{\mathcal{F}}d\mathbf{e}_t}{dt} = {}^{\mathcal{F}}v\kappa\mathbf{e}_n \tag{2-377}$$

Computing the rate of change of $\mathbf{e}_t$ in reference frame $\mathcal{F}$, we have

$$\frac{{}^{\mathcal{F}}d\mathbf{e}_t}{dt} = \frac{{}^{\mathcal{A}}d\mathbf{e}_t}{dt} + {}^{\mathcal{F}}\boldsymbol{\omega}^{\mathcal{A}} \times \mathbf{e}_t \tag{2-378}$$

Now we have

$$\begin{aligned} \frac{{}^{\mathcal{A}}d\mathbf{e}_t}{dt} &= -\tfrac{1}{2}\dot{\theta}\sin(\theta/2)\mathbf{e}_r + \tfrac{1}{2}\dot{\theta}\cos(\theta/2)\mathbf{e}_\theta && (2\text{-}379) \\ {}^{\mathcal{F}}\boldsymbol{\omega}^{\mathcal{A}} \times \mathbf{e}_t &= \dot{\theta}\mathbf{e}_z \times [\cos(\theta/2)\mathbf{e}_r + \sin(\theta/2)\mathbf{e}_\theta] \\ &= -\dot{\theta}\sin(\theta/2)\mathbf{e}_r + \dot{\theta}\cos(\theta/2)\mathbf{e}_\theta && (2\text{-}380) \end{aligned}$$

Adding the expressions in Eqs. (2-379) and (2-380), we obtain

$$\frac{{}^{\mathcal{F}}d\mathbf{e}_t}{dt} = -\tfrac{3}{2}\dot{\theta}\sin(\theta/2)\mathbf{e}_r + \tfrac{3}{2}\dot{\theta}\cos(\theta/2)\mathbf{e}_\theta \tag{2-381}$$

It can be seen from Eq. (2-381) that

$$\left\| \frac{{}^{\mathcal{F}}d\mathbf{e}_t}{dt} \right\| = \tfrac{3}{2}\dot{\theta} = {}^{\mathcal{F}}v\kappa \tag{2-382}$$

Then, using ${}^{\mathcal{F}}v$ from Eq. (2-362), we obtain κ as

$$\kappa = \frac{3\dot{\theta}/2}{2a\dot{\theta}\sin(\theta/2)} = \frac{3}{4a\sin(\theta/2)} \tag{2-383}$$

Now, because $\mathbf{e}_n = \left({}^{\mathcal{F}}d\mathbf{e}_t/dt\right)/\|d\mathbf{e}_t/dt\|$, we obtain $\mathbf{e}_n$ as

$$\mathbf{e}_n = -\sin(\theta/2)\mathbf{e}_r + \cos(\theta/2)\mathbf{e}_\theta \tag{2-384}$$

Finally, the principal unit bi-normal vector is computed using $\mathbf{e}_t$ from Eq. (2-376) and $\mathbf{e}_n$ from Eq. (2-384) as

$$\begin{aligned} \mathbf{e}_b &= \mathbf{e}_t \times \mathbf{e}_n \\ &= [\cos(\theta/2)\mathbf{e}_r + \sin(\theta/2)\mathbf{e}_\theta] \times [-\sin(\theta/2)\,\mathbf{e}_r + \cos(\theta/2)\,\mathbf{e}_\theta] && (2\text{-}385) \\ &= \mathbf{e}_z \end{aligned}$$

(c) Position, Velocity, and Acceleration of Particle in Terms of Intrinsic Basis

First, in order to express $\mathbf{r}$ (as given in Eq. (2-351)) in terms of $\{\mathbf{e}_t, \mathbf{e}_n, \mathbf{e}_b\}$, it is necessary to solve for $\mathbf{e}_r$ in terms of $\{\mathbf{e}_t, \mathbf{e}_n, \mathbf{e}_b\}$. We have from Eqs. (2-376) and (2-384)

$$\begin{aligned} \mathbf{e}_t &= \cos(\theta/2)\mathbf{e}_r + \sin(\theta/2)\mathbf{e}_\theta && (2\text{-}386) \\ \mathbf{e}_n &= -\sin(\theta/2)\,\mathbf{e}_r + \cos(\theta/2)\,\mathbf{e}_\theta && (2\text{-}387) \end{aligned}$$

Multiplying Eq. (2-386) and (2-387) by $\cos(\theta/2)$ and $\sin(\theta/2)$, respectively, and subtracting gives

$$\cos(\theta/2)\,\mathbf{e}_t - \sin(\theta/2)\,\mathbf{e}_n = \cos^2(\theta/2)\mathbf{e}_r + \sin^2(\theta/2)\mathbf{e}_r = \mathbf{e}_r \tag{2-388}$$

Substituting $\mathbf{e}_r$ from Eq. (2-388) into (2-351), the position of the particle is obtained in terms of the intrinsic basis $\{\mathbf{e}_t, \mathbf{e}_n, \mathbf{e}_b\}$ as

$$\mathbf{r} = a(1 - \cos\theta)\left[\cos(\theta/2)\,\mathbf{e}_t - \sin(\theta/2)\,\mathbf{e}_n\right] \tag{2-389}$$

Now the velocity in reference frame $\mathcal{F}$ is obtained using the speed in Eq. (2-362) as

$$^{\mathcal{F}}\mathbf{v} = {}^{\mathcal{F}}v\mathbf{e}_t = 2a\dot{\theta}\sin(\theta/2)\mathbf{e}_t \tag{2-390}$$

The acceleration in reference frame $\mathcal{F}$ is given in terms of the intrinsic basis $\{\mathbf{e}_t, \mathbf{e}_n, \mathbf{e}_b\}$ as

$$^{\mathcal{F}}\mathbf{a} = \frac{d}{dt}\left({}^{\mathcal{F}}v\right)\mathbf{e}_t + \kappa\left({}^{\mathcal{F}}v\right)^2\mathbf{e}_n \tag{2-391}$$

Differentiating $^{\mathcal{F}}v$ in Eq. (2-362), we obtain

$$\frac{d}{dt}\left({}^{\mathcal{F}}v\right) = 2a\ddot{\theta}\sin(\theta/2) + a\dot{\theta}^2\cos(\theta/2) \tag{2-392}$$

Next, computing $\kappa\left({}^{\mathcal{F}}v\right)^2$ using κ from Eq. (2-383), we obtain

$$\kappa\left({}^{\mathcal{F}}v\right)^2 = \frac{3}{4a\sin(\theta/2)}\left[2a\dot{\theta}\sin(\theta/2)\right]^2 = 3a\dot{\theta}^2\sin(\theta/2) \tag{2-393}$$

Substituting the results of Eqs. (2-392) and (2-393) into Eq. (2-391), the acceleration of the particle in given in terms of the intrinsic basis as

$$^{\mathcal{F}}\mathbf{a} = \left[2a\ddot{\theta}\sin(\theta/2) + a\dot{\theta}^2\cos(\theta/2)\right]\mathbf{e}_t + 3a\dot{\theta}^2\sin(\theta/2)\mathbf{e}_n \tag{2-394}$$

■

2.12 Kinematics in a Rotating and Translating Reference Frame

In many dynamics problems it is most convenient to describe the motion of a point using a reference frame that both translates and rotates relative to another reference frame. In this Section we describe the kinematics of a particle using a rotating and translating reference frame.

Let $\mathcal{A}$ and $\mathcal{B}$ be two reference frames such that $\mathcal{B}$ rotates and translates relative to $\mathcal{A}$. Furthermore, let O be a point fixed in $\mathcal{A}$ and let Q be a point fixed in $\mathcal{B}$. Then, let $\mathbf{r}_Q$ be the position of a point Q relative to a point O and let $\mathbf{r}_{P/Q}$ be the position of a point P relative to point Q. Finally, let $^{\mathcal{A}}\boldsymbol{\omega}^{\mathcal{B}}$ be the angular velocity of reference frame $\mathcal{B}$ as viewed by an observer in reference frame $\mathcal{A}$. The points O, Q, and P and the reference frames $\mathcal{A}$ and $\mathcal{B}$ are shown in Fig. 2-28, where it can be seen that reference frame $\mathcal{B}$ simultaneously rotates and translates relative to reference frame $\mathcal{A}$. Using Fig. 2-28, the position of point P relative to point O can then be written as

$$\mathbf{r}_P = \mathbf{r}_Q + \mathbf{r}_{P/Q} \tag{2-395}$$

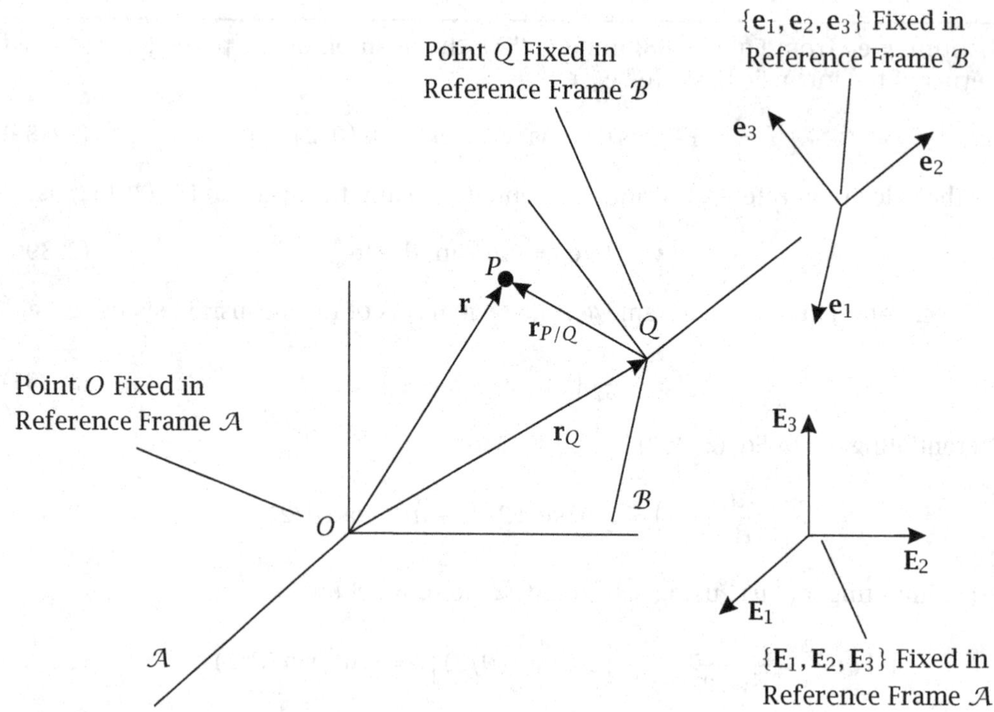

Figure 2–28 Position of a point P relative to a point O fixed in reference frame $\mathcal{A}$ decomposed into the position of a point Q fixed in reference frame $\mathcal{B}$ and the position of point P relative to Q.

The rate of change of $\mathbf{r}_P$ relative to an observer in reference frame $\mathcal{A}$ is then given as

$$^{\mathcal{A}}\mathbf{v}_P = \frac{^{\mathcal{A}}d}{dt}(\mathbf{r}_P) = \frac{^{\mathcal{A}}d}{dt}(\mathbf{r}_Q) + \frac{^{\mathcal{A}}d}{dt}(\mathbf{r}_{P/Q}) \tag{2–396}$$

Noting that $^{\mathcal{A}}d\mathbf{r}_Q/dt = {}^{\mathcal{A}}\mathbf{v}_Q$, Eq. (2–396) can be written as

$$^{\mathcal{A}}\mathbf{v}_P = {}^{\mathcal{A}}\mathbf{v}_Q + \frac{^{\mathcal{A}}d}{dt}(\mathbf{r}_{P/Q}) \tag{2–397}$$

Applying the rate of change transport theorem of Eq. (2–128) on page 47, $^{\mathcal{A}}d(\mathbf{r}_{P/Q})/dt$ is given as

$$\frac{^{\mathcal{A}}d}{dt}(\mathbf{r}_{P/Q}) = \frac{^{\mathcal{B}}d}{dt}(\mathbf{r}_{P/Q}) + {}^{\mathcal{A}}\boldsymbol{\omega}^{\mathcal{B}} \times \mathbf{r}_{P/Q} = {}^{\mathcal{B}}\mathbf{v}_{P/Q} + {}^{\mathcal{A}}\boldsymbol{\omega}^{\mathcal{B}} \times \mathbf{r}_{P/Q} \tag{2–398}$$

Substituting the result of Eq. (2–398) into (2–397), we obtain $^{\mathcal{A}}\mathbf{v}_P$ as

$$\boxed{^{\mathcal{A}}\mathbf{v}_P = {}^{\mathcal{A}}\mathbf{v}_Q + {}^{\mathcal{B}}\mathbf{v}_{P/Q} + {}^{\mathcal{A}}\boldsymbol{\omega}^{\mathcal{B}} \times \mathbf{r}_{P/Q}} \tag{2–399}$$

Equation (2–399) provides a way to compute the velocity of a point P in a reference frame $\mathcal{A}$ given observations of the motion of the point in a reference frame $\mathcal{B}$ where $\mathcal{B}$ simultaneously rotates and translates relative to reference frame $\mathcal{A}$.

Using the expression for ${}^{\mathcal{A}}\mathbf{v}_P$ from Eq. (2–399), the acceleration of point P in reference frame $\mathcal{A}$ is obtained as

$$ {}^{\mathcal{A}}\mathbf{a}_P = \frac{{}^{\mathcal{A}}d}{dt}\left({}^{\mathcal{A}}\mathbf{v}_P\right) = \frac{{}^{\mathcal{A}}d}{dt}\left({}^{\mathcal{A}}\mathbf{v}_Q + {}^{\mathcal{B}}\mathbf{v}_{P/Q} + {}^{\mathcal{A}}\boldsymbol{\omega}^{\mathcal{B}} \times \mathbf{r}_{P/Q}\right) \tag{2–400} $$

Expanding Eq. (2–400) gives

$$ {}^{\mathcal{A}}\mathbf{a}_P = \frac{{}^{\mathcal{A}}d}{dt}\left({}^{\mathcal{A}}\mathbf{v}_Q\right) + \frac{{}^{\mathcal{A}}d}{dt}\left({}^{\mathcal{B}}\mathbf{v}_{P/Q}\right) + \frac{{}^{\mathcal{A}}d}{dt}\left({}^{\mathcal{A}}\boldsymbol{\omega}^{\mathcal{B}} \times \mathbf{r}_{P/Q}\right) \tag{2–401} $$

Now we note that

$$ \frac{{}^{\mathcal{A}}d}{dt}\left({}^{\mathcal{A}}\mathbf{v}_Q\right) = {}^{\mathcal{A}}\mathbf{a}_Q \tag{2–402} $$

where ${}^{\mathcal{A}}\mathbf{a}_Q$ is the acceleration of point Q in reference frame $\mathcal{A}$. Furthermore, from the rate of change transport theorem of Eq. (2–128) on page 47, we have

$$ \frac{{}^{\mathcal{A}}d}{dt}\left({}^{\mathcal{B}}\mathbf{v}_{P/Q}\right) = \frac{{}^{\mathcal{B}}d}{dt}\left({}^{\mathcal{B}}\mathbf{v}_{P/Q}\right) + {}^{\mathcal{A}}\boldsymbol{\omega}^{\mathcal{B}} \times {}^{\mathcal{B}}\mathbf{v}_{P/Q} \tag{2–403} $$

Now we note that

$$ \frac{{}^{\mathcal{B}}d}{dt}\left({}^{\mathcal{B}}\mathbf{v}_{P/Q}\right) = {}^{\mathcal{B}}\mathbf{a}_{P/Q} \tag{2–404} $$

Consequently, Eq. (2–401) can be written as

$$ {}^{\mathcal{A}}\mathbf{a}_P = {}^{\mathcal{A}}\mathbf{a}_Q + {}^{\mathcal{B}}\mathbf{a}_{P/Q} + {}^{\mathcal{A}}\boldsymbol{\omega}^{\mathcal{B}} \times {}^{\mathcal{B}}\mathbf{v}_{P/Q} + \frac{{}^{\mathcal{A}}d}{dt}\left({}^{\mathcal{A}}\boldsymbol{\omega}^{\mathcal{B}} \times \mathbf{r}_{P/Q}\right) \tag{2–405} $$

Next, we have

$$ \frac{{}^{\mathcal{A}}d}{dt}\left({}^{\mathcal{A}}\boldsymbol{\omega}^{\mathcal{B}} \times \mathbf{r}_{P/Q}\right) = \frac{{}^{\mathcal{A}}d}{dt}\left({}^{\mathcal{A}}\boldsymbol{\omega}^{\mathcal{B}}\right) \times \mathbf{r}_{P/Q} + {}^{\mathcal{A}}\boldsymbol{\omega}^{\mathcal{B}} \times \frac{{}^{\mathcal{A}}d}{dt}\left(\mathbf{r}_{P/Q}\right) \tag{2–406} $$

Now, we know that

$$ \begin{aligned} \frac{{}^{\mathcal{A}}d}{dt}\left({}^{\mathcal{A}}\boldsymbol{\omega}^{\mathcal{B}}\right) &= {}^{\mathcal{A}}\boldsymbol{\alpha}^{\mathcal{B}} \\ \frac{{}^{\mathcal{A}}d}{dt}\left(\mathbf{r}_{P/Q}\right) &= {}^{\mathcal{A}}\mathbf{v}_{P/Q} \end{aligned} \tag{2–407} $$

Therefore, Eq. (2–406) can be written as

$$ \frac{{}^{\mathcal{A}}d}{dt}\left({}^{\mathcal{A}}\boldsymbol{\omega}^{\mathcal{B}} \times \mathbf{r}_{P/Q}\right) = {}^{\mathcal{A}}\boldsymbol{\alpha}^{\mathcal{B}} \times \mathbf{r}_{P/Q} + {}^{\mathcal{A}}\boldsymbol{\omega}^{\mathcal{B}} \times {}^{\mathcal{A}}\mathbf{v}_{P/Q} \tag{2–408} $$

Furthermore, we have from the rate of change transport theorem that

$$ {}^{\mathcal{A}}\mathbf{v}_{P/Q} = \frac{{}^{\mathcal{A}}d}{dt}\left(\mathbf{r}_{P/Q}\right) = \frac{{}^{\mathcal{B}}d}{dt}\left(\mathbf{r}_{P/Q}\right) + {}^{\mathcal{A}}\boldsymbol{\omega}^{\mathcal{B}} \times \mathbf{r}_{P/Q} = {}^{\mathcal{B}}\mathbf{v}_{P/Q} + {}^{\mathcal{A}}\boldsymbol{\omega}^{\mathcal{B}} \times \mathbf{r}_{P/Q} \tag{2–409} $$

Substituting the result of Eq. (2–409) into (2–408), we have

$$ \begin{aligned} \frac{{}^{\mathcal{A}}d}{dt}\left({}^{\mathcal{A}}\boldsymbol{\omega}^{\mathcal{B}} \times \mathbf{r}_{P/Q}\right) &= {}^{\mathcal{A}}\boldsymbol{\alpha}^{\mathcal{B}} \times \mathbf{r}_{P/Q} + {}^{\mathcal{A}}\boldsymbol{\omega}^{\mathcal{B}} \times \left({}^{\mathcal{B}}\mathbf{v}_{P/Q} + {}^{\mathcal{A}}\boldsymbol{\omega}^{\mathcal{B}} \times \mathbf{r}_{P/Q}\right) \\ &= {}^{\mathcal{A}}\boldsymbol{\alpha}^{\mathcal{B}} \times \mathbf{r}_{P/Q} + {}^{\mathcal{A}}\boldsymbol{\omega}^{\mathcal{B}} \times {}^{\mathcal{B}}\mathbf{v}_{P/Q} + {}^{\mathcal{A}}\boldsymbol{\omega}^{\mathcal{B}} \times \left({}^{\mathcal{A}}\boldsymbol{\omega}^{\mathcal{B}} \times \mathbf{r}_{P/Q}\right) \end{aligned} \tag{2–410} $$

Then, substituting the expression from Eq. (2–410) into (2–405), we obtain

$$
\begin{aligned}
{}^{\mathcal{A}}\mathbf{a}_P = {}^{\mathcal{A}}\mathbf{a}_Q + {}^{\mathcal{B}}\mathbf{a}_{P/Q} + {}^{\mathcal{A}}\boldsymbol{\omega}^{\mathcal{B}} \times {}^{\mathcal{B}}\mathbf{v}_{P/Q} \\
+ {}^{\mathcal{A}}\boldsymbol{\alpha}^{\mathcal{B}} \times \mathbf{r}_{P/Q} + {}^{\mathcal{A}}\boldsymbol{\omega}^{\mathcal{B}} \times {}^{\mathcal{B}}\mathbf{v}_{P/Q} + {}^{\mathcal{A}}\boldsymbol{\omega}^{\mathcal{B}} \times \left({}^{\mathcal{A}}\boldsymbol{\omega}^{\mathcal{B}} \times \mathbf{r}_{P/Q}\right)
\end{aligned}
\tag{2–411}
$$

Equation (2–411) simplifies to

$$
\boxed{
\begin{aligned}
{}^{\mathcal{A}}\mathbf{a}_P {}^{\mathcal{A}}\mathbf{a}_Q + {}^{\mathcal{B}}\mathbf{a}_{P/Q} + {}^{\mathcal{A}}\boldsymbol{\alpha}^{\mathcal{B}} \times \mathbf{r}_{P/Q} \\
+ 2{}^{\mathcal{A}}\boldsymbol{\omega}^{\mathcal{B}} \times {}^{\mathcal{B}}\mathbf{v}_{P/Q} + {}^{\mathcal{A}}\boldsymbol{\omega}^{\mathcal{B}} \times \left({}^{\mathcal{A}}\boldsymbol{\omega}^{\mathcal{B}} \times \mathbf{r}_{P/Q}\right)
\end{aligned}
}
\tag{2–412}
$$

Equation (2–412) provides a way to compute the acceleration of a point P in a reference frame $\mathcal{A}$ given observations of the motion of the point in a reference frame $\mathcal{B}$, where $\mathcal{B}$ simultaneously rotates and translates relative to reference frame $\mathcal{A}$. It is noted that, for historical reasons, the third, fourth, and fifth terms in Eq. (2–412) have the following names:

$$
\begin{aligned}
{}^{\mathcal{A}}\boldsymbol{\alpha}^{\mathcal{B}} \times \mathbf{r}_{P/Q} &= \text{Euler acceleration} \\
2{}^{\mathcal{A}}\boldsymbol{\omega}^{\mathcal{B}} \times {}^{\mathcal{B}}\mathbf{v}_{P/Q} &= \text{Coriolis acceleration} \\
{}^{\mathcal{A}}\boldsymbol{\omega}^{\mathcal{B}} \times ({}^{\mathcal{A}}\boldsymbol{\omega}^{\mathcal{B}} \times \mathbf{r}_{P/Q}) &= \text{Centripetal acceleration}
\end{aligned}
$$

2.13 Practical Approach to Computing Velocity and Acceleration

In Sections 2.9 and 2.12 we derived expressions for the velocity and acceleration of a point in a reference frame $\mathcal{A}$ given observations about the motion of the point in a reference frame $\mathcal{B}$, where reference frame $\mathcal{B}$ either rotates, or rotates and translates, relative to reference frame $\mathcal{A}$. While in principle these results can be used directly to compute velocity and acceleration, in practice it is often difficult to implement these results because it is necessary to keep track of many quantities simultaneously.[6] Consequently, unless one has a complete understanding of each term in these equations, it is preferable to start at a more fundamental level and apply the rate of change transport theorem of Eq. (2–128) twice. More specifically, the rate of change transport theorem is first applied to position to obtain velocity and is applied a second time to velocity to obtain acceleration. While two applications of the rate of change transport theorem are less direct than a single application of either Eq. (2–158) in Section 2.9 or Eq. (2–412) in Section 2.12, using a two-step approach is more straightforward, thereby minimizing the possibility for error.

Example 2–8

Two rigid rods of lengths l_1 and l_2 are hinged in tandem as shown in Fig. 2–29. Rod OA is hinged at one of its ends to the fixed point O while its other end is hinged at point A, where A is one end of the second rod AP. Point P is located at the free end of rod AP. Knowing that θ is the angle between rod OA and the horizontal and that ϕ is

[6] It is particularly difficult to implement the results for acceleration as given in Eqs. (2–158) and (2–412).

the angle between rod AP and rod OA, determine (a) the velocity and acceleration of point P as viewed by an observer fixed to rod OA and (b) the velocity and acceleration of point P as viewed by an observer fixed to the ground.

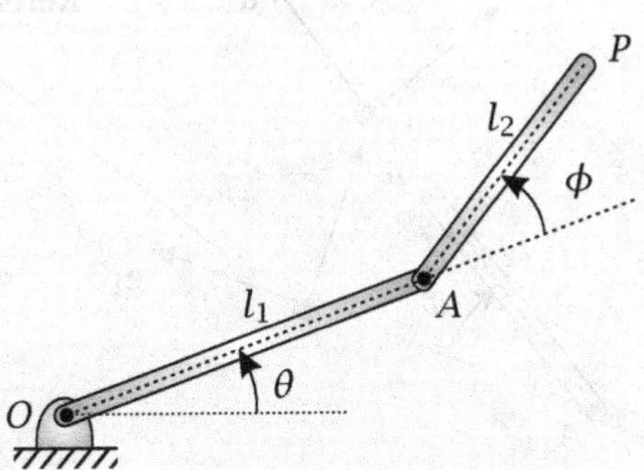

Figure 2–29 Double rod system corresponding to Example 2–8.

Solution to Example 2–8

First, let $\mathcal{F}$ be an inertial reference frame fixed to the ground. Next, for this problem it is useful to use two rotating reference frames to describe the motion. The first rotating reference frame, $\mathcal{A}$, is fixed to rod OA while the second rotating reference frame, $\mathcal{B}$, is fixed to rod AP. Corresponding to reference frame $\mathcal{A}$, we choose the following coordinate system:

$$
\begin{array}{ccc}
 & \text{Origin at point } O & \\
\mathbf{e}_1 & = & \text{Along } OA \\
\mathbf{e}_3 & = & \text{Out of page} \\
\mathbf{e}_2 & = & \mathbf{e}_3 \times \mathbf{e}_1
\end{array}
$$

Corresponding to reference frame $\mathcal{B}$, we choose the following coordinate system:

$$
\begin{array}{ccc}
 & \text{Origin at point } A & \\
\mathbf{u}_1 & = & \text{Along } AP \\
\mathbf{u}_3 & = & \text{Out of page } (= \mathbf{e}_3) \\
\mathbf{u}_2 & = & \mathbf{u}_3 \times \mathbf{u}_1
\end{array}
$$

The geometry of the bases $\{\mathbf{e}_1, \mathbf{e}_2, \mathbf{e}_3\}$ and $\{\mathbf{u}_1, \mathbf{u}_2, \mathbf{u}_3\}$ is shown in Fig. 2–30. In particular, using Fig. 2–30 we have

$$\mathbf{u}_1 = \cos\phi\mathbf{e}_1 + \sin\phi\mathbf{e}_2 \tag{2–413}$$

$$\mathbf{u}_2 = -\sin\phi\mathbf{e}_1 + \cos\phi\mathbf{e}_2 \tag{2–414}$$

$$\mathbf{e}_1 = \cos\phi\mathbf{u}_1 - \sin\phi\mathbf{u}_2 \tag{2–415}$$

$$\mathbf{e}_2 = \sin\phi\mathbf{u}_1 + \cos\phi\mathbf{u}_2 \tag{2–416}$$

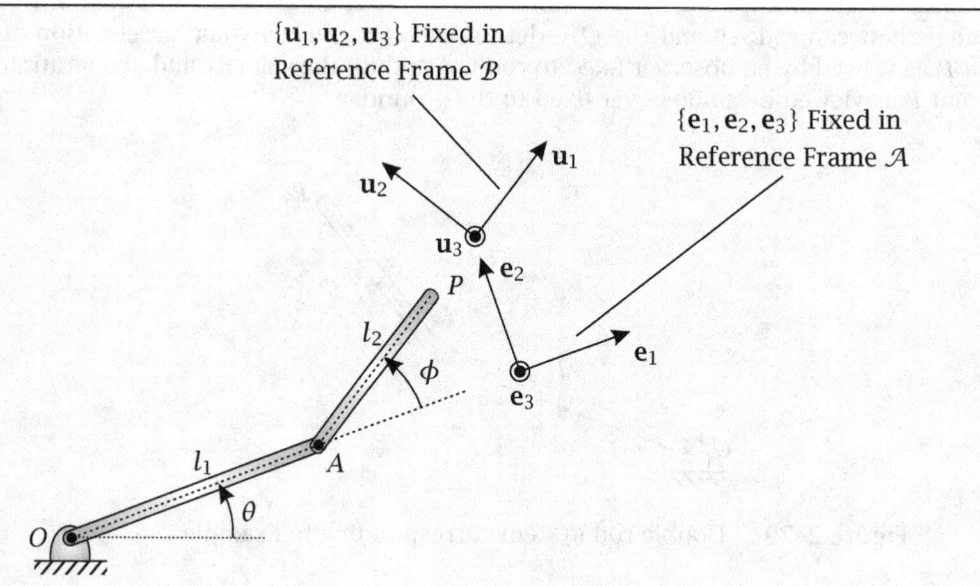

Figure 2-30 Bases $\{\mathbf{e}_1, \mathbf{e}_2, \mathbf{e}_3\}$ and $\{\mathbf{u}_1, \mathbf{u}_2, \mathbf{u}_3\}$ fixed in reference frames $\mathcal{A}$ and $\mathcal{B}$, respectively, for Example 2-8.

(a) Velocity and Acceleration of Point P as Viewed by an Observer Fixed to Rod OA

We note that the position of point P can be expressed as

$$\mathbf{r}_P = \mathbf{r}_A + \mathbf{r}_{P/A} \tag{2-417}$$

Now, in terms of the basis $\{\mathbf{e}_1, \mathbf{e}_2, \mathbf{e}_3\}$, the position of point A is given as

$$\mathbf{r}_A = l_1\mathbf{e}_1 \tag{2-418}$$

Furthermore, in terms of the basis $\{\mathbf{u}_1, \mathbf{u}_2, \mathbf{u}_3\}$, the position of P relative to A is given as

$$\mathbf{r}_{P/A} = l_2\mathbf{u}_1 \tag{2-419}$$

Substituting the expressions from Eqs. (2-418) and (2-419) into Eq. (2-417), we obtain

$$\mathbf{r}_P = l_1\mathbf{e}_1 + l_2\mathbf{u}_1 \tag{2-420}$$

The velocity of point P as viewed by an observer in reference frame $\mathcal{A}$ is then obtained as

$${}^{\mathcal{A}}\mathbf{v}_P = \frac{{}^{\mathcal{A}}d}{dt}(\mathbf{r}_P) = \frac{{}^{\mathcal{A}}d}{dt}(\mathbf{r}_A) + \frac{{}^{\mathcal{A}}d}{dt}(\mathbf{r}_{P/A}) = {}^{\mathcal{A}}\mathbf{v}_A + {}^{\mathcal{A}}\mathbf{v}_{P/A} \tag{2-421}$$

Now because the basis $\{\mathbf{e}_1, \mathbf{e}_2, \mathbf{e}_3\}$ is fixed in reference frame $\mathcal{A}$, we have

$${}^{\mathcal{A}}\mathbf{v}_A = \frac{{}^{\mathcal{A}}d}{dt}(\mathbf{r}_A) = \mathbf{0} \tag{2-422}$$

Furthermore, using the rate of change transport theorem of Eq. (2-128) on page 47 between reference frames $\mathcal{B}$ and $\mathcal{A}$, we have

$${}^{\mathcal{A}}\mathbf{v}_{P/A} = \frac{{}^{\mathcal{A}}d}{dt}(\mathbf{r}_{P/A}) = \frac{{}^{\mathcal{B}}d}{dt}(\mathbf{r}_{P/A}) + {}^{\mathcal{A}}\boldsymbol{\omega}^{\mathcal{B}} \times \mathbf{r}_{P/A} \tag{2-423}$$

Then, because $\mathbf{r}_{P/A}$ is fixed in reference frame $\mathcal{B}$, we have

$$\frac{{}^{\mathcal{B}}d}{dt}(\mathbf{r}_{P/A}) = \mathbf{0} \tag{2-424}$$

Furthermore, because reference frame $\mathcal{B}$ is fixed in rod AP and the angle ϕ is measured relative to rod OA, we have ${}^{\mathcal{A}}\boldsymbol{\omega}^{\mathcal{B}}$ as

$${}^{\mathcal{A}}\boldsymbol{\omega}^{\mathcal{B}} = \dot{\phi}\mathbf{u}_3 \tag{2-425}$$

Consequently,

$${}^{\mathcal{A}}\boldsymbol{\omega}^{\mathcal{B}} \times \mathbf{r}_{P/A} = \dot{\phi}\mathbf{u}_3 \times l_2\mathbf{u}_1 = l_2\dot{\phi}\mathbf{u}_2 \tag{2-426}$$

Adding Eqs. (2-424) and (2-426), we obtain We then obtain ${}^{\mathcal{A}}\mathbf{v}_P$ as

$${}^{\mathcal{A}}\mathbf{v}_P = l_2\dot{\phi}\mathbf{u}_2 \tag{2-427}$$

The acceleration of point P as viewed by an observer in reference frame $\mathcal{A}$, denoted ${}^{\mathcal{A}}\mathbf{a}_P$, is obtained from Eq. (2-128) on page 47 as

$${}^{\mathcal{A}}\mathbf{a}_P = \frac{{}^{\mathcal{A}}d}{dt}\left({}^{\mathcal{A}}\mathbf{v}_P\right) = \frac{{}^{\mathcal{B}}d}{dt}\left({}^{\mathcal{A}}\mathbf{v}_P\right) + {}^{\mathcal{A}}\boldsymbol{\omega}^{\mathcal{B}} \times {}^{\mathcal{A}}\mathbf{v}_P \tag{2-428}$$

Using the expression for ${}^{\mathcal{A}}\mathbf{v}_P$ from Eq. (2-427), we see that

$$\frac{{}^{\mathcal{B}}d}{dt}\left({}^{\mathcal{A}}\mathbf{v}_P\right) = l_2\ddot{\phi}\mathbf{u}_2 \tag{2-429}$$

Furthermore, using ${}^{\mathcal{A}}\boldsymbol{\omega}^{\mathcal{B}}$ from Eq. (2-425), we obtain

$${}^{\mathcal{A}}\boldsymbol{\omega}^{\mathcal{B}} \times {}^{\mathcal{A}}\mathbf{v}_P = \dot{\phi}\mathbf{u}_3 \times l_2\dot{\phi}\mathbf{u}_2 = -l_2\dot{\phi}^2\mathbf{u}_1 \tag{2-430}$$

Consequently, we obtain ${}^{\mathcal{A}}\mathbf{a}_P$ as

$${}^{\mathcal{A}}\mathbf{a}_P = -l_2\dot{\phi}^2\mathbf{u}_1 + l_2\ddot{\phi}\mathbf{u}_2 \tag{2-431}$$

(b) Velocity and Acceleration of Point P as Viewed by an Observer Fixed to Ground

Using the expression for $\mathbf{r}_P$ from Eq. (2-417), the velocity of point P as viewed by an observer fixed to the ground is given as

$${}^{\mathcal{F}}\mathbf{v}_P = \frac{{}^{\mathcal{F}}d}{dt}(\mathbf{r}_P) = \frac{{}^{\mathcal{F}}d}{dt}(\mathbf{r}_A) + \frac{{}^{\mathcal{F}}d}{dt}(\mathbf{r}_{P/A}) = {}^{\mathcal{F}}\mathbf{v}_A + {}^{\mathcal{F}}\mathbf{v}_{P/A} \tag{2-432}$$

Observing from Eq. (2-418) that $\mathbf{r}_A$ is expressed in the basis $\{\mathbf{e}_1, \mathbf{e}_2, \mathbf{e}_3\}$ and $\{\mathbf{e}_1, \mathbf{e}_2, \mathbf{e}_3\}$ is fixed in reference frame $\mathcal{A}$, the first term in Eq. (2-432) is obtained by applying the rate of change transport theorem of Eq. (2-128) on page 47 between reference frames $\mathcal{A}$ and $\mathcal{F}$ as

$${}^{\mathcal{F}}\mathbf{v}_A = \frac{{}^{\mathcal{A}}d}{dt}(\mathbf{r}_A) + {}^{\mathcal{F}}\boldsymbol{\omega}^{\mathcal{A}} \times \mathbf{r}_A \tag{2-433}$$

Because reference frame $\mathcal{A}$ is fixed to rod OA and rod OA rotates with angular rate $\dot{\theta}$ relative to the ground, ${}^{\mathcal{F}}\boldsymbol{\omega}^{\mathcal{A}}$ is given as

$${}^{\mathcal{F}}\boldsymbol{\omega}^{\mathcal{A}} = \dot{\theta}\mathbf{e}_3 \tag{2-434}$$

Furthermore, we have

$$\begin{aligned}\frac{{}^{\mathcal{A}}d}{dt}(\mathbf{r}_A) &= \mathbf{0} && \text{(2-435)}\\ {}^{\mathcal{F}}\boldsymbol{\omega}^{\mathcal{A}}\times\mathbf{r}_A &= \dot{\theta}\mathbf{e}_3\times l_1\mathbf{e}_1 = l_1\dot{\theta}\mathbf{e}_2 && \text{(2-436)}\end{aligned}$$

Adding Eqs. (2-435) and (2-436) gives

$$^{\mathcal{F}}\mathbf{v}_A = l_1\dot{\theta}\mathbf{e}_2 \tag{2-437}$$

Next, observing from Eq. (2-419) that $\mathbf{r}_{P/A}$ is expressed in terms of the basis $\{\mathbf{u}_1,\mathbf{u}_2,\mathbf{u}_3\}$ and $\{\mathbf{u}_1,\mathbf{u}_2,\mathbf{u}_3\}$ is fixed in reference frame $\mathcal{B}$, the second term in Eq. (2-432) is obtained by applying the rate of change transport theorem of Eq. (2-128) on page 47 between reference frames $\mathcal{B}$ and $\mathcal{F}$ as

$$^{\mathcal{F}}\mathbf{v}_{P/A} = \frac{{}^{\mathcal{B}}d}{dt}(\mathbf{r}_A) + {}^{\mathcal{F}}\boldsymbol{\omega}^{\mathcal{B}}\times\mathbf{r}_{P/A} \tag{2-438}$$

Now from the theorem of angular velocity addition as given in Eq. (2-136) on page 48, we have

$$^{\mathcal{F}}\boldsymbol{\omega}^{\mathcal{B}} = {}^{\mathcal{F}}\boldsymbol{\omega}^{\mathcal{A}} + {}^{\mathcal{A}}\boldsymbol{\omega}^{\mathcal{B}} \tag{2-439}$$

Using the expressions for ${}^{\mathcal{A}}\boldsymbol{\omega}^{\mathcal{B}}$ and ${}^{\mathcal{F}}\boldsymbol{\omega}^{\mathcal{A}}$ from Eqs. (2-425) and (2-434), respectively, and noting that $\mathbf{e}_3 = \mathbf{u}_3$, we obtain ${}^{\mathcal{F}}\boldsymbol{\omega}^{\mathcal{B}}$ as

$$^{\mathcal{F}}\boldsymbol{\omega}^{\mathcal{B}} = \dot{\theta}\mathbf{e}_3 + \dot{\phi}\mathbf{u}_3 = (\dot{\theta}+\dot{\phi})\mathbf{u}_3 \tag{2-440}$$

Now we have

$$\begin{aligned}\frac{{}^{\mathcal{B}}d}{dt}(\mathbf{r}_A) &= \mathbf{0} && \text{(2-441)}\\ {}^{\mathcal{F}}\boldsymbol{\omega}^{\mathcal{B}}\times\mathbf{r}_{P/A} &= (\dot{\theta}+\dot{\phi})\mathbf{u}_3\times l_2\mathbf{u}_1 = l_2(\dot{\theta}+\dot{\phi})\mathbf{u}_2 && \text{(2-442)}\end{aligned}$$

Adding Eqs. (2-441) and (2-442), we obtain ${}^{\mathcal{F}}\mathbf{v}_{P/A}$ as

$$^{\mathcal{F}}\mathbf{v}_{P/A} = l_2(\dot{\theta}+\dot{\phi})\mathbf{u}_2 \tag{2-443}$$

Then, adding Eqs. (2-437) and (2-443), we obtain ${}^{\mathcal{F}}\mathbf{v}_P$ as

$$^{\mathcal{F}}\mathbf{v}_P = l_1\dot{\theta}\mathbf{e}_2 + l_2(\dot{\theta}+\dot{\phi})\mathbf{u}_2 \tag{2-444}$$

Using the expression for ${}^{\mathcal{F}}\mathbf{v}_P$ from Eq. (2-444), the acceleration of point P as viewed by an observer in reference frame $\mathcal{F}$ is obtained as

$$^{\mathcal{F}}\mathbf{a}_P = \frac{{}^{\mathcal{F}}d}{dt}\left({}^{\mathcal{F}}\mathbf{v}_P\right) = \frac{{}^{\mathcal{F}}d}{dt}\left({}^{\mathcal{F}}\mathbf{v}_A\right) + \frac{{}^{\mathcal{F}}d}{dt}\left({}^{\mathcal{F}}\mathbf{v}_{P/A}\right) = {}^{\mathcal{F}}\mathbf{a}_A + {}^{\mathcal{F}}\mathbf{a}_{P/A} \tag{2-445}$$

where

$$\begin{aligned}{}^{\mathcal{F}}\mathbf{v}_A &= l_1\dot{\theta}\mathbf{e}_2 && \text{(2-446)}\\ {}^{\mathcal{F}}\mathbf{v}_{P/A} &= l_2(\dot{\theta}+\dot{\phi})\mathbf{u}_2 && \text{(2-447)}\end{aligned}$$

Now since ${}^{\mathcal{F}}\mathbf{v}_A$ is expressed in terms of the basis $\{\mathbf{e}_1, \mathbf{e}_2, \mathbf{e}_3\}$ and $\{\mathbf{e}_1, \mathbf{e}_2, \mathbf{e}_3\}$ is fixed in reference frame $\mathcal{A}$, the first term in Eq. (2-445) is obtained using the rate of change transport theorem of Eq. (2-128) on page 47 between reference frames $\mathcal{A}$ and $\mathcal{F}$ as

$$ {}^{\mathcal{F}}\mathbf{a}_A = \frac{{}^{\mathcal{F}}d}{dt}\left({}^{\mathcal{F}}\mathbf{v}_A\right) = \frac{{}^{\mathcal{A}}d}{dt}\left({}^{\mathcal{F}}\mathbf{v}_A\right) + {}^{\mathcal{F}}\boldsymbol{\omega}^{\mathcal{A}} \times {}^{\mathcal{F}}\mathbf{v}_A \tag{2-448} $$

Now we have

$$ \frac{{}^{\mathcal{A}}d}{dt}\left({}^{\mathcal{F}}\mathbf{v}_A\right) = l_l\ddot{\theta}\mathbf{e}_2 \tag{2-449} $$

$$ {}^{\mathcal{F}}\boldsymbol{\omega}^{\mathcal{A}} \times {}^{\mathcal{F}}\mathbf{v}_A = \dot{\theta}\mathbf{e}_3 \times l_l\dot{\theta}\mathbf{e}_2 = -l_1\dot{\theta}^2\mathbf{e}_1 \tag{2-450} $$

Adding Eqs. (2-449) and (2-450), we obtain

$$ {}^{\mathcal{F}}\mathbf{a}_A = -l_1\dot{\theta}^2\mathbf{e}_1 + l_l\ddot{\theta}\mathbf{e}_2 \tag{2-451} $$

Next, observing that ${}^{\mathcal{F}}\mathbf{v}_{P/A}$ is expressed in terms of the basis $\{\mathbf{u}_1, \mathbf{u}_2, \mathbf{u}_3\}$ and $\{\mathbf{u}_1, \mathbf{u}_2, \mathbf{u}_3\}$ is fixed in reference frame $\mathcal{B}$, the second term in Eq. (2-445) is obtained by applying the rate of change transport theorem of Eq. (2-128) on page 47 between reference frames $\mathcal{B}$ and $\mathcal{F}$ as

$$ {}^{\mathcal{F}}\mathbf{a}_{P/A} = \frac{{}^{\mathcal{F}}d}{dt}\left({}^{\mathcal{F}}\mathbf{v}_{P/A}\right) = \frac{{}^{\mathcal{B}}d}{dt}\left({}^{\mathcal{F}}\mathbf{v}_{P/A}\right) + {}^{\mathcal{F}}\boldsymbol{\omega}^{\mathcal{B}} \times {}^{\mathcal{F}}\mathbf{v}_{P/A} \tag{2-452} $$

Now we have

$$ \frac{{}^{\mathcal{B}}d}{dt}\left({}^{\mathcal{F}}\mathbf{v}_{P/A}\right) = l_2(\ddot{\theta} + \ddot{\phi})\mathbf{u}_2 \tag{2-453} $$

$$ {}^{\mathcal{F}}\boldsymbol{\omega}^{\mathcal{B}} \times {}^{\mathcal{F}}\mathbf{v}_{P/A} = (\dot{\theta} + \dot{\phi})\mathbf{u}_3 \times l_2(\dot{\theta} + \dot{\phi})\mathbf{u}_2 = -l_2(\dot{\theta} + \dot{\phi})^2\mathbf{u}_1 \tag{2-454} $$

where in Eq. (2-454) we have used the expression for ${}^{\mathcal{F}}\boldsymbol{\omega}^{\mathcal{B}}$ from Eq. (2-440). Adding Eqs. (2-453) and (2-454), we obtain

$$ {}^{\mathcal{F}}\mathbf{a}_{P/A} = -l_2(\dot{\theta} + \dot{\phi})^2\mathbf{u}_1 + l_2(\ddot{\theta} + \ddot{\phi})\mathbf{u}_2 \tag{2-455} $$

Then, adding Eqs. (2-451) and (2-455), we obtain the acceleration of point P in reference frame $\mathcal{F}$ as

$$ {}^{\mathcal{F}}\mathbf{a}_P = -l_1\dot{\theta}^2\mathbf{e}_1 + l_1\ddot{\theta}\mathbf{e}_2 - l_2(\dot{\theta} + \dot{\phi})^2\mathbf{u}_1 + l_2(\ddot{\theta} + \ddot{\phi})\mathbf{u}_2 \tag{2-456} $$

■

Example 2-9

A circular disk of radius r rotates about shaft OQ as shown in Fig. 2-31. Shaft OQ has a length L, is oriented horizontally, and is rigidly attached to shaft BC, where shaft BC rotates with constant angular velocity $\boldsymbol{\Omega}$ (where $\|\boldsymbol{\Omega}\| = \Omega$) relative to the ground about the vertical direction. Finally, point P is located on the edge of the disk. Knowing that ϕ is the angle formed by the vertical and the direction of QP, determine the velocity and acceleration of point P as viewed by an observer fixed to the ground.

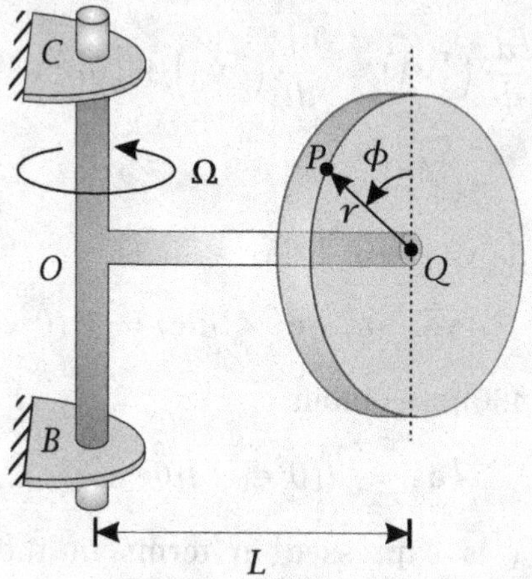

Figure 2-31 Disk on rotating shaft.

Solution to Example 2-9

For this problem, it is convenient to choose one inertial and two noninertial reference frames. The inertial reference frame, denoted $\mathcal{F}$, is fixed to the ground, i.e., $\mathcal{F}$ is fixed to the supports that hold the vertical shaft in place. Corresponding to reference frame $\mathcal{F}$, we choose the following coordinate system:

Origin at point O

$$\begin{array}{lcl} \mathbf{E}_1 & = & \text{Along } OC \\ \mathbf{E}_2 & = & \text{Along } OQ \text{ at } t = 0 \\ \mathbf{E}_3 & = & \mathbf{E}_1 \times \mathbf{E}_2 \end{array}$$

Next, the first noninertial reference frame is denoted $\mathcal{A}$ and is fixed in the horizontal shaft. Then, the following coordinate system is chosen that is fixed in reference frame $\mathcal{A}$:

Origin at point O

$$\begin{array}{lcl} \mathbf{e}_1 & = & \text{Along } OC \\ \mathbf{e}_2 & = & \text{Along } OQ \\ \mathbf{e}_3 & = & \mathbf{e}_1 \times \mathbf{e}_2 \end{array}$$

The second noninertial reference frame, denoted $\mathcal{B}$, is fixed in the disk. Then, the following coordinate system is chosen that is fixed in reference frame $\mathcal{B}$:

Origin at point Q

$$\begin{array}{lcl} \mathbf{u}_1 & = & \text{Along } QP \\ \mathbf{u}_2 & = & \text{Along } OQ \\ \mathbf{u}_3 & = & \mathbf{u}_1 \times \mathbf{u}_2 \end{array}$$

A three-dimensional perspective of the bases $\{\mathbf{e}_1, \mathbf{e}_2, \mathbf{e}_3\}$ and $\{\mathbf{u}_1, \mathbf{u}_2, \mathbf{u}_3\}$ is shown in Fig. 2-32 while a two-dimensional perspective is shown in Fig. 2-33.

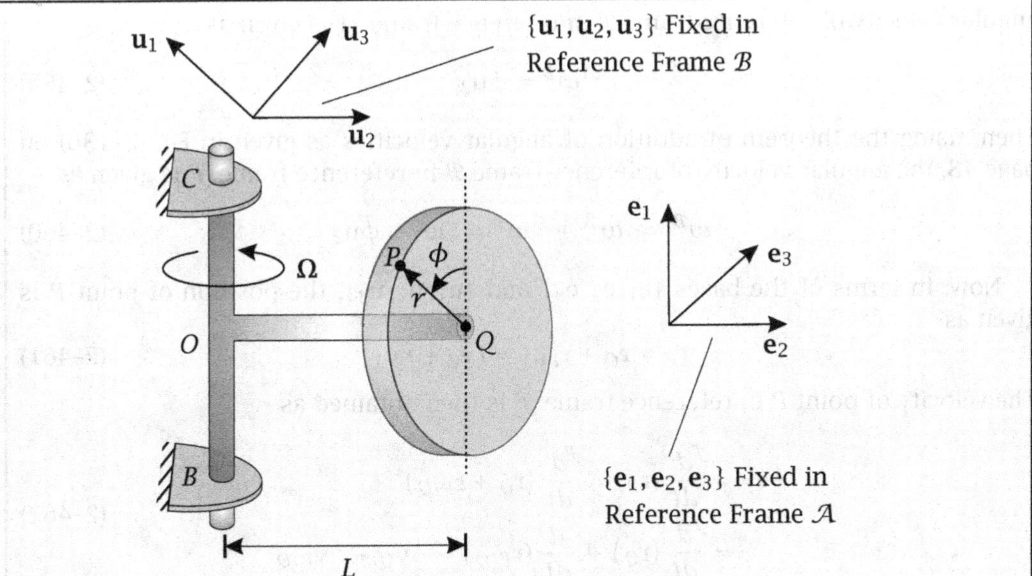

Figure 2-32 Three-dimensional perspective of bases $\{\mathbf{e}_1, \mathbf{e}_2, \mathbf{e}_3\}$ and $\{\mathbf{u}_1, \mathbf{u}_2, \mathbf{u}_3\}$ fixed in reference frames $\mathcal{A}$ and $\mathcal{B}$, respectively, for Example 2-9.

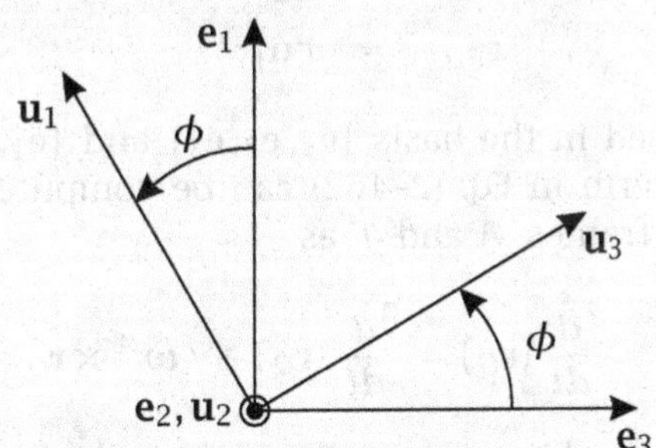

Figure 2-33 Two-dimensional perspective of bases $\{\mathbf{e}_1, \mathbf{e}_2, \mathbf{e}_3\}$ and $\{\mathbf{u}_1, \mathbf{u}_2, \mathbf{u}_3\}$ for Example 2-9.

Using Fig. 2-33, we have

$$\begin{array}{rcl} \mathbf{e}_1 & = & \cos\phi\mathbf{u}_1 + \sin\phi\mathbf{u}_3 \\ \mathbf{e}_3 & = & -\sin\phi\mathbf{u}_1 + \cos\phi\mathbf{u}_3 \end{array} \tag{2-457}$$

The next step in solving this problem is to determine the angular velocity of each reference frame. Because the vertical shaft rotates with angular velocity $\boldsymbol{\Omega}$ relative to the supports and the supports are fixed in reference frame $\mathcal{F}$, the angular velocity of reference frame $\mathcal{A}$ in reference frame $\mathcal{F}$ is given as

$$^{\mathcal{F}}\boldsymbol{\omega}^{\mathcal{A}} = \boldsymbol{\Omega} = \Omega\mathbf{e}_1 \tag{2-458}$$

Next, it is seen that reference frame $\mathcal{B}$ is fixed to the disk and the disk rotates with angular rate $\dot{\phi}$ relative to reference frame $\mathcal{A}$ about the direction $\mathbf{u}_2$. Therefore, the angular velocity of reference frame $\mathcal{B}$ in reference frame $\mathcal{A}$ is given as

$$ {}^{\mathcal{A}}\boldsymbol{\omega}^{\mathcal{B}} = \dot{\phi}\mathbf{u}_2 \tag{2-459}$$

Then, using the theorem of addition of angular velocities as given in Eq. (2-136) on page 48, the angular velocity of reference frame $\mathcal{B}$ in reference frame $\mathcal{F}$ is given as

$$ {}^{\mathcal{F}}\boldsymbol{\omega}^{\mathcal{B}} = {}^{\mathcal{F}}\boldsymbol{\omega}^{\mathcal{A}} + {}^{\mathcal{A}}\boldsymbol{\omega}^{\mathcal{B}} = \Omega\mathbf{e}_1 + \dot{\phi}\mathbf{u}_2 \tag{2-460}$$

Now, in terms of the bases $\{\mathbf{e}_1, \mathbf{e}_2, \mathbf{e}_3\}$ and $\{\mathbf{u}_1, \mathbf{u}_2, \mathbf{u}_3\}$, the position of point P is given as

$$ \mathbf{r}_P = \mathbf{r}_Q + \mathbf{r}_{P/Q} = L\mathbf{e}_2 + r\mathbf{u}_1 \tag{2-461}$$

The velocity of point P in reference frame $\mathcal{F}$ is then obtained as

$$\begin{aligned} {}^{\mathcal{F}}\mathbf{v}_P &= \frac{{}^{\mathcal{F}}d}{dt}(\mathbf{r}_P) = \frac{{}^{\mathcal{F}}d}{dt}(\mathbf{r}_Q + \mathbf{r}_{P/Q}) \\ &= \frac{{}^{\mathcal{F}}d}{dt}(\mathbf{r}_Q) + \frac{{}^{\mathcal{F}}d}{dt}(\mathbf{r}_{P/Q}) = {}^{\mathcal{F}}\mathbf{v}_Q + {}^{\mathcal{F}}\mathbf{v}_{P/Q} \end{aligned} \tag{2-462}$$

where

$$ \mathbf{r}_Q = L\mathbf{e}_2 \tag{2-463}$$

$$ \mathbf{r}_{P/Q} = r\mathbf{u}_1 \tag{2-464}$$

Now, because $\mathbf{r}_Q$ is expressed in the basis $\{\mathbf{e}_1, \mathbf{e}_2, \mathbf{e}_3\}$ and $\{\mathbf{e}_1, \mathbf{e}_2, \mathbf{e}_3\}$ is fixed in reference frame $\mathcal{A}$, the first term in Eq. (2-462) can be computed using the transport theorem between reference frames $\mathcal{A}$ and $\mathcal{F}$ as

$$ {}^{\mathcal{F}}\mathbf{v}_Q = \frac{{}^{\mathcal{F}}d}{dt}(\mathbf{r}_Q) = \frac{{}^{\mathcal{A}}d}{dt}(\mathbf{r}_Q) + {}^{\mathcal{F}}\boldsymbol{\omega}^{\mathcal{A}} \times \mathbf{r}_Q \tag{2-465}$$

Then, because $\{\mathbf{e}_1, \mathbf{e}_2, \mathbf{e}_3\}$ is fixed in reference frame $\mathcal{A}$ and L is constant, the first term in Eq. (2-465) is zero, i.e.,

$$ \frac{{}^{\mathcal{A}}d}{dt}(\mathbf{r}_Q) = \frac{{}^{\mathcal{A}}d}{dt}(L\mathbf{e}_2) = \mathbf{0} \tag{2-466}$$

Furthermore, using ${}^{\mathcal{F}}\boldsymbol{\omega}^{\mathcal{A}}$ from Eq. (2-458), the second term in Eq. (2-465) is obtained as

$$ {}^{\mathcal{F}}\boldsymbol{\omega}^{\mathcal{A}} \times \mathbf{r}_Q = \Omega\mathbf{e}_1 \times (L\mathbf{e}_2) = L\Omega\mathbf{e}_3 \tag{2-467}$$

Adding the results of Eqs. (2-466) and (2-467), we obtain the velocity of point Q in reference frame $\mathcal{F}$ as

$$ {}^{\mathcal{F}}\mathbf{v}_Q = L\Omega\mathbf{e}_3 \tag{2-468}$$

Next, because $\mathbf{r}_{P/Q}$ is expressed in the basis $\{\mathbf{u}_1, \mathbf{u}_2, \mathbf{u}_3\}$ and $\{\mathbf{u}_1, \mathbf{u}_2, \mathbf{u}_3\}$ is fixed in reference frame $\mathcal{B}$, the second term in Eq. (2-462) is obtained using the rate of change transport theorem of Eq. (2-128) on page 47 between reference frames $\mathcal{B}$ and $\mathcal{F}$ as

$$ {}^{\mathcal{F}}\mathbf{v}_{P/Q} = \frac{{}^{\mathcal{F}}d}{dt}(\mathbf{r}_{P/Q}) = \frac{{}^{\mathcal{B}}d}{dt}(\mathbf{r}_{P/Q}) + {}^{\mathcal{F}}\boldsymbol{\omega}^{\mathcal{B}} \times \mathbf{r}_{P/Q} \tag{2-469}$$

Because $\{\mathbf{u}_1, \mathbf{u}_2, \mathbf{u}_3\}$ is fixed in reference frame $\mathcal{B}$ and r is constant, the first term in Eq. (2–469) is zero, i.e.,

$$\frac{^{\mathcal{B}}d}{dt}(\mathbf{r}_{P/Q}) = \frac{^{\mathcal{B}}d}{dt}(r\mathbf{u}_1) = \mathbf{0} \tag{2–470}$$

Furthermore, using $^{\mathcal{F}}\boldsymbol{\omega}^{\mathcal{B}}$ from Eq. (2–460), the second term in Eq. (2–469) is obtained as

$$\begin{aligned} ^{\mathcal{F}}\boldsymbol{\omega}^{\mathcal{B}} \times \mathbf{r}_{P/Q} &= \left(\Omega\mathbf{e}_1 + \dot{\phi}\mathbf{u}_2\right) \times (r\mathbf{u}_1) \\ &= r\Omega\mathbf{e}_1 \times \mathbf{u}_1 + r\dot{\phi}\mathbf{u}_2 \times \mathbf{u}_1 \\ &= r\Omega\mathbf{e}_1 \times \mathbf{u}_1 - r\dot{\phi}\mathbf{u}_3 \end{aligned} \tag{2–471}$$

Then, using Eq. (2–457), we have

$$\mathbf{e}_1 \times \mathbf{u}_1 = (\cos\phi\mathbf{u}_1 + \sin\phi\mathbf{u}_3) \times \mathbf{u}_1 = \sin\phi\mathbf{u}_2 \tag{2–472}$$

Substituting the result of Eq. (2–472) into (2–471), we obtain

$$^{\mathcal{F}}\boldsymbol{\omega}^{\mathcal{B}} \times \mathbf{r}_{P/Q} = r\Omega\sin\phi\mathbf{u}_2 - r\dot{\phi}\mathbf{u}_3 \tag{2–473}$$

Adding the results of Eqs. (2–470) and (2–473), we obtain the velocity of point P relative to point Q in reference frame $\mathcal{F}$ as

$$^{\mathcal{F}}\mathbf{v}_{P/Q} = \frac{^{\mathcal{F}}d}{dt}(\mathbf{r}_{P/Q}) = r\Omega\sin\phi\mathbf{u}_2 - r\dot{\phi}\mathbf{u}_3 \tag{2–474}$$

Finally, adding the results of Eqs. (2–468) and (2–474), we obtain the velocity of point P in reference frame $\mathcal{F}$ as

$$^{\mathcal{F}}\mathbf{v}_P = L\Omega\mathbf{e}_3 + r\Omega\sin\phi\mathbf{u}_2 - r\dot{\phi}\mathbf{u}_3 \tag{2–475}$$

It is noted that $^{\mathcal{F}}\mathbf{v}_P$ can be expressed completely in terms of the basis $\{\mathbf{e}_1, \mathbf{e}_2, \mathbf{e}_3\}$ or $\{\mathbf{u}_1, \mathbf{u}_2, \mathbf{u}_3\}$ by applying the relationship of Eq. (2–457).

Now, the acceleration of point P in reference frame $\mathcal{F}$ is given as

$$\begin{aligned} ^{\mathcal{F}}\mathbf{a}_P &= \frac{^{\mathcal{F}}d}{dt}\left(^{\mathcal{F}}\mathbf{v}_P\right) = \frac{^{\mathcal{F}}d}{dt}\left(^{\mathcal{F}}\mathbf{v}_Q + {}^{\mathcal{F}}\mathbf{v}_{P/Q}\right) \\ &= \frac{^{\mathcal{F}}d}{dt}\left(^{\mathcal{F}}\mathbf{v}_Q\right) + \frac{^{\mathcal{F}}d}{dt}\left(^{\mathcal{F}}\mathbf{v}_{P/Q}\right) = {}^{\mathcal{F}}\mathbf{a}_Q + {}^{\mathcal{F}}\mathbf{a}_{P/Q} \end{aligned} \tag{2–476}$$

Restating Eqs. (2–468) and (2–474), the quantities $^{\mathcal{F}}\mathbf{v}_Q$ and $^{\mathcal{F}}\mathbf{v}_{P/Q}$ are given as

$$^{\mathcal{F}}\mathbf{v}_Q = L\Omega\mathbf{e}_3 \tag{2–477}$$

$$^{\mathcal{F}}\mathbf{v}_{P/Q} = r\Omega\sin\phi\mathbf{u}_2 - r\dot{\phi}\mathbf{u}_3 \tag{2–478}$$

Once again, because $^{\mathcal{F}}\mathbf{v}_Q$ in Eq. (2–477) is expressed in the basis $\{\mathbf{e}_1, \mathbf{e}_2, \mathbf{e}_3\}$ and $\{\mathbf{e}_1, \mathbf{e}_2, \mathbf{e}_3\}$ is fixed in reference frame $\mathcal{A}$, the first term in Eq. (2–476) can be obtained by applying the transport theorem between reference frames $\mathcal{A}$ and $\mathcal{F}$ as

$$^{\mathcal{F}}\mathbf{a}_Q = \frac{^{\mathcal{F}}d}{dt}\left(^{\mathcal{F}}\mathbf{v}_Q\right) = \frac{^{\mathcal{A}}d}{dt}\left(^{\mathcal{F}}\mathbf{v}_Q\right) + {}^{\mathcal{F}}\boldsymbol{\omega}^{\mathcal{A}} \times {}^{\mathcal{F}}\mathbf{v}_Q \tag{2–479}$$

Now, because L and Ω are constant, we have from Eq. (2–477) that

$$\frac{{}^{\mathcal{A}}d}{dt}\left({}^{\mathcal{F}}\mathbf{v}_Q\right) = \frac{{}^{\mathcal{A}}d}{dt}\left(L\Omega\mathbf{e}_3\right) = \mathbf{0} \tag{2–480}$$

Next, using the expression for ${}^{\mathcal{F}}\boldsymbol{\omega}^{\mathcal{A}}$ from Eq. (2–458), the second term in Eq. (2–479) is obtained as

$${}^{\mathcal{F}}\boldsymbol{\omega}^{\mathcal{A}} \times {}^{\mathcal{F}}\mathbf{v}_Q = \Omega\mathbf{e}_1 \times L\Omega\mathbf{e}_3 = -L\Omega^2\mathbf{e}_2 \tag{2–481}$$

Adding the results of Eqs. (2–480) and (2–481), we obtain the acceleration of point Q in reference frame $\mathcal{F}$ as

$${}^{\mathcal{F}}\mathbf{a}_Q = -L\Omega^2\mathbf{e}_2 \tag{2–482}$$

Then, because ${}^{\mathcal{F}}\mathbf{v}_{P/Q}$ in Eq. (2–478) is expressed in the basis $\{\mathbf{u}_1, \mathbf{u}_2, \mathbf{u}_3\}$ and $\{\mathbf{u}_1, \mathbf{u}_2, \mathbf{u}_3\}$ is fixed in reference frame $\mathcal{B}$, the second term in Eq. (2–476) can be obtained by applying the rate of change transport theorem between reference frames $\mathcal{B}$ and $\mathcal{F}$ as

$${}^{\mathcal{F}}\mathbf{a}_{P/Q} = \frac{{}^{\mathcal{F}}d}{dt}\left({}^{\mathcal{F}}\mathbf{v}_{P/Q}\right) = \frac{{}^{\mathcal{B}}d}{dt}\left({}^{\mathcal{F}}\mathbf{v}_{P/Q}\right) + {}^{\mathcal{F}}\boldsymbol{\omega}^{\mathcal{B}} \times {}^{\mathcal{F}}\mathbf{v}_{P/Q} \tag{2–483}$$

Using the expression for ${}^{\mathcal{F}}\mathbf{v}_{P/Q}$ from Eq. (2–478), the first term in Eq. (2–483) becomes

$$\frac{{}^{\mathcal{B}}d}{dt}\left({}^{\mathcal{F}}\mathbf{v}_{P/Q}\right) = r\Omega\dot{\phi}\cos\phi\mathbf{u}_2 - r\ddot{\phi}\mathbf{u}_3 \tag{2–484}$$

Next, using ${}^{\mathcal{F}}\boldsymbol{\omega}^{\mathcal{B}}$ from Eq. (2–460), the second term in Eq. (2–483) is obtained as

$$\begin{aligned} {}^{\mathcal{F}}\boldsymbol{\omega}^{\mathcal{B}} \times {}^{\mathcal{F}}\mathbf{v}_{P/Q} &= (\Omega\mathbf{e}_1 + \dot{\phi}\mathbf{u}_2) \times \left(r\Omega\sin\phi\mathbf{u}_2 - r\dot{\phi}\mathbf{u}_3\right) \\ &= r\Omega^2\sin\phi\mathbf{e}_1 \times \mathbf{u}_2 - r\Omega\dot{\phi}\mathbf{e}_1 \times \mathbf{u}_3 - r\dot{\phi}^2\mathbf{u}_1 \end{aligned} \tag{2–485}$$

Applying Eq. (2–457), we have

$$\mathbf{e}_1 \times \mathbf{u}_2 = (\cos\phi\mathbf{u}_1 + \sin\phi\mathbf{u}_3) \times \mathbf{u}_2 = \cos\phi\mathbf{u}_3 - \sin\phi\mathbf{u}_1 \tag{2–486}$$

$$\mathbf{e}_1 \times \mathbf{u}_3 = (\cos\phi\mathbf{u}_1 + \sin\phi\mathbf{u}_3) \times \mathbf{u}_3 = -\cos\phi\mathbf{u}_2 \tag{2–487}$$

Substituting the results of Eqs. (2–486) and (2–487) into Eq. (2–485), we obtain

$$\begin{aligned} {}^{\mathcal{F}}\boldsymbol{\omega}^{\mathcal{B}} \times {}^{\mathcal{F}}\mathbf{v}_{P/Q} &= r\Omega^2\sin\phi(\cos\phi\mathbf{u}_3 - \sin\phi\mathbf{u}_1) + r\Omega\dot{\phi}\cos\phi\mathbf{u}_2 - r\dot{\phi}^2\mathbf{u}_1 \\ &= r\Omega^2\cos\phi\sin\phi\mathbf{u}_3 - r\Omega^2\sin^2\phi\mathbf{u}_1 \\ &\quad + r\Omega\dot{\phi}\cos\phi\mathbf{u}_2 - r\dot{\phi}^2\mathbf{u}_1 \end{aligned} \tag{2–488}$$

Then, adding the results of Eqs. (2–484) and (2–488), the acceleration of point P relative to point Q in reference frame $\mathcal{F}$ is obtained as

$$\begin{aligned} {}^{\mathcal{F}}\mathbf{a}_{P/Q} &= r\Omega\dot{\phi}\cos\phi\mathbf{u}_2 - r\ddot{\phi}\mathbf{u}_3 + r\Omega^2\cos\phi\sin\phi\mathbf{u}_3 \\ &\quad - r\Omega^2\sin^2\phi\mathbf{u}_1 + r\Omega\dot{\phi}\cos\phi\mathbf{u}_2 - r\dot{\phi}^2\mathbf{u}_1 \end{aligned} \tag{2–489}$$

Finally, adding the results of Eqs. (2–482) and (2–489), we obtain ${}^{\mathcal{F}}\mathbf{a}_P$ as

$$\begin{aligned} {}^{\mathcal{F}}\mathbf{a}_P &= -L\Omega^2\mathbf{e}_2 + r\Omega\dot{\phi}\cos\phi\mathbf{u}_2 - r\ddot{\phi}\mathbf{u}_3 + r\Omega^2\cos\phi\sin\phi\mathbf{u}_3 \\ &\quad - r\Omega^2\sin^2\phi\mathbf{u}_1 + r\Omega\dot{\phi}\cos\phi\mathbf{u}_2 - r\dot{\phi}^2\mathbf{u}_1 \end{aligned} \tag{2–490}$$

Simplifying Eq. (2–490), we obtain the acceleration of point P in reference frame $\mathcal{F}$ as

$$\begin{aligned}{}^{\mathcal{F}}\mathbf{a}_P = & -L\Omega^2\mathbf{e}_2 - (r\Omega^2\sin^2\phi + r\dot{\phi}^2)\mathbf{u}_1 \\ & + 2r\Omega\dot{\phi}\cos\phi\mathbf{u}_2 + (r\Omega^2\cos\phi\sin\phi - r\ddot{\phi})\mathbf{u}_3\end{aligned} \tag{2–491}$$

■

2.14 Kinematics of a Particle in Continuous Contact with a Surface

A special case of particle kinematics that occurs frequently and deserves special attention is that of a particle sliding while in continuous contact with a surface. Consider now a particle that moves in an arbitrary reference frame $\mathcal{A}$ while constrained to slide on a rigid surface S (i.e., S defines a reference frame) as shown in Fig. 2–34. Furthermore, let C denote the point on S that is instantaneously in contact with the particle and let $\mathbf{r}$ and $\mathbf{r}_C^S$ be the position of the particle and the position point C on S, respectively, measured relative to a point O, where O is fixed in reference frame $\mathcal{A}$. Then, assuming that S can be described by a differentiable function, there exists a well-defined plane, denoted $\mathcal{T}_C^S$, that is tangent to S at C. Correspondingly, let $\mathbf{n}$ be the unit vector orthogonal to $\mathcal{T}_C^S$ at C, and let $\mathbf{u}$ and $\mathbf{w}$ be two unit vectors that lie in $\mathcal{T}_C^S$ such that $\{\mathbf{u}, \mathbf{w}, \mathbf{n}\}$ forms a right-handed basis.

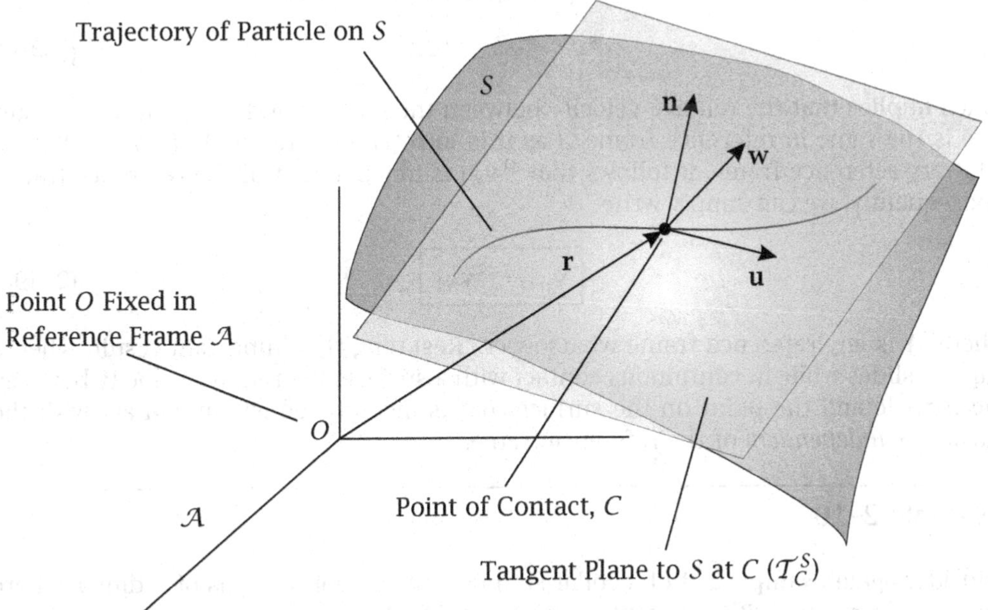

Figure 2–34 Particle moving while in continuous contact with a surface S.

Suppose now that we let ${}^{\mathcal{A}}\mathbf{v}$ and ${}^{\mathcal{A}}\mathbf{v}_C^S$ be the velocity of the particle and velocity of the point C on S, respectively, in reference frame $\mathcal{A}$. Then the velocity of the particle relative to the instantaneous point of contact C in reference frame $\mathcal{A}$ is given as

$$\boxed{{}^{\mathcal{A}}\mathbf{v}_{\text{rel}} = {}^{\mathcal{A}}\mathbf{v} - {}^{\mathcal{A}}\mathbf{v}_C^S} \tag{2-492}$$

It is important to note that while ${}^{\mathcal{A}}\mathbf{v}$ and ${}^{\mathcal{A}}\mathbf{v}_C^S$ depend on the choice of the reference frame $\mathcal{A}$, the relative velocity ${}^{\mathcal{A}}\mathbf{v}_{\text{rel}}$ as defined in Eq. (2-492) is *independent* of the reference frame in which the motion is observed. To verify that ${}^{\mathcal{A}}\mathbf{v}_{\text{rel}}$ is, in fact, independent of the reference frame in which the observations are made, let $\mathcal{B}$ be any reference frame that is *different* from $\mathcal{A}$. We then have

$$\begin{aligned} {}^{\mathcal{A}}\mathbf{v} &= {}^{\mathcal{B}}\mathbf{v} + {}^{\mathcal{A}}\boldsymbol{\omega}^{\mathcal{B}} \times \mathbf{r} && (2\text{-}493) \\ {}^{\mathcal{A}}\mathbf{v}_C^S &= {}^{\mathcal{B}}\mathbf{v}_C^S + {}^{\mathcal{A}}\boldsymbol{\omega}^{\mathcal{B}} \times \mathbf{r}_C^S && (2\text{-}494) \end{aligned}$$

Now, because the particle is moving while in continuous contact with the surface S, we see that the position of the particle and the position of the point on S that is instantaneously in contact with the particle are the same, i.e.,

$$\mathbf{r} = \mathbf{r}_C^S \tag{2-495}$$

Subtracting Eq. (2-494) from (2-493) and applying the result of Eq. (2-495), we obtain

$${}^{\mathcal{A}}\mathbf{v} - {}^{\mathcal{A}}\mathbf{v}_C^S = {}^{\mathcal{B}}\mathbf{v} - {}^{\mathcal{B}}\mathbf{v}_C^S \tag{2-496}$$

Observing that the quantity ${}^{\mathcal{A}}\mathbf{v} - {}^{\mathcal{A}}\mathbf{v}_C^S$ is equal to ${}^{\mathcal{A}}\mathbf{v}_{\text{rel}}$, we have

$${}^{\mathcal{B}}\mathbf{v} - {}^{\mathcal{B}}\mathbf{v}_C^S \equiv {}^{\mathcal{A}}\mathbf{v}_{\text{rel}} \tag{2-497}$$

which implies that the relative velocity between the particle and the point on the surface is the same in reference frame $\mathcal{B}$ as it is in reference frame $\mathcal{A}$. Because $\mathcal{B}$ is an arbitrary reference frame, it follows that ${}^{\mathcal{A}}\mathbf{v}_{\text{rel}}$ is independent of the reference frame. Consequently, we can simply write

$$\boxed{\mathbf{v}_{\text{rel}} = {}^{\mathcal{A}}\mathbf{v}_{\text{rel}}} \tag{2-498}$$

where $\mathcal{A}$ is any reference frame whatsoever. Restating this important result, when a particle slides while in continuous contact with a surface, the relative velocity between the particle and the point on the surface that is instantaneously in contact with the particle is *independent* of the reference frame.

Example 2–10

Consider again Example 2–4 of a collar sliding on a circular annulus of radius r where the annulus rotates with constant angular velocity Ω relative to the ground, as shown in Fig. 2–35. Determine the velocity of the collar relative to the instantaneous point on the annulus that is in contact with the collar using the following reference frames: (a) the annulus and (b) the ground.

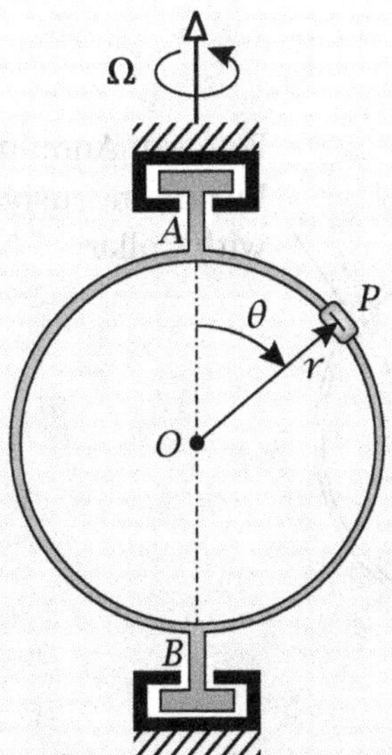

Figure 2–35 Collar sliding on rotating annulus.

Solution to Example 2–10

Recall from Example 2-4 that the following three reference frames were used: (1) reference frame $\mathcal{F}$, fixed to the ground; (2) reference frame $\mathcal{A}$, fixed to the annulus; and (3) reference frame $\mathcal{B}$, fixed to the motion of the particle. Furthermore, the following coordinate system was fixed in reference frame $\mathcal{A}$:

	Origin at point O	
$\mathbf{u}_x$	=	Along OA
$\mathbf{u}_z$	=	Orthogonal to annulus (into page)
$\mathbf{u}_y$	=	$\mathbf{u}_z \times \mathbf{u}_x$

Finally, the following coordinate system was fixed in reference frame $\mathcal{B}$:

	Origin at point O	
$\mathbf{e}_r$	=	Along OP
$\mathbf{e}_z$	=	Orthogonal to annulus (into page)
$\mathbf{e}_\theta$	=	$\mathbf{E}_z \times \mathbf{e}_r$

(a) Relative Velocity Using Annulus as Reference Frame

Consider the point $\bar{P}$ *fixed to the annulus* that is instantaneously in contact with the collar as shown in Fig. 2-36.

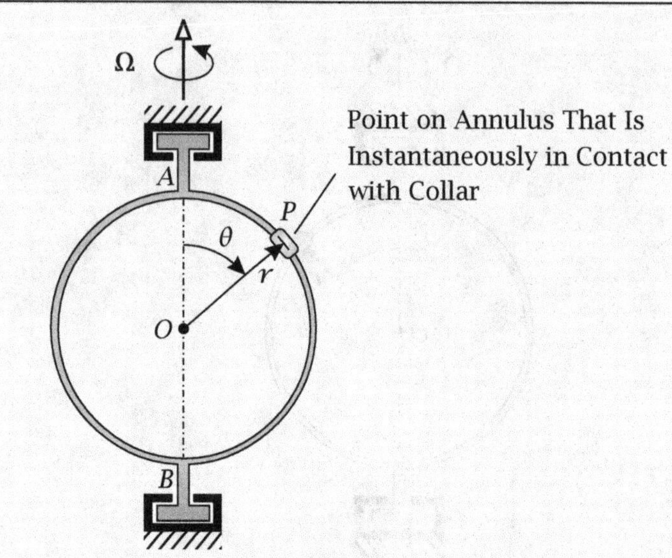

Figure 2-36 Point P fixed to annulus that is instantaneously in contact with collar for Example 2-10.

The position of the point P fixed to the annulus that is instantaneously in contact with the collar is given as

$$\mathbf{r}_P^{\mathcal{A}} = r\mathbf{e}_r \tag{2-499}$$

Now, because in this case we are considering the point P that is fixed to the annulus (i.e., reference frame $\mathcal{A}$), we know that the velocity of point P as viewed by an observer fixed to the annulus must be zero, i.e.,

$${}^{\mathcal{A}}\mathbf{v}_P^{\mathcal{A}} = \mathbf{0} \tag{2-500}$$

Next, from Eq. (2-207) on page 61, the velocity of the *collar* as viewed by an observer in reference frame $\mathcal{A}$ was given as

$${}^{\mathcal{A}}\mathbf{v} = r\dot{\theta}\mathbf{e}_\theta \tag{2-501}$$

Subtracting Eq. (2-500) from (2-501), we obtain the velocity of the collar relative to the point on the annulus that is instantaneously in contact with the collar as

$${}^{\mathcal{A}}\mathbf{v}_{\text{rel}} = r\dot{\theta}\mathbf{e}_\theta \tag{2-502}$$

Finally, recall from its definition that $\mathbf{v}_{\text{rel}}$ is independent of reference frame. Consequently, the dependence on reference frame in Eq. (2-502) can be dropped to give $\mathbf{v}_{\text{rel}}$ as

$$\mathbf{v}_{\text{rel}} = r\dot{\theta}\mathbf{e}_\theta \tag{2-503}$$

(b) Relative Velocity Using Ground as Reference Frame

The relative velocity between the collar and the point on the annulus that is instantaneously in contact with the collar can also be determined using reference frame $\mathcal{F}$

instead of reference frame $\mathcal{A}$. First, the velocity of the point P that is fixed to the collar and is instantaneously in contact with the annulus, *as viewed by an observer in reference frame $\mathcal{F}$*, is given as

$$^{\mathcal{F}}\mathbf{v}_P^{\mathcal{A}} = \frac{^{\mathcal{F}}d}{dt}\left(\mathbf{r}_P^{\mathcal{A}}\right) = \frac{^{\mathcal{A}}d}{dt}\left(\mathbf{r}_P^{\mathcal{A}}\right) + {}^{\mathcal{F}}\boldsymbol{\omega}^{\mathcal{A}} \times \mathbf{r}_P^{\mathcal{A}} \tag{2-504}$$

It is emphasized in Eq. (2-504) that, because we are considering the point P as fixed to the annulus (i.e., reference frame $\mathcal{A}$), the velocity of point P in reference frame $\mathcal{F}$ is computed using the rate of change transport theorem between reference frame $\mathcal{A}$ and reference frame $\mathcal{F}$. Then, because r is constant and $\mathbf{e}_r$ is now a direction that is fixed in reference frame $\mathcal{A}$ (because P is fixed in reference frame $\mathcal{A}$), we have

$$\frac{^{\mathcal{A}}d}{dt}\left(\mathbf{r}_P^{\mathcal{A}}\right) = \mathbf{0} \tag{2-505}$$

Next, using the expression for ${}^{\mathcal{F}}\boldsymbol{\omega}^{\mathcal{A}}$ from Eq. (2-213) on page 62, we have

$$^{\mathcal{F}}\boldsymbol{\omega}^{\mathcal{A}} \times \mathbf{r}_P^{\mathcal{A}} = \Omega\mathbf{u}_x \times r\mathbf{e}_r \tag{2-506}$$

Then, substituting the expression for $\mathbf{u}_x$ from Eq. (2-200) on page 61 into Eq. (2-506), we obtain

$$^{\mathcal{F}}\boldsymbol{\omega}^{\mathcal{A}} \times \mathbf{r}_P^{\mathcal{A}} = \Omega(\cos\theta\,\mathbf{e}_r - \sin\theta\,\mathbf{e}_\theta) \times r\mathbf{e}_r = r\Omega\sin\theta\,\mathbf{e}_z \tag{2-507}$$

Adding Eqs. (2-507) and (2-505) gives

$$^{\mathcal{F}}\mathbf{v}_P^{\mathcal{A}} = r\Omega\sin\theta\,\mathbf{e}_z \tag{2-508}$$

Next, we have the velocity of the collar in reference frame $\mathcal{F}$ from Eq. (2-216) on page 62 as

$$^{\mathcal{F}}\mathbf{v} = r\dot{\theta}\mathbf{e}_\theta + r\Omega\sin\theta\,\mathbf{e}_z \tag{2-509}$$

Subtracting Eq. (2-508) from (2-509), we obtain the velocity of the collar relative to the point on the annulus that is instantaneously in contact with the collar as

$$^{\mathcal{F}}\mathbf{v}_{\text{rel}} = r\dot{\theta}\mathbf{e}_\theta + r\Omega\sin\theta\,\mathbf{e}_z - r\Omega\sin\theta\,\mathbf{e}_z = r\dot{\theta}\mathbf{e}_\theta \tag{2-510}$$

It is seen that the result of Eq. (2-510) is identical to that of Eq. (2-503), i.e., the quantity $\mathbf{v}_{\text{rel}}$ is the same regardless of whether the relative velocity is computed in reference frame $\mathcal{A}$ or reference frame $\mathcal{F}$. Eliminating the dependence in Eq. (2-510) on the reference frame, we obtain $\mathbf{v}_{\text{rel}}$ as

$$\mathbf{v}_{\text{rel}} = r\dot{\theta}\mathbf{e}_\theta \tag{2-511}$$

■

2.15 Kinematics of Rigid Bodies

A *rigid body*, denoted $\mathcal{R}$, is a collection of points, each of which may or may not have mass, such that the distance between any two points on the body is a constant. Given this definition, it is seen that a rigid body is kinematically identical to a reference frame (the only difference being that a reference frame consists of a collection of points in space and the points are massless). Thus, from the standpoint of kinematics, a rigid body and a reference frame are interchangeable, i.e., kinematically, a rigid body *is* a reference frame. Now, because a rigid body $\mathcal{R}$ defines a reference frame, it follows directly that the angular velocity of a rigid body $\mathcal{R}$ as viewed by an observer in an arbitrary reference frame $\mathcal{A}$ is the same as the angular velocity of the reference frame defined by the rigid body. Consequently, when referring to the angular velocity of a rigid body, the notation ${}^{\mathcal{A}}\boldsymbol{\omega}^{\mathcal{R}}$ will mean the angular velocity of the rigid body $\mathcal{R}$ (i.e., the angular velocity of reference frame $\mathcal{R}$) as viewed by an observer in reference frame $\mathcal{A}$. Finally, any collection of points that are not part of the rigid body but are fixed to the reference frame defined by the rigid body is called an *extension* of the rigid body.

2.15.1 Configuration and Degrees of Freedom of a Rigid Body

In Section 2.4 we discussed the three scalar quantities that are required to describe the configuration (i.e., the position) of a point moving in $\mathbb{E}^3$ without any constraints. Consequently, a point (i.e., a particle) moving in $\mathbb{E}^3$ without any constraints has *three degrees of freedom*. However, in order to specify the configuration of a rigid body, *more* quantities are required than are required to specify the configuration of a point. In other words, the number of degrees of freedom for a rigid body is *greater* than the number of degrees of freedom of a point. We now discuss the configuration of a rigid body moving without constraints in $\mathbb{E}^3$.

Using the schematic of the configuration of a rigid body $\mathcal{R}$ as given in Fig. 2-37, we can determine the number of degrees of freedom of an unconstrained rigid body as follows. First, as already discussed, three quantities are required to describe the configuration of an arbitrary point Q fixed in the rigid body $\mathcal{R}$. In addition, we know that a rigid body defines a reference frame and that the rate of change of the orientation of the rigid body $\mathcal{R}$ in an arbitrary reference frame $\mathcal{A}$ is obtained from the angular velocity of $\mathcal{R}$ in reference frame $\mathcal{A}$, ${}^{\mathcal{A}}\boldsymbol{\omega}^{\mathcal{R}}$. Furthermore, we know that the angular velocity of a rigid body is specified by three scalar quantities. Consequently, it must be the case that the orientation of a rigid body can be described using three scalar quantities. Combining the three quantities required to describe the configuration of an arbitrary point on the body with the three quantities required to describe the orientation of the entire body relative to an arbitrary reference frame $\mathcal{A}$, we see that a total of *six* scalar quantities are required to specify completely the configuration of an unconstrained rigid body $\mathcal{R}$. An unconstrained rigid body is thus said to have *six degrees of freedom*.

2.15.2 Rigid Body Motion Using Body-Fixed Coordinate Systems

While in principle the motion of a rigid body can be described using a coordinate system fixed in any reference frame whatsoever, in practice it is inconvenient to choose the coordinate system in such an arbitrary manner. Instead, it is useful to use the

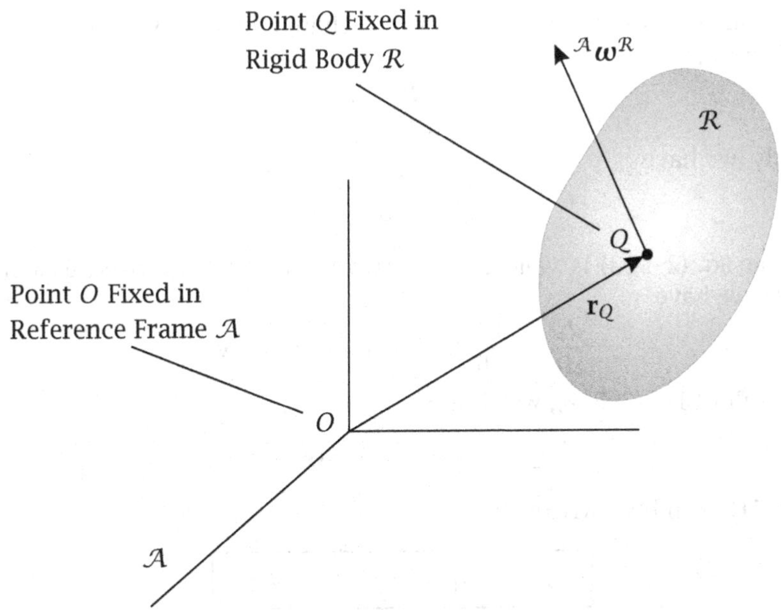

Figure 2-37 Configuration of a rigid body $\mathcal{R}$ in an arbitrary reference frame $\mathcal{A}$. The position of an arbitrary Point Q fixed in $\mathcal{R}$ requires the specification of three independent scalar quantities while the rate of change of the orientation of the rigid body relative to reference frame $\mathcal{A}$ requires the specification of an additional three independent quantities. Consequently, an unconstrained rigid body has six degrees of freedom.

reference frame defined by the rigid body as the observation reference frame. In other words, it is convenient to observe the translational and rotational motion of a rigid body $\mathcal{R}$ in the rigid body reference frame $\mathcal{R}$ itself. Correspondingly, it is convenient to choose a coordinate system that is fixed in the rigid body. Any coordinate system that is fixed in a rigid body is called a *body-fixed coordinate system.* Commensurate with the definition of a body-fixed coordinate system, any basis $\{\mathbf{e}_1, \mathbf{e}_2, \mathbf{e}_3\}$ that is fixed in a rigid body $\mathcal{R}$ is called a *body-fixed basis.*

2.15.3 Translational Kinematics of Points on a Rigid Body

Let P_1 and P_2 be two points on a rigid body $\mathcal{R}$ with positions $\mathbf{r}_1$ and $\mathbf{r}_2$, respectively, relative to a point O fixed in an arbitrary reference frame $\mathcal{A}$. Next, let $\boldsymbol{\rho} = \mathbf{r}_2 - \mathbf{r}_1$ be the position of P_2 relative to P_1. Now, because P_1 and P_2 are fixed to the rigid body, $\boldsymbol{\rho}$ is a vector that is fixed in the rigid body. Furthermore, because P_1 and P_2 are points on the rigid body $\mathcal{R}$, the distance between P_1 and P_2 is constant which implies that $\|\boldsymbol{\rho}\|$ is also constant. Consequently, from Eq. (2-128) on page 47 we have

$$\frac{{}^{\mathcal{A}}d\boldsymbol{\rho}}{dt} = \frac{{}^{\mathcal{R}}d\boldsymbol{\rho}}{dt} + {}^{\mathcal{A}}\boldsymbol{\omega}^{\mathcal{R}} \times \boldsymbol{\rho} \tag{2-512}$$

However, because $\boldsymbol{\rho}$ is fixed in $\mathcal{R}$, the rate of change of $\boldsymbol{\rho}$ as viewed by an observer fixed to $\mathcal{R}$ is zero, i.e.,

$$\frac{^{\mathcal{R}}d\boldsymbol{\rho}}{dt} = \mathbf{0} \tag{2-513}$$

Consequently, we have

$$\frac{^{\mathcal{A}}d\boldsymbol{\rho}}{dt} = {}^{\mathcal{A}}\boldsymbol{\omega}^{\mathcal{R}} \times \boldsymbol{\rho} \tag{2-514}$$

It is seen that Eq. (2–513) is valid for *any* two points that are fixed in a rigid body. Furthermore, we have

$$\frac{^{\mathcal{A}}d\boldsymbol{\rho}}{dt} = \frac{^{\mathcal{A}}d}{dt}(\mathbf{r}_2 - \mathbf{r}_1) = {}^{\mathcal{A}}\mathbf{v}_2 - {}^{\mathcal{A}}\mathbf{v}_1 \tag{2-515}$$

Using the result of Eq. (2–514), we obtain

$${}^{\mathcal{A}}\mathbf{v}_2 - {}^{\mathcal{A}}\mathbf{v}_1 = {}^{\mathcal{A}}\boldsymbol{\omega}^{\mathcal{R}} \times (\mathbf{r}_2 - \mathbf{r}_1) \tag{2-516}$$

Equation (2–516) can be rewritten as

$$\boxed{{}^{\mathcal{A}}\mathbf{v}_2 = {}^{\mathcal{A}}\mathbf{v}_1 + {}^{\mathcal{A}}\boldsymbol{\omega}^{\mathcal{R}} \times (\mathbf{r}_2 - \mathbf{r}_1)} \tag{2-517}$$

Equation (2–517) relates the velocity between any two points on a rigid body as viewed by an observer in reference frame $\mathcal{A}$.

The acceleration between any two points on a rigid body as viewed by an observer in reference frame $\mathcal{A}$ is obtained by applying Eq. (2–128) on page 47 to Eq. (2–517). We then have

$$\frac{^{\mathcal{A}}d}{dt}\left({}^{\mathcal{A}}\mathbf{v}_2\right) = \frac{^{\mathcal{A}}d}{dt}\left({}^{\mathcal{A}}\mathbf{v}_1\right) + \frac{^{\mathcal{A}}d}{dt}\left({}^{\mathcal{A}}\boldsymbol{\omega}^{\mathcal{R}} \times (\mathbf{r}_2 - \mathbf{r}_1)\right) \tag{2-518}$$

Now we note that

$$\begin{aligned} \frac{^{\mathcal{A}}d}{dt}\left({}^{\mathcal{A}}\mathbf{v}_1\right) &= {}^{\mathcal{A}}\mathbf{a}_1 \\ \frac{^{\mathcal{A}}d}{dt}\left({}^{\mathcal{A}}\mathbf{v}_2\right) &= {}^{\mathcal{A}}\mathbf{a}_2 \end{aligned} \tag{2-519}$$

Consequently, Eq. (2–518) can be rewritten as

$${}^{\mathcal{A}}\mathbf{a}_2 = {}^{\mathcal{A}}\mathbf{a}_1 + \frac{^{\mathcal{A}}d}{dt}\left({}^{\mathcal{A}}\boldsymbol{\omega}^{\mathcal{R}} \times (\mathbf{r}_2 - \mathbf{r}_1)\right) \tag{2-520}$$

The second term in Eq. (2–520) can be written as

$$\frac{^{\mathcal{A}}d}{dt}\left({}^{\mathcal{A}}\boldsymbol{\omega}^{\mathcal{R}} \times (\mathbf{r}_2 - \mathbf{r}_1)\right) = \frac{^{\mathcal{A}}d}{dt}\left({}^{\mathcal{A}}\boldsymbol{\omega}^{\mathcal{R}}\right) \times (\mathbf{r}_2 - \mathbf{r}_1) + {}^{\mathcal{A}}\boldsymbol{\omega}^{\mathcal{R}} \times \frac{^{\mathcal{A}}d}{dt}(\mathbf{r}_2 - \mathbf{r}_1) \tag{2-521}$$

Now we recall that

$$\frac{^{\mathcal{A}}d}{dt}\left({}^{\mathcal{A}}\boldsymbol{\omega}^{\mathcal{R}}\right) = {}^{\mathcal{A}}\boldsymbol{\alpha}^{\mathcal{R}} \tag{2-522}$$

where ${}^{\mathcal{A}}\boldsymbol{\alpha}^{\mathcal{R}}$ is the angular acceleration of $\mathcal{R}$ as viewed by an observer in reference frame $\mathcal{A}$. Also, using Eq. (2–516), we have

$$\frac{^{\mathcal{A}}d}{dt}\left({}^{\mathcal{A}}\boldsymbol{\omega}^{\mathcal{R}} \times (\mathbf{r}_2 - \mathbf{r}_1)\right) = {}^{\mathcal{A}}\boldsymbol{\alpha}^{\mathcal{R}} \times (\mathbf{r}_2 - \mathbf{r}_1) + {}^{\mathcal{A}}\boldsymbol{\omega}^{\mathcal{R}} \times \left[{}^{\mathcal{A}}\boldsymbol{\omega}^{\mathcal{R}} \times (\mathbf{r}_2 - \mathbf{r}_1)\right] \tag{2-523}$$

Equation (2-520) then simplifies to

$$\boxed{{}^{\mathcal{A}}\mathbf{a}_2 = {}^{\mathcal{A}}\mathbf{a}_1 + {}^{\mathcal{A}}\boldsymbol{\alpha}^{\mathcal{R}} \times (\mathbf{r}_2 - \mathbf{r}_1) + {}^{\mathcal{A}}\boldsymbol{\omega}^{\mathcal{R}} \times \left[{}^{\mathcal{A}}\boldsymbol{\omega}^{\mathcal{R}} \times (\mathbf{r}_2 - \mathbf{r}_1)\right]} \tag{2-524}$$

Equation (2-524) relates the acceleration between any two points on a rigid body as viewed by an observer in reference frame $\mathcal{A}$.

2.15.4 Kinematics of Rolling and Sliding Rigid Bodies

Many applications of rigid body motion involve two rigid bodies moving while maintaining contact with one another. The two important forms of contact between rigid bodies are *rolling* and *sliding*. We now discuss the kinematics associated with two rigid bodies moving while either rolling or sliding relative to one another.[7]

Rolling and Sliding Between Two Moving Rigid Bodies

Consider a rigid body $\mathcal{R}$, called the *primary* rigid body, moving in an arbitrary reference frame $\mathcal{A}$ while maintaining continuous contact over a nonzero time interval $t \in [t_1, t_2]$ with a rigid body $\mathcal{S}$, called the *secondary* rigid body, as shown in Fig. 2-38. Furthermore, let O be a point fixed in reference frame $\mathcal{A}$ and let Q be a point fixed in the reference frame defined by $\mathcal{R}$.[8] Finally, assume that the motion of $\mathcal{S}$ in reference frame $\mathcal{A}$ is *known* (i.e., the velocity and acceleration of every point on $\mathcal{S}$ in reference frame $\mathcal{A}$ is known).

Now, because $\mathcal{R}$ is in contact with $\mathcal{S}$, at *every instant* of the contacting motion there exists a point C, called the *instantaneous point of contact*, that is common to both $\mathcal{R}$ and $\mathcal{S}$. Moreover, because both bodies are moving, the instantaneous point of contact C changes continuously during the contacting motion. Next, assuming that the surfaces of $\mathcal{R}$ and $\mathcal{S}$ can each be described by differentiable functions, there exists a well-defined *plane of contact*, denoted $\mathcal{T}_C^{\mathcal{R}} = \mathcal{T}_C^{\mathcal{S}} \equiv \mathcal{T}_C$, that is tangent to both $\mathcal{R}$ and $\mathcal{S}$ at the instantaneous point of contact C. Then, let $\mathbf{n}$ be the unit vector in the direction orthogonal to $\mathcal{T}_C$.

Suppose now that we let ${}^{\mathcal{A}}\mathbf{v}_C^{\mathcal{R}}$ and ${}^{\mathcal{A}}\mathbf{v}_C^{\mathcal{S}}$ be the velocity of point C on the rigid body $\mathcal{R}$ and the velocity of point C on the rigid body $\mathcal{S}$, respectively, in reference frame $\mathcal{A}$.[9] Then, the rigid body $\mathcal{R}$ is said to instantaneously *roll* or *roll without slip* on $\mathcal{S}$ if the velocity of point C on $\mathcal{R}$ as viewed by an observer in reference frame $\mathcal{A}$ is equal to the velocity of point C on $\mathcal{S}$ as viewed by an observer in reference frame $\mathcal{A}$, i.e., the condition for $\mathcal{R}$ to roll on $\mathcal{S}$ is given as

$$\boxed{{}^{\mathcal{A}}\mathbf{v}_C^{\mathcal{R}} = {}^{\mathcal{A}}\mathbf{v}_C^{\mathcal{S}}} \tag{2-525}$$

[7]The approach used in this section is inspired by the description of rolling and sliding as given in O'Reilly (2001), the key difference being that in the description given in O'Reilly (2001) the secondary rigid body is a *fixed* surface whereas in the description given here the secondary rigid body may be *moving*. The inclusion of a moving secondary rigid body is important because such situations are encountered in practice (e.g., Problem 2-14 on page 136 is an example of a system of two rigid bodies that are both moving while maintaining continuous contact with one another).

[8]While point Q in Fig. 2-38 is depicted as being a point on the rigid body $\mathcal{R}$, in general the point Q may be a point either on or off the rigid body. In the case that point Q lies *off* $\mathcal{R}$, we assume in this analysis that Q lies in an *extension* of $\mathcal{R}$ and thus is fixed in $\mathcal{R}$.

[9]It is emphasized that when using the notation ${}^{\mathcal{A}}\mathbf{v}_C^{\mathcal{R}}$ we are considering point C to be a point that is fixed in the rigid body $\mathcal{R}$.

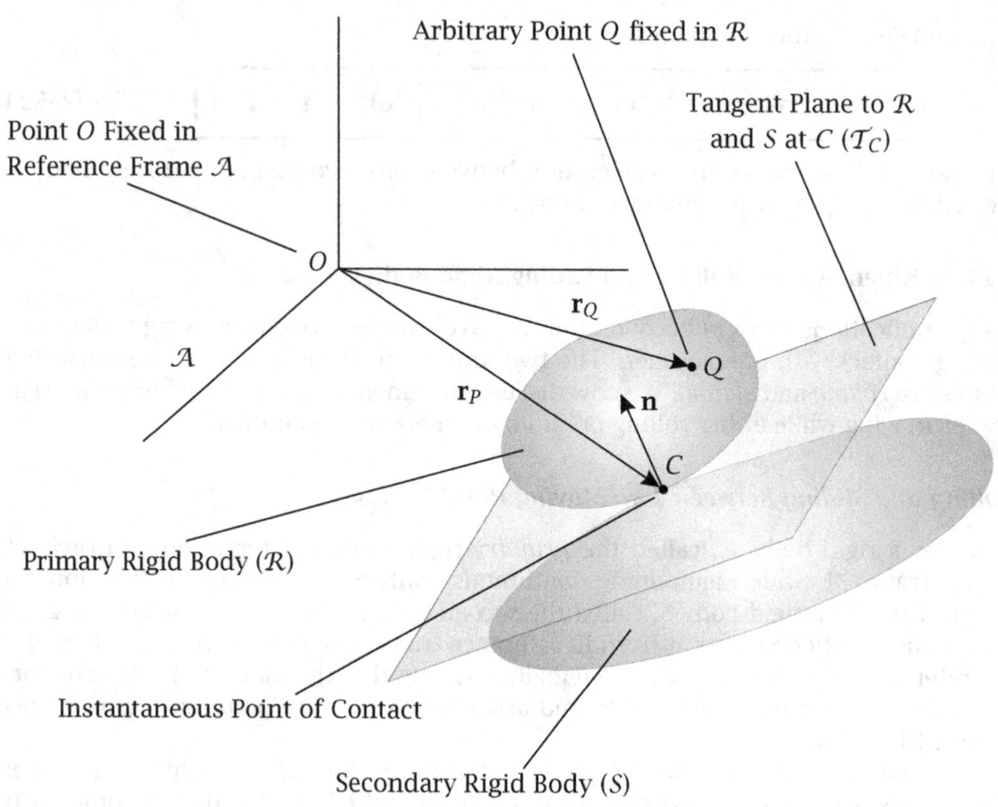

Figure 2–38 Primary rigid body $\mathcal{R}$ moving while in continuous contact with a secondary rigid Body $\mathcal{S}$.

The rigid body $\mathcal{R}$ is said to *slide* on $\mathcal{S}$ if ${}^{\mathcal{A}}\mathbf{v}_C^{\mathcal{R}} \neq {}^{\mathcal{A}}\mathbf{v}_C^{\mathcal{S}}$, but the component of ${}^{\mathcal{A}}\mathbf{v}_C^{\mathcal{S}}$ that lies in the direction of $\mathbf{n}$ is equal to the component of ${}^{\mathcal{A}}\mathbf{v}_C^{\mathcal{R}}$ in the direction of $\mathbf{n}$, i.e., the condition for $\mathcal{R}$ to slide on $\mathcal{S}$ is given as

$$\boxed{{}^{\mathcal{A}}\mathbf{v}_C^{\mathcal{R}} \cdot \mathbf{n} = {}^{\mathcal{A}}\mathbf{v}_C^{\mathcal{S}} \cdot \mathbf{n}\,,\ ({}^{\mathcal{A}}\mathbf{v}_C^{\mathcal{R}} \neq {}^{\mathcal{A}}\mathbf{v}_C^{\mathcal{S}})} \tag{2–526}$$

By comparing Eqs. (2–525) and (2–526) it is seen that rolling is a condition that is *distinct* from sliding because when a rigid body rolls on another rigid body the quantities ${}^{\mathcal{A}}\mathbf{v}_C^{\mathcal{R}}$ and ${}^{\mathcal{A}}\mathbf{v}_C^{\mathcal{S}}$ are equal, whereas when a rigid body slides along another rigid body only the component of ${}^{\mathcal{A}}\mathbf{v}_C^{\mathcal{R}}$ and the component of ${}^{\mathcal{A}}\mathbf{v}_C^{\mathcal{S}}$ in the direction of $\mathbf{n}$ are equal.[10]

Suppose now that we let $\mathbf{r}_Q$ be the position of point Q relative to point O, and let ${}^{\mathcal{A}}\mathbf{v}_Q$ and ${}^{\mathcal{A}}\mathbf{a}_Q$ be the velocity and acceleration, respectively, of point Q in reference frame $\mathcal{A}$. Then, because Q is fixed in $\mathcal{R}$, from Eq. (2–517) the velocity of point C on $\mathcal{R}$

[10]Some textbooks, such as Baruh (1999), use the general term "rolling" to denote any situation where contact between two bodies is maintained. Then, a distinction between rolling and sliding is made by using the terms "rolling without sliding" and "rolling with sliding," respectively. However, in this book we use the terminology *rolling* or *rolling without slip* to denote a rolling rigid body and use the terminology *sliding* to denote a rigid body that is in contact with another rigid body but is not rolling.

in reference frame $\mathcal{A}$ is given as

$$ {}^{\mathcal{A}}\mathbf{v}_C^{\mathcal{R}} = {}^{\mathcal{A}}\mathbf{v}_Q + {}^{\mathcal{A}}\boldsymbol{\omega}^{\mathcal{R}} \times (\mathbf{r}_C - \mathbf{r}_Q) \tag{2-527} $$

Substituting ${}^{\mathcal{A}}\mathbf{v}_C^{\mathcal{R}}$ from Eq. (2–527) into (2–525), we have

$$ {}^{\mathcal{A}}\mathbf{v}_Q + {}^{\mathcal{A}}\boldsymbol{\omega}^{\mathcal{R}} \times (\mathbf{r}_C - \mathbf{r}_Q) = {}^{\mathcal{A}}\mathbf{v}_C^{S} \tag{2-528} $$

Equation (2–528) provides an alternate expression for the case where the rigid body $\mathcal{R}$ rolls on the rigid body S. Similarly, substituting ${}^{\mathcal{A}}\mathbf{v}_C^{\mathcal{R}}$ from Eq. (2–527) into (2–526), we have

$$ \left[{}^{\mathcal{A}}\mathbf{v}_Q + {}^{\mathcal{A}}\boldsymbol{\omega}^{\mathcal{R}} \times (\mathbf{r}_C - \mathbf{r}_Q)\right] \cdot \mathbf{n} = {}^{\mathcal{A}}\mathbf{v}_C^{S} \cdot \mathbf{n} \tag{2-529} $$

Equation (2–529) provides an alternate expression for the case where the rigid body $\mathcal{R}$ slides on the rigid body S.

Rolling and Sliding of a Rigid Body Along a Fixed Surface

An important special case of rolling and sliding that occurs frequently is that of a rigid body rolling or sliding on an absolutely *fixed* body. Furthermore, suppose that observations of the motion are made in a fixed inertial reference frame $\mathcal{F}$. Then, because S is absolutely fixed, S itself is a fixed inertial reference frame (i.e., $S \equiv \mathcal{F}$). Therefore, the velocity of the instantaneous point of contact in reference frame $\mathcal{F}$ is zero, i.e.,

$$ {}^{\mathcal{F}}\mathbf{v}_C^{S} = \mathbf{0} \tag{2-530} $$

Consequently, the rolling condition of Eq. (2–525) simplifies to (O'Reilly, 2001)

$$ \boxed{{}^{\mathcal{F}}\mathbf{v}_C^{\mathcal{R}} = {}^{\mathcal{F}}\mathbf{v}_C^{S} = \mathbf{0}} \tag{2-531} $$

Furthermore, the sliding condition of Eq. (2–526) simplifies to (O'Reilly, 2001)

$$ \boxed{{}^{\mathcal{F}}\mathbf{v}_C^{\mathcal{R}} \cdot \mathbf{n} = 0} \tag{2-532} $$

In other words, when a rigid body rolls on a fixed rigid body, the velocity of the instantaneous point of contact C on $\mathcal{R}$ is zero when viewed by an observer in the fixed inertial reference frame $\mathcal{F}$. Now, it is important to realize that, while ${}^{\mathcal{F}}\mathbf{v}_C^{\mathcal{R}}$ may be zero, the acceleration of C on $\mathcal{R}$ in $\mathcal{F}$, ${}^{\mathcal{F}}\mathbf{a}_C^{\mathcal{R}}$, is *not* zero. The fact that ${}^{\mathcal{F}}\mathbf{a}_C^{\mathcal{R}}$ is not zero is seen by applying Eq. (2–524) as

$$ {}^{\mathcal{A}}\mathbf{a}_C^{\mathcal{R}} = {}^{\mathcal{A}}\mathbf{a}_Q + {}^{\mathcal{A}}\boldsymbol{\alpha}^{\mathcal{R}} \times (\mathbf{r}_C - \mathbf{r}_Q) + {}^{\mathcal{A}}\boldsymbol{\omega}^{\mathcal{R}} \times \left[{}^{\mathcal{A}}\boldsymbol{\omega}^{\mathcal{R}} \times (\mathbf{r}_C - \mathbf{r}_Q)\right] \tag{2-533} $$

Equation (2–533) shows that, in general, the acceleration of the point of contact, ${}^{\mathcal{A}}\mathbf{a}_C^{\mathcal{R}}$, is *not* zero for the case of a rigid body rolling on an absolutely fixed surface. Lastly, it is noted that Eqs. (2–531) and (2–532) are valid *only* when the rigid body S defines a fixed inertial reference frame; in the *general* case of a rigid body in contact with a *moving* rigid body, either Eq. (2–525) or (2–526) must be applied.

Independence of Relative Velocity with Respect to Reference Frame During Sliding

A final interesting point is that, when a rigid body slides along another rigid body, the relative velocity between the two bodies at the instantaneous point of contact is independent of the choice of the reference frame in which the observations are made. To see this last fact, suppose that we choose to observe the motion of $\mathcal{R}$ and $\mathcal{S}$ in a reference frame $\mathcal{B}$ that is different from reference frame $\mathcal{A}$. Then, we have from Eq. (2–399) that

$$ {}^{\mathcal{B}}\mathbf{v}_C^{\mathcal{R}} = {}^{\mathcal{B}}\mathbf{v}_O + {}^{\mathcal{A}}\mathbf{v}_C^{\mathcal{R}} + {}^{\mathcal{B}}\boldsymbol{\omega}^{\mathcal{A}} \times \mathbf{r}_C \tag{2-534} $$

$$ {}^{\mathcal{B}}\mathbf{v}_C^{\mathcal{S}} = {}^{\mathcal{B}}\mathbf{v}_O + {}^{\mathcal{A}}\mathbf{v}_C^{\mathcal{S}} + {}^{\mathcal{B}}\boldsymbol{\omega}^{\mathcal{A}} \times \mathbf{r}_C \tag{2-535} $$

where we recall again that the position of the instantaneous point of contact on both rigid bodies is the same. Subtracting Eq. (2–535) from (2–534), we obtain

$$ {}^{\mathcal{B}}\mathbf{v}_C^{\mathcal{R}} - {}^{\mathcal{B}}\mathbf{v}_C^{\mathcal{S}} = {}^{\mathcal{A}}\mathbf{v}_C^{\mathcal{R}} - {}^{\mathcal{A}}\mathbf{v}_C^{\mathcal{S}} \tag{2-536} $$

Because $\mathcal{B}$ can be any reference frame whatsoever, Eq. (2–536) shows that the relative velocity ${}^{\mathcal{A}}\mathbf{v}_C^{\mathcal{R}} - {}^{\mathcal{A}}\mathbf{v}_C^{\mathcal{S}}$ is the same *regardless* of the reference frame in which the motion is observed. It is noted that the fact that ${}^{\mathcal{A}}\mathbf{v}_C^{\mathcal{R}} - {}^{\mathcal{A}}\mathbf{v}_C^{\mathcal{S}}$ is independent of the reference frame is useful because in some problems it may be most convenient to compute ${}^{\mathcal{A}}\mathbf{v}_C^{\mathcal{R}} - {}^{\mathcal{A}}\mathbf{v}_C^{\mathcal{S}}$ in a particular reference frame.

Example 2–11

A disk of radius r rolls without slip along a fixed inclined plane with inclination angle β as shown in Fig. 2-39. Knowing that the center of the disk is denoted by O, point C is the instantaneous point of contact of the disk with the incline, and point P is located on the edge of the disk such that θ is the angle between OC and OP, determine (a) the velocity and acceleration of point O as viewed by an observer fixed to the ground and (b) the velocity and acceleration of point P as viewed by an observer fixed to the ground.

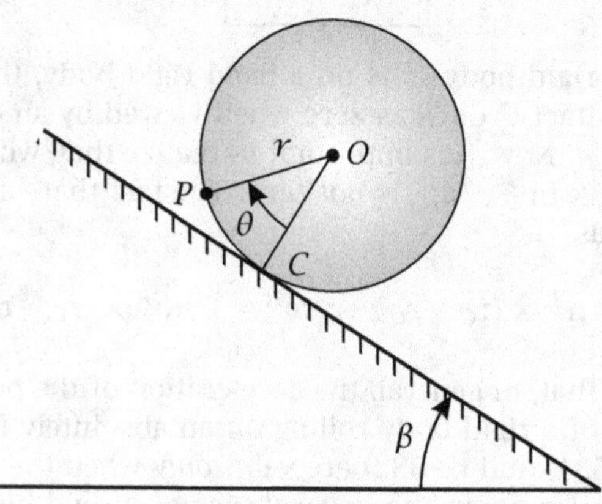

Figure 2–39 Disk of radius r rolling on a fixed incline.

Solution to Example 2–11

First, let $\mathcal{F}$ be a fixed reference frame. Then, choose the following coordinate system fixed in reference frame $\mathcal{F}$:

	Origin at point O at time $t = 0$	
$\mathbf{E}_x$	=	Down incline
$\mathbf{E}_z$	=	Into page
$\mathbf{E}_y$	=	$\mathbf{E}_z \times \mathbf{E}_x$

Next, let $\mathcal{R}$ be a reference frame fixed to the disk. Then, choose the following coordinate system fixed in reference frame $\mathcal{R}$:

	Origin at point O Moving with disk	
$\mathbf{e}_r$	=	Along OP
$\mathbf{e}_z$	=	Into page = $\mathbf{E}_z$
$\mathbf{e}_\theta$	=	$\mathbf{e}_z \times \mathbf{e}_r$

The geometry of the bases $\{\mathbf{E}_x, \mathbf{E}_y, \mathbf{E}_z\}$ and $\{\mathbf{e}_r, \mathbf{e}_\theta, \mathbf{e}_z\}$ is shown in Fig. 2–40.

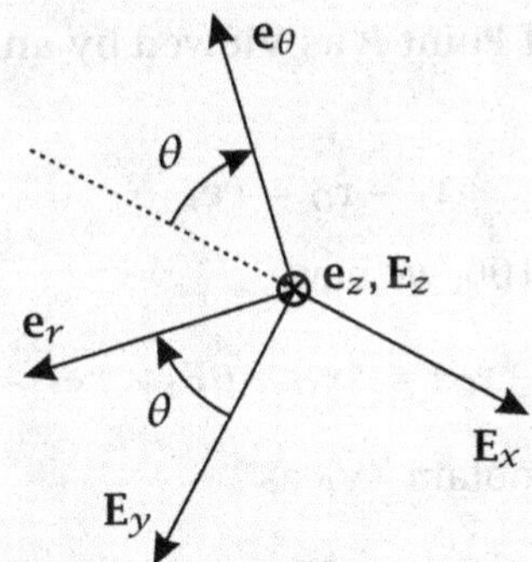

Figure 2–40 Geometry of bases $\{\mathbf{e}_r, \mathbf{e}_\theta, \mathbf{e}_z\}$ and $\{\mathbf{E}_x, \mathbf{E}_y, \mathbf{E}_z\}$ for Example 2–11.

(a) Velocity and Acceleration of Point O as Viewed by an Observer Fixed to Ground

Because the direction $\mathbf{E}_y$ is fixed and θ is measured relative to $\mathbf{E}_y$, we see that the angular velocity of the disk as viewed by an observer in reference frame $\mathcal{F}$ is given as

$$^{\mathcal{F}}\boldsymbol{\omega}^{\mathcal{R}} = \dot{\theta}\mathbf{e}_z = \dot{\theta}\mathbf{E}_z \tag{2–537}$$

Computing the rate of change of $^{\mathcal{F}}\boldsymbol{\omega}^{\mathcal{R}}$ in Eq. (2–537), we obtain the angular acceleration of the disk as

$$^{\mathcal{F}}\boldsymbol{\alpha}^{\mathcal{R}} = \frac{^{\mathcal{F}}d}{dt}\left(^{\mathcal{F}}\boldsymbol{\omega}^{\mathcal{R}}\right) = \ddot{\theta}\mathbf{e}_z \tag{2–538}$$

Next, because the disk rolls without slip along the incline S and the incline is fixed in $\mathcal{F}$, we know that the velocity of point P on $\mathcal{R}$ as viewed by an observer in reference frame $\mathcal{F}$ is zero, i.e., ${}^{\mathcal{F}}\mathbf{v}_C = \mathbf{0}$. Then, using the result of Eq. (2–517) on page 106, we have

$$ {}^{\mathcal{F}}\mathbf{v}_O = {}^{\mathcal{F}}\mathbf{v}_C + {}^{\mathcal{F}}\boldsymbol{\omega}^{\mathcal{R}} \times (\mathbf{r}_O - \mathbf{r}_C) \tag{2–539}$$

Now, in terms of the basis $\{\mathbf{E}_x, \mathbf{E}_y, \mathbf{E}_z\}$, $\mathbf{r}_O - \mathbf{r}_C$ is given as

$$ \mathbf{r}_O - \mathbf{r}_C = -r\mathbf{E}_y \tag{2–540}$$

Then, using the angular velocity from Eq. (2–537) and the fact that ${}^{\mathcal{F}}\mathbf{v}_C = \mathbf{0}$, Eq. (2–539) becomes

$$ {}^{\mathcal{F}}\mathbf{v}_O = \dot{\theta}\mathbf{E}_z \times (-r\mathbf{E}_y) = r\dot{\theta}\mathbf{E}_x \tag{2–541}$$

The position of point O is then found by integrating Eq. (2–541) as

$$ \mathbf{r}_O = \int_0^t r\dot{\theta}d\tau\,\mathbf{E}_x = \int_0^\theta r d\nu \mathbf{E}_x = r\theta\mathbf{E}_x \tag{2–542}$$

where τ and ν are dummy variables of integration. Furthermore, using ${}^{\mathcal{F}}\mathbf{v}_O$ from Eq. (2–541), we obtain the acceleration of point O as viewed by an observer in reference frame $\mathcal{F}$ as

$$ {}^{\mathcal{F}}\mathbf{a}_O = r\ddot{\theta}\mathbf{E}_x \tag{2–543}$$

(b) Velocity and Acceleration of Point P as Viewed by an Observer Fixed to Ground

First, we have

$$ \mathbf{r}_P - \mathbf{r}_O = r\mathbf{e}_r \tag{2–544}$$

Then, from Eq. (2–517) on page 106, we have

$$ {}^{\mathcal{F}}\mathbf{v}_P = {}^{\mathcal{F}}\mathbf{v}_O + {}^{\mathcal{F}}\boldsymbol{\omega}^{\mathcal{R}} \times (\mathbf{r}_P - \mathbf{r}_O) = {}^{\mathcal{F}}\mathbf{v}_O + \dot{\theta}\mathbf{E}_z \times r\mathbf{e}_r = {}^{\mathcal{F}}\mathbf{v}_O + r\dot{\theta}\mathbf{e}_\theta \tag{2–545}$$

Using ${}^{\mathcal{F}}\mathbf{v}_O$ from Eq. (2–541), we obtain ${}^{\mathcal{F}}\mathbf{v}_P$ as

$$ {}^{\mathcal{F}}\mathbf{v}_P = r\dot{\theta}\mathbf{E}_x + r\dot{\theta}\mathbf{e}_\theta \tag{2–546}$$

Furthermore, from Eq. (2–524), we have

$$ {}^{\mathcal{F}}\mathbf{a}_P = {}^{\mathcal{F}}\mathbf{a}_O + {}^{\mathcal{F}}\boldsymbol{\alpha}^{\mathcal{R}} \times (\mathbf{r}_P - \mathbf{r}_O) + {}^{\mathcal{F}}\boldsymbol{\omega}^{\mathcal{R}} \times \left[{}^{\mathcal{F}}\boldsymbol{\omega}^{\mathcal{R}} \times (\mathbf{r}_P - \mathbf{r}_O)\right] \tag{2–547}$$

Substituting ${}^{\mathcal{F}}\boldsymbol{\omega}^{\mathcal{R}}$ from Eq. (2–537), ${}^{\mathcal{F}}\boldsymbol{\alpha}^{\mathcal{R}}$ from Eq. (2–538), and $\mathbf{r}_P - \mathbf{r}_O$ from Eq. (2–544), we obtain

$$ {}^{\mathcal{F}}\mathbf{a}_P = {}^{\mathcal{F}}\mathbf{a}_O + \ddot{\theta}\mathbf{E}_z \times r\mathbf{e}_r + \dot{\theta}\mathbf{E}_z \times \left[\dot{\theta}\mathbf{E}_z \times r\mathbf{e}_r\right] \tag{2–548}$$

Simplifying Eq. (2–548) gives

$$ {}^{\mathcal{F}}\mathbf{a}_P = {}^{\mathcal{F}}\mathbf{a}_O + r\ddot{\theta}\mathbf{e}_\theta - r\dot{\theta}^2\mathbf{e}_r \tag{2–549}$$

Using ${}^{\mathcal{F}}\mathbf{a}_O$ from Eq. (2–543), we obtain the acceleration of point P as

$$ {}^{\mathcal{F}}\mathbf{a}_P = r\ddot{\theta}\mathbf{E}_x + r\ddot{\theta}\mathbf{e}_\theta - r\dot{\theta}^2\mathbf{e}_r \tag{2–550}$$

Comments on the Solution

Recall in Section 2.15.4 that we developed the conditions for a rigid body rolling on a fixed surface. In particular, we stated in Section 2.15.4 that, while the velocity of the point of contact with a fixed surface is zero, the acceleration of the point of contact with the surface is *not zero.* For this example, the point P on the edge of the disk will be in contact with the surface when the angle θ is zero. Furthermore, from Fig. 2–40 it is seen that at the instant when $\theta = 0$, the $\mathbf{e}_r$-direction is aligned with the $\mathbf{E}_y$-direction and the $\mathbf{e}_\theta$-direction is aligned with the $-\mathbf{E}_x$-direction. Consequently, we have

$$\begin{aligned} \mathbf{e}_r(\theta = 0) &= \mathbf{E}_y \\ \mathbf{e}_\theta(\theta = 0) &= -\mathbf{E}_x \end{aligned} \tag{2–551}$$

Therefore, the acceleration of point P when $\theta = 0$ is given as

$$^{\mathcal{F}}\mathbf{a}_P(\theta = 0) = r\ddot{\theta}\mathbf{E}_x - r\ddot{\theta}\mathbf{E}_x - r\dot{\theta}^2\mathbf{e}_r = -r\dot{\theta}^2\mathbf{e}_r \tag{2–552}$$

It is seen from Eq. (2–552) that, when point P is in contact with the incline, the acceleration of P is *not* zero.

■

2.15.5 Orientation of a Rigid Body: Eulerian Angles

In Section 2.15.3 we described the translational kinematics of points on a rigid body. We now turn our attention to the *rotational* kinematics of a rigid body. In particular, we now focus our attention on describing the orientation of a rigid body $\mathcal{R}$ as viewed by an observer in an arbitrary reference frame $\mathcal{A}$.

As stated in Section 2.15.1, the *rate of change* of the orientation of a rigid body in an arbitrary reference frame $\mathcal{A}$ is described using the angular velocity of the rigid body $\mathcal{R}$ in a reference frame $\mathcal{A}$, $^{\mathcal{A}}\boldsymbol{\omega}^{\mathcal{R}}$. Now, it would *seem* as if the orientation of the rigid body could be obtained by integrating the angular velocity $^{\mathcal{A}}\boldsymbol{\omega}^{\mathcal{R}}$, i.e., it would appear as if one could find three scalar quantities whose rates of change are the components of the vector $^{\mathcal{A}}\boldsymbol{\omega}^{\mathcal{R}}$ resolved in an arbitrary basis. However, as it turns out, the components of $^{\mathcal{A}}\boldsymbol{\omega}^{\mathcal{R}}$ resolved in an arbitrary basis do not, in general, arise from the rates of change of three scalar quantities where the three scalar quantities describe the orientation of $\mathcal{R}$ in $\mathcal{A}$. Quantitatively, suppose that we choose to express $^{\mathcal{A}}\boldsymbol{\omega}^{\mathcal{R}}$ in a basis $\{\mathbf{e}_1, \mathbf{e}_2, \mathbf{e}_3\}$ that is fixed in the rigid body $\mathcal{R}$. Then the angular velocity of $\mathcal{R}$ in $\mathcal{A}$ can be written as

$$^{\mathcal{A}}\boldsymbol{\omega}^{\mathcal{R}} = \omega_1\mathbf{e}_1 + \omega_2\mathbf{e}_2 + \omega_3\mathbf{e}_3 \tag{2–553}$$

Suppose further that we have three quantities q_1, q_2, and q_3 that can be used to uniquely describe the orientation of the rigid body $\mathcal{R}$ in reference frame $\mathcal{A}$. Then in general it is the case that

$$\begin{aligned} \dot{q}_1 &\neq \omega_1 \\ \dot{q}_2 &\neq \omega_2 \\ \dot{q}_3 &\neq \omega_3 \end{aligned} \tag{2–554}$$

Consequently, it is not possible to find three quantities q_1, q_2, and q_3 that simultaneously describe the orientation of the rigid body and whose rates of change correspond to the components ω_1, ω_2, and ω_3 of ${}^{\mathcal{A}}\boldsymbol{\omega}^{\mathcal{R}}$. Therefore, in order to specify the orientation of $\mathcal{R}$ in an arbitrary reference frame $\mathcal{A}$, it is necessary to find an *alternate* set of three independent scalar quantities.

A set of three quantities that are commonly used to describe the orientation of a rigid body are *Eulerian angles* or, more simply, *Euler angles.* Euler angles arise from a sequence of three single-axis rotations (SAR). These single-axis rotations together rotate the body from an initial orientation to a final orientation. However, because the three rotations can be done in any order, Euler angles are not unique. In particular, 12 different sets of Euler angles exist to specify the orientation of a rigid body relative to an arbitrary reference frame. Of these 12 sequences, 3 Euler angle conventions are commonly used. They are called *Type I, Type II,* and *Type III* Euler angles. While in principle any of these three conventions can be used, this discussion will be limited to Type I Euler angle conventions because this convention is commonly found in other textbooks and is also common in aeronautical engineering. A summary of the rotations for Types I, II, III Euler angles are given in Tables 2–1 to 2–3.

Table 2–1 Angle definitions and sequence of rotations for Type I Eulerian angles.

Axis of Rotation	Angle	Basis Before Rotation	Basis After Rotation
$\mathbf{E}_3 = \mathbf{p}_3$	ψ	$\{\mathbf{E}_1, \mathbf{E}_2, \mathbf{E}_3\}$	$\{\mathbf{p}_1, \mathbf{p}_2, \mathbf{p}_3\}$
$\mathbf{p}_2 = \mathbf{q}_2$	θ	$\{\mathbf{p}_1, \mathbf{p}_2, \mathbf{p}_3\}$	$\{\mathbf{q}_1, \mathbf{q}_2, \mathbf{q}_3\}$
$\mathbf{q}_1 = \mathbf{e}_1$	ϕ	$\{\mathbf{q}_1, \mathbf{q}_2, \mathbf{q}_3\}$	$\{\mathbf{e}_1, \mathbf{e}_2, \mathbf{e}_3\}$

Table 2–2 Angle definitions and sequence of rotations for Type II Eulerian angles.

Axis of Rotation	Angle	Basis Before Rotation	Basis After Rotation
$\mathbf{E}_3 = \mathbf{p}_3$	ϕ	$\{\mathbf{E}_1, \mathbf{E}_2, \mathbf{E}_3\}$	$\{\mathbf{p}_1, \mathbf{p}_2, \mathbf{p}_3\}$
$\mathbf{p}_1 = \mathbf{q}_1$	θ	$\{\mathbf{p}_1, \mathbf{p}_2, \mathbf{p}_3\}$	$\{\mathbf{q}_1, \mathbf{q}_2, \mathbf{q}_3\}$
$\mathbf{q}_3 = \mathbf{e}_3$	ψ	$\{\mathbf{q}_1, \mathbf{q}_2, \mathbf{q}_3\}$	$\{\mathbf{e}_1, \mathbf{e}_2, \mathbf{e}_3\}$

Table 2–3 Angle definitions and sequence of rotations for Type III Eulerian angles.

Axis of Rotation	Angle	Basis Before Rotation	Basis After Rotation
$\mathbf{E}_3 = \mathbf{p}_3$	ϕ	$\{\mathbf{E}_1, \mathbf{E}_2, \mathbf{E}_3\}$	$\{\mathbf{p}_1, \mathbf{p}_2, \mathbf{p}_3\}$
$\mathbf{p}_2 = \mathbf{q}_2$	θ	$\{\mathbf{p}_1, \mathbf{p}_2, \mathbf{p}_3\}$	$\{\mathbf{q}_1, \mathbf{q}_2, \mathbf{q}_3\}$
$\mathbf{q}_3 = \mathbf{e}_3$	ψ	$\{\mathbf{q}_1, \mathbf{q}_2, \mathbf{q}_3\}$	$\{\mathbf{e}_1, \mathbf{e}_2, \mathbf{e}_3\}$

The following assumptions are used for any of the three types of Euler angles. First, let $\{\mathbf{E}_1, \mathbf{E}_2, \mathbf{E}_3\}$ be a basis fixed in an arbitrary reference frame $\mathcal{A}$. Next, let $\mathcal{R}$ be a rigid body and let $\{\mathbf{e}_1, \mathbf{e}_2, \mathbf{e}_3\}$ be a body-fixed basis. Furthermore, because in this Section we are interested only in the rotation of the body, let O be a point fixed in reference frame $\mathcal{A}$ and let Q be a point fixed in the rigid body $\mathcal{R}$ such that O and Q coincide. Finally, assume that, *if no rotations are performed*, the bases $\{\mathbf{E}_1, \mathbf{E}_2, \mathbf{E}_3\}$ and $\{\mathbf{e}_1, \mathbf{e}_2, \mathbf{e}_3\}$ are aligned.

Type I Euler Angles

The three rotations of the rigid body $\mathcal{R}$ in reference frame $\mathcal{A}$ that define the class of Type I Euler angles are shown in Figs. 2-41 to 2-43 and are given as follows: (1) a positive rotation of the basis $\mathbf{E} = \{\mathbf{E}_1, \mathbf{E}_2, \mathbf{E}_3\}$ about the $\mathbf{E}_3$-direction by an angle ψ that results in the basis $\mathbf{P} = \{\mathbf{p}_1, \mathbf{p}_2, \mathbf{p}_3\}$; (2) a positive rotation of the basis $\{\mathbf{p}_1, \mathbf{p}_2, \mathbf{p}_3\}$ about the $\mathbf{p}_2$-direction by an angle θ that results in the basis $\mathbf{Q} = \{\mathbf{q}_1, \mathbf{q}_2, \mathbf{q}_3\}$; and (3) a positive rotation of the basis $\{\mathbf{q}_1, \mathbf{q}_2, \mathbf{q}_3\}$ about the $\mathbf{q}_1$-direction by an angle ϕ that results in the final basis $\mathbf{e} = \{\mathbf{e}_1, \mathbf{e}_2, \mathbf{e}_3\}$. Because of the axes about which the rotations for Type I Euler angles are performed (namely, about the directions $\mathbf{E}_3$, $\mathbf{p}_2$, and $\mathbf{q}_1$, respectively), Type I Euler angles are often called "3-2-1" Euler angles.

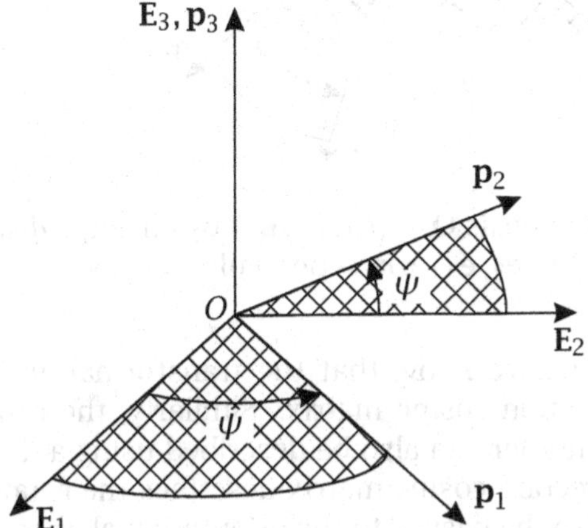

Figure 2-41 Rotation 1 of basis $\mathbf{E} = \{\mathbf{E}_1, \mathbf{E}_2, \mathbf{E}_3\}$ by an angle ψ about the $\mathbf{E}_3$-direction resulting in the basis $\mathbf{P} = \{\mathbf{p}_1, \mathbf{p}_2, \mathbf{p}_3\}$ for Type I Euler angles.

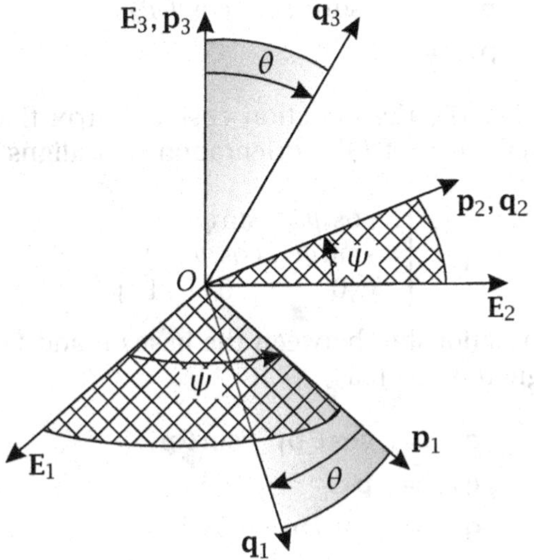

Figure 2-42 Rotation 2 of basis $\mathbf{P} = \{\mathbf{p}_1, \mathbf{p}_2, \mathbf{p}_3\}$ by an angle θ about the $\mathbf{p}_2$-direction resulting in the basis $\mathbf{Q} = \{\mathbf{q}_1, \mathbf{q}_2, \mathbf{q}_3\}$ for Type I Euler angles.

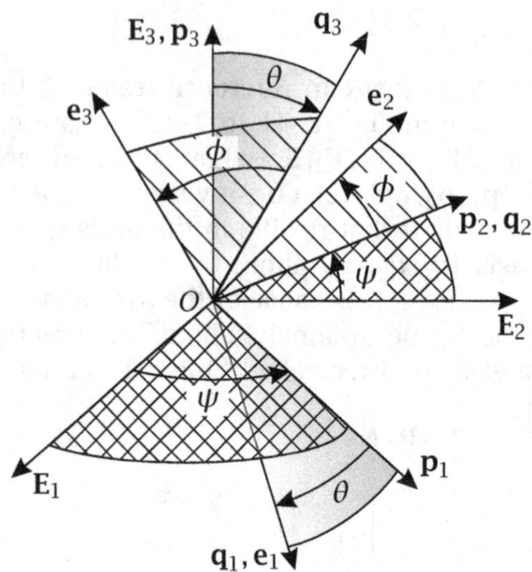

Figure 2–43 Rotation 3 of basis $\mathbf{Q} = \{\mathbf{q}_1, \mathbf{q}_2, \mathbf{q}_3\}$ by an angle ϕ about the $\mathbf{q}_1$-direction resulting in the basis $\mathbf{e} = \{\mathbf{e}_1, \mathbf{e}_2, \mathbf{e}_3\}$ for Type I Euler angles.

Now, from Section 1.4.4 we know that the transformation between any two orthonormal bases is a direction cosine matrix. Similarly, the rotation of a rigid body from one orientation to another can also be described using a direction cosine matrix. In this latter case, the direction cosine matrix describes the rotation of the body from the orientation aligned with the basis **E** to the orientation aligned with the basis **e**. This direction cosine matrix is obtained as follows. First, using Fig. 2–41, the relationship between the bases **E** and **P** resulting from the first rotation by the angle ψ is given as

$$\mathbf{p}_1 = \cos\psi\,\mathbf{E}_1 + \sin\psi\,\mathbf{E}_2 \tag{2–555}$$
$$\mathbf{p}_2 = -\sin\psi\,\mathbf{E}_1 + \cos\psi\,\mathbf{E}_2 \tag{2–556}$$
$$\mathbf{p}_3 = \mathbf{E}_3 \tag{2–557}$$

Then, using Eqs. (2–555)-(2–557), the direction cosine matrix that rotates the body from the orientation aligned with **E** to the orientation that aligns the body with **P** is given as

$$\{\mathbf{C}\}_{\mathrm{E}}^{\mathrm{P}} = \left\{\begin{array}{ccc} \cos\psi & \sin\psi & 0 \\ -\sin\psi & \cos\psi & 0 \\ 0 & 0 & 1 \end{array}\right\} \tag{2–558}$$

Next, using Fig. 2–42, the relationship between the bases **P** and **Q** resulting from the second rotation by the angle θ is given as

$$\mathbf{q}_1 = \cos\theta\,\mathbf{p}_1 - \sin\theta\,\mathbf{p}_3 \tag{2–559}$$
$$\mathbf{q}_2 = \mathbf{p}_2 \tag{2–560}$$
$$\mathbf{q}_3 = \sin\theta\,\mathbf{p}_1 + \cos\theta\,\mathbf{p}_3 \tag{2–561}$$

Then, using Eqs. (2–559)-(2–561), the direction cosine matrix that rotates the body from the orientation aligned with **P** to the orientation that aligns the body with **Q** is given as

$$\{\mathbf{C}\}_{\mathbf{P}}^{\mathbf{Q}} = \begin{Bmatrix} \cos\theta & 0 & -\sin\theta \\ 0 & 1 & 0 \\ \sin\theta & 0 & \cos\theta \end{Bmatrix} \tag{2-562}$$

Finally, using Fig. 2-43, the relationship between the bases **Q** and **e** resulting from the third rotation by ϕ is given as

$$\mathbf{e}_1 = \mathbf{q}_1 \tag{2-563}$$
$$\mathbf{e}_2 = \cos\phi\mathbf{q}_2 + \sin\phi\mathbf{q}_3 \tag{2-564}$$
$$\mathbf{e}_3 = -\sin\phi\mathbf{q}_2 + \cos\phi\mathbf{q}_3 \tag{2-565}$$

Then, using Eqs. (2-563)-(2-565), the direction cosine matrix that rotates the body from the orientation aligned with **Q** to the orientation that aligns the body with **e** is given as

$$\{\mathbf{C}\}_{\mathbf{Q}}^{\mathbf{e}} = \begin{Bmatrix} 1 & 0 & 0 \\ 0 & \cos\phi & \sin\phi \\ 0 & -\sin\phi & \cos\phi \end{Bmatrix} \tag{2-566}$$

Using Eqs. (2-558), (2-562), and (2-566), the rotation of the body from the initial orientation aligned with the basis **E** to the final orientation aligned with the basis **e** is obtained by multiplying the three direction cosine matrices $\{\mathbf{C}\}_{\mathbf{E}}^{\mathbf{P}}$, $\{\mathbf{C}\}_{\mathbf{P}}^{\mathbf{Q}}$, and $\{\mathbf{C}\}_{\mathbf{Q}}^{\mathbf{e}}$ as

$$\{\mathbf{C}\}_{\mathbf{E}}^{\mathbf{e}} = \{\mathbf{C}\}_{\mathbf{Q}}^{\mathbf{e}}\{\mathbf{C}\}_{\mathbf{P}}^{\mathbf{Q}}\{\mathbf{C}\}_{\mathbf{E}}^{\mathbf{P}} \tag{2-567}$$

Now, because the direction cosine matrices $\{\mathbf{C}\}_{\mathbf{E}}^{\mathbf{P}}$, $\{\mathbf{C}\}_{\mathbf{P}}^{\mathbf{Q}}$, and $\{\mathbf{C}\}_{\mathbf{Q}}^{\mathbf{e}}$ correspond to rotations by the angles ψ, θ, and ϕ, respectively, we can write

$$\begin{aligned} \{\mathbf{C}\}_{\mathbf{E}}^{\mathbf{P}} &= \{\mathbf{C}\}_{\psi} \\ \{\mathbf{C}\}_{\mathbf{P}}^{\mathbf{Q}} &= \{\mathbf{C}\}_{\theta} \\ \{\mathbf{C}\}_{\mathbf{Q}}^{\mathbf{e}} &= \{\mathbf{C}\}_{\phi} \end{aligned} \tag{2-568}$$

Using the expressions in Eq. (2-568), the direction cosine matrix that rotates the rigid body from the initial orientation aligned with basis **E** to the final orientation aligned with the basis **e** is given as

$$\{\mathbf{C}\}_{\mathbf{E}}^{\mathbf{e}} = \{\mathbf{C}\}_{\phi}\{\mathbf{C}\}_{\theta}\{\mathbf{C}\}_{\psi} \tag{2-569}$$

Finally, using the expressions for $\{\mathbf{C}\}_{\mathbf{E}}^{\mathbf{P}}$, $\{\mathbf{C}\}_{\mathbf{P}}^{\mathbf{Q}}$, and $\{\mathbf{C}\}_{\mathbf{Q}}^{\mathbf{e}}$ from Eqs. (2-558), (2-562), and (2-566), we obtain the direction cosine matrix from **E** to **e** as

$$\{\mathbf{C}\}_{\mathbf{E}}^{\mathbf{e}} = \begin{Bmatrix} \cos\theta\cos\psi & \cos\theta\sin\psi & -\sin\theta \\ \begin{matrix}(-\cos\phi\sin\psi \\ +\sin\phi\sin\theta\cos\psi)\end{matrix} & \begin{matrix}(\cos\phi\cos\psi \\ +\sin\phi\sin\theta\sin\psi)\end{matrix} & \sin\phi\cos\theta \\ \begin{matrix}(\sin\phi\sin\psi \\ +\cos\phi\sin\theta\cos\psi)\end{matrix} & \begin{matrix}(-\sin\phi\cos\psi \\ +\cos\phi\sin\theta\sin\psi)\end{matrix} & \cos\phi\cos\theta \end{Bmatrix} \tag{2-570}$$

For completeness, it is noted that the direction cosine *tensor* that rotates the rigid body from the orientation aligned with the basis $\mathbf{E}$ to the basis aligned with $\mathbf{e}$ is given as

$$\begin{aligned}\mathbf{C} = {} & \cos\theta\cos\psi\,\mathbf{e}_1\otimes\mathbf{E}_1 + \cos\theta\sin\psi\,\mathbf{e}_1\otimes\mathbf{E}_2 - \sin\theta\,\mathbf{e}_1\otimes\mathbf{E}_3 \\ & + (-\cos\phi\sin\psi + \sin\phi\sin\theta\cos\psi)\mathbf{e}_2\otimes\mathbf{E}_1 \\ & + (\cos\phi\cos\psi + \sin\phi\sin\theta\sin\psi)\mathbf{e}_2\otimes\mathbf{E}_2 \\ & + \sin\phi\cos\theta\,\mathbf{e}_2\otimes\mathbf{E}_3 \\ & + (\sin\phi\sin\psi + \cos\phi\sin\theta\cos\psi)\mathbf{e}_3\otimes\mathbf{E}_1 \\ & + (-\sin\phi\cos\psi + \cos\phi\sin\theta\sin\psi)\mathbf{e}_3\otimes\mathbf{E}_2 \\ & + \cos\phi\cos\theta\,\mathbf{e}_3\otimes\mathbf{E}_3\end{aligned} \tag{2-571}$$

Now we can obtain an expression for the angular velocity of the rigid body $\mathcal{R}$ in the reference frame $\mathcal{A}$ (i.e., ${}^{\mathcal{A}}\boldsymbol{\omega}^{\mathcal{R}}$) as follows. First, suppose that we let $\mathcal{P}$ denote the reference frame in which the basis $\mathbf{P}$ is fixed. Then, because the first Euler angle rotation is performed relative to the basis $\{\mathbf{E}_1, \mathbf{E}_2, \mathbf{E}_3\}$ and $\{\mathbf{E}_1, \mathbf{E}_2, \mathbf{E}_3\}$ is fixed in reference frame $\mathcal{A}$, the angular velocity of reference frame $\mathcal{P}$ in reference frame $\mathcal{A}$ is given as

$${}^{\mathcal{A}}\boldsymbol{\omega}^{\mathcal{P}} = \dot{\psi}\mathbf{E}_3 = \dot{\psi}\mathbf{p}_3 \tag{2-572}$$

Next, suppose that we let $\mathcal{Q}$ denote the reference frame in which the basis $\mathbf{Q}$ is fixed. Then, because the second Euler angle rotation is performed relative to the basis $\mathbf{P}$ and $\mathbf{P}$ is fixed in reference frame $\mathcal{P}$, the angular velocity of reference frame $\mathcal{Q}$ in reference frame $\mathcal{P}$ is given as

$${}^{\mathcal{P}}\boldsymbol{\omega}^{\mathcal{Q}} = \dot{\theta}\mathbf{p}_2 = \dot{\theta}\mathbf{q}_2 \tag{2-573}$$

Finally, observing that the third rotation is performed relative to the basis $\mathbf{Q}$ and $\mathbf{Q}$ is fixed in reference frame $\mathcal{Q}$, the angular velocity of the rigid body $\mathcal{R}$ in the reference frame $\mathcal{Q}$ is given as

$${}^{\mathcal{Q}}\boldsymbol{\omega}^{\mathcal{R}} = \dot{\phi}\mathbf{q}_1 = \dot{\phi}\mathbf{e}_1 \tag{2-574}$$

Then, applying the angular velocity addition theorem to the expressions for ${}^{\mathcal{A}}\boldsymbol{\omega}^{\mathcal{Q}}$, ${}^{\mathcal{Q}}\boldsymbol{\omega}^{\mathcal{P}}$, and ${}^{\mathcal{Q}}\boldsymbol{\omega}^{\mathcal{R}}$ as given, respectively, in Eqs. (2–572), (2–573), and (2–574), we obtain ${}^{\mathcal{A}}\boldsymbol{\omega}^{\mathcal{R}}$ as

$${}^{\mathcal{A}}\boldsymbol{\omega}^{\mathcal{R}} = {}^{\mathcal{A}}\boldsymbol{\omega}^{\mathcal{P}} + {}^{\mathcal{P}}\boldsymbol{\omega}^{\mathcal{Q}} + {}^{\mathcal{Q}}\boldsymbol{\omega}^{\mathcal{R}} = \dot{\psi}\mathbf{p}_3 + \dot{\theta}\mathbf{q}_2 + \dot{\phi}\mathbf{e}_1 \tag{2-575}$$

Now we can obtain an expression for ${}^{\mathcal{A}}\boldsymbol{\omega}^{\mathcal{R}}$ purely in terms of the body-fixed basis $\mathbf{e}$ as follows. First, solving Eqs. (2–559) and (2–561) simultaneously for $\mathbf{p}_3$, we obtain

$$\mathbf{p}_3 = -\sin\theta\,\mathbf{q}_1 + \cos\theta\,\mathbf{q}_3 \tag{2-576}$$

Next, solving Eqs. (2–564) and (2–565) simultaneously for $\mathbf{q}_3$, we obtain

$$\mathbf{q}_3 = \sin\phi\mathbf{e}_2 + \cos\phi\mathbf{e}_3 \tag{2-577}$$

Substituting $\mathbf{q}_3$ from Eq. (2–577) into Eq. (2–576), we obtain

$$\mathbf{p}_3 = -\sin\theta\,\mathbf{q}_1 + \cos\theta\,(\sin\phi\mathbf{e}_2 + \cos\phi\mathbf{e}_3) \tag{2-578}$$

Then, using the fact that $\mathbf{q}_1 = \mathbf{e}_1$, Eq. (2–578) simplifies to

$$\mathbf{p}_3 = -\sin\theta\,\mathbf{e}_1 + \cos\theta\sin\phi\mathbf{e}_2 + \cos\theta\cos\phi\mathbf{e}_3 \tag{2-579}$$

Furthermore, solving Eqs. (2-564) and (2-565) for $\mathbf{q}_2$, we obtain

$$\mathbf{q}_2 = \cos\phi\mathbf{e}_2 - \sin\phi\mathbf{e}_3 \tag{2-580}$$

Substituting the results from Eqs. (2-579) and (2-580) into Eq. (2-575), we obtain

$$^{\mathcal{A}}\boldsymbol{\omega}^{\mathcal{R}} = \dot{\psi}(-\sin\theta\,\mathbf{e}_1+\cos\theta\,\sin\phi\mathbf{e}_2+\cos\theta\,\cos\phi\mathbf{e}_3)+\dot{\theta}(\cos\phi\mathbf{e}_2-\sin\phi\mathbf{e}_3)+\dot{\phi}\mathbf{e}_1 \tag{2-581}$$

Simplifying Eq. (2-581), we obtain the angular velocity of the rigid body in reference frame $\mathcal{A}$ as

$$\boxed{\begin{aligned} ^{\mathcal{A}}\boldsymbol{\omega}^{\mathcal{R}} = {} & (\dot{\phi} - \dot{\psi}\sin\theta)\mathbf{e}_1 + (\dot{\psi}\cos\theta\,\sin\phi + \dot{\theta}\cos\phi)\mathbf{e}_2 \\ & + (\dot{\psi}\cos\theta\,\cos\phi - \dot{\theta}\sin\phi)\mathbf{e}_3 \end{aligned}} \tag{2-582}$$

Then, applying the definition of the angular acceleration of a reference frame from Eq. (2-141), we have

$$^{\mathcal{A}}\boldsymbol{\alpha}^{\mathcal{R}} = \frac{^{\mathcal{A}}d}{dt}\left(^{\mathcal{A}}\boldsymbol{\omega}^{\mathcal{R}}\right) = \frac{^{\mathcal{R}}d}{dt}\left(^{\mathcal{A}}\boldsymbol{\omega}^{\mathcal{R}}\right) \tag{2-583}$$

Because the expression for $^{\mathcal{A}}\boldsymbol{\omega}^{\mathcal{R}}$ in Eq. (2-582) is expressed in a body-fixed basis (i.e., a basis fixed in $\mathcal{R}$), it is most convenient to compute the rate of change of $^{\mathcal{A}}\boldsymbol{\omega}^{\mathcal{R}}$ in reference frame $\mathcal{R}$ as

$$^{\mathcal{A}}\boldsymbol{\alpha}^{\mathcal{R}} = \frac{^{\mathcal{R}}d}{dt}\left(^{\mathcal{A}}\boldsymbol{\omega}^{\mathcal{R}}\right) \tag{2-584}$$

Suppose now that, consistent with Eq. (2-553), we let

$$\omega_1 = \dot{\phi} - \dot{\psi}\sin\theta \tag{2-585}$$

$$\omega_2 = \dot{\psi}\cos\theta\,\sin\phi + \dot{\theta}\cos\phi \tag{2-586}$$

$$\omega_3 = \dot{\psi}\cos\theta\,\cos\phi - \dot{\theta}\sin\phi \tag{2-587}$$

Now from the perspective of an observer fixed to the rigid body, the basis vectors $\mathbf{e}_1$, $\mathbf{e}_2$ and $\mathbf{e}_3$ appear to be fixed. Consequently, we have

$$^{\mathcal{A}}\boldsymbol{\alpha}^{\mathcal{R}} = \dot{\omega}_1\mathbf{e}_1 + \dot{\omega}_2\mathbf{e}_2 + \dot{\omega}_3\mathbf{e}_3 \tag{2-588}$$

Computing the rates of change of ω_1, ω_2, and ω_3 in Eqs. (2-585)-(2-587), gives

$$\dot{\omega}_1 = \ddot{\phi} - \ddot{\psi}\sin\theta - \dot{\psi}\dot{\theta}\cos\theta \tag{2-589}$$

$$\begin{aligned} \dot{\omega}_2 = {} & \ddot{\psi}\cos\theta\,\sin\phi + \dot{\psi}(-\dot{\theta}\sin\theta\,\sin\phi + \dot{\phi}\cos\theta\,\cos\phi) \\ & +\ddot{\theta}\cos\phi - \dot{\theta}\dot{\phi}\sin\phi \end{aligned} \tag{2-590}$$

$$\begin{aligned} \dot{\omega}_3 = {} & \ddot{\psi}\cos\theta\,\cos\phi - \dot{\psi}(\dot{\theta}\sin\theta\,\cos\phi + \dot{\phi}\cos\theta\,\sin\phi) \\ & -\ddot{\theta}\sin\phi - \dot{\theta}\dot{\phi}\cos\phi \end{aligned} \tag{2-591}$$

The angular acceleration of the rigid body in reference frame $\mathcal{A}$ is then given as

$$\boxed{\begin{aligned} ^{\mathcal{A}}\boldsymbol{\alpha}^{\mathcal{R}} = {} & (\ddot{\phi} - \ddot{\psi}\sin\theta - \dot{\psi}\dot{\theta}\cos\theta)\mathbf{e}_1 \\ & + \Big[\ddot{\psi}\cos\theta\,\sin\phi + \dot{\psi}(-\dot{\theta}\sin\theta\,\sin\phi + \dot{\phi}\cos\theta\,\cos\phi) \\ & \quad + \ddot{\theta}\cos\phi - \dot{\theta}\dot{\phi}\sin\phi)\Big]\mathbf{e}_2 \\ & + \Big[\ddot{\psi}\cos\theta\,\cos\phi - \dot{\psi}(\dot{\theta}\sin\theta\,\cos\phi + \dot{\phi}\cos\theta\,\sin\phi) \\ & \quad - \ddot{\theta}\sin\phi - \dot{\theta}\dot{\phi}\cos\phi\Big]\mathbf{e}_3 \end{aligned}} \tag{2-592}$$

Assume now that the Euler angles ψ, θ, and ϕ are limited to the ranges

$$\begin{aligned} 0 &\le \psi \le 2\pi \\ -\pi/2 &\le \theta \le \pi/2 \\ 0 &\le \phi \le 2\pi \end{aligned} \tag{2-593}$$

While any orientation of the rigid body $\mathcal{R}$ relative to the arbitrary reference frame $\mathcal{A}$ can be obtained using the ranges as prescribed in Eq. (2–593), for the orientations where $\theta = \pm\pi/2$ (i.e., the $\mathbf{e}_1$-direction is vertical) no unique values exist for ψ and ϕ. Hence, for $\theta = \pm\pi/2$, the angles ψ and ϕ are undefined. However, it can be shown for $\theta = \pi/2$ that the angle $\psi - \phi$ is well defined while for $\theta = -\pi/2$, the angle $\psi + \phi$ is well defined. The orientations where $\theta = \pm\pi/2$ are called *gimbal lock*. During gimbal lock, the component of ${}^{\mathcal{A}}\boldsymbol{\omega}^{\mathcal{R}}$ in the direction of $\mathbf{p}_1$ is undefined.

Euler Basis and Dual Euler Basis

As seen in this section, Euler angles provide a way to describe the orientation of a rigid body $\mathcal{R}$ relative to an arbitrary reference frame $\mathcal{A}$. Now, it is important to understand that the sequence of rotations is *not* commutative, i.e., two different orders of rotation will lead to different orientations of the rigid body. As a result, every intermediate orientation depends on the order of all previous rotations (e.g., the orientation of the rigid body obtained after applying the first two rotations depends on the order of the first of these two rotations). Furthermore, it is important to understand that the directions about which the rotations are performed *do not* form a mutually orthogonal basis. More specifically, we know that the directions of the three rotations in the Type I Euler angle sequence are $\mathbf{E}_3 = \mathbf{p}_3$, $\mathbf{p}_2 = \mathbf{q}_2$, and $\mathbf{q}_1 = \mathbf{e}_1$. Suppose now that we let

$$\mathbf{k}_1 = \mathbf{p}_3 \tag{2-594}$$

$$\mathbf{k}_2 = \mathbf{q}_2 \tag{2-595}$$

$$\mathbf{k}_3 = \mathbf{e}_1 \tag{2-596}$$

The (nonorthogonal) basis $\mathbf{K} = \{\mathbf{k}_1, \mathbf{k}_2, \mathbf{k}_3\}$ is called the *Euler basis* corresponding to Type I Euler angles. It is important to note that an Euler basis is specific to the Euler angle sequence, i.e., there exists a different Euler basis for every Euler angle convention. We note that the Euler basis can be expressed in the body-fixed basis $\mathbf{e} = \{\mathbf{e}_1, \mathbf{e}_2, \mathbf{e}_3\}$ as

$$\mathbf{k}_1 = -\sin\theta\,\mathbf{e}_1 + \sin\phi\cos\theta\,\mathbf{e}_2 + \cos\phi\cos\theta\,\mathbf{e}_3 \tag{2-597}$$

$$\mathbf{k}_2 = \cos\phi\mathbf{e}_2 - \sin\phi\mathbf{e}_3 \tag{2-598}$$

$$\mathbf{k}_3 = \mathbf{e}_1 \tag{2-599}$$

Using the expression for ${}^{\mathcal{A}}\boldsymbol{\omega}^{\mathcal{R}}$ as given in Eq. (2–575), the angular velocity of the rigid body in an arbitrary reference frame $\mathcal{A}$ can be written in terms of the (Type I) Euler basis as

$${}^{\mathcal{A}}\boldsymbol{\omega}^{\mathcal{R}} = \dot{\psi}\mathbf{k}_1 + \dot{\theta}\mathbf{k}_2 + \dot{\phi}\mathbf{k}_3 \tag{2-600}$$

Suppose now that we define a basis $\mathbf{K}^* = \{\mathbf{k}_1^*, \mathbf{k}_2^*, \mathbf{k}_3^*\}$ that satisfies the following property:

$$\mathbf{k}_i^* \cdot \mathbf{k}_j = \begin{cases} 1 & , \quad i = j \\ 0 & , \quad i \neq j \end{cases} \quad (i, j = 1, 2, 3) \tag{2-601}$$

The basis $\mathbf{K}^*$ is called the *dual Euler basis*[11] (O'Reilly, 2004). Furthermore, suppose we choose to express the dual Euler basis $\mathbf{K}^*$ in terms of the body-fixed basis $\{\mathbf{e}_1, \mathbf{e}_2, \mathbf{e}_3\}$. We then have

$$\mathbf{k}_1^* = k_{11}\mathbf{e}_1 + k_{12}\mathbf{e}_2 + k_{13}\mathbf{e}_3 \qquad (2\text{-}602)$$

$$\mathbf{k}_2^* = k_{21}\mathbf{e}_1 + k_{22}\mathbf{e}_2 + k_{33}\mathbf{e}_3 \qquad (2\text{-}603)$$

$$\mathbf{k}_3^* = k_{31}\mathbf{e}_1 + k_{32}\mathbf{e}_2 + k_{33}\mathbf{e}_3 \qquad (2\text{-}604)$$

where the coefficients k_{ij} $(i, j = 1, 2, 3)$ have to be determined. We can determine these coefficients as follows. Taking the scalar product of each vector in the basis $\mathbf{K}^*$ with each vector in the basis $\mathbf{K}$ and using the property of Eq. (2-601), we obtain the following relationships:

$$\begin{array}{lcl}
\mathbf{k}_1^* \cdot \mathbf{k}_1 = -k_{11}\sin\theta + k_{12}\sin\phi\cos\theta + k_{13}\cos\phi\cos\theta & = & 1 \\
\mathbf{k}_1^* \cdot \mathbf{k}_2 = k_{12}\cos\phi - k_{13}\sin\phi & = & 0 \\
\mathbf{k}_1^* \cdot \mathbf{k}_3 = k_{11} & = & 0 \\
\mathbf{k}_2^* \cdot \mathbf{k}_1 = -k_{21}\sin\theta + k_{22}\sin\phi\cos\theta + k_{23}\cos\phi\cos\theta & = & 0 \\
\mathbf{k}_2^* \cdot \mathbf{k}_2 = k_{22}\cos\phi - k_{23}\sin\phi & = & 1 \\
\mathbf{k}_2^* \cdot \mathbf{k}_3 = k_{21} & = & 0 \\
\mathbf{k}_3^* \cdot \mathbf{k}_1 = -k_{31}\sin\theta + k_{32}\sin\phi\cos\theta + k_{33}\cos\phi\cos\theta & = & 0 \\
\mathbf{k}_3^* \cdot \mathbf{k}_2 = k_{32}\cos\phi - k_{33}\sin\phi & = & 0 \\
\mathbf{k}_3^* \cdot \mathbf{k}_3 = k_{31} & = & 1
\end{array} \qquad (2\text{-}605)$$

Equation (2-605) can be written in matrix form as

$$\left\{\begin{array}{ccc} -\sin\theta & \sin\phi\cos\theta & \cos\phi\cos\theta \\ 0 & \cos\phi & -\sin\phi \\ 1 & 0 & 0 \end{array}\right\}\left\{\begin{array}{ccc} k_{11} & k_{21} & k_{31} \\ k_{12} & k_{22} & k_{32} \\ k_{13} & k_{23} & k_{33} \end{array}\right\} = \left\{\begin{array}{ccc} 1 & 0 & 0 \\ 0 & 1 & 0 \\ 0 & 0 & 1 \end{array}\right\} \qquad (2\text{-}606)$$

Now suppose we let

$$\{\mathbf{K}\}_{\mathbf{e}} = \left\{\begin{array}{ccc} -\sin\theta & \sin\phi\cos\theta & \cos\phi\cos\theta \\ 0 & \cos\phi & -\sin\phi \\ 1 & 0 & 0 \end{array}\right\} \qquad (2\text{-}607)$$

where $\{\mathbf{K}\}_{\mathbf{e}}$ is the matrix representation of the Euler basis in the body-fixed basis $\mathbf{e}$. Multiplying both sides of Eq. (2-606) by $\{\mathbf{K}\}_{\mathbf{e}}^{-1}$, we obtain

$$\left\{\begin{array}{ccc} k_{11} & k_{21} & k_{31} \\ k_{12} & k_{22} & k_{32} \\ k_{13} & k_{23} & k_{33} \end{array}\right\} = \{\mathbf{K}\}_{\mathbf{e}}^{-1} = \left\{\begin{array}{ccc} 0 & 0 & 1 \\ \sin\phi\sec\theta & \cos\phi & \sin\phi\tan\theta \\ \cos\phi\sec\theta & -\sin\phi & \cos\phi\tan\theta \end{array}\right\} \qquad (2\text{-}608)$$

Alternatively, we have

$$\{\mathbf{K}\}_{\mathbf{e}}^{-T} = \left\{\begin{array}{ccc} k_{11} & k_{12} & k_{13} \\ k_{21} & k_{22} & k_{23} \\ k_{31} & k_{32} & k_{33} \end{array}\right\} = \left\{\begin{array}{ccc} 0 & \sin\phi\sec\theta & \cos\phi\sec\theta \\ 0 & \cos\phi & -\sin\phi \\ 1 & \sin\phi\tan\theta & \cos\phi\tan\theta \end{array}\right\} \qquad (2\text{-}609)$$

[11]Dr. Oliver M. O'Reilly is credited as the one who discovered the dual Euler basis (O'Reilly, 2004). The author gratefully acknowledges Dr. O'Reilly for his help in becoming aware of and gaining insight into the importance of the dual Euler basis.

where the notation $\{\cdot\}^{-T}$ means inverse transpose. Using the coefficients k_{ij} $(i, j = 1, 2, 3)$ from either Eq. (2–608) or (2–609), the dual Euler basis vectors are given in terms of the body-fixed basis $\mathbf{e} = \{\mathbf{e}_1, \mathbf{e}_2, \mathbf{e}_3\}$ as

$$\mathbf{k}_1^* = \sin\phi \sec\theta \, \mathbf{e}_2 + \cos\phi \sec\theta \, \mathbf{e}_3 \tag{2–610}$$

$$\mathbf{k}_2^* = \cos\phi \mathbf{e}_2 - \sin\phi \mathbf{e}_3 \tag{2–611}$$

$$\mathbf{k}_3^* = \mathbf{e}_1 + \sin\phi \tan\theta \, \mathbf{e}_2 + \cos\phi \tan\theta \, \mathbf{e}_3 \tag{2–612}$$

It is noted that the Euler basis and the dual Euler basis will become important in Chapter 5 when describing a conservative pure torque.

Example 2–12

A circular disk of radius r rolls without slip along a fixed horizontal surface as shown in Fig. 2–44. Using the angles ψ, θ, and ϕ to describe the orientation of the disk relative to the surface, determine the velocity and acceleration of the center of the disk as viewed by an observer fixed to the surface.

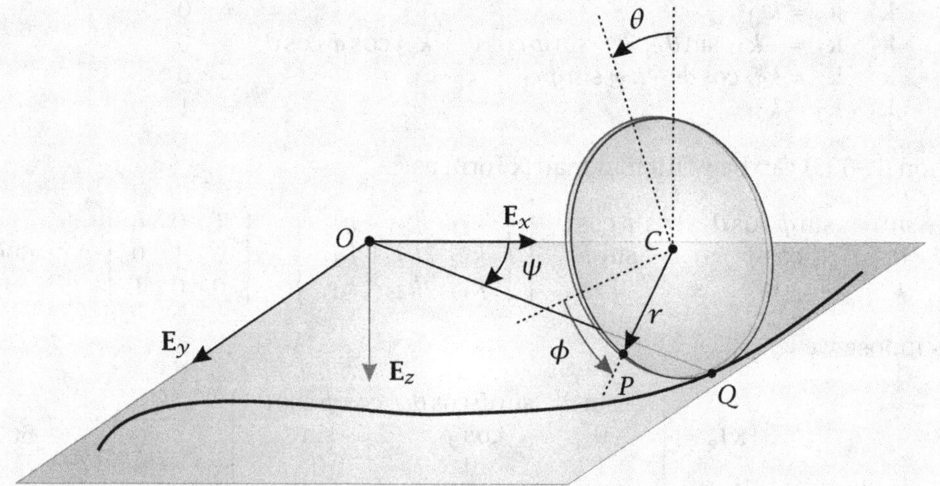

Figure 2–44 Disk rolling on horizontal surface.

Solution to Example 2–12

First, let S be the reference frame of the surface. Then choose the following coordinate system fixed in reference frame S:

Origin at O		
$\mathbf{E}_x$	=	As given
$\mathbf{E}_y$	=	As given
$\mathbf{E}_z$	=	As given

Next, let $\mathcal{P}$ be a reference frame fixed to the direction of OQ. Then choose the following coordinate system fixed in reference frame $\mathcal{P}$:

$$\begin{array}{ccc} & \text{Origin at } O & \\ \mathbf{p}_1 & = & \text{Along } OQ \\ \mathbf{p}_3 & = & \text{Along } \mathbf{E}_z \\ \mathbf{p}_2 & = & \mathbf{p}_3 \times \mathbf{p}_1 \end{array}$$

Next, let $\mathcal{Q}$ be a reference frame fixed to the direction of CQ. Then choose the following coordinate system fixed in reference frame $\mathcal{Q}$:

$$\begin{array}{ccc} & \text{Origin at } O & \\ \mathbf{q}_3 & = & \text{Along } CQ \\ \mathbf{q}_2 & = & \text{Along } \mathbf{p}_2 \\ \mathbf{q}_1 & = & \mathbf{q}_2 \times \mathbf{q}_3 \end{array}$$

Finally, let $\mathcal{R}$ be a reference frame fixed to the disk. Then, choose the following coordinate system fixed in reference frame $\mathcal{R}$:

$$\begin{array}{ccc} & \text{Origin at } O & \\ \mathbf{e}_1 & = & \mathbf{q}_1 \\ \mathbf{e}_2 & = & \text{Along } CP \\ \mathbf{e}_3 & = & \mathbf{q}_2 \times \mathbf{q}_3 \end{array}$$

The bases $\{\mathbf{p}_1, \mathbf{p}_2, \mathbf{p}_3\}$, $\{\mathbf{q}_1, \mathbf{q}_2, \mathbf{q}_3\}$ and $\{\mathbf{e}_1, \mathbf{e}_2, \mathbf{e}_3\}$ are shown in Fig. 2-45.

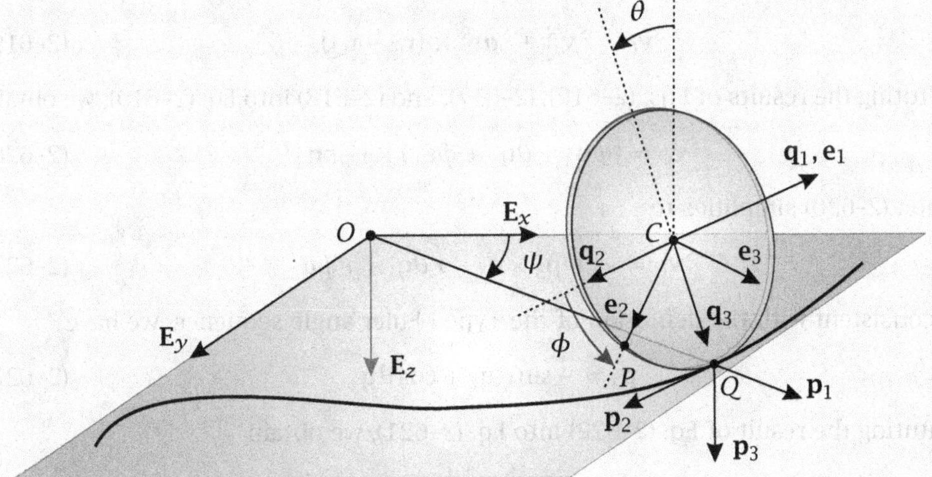

Figure 2-45 Geometry of bases $\{\mathbf{p}_1, \mathbf{p}_2, \mathbf{p}_3\}$, $\{\mathbf{q}_1, \mathbf{q}_2, \mathbf{q}_3\}$, and $\{\mathbf{e}_1, \mathbf{e}_2, \mathbf{e}_3\}$ for Example 2-12.

It is important to observe that the angles ψ, θ, and ϕ form a set of Type I Euler angles. The fact that $\{\psi, \theta, \phi\}$ is a Type I Euler angle set can be verified as follows. First, the angle ψ is measured from the $\mathbf{E}_x$-direction and corresponds to a rotation about the $\mathbf{E}_z$-direction (i.e., the "3" direction). Next, the angle θ is measured from the negative $\mathbf{p}_3$-direction and corresponds to a rotation about the $\mathbf{p}_2$-direction (i.e., the "2" direction). Finally, the angle ϕ is measured from the $\mathbf{q}_2$-direction and corresponds to a rotation about the $\mathbf{q}_1$-direction (i.e., the "1" direction). Therefore, the set $\{\psi, \theta, \phi\}$ is a "3-2-1" Euler angle sequence, i.e., $\{\psi, \theta, \phi\}$ is a set of Type I Euler angles.

Now, because $\{\psi, \theta, \phi\}$ is a set of Type I Euler angles, we have the following:

$$
\begin{aligned}
{}^{\mathcal{S}}\boldsymbol{\omega}^{\mathcal{P}} &= \dot{\psi}\mathbf{p}_3 && (2\text{-}613)\\
{}^{\mathcal{P}}\boldsymbol{\omega}^{\mathcal{Q}} &= \dot{\theta}\mathbf{q}_2 && (2\text{-}614)\\
{}^{\mathcal{Q}}\boldsymbol{\omega}^{\mathcal{R}} &= \dot{\phi}\mathbf{e}_1 && (2\text{-}615)
\end{aligned}
$$

where ${}^{\mathcal{S}}\boldsymbol{\omega}^{\mathcal{P}}$ is the angular velocity of $\mathcal{P}$ in $\mathcal{S}$, ${}^{\mathcal{P}}\boldsymbol{\omega}^{\mathcal{Q}}$ is the angular velocity of $\mathcal{Q}$ in $\mathcal{P}$, and ${}^{\mathcal{Q}}\boldsymbol{\omega}^{\mathcal{R}}$ is the angular velocity of $\mathcal{R}$ in $\mathcal{Q}$. The angular velocity of the disk as viewed by an observer fixed to the surface is then obtained from the angular velocity addition theorem as

$$
{}^{\mathcal{S}}\boldsymbol{\omega}^{\mathcal{R}} = {}^{\mathcal{S}}\boldsymbol{\omega}^{\mathcal{P}} + {}^{\mathcal{P}}\boldsymbol{\omega}^{\mathcal{Q}} + {}^{\mathcal{Q}}\boldsymbol{\omega}^{\mathcal{R}} = \dot{\psi}\mathbf{p}_3 + \dot{\theta}\mathbf{q}_2 + \dot{\phi}\mathbf{e}_1 \qquad (2\text{-}616)
$$

where we note that $\{\mathbf{p}_3, \mathbf{q}_2, \mathbf{e}_1\}$ is the Euler basis. Furthermore, the position of point C relative to point Q is given in terms of the basis $\{\mathbf{q}_1, \mathbf{q}_2, \mathbf{q}_3\}$ as

$$
\mathbf{r}_C - \mathbf{r}_Q = -r\mathbf{q}_3 \qquad (2\text{-}617)
$$

Now, because the disk rolls without slip along a fixed surface, we have

$$
{}^{\mathcal{S}}\mathbf{v}_Q^{\mathcal{R}} = \mathbf{0} \qquad (2\text{-}618)
$$

The velocity of the center of the disk as viewed by an observer fixed to the surface is then given from Eq. (2–517) as

$$
{}^{\mathcal{S}}\mathbf{v}_C = {}^{\mathcal{S}}\mathbf{v}_Q^{\mathcal{R}} + {}^{\mathcal{S}}\boldsymbol{\omega}^{\mathcal{R}} \times (\mathbf{r}_C - \mathbf{r}_Q) \qquad (2\text{-}619)
$$

Substituting the results of Eqs. (2–616), (2–617), and (2–618) into Eq. (2–619), we obtain

$$
{}^{\mathcal{S}}\mathbf{v}_C = (\dot{\psi}\mathbf{p}_3 + \dot{\theta}\mathbf{q}_2 + \dot{\phi}\mathbf{e}_1) \times (-r\mathbf{q}_3) \qquad (2\text{-}620)
$$

Equation (2–620) simplifies to

$$
{}^{\mathcal{S}}\mathbf{v}_C = -r\dot{\psi}\mathbf{p}_3 \times \mathbf{q}_3 - r\dot{\theta}\mathbf{q}_1 + r\dot{\phi}\mathbf{q}_2 \qquad (2\text{-}621)
$$

Now, consistent with the definition of the Type I Euler angle sequence, we have

$$
\mathbf{p}_3 = -\sin\theta\,\mathbf{q}_1 + \cos\theta\,\mathbf{q}_3 \qquad (2\text{-}622)
$$

Substituting the result of Eq. (2–622) into Eq. (2–621), we obtain

$$
\begin{aligned}
{}^{\mathcal{S}}\mathbf{v}_C &= -r\dot{\psi}(-\sin\theta\,\mathbf{q}_1 + \cos\theta\,\mathbf{q}_3) \times \mathbf{q}_3 - r\dot{\theta}\mathbf{q}_1 + r\dot{\phi}\mathbf{q}_2\\
&= -r\dot{\psi}\sin\theta\,\mathbf{q}_2 - r\dot{\theta}\mathbf{q}_1 + r\dot{\phi}\mathbf{q}_2
\end{aligned} \qquad (2\text{-}623)
$$

Rearranging Eq. (2–623), the velocity of the center of the disk as viewed by an observer fixed to the surface is given as

$$
{}^{\mathcal{S}}\mathbf{v}_C = -r\dot{\theta}\mathbf{q}_1 + (r\dot{\phi} - r\dot{\psi}\sin\theta)\mathbf{q}_2 \qquad (2\text{-}624)
$$

The acceleration of the center of the disk as viewed by an observer fixed to the surface is obtained as follows. First, we see from Eq. (2–624) that ${}^{\mathcal{S}}\mathbf{v}_C$ is expressed in the basis $\{\mathbf{q}_1, \mathbf{q}_2, \mathbf{q}_3\}$ and $\{\mathbf{q}_1, \mathbf{q}_2, \mathbf{q}_3\}$ is fixed in reference frame $\mathcal{Q}$. Consequently, it is most convenient to determine ${}^{\mathcal{S}}\mathbf{a}_C$ by applying the rate of change transport theorem between reference frames $\mathcal{Q}$ and $\mathcal{S}$, i.e.,

$$ {}^{\mathcal{S}}\mathbf{a}_C = \frac{{}^{\mathcal{S}}d}{dt}\left({}^{\mathcal{S}}\mathbf{v}_C\right) = \frac{{}^{\mathcal{Q}}d}{dt}\left({}^{\mathcal{S}}\mathbf{v}_C\right) + {}^{\mathcal{S}}\boldsymbol{\omega}^{\mathcal{Q}} \times {}^{\mathcal{S}}\mathbf{v}_C \tag{2-625} $$

Now the angular velocity of $\mathcal{Q}$ in $\mathcal{S}$ is given as

$$ {}^{\mathcal{S}}\boldsymbol{\omega}^{\mathcal{Q}} = {}^{\mathcal{S}}\boldsymbol{\omega}^{\mathcal{P}} + {}^{\mathcal{P}}\boldsymbol{\omega}^{\mathcal{Q}} = \dot{\psi}\mathbf{p}_3 + \dot{\theta}\mathbf{q}_2 \tag{2-626} $$

Next, we have

$$ \frac{{}^{\mathcal{Q}}d}{dt}\left({}^{\mathcal{S}}\mathbf{v}_C\right) = -r\ddot{\theta}\mathbf{q}_1 + (r\ddot{\phi} - r\ddot{\psi}\sin\theta - r\dot{\psi}\dot{\theta}\cos\theta)\mathbf{q}_2 \tag{2-627} $$

$$ {}^{\mathcal{S}}\boldsymbol{\omega}^{\mathcal{Q}} \times {}^{\mathcal{S}}\mathbf{v}_C = (\dot{\psi}\mathbf{p}_3 + \dot{\theta}\mathbf{q}_2) \times \left[-r\dot{\theta}\mathbf{q}_1 + (r\dot{\phi} - r\dot{\psi}\sin\theta)\mathbf{q}_2\right] \tag{2-628} $$

Computing the vector product in Eq. (2-628), we obtain

$$ {}^{\mathcal{S}}\boldsymbol{\omega}^{\mathcal{Q}} \times {}^{\mathcal{S}}\mathbf{v}_C = -r\dot{\psi}\dot{\theta}\mathbf{p}_3 \times \mathbf{q}_1 + r\dot{\psi}(\dot{\phi} - r\dot{\psi}\sin\theta)\mathbf{p}_3 \times \mathbf{q}_2 + r\dot{\theta}^2\mathbf{q}_3 \tag{2-629} $$

Using the expression for $\mathbf{p}_3$ from Eq. (2-622), the quantities $\mathbf{p}_3 \times \mathbf{q}_1$ and $\mathbf{p}_3 \times \mathbf{q}_2$ are given as

$$ \mathbf{p}_3 \times \mathbf{q}_1 = (-\sin\theta\,\mathbf{q}_1 + \cos\theta\,\mathbf{q}_3) \times \mathbf{q}_1 = \cos\theta\,\mathbf{q}_2 \tag{2-630} $$

$$ \mathbf{p}_3 \times \mathbf{q}_2 = (-\sin\theta\,\mathbf{q}_1 + \cos\theta\,\mathbf{q}_3) \times \mathbf{q}_2 = -\sin\theta\,\mathbf{q}_3 - \cos\theta\,\mathbf{q}_1 \tag{2-631} $$

Substituting the results of Eqs. (2-630) and (2-631) into (2-629) gives

$$ {}^{\mathcal{S}}\boldsymbol{\omega}^{\mathcal{Q}} \times {}^{\mathcal{S}}\mathbf{v}_C = -r\dot{\psi}\dot{\theta}\cos\theta\,\mathbf{q}_2 - r\dot{\psi}(\dot{\phi} - \dot{\psi}\sin\theta)(\sin\theta\,\mathbf{q}_3 + \cos\theta\,\mathbf{q}_1) + r\dot{\theta}^2\mathbf{q}_3 \tag{2-632} $$

Simplifying Eq. (2-632), we obtain

$$ \begin{aligned} {}^{\mathcal{S}}\boldsymbol{\omega}^{\mathcal{Q}} \times {}^{\mathcal{S}}\mathbf{v}_C = & -r\dot{\psi}(\dot{\phi} - \dot{\psi}\sin\theta)\cos\theta\,\mathbf{q}_1 - r\dot{\psi}\dot{\theta}\cos\theta\,\mathbf{q}_2 \\ & + \left[r\dot{\theta}^2 - r\dot{\psi}(\dot{\phi} - r\dot{\psi}\sin\theta)\sin\theta\right]\mathbf{q}_3 \end{aligned} \tag{2-633} $$

Adding Eqs. (2-627) and (2-633), the acceleration of the center of the disk as viewed by an observer fixed to the surface is given as

$$ \begin{aligned} {}^{\mathcal{S}}\mathbf{a}_C = & -\left[r\ddot{\theta} + r\dot{\psi}(\dot{\phi} - \dot{\psi}\sin\theta)\cos\theta\right]\mathbf{q}_1 \\ & + (r\ddot{\phi} - r\ddot{\psi}\sin\theta - 2r\dot{\psi}\dot{\theta}\cos\theta)\mathbf{q}_2 \\ & + \left[r\dot{\theta}^2 - r\dot{\psi}(\dot{\phi} - \dot{\psi}\sin\theta)\sin\theta\right]\mathbf{q}_3 \end{aligned} \tag{2-634} $$

■

Summary of Chapter 2

This chapter was devoted to developing a framework for determining the motion of a particle or a rigid body without regard to the forces that cause the motion. We began by defining a *reference frame* $\mathcal{A}$ as a collection of points such that the distance between any two points in $\mathcal{A}$ does not change with time. We then defined the concept of an *observer* in reference frame $\mathcal{A}$ as any device rigidly attached to reference frame $\mathcal{A}$ that makes observations in a particular reference frame. Next, we stated as assumptions of Newtonian mechanics that observations of time and space are the same for observers in all reference frames. Given the definition of a reference frame and the assumptions of the equivalence of space and time in different reference frames, it was discussed that, while observations of a vector are the same in different reference frames, observations of the rate of change of a vector in two different reference frames $\mathcal{A}$ and $\mathcal{B}$ are *not* the same, i.e., in general it is the case that

$$\frac{{}^{\mathcal{A}}d\mathbf{b}}{dt} \neq \frac{{}^{\mathcal{B}}d\mathbf{b}}{dt}$$

Then, given a reference frame $\mathcal{B}$ that rotates with angular velocity ${}^{\mathcal{A}}\boldsymbol{\omega}^{\mathcal{B}}$ relative to a reference frame $\mathcal{A}$, it was shown that the rate of change of an arbitrary vector $\mathbf{b}$ as viewed by an observer in $\mathcal{A}$ is given as

$$\frac{{}^{\mathcal{A}}d\mathbf{b}}{dt} = \frac{{}^{\mathcal{B}}d\mathbf{b}}{dt} + {}^{\mathcal{A}}\boldsymbol{\omega}^{\mathcal{B}} \times \mathbf{b} \tag{2-128}$$

Equation (2-128) was referred to as the *rate of change transport theorem* or, more simply, the *transport theorem*, and provides a mechanism for determining the rate of change of a vector in a desired reference frame given observations of the rate of change of that same vector in another reference frame. Using the rate of change transport theorem, the velocity and acceleration of a point P as viewed by an observer in an arbitrary reference frame $\mathcal{A}$ were determined, respectively, as

$${}^{\mathcal{A}}\mathbf{v} = \frac{{}^{\mathcal{B}}d\mathbf{r}}{dt} + {}^{\mathcal{A}}\boldsymbol{\omega}^{\mathcal{B}} \times \mathbf{r} \tag{2-151}$$

and

$${}^{\mathcal{A}}\mathbf{a} = {}^{\mathcal{B}}\mathbf{a} + {}^{\mathcal{A}}\boldsymbol{\alpha}^{\mathcal{B}} \times \mathbf{r} + 2\,{}^{\mathcal{A}}\boldsymbol{\omega}^{\mathcal{B}} \times {}^{\mathcal{B}}\mathbf{v} + {}^{\mathcal{A}}\boldsymbol{\omega}^{\mathcal{B}} \times \left({}^{\mathcal{A}}\boldsymbol{\omega}^{\mathcal{B}} \times \mathbf{r}\right) \tag{2-158}$$

where ${}^{\mathcal{A}}\boldsymbol{\alpha}^{\mathcal{B}}$ is the *angular acceleration* of reference frame $\mathcal{B}$ as viewed by an observer in reference frame $\mathcal{A}$.

Using the general kinematic results for velocity and acceleration in a reference frame $\mathcal{A}$, the particular representations of velocity and acceleration for several basic coordinate systems were derived. In particular, expressions for velocity and acceleration were derived in terms of Cartesian, cylindrical, spherical, and intrinsic bases. For a Cartesian basis $\{\mathbf{e}_x, \mathbf{e}_y, \mathbf{e}_z\}$ attached to a reference frame $\mathcal{A}$, the velocity and acceleration were given, respectively, as

$${}^{\mathcal{A}}\mathbf{v} = \dot{x}\mathbf{e}_x + \dot{y}\mathbf{e}_y + \dot{z}\mathbf{e}_z \tag{2-160}$$

and

$${}^{\mathcal{A}}\mathbf{a} = \ddot{x}\mathbf{e}_x + \ddot{y}\mathbf{e}_y + \ddot{z}\mathbf{e}_z \tag{2-161}$$

For a cylindrical basis $\{\mathbf{e}_r, \mathbf{e}_\theta, \mathbf{e}_z\}$ attached to reference frame $\mathcal{B}$, the angular velocity of reference frame $\mathcal{B}$ in reference frame $\mathcal{A}$ was derived as

$$^{\mathcal{A}}\boldsymbol{\omega}^{\mathcal{B}} = \dot{\theta}\mathbf{e}_z \tag{2-180}$$

The velocity and acceleration as viewed by an observer in reference frame $\mathcal{A}$ were then derived, respectively, in terms of a cylindrical basis as

$$^{\mathcal{A}}\mathbf{v} = \dot{r}\mathbf{e}_r + r\dot{\theta}\mathbf{e}_\theta + \dot{z}\mathbf{e}_z \tag{2-184}$$

and

$$^{\mathcal{A}}\mathbf{a} = (\ddot{r} - r\dot{\theta}^2)\mathbf{e}_r + (2\dot{r}\dot{\theta} + r\ddot{\theta})\mathbf{e}_\theta + \ddot{z}\mathbf{e}_z \tag{2-188}$$

For a spherical basis $\{\mathbf{u}_r, \mathbf{u}_\phi, \mathbf{u}_\theta\}$ attached to reference frame C, the angular velocity of reference frame C in reference frame $\mathcal{A}$ was derived as

$$^{\mathcal{A}}\boldsymbol{\omega}^{C} = \dot{\theta}\cos\phi\mathbf{u}_r - \dot{\theta}\sin\phi\mathbf{u}_\phi + \dot{\phi}\mathbf{u}_\theta \tag{2-252}$$

The velocity and acceleration as viewed by an observer in reference frame $\mathcal{A}$ were then derived, respectively, in terms of a spherical basis as

$$^{\mathcal{A}}\mathbf{v} = \dot{r}\mathbf{u}_r + r\dot{\phi}\mathbf{u}_\phi + r\dot{\theta}\sin\phi\mathbf{u}_\theta \tag{2-257}$$

and

$$\begin{aligned} ^{\mathcal{A}}\mathbf{a} &= (\ddot{r} - r\dot{\phi}^2 - r\dot{\theta}^2\sin^2\phi)\mathbf{u}_r \\ &+ (2\dot{r}\dot{\phi} + r\ddot{\phi} - r\dot{\theta}^2\cos\phi\sin\phi)\mathbf{u}_\phi \\ &+ (r\ddot{\theta}\sin\phi + 2r\dot{\phi}\dot{\theta}\cos\phi + 2\dot{r}\dot{\theta}\sin\phi)\mathbf{u}_\theta \end{aligned} \tag{2-261}$$

Finally, for an intrinsic basis $\{\mathbf{e}_t, \mathbf{e}_n, \mathbf{e}_b\}$ attached to reference frame $\mathcal{B}$, the velocity and acceleration were derived as

$$^{\mathcal{A}}\mathbf{v} = {}^{\mathcal{A}}v\mathbf{e}_t \tag{2-282}$$

and

$$^{\mathcal{A}}\mathbf{a} = \frac{d}{dt}\left({}^{\mathcal{A}}v\right)\mathbf{e}_t + \kappa\left({}^{\mathcal{A}}v\right)^2\mathbf{e}_n \tag{2-308}$$

Furthermore, the angular velocity of reference frame $\mathcal{B}$ as viewed by an observer in reference frame $\mathcal{A}$ was derived as

$$^{\mathcal{A}}\boldsymbol{\omega}^{\mathcal{B}} = {}^{\mathcal{A}}v(\tau\mathbf{e}_t + \kappa\mathbf{e}_b) \tag{2-302}$$

where ${}^{\mathcal{A}}v = \|{}^{\mathcal{A}}\mathbf{v}\|$ is the speed as viewed by an observer in reference frame $\mathcal{A}$, and κ and τ are the curvature and torsion, respectively, as viewed by an observer in reference frame $\mathcal{A}$. The curvature and torsion were derived, respectively, as

$$\kappa = \frac{1}{{}^{\mathcal{A}}v}\left\|\frac{^{\mathcal{A}}d\mathbf{e}_t}{dt}\right\| \tag{2-306}$$

$$\tau = \frac{1}{{}^{\mathcal{A}}v}\left\|\frac{^{\mathcal{A}}d\mathbf{e}_b}{dt}\right\| \tag{2-307}$$

The next topic covered in this chapter was the kinematics of a point in a rotating and translating reference frame. In particular, for the case of a reference frame $\mathcal{B}$ that rotates and translates relative to a reference frame $\mathcal{A}$, the velocity and acceleration of a point P as viewed by an observer in reference frame $\mathcal{A}$ were obtained, respectively, as

$$ {}^{\mathcal{A}}\mathbf{v}_P = {}^{\mathcal{A}}\mathbf{v}_Q + {}^{\mathcal{B}}\mathbf{v}_{P/Q} + {}^{\mathcal{A}}\boldsymbol{\omega}^{\mathcal{B}} \times \mathbf{r}_{P/Q} \tag{2-399} $$

and

$$ \begin{aligned} {}^{\mathcal{A}}\mathbf{a}_P = {}^{\mathcal{A}}\mathbf{a}_Q + {}^{\mathcal{B}}\mathbf{a}_{P/Q} + {}^{\mathcal{A}}\boldsymbol{\alpha}^{\mathcal{B}} \times \mathbf{r}_{P/Q} \\ + 2\,{}^{\mathcal{A}}\boldsymbol{\omega}^{\mathcal{B}} \times {}^{\mathcal{B}}\mathbf{v}_{P/Q} + {}^{\mathcal{A}}\boldsymbol{\omega}^{\mathcal{B}} \times \left({}^{\mathcal{A}}\boldsymbol{\omega}^{\mathcal{B}} \times \mathbf{r}_{P/Q}\right) \end{aligned} \tag{2-412} $$

The next topic covered in this chapter was kinematics of rigid bodies. A rigid body was defined as a collection of material points in $\mathbb{E}^3$ such that the distance between any two points in the collection is a constant. From this definition of a rigid body, it was discussed that, kinematically, a rigid body and a reference frame are interchangeable, i.e., a rigid body is a reference frame. Consequently, the angular velocity of a rigid body $\mathcal{R}$ as viewed by an observer in reference frame $\mathcal{A}$, ${}^{\mathcal{A}}\boldsymbol{\omega}^{\mathcal{R}}$, is the angular velocity of the reference frame defined by the rigid body as viewed by an observer in reference frame $\mathcal{A}$. A reference frame attached to a rigid body was called a *body-fixed reference frame* and a coordinate system attached to a body-fixed reference frame was called a *body-fixed coordinate system.* In terms of a body-fixed reference frame, the following two key results were derived that relate the velocity and acceleration between any two points on the rigid body $\mathcal{R}$:

$$ {}^{\mathcal{A}}\mathbf{v}_2 = {}^{\mathcal{A}}\mathbf{v}_1 + {}^{\mathcal{A}}\boldsymbol{\omega}^{\mathcal{R}} \times (\mathbf{r}_2 - \mathbf{r}_1) \tag{2-517} $$

$$ {}^{\mathcal{A}}\mathbf{a}_2 = {}^{\mathcal{A}}\mathbf{a}_1 + {}^{\mathcal{A}}\boldsymbol{\alpha}^{\mathcal{R}} \times (\mathbf{r}_2 - \mathbf{r}_1) + {}^{\mathcal{A}}\boldsymbol{\omega}^{\mathcal{R}} \times \left[{}^{\mathcal{A}}\boldsymbol{\omega}^{\mathcal{R}} \times (\mathbf{r}_2 - \mathbf{r}_1)\right] \tag{2-524} $$

where ${}^{\mathcal{A}}\boldsymbol{\alpha}^{\mathcal{R}}$ is the *angular acceleration* of $\mathcal{R}$ as viewed by an observer in reference frame $\mathcal{A}$.

Using the relationships that govern the kinematics of a rigid body, the kinematics of a rigid body rolling or sliding along another rigid body S, was described. First, the condition for a rigid body $\mathcal{R}$ to roll on S was given as

$$ {}^{\mathcal{A}}\mathbf{v}_C^{\mathcal{R}} = {}^{\mathcal{A}}\mathbf{v}_C^{S} \tag{2-525} $$

where C is the instantaneous point of contact between $\mathcal{R}$ and S, and ${}^{\mathcal{A}}\mathbf{v}_C^{\mathcal{R}}$ and ${}^{\mathcal{A}}\mathbf{v}_C^{S}$ are the velocity of point C on $\mathcal{R}$ and the velocity of point C on S, respectively, in an arbitrary reference frame $\mathcal{A}$. Next, the condition for $\mathcal{R}$ to slide on S was given as

$$ {}^{\mathcal{A}}\mathbf{v}_C^{\mathcal{R}} \cdot \mathbf{n} = {}^{\mathcal{A}}\mathbf{v}_C^{S} \cdot \mathbf{n} \; ({}^{\mathcal{A}}\mathbf{v}_C^{\mathcal{R}} \neq {}^{\mathcal{A}}\mathbf{v}_C^{S}) \tag{2-526} $$

where $\mathbf{n}$ is the unit vector in the direction orthogonal to S at point C. Finally, the commonly encountered special case of rolling and sliding on a fixed surface was discussed.

The final topic covered in this chapter was the description of the orientation of a rigid body. In particular, it was discussed that the angular velocity of a rigid body $\mathcal{R}$ in an arbitrary reference frame $\mathcal{A}$ is, in general, *not* the rate of change of a vector that itself describes the orientation of the rigid body of $\mathcal{R}$ in $\mathcal{A}$. Consequently, it was necessary to develop an alternate means to describe the orientation of a rigid body. It was then discussed that a set of quantities that are commonly used to describe the orientation of a rigid body are the *Eulerian angles.* While many different conventions for Euler angles can be used, in this Chapter we described the *Type I* Euler angle convention. Using Type I Euler angles ψ, θ, and ϕ, it was shown that the direction cosine matrix from a basis $\mathbf{E} = \{\mathbf{E}_1, \mathbf{E}_2, \mathbf{E}_3\}$ (where $\mathbf{E}$ is fixed in reference frame $\mathcal{A}$) to a basis $\mathbf{e} = \{\mathbf{e}_1, \mathbf{e}_2, \mathbf{e}_3\}$ (where $\mathbf{e}$ is fixed in the rigid body $\mathcal{R}$) is given as

$$\{\mathbf{C}\}_{\mathbf{E}}^{\mathbf{e}} = \left\{ \begin{array}{ccc} \cos\theta\cos\psi & \cos\theta\sin\psi & -\sin\theta \\ (-\cos\phi\sin\psi & (\cos\phi\cos\psi & \sin\phi\cos\theta \\ +\sin\phi\sin\theta\cos\psi) & +\sin\phi\sin\theta\sin\psi) & \\ (\sin\phi\sin\psi & (-\sin\phi\cos\psi & \cos\phi\cos\theta \\ +\cos\phi\sin\theta\cos\psi) & +\cos\phi\sin\theta\sin\psi & \end{array} \right\} \tag{2-570}$$

Furthermore, the angular velocity of the rigid body $\mathcal{R}$ in the arbitrary reference frame $\mathcal{A}$ can be expressed in the body-fixed basis $\{\mathbf{e}_1, \mathbf{e}_2, \mathbf{e}_3\}$ as

$$\begin{aligned} {}^{\mathcal{A}}\boldsymbol{\omega}^{\mathcal{R}} &= (\dot\phi - \dot\psi\sin\theta)\mathbf{e}_1 + (\dot\psi\cos\theta\sin\phi + \dot\theta\cos\phi)\mathbf{e}_2 \\ &+ (\dot\psi\cos\theta\cos\phi - \dot\theta\sin\phi)\mathbf{e}_3 \end{aligned} \tag{2-582}$$

Finally, two important nonorthogonal bases called the *Euler basis* and *dual Euler basis* were derived. The Euler basis is given in terms of the body-fixed basis $\mathbf{E} = \{\mathbf{e}_1, \mathbf{e}_2, \mathbf{e}_3\}$ as

$$\mathbf{k}_1 = -\sin\theta\,\mathbf{e}_1 + \sin\phi\cos\theta\,\mathbf{e}_2 + \cos\phi\cos\theta\,\mathbf{e}_3 \tag{2-597}$$

$$\mathbf{k}_2 = \cos\phi\mathbf{e}_2 - \sin\phi\mathbf{e}_3 \tag{2-598}$$

$$\mathbf{k}_3 = \mathbf{e}_1 \tag{2-599}$$

Furthermore, the dual Euler basis, $\mathbf{K}^* = \{\mathbf{k}_1^*, \mathbf{k}_2^*, \mathbf{k}_3^*\}$, is given in terms of the body-fixed basis $\mathbf{e} = \{\mathbf{e}_1, \mathbf{e}_2, \mathbf{e}_3\}$ as

$$\mathbf{k}_1^* = \sin\phi\sec\theta\,\mathbf{e}_2 + \cos\phi\sec\theta\,\mathbf{e}_3 \tag{2-610}$$

$$\mathbf{k}_2^* = \cos\phi\mathbf{e}_2 - \sin\phi\mathbf{e}_3 \tag{2-611}$$

$$\mathbf{k}_3^* = \mathbf{e}_1 + \sin\phi\tan\theta\,\mathbf{e}_2 + \cos\phi\tan\theta\,\mathbf{e}_3 \tag{2-612}$$

In terms of the Euler basis, the angular velocity of a rigid body $\mathcal{R}$ in an arbitrary reference frame $\mathcal{A}$ is given as

$${}^{\mathcal{A}}\boldsymbol{\omega}^{\mathcal{R}} = \dot\psi\mathbf{k}_1 + \dot\theta\mathbf{k}_2 + \dot\phi\mathbf{k}_3 \tag{2-600}$$

Problems for Chapter 2

2–1 A bug B crawls radially outward at constant speed v_0 from the center of a rotating disk as shown in Fig. P2-1. Knowing that the disk rotates about its center O with constant absolute angular velocity $\mathbf{\Omega}$ relative to the ground (where $\|\mathbf{\Omega}\| = \Omega$), determine the velocity and acceleration of the bug as viewed by an observer fixed to the ground.

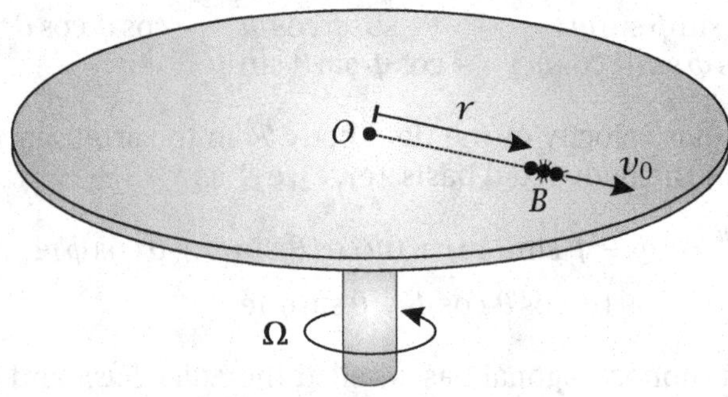

Figure P2-1

2–2 A particle, denoted by P, slides on a circular table as shown in Fig. P2-2. The position of the particle is known in terms of the radius r and the angle θ, where r is measured from the center of the table at point O and θ is measured relative to the direction of OQ, where Q is a point on the circumference of the table. Knowing that the table rotates with constant angular rate Ω, determine the velocity and acceleration of the particle as viewed by an observer in a fixed reference frame.

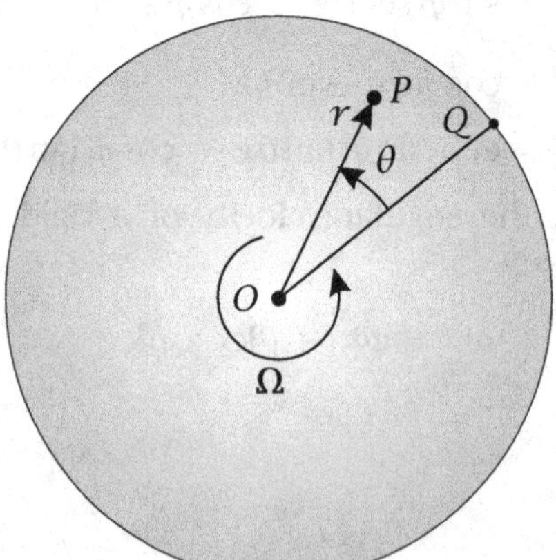

Figure P2-2

2–3 A collar slides along a rod as shown in Fig. P2-3. The rod is free to rotate about a hinge at the fixed point O. Simultaneously, the rod rotates about the vertical direction with constant angular velocity $\mathbf{\Omega}$ relative to the ground. Knowing that r describes the location of the collar along the rod, θ is the angle measured from the vertical, and $\Omega = \|\boldsymbol{\Omega}\|$, determine the velocity and acceleration of the collar as viewed by an observer fixed to the ground.

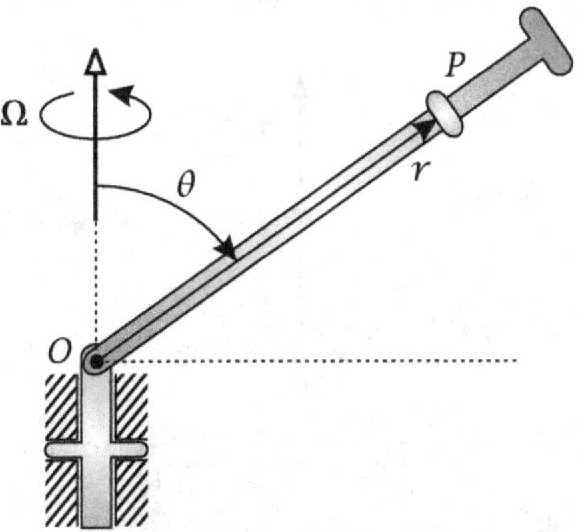

Figure P2-3

2-4 A particle slides along a track in the form of a parabola $y = x^2/a$ as shown in Fig. P2-4. The parabola rotates about the vertical with a constant angular velocity $\boldsymbol{\Omega}$ relative to a fixed reference frame (where $\Omega = \|\boldsymbol{\Omega}\|$). Determine the velocity and acceleration of the particle as viewed by an observer in a fixed reference frame.

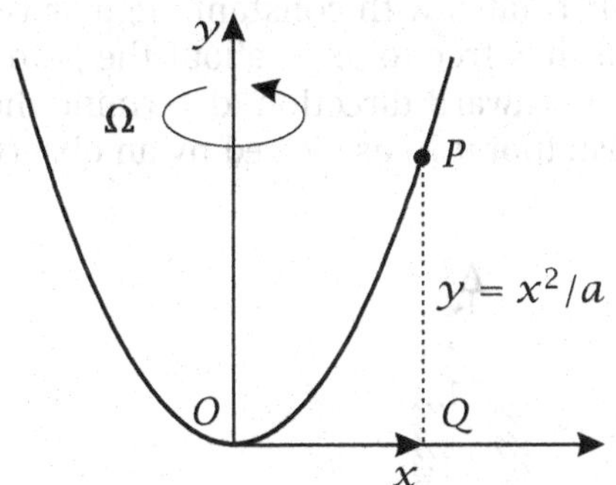

Figure P2-4

2-5 A satellite is in motion over the Earth as shown in Fig. P2-5. The Earth is modeled as a sphere of radius R that rotates with constant angular velocity $\boldsymbol{\Omega}$ in a direction $\mathbf{e}_z$, where $\mathbf{e}_z$ lies along a radial line that lies in the direction from the center of the Earth at point O to the North Pole of the Earth at point N. Furthermore, the center of the Earth is assumed to be an absolutely *fixed point*. The position of the satellite is known in terms of an *Earth-centered Earth-fixed* Cartesian coordinate system whose right-handed basis $\{\mathbf{e}_x, \mathbf{e}_y, \mathbf{e}_z\}$ is defined as follows:

- The direction $\mathbf{e}_x$ lies orthogonal to $\mathbf{e}_z$ in the equatorial plane of the Earth along the line from O to P, where P lies at the intersection of the equator with the great circle called the *Prime Meridian*
- The direction $\mathbf{e}_y$ lies orthogonal to both $\mathbf{e}_x$ and $\mathbf{e}_z$ in the equatorial plane of the Earth such that $\mathbf{e}_y = \mathbf{e}_z \times \mathbf{e}_x$

Using the basis $\{\mathbf{e}_x, \mathbf{e}_y, \mathbf{e}_z\}$ to express all quantities, determine the velocity and acceleration of the spacecraft (a) as viewed by an observer fixed to the Earth and (b) as viewed by an observer in a fixed inertial reference frame.

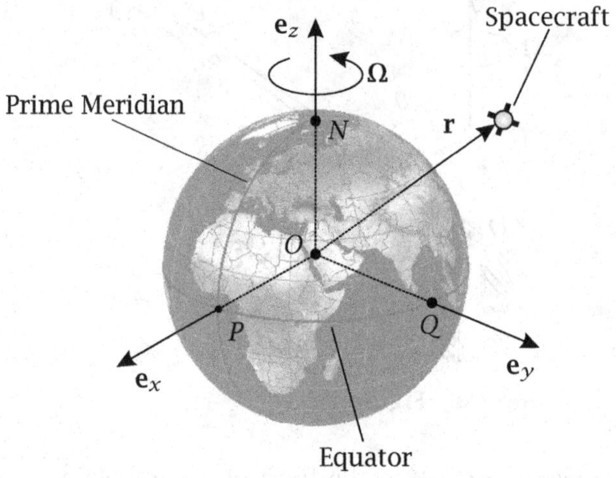

Figure P2-5

2–6 An arm of length L is hinged at one of its ends at point O to a vertical shaft as shown in Fig. P2-6. The shaft rotates with constant angular velocity $\mathbf{\Omega}$ relative to the ground. Assuming that the arm is free to pivot about the point O and θ describes the angle of the shaft from the downward direction, determine the velocity and acceleration of the free end of the arm (point P) as viewed by an observer fixed to the ground.

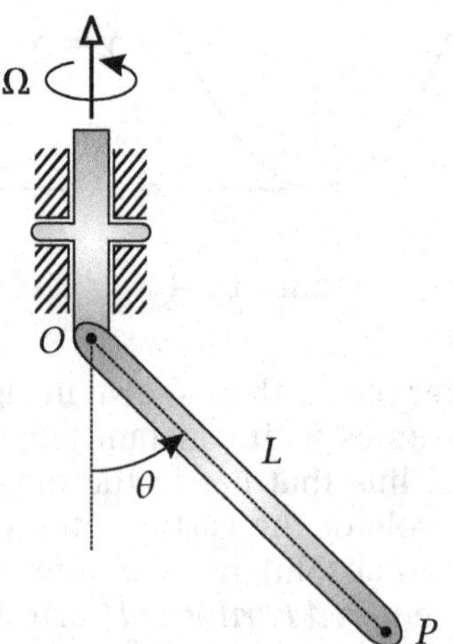

Figure P2-6

2–7 A circular disk of radius R rolls without slip along a horizontal surface as shown in Fig. P2-7. Knowing that point P is located on the edge of the disk and θ is the angle formed by the vertical and the direction OP (where O is the center of the disk), determine the following quantities relative to an observer fixed in the ground: (a) the velocity of point P and (b) the acceleration of point P.

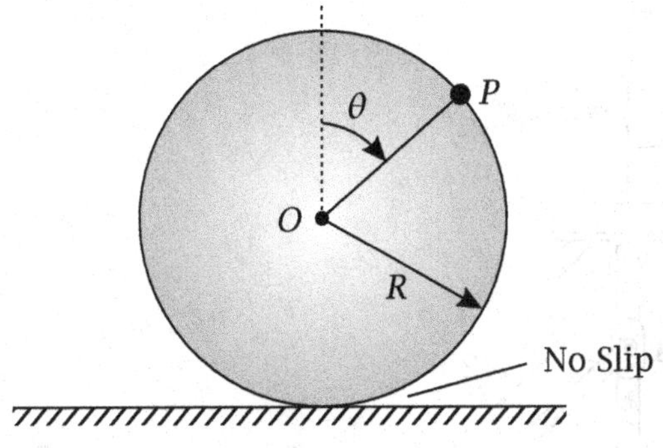

Figure P2-7

2–8 A bead slides along a fixed circular helix of radius R and helical inclination angle ϕ as shown in Fig. P2-8. Knowing that the angle θ measures the position of the bead and is equal to zero when the bead is at the base of the helix, determine the following quantities relative to an observer fixed to the helix: (a) the arc-length parameter s as a function of the angle θ; (b) the intrinsic basis $\{\mathbf{e}_t, \mathbf{e}_n, \mathbf{e}_b\}$ and the curvature of the trajectory as a function of the angle θ; and (c) the velocity and acceleration of the particle in terms of the intrinsic basis $\{\mathbf{e}_t, \mathbf{e}_n, \mathbf{e}_b\}$.

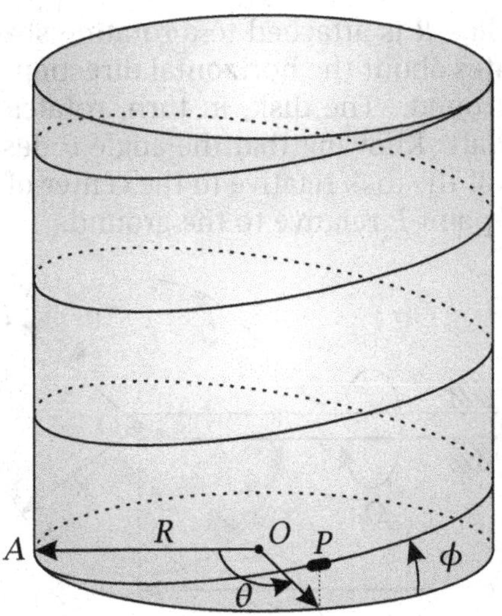

Figure P2-8

2–9 Arm AB is hinged at points A and B to collars that slide along vertical and horizontal shafts, respectively, as shown in Fig. P2-9. The vertical shaft rotates with angular velocity $\mathbf{\Omega}$ relative to a fixed reference frame (where $\Omega = \|\mathbf{\Omega}\|$), and point B moves with constant velocity $\mathbf{v}_0$ relative to the horizontal shaft. Knowing that point P is located at the center of the arm and the angle θ describes the orientation of the arm with respect to the vertical shaft, determine the velocity and acceleration of point P

as viewed by an observer fixed to the ground. In simplifying your answers, find an expression for $\dot{\theta}$ in terms of v_0 and l and express your answers in terms of only l, Ω, $\dot{\Omega}$, θ, and v_0.

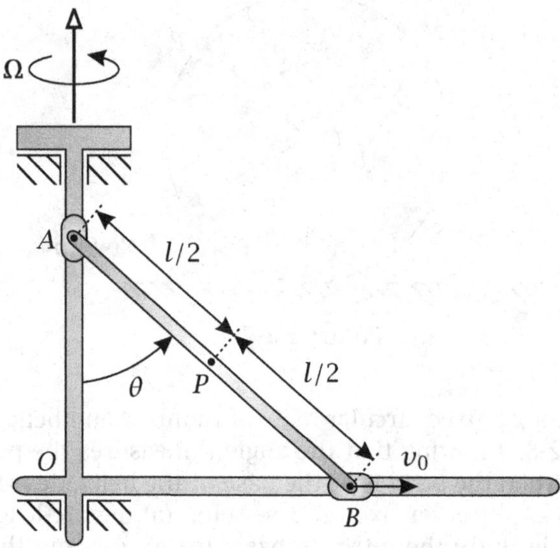

Figure P2-9

2–10 A circular disk of radius R is attached to a rotating shaft of length L as shown in Fig. P2-10. The shaft rotates about the horizontal direction with a constant angular velocity Ω relative to the ground. The disk, in turn, rotates about its center about an axis orthogonal to the shaft. Knowing that the angle θ describes the position of a point P located on the edge of the disk relative to the center of the disk, determine the velocity and acceleration of point P relative to the ground.

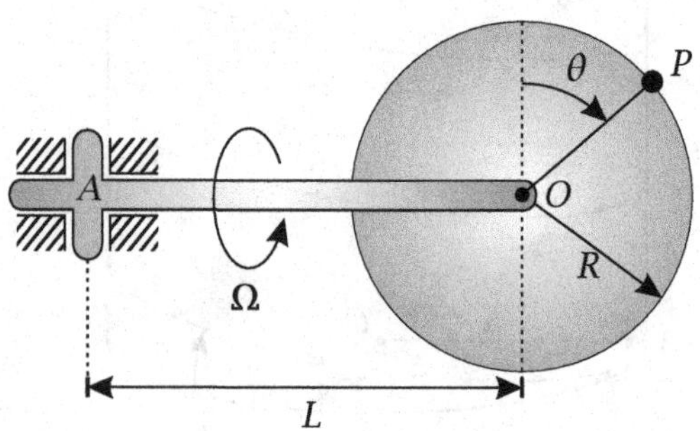

Figure P2-10

2–11 A rod of length L with a wheel of radius R attached to one of its ends is rotating about the vertical axis OA with a constant angular velocity Ω relative to a fixed reference frame as shown in Fig. P2-11. The wheel is vertical and rolls without slip along a fixed horizontal surface. Determine the angular velocity and angular acceleration of the wheel as viewed by an observer in a fixed reference frame.

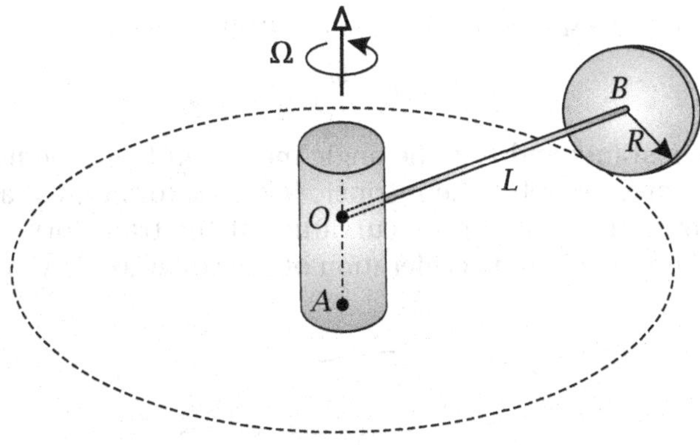

Figure P2-11

2–12 A disk of radius r rolls without slip along the inside of a fixed circular track of radius $R > r$ as shown in Fig. P2-12. The center of the circular track is denoted by point O while the center of the disk is denoted by point P. The angle θ is measured from the fixed vertically downward direction to the center of the disk while the angle ϕ is measured from the OP-direction to the OQ-direction, where point Q is located on the edge of the disk. Knowing that point Q coincides with point A (where point A is located at the bottom of the track) when θ and ϕ are zero, determine the following quantities relative to an observer in a fixed reference frame: (a) the angular velocity and angular acceleration of the disk and (b) the velocity and acceleration of point Q. Express your answers in terms of the angle ϕ (and the time derivatives of ϕ).

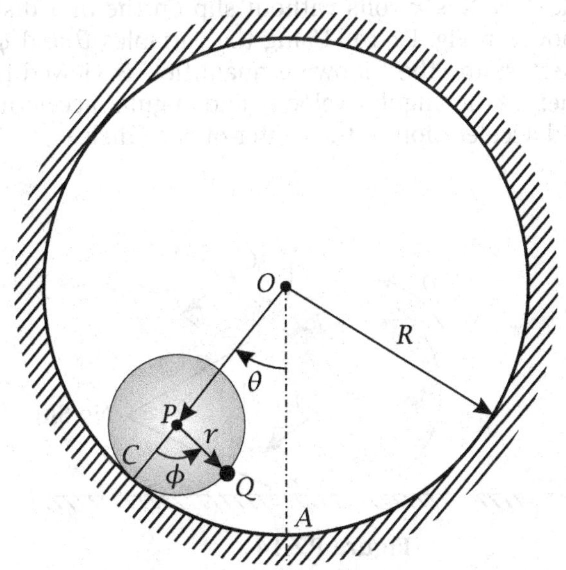

Figure P2-12

2–13 A collar is constrained to slide along a track in the form of a logarithmic spiral as shown in Fig. P2-13. The equation for the spiral is given as

$$r = r_0 e^{-a\theta}$$

where r_0 and a are constants and θ is the angle measured from the horizontal direction. Determine (a) expressions for the intrinsic basis vectors $\mathbf{e}_t$, $\mathbf{e}_n$, and $\mathbf{e}_b$ in terms any other basis of your choosing; (b) the curvature of the trajectory as a function of the angle θ; and (c) the velocity and acceleration of the collar as viewed by an observer fixed to the track.

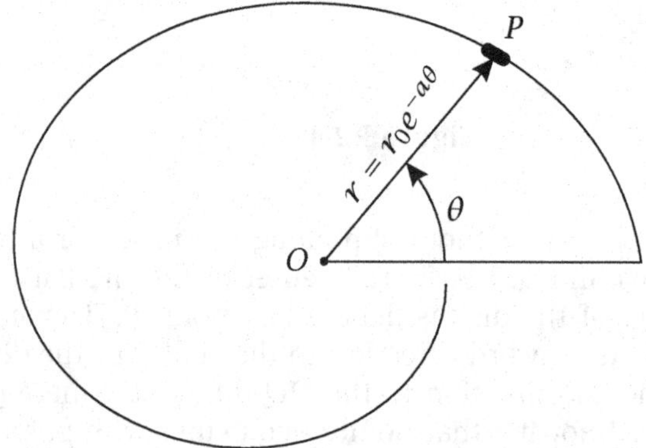

Figure P2-13

2–14 A circular disk of radius R rolls without slip along a fixed horizontal surface. A second circular disk of radius r rolls without slip on the first disk. The geometry of the two disks is shown in Fig. P2-14. Using the variables θ and ϕ to describe the motion of the disks, determine the following quantities as viewed by an observer in a fixed reference frame: (a) the angular velocity and angular acceleration of each disk and (b) the velocity and acceleration of the center of each disk.

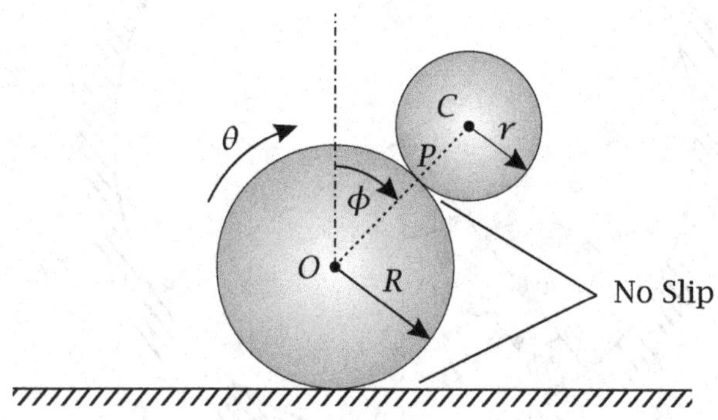

Figure P2-14

2–15 A circular disk of radius R is attached to a rotating shaft of length L as shown in Fig. P2-15. The shaft rotates about the vertical direction with a constant angular velocity Ω relative to the ground. The disk, in turn, rotates about its center about an axis orthogonal to the shaft. Knowing that the angle θ describes the position of a point P located on the edge of the disk relative to the center of the disk, determine the following quantities as viewed by an observer fixed to the ground: (a) the angular velocity of the disk and (b) the velocity and acceleration of point P.

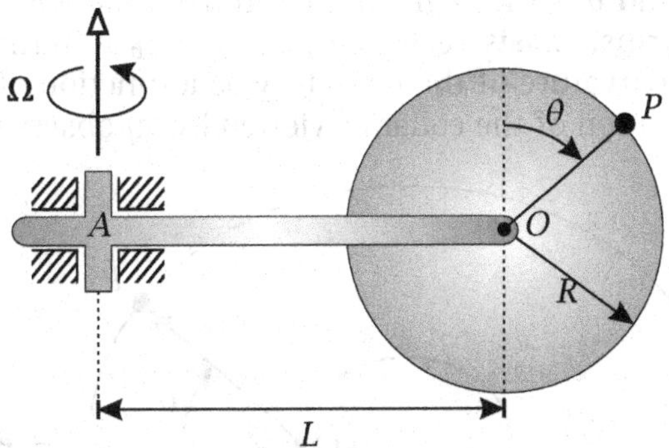

Figure P2-15

2–16 A disk of radius R rotates freely about its center at a point located on the end of an arm of length L as shown in Fig. P2-16. The arm itself pivots freely at its other end at point O to a vertical shaft. Finally, the shaft rotates with constant angular velocity Ω relative to the ground. Knowing that ϕ describes the location of a point P on the edge of the disk relative to the direction OQ and that θ is formed by the arm with the downward direction, determine the following quantities as viewed by an observer fixed to the ground: (a) the angular velocity of the disk and (b) the velocity and acceleration of point P.

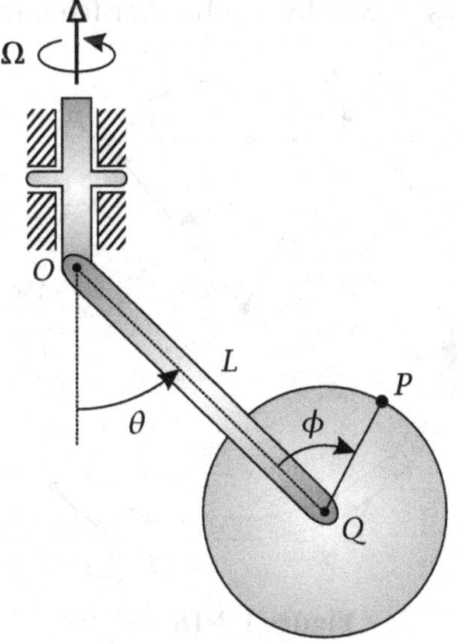

Figure P2-16

2–17 A particle slides along a track in the form of a spiral as shown in Fig. P2-17. The equation for the spiral is

$$r = a\theta$$

where a is a constant and θ is the angle measured from the horizontal. Determine (a) expressions for the intrinsic basis vectors $\mathbf{e}_t$, $\mathbf{e}_n$, and $\mathbf{e}_b$ in terms any other basis of your choosing; (b) the curvature of the trajectory as a function of the angle θ; and (c) the velocity and acceleration of the collar as viewed by an observer fixed to the track.

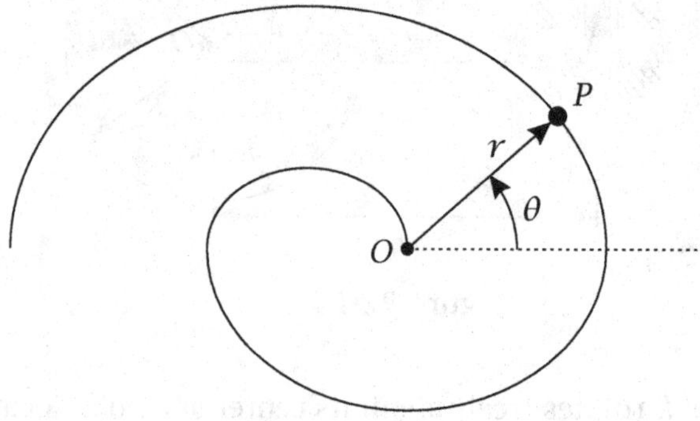

Figure P2-17

2–18 A particle slides without friction inside a circular tube of radius R as shown in Fig. P2-18. The tube is hinged at a point on its diameter to a fixed hinge at point O such that the tube rotates with constant angular velocity Ω in the vertical plane about point O. Knowing that the angle θ describes the location of the particle relative to the direction from O to the center of the tube at point Q, determine the velocity and acceleration of the particle as viewed by an observer fixed to the ground.

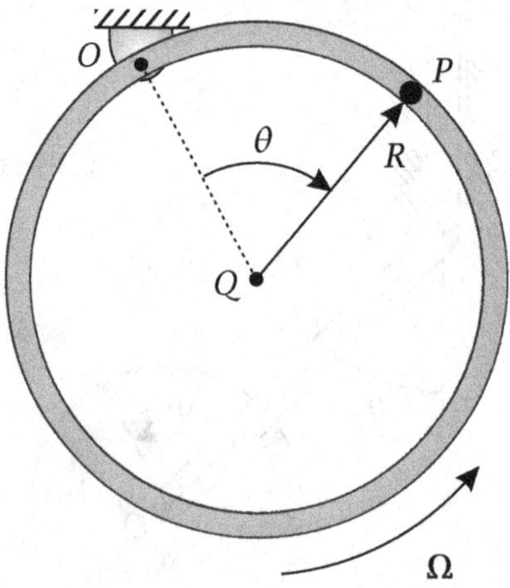

Figure P2-18

2–19 A particle P slides without friction along the inside of a fixed hemispherical bowl of radius R as shown in Fig. P2-19. The basis $\{\mathbf{E}_x, \mathbf{E}_y, \mathbf{E}_z\}$ is fixed to the bowl. Furthermore, the angle θ is measured from the $\mathbf{E}_x$-direction to the direction OQ, where point Q lies on the rim of the bowl while the angle ϕ is measured from the OQ-direction to the position of the particle. Determine the velocity and acceleration of the particle as viewed by an observer fixed to the bowl. **Hint:** Express the position in terms of a spherical basis that is fixed to the direction OP; then determine the velocity and acceleration as viewed by an observer fixed to the bowl in terms of this spherical basis.

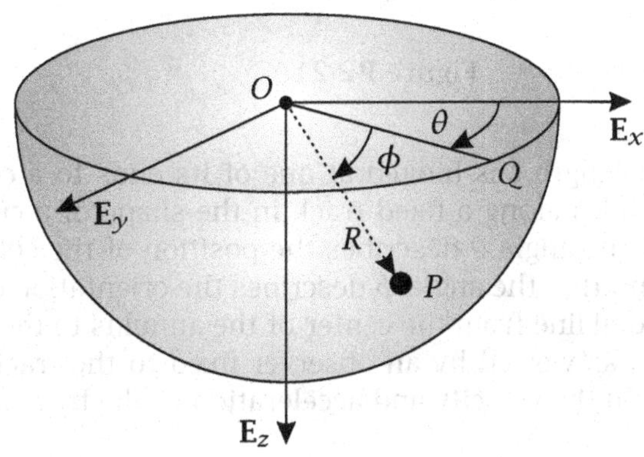

Figure P2-19

2–20 A particle P slides along a circular table as shown in Fig. P2-20. The table is rigidly attached to two shafts such that the shafts and table rotate with angular velocity Ω about an axis along the direction of the shafts. Knowing that the position of the particle is given in terms of a polar coordinate system relative to the table, determine (a) the angular velocity of the table as viewed by an observer fixed to the ground, (b) the velocity and acceleration of the particle as viewed by an observer fixed to the table, and (c) the velocity and acceleration of the particle as viewed by an observer fixed to the ground.

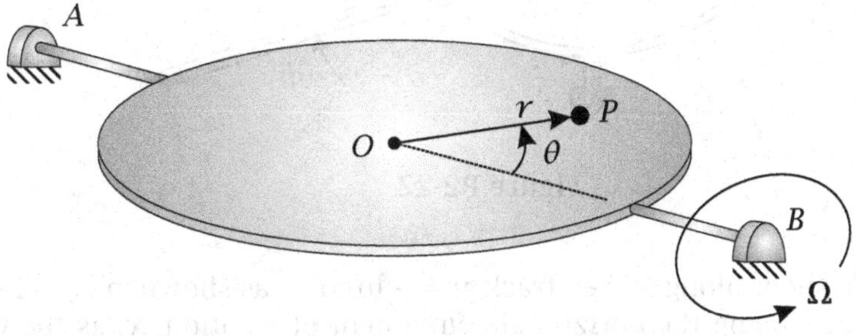

Figure P2-20

2–21 A slender rod of length l is hinged to a collar as shown in Fig. P2-21. The collar slides freely along a fixed horizontal track. Knowing that x is the horizontal displacement of the collar and that θ describes the orientation of the rod relative to the vertical direction, determine the velocity and acceleration of the free end of the rod as viewed by an observer fixed to the track.

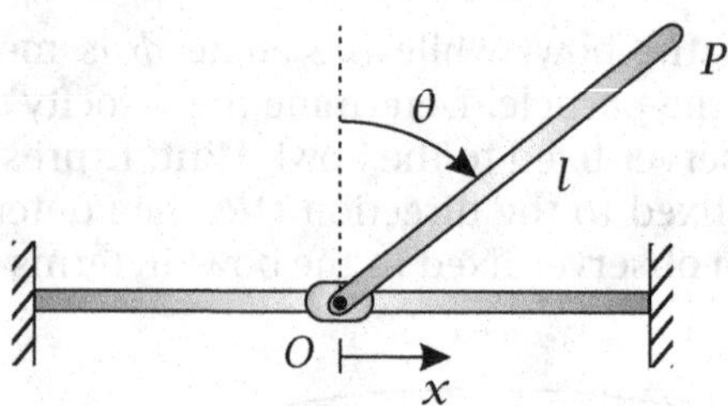

Figure P2-21

2–22 A slender rod of length l is hinged at one of its ends to a collar as shown in Fig. P2-22. The collar slides along a fixed track in the shape of a circular annulus of radius R. Knowing that the angle θ describes the position of the collar relative to the center of the annulus and that the angle ϕ describes the orientation of the rod relative to the instantaneous radial line from the center of the annulus to the collar, determine the following quantities as viewed by an observer fixed to the track: (a) the angular velocity of the rod and (b) the velocity and acceleration of the free end of the rod.

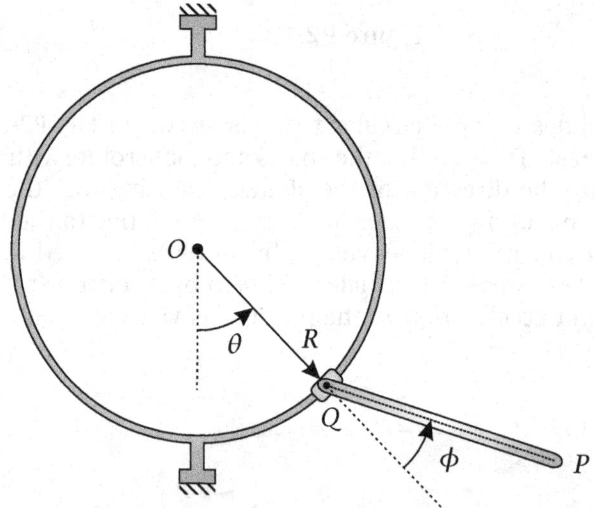

Figure P2-22

2–23 A particle slides along a fixed track $y = -\ln \cos x$ as shown in Fig. P2-23 (where $-\pi/2 < x < \pi/2$). Using the horizontal component of position, x, as the variable to describe the motion and the initial condition $x(t = 0) = x_0$, determine the following quantities as viewed by an observer fixed to the track: (a) the arc-length parameter s as a function of x, (b) the intrinsic basis $\{\mathbf{e}_t, \mathbf{e}_n, \mathbf{e}_b\}$ and the curvature κ, and (c) the velocity and acceleration of the particle.

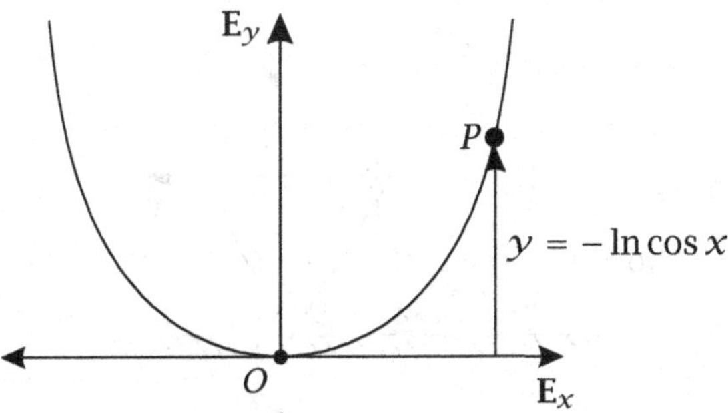

Figure P2-23

2–24 A system consists of a particle P and two disks, $\mathcal{D}_1$ and $\mathcal{D}_2$, of radius R and $2R$, respectively, as shown in Fig. P2-24. Disk $\mathcal{D}_1$ rotates freely about a point A that is located a distance R from the center of disk $\mathcal{D}_2$ with angular velocity Ω_1 relative to disk $\mathcal{D}_2$, while disk $\mathcal{D}_2$ rotates about its center at point O with angular velocity Ω_2 relative to the ground. Knowing that the particle slides along disk $\mathcal{D}_1$, determine (a) the angular velocity of disk $\mathcal{D}_1$ as viewed by an observer fixed to the ground and (b) the velocity and acceleration of the particle as viewed by an observer fixed to the ground.

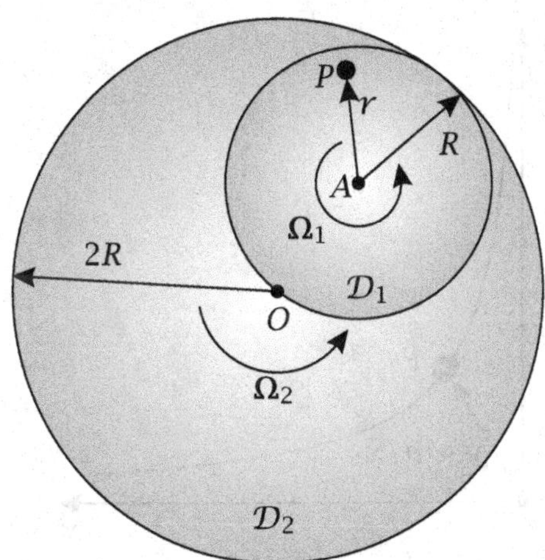

Figure P2-24

2–25 An arm of length l is hinged at one of its ends to the center of a circular disk of radius r as shown in Fig. P2-25. The disk rolls without slip along a fixed horizontal surface. Using the variable x to describe the position of the center of the disk and the variable θ to describe the orientation of the arm relative to the vertically downward direction, determine the following quantities as viewed by an observer fixed to the ground: (a) the velocity and acceleration of the center of the disk and (b) the velocity and acceleration of the free end of the arm.

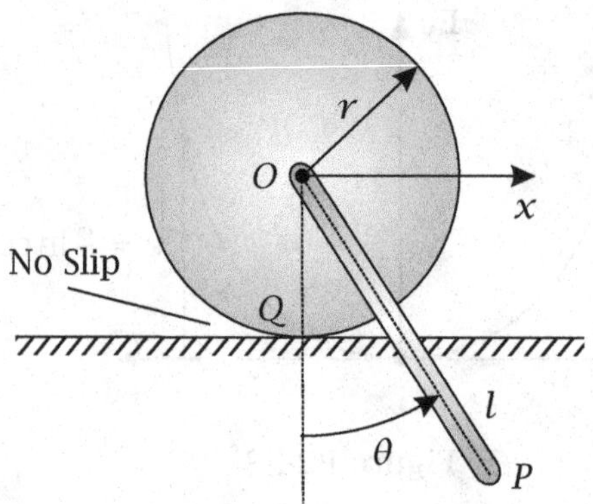

Figure P2-25

2–26 A particle slides along a track in the form of a hyperbola $y = a/x$ (where a is a constant and $x > 0$) as shown in Fig. P3-16. Using the initial condition $x(t = 0) = x_0$, determine the following quantities as viewed by an observer fixed to the track: (a) the arc-length parameter s as a function of x; (b) the intrinsic basis $\{\mathbf{e}_t, \mathbf{e}_n, \mathbf{e}_b\}$ and the curvature κ; and (c) the velocity and acceleration of the particle.

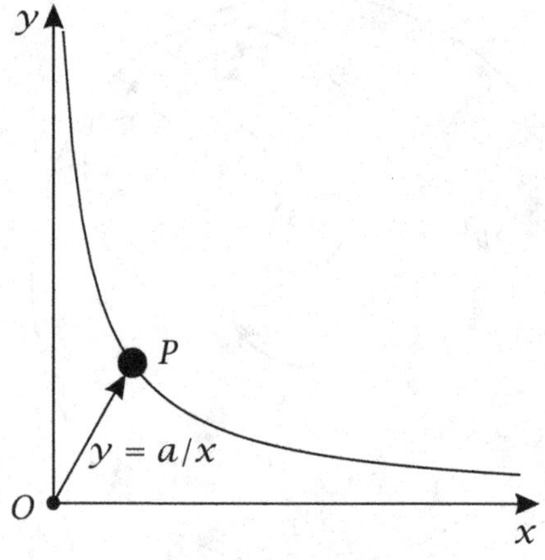

Figure P2-26

2–27 A satellite is in motion over the Earth as shown in Fig. P2-27. The Earth is modeled as a sphere of radius R that rotates with constant angular velocity Ω in a direction $\mathbf{e}_z$, where $\mathbf{e}_z$ lies in the direction from the center of the Earth at the fixed point O to the North Pole of the Earth at point N. The position of the satellite is known geographically in terms of its radial distance, r, from the center of the Earth, its *Earth-relative longitude*, θ, where θ is the angle measured from direction $\mathbf{e}_x$, where $\mathbf{e}_x$ lies along the line from the center of the Earth to the intersection of the Equator with the Prime Meridian, and its *latitude*, ϕ, where ϕ is measured from the line that lies along the projection of the position into the equatorial plane. Using the spherical basis $\{\mathbf{e}_r, \mathbf{e}_\theta, \mathbf{e}_\phi\}$ to describe the position of the spacecraft (where $\mathbf{e}_r = \mathbf{r}/r$, $\mathbf{e}_\theta = (\mathbf{e}_z \times \mathbf{e}_r)/\|\mathbf{e}_z \times \mathbf{e}_r\|$, and $\mathbf{e}_\phi = \mathbf{e}_r \times \mathbf{e}_\theta$), determine the velocity and acceleration of the satellite (a) as viewed by an observer fixed to the Earth and (b) as viewed by an observer in a fixed inertial reference frame.

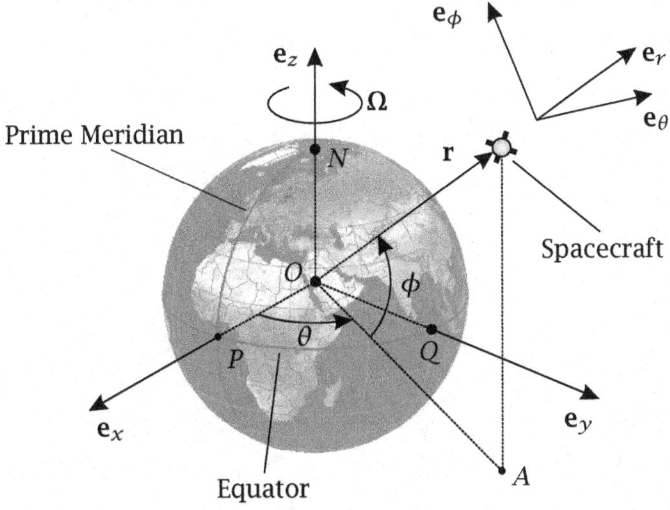

Figure P2-27

Chapter 3

Kinetics of Particles

If I have seen further it is by standing upon the shoulders of giants.

- Sir Isaac Newton (1642–1727)
British Physicist and Mathematician

Until now, we have been concerned with the kinematics of particles where the objective has been to determine the motion of a particle or rigid body *without* regard to the cause of the motion. Clearly in any physical system motion cannot occur without the application of some kind of external stimulus. In particular, in order for a particle to accelerate, it is necessary to apply a force to the particle. In order to study the motion that results from the application of a force (or, in general, the application of multiple forces) to a particle, it is necessary to study the *kinetics* of the particle. The objective of kinetics is threefold: (1) to describe quantitatively the forces that act on a particle; (2) to determine the motion that results from the application of these forces using postulated laws of physics; and (3) to analyze the motion.

The first topic in this chapter is the development of models for forces that are commonly used in dynamics. In particular, models are developed for contact forces, spring forces, and gravitational forces. These models will be used throughout the remainder of this book when solving problems.

The next topic in this chapter covers *Newton's laws*, which are the fundamental postulates that govern the nonrelativistic motion of particles. A framework is then established for the systematic application of Newton's laws. Several example problems are then solved using Newton's laws.

The next topics in this chapter are linear momentum and angular momentum of a particle. First, the definition of linear momentum of a particle is established. Using the definition of linear momentum, the principle of linear impulse and linear momentum for a particle is derived. Next, the definition of the moment of a force and the angular momentum of a particle are established. Using the definition of angular momentum, the principle of angular impulse and angular momentum for a particle is derived.

The final topics covered in this chapter are the power, work, and energy of a particle. First, the definitions of kinetic energy and work of a force are stated. These definitions lead to the work-energy theorem for a particle. A special class of forces, called conservative forces, are then defined. Using the work-energy theorem for a particle, the principle of work and energy for a particle is derived. Then, using the definition of a conservative force, an alternate form of the work-energy theorem for a particle is derived. Finally, using the alternate form of the work-energy theorem for a particle, an alternate form of the principle of work and energy for a particle is derived.

3.1 Forces Commonly Used in Dynamics

A fundamental component in the study of kinetics of particles is the resultant force that acts on a particle. In general, this resultant force is made up of forces that have significantly different characteristics. In particular, three distinctly different types of forces that commonly arise in dynamics are friction forces, spring forces, and gravitational forces. In this section we develop models for each of these forces. These models will be used in the remainder of the book in the study of kinetics of particles, systems of particles, and rigid bodies.

3.1.1 Contact Forces

Many dynamics problems involve the motion of a particle sliding along a surface in $\mathbb{E}^3$. In this section we discuss the kinematics and kinetics associated with the motion of a particle in continuous contact with a surface.

Decomposition of a Contact Force Applied by a Surface

Consider a particle of mass m that moves in an arbitrary reference frame $\mathcal{A}$ while constrained to slide on a rigid surface S (i.e., S defines a reference frame) as shown in Fig. 3–1. Furthermore, let C denote the point on S that is instantaneously in contact with the particle, and let $\mathbf{r}$ and $\mathbf{r}_C^S$ be the position of the particle and the position of C on S, respectively, measured relative to a point O, where O is fixed in reference frame $\mathcal{A}$. Then, assuming that S can be described by a differentiable function, there exists a well-defined plane, denoted $\mathcal{T}_C^S$, that is tangent to S at C. Correspondingly, let $\mathbf{n}$ be the unit vector orthogonal to $\mathcal{T}_C^S$ at C, and let $\mathbf{u}$ and $\mathbf{w}$ be two unit vectors that lie in $\mathcal{T}_C^S$ such that $\{\mathbf{u}, \mathbf{w}, \mathbf{n}\}$ forms a right-handed basis.

In general, the force exerted by a surface S on the particle can be decomposed into the following two parts: (1) a force $\mathbf{N}$ that is orthogonal to the tangent plane $\mathcal{T}_C^S$ and (2) a force $\mathbf{N}$ that lies in the tangent plane $\mathcal{T}_C^S$. Then the total force exerted by S on the particle, denoted $\mathbf{F}_S$, is given as

$$\mathbf{F}_S = \mathbf{N} + \mathbf{F}_f \tag{3–1}$$

The force $\mathbf{N}$ is called the *reaction force* or *normal force* exerted by the surface S on the particle, while the force $\mathbf{F}_f$ is called the *friction force* exerted by the surface on the particle. Now, because $\mathbf{N}$ lies orthogonal to $\mathcal{T}_C^S$ while $\mathbf{F}_f$ lies in $\mathcal{T}_C^S$, these forces can be expressed in terms of the basis $\{\mathbf{u}, \mathbf{w}, \mathbf{n}\}$ as

$$\begin{aligned} \mathbf{N} &= N\mathbf{n} \\ \mathbf{F}_f &= F_{f,u}\mathbf{u} + F_{f,w}\mathbf{w} \end{aligned} \tag{3–2}$$

Then, by assumption it is seen that the reaction force satisfies the property that

$$\mathbf{N} \cdot \mathbf{b} = 0 \tag{3–3}$$

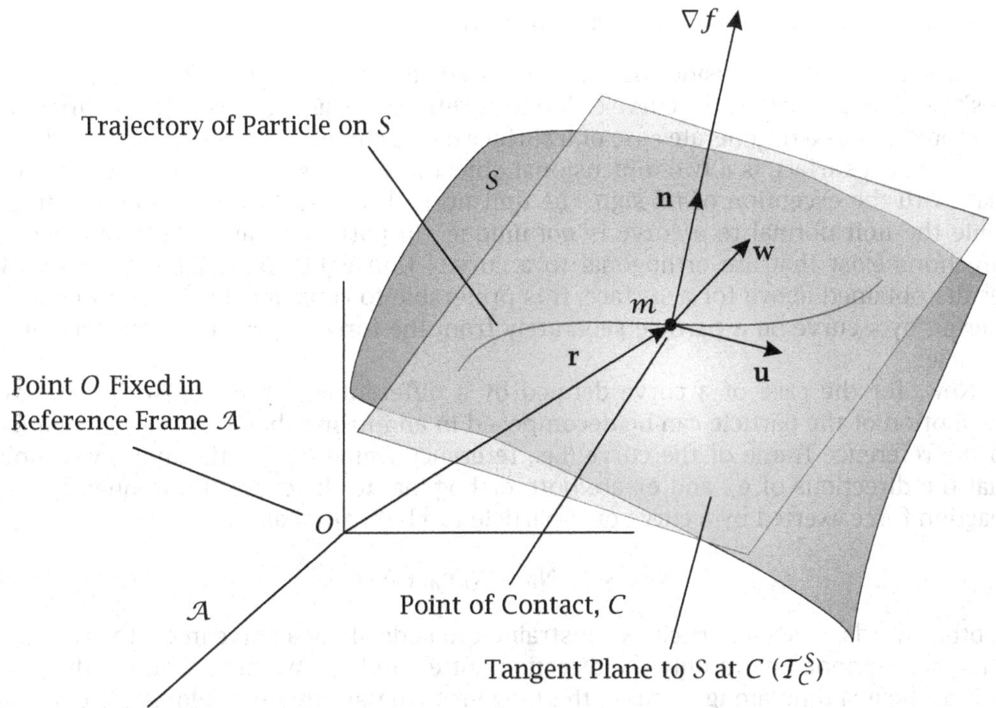

Figure 3–1 Particle sliding with friction along a surface S.

for any vector $\mathbf{b} \in \mathcal{T}_C^S$. We now develop models for both reaction forces and friction forces.

Reaction Force Exerted by a Surface on a Particle

It is seen from Eq. (3–1) that the force of reaction of a surface S on a particle lies in the direction orthogonal to the tangent plane at the point of contact between the particle and the surface. Suppose now that the equation for the surface S can be described as

$$f(\mathbf{r}) = 0 \tag{3-4}$$

where $\mathbf{r}$ is the position of a particular point on S. Then from calculus we know that the normal to S at $\mathbf{r}$ lies in the direction of the *gradient* of S at $\mathbf{r}$. Denoting the gradient of the function f by ∇f, the unit vector normal to S at $\mathbf{r}$ is given as

$$\mathbf{n} = \frac{\nabla f}{\|\nabla f\|} \tag{3-5}$$

In general, the function f will be most conveniently described in either Cartesian coordinates, cylindrical coordinates, or spherical coordinates relative to an arbitrary reference frame $\mathcal{A}$. Expressions for the gradient of a scalar function $f(\mathbf{r})$ in these coordinate systems are given in Appendix B.

Reaction Force Exerted by a Curve on a Particle

Suppose now that we consider the case of a particle sliding along a rigid *curve*, C, in $\mathbb{E}^3$ as shown in Fig. 3–2 (i.e., the curve C defines a reference frame). Now, while a curve can be thought of as a degenerate case of a surface (i.e., a curve is a one-dimensional object in $\mathbb{E}^3$ while a surface is a two-dimensional object in $\mathbb{E}^3$), it is important to understand that, with the exception of its sign, the unit normal to a surface is uniquely defined while the unit normal to a curve is *not* unique. In particular, an infinite number of directions exist that are orthogonal to a curve. Consequently, rather than use the results obtained above for a surface, it is preferable to consider the forces of contact exerted by a curve on a particle separately from the forces exerted by a surface on a particle.

Now, for the case of a curve defined by a differentiable function we know that the motion of the particle can be decomposed in an intrinsic basis $\{\mathbf{e}_t, \mathbf{e}_n, \mathbf{e}_b\}$ relative to the reference frame of the curve (i.e., reference frame C). Furthermore, we know that the directions of $\mathbf{e}_n$ and $\mathbf{e}_b$ are both orthogonal to the curve. Consequently, the reaction force exerted by a curve on a particle can be written as

$$\mathbf{N} = \mathbf{N}_n + \mathbf{N}_b = N_n\mathbf{e}_n + N_b\mathbf{e}_b \tag{3-6}$$

In other words, when a particle is constrained to slide along a curve in $\mathbb{E}^3$, the reaction force has components in only the directions of $\mathbf{e}_n$ and $\mathbf{e}_b$. We note that, for the case of a particle sliding along a curve, the tangent plane at any point along the curve is actually only a line, i.e., only *one* direction is required to define the plane tangent to a curve (this direction being $\mathbf{e}_t$).

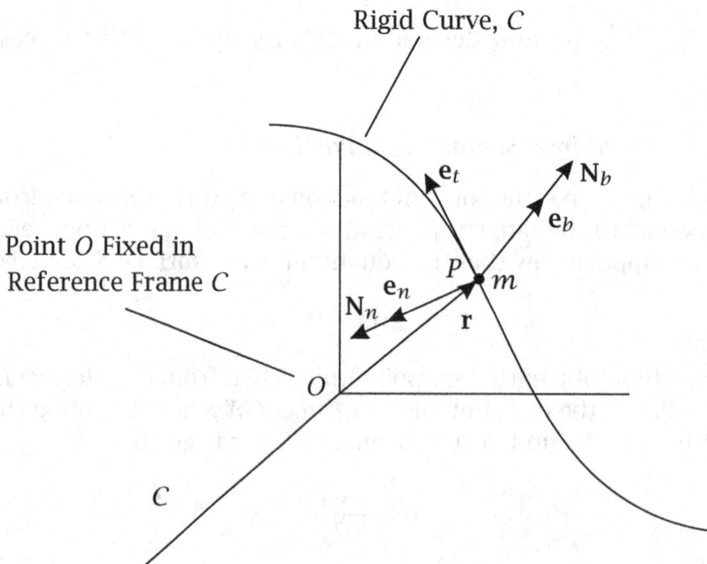

Figure 3–2 Reaction force $\mathbf{N}$ exerted by a rigid curve, C, on a particle. The reaction force $\mathbf{N}$ has components in only the directions of $\mathbf{e}_n$ and $\mathbf{e}_b$.

Friction Force Exerted by a Surface on a Particle

With regard to the force of friction exerted by a surface on a particle, the two most common types of friction forces are *Coulomb friction* (Coulomb, 1785) and *viscous friction.* We now derive models for a Coulomb friction force and a viscous friction force.

Coulomb Friction

The fundamental assumption in Coulomb friction is that the friction force $\mathbf{F}_f$ is proportional to the normal force. Furthermore, the constant of proportionality is called the *coefficient of Coulomb friction.*[1] There are two types of Coulomb friction: static and dynamic. Static Coulomb friction arises when the relative velocity between the particle and the instantaneous point of contact of the particle with the surface is zero, i.e., $\mathbf{v}_{\text{rel}} = \mathbf{0}$. For the case of static Coulomb friction we have

$$\boxed{\|\mathbf{F}_f\| \le \mu_s \|\mathbf{N}\|} \tag{3-7}$$

where the constant μ_s is called the *coefficient of static Coulomb friction.* Examining Eq. (3-7), it is seen that, in the case of static Coulomb friction, the actual friction force is indeterminate because the magnitude of $\mathbf{F}_f$ is only known to be less than the product of the coefficient of static friction and the normal force. Dynamic Coulomb friction arises when $\mathbf{v}_{\text{rel}} \neq \mathbf{0}$. For the case of dynamic Coulomb friction we have

$$\boxed{\mathbf{F}_f = -\mu_d \|\mathbf{N}\| \frac{\mathbf{v}_{\text{rel}}}{\|\mathbf{v}_{\text{rel}}\|}} \tag{3-8}$$

where the constant μ_d is called the *coefficient of dynamic Coulomb friction* and $\mathbf{v}_{\text{rel}}$ is the velocity of the particle relative to the point on S that is instantaneously in contact with the particle (see Section 2.14 for the definition of $\mathbf{v}_{\text{rel}}$). It is noted that the dynamic Coulomb friction force is obtained using the *relative* velocity between the particle and the surface, *not* the velocity of the particle.

Viscous Friction

The fundamental assumption in viscous friction is that the friction force is proportional to $\mathbf{v}_{\text{rel}}$ and has a constant of proportionality c called the *coefficient of viscous friction.* The viscous friction force is then given as

$$\boxed{\mathbf{F}_f = -c\mathbf{v}_{\text{rel}}} \tag{3-9}$$

As with dynamic Coulomb friction, it is noted that the viscous friction force is obtained using the *relative* velocity between the particle and the surface, *not* the velocity of the particle (again, see Section 2.14 for the definition of $\mathbf{v}_{\text{rel}}$).

[1]Models for Coulomb friction can be found in many books on dynamics. See Beer and Johnston (1997), Bedford and Fowler (2005), or Greenwood (1988) for other descriptions of Coulomb friction.

3.1.2 Spring Forces

Another force that arises in the study of dynamics is a *spring force.* Two particular spring forces of interest are those due to a *linear spring* and a *curvilinear spring.* In this section we develop a model for the force exerted by each of these types of springs.

Linear Spring Forces

Consider a particle of mass m attached to a massless spring and moving in an arbitrary reference frame $\mathcal{A}$ as shown in Fig. 3-3 O'Reilly (2001).

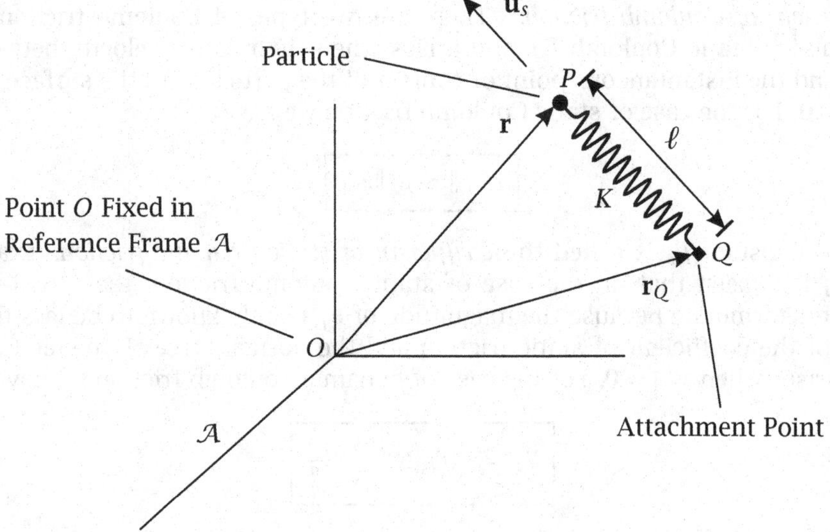

Figure 3-3 Particle attached to a linear spring with spring constant K and unstretched length ℓ_0.

Furthermore, let O be a point fixed in reference frame $\mathcal{A}$, and let $\mathbf{r}_Q$ and $\mathbf{r}$ be the position of the attachment point of the spring and the position of the particle, respectively, relative to point O (it is noted that the attachment point Q may either be fixed or moving in reference frame $\mathcal{A}$). Next, assume that the spring force satisfies the following properties: (1) the spring force is proportional to the *compressed or stretched length* of the spring; (2) the spring force has a positive constant of proportionality K; (3) the spring force lies along the line connecting Q and P. Assumption (1) is called *Hooke's law.*[2] Furthermore, under assumptions (1) through (3), the spring is said to be a *linear spring.* Next, suppose we denote the *length* of the spring by ℓ. Then it is seen that ℓ is given as

$$\ell = \|\mathbf{r} - \mathbf{r}_Q\| \tag{3-10}$$

Suppose we denote the *unstretched length* or the *zero-force length* of the spring by ℓ_0. Then the *stretched length* of the spring, denoted $\Delta\ell$, is given as

$$\Delta\ell = \ell - \ell_0 \tag{3-11}$$

[2]Hooke's law was originally stated in 1676 by Robert Hooke (1635–1703) in the form of a Latin cryptogram (Thornton and Marion, 2004). Hooke later provided a translation of the Latin cryptogram as *ut tensio sic vis (the stretch is proportional to the force)* (O'Reilly, 2001).

Applying assumptions (1) and (2), the *magnitude* of the force generated by the spring, denoted $\|\mathbf{F}_s\|$, is given as

$$\|\mathbf{F}_s\| = K\,|(\ell - \ell_0)| \tag{3-12}$$

Also, applying assumption (3) and denoting the unit vector from $\mathbf{r}_Q$ to $\mathbf{r}$ as $\mathbf{u}_s$, the direction of the spring force is given as

$$\mathbf{u}_s = \frac{\mathbf{r} - \mathbf{r}_Q}{\|\mathbf{r} - \mathbf{r}_Q\|} \tag{3-13}$$

It can be seen from Eq. (3–11) that the quantity $\ell - \ell_0$ can be either positive or negative. Therefore, we need to examine two cases.

Case 1: $\ell - \ell_0 < 0$ ***(Compression)***

In the case where $\ell - \ell_0 < 0$, the spring is said to be *compressed.* In the case of compression, the spring exerts a force on the particle in the direction from $\mathbf{r}_Q$ to $\mathbf{r}$, i.e., the spring force is in the direction of $\mathbf{u}_s$. However, because $\ell - \ell_0 < 0$ when the spring is compressed, we have

$$|\ell - \ell_0| = -(\ell - \ell_0) \tag{3-14}$$

The force of the spring in compression is then given as

$$\mathbf{F}_s = -K(\ell - \ell_0)\mathbf{u}_s \tag{3-15}$$

Case 2: $\ell - \ell_0 > 0$ ***(Extension)***

In the case where $\ell - \ell_0 > 0$, the spring is said to be *extended* or *stretched.* In the case of extension, the spring exerts a force on the particle in the direction $\mathbf{r}$ to $\mathbf{r}_Q$, i.e., the spring force is in the direction of $-\mathbf{u}_s$. However, because $\ell - \ell_0 > 0$ when the spring is compressed, we have

$$|\ell - \ell_0| = (\ell - \ell_0) \tag{3-16}$$

The force of the spring in extension is then given as

$$\mathbf{F}_s = K(\ell - \ell_0)\,[-\mathbf{u}_s] \tag{3-17}$$

Equation (3–17) can be rewritten as

$$\mathbf{F}_s = -K(\ell - \ell_0)\mathbf{u}_s \tag{3-18}$$

The results of the two cases show that, *regardless* of whether the spring is compressed or extended, the force exerted by the spring on the particle is given as

$$\boxed{\mathbf{F}_s = -K(\ell - \ell_0)\mathbf{u}_s} \tag{3-19}$$

where ℓ is the length of the spring and is obtained from Eq. (3–10), ℓ_0 is the unstretched length of the spring, and $\mathbf{u}_s$ is obtained from Eq. (3–13).

It is important to note that the aforementioned model for spring force is an approximation based on experimental evidence. Consequently, Hooke's law is not always valid. For example, it is clear that a *real* spring cannot be stretched to an arbitrarily large length (since the spring would break or permanently deform so that it would not return to its original shape) nor can a real spring be compressed to a length of zero (since the spring has mass and, thus, occupies a nonzero amount of space). Nevertheless, for many engineering applications where the spring stretches or compresses to a "reasonable" length, the model for a linear spring obtained in Eq. (3–19) is an excellent approximation.

Curvilinear Spring

Consider a particle attached to a massless spring and moving in an arbitrary reference frame $\mathcal{A}$ such that the particle slides along a rigid curve C (i.e., C defines a reference frame) and the spring conforms to the shape of C as shown in Fig. 3–4. Furthermore, suppose that one end of the spring is located at a point Q on C while the other end of the spring is attached to the particle (it is noted that point Q may either be fixed or moving relative to C). Finally, let the force generated by the spring have the form

$$\boxed{\mathbf{F}_s = -K(\ell - \ell_0)\mathbf{e}_t} \tag{3–20}$$

where K is the spring constant, ℓ is the stretched length of the spring along the curve C, ℓ_0 is the unstretched length of the spring, and $\mathbf{e}_t$ is the tangent vector at point P as viewed by an observer fixed to C. A spring whose force satisfies Eq. (3–20) is called a *curvilinear spring*; a schematic of a curvilinear spring is shown in Fig. 3–4.

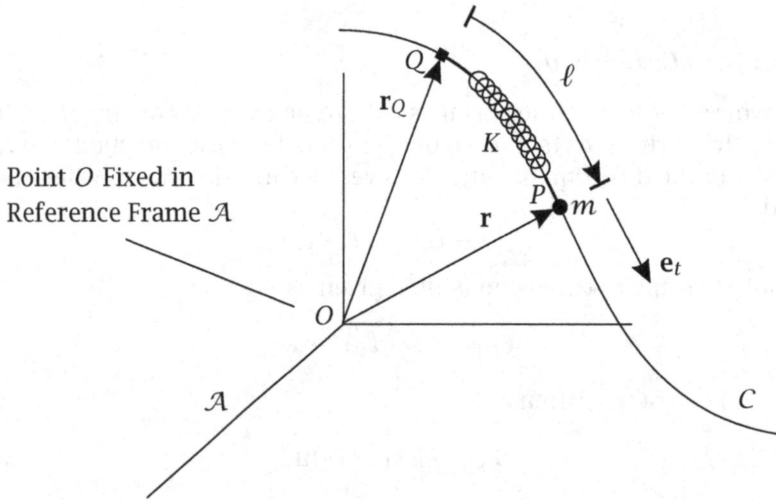

Figure 3–4 Particle sliding on a curve C while attached to a curvilinear spring with spring constant K and unstretched length ℓ_0.

As with a linear spring, in the case of either compression or extension, the force generated by a curvilinear spring satisfies Eq. (3–20). Moreover, it is seen from Eq. (3–20) that a curvilinear spring is a generalization of a linear spring to the case where the spring bends to conform to the shape of an arbitrary curve.

3.1.3 Central Forces

Let $\mathcal{A}$ be an arbitrary reference frame and let O be a point fixed in reference frame $\mathcal{A}$. Furthermore, let $\mathbf{r}$ be the position of a particle of mass m where $\mathbf{r}$ is measured relative to point O as shown in Fig. 3-5. Then a force $\mathbf{F}_c$ acting on m is said to be a *central force* relative to point O if $\mathbf{F}_c$ acts along the line connecting the particle and point O. A central force can be written as

$$\mathbf{F}_c = F_c \mathbf{e}_r \tag{3-21}$$

where $\mathbf{e}_r$ is the unit vector in the direction of $\mathbf{r}$, i.e.,

$$\mathbf{e}_r = \frac{\mathbf{r}}{\|\mathbf{r}\|} \tag{3-22}$$

Using Eq. (3-21) together with Eq. (3-22), a central force can be written as

$$\mathbf{F}_c = F_c \frac{\mathbf{r}}{\|\mathbf{r}\|} \tag{3-23}$$

It is important to note that, because the line of action of $\mathbf{F}_c$ matters (i.e., the force $\mathbf{F}_c$ lies along the particular line between the particle and point O), a central force is a *sliding vector*.

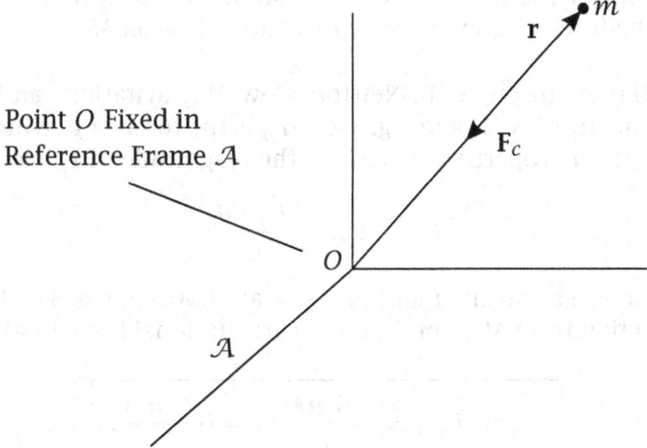

Figure 3-5 Particle under the influence of a central force $\mathbf{F}_c$ relative to a point O.

3.1.4 Gravitational Forces

Newton's Law of Gravitation

A particular case of a central force is a so-called gravitational force. *Gravitational forces* arise from the attraction of one body on another. This law is commonly known as *Newton's law of gravitation*. Consider two particles of mass M and m, respectively. Then Newton's law of gravitation states the following:

(1) The force of gravitational attraction of M on m, denoted $\mathbf{F}_g$, is equal and opposite to the force of attraction of m on M.

(2) The force of gravitational attraction between M and m is inversely proportional to the square of the distance between M and m.

(3) The force of gravitational attraction is proportional to the product of M and m.

(4) The force of gravitational attraction of M on m lies in the direction from m to M while the force of gravitational attraction of m on M lies in the direction from M to m.

Figure 3-6 shows a schematic of Newton's law of gravitation.

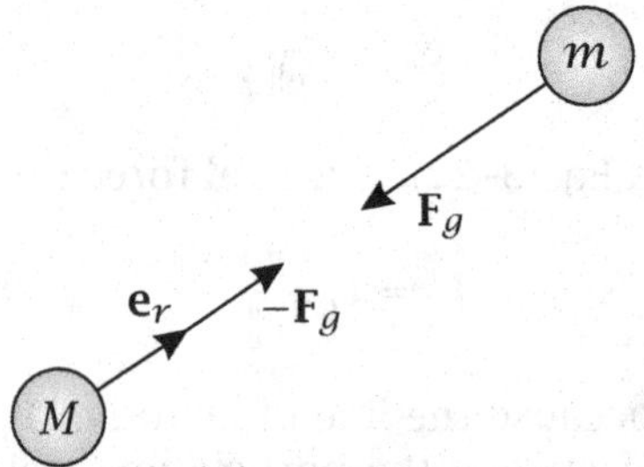

Figure 3-6 Schematic of Newton's law of gravitation between particles of mass M and m. The quantity $\mathbf{F}_g$ is the force of gravitational attraction of M on m while the quantity $-\mathbf{F}_g$ is the force of gravitational attraction of m on M.

Using Properties (1) through (4), Newton's law of gravitation can be quantified as follows. First, from Property (1) and Fig. 3-6, $-\mathbf{F}_g$ is the force of gravitational attraction of m on M. Next, from Properties (2) and (3), the magnitude of $\mathbf{F}_g$ can be written as

$$F_g = \|\mathbf{F}_g\| = \frac{GmM}{r^2} \tag{3-24}$$

where r is the distance between M and m. Now, as shown in Fig. 3-6, let $\mathbf{e}_r$ be the unit vector in the direction from M to m. Then the gravitational force of attraction of M on m is given as

$$\boxed{\mathbf{F}_g = -F_g\mathbf{e}_r = -\frac{GmM}{r^2}\mathbf{e}_r = -\frac{GmM}{r^3}\mathbf{r}} \tag{3-25}$$

Similarly, the gravitational force of attraction of m on M is given as

$$-\mathbf{F}_g = F_g\mathbf{e}_r = \frac{GmM}{r^2}\mathbf{e}_r = \frac{GmM}{r^3}\mathbf{r} \tag{3-26}$$

The quantity G is called the *constant of gravitation.*[3]

3.1.5 Force of Gravity

A special case of a gravitational force is the gravitational force exerted by a large body on a small body when the small body is in close proximity to the large body. An example of such a situation is a small mass near a planet. Suppose we let M be

[3]The approximate numerical values of G in SI units and English units are $G = 6.673 \times 10^{-11}\text{m}^3/(\text{kg s}^2)$ and $G = 3.44 \times 10^{-8}\ \text{ft}^4/(\text{lb s}^4)$, respectively.

the mass of the large body and let m be the mass of the small body where $m \ll M$. Furthermore, let the large body be a homogeneous sphere of radius R and center at point O. Consequently, it can be proved that the force of gravitational attraction of the large body on the small body is equivalent to that of a particle of mass M located at the center of the sphere.[4]). Finally, let the small body of mass m be located a distance h from the surface of the large spherical body (where $h \ll R$) as shown in Fig. 3-7.

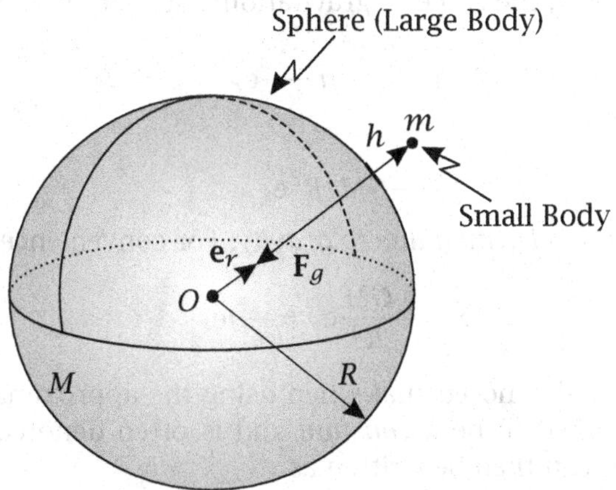

Figure 3-7 Schematic of force of gravitational attraction, $\mathbf{F}_g$, of a large homogeneous sphere of mass M on a small body of mass $m \ll M$.

Then, from Newton's law of gravitation, the force of gravitational attraction of M on m is given as

$$\mathbf{F}_g = -\frac{GmM}{r^2}\mathbf{e}_r = -\frac{GmM}{(R+h)^2}\mathbf{e}_r \tag{3-27}$$

where h is the height above the sphere in the direction of $\mathbf{e}_r$, where $\mathbf{e}_r$ is the unit vector in the direction from O to m. We can see from Eq. (3-27) that the magnitude of $\mathbf{F}_g$ is given as

$$\|\mathbf{F}_g\| = F_g = \frac{GmM}{(R+h)^2} \tag{3-28}$$

Now we can rewrite the term $R + h$ as

$$R + h = R\left(1 + \frac{h}{R}\right) \tag{3-29}$$

Substituting $R + h$ from Eq. (3-29) into (3-28), we obtain

$$F_g = \frac{GmM}{R^2\left(1 + \frac{h}{R}\right)^2} \tag{3-30}$$

Now, because $h \ll R$, we have

$$h/R \ll 1 \tag{3-31}$$

[4]See Greenwood (1988) for a proof that the force of gravitational attraction of a homogeneous of mass M and radius R is equivalent to the force exerted by a particle of mass M located at the center of the sphere.

Consequently,

$$\left(1+\frac{h}{R}\right)^2 \approx 1 \tag{3-32}$$

which implies that

$$F_g \approx \frac{GmM}{R^2} = m\frac{GM}{R^2} \tag{3-33}$$

Then, for small values of h, the force of gravitational attraction of M on m is given as

$$\mathbf{F}_g \approx -m\frac{GM}{R^2}\mathbf{e}_r \tag{3-34}$$

The quantity

$$-GM/R^2\mathbf{e}_r \tag{3-35}$$

is often called the *local acceleration due to gravity.* For convenience, we write

$$\mathbf{g} = -\frac{GM}{R^2}\mathbf{e}_r = -g\mathbf{e}_r \tag{3-36}$$

where $g = GM/R^2$. It is also noted that when using the approximation of Eq. (3-36), the direction $\mathbf{e}_r$ is assumed to be a *constant* and is often denoted $\mathbf{e}_v$. The force of gravitation in Eq. (3-34) can then be written as

$$\mathbf{F}_g = -mg\mathbf{e}_v \tag{3-37}$$

For convenience, the direction $\mathbf{e}_v$ is called the *local vertical* direction or simply the *vertical* direction. Furthermore, the acceleration due to gravity, denoted $\mathbf{g}$, is given as

$$\mathbf{g} = -g\mathbf{e}_v \tag{3-38}$$

Then the force of gravity can be written as

$$\boxed{\mathbf{F}_g = m\mathbf{g}} \tag{3-39}$$

It is noted that for some problems the direction of $\mathbf{F}_g$ in Eq. (3-39) may be more conveniently specified using the direction *opposite* the local vertical, in which case the force of gravity is given as

$$\mathbf{F}_g = mg\mathbf{u}_v \tag{3-40}$$

where $\mathbf{u}_v = -\mathbf{e}_v$.

3.2 Inertial Reference Frames

Before proceeding to describe Newton's laws, it is important to emphasize that, while motion can be observed in an arbitrary (noninertial) reference frame, the laws of motion for a particle are valid only in an inertial reference frame. Therefore, from this point forth, with only a few exceptions, all kinematic quantities will be defined relative to an inertial reference frame. For clarity, in the development of the theory, the left superscript $\mathcal{N}$ will always denote motion *as viewed by an observer in an inertial reference frame.* Finally, when solving problems, the left superscript $\mathcal{F}$ will denote a fixed inertial reference frame while any other calligraphic letter (e.g., $\mathcal{A}$, $\mathcal{B}$) will denote a general (noninertial) reference frame.

3.3 Newton's Laws for a Particle

We now state the axioms that govern the motion of particles. These axioms are called *Newton's laws* and are given as follows:

Newton's 1st Law:

An object in motion tends to remain in motion unless acted upon by an external force. An object at rest tends to remain at rest unless acted upon by an outside force.

Newton's 2nd Law:[5]

Let **F** *be the* total *or* resultant *force acting on a particle of mass* m*. Then*

$$\boxed{\mathbf{F} = m\,{}^{\mathcal{N}}\mathbf{a}} \tag{3-41}$$

where ${}^{\mathcal{N}}\mathbf{a} \equiv \mathbf{a}$ *is the acceleration of the particle as viewed by an observer in any inertial reference frame* $\mathcal{N}$ (Newton, 1687).[6]

Newton's 3rd Law:

For every action, there is an equal and opposite reaction.

3.4 Comments on Newton's Laws

It can be seen that the statements of Newton's laws are extremely simple. However, it is important for the reader to understand that the application of Newton's laws is often a highly nontrivial matter. First, because Newton's laws are valid only in an inertial reference frame, it is necessary to determine the acceleration of the particle in an inertial reference frame using observations of the motion in a noninertial reference frame. Next, once an observation reference frame has been chosen, it is necessary to specify an appropriate coordinate system in which to quantify the motion. While in principle *any* coordinate system may be used, choosing an inappropriate coordinate system can result in extremely tedious algebra and can distort the key aspects of a particular problem. Consequently, it is extremely important to think carefully and to choose a coordinate system that is well suited to the geometry of the problem being solved. The acceleration of the particle in an inertial reference frame must then be computed in terms of the chosen coordinate system. Finally, the forces acting on the particle must be resolved in the chosen coordinate system. While it may seem obvious, it is not only necessary to take great care to ensure that each force is specified in the proper direction, but it is also necessary to examine carefully the problem to ensure that no forces are omitted.

[5] Newton's 2nd law is an example of *Galilean invariance* or *Newtonian relativity*.

[6] Isaac Newton (1642–1727) wrote his 2nd law in Volume 1 of his famous *Philosophiae Naturalis Principia Mathematica* (better known as *Principia*). Newton's *Principia* was originally published in 1687. An English translation of *Principia* is given in Newton (1687).

3.5 Examples of Application of Newton's Laws in Particle Dynamics

In this section we apply Newton's laws to some representative problems in particle dynamics. In order to illustrate the concepts in a clear manner, the examples are solved using an approach that is completely consistent with the previously developed theory.

Example 3–1

A rigid massless arm of length l is hinged at the fixed point O as shown in Fig. 3-8. Attached to the arm at its other end is a particle of mass m. Knowing that the particle is under the influence of gravity (where gravity acts vertically downward), determine (a) the differential equation of motion for the particle in terms of the angle θ and (b) the tension in the arm as a function of the angle θ.

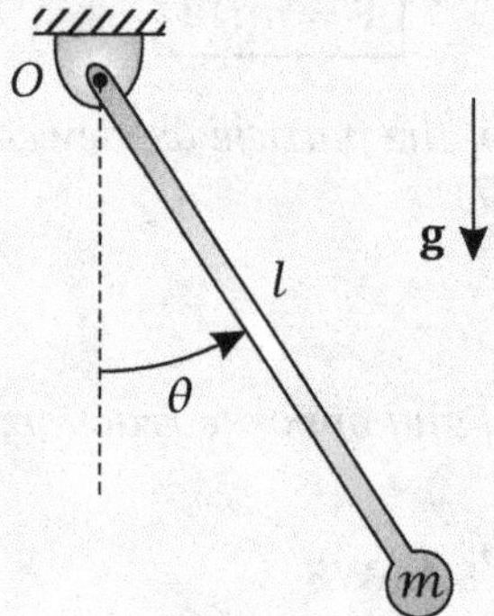

Figure 3–8 Particle of mass m connected to a rigid massless arm of length l.

Solution to Example 3–1

Kinematics

First, let $\mathcal{F}$ be a fixed inertial reference frame. Furthermore, let us choose the following coordinate system that is fixed in $\mathcal{F}$:

$$\begin{array}{ccc} & \text{Origin at point } O & \\ \mathbf{E}_x & = & \text{Along } Om \text{ at } \theta = 0 \\ \mathbf{E}_z & = & \text{Out of page} \\ \mathbf{E}_y & = & \mathbf{E}_z \times \mathbf{E}_x \end{array}$$

Next, for this problem it is useful to make observations regarding the motion of the system using a noninertial reference frame, denoted $\mathcal{R}$, that is fixed in the arm and, thus, rotates relative to $\mathcal{F}$. A convenient coordinate system that is fixed in reference frame $\mathcal{R}$ is as follows:

$$\begin{array}{ccc} & \text{Origin at point } O & \\ \mathbf{e}_r & = & \text{Along } Om \\ \mathbf{e}_z & = & \text{Out of page} \\ \mathbf{e}_\theta & = & \mathbf{e}_z \times \mathbf{e}_r \end{array}$$

The geometry of the bases $\{\mathbf{E}_x, \mathbf{E}_y, \mathbf{E}_z\}$ and $\{\mathbf{e}_r, \mathbf{e}_\theta, \mathbf{E}_z\}$ is shown in Fig. 3-9.

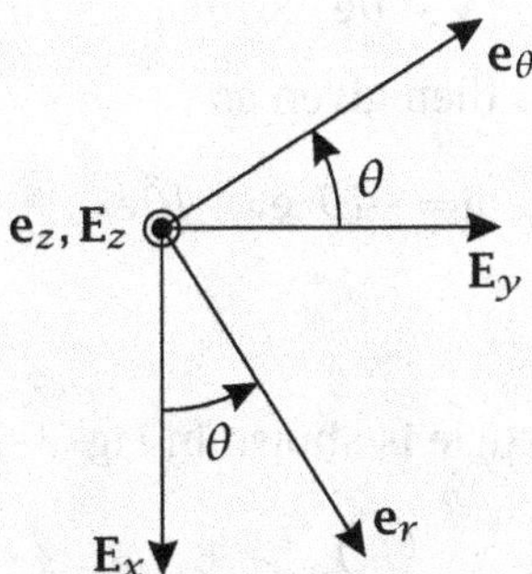

Figure 3-9 Geometry of bases $\{\mathbf{E}_x, \mathbf{E}_y, \mathbf{E}_z\}$ and $\{\mathbf{e}_r, \mathbf{e}_\theta, \mathbf{e}_z\}$ for Example 3-1.

Using Fig. 3-9, we have

$$\mathbf{E}_x = \cos\theta\,\mathbf{e}_r - \sin\theta\,\mathbf{e}_\theta \tag{3-42}$$

$$\mathbf{E}_y = \sin\theta\,\mathbf{e}_r + \cos\theta\,\mathbf{e}_\theta \tag{3-43}$$

The position of the particle in terms of the basis $\{\mathbf{e}_r, \mathbf{e}_\theta, \mathbf{e}_z\}$ is

$$\mathbf{r} = l\mathbf{e}_r \tag{3-44}$$

Because reference frame $\mathcal{R}$ is fixed to the arm and θ is the angle formed by the arm with the (fixed) vertically downward direction, the angular velocity of reference frame $\mathcal{R}$ in $\mathcal{F}$ is given as

$${}^{\mathcal{F}}\boldsymbol{\omega}^{\mathcal{R}} = \dot{\theta}\mathbf{e}_z \tag{3-45}$$

The velocity of the particle in reference frame $\mathcal{F}$ is then given as

$${}^{\mathcal{F}}\mathbf{v} = \frac{{}^{\mathcal{R}}d\mathbf{r}}{dt} + {}^{\mathcal{F}}\boldsymbol{\omega}^{\mathcal{R}} \times \mathbf{r} \tag{3-46}$$

Because the length of the pendulum is constant, we have

$$\frac{{}^{\mathcal{R}}d\mathbf{r}}{dt} = \mathbf{0} \tag{3-47}$$

Furthermore,

$${}^{\mathcal{F}}\boldsymbol{\omega}^{\mathcal{R}} \times \mathbf{r} = \dot{\theta}\mathbf{e}_z \times l\mathbf{e}_r = l\dot{\theta}\mathbf{e}_\theta \tag{3-48}$$

Adding the results of Eqs. (3-47) and (3-48), we obtain the velocity as

$${}^{\mathcal{F}}\mathbf{v} = l\dot{\theta}\mathbf{e}_\theta \tag{3-49}$$

The acceleration is then given as

$${}^{\mathcal{F}}\mathbf{a} = \frac{{}^{\mathcal{F}}d}{dt}\left({}^{\mathcal{F}}\mathbf{v}\right) = \frac{{}^{\mathcal{R}}d}{dt}\left({}^{\mathcal{F}}\mathbf{v}\right) + {}^{\mathcal{F}}\boldsymbol{\omega}^{\mathcal{R}} \times {}^{\mathcal{F}}\mathbf{v} \tag{3-50}$$

Now we have

$$\frac{{}^{\mathcal{R}}d}{dt}\left({}^{\mathcal{F}}\mathbf{v}\right) = l\ddot{\theta}\mathbf{e}_\theta \tag{3-51}$$

and

$$^{\mathcal{F}}\boldsymbol{\omega}^{\mathcal{R}} \times {}^{\mathcal{F}}\mathbf{v} = \dot{\theta}\mathbf{e}_z \times l\dot{\theta}\mathbf{e}_\theta = -l\dot{\theta}^2\mathbf{e}_r \tag{3-52}$$

The acceleration of the particle is then given as

$$^{\mathcal{F}}\mathbf{a} = -l\dot{\theta}^2\mathbf{e}_r + l\ddot{\theta}\mathbf{e}_\theta \tag{3-53}$$

Kinetics

The free body diagram of the particle is shown in Fig. 3-10.

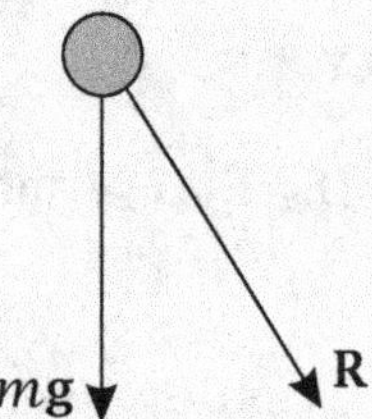

Figure 3-10 Free body diagram of particle for Example 3-1.

Using Fig. 3-10, it can be seen that the following two forces act on the particle:

$m\mathbf{g}$ = Force of gravity
$\mathbf{R}$ = Reaction force exerted by arm on particle

Now, because the force of gravity acts vertically downward, we have

$$m\mathbf{g} = mg\mathbf{E}_x \tag{3-54}$$

Then, substituting the result of Eq. (3-42) into (3-54), the force of gravity is obtained as

$$m\mathbf{g} = mg\cos\theta\,\mathbf{e}_r - mg\sin\theta\,\mathbf{e}_\theta \tag{3-55}$$

Next, the reaction force of the arm on the particle, **R**, must lie in along the arm, i.e., **R** must lie in the $\mathbf{e}_r$-direction. However, because it is not known at this point whether **R** lies in the positive or negative $\mathbf{e}_r$-direction, the *sign* of the reaction force is not known. Therefore, we must specify **R** as

$$\mathbf{R} = R\mathbf{e}_r \tag{3-56}$$

where the scalar R must be determined. Using the expressions for $m\mathbf{g}$ from Eq. (3-55) and **R** from Eq. (3-56), the resultant force acting on the particle can be written as

$$\mathbf{F} = m\mathbf{g} + \mathbf{R} = mg\cos\theta\,\mathbf{e}_r - mg\sin\theta\,\mathbf{e}_\theta + R\mathbf{e}_r \tag{3-57}$$

Combining $\mathbf{e}_r$ and $\mathbf{e}_\theta$ components, Eq. (3-57) can be rewritten as

$$\mathbf{F} = (mg\cos\theta + R)\mathbf{e}_r - mg\sin\theta\,\mathbf{e}_\theta \tag{3-58}$$

Then, applying Newton's 2nd, we have

$$\mathbf{F} = m\,{}^{\mathcal{N}}\mathbf{a} \tag{3-59}$$

where $\mathcal{N}$ is any inertial reference frame. Noting that $\mathcal{F}$ is an inertial reference frame, we can substitute **F** from Eq. (3–58) and ${}^{\mathcal{F}}\mathbf{a}$ from Eq. (3–53) to give

$$(mg\cos\theta + R)\mathbf{e}_r - mg\sin\theta\,\mathbf{e}_\theta = m(-l\dot{\theta}^2\mathbf{e}_r + l\ddot{\theta}\mathbf{e}_\theta) = -ml\dot{\theta}^2\mathbf{e}_r + ml\ddot{\theta}\mathbf{e}_\theta \tag{3-60}$$

Equating the $\mathbf{e}_r$ and $\mathbf{e}_\theta$ components, we obtain the following two scalar equations:

$$mg\cos\theta + R = -ml\dot{\theta}^2 \tag{3-61}$$

$$-mg\sin\theta = ml\ddot{\theta} \tag{3-62}$$

(a) Differential Equation of Motion in Terms of θ

In this example there is only one variable used to describe the motion, namely the angle θ. Consequently, only one differential equation is required to describe the motion. In order to obtain the differential equation, we need to determine an expression that is a function of θ and time derivatives of θ along with any constants given in the problem. It is noted that the differential equation *cannot* contain any reaction forces because the reaction forces are unknown. In this example, the differential equation arises directly from the equations that result from the application of Newton's 2nd law. In particular, Eq. (3–62) is the differential equation of motion. Consequently, we have

$$ml\ddot{\theta} = -mg\sin\theta \tag{3-63}$$

Rearranging Eq. (3–63) and dropping m from both sides, we obtain

$$\ddot{\theta} = -\frac{g}{l}\sin\theta \tag{3-64}$$

(b) Tension in Arm as a Function of Angle θ

Solving Eq. (3–61) for R, we obtain

$$R = -mg\cos\theta - ml\dot{\theta}^2 \tag{3-65}$$

It can be seen from Eq. (3–56) that R is a function of $\dot{\theta}^2$. Consequently, in order to obtain an expression for the tension in the arm as a function of the angle θ, it is necessary to obtain an expression for $\dot{\theta}^2$ in terms of θ. An expression for $\dot{\theta}^2$ in terms of θ can be obtained from Eq. (3–64). In particular, we see from Eq. (3–64) that the motion is a function of the single variable θ. Consequently, the results of rectilinear motion from Section 2.7 can be applied. In this case, the *angular acceleration* $\ddot{\theta}$ is a function of the *angular position* θ. Hence we are in Case 3 as described in Section 2.7 on page 41. First, we can apply Eq. (2–57) by making the following substitutions:

$$\begin{array}{ccc} x & \longrightarrow & \theta \\ v & \longrightarrow & \dot{\theta} \\ a & \longrightarrow & \ddot{\theta} \end{array}$$

We then have

$$\ddot{\theta} = \frac{d}{dt}(\dot{\theta}) = \frac{d\dot{\theta}}{d\theta}\frac{d\theta}{dt} = \dot{\theta}\frac{d\dot{\theta}}{d\theta} \tag{3-66}$$

Next, substituting the expression for $\ddot{\theta}$ into Eq. (3-64), we obtain

$$\dot{\theta}\frac{d\dot{\theta}}{d\theta} = -\frac{g}{l}\sin\theta \tag{3-67}$$

Using the method of separation of variables (Kreyszig, 1988), we have

$$\dot{\theta}d\dot{\theta} = -\frac{g}{l}\sin\theta\, d\theta \tag{3-68}$$

Then, integrating both sides of Eq. (3-68) gives

$$\int_{\dot{\theta}_0}^{\dot{\theta}} v\,dv = \int_{\theta_0}^{\theta} -\frac{g}{l}\sin\eta\, d\eta \tag{3-69}$$

where v and η are dummy variables of integration. Computing the integrals on both sides of Eq. (3-69), we obtain

$$\left.\frac{v^2}{2}\right|_{\dot{\theta}_0}^{\dot{\theta}} = \left.\frac{g}{l}\cos\eta\right|_{\theta_0}^{\theta} \tag{3-70}$$

Simplifying Eq. (3-70), we obtain

$$\dot{\theta}^2 - \dot{\theta}_0^2 = \frac{2g}{l}(\cos\theta - \cos\theta_0) \tag{3-71}$$

Therefore,

$$\dot{\theta}^2 = \dot{\theta}_0^2 + \frac{2g}{l}(\cos\theta - \cos\theta_0) \tag{3-72}$$

Then, substituting the expression for $\dot{\theta}^2$ from Eq. (3-72) into (3-65), we obtain

$$R = -mg\cos\theta - ml\left[\dot{\theta}_0^2 + \frac{2g}{l}(\cos\theta - \cos\theta_0)\right] \tag{3-73}$$

Simplifying Eq. (3-73), we obtain

$$R = -3mg\cos\theta - ml\dot{\theta}_0^2 + 2mg\cos\theta_0 \tag{3-74}$$

The reaction force as a function of θ is then given as

$$\mathbf{R} = R\mathbf{e}_r = \left[-3mg\cos\theta - ml\dot{\theta}_0^2 + 2mg\cos\theta_0\right]\mathbf{e}_r \tag{3-75}$$

Comments on Solution Approach and Results Obtained

The current example demonstrates several important components that are common to all engineering dynamics problems. The first important component is *kinematics*. In Chapter 2 we studied methods to derive velocity and acceleration in various coordinate systems. The current example demonstrates the importance of choosing an appropriate coordinate system in which to solve the problem. Had a coordinate system different from cylindrical coordinates been chosen (e.g., a fixed Cartesian coordinate system), the equations resulting from the application of Newton's 2nd law would have been more complicated. Second, the choice of cylindrical coordinates made it easier to understand the motion. In particular, the motion is naturally decoupled into radial and transverse motion using cylindrical coordinates. Therefore, had a different coordinate system been chosen, the equations resulting from Newton's 2nd law would have been more difficult to interpret.

The second important component is the application of Newton's 2nd law. In this example we had two forces acting on the particle. The first force, namely that of gravity, was known. Consequently, it was necessary to properly specify both its magnitude *and* direction. In particular, for this example gravity acted vertically downward. Therefore, it was necessary to ensure that the force of gravity was placed in the *positive* $\mathbf{E}_x$-direction. The second force, namely that of the reaction force of the arm, was unknown in the sense that its sign was not known. Therefore, it was not important whether this force was initially chosen to be in the $\mathbf{e}_r$-direction or in the negative $\mathbf{e}_r$-direction; the proper sign of R was determined from the algebra.

The third important component was the ability to apply engineering mathematics. In this example we were asked specifically to determine the tension in the arm as a function of the angle θ. Once the kinematics and kinetics were completed, the original dynamics problem was turned into a math problem. In order to solve this math problem, it was necessary to recall fundamental concepts from algebra and calculus. While, strictly speaking, dynamics is not mathematics, it is virtually impossible to solve any dynamics problem without sufficient knowledge of mathematics. Consequently, the reader should keep in mind the importance of mathematics in the formulation and solution of any dynamics problem.

■

Example 3–2

A particle of mass m slides without friction along a fixed track in the form of a parabola as shown in Fig. 3–11. The equation for the parabola is

$$y = \frac{r^2}{2R} \tag{3–76}$$

where R is a constant, r is the distance from point O to point Q, point Q is the projection of point P onto the horizontal direction, and y is the vertical distance. Knowing that gravity acts downward and that the initial conditions are $r(t = 0) = r_0$ and $\dot{r}(t = 0) = 0$, determine (a) the differential equation of motion for the particle in terms of the variable r, and (b) an expression for $\dot{r}^2$ as a function of r.

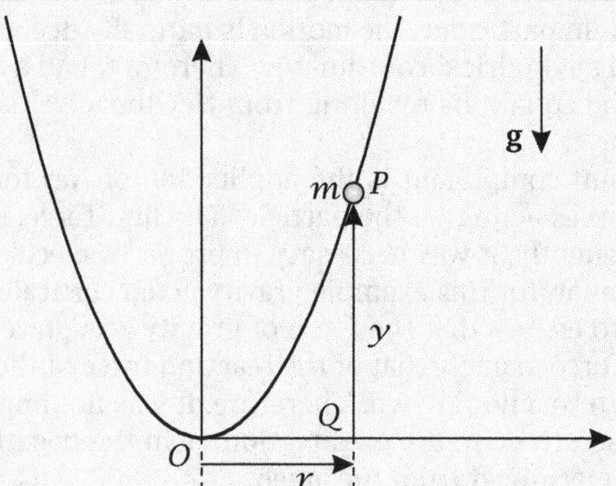

Figure 3–11 Particle sliding without friction along a track in the shape of a parabola.

Solution to Example 3–2

Kinematics

For this problem it is not necessary to use a noninertial reference frame to describe the motion of the particle; an inertial reference frame fixed to the parabola, denoted $\mathcal{F}$, will suffice. Corresponding to $\mathcal{F}$, we choose the following Cartesian coordinate system to express the motion of the particle:

Origin at point O		
$\mathbf{E}_x$	=	Along OQ
$\mathbf{E}_z$	=	Out of page
$\mathbf{E}_y$	=	$\mathbf{E}_z \times \mathbf{E}_x$

The position of the particle is given in terms of the coordinate system $\{\mathbf{E}_x, \mathbf{E}_y, \mathbf{E}_z\}$ as

$$\mathbf{r} = r\mathbf{E}_x + y\mathbf{E}_y \tag{3–77}$$

Substituting Eq. (3–76) into (3–77), the position can be written as

$$\mathbf{r} = r\mathbf{E}_x + \frac{r^2}{2R}\mathbf{E}_y \tag{3–78}$$

Now since we have chosen a fixed (inertial) reference frame in which to observe, the motion of the particle, we have, trivially,

$${}^{\mathcal{F}}\boldsymbol{\omega}^{\mathcal{F}} = \mathbf{0} \tag{3–79}$$

Then, computing the rate of change of $\mathbf{r}$ in reference frame $\mathcal{F}$, the velocity of the particle in reference frame $\mathcal{F}$ is obtained as

$${}^{\mathcal{F}}\mathbf{v} = \frac{{}^{\mathcal{F}}d\mathbf{r}}{dt} = \dot{r}\mathbf{E}_x + \frac{r\dot{r}}{R}\mathbf{E}_y \tag{3–80}$$

Finally, computing the rate of change of ${}^{\mathcal{F}}\mathbf{v}$ in reference frame $\mathcal{F}$, the acceleration of the particle in reference frame $\mathcal{F}$ is obtained as

$$ {}^{\mathcal{F}}\mathbf{a} = \frac{{}^{\mathcal{F}}d}{dt}\left({}^{\mathcal{F}}\mathbf{v}\right) = \ddot{r}\mathbf{E}_x + \frac{\dot{r}^2 + r\ddot{r}}{R}\mathbf{E}_y \tag{3-81} $$

Kinetics

The free body diagram of the particle is shown in Fig. 3-12.

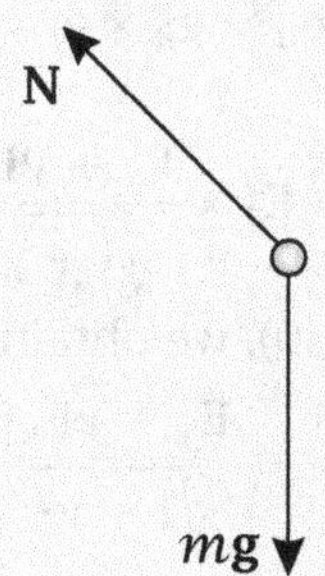

Figure 3-12 Free body diagram of particle for Example 3-2.

Using Fig. 3-12, it is seen that the following two forces act on the particle:

$\mathbf{N}$	=	Reaction force of track on particle
$m\mathbf{g}$	=	Force of gravity

Given the geometry, the forces $\mathbf{N}$ and $m\mathbf{g}$ can be expressed as

$$ \begin{aligned} \mathbf{N} &= N\mathbf{e}_n \\ m\mathbf{g} &= -mg\mathbf{E}_y \end{aligned} \tag{3-82} $$

where $\mathbf{e}_n$ is the inward pointing normal to the track at $\mathbf{r}$. The direction $\mathbf{e}_n$ can be determined by constructing an intrinsic basis in terms of the basis $\{\mathbf{E}_x, \mathbf{E}_y, \mathbf{E}_z\}$ as follows. First, we know that the tangent vector $\mathbf{e}_t$ in reference frame $\mathcal{F}$ is given as

$$ \mathbf{e}_t = \frac{{}^{\mathcal{F}}\mathbf{v}}{\|{}^{\mathcal{F}}\mathbf{v}\|} = \frac{{}^{\mathcal{F}}\mathbf{v}}{{}^{\mathcal{F}}v} \tag{3-83} $$

Then, using ${}^{\mathcal{F}}\mathbf{v}$ from Eq. (3-80), we obtain the speed in reference frame $\mathcal{F}$, ${}^{\mathcal{F}}v$ as

$$ {}^{\mathcal{F}}v = \left\|\dot{r}\mathbf{E}_x + \frac{r\dot{r}}{R}\mathbf{E}_y\right\| = \dot{r}\sqrt{1 + \left(\frac{r}{R}\right)^2} \tag{3-84} $$

Suppose now that we let

$$ y = \frac{r}{R} \tag{3-85} $$

Then the tangent vector can be written as

$$ \mathbf{e}_t = \frac{\dot{r}\mathbf{E}_x + \dot{r}y\mathbf{E}_y}{\dot{r}\sqrt{1 + y^2}} \tag{3-86} $$

Equation (3–86) simplifies to

$$\mathbf{e}_t = \frac{\mathbf{E}_x + y\mathbf{E}_y}{\sqrt{1+y^2}} \tag{3-87}$$

Next, because the motion is two-dimensional, the principal unit bi-normal vector, $\mathbf{e}_b$, must lie in the direction orthogonal to the plane of motion. Consequently, we have

$$\mathbf{e}_b = \mathbf{E}_z \tag{3-88}$$

The unit normal vector, $\mathbf{e}_n$, is then given as

$$\mathbf{e}_n = \mathbf{e}_b \times \mathbf{e}_t \tag{3-89}$$

Therefore,

$$\mathbf{e}_n = \mathbf{E}_z \times \frac{\mathbf{E}_x + y\mathbf{E}_y}{\sqrt{1+y^2}} \tag{3-90}$$

Taking the vector products in Eq. (3–90), we obtain

$$\mathbf{e}_n = \frac{\mathbf{E}_y - y\mathbf{E}_x}{\sqrt{1+y^2}} \tag{3-91}$$

Rearranging Eq. (3–91), we have

$$\mathbf{e}_n = \frac{-y\mathbf{E}_x + \mathbf{E}_y}{\sqrt{1+y^2}} \tag{3-92}$$

Then the reaction force $\mathbf{N}$ is given as

$$\mathbf{N} = N\mathbf{e}_n = N\left[\frac{-y\mathbf{E}_x + \mathbf{E}_y}{\sqrt{1+y^2}}\right] \tag{3-93}$$

The resultant force, $\mathbf{F}$, on the particle is then given as

$$\mathbf{F} = \mathbf{N} + m\mathbf{g} = N\left[\frac{-y\mathbf{E}_x + \mathbf{E}_y}{\sqrt{1+y^2}}\right] - mg\mathbf{E}_y \tag{3-94}$$

Rearranging Eq. (3–94), we have

$$\mathbf{F} = -N\frac{y}{\sqrt{1+y^2}}\mathbf{E}_x + \left[\frac{N}{\sqrt{1+y^2}} - mg\right]\mathbf{E}_y \tag{3-95}$$

Then, applying Newton's 2nd law by setting $\mathbf{F}$ equal to $m\,{}^{\mathcal{F}}\mathbf{a}$ using ${}^{\mathcal{F}}\mathbf{a}$ from Eq. (3–81), we obtain

$$-N\frac{y}{\sqrt{1+y^2}}\mathbf{E}_x + \left[\frac{N}{\sqrt{1+y^2}} - mg\right]\mathbf{E}_y = m\ddot{r}\mathbf{E}_x + m\frac{\dot{r}^2 + r\ddot{r}}{R}\mathbf{E}_y \tag{3-96}$$

Equating components yields the following two scalar equations:

$$\begin{aligned} -N\frac{y}{\sqrt{1+y^2}} &= m\ddot{r} \\ \frac{N}{\sqrt{1+y^2}} - mg &= m\frac{\dot{r}^2 + r\ddot{r}}{R} \end{aligned} \tag{3-97}$$

(a) Differential Equation of Motion in Terms of r

Equation (3-97) can be rewritten as

$$\begin{aligned} -N\frac{y}{\sqrt{1+y^2}} &= m\ddot{r} \\ \frac{N}{\sqrt{1+y^2}} &= mg + m\frac{\dot{r}^2 + r\ddot{r}}{R} \end{aligned} \tag{3-98}$$

Dividing the two expressions in Eq. (3-98), we obtain

$$-y = \frac{m\ddot{r}}{mg + m\dfrac{\dot{r}^2 + r\ddot{r}}{R}} \tag{3-99}$$

Simplifying Eq. (3-99) by dropping m, we obtain

$$-y = \frac{\ddot{r}}{g + \dfrac{\dot{r}^2 + r\ddot{r}}{R}} \tag{3-100}$$

Recalling from Eq. (3-85) that $y = r/R$, Eq. (3-100) can be written as

$$-\frac{r}{R} = \frac{\ddot{r}}{g + \dfrac{\dot{r}^2 + r\ddot{r}}{R}} \tag{3-101}$$

Rearranging Eq. (3-101), we obtain

$$-r\left[g + \frac{\dot{r}^2 + r\ddot{r}}{R}\right] = R\ddot{r} \tag{3-102}$$

We then obtain

$$-rg - \frac{r\dot{r}^2 + r^2\ddot{r}}{R} = R\ddot{r} \tag{3-103}$$

Multiplying Eq. (3-103) by R, we obtain

$$-Rrg - r\dot{r}^2 - r^2\ddot{r} = R^2\ddot{r} \tag{3-104}$$

Rearranging Eq. (3-104), we obtain the differential equation as

$$(r^2 + R^2)\ddot{r} + r(\dot{r}^2 + Rg) = 0 \tag{3-105}$$

(b) Expression for $\dot{r}^2$ as a Function of r

We start with the result from part (c), i.e., we start with Eq. (3-105). We know that

$$\ddot{r} = \frac{d\dot{r}}{dt} = \frac{d\dot{r}}{dr}\frac{dr}{dt} = \dot{r}\frac{d\dot{r}}{dr} \tag{3-106}$$

The expression for $\ddot{r}$ from Eq. (3-106) can then be substituted into Eq. (3-105). This gives

$$(r^2 + R^2)\dot{r}\frac{d\dot{r}}{dr} + r(\dot{r}^2 + Rg) = 0 \tag{3-107}$$

Rearranging Eq. (3–107), we obtain

$$(r^2 + R^2)\dot{r}\frac{d\dot{r}}{dr} = -r(\dot{r}^2 + Rg) \tag{3–108}$$

Separating r and $\dot{r}$ in Eq. (3–108), we have

$$\frac{\dot{r}}{\dot{r}^2 + Rg}d\dot{r} = -\frac{r}{r^2 + R^2}dr \tag{3–109}$$

Integrating both sides, we obtain

$$\int_{\dot{r}_0}^{\dot{r}} \frac{\nu}{\nu^2 + Rg}d\nu = \int_{r_0}^{r} -\frac{\eta}{\eta^2 + R^2}d\eta \tag{3–110}$$

where ν and η are dummy variables of integration. Then, from Appendix B, we have

$$\int \frac{\nu}{\nu^2 + a^2}d\nu = \tfrac{1}{2}\ln|\nu^2 + a^2| + C \tag{3–111}$$

Applying the result of Eq. (3–111) in (3–110), we obtain

$$\tfrac{1}{2}\left[\ln|\nu^2 + Rg|\right]_{\dot{r}_0}^{\dot{r}} = -\tfrac{1}{2}\left[\ln|\eta^2 + R^2|\right]_{r_0}^{r} \tag{3–112}$$

Equation (3–112) simplifies to

$$\left[\ln|\dot{r}^2 + Rg|\right]_{\dot{r}_0}^{\dot{r}} = -\left[\ln|r^2 + R^2|\right]_{r_0}^{r} \tag{3–113}$$

We then obtain

$$\ln|\dot{r}^2 + Rg| - \ln|\dot{r}_0^2 + Rg| = \ln|r_0^2 + R^2| - \ln|r^2 + R^2| \tag{3–114}$$

Simplifying Eq. (3–114), we have

$$\ln\left|\frac{\dot{r}^2 + Rg}{\dot{r}_0^2 + Rg}\right| = \ln\left|\frac{r_0^2 + R^2}{r^2 + R^2}\right| \tag{3–115}$$

Taking exponentials on both sides of Eq. (3–115), we obtain

$$\frac{\dot{r}^2 + Rg}{\dot{r}_0^2 + Rg} = \frac{r_0^2 + R^2}{r^2 + R^2} \tag{3–116}$$

Rearranging Eq. (3–116), we obtain

$$\dot{r}^2 = -Rg + (\dot{r}_0^2 + Rg)\frac{r_0^2 + R^2}{r^2 + R^2} \tag{3–117}$$

Using the given initial condition $\dot{r}(t = 0) = 0$, we obtain

$$\dot{r}^2 = -Rg + Rg\frac{r_0^2 + R^2}{r^2 + R^2} \tag{3–118}$$

This last expression simplifies to

$$\dot{r}^2 = \frac{r_0^2 - r^2}{r^2 + R^2}Rg \tag{3–119}$$

■

Example 3–3

A particle of mass m slides inside a tube as shown in Fig. 3–13. The tube rotates in the horizontal plane with constant angular velocity Ω in a counterclockwise sense about the fixed point O. Knowing that a viscous friction force with coefficient of friction c is exerted on the particle and that r is the displacement of the particle from point O, determine the differential equation of motion for the particle in terms of r.

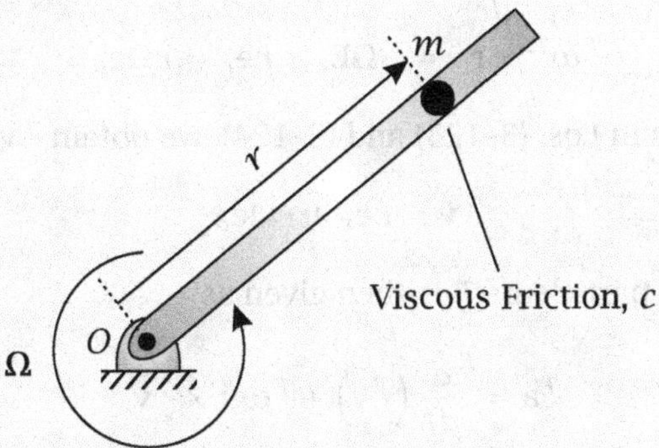

Figure 3–13 Particle sliding with friction inside a whirling tube.

Solution to Example 3–3

Kinematics

First, let $\mathcal{F}$ be a fixed inertial reference frame. Then, choose the following coordinate system fixed in reference frame $\mathcal{F}$:

	Origin at point O	
$\mathbf{E}_x$	=	Along Om at $t = 0$
$\mathbf{E}_z$	=	Out of page
$\mathbf{E}_y$	=	$\mathbf{E}_z \times \mathbf{E}_x$

Next, let $\mathcal{R}$ be a reference frame fixed to the tube. Then, choose the following coordinate system fixed in reference frame $\mathcal{R}$:

	Origin at O	
$\mathbf{e}_r$	=	Along Om
$\mathbf{E}_z$	=	Out of page
$\mathbf{e}_\theta$	=	$\mathbf{E}_z \times \mathbf{e}_r$

Now, since $\mathcal{R}$ is fixed in the tube and the tube rotates with angular rate Ω relative to the fixed reference frame $\mathcal{F}$, the angular velocity of $\mathcal{R}$ in $\mathcal{F}$ is given as

$$ {}^{\mathcal{F}}\boldsymbol{\omega}^{\mathcal{R}} = \boldsymbol{\Omega} = \Omega \mathbf{E}_z \tag{3-120} $$

Furthermore, the position of the particle is given as

$$ \mathbf{r} = r\mathbf{e}_r \tag{3-121} $$

Computing the rate of change of **r** in Eq. (3–121), the velocity of the particle in $\mathcal{F}$ is given as

$$^{\mathcal{F}}\mathbf{v} = \frac{^{\mathcal{R}}d\mathbf{r}}{dt} + {}^{\mathcal{F}}\boldsymbol{\omega}^{\mathcal{R}} \times \mathbf{r} \tag{3–122}$$

Now we have

$$\begin{aligned} \frac{^{\mathcal{R}}d\mathbf{r}}{dt} &= \dot{r}\mathbf{e}_r && (3–123) \\ {}^{\mathcal{F}}\boldsymbol{\omega}^{\mathcal{R}} \times \mathbf{r} &= \Omega\mathbf{E}_z \times r\mathbf{e}_r = r\Omega\mathbf{e}_\theta && (3–124) \end{aligned}$$

Adding the expressions in Eqs. (3–123) and (3–124), we obtain $^{\mathcal{F}}\mathbf{v}$ as

$$^{\mathcal{F}}\mathbf{v} = \dot{r}\mathbf{e}_r + r\Omega\mathbf{e}_\theta \tag{3–125}$$

The acceleration of the particle in $\mathcal{F}$ is then given as

$$^{\mathcal{F}}\mathbf{a} = \frac{^{\mathcal{R}}d}{dt}\left({}^{\mathcal{F}}\mathbf{v}\right) + {}^{\mathcal{F}}\boldsymbol{\omega}^{\mathcal{R}} \times {}^{\mathcal{F}}\mathbf{v} \tag{3–126}$$

Noting that Ω is a constant, we have

$$\begin{aligned} \frac{^{\mathcal{R}}d}{dt}\left({}^{\mathcal{F}}\mathbf{v}\right) &= \ddot{r}\mathbf{e}_r + \dot{r}\Omega\mathbf{e}_\theta && (3–127) \\ {}^{\mathcal{F}}\boldsymbol{\omega}^{\mathcal{R}} \times {}^{\mathcal{F}}\mathbf{v} &= \Omega\mathbf{E}_z \times (\dot{r}\mathbf{e}_r + r\Omega\mathbf{e}_\theta) \\ &= \dot{r}\Omega\mathbf{e}_\theta - r\Omega^2\mathbf{e}_r && (3–128) \end{aligned}$$

Adding Eqs. (3–127) and (3–128), we obtain $^{\mathcal{F}}\mathbf{a}$ as

$$^{\mathcal{F}}\mathbf{a} = (\ddot{r} - \Omega^2 r)\mathbf{e}_r + 2\dot{r}\Omega\mathbf{e}_\theta \tag{3–129}$$

Kinetics

The free body diagram of the particle is shown in Fig. 3–14.

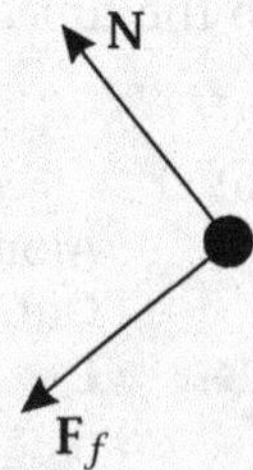

Figure 3–14 Free body diagram for Example 3–3.

Using Fig. 3–14, it is seen that the forces acting on the particle are

$$\begin{aligned} \mathbf{N} &= \text{Reaction force of tube on particle} \\ \mathbf{F}_f &= \text{Force of friction} \end{aligned}$$

Now we know that the reaction force $\mathbf{N}$ must act in the direction orthogonal to the tube. Furthermore, because the motion is planar, the force $\mathbf{N}$ must lie in the plane of motion. Consequently, $\mathbf{N}$ must lie in the direction of $\mathbf{e}_\theta$ and can be expressed as

$$\mathbf{N} = N\mathbf{e}_\theta \tag{3-130}$$

Next, because the friction force is viscous, we have

$$\mathbf{F}_f = -c\mathbf{v}_{\text{rel}} \tag{3-131}$$

Now recall that $\mathbf{v}_{\text{rel}}$ is the velocity of the particle relative to the point fixed to the tube that is instantaneously in contact with the particle. Furthermore, recalling that $\mathbf{v}_{\text{rel}}$ is independent of reference frame, we can arbitrarily compute $\mathbf{v}_{\text{rel}}$ in reference frame of the tube, $\mathcal{R}$, as

$$\mathbf{v}_{\text{rel}} = {}^{\mathcal{R}}\mathbf{v} - {}^{\mathcal{R}}\mathbf{v}^{\mathcal{R}} \tag{3-132}$$

Now we note that, *as viewed by an observer fixed to the tube*, the velocity of the point on the tube that is instantaneously in contact with the particle is *zero*, i.e.,

$${}^{\mathcal{R}}\mathbf{v}^{\mathcal{R}} = \mathbf{0} \tag{3-133}$$

Consequently, $\mathbf{v}_{\text{rel}}$ simplifies to

$$\mathbf{v}_{\text{rel}} = {}^{\mathcal{R}}\mathbf{v} \tag{3-134}$$

Now we have from Eq. (3-123) that

$$\frac{{}^{\mathcal{R}}d\mathbf{r}}{dt} = {}^{\mathcal{R}}\mathbf{v} = \dot{r}\mathbf{e}_r \tag{3-135}$$

Substituting the result of Eq. (3-135) into (3-134), we obtain $\mathbf{v}_{\text{rel}}$ as

$$\mathbf{v}_{\text{rel}} = \dot{r}\mathbf{e}_r \tag{3-136}$$

Substituting the result of Eq. (3-136) into (3-131), the force of viscous friction is obtained as

$$\mathbf{F}_f = -c\dot{r}\mathbf{e}_r \tag{3-137}$$

Then, adding Eqs. (3-130) and (3-137), the resultant force acting on the particle is given as

$$\mathbf{F} = \mathbf{N} = N\mathbf{e}_\theta - c\dot{r}\mathbf{e}_r = -c\dot{r}\mathbf{e}_r + N\mathbf{e}_\theta \tag{3-138}$$

Setting $\mathbf{F}$ in Eq. (3-138) equal to $m^{\mathcal{F}}\mathbf{a}$ using the expression for ${}^{\mathcal{F}}\mathbf{a}$ from Eq. (3-129), we obtain

$$-c\dot{r}\mathbf{e}_r + N\mathbf{e}_\theta = m(\ddot{r} - \Omega^2 r)\mathbf{e}_r + 2m\dot{r}\Omega\mathbf{e}_\theta \tag{3-139}$$

Equating components in Eq. (3-139), we obtain the following two scalar equations:

$$\begin{aligned} m(\ddot{r} - \Omega^2 r) &= -c\dot{r} && (3\text{-}140) \\ 2m\dot{r}\Omega &= N && (3\text{-}141) \end{aligned}$$

Differential Equation of Motion for Particle

It is observed that Eq. (3–140) neither has any unknown reaction forces nor has any unknown parameters. Furthermore, because r is the only variable required to describe the motion of the particle, the differential equation of motion is given by Eq. (3–140). Rearranging Eq. (3–140), we obtain the differential equation of motion as

$$m\ddot{r} + c\dot{r} - m\Omega^2 r = 0 \tag{3–142}$$

It is noted that Eq. (3–142) is a linear constant coefficient ordinary differential equation in the variable r and, thus, has an analytic solution.

■

Example 3–4

A ball bearing slides without friction in the vertical plane along the surface of a semicircular cylinder of radius r as shown in Fig. 3–15. Assuming that the ball bearing is modeled as a particle of mass m, that gravity acts vertically downward, and that the initial conditions are $\theta(t = 0) = 0$ and $\dot{\theta}(t = 0) = 0$, determine (a) the differential equation of motion while the ball bearing maintains contact with the cylinder and (b) the angular displacement θ at which the ball bearing loses contact with the cylinder.

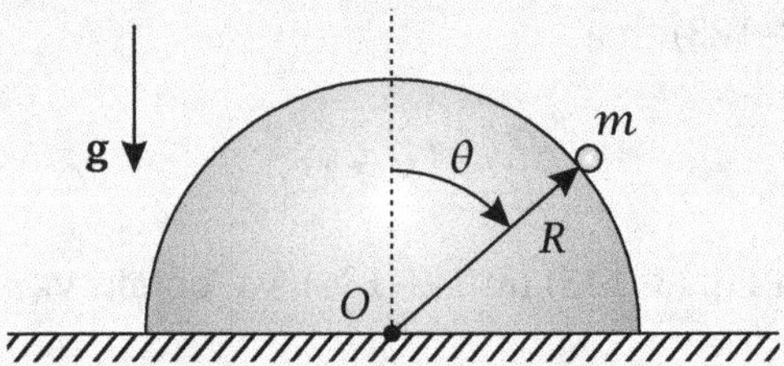

Figure 3–15 Ball bearing sliding without friction on a semicircular cylinder.

Solution to Example 3–4

Kinematics

First, let $\mathcal{F}$ be a reference frame fixed to the cylinder. Then, choose the following coordinate system fixed in reference frame $\mathcal{F}$:

Origin at O		
$\mathbf{E}_x$	=	Along Om when $\theta = 0$
$\mathbf{E}_z$	=	Into page
$\mathbf{E}_y$	=	$\mathbf{E}_z \times \mathbf{E}_x$

Next, let $\mathcal{A}$ be a reference frame fixed to the direction along Om. Then, choose the following coordinate system fixed in reference frame $\mathcal{A}$:

$$\begin{array}{ccc} & \text{Origin at } O & \\ \mathbf{e}_r & = & \text{Along } Om \\ \mathbf{e}_z & = & \text{Into page } (= \mathbf{E}_z) \\ \mathbf{e}_\theta & = & \mathbf{e}_z \times \mathbf{e}_r \end{array}$$

Then, using Fig. 3-16, the relationship between the bases $\{\mathbf{E}_x, \mathbf{E}_y, \mathbf{E}_z\}$ and $\{\mathbf{e}_r, \mathbf{e}_\theta, \mathbf{e}_z\}$ is given as

$$\mathbf{E}_x = \cos\theta\, \mathbf{e}_r - \sin\theta\, \mathbf{e}_\theta \tag{3-143}$$

$$\mathbf{E}_y = \sin\theta\, \mathbf{e}_r + \cos\theta\, \mathbf{e}_\theta \tag{3-144}$$

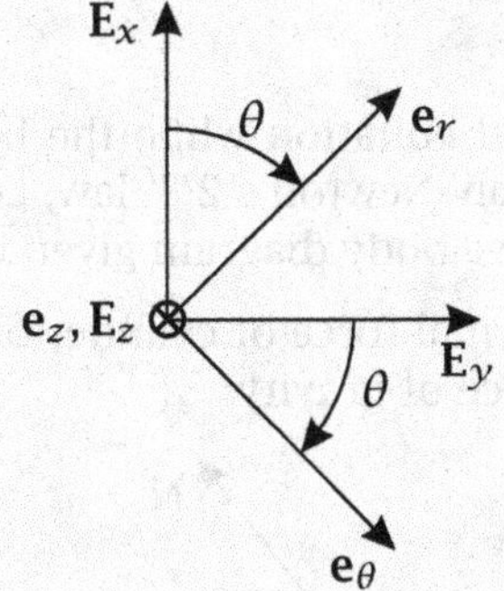

Figure 3-16 Geometry of bases $\{\mathbf{E}_x, \mathbf{E}_y, \mathbf{E}_z\}$ and $\{\mathbf{e}_r, \mathbf{e}_\theta, \mathbf{e}_z\}$ for Example 3-4.

The position of the ball bearing is then given in terms of the basis $\{\mathbf{e}_r, \mathbf{e}_\theta, \mathbf{e}_z\}$ as

$$\mathbf{r} = R\mathbf{e}_r \tag{3-145}$$

Furthermore, the angular velocity of reference frame $\mathcal{A}$ in reference frame $\mathcal{F}$ is given as

$${}^{\mathcal{F}}\boldsymbol{\omega}^{\mathcal{A}} = \dot{\theta}\mathbf{e}_z \tag{3-146}$$

Then, using the rate of change transport theorem between reference frames $\mathcal{A}$ and $\mathcal{F}$, the velocity of the ball bearing in reference frame $\mathcal{F}$ is given as

$${}^{\mathcal{F}}\mathbf{v} = \frac{{}^{\mathcal{F}}d\mathbf{r}}{dt} = \frac{{}^{\mathcal{A}}d\mathbf{r}}{dt} + {}^{\mathcal{F}}\boldsymbol{\omega}^{\mathcal{A}} \times \mathbf{r} \tag{3-147}$$

Now we have

$$\frac{{}^{\mathcal{A}}d\mathbf{r}}{dt} = \mathbf{0} \tag{3-148}$$

$${}^{\mathcal{F}}\boldsymbol{\omega}^{\mathcal{A}} \times \mathbf{r} = \dot{\theta}\mathbf{e}_z \times R\mathbf{e}_r = R\dot{\theta}\mathbf{e}_\theta \tag{3-149}$$

Adding the results of Eqs. (3-148) and (3-149), we obtain the velocity of the ball bearing in reference frame $\mathcal{F}$ as

$${}^{\mathcal{F}}\mathbf{v} = R\dot{\theta}\mathbf{e}_\theta \tag{3-150}$$

Then, applying the rate of change transport theorem to ${}^{\mathcal{F}}\mathbf{v}$ between reference frames $\mathcal{A}$ and $\mathcal{F}$, the acceleration of the ball bearing in reference frame $\mathcal{F}$ is given as

$$ {}^{\mathcal{F}}\mathbf{a} = \frac{{}^{\mathcal{F}}d}{dt}\left({}^{\mathcal{F}}\mathbf{v}\right) = \frac{{}^{\mathcal{A}}d}{dt}\left({}^{\mathcal{F}}\mathbf{v}\right) + {}^{\mathcal{F}}\boldsymbol{\omega}^{\mathcal{A}} \times {}^{\mathcal{F}}\mathbf{v} \tag{3-151} $$

where

$$ \frac{{}^{\mathcal{A}}d}{dt}\left({}^{\mathcal{F}}\mathbf{v}\right) = R\ddot{\theta}\mathbf{e}_\theta \tag{3-152} $$

$$ {}^{\mathcal{F}}\boldsymbol{\omega}^{\mathcal{A}} \times {}^{\mathcal{F}}\mathbf{v} = \dot{\theta}\mathbf{e}_z \times (R\dot{\theta}\mathbf{e}_\theta) = -R\dot{\theta}^2\mathbf{e}_r \tag{3-153} $$

Adding the expressions in Eqs. (3-152) and (3-153), we obtain ${}^{\mathcal{F}}\mathbf{a}$ as

$$ {}^{\mathcal{F}}\mathbf{a} = -R\dot{\theta}^2\mathbf{e}_r + R\ddot{\theta}\mathbf{e}_\theta \tag{3-154} $$

Kinetics

In order to obtain the differential equation while the ball bearing maintains contact with the cylinder, we need to apply Newton's 2nd law, i.e., $\mathbf{F} = m{}^{\mathcal{F}}\mathbf{a}$. We already have ${}^{\mathcal{F}}\mathbf{a}$ from Eq. (3-154). Using the free body diagram given in Fig. 3-17, we have

$$ \begin{array}{rcl} \mathbf{N} & = & \text{Normal force of cylinder on particle} \\ m\mathbf{g} & = & \text{Force of gravity} \end{array} $$

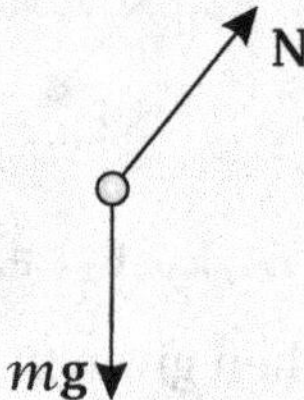

Figure 3-17 Free body diagram for Example 3-4.

Now from the geometry we have

$$ \begin{array}{rcl} \mathbf{N} & = & N\mathbf{e}_r \\ m\mathbf{g} & = & -mg\mathbf{E}_x \end{array} \tag{3-155} $$

Then, substituting the expression for $\mathbf{E}_y$ from Eq. (3-144), the force of gravity can be written as

$$ m\mathbf{g} = -mg(\cos\theta\,\mathbf{e}_r - \sin\theta\,\mathbf{e}_\theta) = -mg\cos\theta\,\mathbf{e}_r + mg\sin\theta\,\mathbf{e}_\theta \tag{3-156} $$

The resultant force on the ball bearing is then given as

$$ \mathbf{F} = N\mathbf{e}_r - mg\cos\theta\,\mathbf{e}_r + mg\sin\theta\,\mathbf{e}_\theta = (N - mg\cos\theta\,)\mathbf{e}_r + mg\sin\theta\,\mathbf{e}_\theta \tag{3-157} $$

Setting $\mathbf{F}$ equal to $m\mathbf{a}$ using ${}^{\mathcal{F}}\mathbf{a}$ from Eq. (3-154), we obtain

$$ (N - mg\cos\theta\,)\mathbf{e}_r + mg\sin\theta\,\mathbf{e}_\theta = -mR\dot{\theta}^2\mathbf{e}_r + mR\ddot{\theta}\mathbf{e}_\theta \tag{3-158} $$

We then obtain the following two scalar equations:

$$ N - mg\cos\theta = -mR\dot{\theta}^2 \tag{3-159} $$

$$ mg\sin\theta = mR\ddot{\theta} \tag{3-160} $$

(a) Differential Equation While Ball Bearing Maintains Contact with Cylinder

Noticing that Eq. (3–160) has no reaction forces, we obtain the differential equation of motion for the ball bearing while it maintains contact with the cylinder as

$$mR\ddot{\theta} = mg\sin\theta \tag{3-161}$$

Simplifying Eq. (3–161), we obtain

$$\ddot{\theta} = \frac{g}{R}\sin\theta \tag{3-162}$$

(b) Angular Displacement at Which Ball Bearing Loses Contact with Cylinder

We note that the ball bearing will lose contact with the cylinder when the reaction force is zero, i.e., when $N = 0$. Setting N equal to zero in Eq. (3–159), we obtain

$$-mg\cos\theta = -mR\dot{\theta}^2 \tag{3-163}$$

which implies that

$$\dot{\theta}^2 = \frac{g}{R}\cos\theta \tag{3-164}$$

Now it is observed that Eq. (3–164) is *not* an algebraic equation in θ because it contains a term of $\dot{\theta}^2$. Therefore, we need to find an expression for $\dot{\theta}^2$ in terms of θ. Such an expression is obtained by solving the differential equation of Eq. (3–162). First, we note that

$$\ddot{\theta} = \frac{d\dot{\theta}}{dt} = \frac{d\dot{\theta}}{d\theta}\frac{d\theta}{dt} = \dot{\theta}\frac{d\dot{\theta}}{d\theta} \tag{3-165}$$

Substituting $\ddot{\theta}$ into Eq. (3–162), we obtain

$$\dot{\theta}\frac{d\dot{\theta}}{d\theta} = \frac{g}{R}\sin\theta \tag{3-166}$$

Separating variables in Eq. (3–166), we obtain

$$\dot{\theta}d\dot{\theta} = \frac{g}{R}\sin\theta\, d\theta \tag{3-167}$$

Integrating both sides of Eq. (3–167) gives

$$\int_{\dot{\theta}_0}^{\dot{\theta}} \eta d\eta = \int_{\theta_0}^{\theta} \frac{g}{R}\sin\nu d\nu \tag{3-168}$$

where η and ν are dummy variables of integration. Using the initial conditions $\theta(t = 0) = 0$ and $\dot{\theta}(t = 0) = 0$, Eq. (3–168) simplifies to

$$\frac{\dot{\theta}^2}{2} = \frac{g}{R}(1 - \cos\theta) \tag{3-169}$$

from which we obtain

$$\dot{\theta}^2 = \frac{2g}{R}(1 - \cos\theta) \tag{3-170}$$

Substituting $\dot{\theta}^2$ from Eq. (3–170) into (3–164), we obtain

$$\frac{2g}{R}(1-\cos\theta) = \frac{g}{R}\cos\theta \tag{3–171}$$

Dropping the common factor of g/R and rearranging Eq. (3–171) to solve for $\cos\theta$, we obtain

$$\cos\theta = 2/3 \tag{3–172}$$

Consequently, the angular displacement at which the ball bearing loses contact with the cylinder is

$$\theta = \cos^{-1}(2/3) \approx 48.1897 \text{ deg} \tag{3–173}$$

■

3.6 Linear Momentum and Linear Impulse for a Particle

3.6.1 Linear Momentum of a Particle

Let P be a particle of mass m moving in an inertial reference frame $\mathcal{N}$. Furthermore, let $\mathbf{r}$ be the position of P and let ${}^{\mathcal{N}}\mathbf{v}$ be the velocity of the particle in $\mathcal{N}$. Then the *linear momentum* of the particle in reference frame $\mathcal{N}$, denoted ${}^{\mathcal{N}}\mathbf{G}$, is defined as

$${}^{\mathcal{N}}\mathbf{G} \equiv \mathbf{G} = m\,{}^{\mathcal{N}}\mathbf{v} \tag{3–174}$$

Computing the rate of change of ${}^{\mathcal{N}}\mathbf{G}$ in reference frame $\mathcal{N}$, we have

$$\frac{{}^{\mathcal{N}}d}{dt}\left({}^{\mathcal{N}}\mathbf{G}\right) = \frac{{}^{\mathcal{N}}d}{dt}\left(m\,{}^{\mathcal{N}}\mathbf{v}\right) = m\frac{{}^{\mathcal{N}}d}{dt}\left({}^{\mathcal{N}}\mathbf{v}\right) = m\,{}^{\mathcal{N}}\mathbf{a} \tag{3–175}$$

Then, applying Newton's 2nd law of Eq. (3–41), we have

$$\boxed{\mathbf{F} = \frac{{}^{\mathcal{N}}d}{dt}\left({}^{\mathcal{N}}\mathbf{G}\right)} \tag{3–176}$$

Equation (3–176) states that the resultant force acting on a particle of mass m is equal to the rate of change of linear momentum in an inertial reference frame $\mathcal{N}$. We note that Eq. (3–176) is an alternate form of Newton's 2nd law.

3.6.2 Principle of Linear Impulse and Momentum for a Particle

Let P be a particle of mass m moving in an inertial reference frame $\mathcal{N}$. Furthermore, let $\mathbf{F}$ be a force acting on P. Suppose now that we consider the effect of the force on a given time interval $t \in [t_1, t_2]$. Then the integral of the force from t_1 to $t_2 > t_1$, denoted $\hat{\mathbf{F}}$, is given as

$$\hat{\mathbf{F}} = \int_{t_1}^{t_2} \mathbf{F}\,dt \tag{3–177}$$

The quantity $\hat{\mathbf{F}}$ is called the *linear impulse* of the force $\mathbf{F}$ on the time interval $t \in [t_1, t_2]$. Then, integrating both sides of Eq. (3–176) from t_1 to t_2, we have

$$\int_{t_1}^{t_2} \mathbf{F} dt = \int_{t_1}^{t_2} \frac{^{\mathcal{N}}d}{dt}\left({}^{\mathcal{N}}\mathbf{G}\right) dt = {}^{\mathcal{N}}\mathbf{G}(t_2) - {}^{\mathcal{N}}\mathbf{G}(t_1) \tag{3–178}$$

Furthermore, applying the definition of linear impulse using Eq. (3–177), Eq. (3–178) can be written as

$$\boxed{\hat{\mathbf{F}} = {}^{\mathcal{N}}\mathbf{G}(t_2) - {}^{\mathcal{N}}\mathbf{G}(t_1)} \tag{3–179}$$

Equation (3–179) is called the *principle of linear impulse and linear momentum* and states that the change in linear momentum of a particle over a specified time interval is equal to the total linear impulse of the external forces acting on the particle during the same time interval. In terms of the velocity of the particle in the inertial reference frame $\mathcal{N}$, Eq. (3–179) can be written as

$$\boxed{\hat{\mathbf{F}} = m{}^{\mathcal{N}}\mathbf{v}(t_2) - m{}^{\mathcal{N}}\mathbf{v}(t_1)} \tag{3–180}$$

3.6.3 Conservation of Linear Momentum

Let ${}^{\mathcal{N}}\mathbf{G}$ be the linear momentum of a particle in an inertial reference frame $\mathcal{N}$ and let $\mathbf{u}$ be a unit vector in $\mathbb{E}^3$. Then ${}^{\mathcal{N}}\mathbf{G}$ is said to be *conserved* in the direction of $\mathbf{u}$ if

$${}^{\mathcal{N}}\mathbf{G} \cdot \mathbf{u} = \text{constant} \tag{3–181}$$

Observing that ${}^{\mathcal{N}}\mathbf{G} \cdot \mathbf{u}$ is a scalar function, we can differentiate Eq. (3–181) to obtain

$$\frac{d}{dt}\left({}^{\mathcal{N}}\mathbf{G} \cdot \mathbf{u}\right) = 0 \tag{3–182}$$

Again, because ${}^{\mathcal{N}}\mathbf{G} \cdot \mathbf{u}$ is a scalar function, we can compute its rate of change arbitrarily in reference frame $\mathcal{N}$ using the product rule for differentiation to obtain

$$\frac{d}{dt}\left({}^{\mathcal{N}}\mathbf{G} \cdot \mathbf{u}\right) = \frac{^{\mathcal{N}}d}{dt}\left({}^{\mathcal{N}}\mathbf{G}\right) \cdot \mathbf{u} + {}^{\mathcal{N}}\mathbf{G} \cdot \frac{^{\mathcal{N}}d\mathbf{u}}{dt} = 0 \tag{3–183}$$

Substituting the result of Eq. (3–176) into (3–183), the linear momentum of a particle will be conserved in the direction of $\mathbf{u}$ if and only if

$$\boxed{\mathbf{F} \cdot \mathbf{u} + {}^{\mathcal{N}}\mathbf{G} \cdot \frac{^{\mathcal{N}}d\mathbf{u}}{dt} = 0} \tag{3–184}$$

Finally, we say that the linear momentum of a particle ${}^{\mathcal{N}}\mathbf{G}$ is *conserved in all directions* or, more simply, the linear momentum ${}^{\mathcal{N}}\mathbf{G}$ is *conserved*, if

$${}^{\mathcal{N}}\mathbf{G} = \text{constant} \tag{3–185}$$

where in this case the constant is a vector. Differentiating Eq. (3–185) in the inertial reference frame $\mathcal{N}$, we obtain

$$\frac{^{\mathcal{N}}d}{dt}\left({}^{\mathcal{N}}\mathbf{G}\right) = \mathbf{0} \tag{3–186}$$

It is seen that if Eq. (3-186) is satisfied, then from Eq. (3-176) we have that linear momentum is conserved if

$$\mathbf{F} = \mathbf{0} \tag{3-187}$$

In other words, the linear momentum of a particle will be conserved (in all directions) if the resultant force acting on the particle is *zero*. It is important to keep in mind that for most problems there may only exist certain directions along which the linear momentum is conserved. In such cases these directions must be determined by applying Eq. (3-184). However, for some problems linear momentum may be conserved in all directions. Determining if linear momentum is conserved in all directions is most conveniently determined by applying Eq. (3-187).

Suppose now that we consider the case where $\mathbf{u}$ is a direction that is fixed in the inertial reference frame $\mathcal{N}$, i.e.,

$$\frac{{}^{\mathcal{N}}d\mathbf{u}}{dt} = \mathbf{0} \tag{3-188}$$

Applying Eq. (3-188), Eq. (3-184) reduces to

$$\mathbf{F} \cdot \mathbf{u} = 0 \tag{3-189}$$

Equation (3-189) states that the linear momentum of a particle in an inertial reference frame $\mathcal{N}$ will be conserved in the inertially fixed direction $\mathbf{u}$ if the component of force acting on the particle in the direction of $\mathbf{u}$ is zero.

3.7 Moment of a Force and Moment Transport Theorem for a Particle

Let $\mathbf{F}$ be a force acting on a particle P of mass m. Furthermore, let $\mathbf{r}$ be the position of the particle and let Q be an arbitrary point. Then the *moment of the force* $\mathbf{F}$ *relative to point* Q, denoted $\mathbf{M}_Q$, is defined as

$$\mathbf{M}_Q = (\mathbf{r} - \mathbf{r}_Q) \times \mathbf{F} \tag{3-190}$$

Suppose now that we consider another arbitrary point Q'. Then, from Eq. (3-190), the moment applied by the force $\mathbf{F}$ relative to Q' is

$$\mathbf{M}_{Q'} = (\mathbf{r} - \mathbf{r}_{Q'}) \times \mathbf{F} \tag{3-191}$$

Then, subtracting Eq. (3-191) from (3-190), we have

$$\mathbf{M}_Q - \mathbf{M}_{Q'} = (\mathbf{r} - \mathbf{r}_Q) \times \mathbf{F} - (\mathbf{r} - \mathbf{r}_{Q'}) \times \mathbf{F} \tag{3-192}$$

Equation (3-192) simplifies to

$$\mathbf{M}_Q - \mathbf{M}_{Q'} = (\mathbf{r}_{Q'} - \mathbf{r}_Q) \times \mathbf{F} \tag{3-193}$$

Rearranging Eq. (3-193), we obtain

$$\mathbf{M}_Q = \mathbf{M}_{Q'} + (\mathbf{r}_{Q'} - \mathbf{r}_Q) \times \mathbf{F} \tag{3-194}$$

Equation (3-194) is called the *moment transport theorem* for a particle and relates the moment due to a force $\mathbf{F}$ relative to an arbitrary point Q to the moment of that same force relative to any other point Q'.

3.8 Angular Momentum and Angular Impulse for a Particle

3.8.1 Angular Momentum and Its Rate of Change of a Particle

Let P be a particle of mass m moving in an inertial reference frame $\mathcal{N}$. Furthermore, let $\mathbf{r}$ be the position of the particle, let ${}^{\mathcal{N}}\mathbf{v}$ be the velocity of the particle in $\mathcal{N}$, and let Q be an arbitrary point. Then the *angular momentum* of the particle in an inertial reference frame $\mathcal{N}$ relative to point Q is defined as[7]

$$ {}^{\mathcal{N}}\mathbf{H}_Q = (\mathbf{r} - \mathbf{r}_Q) \times m({}^{\mathcal{N}}\mathbf{v} - {}^{\mathcal{N}}\mathbf{v}_Q) \tag{3-195} $$

Computing the rate of change of ${}^{\mathcal{N}}\mathbf{H}_Q$ in the inertial reference frame $\mathcal{N}$, we have

$$ \frac{{}^{\mathcal{N}}d}{dt}\left({}^{\mathcal{N}}\mathbf{H}_Q\right) = \frac{{}^{\mathcal{N}}d}{dt}\left[(\mathbf{r} - \mathbf{r}_Q) \times m({}^{\mathcal{N}}\mathbf{v} - {}^{\mathcal{N}}\mathbf{v}_Q)\right] \tag{3-196} $$

Expanding Eq. (3-196), we obtain

$$ \begin{aligned} \frac{{}^{\mathcal{N}}d}{dt}\left({}^{\mathcal{N}}\mathbf{H}_Q\right) = & \frac{{}^{\mathcal{N}}d}{dt}(\mathbf{r} - \mathbf{r}_Q) \times m({}^{\mathcal{N}}\mathbf{v} - {}^{\mathcal{N}}\mathbf{v}_Q) \\ & + (\mathbf{r} - \mathbf{r}_Q) \times m \frac{{}^{\mathcal{N}}d}{dt}\left({}^{\mathcal{N}}\mathbf{v} - {}^{\mathcal{N}}\mathbf{v}_Q\right) \end{aligned} \tag{3-197} $$

Now we note that

$$ \begin{aligned} \frac{{}^{\mathcal{N}}d}{dt}(\mathbf{r} - \mathbf{r}_Q) &= {}^{\mathcal{N}}\mathbf{v} - {}^{\mathcal{N}}\mathbf{v}_Q \\ \frac{{}^{\mathcal{N}}d}{dt}\left({}^{\mathcal{N}}\mathbf{v} - {}^{\mathcal{N}}\mathbf{v}_Q\right) &= {}^{\mathcal{N}}\mathbf{a} - {}^{\mathcal{N}}\mathbf{a}_Q \end{aligned} \tag{3-198} $$

Furthermore,

$$ ({}^{\mathcal{N}}\mathbf{v} - {}^{\mathcal{N}}\mathbf{v}_Q) \times m({}^{\mathcal{N}}\mathbf{v} - {}^{\mathcal{N}}\mathbf{v}_Q) = \mathbf{0} \tag{3-199} $$

Equation (3-197) then simplifies to

$$ \frac{{}^{\mathcal{N}}d}{dt}\left({}^{\mathcal{N}}\mathbf{H}_Q\right) = (\mathbf{r} - \mathbf{r}_Q) \times m\left({}^{\mathcal{N}}\mathbf{a} - {}^{\mathcal{N}}\mathbf{a}_Q\right) \tag{3-200} $$

Then, applying Newton's 2nd law of Eq. (3-41), Eq. (3-200) can be rewritten as

$$ \frac{{}^{\mathcal{N}}d}{dt}\left({}^{\mathcal{N}}\mathbf{H}_Q\right) = (\mathbf{r} - \mathbf{r}_Q) \times \mathbf{F} - (\mathbf{r} - \mathbf{r}_Q) \times m\,{}^{\mathcal{N}}\mathbf{a}_Q \tag{3-201} $$

Using the definition of the moment of a force from Eq. (3-190), we see that

$$ (\mathbf{r} - \mathbf{r}_Q) \times \mathbf{F} = \mathbf{M}_Q \tag{3-202} $$

[7]There are two conventions for the definition of angular momentum. The first convention, shown in Eq. (3-195), uses the *relative velocity* ${}^{\mathcal{N}}\mathbf{v} - {}^{\mathcal{N}}\mathbf{v}_Q$ between the particle and the reference point while the second convention uses the (absolute) velocity of the particle ${}^{\mathcal{N}}\mathbf{v}$. Either of the aforementioned conventions is perfectly valid, but the convention using relative velocity is adopted here because the author feels that the results obtained using relative velocity are more intuitive than the results obtained using absolute velocity. Examples of previous works that define angular momentum using the relative velocity convention include Greenwood (1977; 1988) while examples of works that define angular momentum using the absolute velocity convention include Synge and Griffith (1959), O'Reilly (2001), and Thornton and Marion (2004).

where $\mathbf{M}_Q$ is the moment of the resultant force $\mathbf{F}$ acting on the particle relative to point Q. Consequently, Eq. (3–201) can be written as

$$\boxed{\frac{{}^{\mathcal{N}}d}{dt}\left({}^{\mathcal{N}}\mathbf{H}_Q\right) = \mathbf{M}_Q - (\mathbf{r} - \mathbf{r}_Q) \times m{}^{\mathcal{N}}\mathbf{a}_Q} \tag{3–203}$$

The quantity

$$-(\mathbf{r} - \mathbf{r}_Q) \times m{}^{\mathcal{N}}\mathbf{a}_Q \tag{3–204}$$

in Eq. (3–203) is called the *inertial moment* of point Q relative to point P.

Now suppose that we choose a reference point O that is fixed in an inertial reference frame $\mathcal{N}$. Then the acceleration of point O in $\mathcal{N}$ is zero, i.e., ${}^{\mathcal{N}}\mathbf{a}_O = \mathbf{0}$, which implies that the inertial moment, $-(\mathbf{r} - \mathbf{r}_O) \times m{}^{\mathcal{N}}\mathbf{a}_O$, is zero. Consequently, *for a reference point O fixed in an inertial reference frame $\mathcal{N}$*, Eq. (3–203) reduces to

$$\boxed{\frac{{}^{\mathcal{N}}d}{dt}\left({}^{\mathcal{N}}\mathbf{H}_O\right) = \mathbf{M}_O} \tag{3–205}$$

Equation (3–205) states that the rate of change of angular momentum of a particle relative to a reference point O that is fixed in an inertial reference frame is equal to the moment due to all external forces relative to point O.

3.8.2 Principle of Angular Impulse and Angular Momentum of a Particle

Now consider an interval of time $t \in [t_1, t_2]$. Then the integral of the moment $\mathbf{M}_Q$ from t_1 to t_2, denoted $\hat{\mathbf{M}}_Q$, is given as

$$\hat{\mathbf{M}}_Q = \int_{t_1}^{t_2} \mathbf{M}_Q dt \tag{3–206}$$

The quantity $\hat{\mathbf{M}}_Q$ is called the *angular impulse* of the moment $\mathbf{M}_Q$ on the time interval $t \in [t_1, t_2]$. Integrating both sides of Eq. (3–203) from t_1 to t_2, we obtain

$$\int_{t_1}^{t_2} \left[\mathbf{M}_Q - (\mathbf{r} - \mathbf{r}_Q) \times m{}^{\mathcal{N}}\mathbf{a}_Q\right] dt = \int_{t_1}^{t_2} \left[\frac{{}^{\mathcal{N}}d}{dt}\left({}^{\mathcal{N}}\mathbf{H}_Q\right)\right] dt \tag{3–207}$$

Equation (3–207) can be rewritten as

$$\int_{t_1}^{t_2} \mathbf{M}_Q dt - \int_{t_1}^{t_2} (\mathbf{r} - \mathbf{r}_Q) \times m{}^{\mathcal{N}}\mathbf{a}_Q dt = \int_{t_1}^{t_2} \left[\frac{{}^{\mathcal{N}}d}{dt}\left({}^{\mathcal{N}}\mathbf{H}_Q\right)\right] dt \tag{3–208}$$

Then, using Eqs. (3–206) and (3–207) and observing that

$$\int_{t_1}^{t_2} \frac{{}^{\mathcal{N}}d}{dt}\left({}^{\mathcal{N}}\mathbf{H}_Q\right) dt = {}^{\mathcal{N}}\mathbf{H}_Q(t_2) - {}^{\mathcal{N}}\mathbf{H}_Q(t_1) \tag{3–209}$$

Eq. (3–207) simplifies to

$$\boxed{\hat{\mathbf{M}}_Q - \int_{t_1}^{t_2} \left[(\mathbf{r} - \mathbf{r}_Q) \times m{}^{\mathcal{N}}\mathbf{a}_Q\right] dt = {}^{\mathcal{N}}\mathbf{H}_Q(t_2) - {}^{\mathcal{N}}\mathbf{H}_Q(t_1)} \tag{3–210}$$

Equation (3-210) is called the *principle of angular impulse and angular momentum relative to an arbitrary reference point Q* and states that the change in angular momentum of a particle relative to an arbitrary reference point Q over a time interval $t \in [t_1, t_2]$ is equal to the sum of the angular impulse due to all external forces applied to the particle relative to point Q and the angular impulse due to the inertial moment of point Q relative to point P on the time interval $t \in [t_1, t_2]$.

Suppose now that we choose a reference point O that is fixed in an inertial reference frame. Then the acceleration of point O is zero, i.e., ${}^{\mathcal{N}}\mathbf{a}_O = \mathbf{0}$. Moreover, the second term in Eq. (3-210) is zero and Eq. (3-210) reduces to

$$\boxed{\hat{\mathbf{M}}_O = {}^{\mathcal{N}}\mathbf{H}_O(t_2) - {}^{\mathcal{N}}\mathbf{H}_O(t_1)} \tag{3-211}$$

Equation (3-211) is called the *principle of angular impulse and angular momentum relative to a point O fixed in an inertial reference frame* and states that the change in angular momentum of a particle relative to a point O fixed in an inertial reference frame is equal to the angular impulse due to all external forces applied to the particle relative to point O.

3.8.3 Conservation of Angular Momentum

Let ${}^{\mathcal{N}}\mathbf{H}_Q$ be the angular momentum of a particle P in an inertial reference frame $\mathcal{N}$ relative to an arbitrary reference point Q. Furthermore, let $\mathbf{u}$ be an arbitrary unit vector in $\mathbb{E}^3$. Then the angular momentum ${}^{\mathcal{N}}\mathbf{H}_Q$ is said to be *conserved* in the direction of $\mathbf{u}$ if

$${}^{\mathcal{N}}\mathbf{H}_Q \cdot \mathbf{u} = \text{constant} \tag{3-212}$$

Observing that ${}^{\mathcal{N}}\mathbf{H}_Q \cdot \mathbf{u}$ is a scalar function, we can differentiate Eq. (3-212) to obtain

$$\frac{d}{dt}\left({}^{\mathcal{N}}\mathbf{H}_Q \cdot \mathbf{u}\right) = 0 \tag{3-213}$$

Again, because ${}^{\mathcal{N}}\mathbf{H}_Q \cdot \mathbf{u}$ is a scalar function, its rate of change is independent of reference frame. Consequently, ${}^{\mathcal{N}}\mathbf{H}_Q \cdot \mathbf{u}$ can be differentiated arbitrarily in the inertial reference frame $\mathcal{N}$ to give

$$\frac{{}^{\mathcal{N}}d}{dt}\left({}^{\mathcal{N}}\mathbf{H}_Q \cdot \mathbf{u}\right) = \frac{{}^{\mathcal{N}}d}{dt}\left({}^{\mathcal{N}}\mathbf{H}_Q\right) \cdot \mathbf{u} + {}^{\mathcal{N}}\mathbf{H}_Q \cdot \frac{{}^{\mathcal{N}}d\mathbf{u}}{dt} \tag{3-214}$$

Then, using the result of Eq. (3-203), Eq. (3-214) can be written as

$$\boxed{\left(\mathbf{M}_Q - (\mathbf{r} - \mathbf{r}_Q) \times m\,{}^{\mathcal{N}}\mathbf{a}_Q\right) \cdot \mathbf{u} + {}^{\mathcal{N}}\mathbf{H}_Q \cdot \frac{{}^{\mathcal{N}}d\mathbf{u}}{dt} = 0} \tag{3-215}$$

Consequently, the angular momentum ${}^{\mathcal{N}}\mathbf{H}_Q$ in an inertial reference frame $\mathcal{N}$ will be conserved in the direction of $\mathbf{u}$ if and only if Eq. (3-215) is satisfied. Finally, we say that the angular momentum of a particle relative to an arbitrary point Q, ${}^{\mathcal{N}}\mathbf{H}_Q$, is *conserved in all directions* or, more simply, is *conserved,* if

$${}^{\mathcal{N}}\mathbf{H}_Q = \text{constant} \tag{3-216}$$

where in this case the constant is a vector. Differentiating Eq. (3-216) in the inertial reference frame $\mathcal{N}$, we obtain

$$\frac{{}^{\mathcal{N}}d}{dt}\left({}^{\mathcal{N}}\mathbf{H}_Q\right) = \mathbf{0} \tag{3-217}$$

It is seen that if Eq. (3-217) is satisfied, then from Eq. (3-203) we have that angular momentum ${}^{\mathcal{N}}\mathbf{H}_Q$ is conserved if

$$\mathbf{M}_Q - (\mathbf{r} - \mathbf{r}_Q) \times m\,{}^{\mathcal{N}}\mathbf{a}_Q = \mathbf{0} \tag{3-218}$$

As with linear momentum, it is important to keep in mind that for most problems there may only exist certain directions along which the angular momentum is conserved. In such cases these directions must be determined by applying Eq. (3-215). However, for some problems angular momentum may be conserved in all directions. Determining if angular momentum is conserved in all directions is most conveniently determined by showing that ${}^{\mathcal{N}}d({}^{\mathcal{N}}\mathbf{H}_Q)/dt = \mathbf{0}$ or by showing that $\mathbf{M}_Q - (\mathbf{r} - \mathbf{r}_Q) \times m\,{}^{\mathcal{N}}\mathbf{a}_Q = \mathbf{0}$

Suppose now that we choose a reference point O that is fixed in an inertial reference frame. Then the inertial moment $-(\mathbf{r} - \mathbf{r}_O) \times m\,{}^{\mathcal{N}}\mathbf{a}_O = \mathbf{0}$. Equation (3-215) then simplifies to

$$\boxed{\mathbf{M}_O \cdot \mathbf{u} + {}^{\mathcal{N}}\mathbf{H}_O \cdot \frac{{}^{\mathcal{N}}d\mathbf{u}}{dt} = 0} \tag{3-219}$$

Consequently, the angular momentum ${}^{\mathcal{N}}\mathbf{H}_O$ relative to an *inertially fixed point O* will be conserved in the direction of $\mathbf{u}$ if Eq. (3-219) is satisfied.

Example 3-5

A particle of mass m is suspended from point P located at one end of a slender rigid massless rod as shown in Fig. 3-18. The other end of the rod is hinged at a point Q, where point Q is attached to a massless collar. The collar slides with a specified displacement $x(t)$ along a horizontal track. Knowing that θ describes the orientation of the rod with the vertically downward direction and that gravity acts downward, determine the differential equation of motion for the particle using (a) Newton's 2nd law and (b) a balance of angular momentum relative to point Q.

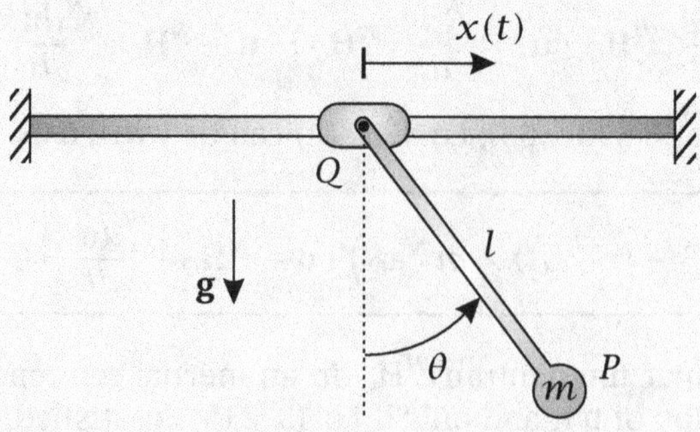

Figure 3-18 Pendulum attached to a rod sliding on a massless collar.

Solution to Example 3–5

Kinematics

First, let $\mathcal{F}$ be a reference frame fixed to the track. Then, choose the following coordinate system fixed in reference frame $\mathcal{F}$:

$$
\begin{array}{ccc}
 & \text{Origin at } Q \text{ when } t = 0 & \\
\mathbf{E}_x & = & \text{Along } AB \\
\mathbf{E}_z & = & \text{Out of page} \\
\mathbf{E}_y & = & \mathbf{E}_z \times \mathbf{E}_x
\end{array}
$$

Next, let $\mathcal{A}$ be a reference frame fixed to the rod. Then, choose the following coordinate system fixed in reference frame $\mathcal{A}$:

$$
\begin{array}{ccc}
 & \text{Origin at } Q & \\
\mathbf{e}_r & = & \text{Along } QP \\
\mathbf{e}_z & = & \text{Out of page} \\
\mathbf{e}_\theta & = & \mathbf{E}_z \times \mathbf{e}_r
\end{array}
$$

The relationship between the bases $\{\mathbf{E}_x, \mathbf{E}_y, \mathbf{E}_z\}$ and $\{\mathbf{e}_r, \mathbf{e}_\theta, \mathbf{e}_z\}$ is shown in Fig. 3–19.

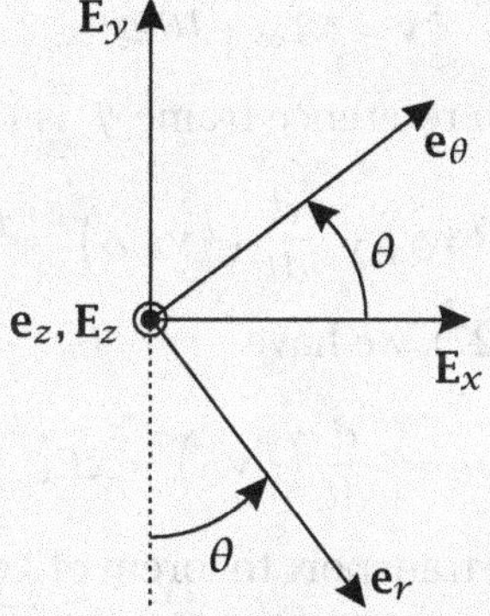

Figure 3–19 Geometry of bases $\{\mathbf{E}_x, \mathbf{E}_y, \mathbf{E}_z\}$ and $\{\mathbf{e}_r, \mathbf{e}_\theta, \mathbf{E}_z\}$ for Example 3–5.

In particular, using Fig. 3–19, the relationship between the basis $\{\mathbf{E}_x, \mathbf{E}_y, \mathbf{E}_z\}$ and $\{\mathbf{e}_r, \mathbf{e}_\theta, \mathbf{e}_z\}$ is given as

$$\mathbf{E}_x = \sin\theta\,\mathbf{e}_r + \cos\theta\,\mathbf{e}_\theta \tag{3-220}$$

$$\mathbf{E}_y = -\cos\theta\,\mathbf{e}_r + \sin\theta\,\mathbf{e}_\theta \tag{3-221}$$

Furthermore, it is noted that the angular velocity of reference frame $\mathcal{F}$ is zero (because $\mathcal{F}$ is a fixed inertial reference frame) while the angular velocity of reference frame $\mathcal{A}$ in reference frame $\mathcal{F}$ is

$${}^{\mathcal{F}}\boldsymbol{\omega}^{\mathcal{A}} = \dot{\theta}\mathbf{e}_z \tag{3-222}$$

The position of the particle can then be expressed in terms of the basis $\{\mathbf{E}_x, \mathbf{E}_y, \mathbf{E}_z\}$ fixed in reference frame $\mathcal{F}$ and the basis $\{\mathbf{e}_r, \mathbf{e}_\theta, \mathbf{e}_z\}$ fixed in reference frame $\mathcal{A}$ as

$$\mathbf{r} = x\mathbf{E}_x + l\mathbf{e}_r \tag{3-223}$$

Now let

$$\begin{aligned} \mathbf{r}_Q &= x\mathbf{E}_x \\ \mathbf{r}_{P/Q} &= l\mathbf{e}_r \end{aligned} \tag{3-224}$$

Then the position of the particle can be written as

$$\mathbf{r} = \mathbf{r}_Q + \mathbf{r}_{P/Q} \tag{3-225}$$

Computing the rate of change of $\mathbf{r}$ in the inertial reference frame $\mathcal{F}$, we have

$${}^{\mathcal{F}}\mathbf{v} = \frac{{}^{\mathcal{F}}d\mathbf{r}}{dt} = \frac{{}^{\mathcal{F}}d}{dt}(\mathbf{r}_Q) + \frac{{}^{\mathcal{F}}d}{dt}(\mathbf{r}_{P/Q}) = {}^{\mathcal{F}}\mathbf{v}_Q + {}^{\mathcal{F}}\mathbf{v}_{P/Q} \tag{3-226}$$

Now, because $\mathcal{F}$ is fixed, we have

$${}^{\mathcal{F}}\mathbf{v}_Q = \frac{{}^{\mathcal{F}}d}{dt}(\mathbf{r}_Q) = \dot{x}\mathbf{E}_x \tag{3-227}$$

Furthermore,

$${}^{\mathcal{F}}\mathbf{v}_{P/Q} = \frac{{}^{\mathcal{F}}d}{dt}(\mathbf{r}_{P/Q}) = \frac{{}^{\mathcal{A}}d}{dt}(\mathbf{r}_{P/Q}) + {}^{\mathcal{F}}\boldsymbol{\omega}^{\mathcal{A}} \times \mathbf{r}_Q = \dot{\theta}\mathbf{e}_z \times l\mathbf{e}_r = l\dot{\theta}\mathbf{e}_\theta \tag{3-228}$$

Adding the expressions in Eqs. (3-227) and (3-228), we obtain

$${}^{\mathcal{F}}\mathbf{v} = \dot{x}\mathbf{E}_x + l\dot{\theta}\mathbf{e}_\theta \tag{3-229}$$

The acceleration of the particle in reference frame $\mathcal{F}$ is then obtained as

$${}^{\mathcal{F}}\mathbf{a} = \frac{{}^{\mathcal{F}}d}{dt}\left({}^{\mathcal{F}}\mathbf{v}\right) = \frac{{}^{\mathcal{F}}d}{dt}\left({}^{\mathcal{F}}\mathbf{v}_Q\right) + \frac{{}^{\mathcal{F}}d}{dt}\left({}^{\mathcal{F}}\mathbf{v}_{P/Q}\right) = {}^{\mathcal{F}}\mathbf{a}_Q + {}^{\mathcal{F}}\mathbf{a}_{P/Q} \tag{3-230}$$

First, using the result of Eq. (3-227), we have

$${}^{\mathcal{F}}\mathbf{a}_Q = \frac{{}^{\mathcal{F}}d}{dt}\left({}^{\mathcal{F}}\mathbf{v}_Q\right) = \ddot{x}\mathbf{E}_x \tag{3-231}$$

Next, applying the rate of change transport theorem of Eq. (2-128) on page 47 between reference frames $\mathcal{A}$ and $\mathcal{F}$ to ${}^{\mathcal{F}}\mathbf{v}_{P/Q}$ using the result of Eq. (3-228), we obtain

$$\begin{aligned} {}^{\mathcal{F}}\mathbf{a}_{P/Q} &= \frac{{}^{\mathcal{F}}d}{dt}\left({}^{\mathcal{F}}\mathbf{v}_{P/Q}\right) = \frac{{}^{\mathcal{A}}d}{dt}\left({}^{\mathcal{F}}\mathbf{v}_{P/Q}\right) + {}^{\mathcal{F}}\boldsymbol{\omega}^{\mathcal{A}} \times {}^{\mathcal{F}}\mathbf{v}_{P/Q} \\ &= l\ddot{\theta}\mathbf{e}_\theta + \dot{\theta}\mathbf{e}_z \times l\dot{\theta}\mathbf{e}_\theta = -l\dot{\theta}^2\mathbf{e}_r + l\ddot{\theta}\mathbf{e}_\theta \end{aligned} \tag{3-232}$$

Adding the results of Eqs. (3-231) and (3-232), we obtain the acceleration of the particle in reference frame $\mathcal{F}$ as

$${}^{\mathcal{F}}\mathbf{a} = \ddot{x}\mathbf{E}_x - l\dot{\theta}^2\mathbf{e}_r + l\ddot{\theta}\mathbf{e}_\theta \tag{3-233}$$

Finally, substituting the expression for $\mathbf{E}_x$ in terms of $\mathbf{e}_r$ and $\mathbf{e}_\theta$ from Eq. (3-220), we obtain ${}^{\mathcal{F}}\mathbf{a}$ in terms of the basis $\{\mathbf{e}_r, \mathbf{e}_\theta, \mathbf{e}_z\}$ as

$${}^{\mathcal{F}}\mathbf{a} = \ddot{x}(\sin\theta\,\mathbf{e}_r + \cos\theta\,\mathbf{e}_\theta) - l\dot{\theta}^2\mathbf{e}_r + l\ddot{\theta}\mathbf{e}_\theta \tag{3-234}$$

Simplifying Eq. (3-234) gives

$${}^{\mathcal{F}}\mathbf{a} = (\ddot{x}\sin\theta - l\dot{\theta}^2)\mathbf{e}_r + (\ddot{x}\cos\theta + l\ddot{\theta})\mathbf{e}_\theta \tag{3-235}$$

Kinetics

The free body diagram of the particle is shown in Fig. 3-20.

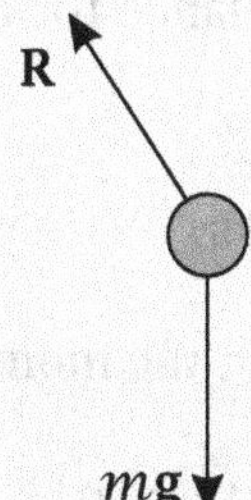

Figure 3-20 Free body diagram for Example 3-5.

Using Fig. 3-20, it is seen that the following forces act on the particle:

$$\begin{aligned} \mathbf{R} &= \text{Reaction force of rod on particle} \\ m\mathbf{g} &= \text{Force of gravity} \end{aligned}$$

In examining the forces in this problem, it is important to note that the reaction force **R** must lie along the direction of the rod. Consequently, we have

$$\mathbf{R} = R\mathbf{e}_r \tag{3-236}$$

$$m\mathbf{g} = -mg\mathbf{E}_y \tag{3-237}$$

The resultant force acting on the particle is then given as

$$\mathbf{F} = \mathbf{R} + m\mathbf{g} = R\mathbf{e}_r - mg\mathbf{E}_y \tag{3-238}$$

Then, using the expression for $\mathbf{E}_y$ from Eq. (3-221), the resultant force is obtained in terms of the basis $\{\mathbf{e}_r, \mathbf{e}_\theta, \mathbf{e}_z\}$ as

$$\mathbf{F} = R\mathbf{e}_r - mg(-\cos\theta\,\mathbf{e}_r + \sin\theta\,\mathbf{e}_\theta) = (R + mg\cos\theta)\mathbf{e}_r - mg\sin\theta\,\mathbf{e}_\theta \tag{3-239}$$

(a) Differential Equation of Motion via Newton's 2nd Law

Applying Newton's 2nd law to the particle using the resultant force in Eq. (3-239) and the acceleration in Eq. (3-233), we have

$$(R + mg\cos\theta)\mathbf{e}_r - mg\sin\theta\,\mathbf{e}_\theta = m(\ddot{x}\sin\theta - l\dot{\theta}^2)\mathbf{e}_r + m(\ddot{x}\cos\theta + l\ddot{\theta})\mathbf{e}_\theta \tag{3-240}$$

Equating components in Eq. (3-240), we obtain the following two scalar equations:

$$m(\ddot{x}\sin\theta - l\dot{\theta}^2) = R + mg\cos\theta \tag{3-241}$$

$$m(\ddot{x}\cos\theta + l\ddot{\theta}) = -mg\sin\theta \tag{3-242}$$

It is seen that Eq. (3-242) has no unknown forces. Consequently, the differential equation of motion is given as

$$m(\ddot{x}\cos\theta + l\ddot{\theta}) = -mg\sin\theta \tag{3-243}$$

Rearranging Eq. (3-244), we obtain

$$ml\ddot{\theta} + mg\sin\theta + m\ddot{x}\cos\theta = 0 \tag{3-244}$$

(b) Differential Equation via Balance of Angular Momentum Relative to Point Q

We note that point Q is not inertially fixed and, thus, is an *arbitrary* point. Therefore, it is necessary to apply the moment balance for an arbitrary reference point as given in Eq. (3–203), i.e., we need to apply

$$\frac{^{\mathcal{F}}d}{dt}\left({^{\mathcal{F}}}\mathbf{H}_Q\right) = \mathbf{M}_Q - (\mathbf{r} - \mathbf{r}_Q) \times m\,{^{\mathcal{F}}}\mathbf{a}_Q \tag{3-245}$$

Now, because $\mathbf{R}$ passes through point Q, the moment relative to point Q is due purely to gravity and is given as

$$\mathbf{M}_Q = (\mathbf{r}_g - \mathbf{r}_Q) \times m\mathbf{g} \tag{3-246}$$

Furthermore, because the force of gravity acts at the location of the particle, we have that

$$\mathbf{r}_g = \mathbf{r} = x\mathbf{E}_x + l\mathbf{e}_r \tag{3-247}$$

Now, because $\mathbf{r}_Q = x\mathbf{E}_x$, we have

$$\mathbf{r}_g - \mathbf{r}_Q = l\mathbf{e}_r \tag{3-248}$$

The moment relative to point Q is then given as

$$\mathbf{M}_Q = l\mathbf{e}_r \times (-mg\mathbf{E}_y) = -mgl\mathbf{e}_r \times \mathbf{E}_y \tag{3-249}$$

Then, using Eq. (3–221), we have $\mathbf{e}_r \times \mathbf{E}_y$ as

$$\mathbf{e}_r \times \mathbf{E}_y = \mathbf{e}_r \times (-\cos\theta\,\mathbf{e}_r + \sin\theta\,\mathbf{e}_\theta) = \sin\theta\,\mathbf{e}_z \tag{3-250}$$

Using Eq. (3–250), the moment relative to point Q simplifies to

$$\mathbf{M}_Q = -mgl\sin\theta\,\mathbf{e}_z \tag{3-251}$$

Next, using $\mathbf{r}_Q$ from Eq. (3–224), we have

$$\mathbf{r} - \mathbf{r}_Q = \mathbf{r}_{P/Q} = l\mathbf{e}_r \tag{3-252}$$

Then, since ${^{\mathcal{F}}}\mathbf{a}_Q = \ddot{\mathbf{r}}_Q = \ddot{x}\mathbf{E}_x$, the inertial moment is given as

$$-(\mathbf{r} - \mathbf{r}_Q) \times m\,{^{\mathcal{F}}}\mathbf{a}_Q = -l\mathbf{e}_r \times m\ddot{x}\mathbf{E}_x = -ml\ddot{x}\mathbf{e}_r \times \mathbf{E}_x \tag{3-253}$$

Using Eq. (3–220), we obtain $\mathbf{e}_r \times \mathbf{E}_x$ as

$$\mathbf{e}_r \times \mathbf{E}_x = \mathbf{e}_r \times (\sin\theta\,\mathbf{e}_r + \cos\theta\,\mathbf{e}_\theta) = \cos\theta\,\mathbf{e}_z \tag{3-254}$$

The inertial moment relative to point Q then simplifies to

$$-(\mathbf{r} - \mathbf{r}_Q) \times m\,{^{\mathcal{F}}}\mathbf{a}_Q = -ml\ddot{x}\cos\theta\,\mathbf{e}_z \tag{3-255}$$

Finally, the rate of change of angular momentum relative to point Q is given as

$$\frac{^{\mathcal{F}}d}{dt}\left({^{\mathcal{F}}}\mathbf{H}_Q\right) = (\mathbf{r} - \mathbf{r}_Q) \times m(\mathbf{a} - \mathbf{a}_Q) \tag{3-256}$$

Noting that

$$^{\mathcal{F}}\mathbf{a} - {}^{\mathcal{F}}\mathbf{a}_Q = -l\dot{\theta}^2\mathbf{e}_r + l\ddot{\theta}\mathbf{e}_\theta \quad (3\text{-}257)$$

and using the expression for $\mathbf{r} - \mathbf{r}_Q$ from Eq. (3-252), we obtain the rate of change of angular momentum relative to point Q as

$$\frac{^{\mathcal{F}}d}{dt}\left({}^{\mathcal{F}}\mathbf{H}_Q\right) = l\mathbf{e}_r \times m(-l\dot{\theta}^2\mathbf{e}_r + l\ddot{\theta}\mathbf{e}_\theta) = ml^2\ddot{\theta}\mathbf{e}_z \quad (3\text{-}258)$$

Then, substituting the results of Eqs. (3-251), (3-255), and (3-258) into (3-245), we obtain

$$ml^2\ddot{\theta}\mathbf{e}_z = -mgl\sin\theta\,\mathbf{e}_z - ml\ddot{x}\cos\theta\,\mathbf{e}_z \quad (3\text{-}259)$$

Rearranging and simplifying Eq. (3-259), we obtain the differential equation as

$$ml\ddot{\theta} + mg\sin\theta + m\ddot{x}\cos\theta = 0 \quad (3\text{-}260)$$

It is noted that the result of Eq. (3-260) obtained using a balance of angular momentum relative to point Q is identical to that of Eq. (3-244) obtained using Newton's 2nd law.

■

3.9 Instantaneous Linear and Angular Impulse

In many applications in dynamics, impulses are assumed to occur without the passage of time. In order for an impulse to have a nonzero value over a zero time interval, it is necessary that the force be infinite. While strictly speaking it is not possible to apply an infinite force, such an assumption is often a good approximation to reality. Modeling an infinite force over a zero time interval is accomplished by introducing a special function called the *Dirac delta function*. The Dirac delta function, denoted $\delta(t-s)$, is defined as follows:

$$\delta(t-s) = \begin{cases} 0 & , \quad t \neq s \\ \infty & , \quad t = s \end{cases} \quad (3\text{-}261)$$

Furthermore, $\delta(t-s)$ has the following property:

$$\int_{-\infty}^{\infty} f(t)\delta(t-s)dt = f(s) \quad (3\text{-}262)$$

From Eq. (3-262) it is seen that if $f(t) = 1$, then

$$\int_{-\infty}^{\infty} \delta(t-s)dt = 1 \quad (3\text{-}263)$$

where $f(t)$ is an arbitrary function of t. Suppose now that we have an infinite force $\mathbf{F}$ modeled as follows:

$$\mathbf{F}(t) = \hat{\mathbf{F}}\delta(t-\tau) \quad (3\text{-}264)$$

where $\hat{\mathbf{F}}$ is finite. Suppose now that we consider the instant of time $t = \tau$. Applying Eq. (3-177), the linear impulse of the force $\mathbf{F}$ is given as

$$\hat{\mathbf{F}} = \int_{t=\tau^-}^{t=\tau^+} \mathbf{F}(t)dt = \int_{-\infty}^{\infty} \mathbf{F}(t)dt \quad (3\text{-}265)$$

A linear impulse of the form of Eq. (3–265) is called an *instantaneous linear impulse.* Furthermore, suppose that we consider the application of the force $\mathbf{F}$ to a particle P such that $\mathbf{F}$ is located at the position $\mathbf{r}$ and relative to a point Q. Then the angular impulse of the force $\mathbf{F}$ relative to point Q is given from Eq. (3–206) as

$$\hat{\mathbf{M}}_Q = \int_{t=\tau^-}^{t=\tau^+} (\mathbf{r} - \mathbf{r}_Q) \times \mathbf{F} dt = \int_{-\infty}^{\infty} (\mathbf{r} - \mathbf{r}_Q) \times \mathbf{F}(t) dt = (\mathbf{r} - \mathbf{r}_Q) \times \hat{\mathbf{F}} \tag{3-266}$$

An angular impulse that arises from an instantaneous linear impulse is called an *instantaneous angular impulse.*

3.10 Power, Work, and Energy for a Particle

3.10.1 Power and Work of a Force

Consider a particle of mass m moving in an inertial reference frame $\mathcal{N}$ with velocity ${}^{\mathcal{N}}\mathbf{v}$. Furthermore, let $\mathbf{F}$ be a force acting on the particle. Then the *power* of the force $\mathbf{F}$ in reference frame $\mathcal{N}$, denoted ${}^{\mathcal{N}}P$, is defined as

$${}^{\mathcal{N}}P = \mathbf{F} \cdot {}^{\mathcal{N}}\mathbf{v} \tag{3-267}$$

Consider now an interval of time $t \in [t_1, t_2]$ over which the force $\mathbf{F}$ acts, and let $\mathbf{r}_1 = \mathbf{r}(t_1)$ and $\mathbf{r}_2 = \mathbf{r}(t_2)$ be the positions of the particle P at t_1 and t_2, respectively, as shown in Fig. 3-21. Then the *work* done by the force $\mathbf{F}$ in reference frame $\mathcal{N}$ from t_1

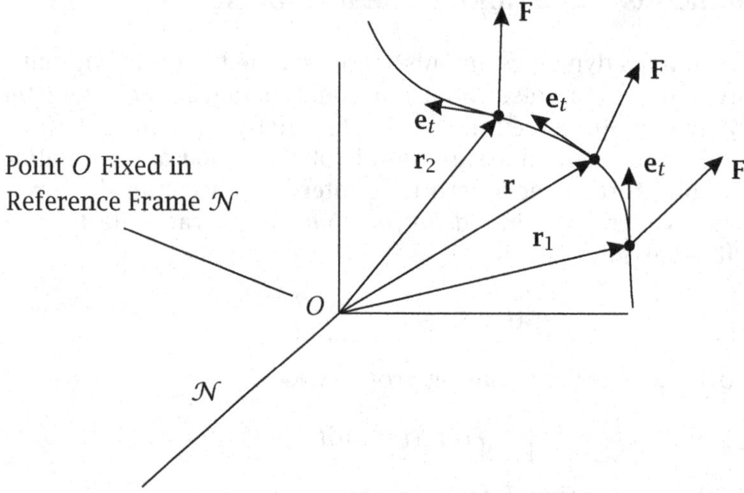

Figure 3–21 Work done by a force $\mathbf{F}$ in an inertial reference frame $\mathcal{N}$ on a time interval $t \in [t_1, t_2]$ in moving a particle of mass m from a position $\mathbf{r}(t_1) = \mathbf{r}_1$ to a position $\mathbf{r}(t_2) = \mathbf{r}_2$.

to t_2 is defined as

$${}^{\mathcal{N}}W_{12} = \int_{t_1}^{t_2} {}^{\mathcal{N}}P dt = \int_{t_1}^{t_2} \mathbf{F} \cdot {}^{\mathcal{N}}\mathbf{v} dt \tag{3-268}$$

Now from Eq. (2–29) we have

$${}^{\mathcal{N}}\mathbf{v} = \frac{{}^{\mathcal{N}}d\mathbf{r}}{dt}$$

Consequently, the work ${}^{\mathcal{N}}W_{12}$ can be written as

$$ {}^{\mathcal{N}}W_{12} = \int_{t_1}^{t_2} \mathbf{F} \cdot \frac{{}^{\mathcal{N}}d\mathbf{r}}{dt} dt = \int_{\mathbf{r}_1}^{\mathbf{r}_2} \mathbf{F} \cdot {}^{\mathcal{N}}d\mathbf{r} \tag{3-269} $$

where ${}^{\mathcal{N}}d\mathbf{r}$ is a differential change in the position of the particle in reference frame $\mathcal{N}$. Now recall from Section 2.11.4 that the velocity of a particle can be written in terms of the tangent vector $\mathbf{e}_t$ as

$$ {}^{\mathcal{N}}\mathbf{v} = {}^{\mathcal{N}}v\mathbf{e}_t $$

where ${}^{\mathcal{N}}v$ is the speed of the particle in reference frame $\mathcal{N}$. The work in reference frame $\mathcal{N}$ can then be written as

$$ {}^{\mathcal{N}}W_{12} = \int_{t_1}^{t_2} \mathbf{F} \cdot {}^{\mathcal{N}}v\mathbf{e}_t dt \tag{3-270} $$

It can be seen from Eq. (3–270) that only the component of the force $\mathbf{F}$ that lies in the direction of ${}^{\mathcal{N}}\mathbf{v}$ does any work; any component of $\mathbf{F}$ orthogonal to ${}^{\mathcal{N}}\mathbf{v}$ does not contribute to the work. Furthermore, it is seen from Eq. (3–268) that, due to its dependence on ${}^{\mathcal{N}}\mathbf{v}$, the work of a force $\mathbf{F}$ depends on the choice of the reference frame.

3.10.2 Kinetic Energy and Work-Energy Theorem for a Particle

The *kinetic energy* of a particle of mass m in an inertial reference frame $\mathcal{N}$, denoted ${}^{\mathcal{N}}T$, is defined as

$$ {}^{\mathcal{N}}T = \tfrac{1}{2}m{}^{\mathcal{N}}\mathbf{v} \cdot {}^{\mathcal{N}}\mathbf{v} \tag{3-271} $$

where ${}^{\mathcal{N}}\mathbf{v}$ is the velocity of the particle in reference frame $\mathcal{N}$. Noting that ${}^{\mathcal{N}}T$ is a scalar function and, hence, is independent of reference frame, its rate of change is given as

$$ \frac{d}{dt}\left({}^{\mathcal{N}}T\right) = \frac{1}{2}m\frac{d}{dt}\left[{}^{\mathcal{N}}\mathbf{v} \cdot {}^{\mathcal{N}}\mathbf{v}\right] \tag{3-272} $$

Differentiating the vectors on the right-hand side of Eq. (3–272) in reference frame $\mathcal{N}$, we obtain

$$ \frac{d}{dt}\left({}^{\mathcal{N}}T\right) = \frac{1}{2}m\left[\frac{{}^{\mathcal{N}}d}{dt}\left({}^{\mathcal{N}}\mathbf{v}\right) \cdot {}^{\mathcal{N}}\mathbf{v} + {}^{\mathcal{N}}\mathbf{v} \cdot \frac{{}^{\mathcal{N}}d}{dt}\left({}^{\mathcal{N}}\mathbf{v}\right)\right] \tag{3-273} $$

Furthermore, noting that ${}^{\mathcal{N}}d({}^{\mathcal{N}}\mathbf{v})/dt = {}^{\mathcal{N}}\mathbf{a}$ and using the fact that the scalar product is commutative, we obtain

$$ \frac{d}{dt}\left({}^{\mathcal{N}}T\right) = m{}^{\mathcal{N}}\mathbf{a} \cdot {}^{\mathcal{N}}\mathbf{v} \tag{3-274} $$

Then, applying Newton's 2nd law of Eq. (3–41), Eq. (3–274) becomes

$$ \boxed{\frac{d}{dt}\left({}^{\mathcal{N}}T\right) = \mathbf{F} \cdot {}^{\mathcal{N}}\mathbf{v} = {}^{\mathcal{N}}P} \tag{3-275} $$

where ${}^{\mathcal{N}}P$ is the power of the force $\mathbf{F}$ in an inertial reference frame $\mathcal{N}$ as defined in Eq. (3–267). Equation (3–275) is called the *work-energy theorem for a particle* and relates the rate of change of kinetic energy of a particle in an inertial reference frame $\mathcal{N}$ to the power of the resultant force acting on the particle. It is important to understand that Eq. (3–275) is valid only in an *inertial reference frame*.

3.10.3 Principle of Work and Energy for a Particle

Now suppose that we consider motion over a time interval $t \in [t_1, t_2]$. Integrating Eq. (3–275) from t_1 to t_2, we have

$$\int_{t_1}^{t_2} \frac{d}{dt}\left({}^{\mathcal{N}}T\right) dt = \int_{t_1}^{t_2} \mathbf{F} \cdot {}^{\mathcal{N}}\mathbf{v} dt \tag{3–276}$$

Now we know from Eq. (3–268) that the work done by all forces acting on the particle on the interval $t \in [t_1, t_2]$ is given as

$${}^{\mathcal{N}}W_{12} = \int_{t_1}^{t_2} \mathbf{F} \cdot {}^{\mathcal{N}}\mathbf{v} dt \tag{3–277}$$

Furthermore, we have

$$\int_{t_1}^{t_2} \frac{d}{dt}\left({}^{\mathcal{N}}T\right) dt = {}^{\mathcal{N}}T(t_2) - {}^{\mathcal{N}}T(t_1) = {}^{\mathcal{N}}T_2 - {}^{\mathcal{N}}T_1 \tag{3–278}$$

Using Eq. (3–277) together with Eq. (3–278), we obtain

$$\boxed{{}^{\mathcal{N}}T_2 - {}^{\mathcal{N}}T_1 = {}^{\mathcal{N}}W_{12}} \tag{3–279}$$

Equation (3–279) is called the *principle of work and energy* for a particle and states that the change in kinetic energy between any two points along the trajectory of a particle in an inertial reference frame is equal to the work done by all forces acting on the particle between those same two points.

3.10.4 Conservative Forces and Potential Energy

It is seen from Eq. (3–268) that, in general, the work done by a force $\mathbf{F}$ over a time interval $t \in [t_1, t_2]$ depends on both the trajectory taken by the particle on $[t_1, t_2]$ and the position of the particle at the endpoints t_1 and t_2. Consider now a force $\mathbf{F}^c$ whose work in moving a particle from an initial position $\mathbf{r}(t_1) = \mathbf{r}_1$ to a final position $\mathbf{r}_2(t_2) = \mathbf{r}_2$ *does not* depend on the trajectory taken by the particle but, instead, depends on *only* the endpoints $\mathbf{r}_1$ and $\mathbf{r}_2$. Then the force $\mathbf{F}^c$ is said to be a *conservative force.* A schematic of a conservative force $\mathbf{F}^c$ is shown in Fig. 3–22.
Suppose now that ${}^{\mathcal{N}}\mathbf{v}^{(1)}, \ldots, {}^{\mathcal{N}}\mathbf{v}^{(n)}$ are the velocities in an inertial reference frame $\mathcal{N}$ that correspond to an arbitrary set of trajectories that start at $\mathbf{r}_1$ and end at $\mathbf{r}_2$. Then the force $\mathbf{F}^c$ is conservative if the work done from $\mathbf{r}_1$ to $\mathbf{r}_2$ is the same for all trajectories ${}^{\mathcal{N}}\mathbf{v}^{(i)}$ $i = 1, 2, \ldots, n$, i.e.,

$${}^{\mathcal{N}}W_{12}^c = \int_{t_1}^{t_2} \mathbf{F}^c \cdot {}^{\mathcal{N}}\mathbf{v}^{(1)} = \cdots = \int_{t_1}^{t_2} \mathbf{F}^c \cdot {}^{\mathcal{N}}\mathbf{v}^{(n)} \tag{3–280}$$

A consequence of the fact that a conservative force is independent of the trajectory is that there exists a scalar function ${}^{\mathcal{N}}U = {}^{\mathcal{N}}U(\mathbf{r})$ such that

$$\boxed{\mathbf{F}^c = -\nabla_{\mathbf{r}} {}^{\mathcal{N}}U = -\frac{\partial}{\partial \mathbf{r}}\left({}^{\mathcal{N}}U\right)} \tag{3–281}$$

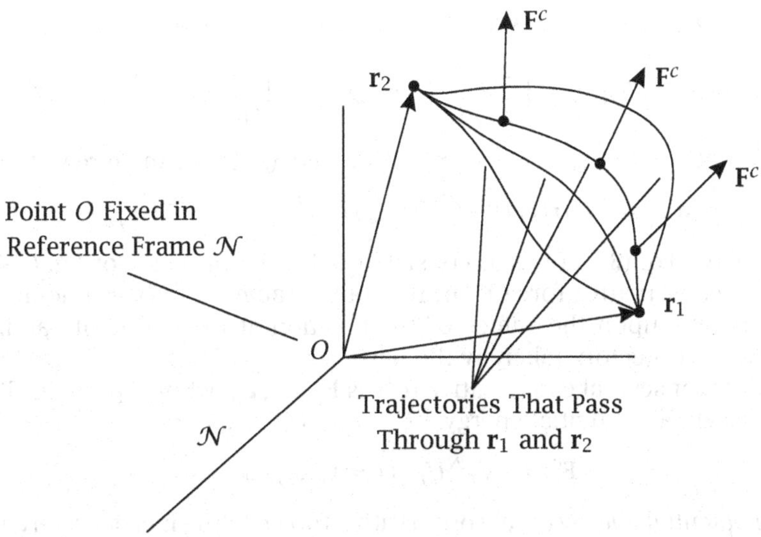

Figure 3-22 Work done by a conservative force $\mathbf{F}^c$ in moving a particle from a position $\mathbf{r}(t_1) = \mathbf{r}_1$ to a position $\mathbf{r}(t_2) = \mathbf{r}_2$. The work of the conservative force depends only on the endpoints $\mathbf{r}_1$ and $\mathbf{r}_2$ and does not depend on the trajectory subtended by the particle in moving from $\mathbf{r}_1$ to $\mathbf{r}_2$.

where $\mathbf{r}$ is the vector that defines the location in $\mathbb{E}^3$ relative to a point O fixed in reference frame $\mathcal{N}$ at which the force $\mathbf{F}^c$ acts. The scalar function ${}^{\mathcal{N}}U$ in Eq. (3-281) is called a *potential energy of the force* $\mathbf{F}^c$ *in the inertial reference frame* $\mathcal{N}$ and the operator $\nabla_{\mathbf{r}}$ is the *gradient with respect to* $\mathbf{r}$.[8] Then, using the definition of power from Eq. (3-267), the power of the conservative force $\mathbf{F}^c$ in the inertial reference frame $\mathcal{N}$, denoted ${}^{\mathcal{N}}p^c$, is given as

$$ {}^{\mathcal{N}}p^c = \mathbf{F}^c \cdot {}^{\mathcal{N}}\mathbf{v} = -\nabla_{\mathbf{r}}{}^{\mathcal{N}}U \cdot {}^{\mathcal{N}}\mathbf{v} = -\frac{\partial}{\partial \mathbf{r}}\left({}^{\mathcal{N}}U\right) \cdot {}^{\mathcal{N}}\mathbf{v} \tag{3-282} $$

Next, applying Eq. (2-22) to (3-282), we see that

$$ \nabla_{\mathbf{r}}{}^{\mathcal{N}}U \cdot {}^{\mathcal{N}}\mathbf{v} = \frac{d}{dt}\left({}^{\mathcal{N}}U\right) \tag{3-283} $$

Substituting the identity of Eq. (3-283) into (3-282), we obtain ${}^{\mathcal{N}}p^c$ as

$$ \boxed{{}^{\mathcal{N}}p^c = -\frac{d}{dt}\left({}^{\mathcal{N}}U\right)} \tag{3-284} $$

Consequently, the power of a conservative force in an inertial reference frame $\mathcal{N}$ is the negative of the rate of change of a potential energy of the conservative force where potential energy is computed in the inertial reference frame $\mathcal{N}$.

Now consider a time interval $t \in [t_1, t_2]$ over which the conservative force $\mathbf{F}^c$ acts. Using Eq. (3-282), the work done by the conservative force $\mathbf{F}^c$ in reference frame $\mathcal{N}$

[8]We use the terminology "a potential energy" instead of "the potential energy" because all potential energies, up to an arbitrary function of time, have the same gradient with respect to position.

from t_1 to t_2, denoted ${}^{\mathcal{N}}W_{12}^c$, is given as

$$ {}^{\mathcal{N}}W_{12}^c = \int_{t_1}^{t_2} -\frac{d}{dt}\left({}^{\mathcal{N}}U\right) dt = -\int_{{}^{\mathcal{N}}U_1}^{{}^{\mathcal{N}}U_2} dU \tag{3-285} $$

Then, using the fact that $\mathbf{r}(t_1) = \mathbf{r}_1$ and $\mathbf{r}(t_2) = \mathbf{r}_2$, Eq. (3–285) can be rewritten as

$$ {}^{\mathcal{N}}W_{12}^c = {}^{\mathcal{N}}U(\mathbf{r}(t_1)) - {}^{\mathcal{N}}U(\mathbf{r}(t_2)) = {}^{\mathcal{N}}U(\mathbf{r}_1) - {}^{\mathcal{N}}U(\mathbf{r}_2) \tag{3-286} $$

It can be seen from Eq. (3–286) that, consistent with the property of Eq. (3–280), the work done by a conservative force $\mathbf{F}^c$ in reference frame $\mathcal{N}$ over a time interval $t \in [t_1, t_2]$ depends only upon the values of the position at the endpoints and, thus, is independent of the trajectory taken by the particle.

Suppose now that several conservative forces $\mathbf{F}_1^c, \ldots, \mathbf{F}_n^c$ act on a particle. Then each force has an associated potential energy ${}^{\mathcal{N}}U_1, \ldots, {}^{\mathcal{N}}U_n$, i.e.,

$$ \mathbf{F}_i^c = -\nabla_{\mathbf{r}}{}^{\mathcal{N}}U_i \quad (i = 1, \ldots, n) \tag{3-287} $$

Then the *total potential energy* of all conservative forces in the inertial reference frame $\mathcal{N}$, denoted U, is defined as

$$ {}^{\mathcal{N}}U = \sum_{i=1}^{n} {}^{\mathcal{N}}U_i \tag{3-288} $$

Correspondingly, the total conservative force acting on the particle is given as

$$ \mathbf{F}^c = \sum_{i=1}^{n} \mathbf{F}_i^c = -\sum_{i=1}^{n} \nabla_{\mathbf{r}}{}^{\mathcal{N}}U_i = -\nabla_{\mathbf{r}}{}^{\mathcal{N}}U \tag{3-289} $$

3.10.5 Examples of Conservative Forces

In this section we provide examples of several commonly encountered conservative forces. These forces will be used extensively in the analysis of dynamics problems.

Constant Forces

Let $\mathbf{c}$ be a constant force. Then the potential energy of $\mathbf{c}$ in reference frame $\mathcal{N}$, denoted ${}^{\mathcal{N}}U_c$, is given as

$$ {}^{\mathcal{N}}U_c = -\mathbf{c} \cdot \mathbf{r} \tag{3-290} $$

To verify that ${}^{\mathcal{N}}U_c$ is in fact the potential energy of the constant force $\mathbf{c}$, we can compute the rate of change of ${}^{\mathcal{N}}U_c$ as

$$ \frac{d}{dt}\left({}^{\mathcal{N}}U_c\right) = \nabla_{\mathbf{r}}(-\mathbf{c} \cdot \mathbf{r}) \cdot \frac{{}^{\mathcal{N}}d\mathbf{r}}{dt} = -\mathbf{c} \cdot {}^{\mathcal{N}}\mathbf{v} \tag{3-291} $$

where we note that, since $\mathbf{c}$ is constant, we have $\nabla(\mathbf{c} \cdot \mathbf{r}) \equiv \mathbf{c}$. Then, observing that $-\mathbf{c} \cdot {}^{\mathcal{N}}\mathbf{v}$ is the negative of the power of the force $\mathbf{c}$ in the inertial reference frame $\mathcal{N}$, we obtain

$$ {}^{\mathcal{N}}P_c = \mathbf{c} \cdot {}^{\mathcal{N}}\mathbf{v} \tag{3-292} $$

Therefore,

$$ {}^{\mathcal{N}}P_c = -\frac{d}{dt}\left({}^{\mathcal{N}}U_c\right) \tag{3-293} $$

which implies that $\mathbf{c}$ is conservative.

Spring Forces

Recall from Eq. (3–19) that the force applied to a particle by a linear spring with spring constant K is given as

$$\mathbf{F}_s = -K(\ell - \ell_0)\mathbf{u}_s \tag{3-294}$$

where $\ell = \|\mathbf{r} - \mathbf{r}_Q\|$ is the length of the spring, $\mathbf{r}$ is the position of the particle, and $\mathbf{r}_Q$ is the attachment point of the spring. Now let $\mathcal{N}$ be an inertial reference frame and assume that the attachment point is fixed in $\mathcal{N}$, i.e., assume that

$$\begin{aligned} {}^{\mathcal{N}}\mathbf{v}_Q &= \mathbf{0} \\ {}^{\mathcal{N}}\mathbf{a}_Q &= \mathbf{0} \end{aligned} \tag{3-295}$$

Then the potential energy of a linear spring in reference frame $\mathcal{N}$, denoted ${}^{\mathcal{N}}U_s$, is given as

$${}^{\mathcal{N}}U_s = \frac{K}{2}(\ell - \ell_0)^2 \tag{3-296}$$

where $\ell = \|\mathbf{r} - \mathbf{r}_Q\|$ and ℓ_0 are the stretched and unstretched length of the spring, respectively. To verify that ${}^{\mathcal{N}}U_s$ is the potential energy of $\mathbf{F}_s$, we need to compute $d({}^{\mathcal{N}}U_s)/dt$. Computing the rate of change of ${}^{\mathcal{N}}U_s$, we have

$$\frac{d}{dt}\left({}^{\mathcal{N}}U_s\right) = \frac{d}{dt}\left[\frac{K}{2}(\ell - \ell_0)^2\right] = K(\ell - \ell_0)\frac{d}{dt}(\ell - \ell_0) \tag{3-297}$$

Using the fact that $\ell = \|\mathbf{r} - \mathbf{r}_Q\|$, we obtain

$$\frac{d}{dt}(\ell - \ell_0) = \nabla_{\mathbf{r}-\mathbf{r}_Q}(\ell - \ell_0) \cdot \frac{{}^{\mathcal{N}}d}{dt}(\mathbf{r} - \mathbf{r}_Q) \tag{3-298}$$

where the gradient is evaluated with respect to $\mathbf{r} - \mathbf{r}_Q$. The gradient is given as

$$\begin{aligned} \nabla_{\mathbf{r}-\mathbf{r}_Q}(\ell - \ell_0) &= \nabla_{\mathbf{r}-\mathbf{r}_Q}\|\mathbf{r} - \mathbf{r}_Q\| \\ &= {}^{\mathcal{N}}\nabla_{\mathbf{r}-\mathbf{r}_Q}\left[(\mathbf{r} - \mathbf{r}_Q) \cdot (\mathbf{r} - \mathbf{r}_Q)\right]^{1/2} \\ &= \frac{\mathbf{r} - \mathbf{r}_Q}{\|\mathbf{r} - \mathbf{r}_Q\|} \\ &= \frac{\mathbf{r} - \mathbf{r}_Q}{\ell} \end{aligned} \tag{3-299}$$

Consequently,

$$\frac{d}{dt}(\ell - \ell_0) = \frac{\mathbf{r} - \mathbf{r}_Q}{\ell} \cdot \frac{{}^{\mathcal{N}}d}{dt}(\mathbf{r} - \mathbf{r}_Q) \tag{3-300}$$

The quantity $d({}^{\mathcal{N}}U_s)/dt$ then becomes

$$\frac{d}{dt}\left({}^{\mathcal{N}}U_s\right) = K(\ell - \ell_0)\frac{\mathbf{r} - \mathbf{r}_Q}{\ell} \cdot \frac{{}^{\mathcal{N}}d}{dt}(\mathbf{r} - \mathbf{r}_Q) \tag{3-301}$$

Now, because the attachment point is fixed in the inertial reference frame $\mathcal{N}$, we have

$$\frac{{}^{\mathcal{N}}d}{dt}(\mathbf{r} - \mathbf{r}_Q) = {}^{\mathcal{N}}\mathbf{v} \tag{3-302}$$

Consequently,

$$\frac{d}{dt}\left({}^{\mathcal{N}}U_s\right) = K\left(\ell - \ell_0\right)\frac{\mathbf{r} - \mathbf{r}_Q}{\ell}\cdot{}^{\mathcal{N}}\mathbf{v} \tag{3-303}$$

Using the fact that $(\mathbf{r} - \mathbf{r}_Q)/\ell = \mathbf{u}_s$, Eq. (3-303) can be rewritten as

$$\frac{d}{dt}\left({}^{\mathcal{N}}U_s\right) = -K\left(\ell - \ell_0\right)\mathbf{u}_s\cdot{}^{\mathcal{N}}\mathbf{v} \tag{3-304}$$

Now we observe that $\mathbf{F}_s = -K\left(\ell - \ell_0\right)\mathbf{u}_s$. Therefore, $d({}^{\mathcal{N}}U_s)/dt$ simplifies to

$$\frac{d}{dt}\left({}^{\mathcal{N}}U_s\right) = -\mathbf{F}_s\cdot{}^{\mathcal{N}}\mathbf{v} \tag{3-305}$$

It is seen that the quantity $\mathbf{F}_s\cdot{}^{\mathcal{N}}\mathbf{v}$ is the power of $\mathbf{F}_s$ in reference frame $\mathcal{N}$, i.e.,

$${}^{\mathcal{N}}P_s = \mathbf{F}_s\cdot{}^{\mathcal{N}}\mathbf{v} \tag{3-306}$$

Therefore,

$${}^{\mathcal{N}}P_s = -\frac{d}{dt}\left({}^{\mathcal{N}}U_s\right) \tag{3-307}$$

which implies that $\mathbf{F}_s$ is conservative.

Gravitational Forces

Recall from Newton's law of gravitation of Eq. (3-25) that the force of attraction of a particle of mass M on a particle of mass m is given as

$$\mathbf{F}_g = -\frac{GmM}{r^3}\mathbf{r} \tag{3-308}$$

where $\mathbf{r} = \mathbf{r} - \mathbf{r}_M$ is the position of m relative to M and $r = \|\mathbf{r}\|$. Furthermore, assume that the position of M is inertially fixed, i.e., assume that

$$\begin{aligned} {}^{\mathcal{N}}\mathbf{v}_M &= \mathbf{0} \\ {}^{\mathcal{N}}\mathbf{a}_M &= \mathbf{0} \end{aligned} \tag{3-309}$$

Then the potential energy of a gravitational force, denoted U_g, is given as

$${}^{\mathcal{N}}U_g = -\frac{GmM}{r} \tag{3-310}$$

To verify that ${}^{\mathcal{N}}U_g$ is indeed the potential energy of a gravitational force, we compute $d({}^{\mathcal{N}}U_g)/dt$ as

$$\frac{d}{dt}\left({}^{\mathcal{N}}U_g\right) = \frac{GmM}{r^2}\frac{dr}{dt} \tag{3-311}$$

Now we note that, using the scalar product, an alternate expression for r is

$$r = \sqrt{\mathbf{r}\cdot\mathbf{r}} = (\mathbf{r}\cdot\mathbf{r})^{1/2} \tag{3-312}$$

Differentiating r in the inertial reference frame $\mathcal{N}$ using Eq. (3-312), we obtain

$$\frac{dr}{dt} = \dot{r} = \tfrac{1}{2}(\mathbf{r}\cdot\mathbf{r})^{-1/2}\left[\frac{{}^{\mathcal{N}}d\mathbf{r}}{dt}\cdot\mathbf{r} + \mathbf{r}\cdot\frac{{}^{\mathcal{N}}d\mathbf{r}}{dt}\right] \tag{3-313}$$

Furthermore, because the mass M is inertially fixed, we have

$$\frac{{}^{\mathcal{N}}d}{dt}\mathbf{r} = {}^{\mathcal{N}}\mathbf{v} - {}^{\mathcal{N}}\mathbf{v}_M = {}^{\mathcal{N}}\mathbf{v} \tag{3-314}$$

Consequently, Eq. (3–313) simplifies to

$$\dot{r} = \tfrac{1}{2}(\mathbf{r}\cdot\mathbf{r})^{-1/2}\left[2\,{}^{\mathcal{N}}\mathbf{v}\cdot\mathbf{r}\right] = \frac{{}^{\mathcal{N}}\mathbf{v}\cdot\mathbf{r}}{\sqrt{\mathbf{r}\cdot\mathbf{r}}} \tag{3-315}$$

Furthermore, substituting Eq. (3–312) into (3–315), we obtain

$$\dot{r} = \frac{{}^{\mathcal{N}}\mathbf{v}\cdot\mathbf{r}}{r} \tag{3-316}$$

Then, substituting $\dot{r}$ from Eq. (3–316) into (3–311) gives

$$\frac{d}{dt}\left({}^{\mathcal{N}}U_g\right) = \frac{GmM}{r^2}\frac{{}^{\mathcal{N}}\mathbf{v}\cdot\mathbf{r}}{r} = \frac{GmM}{r^3}\mathbf{r}\cdot{}^{\mathcal{N}}\mathbf{v} \tag{3-317}$$

Using the definition of the gravitational force from Eq. (3–308), Eq. (3–317) becomes

$$\frac{d}{dt}\left({}^{\mathcal{N}}U_g\right) = -\mathbf{F}_g\cdot{}^{\mathcal{N}}\mathbf{v} \tag{3-318}$$

Now we note that the power of a gravitational force in reference frame $\mathcal{N}$, denoted P_g, is given as $\mathbf{F}_g\cdot{}^{\mathcal{N}}\mathbf{v}$. Therefore,

$${}^{\mathcal{N}}P_g = -\frac{d}{dt}\left({}^{\mathcal{N}}U_g\right) \tag{3-319}$$

which implies that $\mathbf{F}_g$ is conservative.

3.10.6 Alternate Form of Work-Energy Theorem for a Particle

Let $\mathbf{r}$ be the position of a particle and let ${}^{\mathcal{N}}\mathbf{v}$ be the velocity of the particle in an inertial reference frame $\mathcal{N}$. Consider now that the particle is being acted on by conservative forces $\mathbf{F}_1^c,\ldots,\mathbf{F}_n^c$ and a resultant nonconservative force $\mathbf{F}^{nc}$. Furthermore, let ${}^{\mathcal{N}}U_1(\mathbf{r}),\ldots,{}^{\mathcal{N}}U_n(\mathbf{r})$ be the potential energies of $\mathbf{F}_1^c,\ldots,\mathbf{F}_n^c$, respectively, in the inertial reference frame $\mathcal{N}$. Then

$$\mathbf{F}_i^c = -\nabla_{\mathbf{r}}{}^{\mathcal{N}}U_i \quad (i = 1,\ldots,n) \tag{3-320}$$

The resultant conservative force acting on the particle, denoted $\mathbf{F}_c$, is then given as

$$\mathbf{F}^c = \sum_{i=1}^{n}\mathbf{F}_i^c = \sum_{i=1}^{n}-\nabla_{\mathbf{r}}{}^{\mathcal{N}}U_i \tag{3-321}$$

The total potential energy in the inertial reference frame $\mathcal{N}$ is then obtained from Eq. (3–288) as

$${}^{\mathcal{N}}U = \sum_{i=1}^{n}{}^{\mathcal{N}}U_i \tag{3-322}$$

Consequently, the total conservative force acting on the particle is given as

$$\mathbf{F}^c = -\nabla_{\mathbf{r}} {}^{\mathcal{N}}U \tag{3-323}$$

Then the resultant force acting on the particle is given as

$$\mathbf{F} = \mathbf{F}^c + \mathbf{F}^{nc} = -\nabla_{\mathbf{r}} {}^{\mathcal{N}}U + \mathbf{F}^{nc} \tag{3-324}$$

Now, from the work-energy theorem of Eq. (3-275), we have

$$\frac{d}{dt}\left({}^{\mathcal{N}}T\right) = \mathbf{F} \cdot {}^{\mathcal{N}}\mathbf{v} \tag{3-325}$$

Substituting the expression from Eq. (3-324) into (3-325), we obtain

$$\frac{d}{dt}\left({}^{\mathcal{N}}T\right) = \left(\nabla_{\mathbf{r}} {}^{\mathcal{N}}U + \mathbf{F}^{nc}\right) \cdot {}^{\mathcal{N}}\mathbf{v} \tag{3-326}$$

Consequently,

$$\frac{d}{dt}\left({}^{\mathcal{N}}T\right) = -\nabla_{\mathbf{r}}\left({}^{\mathcal{N}}U\right) \cdot {}^{\mathcal{N}}\mathbf{v} + \mathbf{F}^{nc} \cdot {}^{\mathcal{N}}\mathbf{v} \tag{3-327}$$

Now recall from Eq. (3-282) that

$$\nabla_{\mathbf{r}}\left({}^{\mathcal{N}}U\right) \cdot {}^{\mathcal{N}}\mathbf{v} = \frac{d}{dt}\left({}^{\mathcal{N}}U\right) \tag{3-328}$$

Equation (3-327) can then be written as

$$\frac{d}{dt}\left({}^{\mathcal{N}}T\right) = -\frac{d}{dt}\left({}^{\mathcal{N}}U\right) + \mathbf{F}^{nc} \cdot {}^{\mathcal{N}}\mathbf{v} \tag{3-329}$$

Equation (3-329) can be rearranged to give

$$\frac{d}{dt}\left({}^{\mathcal{N}}T + {}^{\mathcal{N}}U\right) = \mathbf{F}^{nc} \cdot {}^{\mathcal{N}}\mathbf{v} \tag{3-330}$$

The quantity

$${}^{\mathcal{N}}E = {}^{\mathcal{N}}T + {}^{\mathcal{N}}U \tag{3-331}$$

is called the *total energy* in the inertial reference frame $\mathcal{N}$. In terms of the total energy, we have

$$\boxed{\frac{d}{dt}\left({}^{\mathcal{N}}E\right) = \mathbf{F}^{nc} \cdot {}^{\mathcal{N}}\mathbf{v}} \tag{3-332}$$

Equation (3-332) is an alternate form of the work-energy theorem and states that the rate of change of the total energy for a particle in an inertial reference frame is equal to the power of the nonconservative forces in an inertial reference frame. Similar to previous results involving the laws of kinetics, it is noted that Eq. (3-332) is valid only in an *inertial reference frame*.

3.10.7 Conservation of Energy for a Particle

Now suppose that the resultant nonconservative force $\mathbf{F}^{nc}$ does no work, i.e.,

$$\mathbf{F}^{nc} \cdot {}^{\mathcal{N}}\mathbf{v} = 0 \tag{3-333}$$

Then

$$\frac{d}{dt}\left({}^{\mathcal{N}}E\right) = 0 \tag{3-334}$$

which implies that

$${}^{\mathcal{N}}E = \text{constant} \tag{3-335}$$

When the total energy of a particle is a constant in an inertial reference frame $\mathcal{N}$, the total energy of the particle in $\mathcal{N}$ is said to be *conserved.* It is seen from Eqs. (3-333) and (3-334) that the total energy of a particle in an inertial reference frame $\mathcal{N}$ will be conserved only if the power produced by all nonconservative forces in $\mathcal{N}$ is zero.

3.10.8 Alternate Form of the Principle of Work and Energy for a Particle

Now suppose that we consider motion over a time interval $t \in [t_1, t_2]$. Then, integrating Eq. (3-332) from t_1 to t_2, we have

$$\int_{t_1}^{t_2} \frac{d}{dt}\left({}^{\mathcal{N}}E\right) dt = \int_{t_1}^{t_2} \mathbf{F}^{nc} \cdot {}^{\mathcal{N}}\mathbf{v} dt \tag{3-336}$$

Now we know from the definition of Eq. (3-268) that the work done by the nonconservative forces on the interval $t \in [t_1, t_2]$ in the inertial reference frame $\mathcal{N}$ is

$${}^{\mathcal{N}}W_{12}^{nc} = \int_{t_1}^{t_2} \mathbf{F}^{nc} \cdot {}^{\mathcal{N}}\mathbf{v} dt \tag{3-337}$$

Furthermore, we have

$$\int_{t_1}^{t_2} \frac{d}{dt}\left({}^{\mathcal{N}}E\right) dt = {}^{\mathcal{N}}E(t_2) - {}^{\mathcal{N}}E(t_1) = {}^{\mathcal{N}}E_2 - {}^{\mathcal{N}}E_1 \tag{3-338}$$

Using Eq. (3-337) together with Eq. (3-338), we obtain

$$\boxed{{}^{\mathcal{N}}E_2 - {}^{\mathcal{N}}E_1 = {}^{\mathcal{N}}W_{12}^{nc}} \tag{3-339}$$

Equation (3-339) is an alternate form for the principle of work and energy for a particle and states that the change in total energy between any two points along the trajectory of a particle in an inertial reference frame is equal to the work done by all *nonconservative forces* acting on the particle between those same two points. It is noted that, in the case that the nonconservative forces do no work, we have ${}^{\mathcal{N}}W_{12}^{nc} = 0$ which implies that

$${}^{\mathcal{N}}E_2 = {}^{\mathcal{N}}E_1 \tag{3-340}$$

Equation (3-340) states that, if the nonconservative forces acting on a particle do no work, then the total energy of the particle is conserved.

Example 3–6

A particle of mass m slides without friction along a fixed track in the form of a linear spiral. The equation for the spiral is given in cylindrical coordinates as

$$r = a\theta$$

Knowing that gravity acts vertically downward, determine the differential equation of motion for the particle using (a) the work-energy theorem for a particle and (b) the alternate form of the work-energy theorem for a particle.

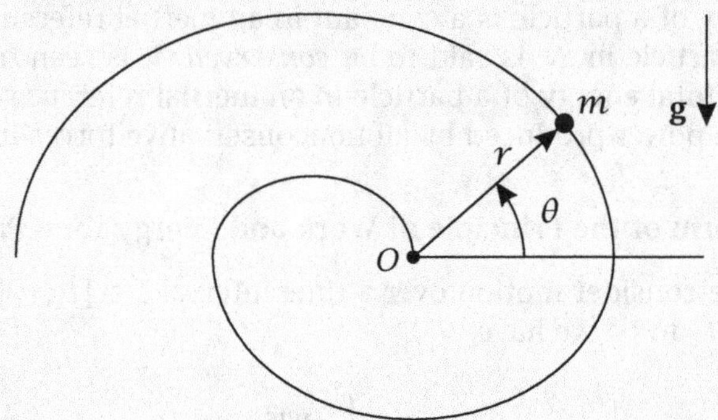

Figure 3–23 Particle of mass m sliding without friction along a track in the shape of a linear spiral.

Solution to Example 3–6

Kinematics

First, let $\mathcal{F}$ be a reference frame fixed to the track. Then, choose the following coordinate system fixed in reference frame $\mathcal{F}$:

$$\begin{array}{lcl} \mathbf{E}_x & = & \text{To the right} \\ \mathbf{E}_z & = & \text{Out of page} \\ \mathbf{E}_y & = & \mathbf{E}_z \times \mathbf{E}_x \end{array}$$

Next, let $\mathcal{A}$ be a reference frame fixed to the direction of OP. Then, choose the following coordinate system fixed in reference frame $\mathcal{A}$:

$$\begin{array}{lcl} \mathbf{e}_r & = & \text{Along } OP \\ \mathbf{e}_z & = & \text{Out of page} \\ \mathbf{e}_\theta & = & \mathbf{e}_z \times \mathbf{e}_r \end{array}$$

Using Fig. 3-24, the relationship between the bases $\{\mathbf{e}_r, \mathbf{e}_\theta, \mathbf{e}_z\}$ and $\{\mathbf{E}_x, \mathbf{E}_y, \mathbf{E}_z\}$ is given as

$$\mathbf{E}_x = \cos\theta\,\mathbf{e}_r - \sin\theta\,\mathbf{e}_\theta \qquad (3\text{–}341)$$

$$\mathbf{E}_y = \sin\theta\,\mathbf{e}_r + \cos\theta\,\mathbf{e}_\theta \qquad (3\text{–}342)$$

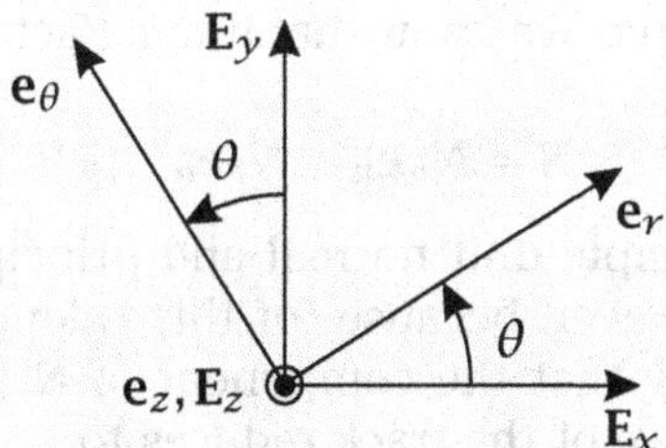

Figure 3-24 Geometry of bases $\{\mathbf{E}_x, \mathbf{E}_y, \mathbf{E}_z\}$ and $\{\mathbf{e}_r, \mathbf{e}_\theta, \mathbf{e}_z\}$ for Example 3-6.

The position of the particle is then given in terms of the basis $\{\mathbf{e}_r, \mathbf{e}_\theta, \mathbf{e}_z\}$ as

$$\mathbf{r} = r\mathbf{e}_r = a\theta\mathbf{e}_r \tag{3-343}$$

Furthermore, the angular velocity of reference frame $\mathcal{A}$ in reference frame $\mathcal{F}$ is given as

$$^{\mathcal{F}}\boldsymbol{\omega}^{\mathcal{A}} = \dot{\theta}\mathbf{e}_z \tag{3-344}$$

The velocity in reference frame $\mathcal{F}$ is then obtained using the rate of change transport theorem as

$$^{\mathcal{F}}\mathbf{v} = \frac{^{\mathcal{F}}d\mathbf{r}}{dt} = \frac{^{\mathcal{A}}d\mathbf{r}}{dt} + {^{\mathcal{F}}\boldsymbol{\omega}^{\mathcal{A}}} \times \mathbf{r} \tag{3-345}$$

Now we have

$$\frac{^{\mathcal{A}}d\mathbf{r}}{dt} = a\dot{\theta}\mathbf{e}_r \tag{3-346}$$

$$^{\mathcal{F}}\boldsymbol{\omega}^{\mathcal{A}} \times \mathbf{r} = \dot{\theta}\mathbf{e}_z \times a\theta\mathbf{e}_r = a\theta\dot{\theta}\mathbf{e}_\theta \tag{3-347}$$

Adding Eqs. (3-346) and (3-347), we obtain $^{\mathcal{F}}\mathbf{v}$ as

$$^{\mathcal{F}}\mathbf{v} = a\dot{\theta}\mathbf{e}_r + a\theta\dot{\theta}\mathbf{e}_\theta \tag{3-348}$$

It is noted that, because in this problem we are asked to derive the differential equation of motion using the work-energy theorem, it is only necessary to compute the velocity of the particle in an inertial reference frame (i.e., we do not need to compute the acceleration of the particle).

Kinetics

The free body diagram of the particle is shown in Fig. 3-25, from which is it seen that the following forces act on the particle:

$$\begin{aligned} \mathbf{N} &= \text{Reaction force of track on particle} \\ m\mathbf{g} &= \text{Force of gravity} \end{aligned}$$

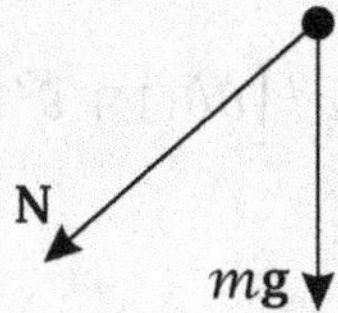

Figure 3-25 Free body diagram for Example 3-6.

Now, because **N** is a contact force, we know that it must act orthogonal to the direction of the track, i.e.,

$$\mathbf{N} = N_n \mathbf{e}_n + N_b \mathbf{e}_b \tag{3-349}$$

where $\mathbf{e}_n$ and $\mathbf{e}_b$ are the principle unit normal and principle unit bi-normal vectors, respectively, to the track. However, because for this example the motion lies entirely in the $\{\mathbf{e}_r, \mathbf{e}_\theta\}$-plane, we know that the component of **N** in the $\mathbf{e}_b$-direction is zero. Consequently, the reaction force of the track reduces to

$$\mathbf{N} = N_n \mathbf{e}_n \equiv N \mathbf{e}_n \tag{3-350}$$

As will become clear shortly, rather than obtaining an expression for **N** in terms of the basis $\{\mathbf{e}_r, \mathbf{e}_\theta, \mathbf{e}_z\}$, it is preferable to leave **N** in the form given in Eq. (3-350). Next, the force of gravity is given as

$$m\mathbf{g} = -mg\mathbf{E}_y \tag{3-351}$$

Then, using the expression for $\mathbf{E}_y$ in terms of $\mathbf{e}_r$ and $\mathbf{e}_\theta$ as given in Eq. (3-342), we can write the force of gravity as

$$m\mathbf{g} = -mg(\sin\theta\,\mathbf{e}_r + \cos\theta\,\mathbf{e}_\theta) = -mg\sin\theta\,\mathbf{e}_r - mg\cos\theta\,\mathbf{e}_\theta \tag{3-352}$$

The resultant force acting on the particle is then given as

$$\mathbf{F} = \mathbf{N} + m\mathbf{g} = N\mathbf{e}_n - mg\sin\theta\,\mathbf{e}_r - mg\cos\theta\,\mathbf{e}_\theta \tag{3-353}$$

We can now proceed to use both forms of the work-energy theorem for a particle.

(a) Differential Equation Using Work-Energy Theorem

The work-energy theorem in the inertial reference frame $\mathcal{F}$ is given from Eq. (3-275) as

$$\frac{d}{dt}\left({}^{\mathcal{F}}T\right) = \mathbf{F} \cdot {}^{\mathcal{F}}\mathbf{v} \tag{3-354}$$

Using the expression for ${}^{\mathcal{F}}\mathbf{v}$ from Eq. (3-348), the kinetic energy in reference frame $\mathcal{F}$ is given as

$$\begin{aligned} {}^{\mathcal{F}}T &= \tfrac{1}{2}m\,{}^{\mathcal{F}}\mathbf{v} \cdot {}^{\mathcal{F}}\mathbf{v} = \tfrac{1}{2}m\left(a\dot{\theta}\mathbf{e}_r + a\theta\dot{\theta}\mathbf{e}_\theta\right) \cdot \left(a\dot{\theta}\mathbf{e}_r + a\theta\dot{\theta}\mathbf{e}_\theta\right) \\ &= \tfrac{1}{2}m\left(a^2\dot{\theta}^2 + a^2\theta^2\dot{\theta}^2\right) = \tfrac{1}{2}ma^2\dot{\theta}^2(1+\theta^2) \end{aligned} \tag{3-355}$$

Computing the rate of change of ${}^{\mathcal{F}}T$ (remembering that ${}^{\mathcal{F}}T$ is a scalar and, thus, its rate of change is independent of reference frame), we have

$$\frac{d}{dt}\left({}^{\mathcal{F}}T\right) = ma^2\left[\dot{\theta}\ddot{\theta}(1+\theta^2) + \dot{\theta}^2(\theta\dot{\theta})\right] \tag{3-356}$$

Simplifying Eq. (3-356) gives

$$\frac{d}{dt}\left({}^{\mathcal{F}}T\right) = ma^2\dot{\theta}\left[\ddot{\theta}(1+\theta^2) + \dot{\theta}^2\theta\right] \tag{3-357}$$

Next, using the expression for $\mathbf{F}$ from Eq. (3-353), the power in the inertial reference frame $\mathcal{F}$ produced by the resultant force $\mathbf{F}$ is

$$\mathbf{F} \cdot {}^{\mathcal{F}}\mathbf{v} = (\mathbf{N} + m\mathbf{g}) \cdot {}^{\mathcal{F}}\mathbf{v} \tag{3-358}$$

Now we know that $\mathbf{N}$ acts in the direction of $\mathbf{e}_n$ while ${}^{\mathcal{F}}\mathbf{v}$ lies in the direction of $\mathbf{e}_t$, where $\mathbf{e}_t$ is tangent to the track. Therefore, we have $\mathbf{N} \cdot {}^{\mathcal{F}}\mathbf{v} = 0$ which implies that

$$\begin{aligned} \mathbf{F} \cdot {}^{\mathcal{F}}\mathbf{v} = m\mathbf{g} \cdot {}^{\mathcal{F}}\mathbf{v} &= (-mg\sin\theta\,\mathbf{e}_r - mg\cos\theta\,\mathbf{e}_\theta) \cdot \left(a\dot\theta\mathbf{e}_r + a\theta\dot\theta\mathbf{e}_\theta\right) \\ &= -mga(\dot\theta\sin\theta + \theta\dot\theta\cos\theta) \end{aligned} \tag{3-359}$$

Then, setting the power in Eq. (3-359) equal to the rate of change of kinetic energy as given in Eq. (3-357), we obtain

$$ma^2\dot\theta\left[\ddot\theta(1+\theta^2) + \dot\theta^2\theta\right] = -mga(\dot\theta\sin\theta + \theta\dot\theta\cos\theta) \tag{3-360}$$

Noting that $\dot\theta \neq 0$ as a function of time (otherwise the particle would not be moving), we can drop $\dot\theta$ from both terms in Eq. (3-360) to obtain the differential equation of motion as

$$ma^2\left[\ddot\theta(1+\theta^2) + \dot\theta^2\theta\right] + mga(\sin\theta + \theta\cos\theta) = 0 \tag{3-361}$$

(b) Differential Equation Using Alternate Form of Work-Energy Theorem

The alternate form of the work-energy theorem is given in the inertial reference frame $\mathcal{F}$ from Eq. (3-332) as

$$\frac{d}{dt}\left({}^{\mathcal{F}}E\right) = \mathbf{F}^{nc} \cdot {}^{\mathcal{F}}\mathbf{v} \tag{3-362}$$

where ${}^{\mathcal{F}}E$ is the total energy in the inertial reference frame $\mathcal{F}$. Now we know that

$${}^{\mathcal{F}}E = {}^{\mathcal{F}}T + {}^{\mathcal{F}}U \tag{3-363}$$

where ${}^{\mathcal{F}}T$ is the kinetic energy in the inertial reference frame $\mathcal{F}$ (as previously derived) while ${}^{\mathcal{F}}U$ is the potential energy in reference frame $\mathcal{F}$. Now in this problem the only conservative force is that due to gravity. Moreover, since gravity is a conservative force, the potential energy in reference frame $\mathcal{F}$ is given as

$${}^{\mathcal{F}}U = {}^{\mathcal{F}}U_g = -m\mathbf{g} \cdot \mathbf{r} \tag{3-364}$$

Substituting the expression for $m\mathbf{g}$ from Eq. (3-352) and the expression for $\mathbf{r}$ from Eq. (3-343) into (3-364), we obtain ${}^{\mathcal{F}}U$ as

$${}^{\mathcal{F}}U = -(-mg\sin\theta\,\mathbf{e}_r - mg\cos\theta\,\mathbf{e}_\theta) \cdot a\theta\mathbf{e}_r = mga\theta\sin\theta \tag{3-365}$$

Then, adding ${}^{\mathcal{F}}T$ in Eq. (3-355) to ${}^{\mathcal{F}}U$ in Eq. (3-365), the total energy in reference frame $\mathcal{F}$ is obtained as

$${}^{\mathcal{F}}E = {}^{\mathcal{F}}T + {}^{\mathcal{F}}U = \tfrac{1}{2}ma^2\dot\theta^2(1+\theta^2) + mga\theta\sin\theta \tag{3-366}$$

Computing the rate of change of ${}^{\mathcal{F}}E$ in Eq. (3-366), we obtain

$$\begin{aligned} \frac{d}{dt}\left({}^{\mathcal{F}}E\right) &= ma^2\left[\dot\theta\ddot\theta(1+\theta^2) + \dot\theta^2(\theta\dot\theta)\right] + mga\left(\dot\theta\sin\theta + \theta\dot\theta\cos\theta\right) \\ &= ma^2\dot\theta\left[\ddot\theta(1+\theta^2) + \dot\theta^2\theta\right] + mga\dot\theta\,(\sin\theta + \theta\cos\theta) \end{aligned} \tag{3-367}$$

Finally, because $\mathbf{N}$ is the only nonconservative force acting on the particle and $\mathbf{N} \cdot {}^{\mathcal{F}}\mathbf{v} = 0$, we have

$$\mathbf{F}^{nc} \cdot {}^{\mathcal{F}}\mathbf{v} = \mathbf{N} \cdot {}^{\mathcal{F}}\mathbf{v} = 0 \tag{3-368}$$

Substituting the results of Eqs. (3-367) and (3-368) into (3-362), we obtain

$$ma^2\dot{\theta}\left[\ddot{\theta}(1+\theta^2) + \dot{\theta}^2\theta\right] + mga\dot{\theta}\,(\sin\theta + \theta\cos\theta) = 0 \tag{3-369}$$

Again, noting that $\dot{\theta}$ is not equal to zero as a function of time, we can drop $\dot{\theta}$ from Eq. (3-369) to obtain the differential equation of motion as

$$ma^2\left[\ddot{\theta}(1+\theta^2) + \dot{\theta}^2\theta\right] + mga\,(\sin\theta + \theta\cos\theta) = 0 \tag{3-370}$$

in agreement with the result from part (a) above.

■

Example 3–7

A collar of mass m slides without friction along a fixed horizontal rigid massless rod as shown in Fig. 3-26. The collar is attached to a linear spring with spring constant K and unstretched length ℓ_0. Knowing that $q(t)$ is the prescribed displacement of the attachment point of the spring, x is the displacement of the collar relative to the spring, and assuming no gravity, determine the differential equation of motion using (a) Newton's 2nd law and (b) the work-energy theorem for a rigid body. Also, (c) show that an *incorrect* result is obtained by assuming that the spring force is conservative.

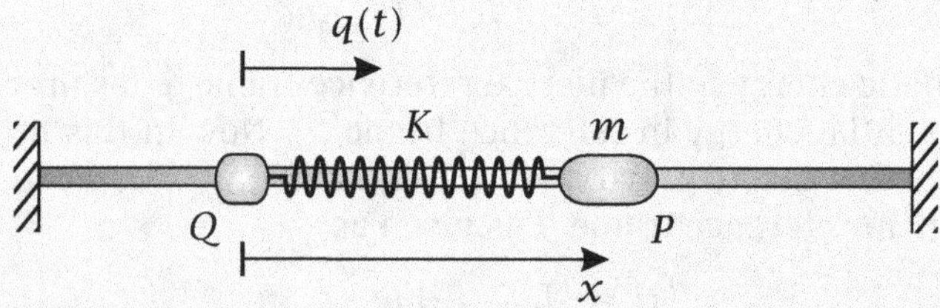

Figure 3–26 Collar sliding on a fixed horizontal track attached to a linear spring with a moving attachment point.

Solution to Example 3–7

Preliminaries

The purpose of this example is to demonstrate the proper application of the linear spring force model and show that the force exerted by a linear spring with a moving attachment point is *not* conservative. To this end, this problem will be solved using Newton's 2nd law and the work-energy theorem for a particle. In addition, in order to demonstrate for this problem that the force exerted by the linear spring is *not* conservative, the alternate form of the work-energy theorem will be applied using the assumption that the force in the linear spring is conservative. It will then be shown that using the assumption that the linear spring force is conservative leads to an incorrect result, i.e., assuming that the spring force is conservative leads to results that are different from the results obtained using Newton's 2nd law and the work-energy theorem.

Kinematics

Let $\mathcal{F}$ be a reference frame fixed to the track. Then, choose the following coordinate system fixed in reference frame $\mathcal{F}$:

$$
\begin{array}{ccc}
 & \text{Origin at point } Q & \\
 & \text{when } t = 0 & \\
\mathbf{E}_x & = & \text{To the right} \\
\mathbf{E}_z & = & \text{Out of page} \\
\mathbf{E}_y & = & \mathbf{E}_z \times \mathbf{E}_x
\end{array}
$$

Then, since q is the displacement of the attachment point of the spring relative to the track, we have

$$\mathbf{r}_Q = q\mathbf{E}_x \tag{3-371}$$

Next, since x is the displacement of the collar relative to the attachment point of the spring, we have

$$\mathbf{r}_{P/Q} = x\mathbf{E}_x \tag{3-372}$$

Consequently, the position of the collar relative to the track is given as

$$\mathbf{r} \equiv \mathbf{r}_P = \mathbf{r}_Q + \mathbf{r}_{P/Q} = q\mathbf{E}_x + x\mathbf{E}_x = (q + x)\mathbf{E}_x = (x + q)\mathbf{E}_x \tag{3-373}$$

The velocity of the collar in reference frame $\mathcal{F}$ is then given as

$${}^{\mathcal{F}}\mathbf{v} = \frac{{}^{\mathcal{F}}d\mathbf{r}}{dt} = (\dot{x} + \dot{q})\mathbf{E}_x \tag{3-374}$$

Finally, the acceleration of the collar in reference frame $\mathcal{F}$ is given as

$${}^{\mathcal{F}}\mathbf{a} = \frac{{}^{\mathcal{F}}d}{dt}\left({}^{\mathcal{F}}\mathbf{v}\right) = (\ddot{x} + \ddot{q})\mathbf{E}_x \tag{3-375}$$

Kinetics

The free body diagram of the collar is shown in Fig. 3-27.

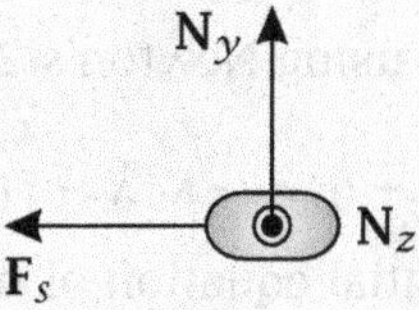

Figure 3-27 Free body diagram of collar for Example 3-7.

It can be seen that the two forces acting on the collar are the force in the linear spring, $\mathbf{F}_s$, and the reaction force of the track on the collar, $\mathbf{N}$. First, from Eq. (3-19) the force exerted by the linear spring on the collar is given as

$$\mathbf{F}_s = -K(\ell - \ell_0)\mathbf{u}_s \tag{3-376}$$

Then, we have

$$\mathbf{r} - \mathbf{r}_Q = (x + q)\mathbf{E}_x - q\mathbf{E}_x = x\mathbf{E}_x \tag{3-377}$$

Substituting the result of Eq. (3-377) into (3-10), we obtain the length of the spring as

$$\ell = \|\mathbf{r} - \mathbf{r}_Q\| = \|(x + q)\mathbf{E}_x - q\mathbf{E}_x\| = \|x\mathbf{E}_x\| = |x| = x \tag{3-378}$$

Using the result of Eq. (3-377) in (3-13), the unit vector in the direction along the spring is given as

$$\mathbf{u}_s = \frac{\mathbf{r} - \mathbf{r}_Q}{\|\mathbf{r} - \mathbf{r}_Q\|} = \frac{x\mathbf{E}_x}{x} = \mathbf{E}_x \tag{3-379}$$

The force exerted by the linear spring is then given as

$$\mathbf{F}_s = -K(x - \ell_0)\mathbf{E}_x \tag{3-380}$$

Next, the reaction force of the track on the collar must lie in a direction orthogonal to the track. Because the directions $\mathbf{E}_y$ and $\mathbf{E}_z$ are both orthogonal to the track, the reaction force of the track on the collar can be written as

$$\mathbf{N} = \mathbf{N}_y + \mathbf{N}_z = N_y\mathbf{E}_y + N_z\mathbf{E}_z \tag{3-381}$$

Then, adding the expression for $\mathbf{F}_s$ from Eq. (3-380) and the expression for $\mathbf{N}$ from Eq. (3-381), the resultant force acting on the collar is given as

$$\mathbf{F} = \mathbf{F}_s + \mathbf{N} = -K(x - \ell_0)\mathbf{E}_x + N_y\mathbf{E}_y + N_z\mathbf{E}_z \tag{3-382}$$

Setting $\mathbf{F}$ in Eq. (3-382) equal to $m\,{}^{\mathcal{F}}\mathbf{a}$ using the expression for ${}^{\mathcal{F}}\mathbf{a}$ from Eq. (3-375), we obtain

$$-K(x - \ell_0)\mathbf{E}_x + N_y\mathbf{E}_y + N_z\mathbf{E}_z = m(\ddot{x} + \ddot{q})\mathbf{E}_x \tag{3-383}$$

Equating components in Eq. (3-383), we obtain the following three scalar equations:

$$m(\ddot{x} + \ddot{q}) = -K(x - \ell_0) \tag{3-384}$$

$$N_y = 0 \tag{3-385}$$

$$N_z = 0 \tag{3-386}$$

(a) Differential Equation Using Newton's 2nd Law

The differential equation of motion using Newton's 2nd law is obtained directly from Eq. (3-384) as

$$m(\ddot{x} + \ddot{q}) = -K(x - \ell_0) \tag{3-387}$$

Rearranging Eq. (3-387), the differential equation of motion can be written as

$$m\ddot{x} + Kx = -m\ddot{q} + K\ell_0 \tag{3-388}$$

(b) Differential Equation Using Work-Energy Theorem

The work-energy theorem for the collar is given from Eq. (3-275) as

$$\frac{d}{dt}\left({}^{\mathcal{F}}T\right) = \mathbf{F}\cdot{}^{\mathcal{F}}\mathbf{v} \tag{3-389}$$

First, using the velocity of the collar in reference frame $\mathcal{F}$ as obtained in Eq. (3-389), the kinetic energy in reference frame $\mathcal{F}$ is given as

$${}^{\mathcal{F}}T = \tfrac{1}{2}m{}^{\mathcal{F}}\mathbf{v}\cdot{}^{\mathcal{F}}\mathbf{v} = \tfrac{1}{2}m(\dot{x}+\dot{q})\mathbf{E}_x\cdot(\dot{x}+\dot{q})\mathbf{E}_x = \tfrac{1}{2}m(\dot{x}+\dot{q})^2 \tag{3-390}$$

The rate of change of ${}^{\mathcal{F}}T$ is then given as

$$\frac{d}{dt}\left({}^{\mathcal{F}}T\right) = m(\dot{x}+\dot{q})(\ddot{x}+\ddot{q}) \tag{3-391}$$

Next, using the resultant force acting on the collar as given in Eq. (3-382), the power produced by all forces acting on the collar is given as

$$\begin{aligned}\mathbf{F}\cdot{}^{\mathcal{F}}\mathbf{v} &= (\mathbf{F}_s+\mathbf{N})\cdot{}^{\mathcal{F}}\mathbf{v}\\ &= (-K(x-\ell_0)\mathbf{E}_x + N_y\mathbf{E}_y + N_z\mathbf{E}_z)\cdot(\dot{x}+\dot{q})\mathbf{E}_x\\ &= -K(x-\ell_0)(\dot{x}+\dot{q})\end{aligned} \tag{3-392}$$

Then, setting $d({}^{\mathcal{F}}T)/dt$ equal to $\mathbf{F}\cdot{}^{\mathcal{F}}\mathbf{v}$, we obtain

$$m(\dot{x}+\dot{q})(\ddot{x}+\ddot{q}) = -K(x-\ell_0)(\dot{x}+\dot{q}) \tag{3-393}$$

Rearranging Eq. (3-393) gives

$$(\dot{x}+\dot{q})\left[m(\ddot{x}+\ddot{q}) + K(x-\ell_0)\right] = 0 \tag{3-394}$$

Now, observing that $(\dot{q}+\dot{x})\mathbf{E}_x$ is the velocity of the collar in reference frame $\mathcal{F}$, the quantity $\dot{x}+\dot{q}$ cannot be equal to zero. Consequently, the term in square brackets in Eq. (3-394) must be zero, i.e.,

$$m(\ddot{x}+\ddot{q}) + K(x-\ell_0) = 0 \tag{3-395}$$

Rearranging Eq. (3-395), we obtain

$$m\ddot{x} + Kx = -m\ddot{q} + K\ell_0 \tag{3-396}$$

It is noted that Eq. (3-396) is identical to Eq. (3-388).

(c) Incorrect Result Obtained by Assuming Spring Force is Conservative

As discussed in Section 3.10.5, the force of a linear spring is only conservative when the attachment point is inertially fixed. Since in this problem it is known that the attachment point moves with velocity ${}^{\mathcal{F}}\mathbf{v}_Q = \dot{q}\mathbf{E}_x \neq \mathbf{0}$ and acceleration ${}^{\mathcal{F}}\mathbf{a}_Q = \ddot{q}\mathbf{E}_x \neq \mathbf{0}$, point Q is *not* inertially fixed. Consequently, the force exerted by the linear spring on the collar is not conservative.

Suppose now that we assume *incorrectly* that the force exerted by the linear spring on the collar is conservative and apply the alternate form of the work-energy theorem using this assumption. Then, from Eq. (3–332) we would have

$$\frac{d}{dt}\left({}^{\mathcal{F}}E\right) = \mathbf{F}^{nc} \cdot {}^{\mathcal{F}}\mathbf{v} \tag{3-397}$$

where $\mathbf{F}^{nc}$ is the resultant nonconservative force acting on the collar. Assuming *incorrectly* that the spring force is conservative, the potential energy associated with the spring force in reference frame $\mathcal{F}$ would be obtained from Eq. (3–296) as

$${}^{\mathcal{F}}U = \tfrac{K}{2}\left(\ell - \ell_0\right)^2 = \tfrac{K}{2}\left(x - \ell_0\right)^2 \tag{3-398}$$

Then, using the expression for the kinetic energy from Eq. (3–390), the total energy in reference frame $\mathcal{F}$ would be obtained as

$${}^{\mathcal{F}}E = {}^{\mathcal{F}}T + {}^{\mathcal{F}}U = \tfrac{1}{2}m(\dot{x} + \dot{q})^2 + \tfrac{K}{2}\left(x - \ell_0\right)^2 \tag{3-399}$$

Next, under the incorrect assumption that the spring force is conservative, the resultant nonconservative force acting on the collar would be given incorrectly as

$$\mathbf{F}^{nc} = \mathbf{N} \tag{3-400}$$

which would imply

$$\mathbf{F}^{nc} \cdot {}^{\mathcal{F}}\mathbf{v} = \mathbf{N} \cdot {}^{\mathcal{F}}\mathbf{v} = (N_y\mathbf{E}_y + N_z\mathbf{E}_z) \cdot (\dot{x} + \dot{q})\mathbf{E}_x = 0 \tag{3-401}$$

Consequently, Eq. (3–401) simplifies to

$$\frac{d}{dt}\left({}^{\mathcal{F}}E\right) = 0 \tag{3-402}$$

Computing the rate of change of ${}^{\mathcal{F}}E$ gives

$$\frac{d}{dt}\left({}^{\mathcal{F}}E\right) = m(\dot{x} + \dot{q})(\ddot{x} + \ddot{q}) + K(x - \ell_0)\dot{x} \tag{3-403}$$

Now, unlike the result obtained using the work-energy theorem, in this case the two terms in Eq. (3–403) *do not* have a common factor of $\dot{x} + \dot{q}$. Consequently, it is not possible to set $d({}^{\mathcal{F}}E)/dt$ equal to zero in Eq. (3–403) and obtain the differential equation of motion. Therefore, it is *incorrect* to assume that the force exerted by the linear spring is conservative. It is noted that, *if the attachment point of the spring was inertially fixed*, then both ${}^{\mathcal{F}}\mathbf{v}_Q$ and ${}^{\mathcal{F}}\mathbf{a}_Q$ would be zero, which would imply that $\dot{q}$ and $\ddot{q}$ would be zero. Then, Eq. (3–403) would reduce to

$$\frac{d}{dt}\left({}^{\mathcal{F}}E\right) = m\dot{x}\ddot{x} + K(x - \ell_0)\dot{x} \tag{3-404}$$

Setting Eq. (3–404) equal to zero and dropping the common factor of $\dot{x}$, we would obtain the correct differential equation of motion for the collar for the case where the attachment point of the spring was inertially fixed.

■

Example 3–8

A block of mass m is connected to a linear spring with spring constant K and unstretched length ℓ_0 as shown in Fig. 3–28. The block is initially at rest and the spring is initially unstressed when a horizontal impulse $\hat{\mathbf{P}}$ is applied. Determine (a) the velocity of the block immediately after the application of the impulse and (b) the maximum compression of the spring after the impulse is applied.

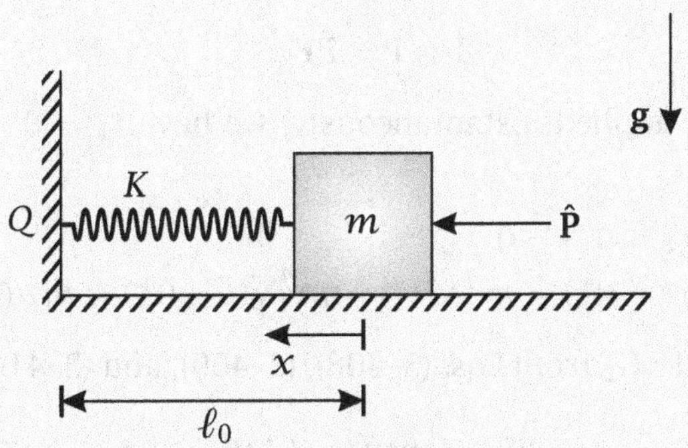

Figure 3–28 Block of mass m connected to linear spring and struck by horizontal impulse $\hat{\mathbf{P}}$.

Solution to Example 3–8

Kinematics

This problem is a simple but excellent example of both the principle of linear impulse and momentum and the principle of work and energy for a particle. First, let $\mathcal{F}$ be a fixed inertial reference frame. Next, choose the following coordinate system fixed in $\mathcal{F}$:

	Origin at block when $x = 0$	
$\mathbf{E}_x$	=	To the left
$\mathbf{E}_z$	=	Into page
$\mathbf{E}_y$	=	$\mathbf{E}_z \times \mathbf{E}_x$

Then, the position of the block is given in terms of the displacement x as

$$\mathbf{r} = x\mathbf{E}_x \tag{3–405}$$

Because $\{\mathbf{E}_x, \mathbf{E}_y, \mathbf{E}_z\}$ is a fixed basis, the velocity of the block in reference frame $\mathcal{F}$ is given as

$$^{\mathcal{F}}\mathbf{v} = \frac{^{\mathcal{F}}d\mathbf{r}}{dt} = \dot{x}\mathbf{E}_x = v\mathbf{E}_x \tag{3–406}$$

Kinetics

Kinetics During Application of Impulse

The velocity of the block the instant after the impulse is applied is found by applying the principle of linear impulse and linear momentum of Eq. (3–179), i.e.,

$$\hat{\mathbf{F}} = {}^{\mathcal{F}}\mathbf{G}(t_2) - {}^{\mathcal{F}}\mathbf{G}(t_1) = {}^{\mathcal{F}}\mathbf{G}_2 - {}^{\mathcal{F}}\mathbf{G}_1 \tag{3-407}$$

Now, because $\hat{\mathbf{P}}$ is the only impulse applied to the particle, we have

$$\hat{\mathbf{F}} = \hat{\mathbf{P}} = \hat{P}\mathbf{E}_x \tag{3-408}$$

Moreover, because $\hat{\mathbf{P}}$ is applied instantaneously, we have $t_1 = 0^-$ and $t_2 = 0^+$ which implies that

$$\begin{aligned} {}^{\mathcal{F}}\mathbf{G}_1 &= {}^{\mathcal{F}}\mathbf{G}(t = 0^-) = \mathbf{0} && (3\text{-}409) \\ {}^{\mathcal{F}}\mathbf{G}_2 &= {}^{\mathcal{F}}\mathbf{G}(t = 0^+) = m\,{}^{\mathcal{F}}\mathbf{v}(t_2) = m\,{}^{\mathcal{F}}\mathbf{v}(t = 0^+) = mv(t = 0^+)\mathbf{E}_x && (3\text{-}410) \end{aligned}$$

Substituting $\hat{\mathbf{F}}$, ${}^{\mathcal{F}}\mathbf{G}_1$, and ${}^{\mathcal{F}}\mathbf{G}_2$ from Eqs. (3–408), (3–409), and (3–410), respectively, into (3–407), we obtain

$$\hat{P}\mathbf{E}_x = mv(t = 0^+)\mathbf{E}_x \tag{3-411}$$

Solving Eq. (3–411) for $v(t = 0^+)$ gives

$$v(t = 0^+) = \frac{\hat{P}}{m} \tag{3-412}$$

We then obtain the velocity of the block the instant after the impulse $\hat{\mathbf{P}}$ is applied as

$${}^{\mathcal{F}}\mathbf{v}(t_2) = \frac{\hat{P}}{m}\mathbf{E}_x \tag{3-413}$$

Kinetics After Application of Impulse

The maximum compression of the spring after the impulse $\hat{\mathbf{P}}$ is applied is obtained using the principle of work and energy of Eq. (3–279) in reference frame $\mathcal{F}$, i.e.,

$${}^{\mathcal{F}}T_3 - {}^{\mathcal{F}}T_2 = {}^{\mathcal{F}}W_{23} \tag{3-414}$$

The free body diagram of the block *after* the impulse is applied is shown in Fig. 3–29.

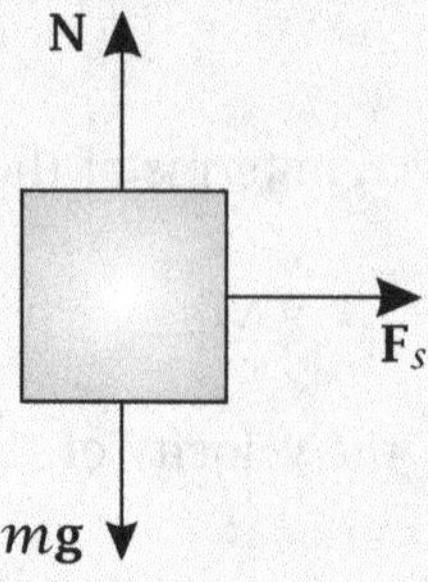

Figure 3–29 Free body diagram of block in Example 3–8 after impulse $\hat{\mathbf{P}}$ is applied.

It is seen that the only forces acting on the block after the impulse $\hat{\mathbf{P}}$ is applied are those of gravity and the spring. First, the spring force is given as

$$\mathbf{F}_s = -K(\ell - \ell_0)\mathbf{u}_s \tag{3-415}$$

Now in this case the direction along the attachment point of the spring to the particle is $\mathbf{u}_s = -\mathbf{E}_x$. Furthermore, the length of the spring is given as

$$\ell = \|\mathbf{r} - \mathbf{r}_Q\| \tag{3-416}$$

Noting that $\mathbf{r} = x\mathbf{E}_x$ and $\mathbf{r}_Q = \ell_0\mathbf{E}_x$, we have

$$\ell = \|x\mathbf{E}_x - \ell_0\mathbf{E}_x\| = |x - \ell_0| \tag{3-417}$$

Now, since $x - \ell_0 < 0$, we have

$$\ell = |x - \ell_0| = -(x - \ell_0) \tag{3-418}$$

We then obtain the spring force in terms of $\{\mathbf{E}_x, \mathbf{E}_y, \mathbf{E}_z\}$ as

$$\mathbf{F}_s = -K(-(x - \ell_0) - \ell_0)(-\mathbf{E}_x) = -Kx\mathbf{E}_x \tag{3-419}$$

Next, the force of gravity is given as

$$m\mathbf{g} = -mg\mathbf{E}_y \tag{3-420}$$

The resultant force acting on the particle is then given as

$$\mathbf{F} = \mathbf{F}_s + m\mathbf{g} = -Kx\mathbf{E}_x - mg\mathbf{E}_y \tag{3-421}$$

Suppose now that we let t_3 be the time when the spring has reached its maximum compression. Then the work done by the resultant force acting on the block between $t_2 = 0^+$ and t_3 is given as

$${}^{\mathcal{F}}W_{23} = \int_{t_2}^{t_3} \mathbf{F} \cdot {}^{\mathcal{F}}\mathbf{v}dt = \int_{t_2}^{t_3} (-Kx\mathbf{E}_x - mg\mathbf{E}_y) \cdot \dot{x}\mathbf{E}_x dt = \int_{x_2}^{x_3} -Kxdx \tag{3-422}$$

where $x_2 = x(t_2)$ and $x_3 = x(t_3)$. Integrating Eq. (3–422) and using the fact that $x_2 = 0$, we obtain

$${}^{\mathcal{F}}W_{23} = -\tfrac{1}{2}Kx_3^2 \tag{3-423}$$

Next, the kinetic energy in reference frame $\mathcal{F}$ at times t_2 and t_3 are given, respectively, as

$$\begin{aligned} {}^{\mathcal{F}}T_2 &= \tfrac{1}{2}m{}^{\mathcal{F}}\mathbf{v}(t_2) \cdot {}^{\mathcal{F}}\mathbf{v}(t_2) = \tfrac{1}{2}mv^2(t_2) = \tfrac{1}{2}mv_2^2 && (3\text{-}424) \\ {}^{\mathcal{F}}T_3 &= \tfrac{1}{2}m{}^{\mathcal{F}}\mathbf{v}(t_3) \cdot {}^{\mathcal{F}}\mathbf{v}(t_3) = \tfrac{1}{2}mv^2(t_3) = \tfrac{1}{2}mv_3^2 && (3\text{-}425) \end{aligned}$$

where $v_2 = v(t_2)$ and $v_3 = v(t_3)$. Then, using ${}^{\mathcal{F}}\mathbf{v}(t_2)$ from Eq. (3–413), we obtain

$${}^{\mathcal{F}}T_2 = \tfrac{1}{2}m\left(\frac{\hat{P}}{m}\right)^2 \tag{3-426}$$

$$ {}^{\mathcal{F}}T_3 = \tfrac{1}{2}mv_3^2 \tag{3-427} $$

Substituting ${}^{\mathcal{F}}T_2$ and ${}^{\mathcal{F}}T_3$ from Eqs. (3–426) and (3–427), respectively, and ${}^{\mathcal{F}}W_{23}$ from Eq. (3–423) into (3–414), we obtain

$$ \tfrac{1}{2}mv_3^2 - \tfrac{1}{2}m\left(\frac{\hat{P}}{m}\right)^2 = -\tfrac{1}{2}Kx_3^2 \tag{3-428} $$

It is known that the spring will attain its maximum compression when the velocity of the block is zero. Therefore, at the point of maximum compression of the spring we have

$$ {}^{\mathcal{F}}\mathbf{v}(t_3) = \mathbf{0} \Longrightarrow v_3 = 0 \tag{3-429} $$

Equation (3–428) simplifies to

$$ -\tfrac{1}{2}m\left(\frac{\hat{P}}{m}\right)^2 = -\tfrac{1}{2}Kx_3^2 \tag{3-430} $$

Solving Eq. (3–430) for x_3, we obtain the maximum compression of the block as

$$ x_3 = \frac{\hat{P}}{\sqrt{Km}} \tag{3-431} $$

■

Example 3–9

Consider Example 3–2 on page 163 of the particle moving under the influence of gravity along a track in the form of a parabola. Determine the differential equation of motion using (a) the work-energy theorem for a particle and (b) the alternate form of the work-energy theorem.

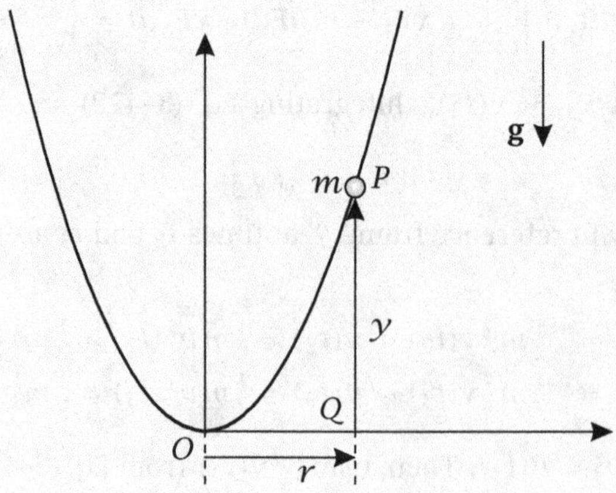

Figure 3–30 Particle sliding without friction along a track in the shape of a parabola.

Solution to Example 3–9

(a) Differential Equation Using Work-Energy Theorem

Using the work-energy theorem from Eq. (3–275), it is necessary to determine both the kinetic energy of the particle in the inertial reference frame $\mathcal{F}$ and the power of the external forces acting on the particle in reference frame $\mathcal{F}$. Recall from Eq. (3–80) on page 164 of Example 3–2 that the velocity of the particle in the fixed inertial reference frame $\mathcal{F}$ is given as

$$ {}^{\mathcal{F}}\mathbf{v} = \dot{r}\mathbf{E}_x + \frac{r\dot{r}}{R}\mathbf{E}_y \tag{3–432} $$

Therefore, the kinetic energy of the particle in $\mathcal{F}$ is given as

$$ {}^{\mathcal{F}}T = \tfrac{1}{2}m\,{}^{\mathcal{F}}\mathbf{v}\cdot{}^{\mathcal{F}}\mathbf{v} = \tfrac{1}{2}m\dot{r}^2\left[1+\frac{r^2}{R^2}\right] \tag{3–433} $$

Recall also that the resultant force acting on the parabola is

$$ \mathbf{F} = \mathbf{N} + m\mathbf{g} = N\mathbf{e}_n - mg\mathbf{E}_y \tag{3–434} $$

where $\mathbf{e}_n$ is defined as in Eq. (3–92) on page 166. Then the power of all forces acting on the particle in reference frame $\mathcal{F}$ is given as

$$ {}^{\mathcal{F}}P = \mathbf{F}\cdot{}^{\mathcal{F}}\mathbf{v} = (N\mathbf{e}_n - mg\mathbf{E}_y)\cdot{}^{\mathcal{F}}\mathbf{v} \tag{3–435} $$

Now, because ${}^{\mathcal{F}}\mathbf{v}$ lies in the direction of $\mathbf{e}_t$, we have

$$ {}^{\mathcal{F}}P = (N\mathbf{e}_n - mg\mathbf{E}_y)\cdot{}^{\mathcal{F}}v\mathbf{e}_t = N\mathbf{e}_n\cdot\mathbf{e}_t - mg\mathbf{E}_y\cdot{}^{\mathcal{F}}v\mathbf{e}_t \tag{3–436} $$

Furthermore, because $\mathbf{e}_n\cdot\mathbf{e}_t = 0$, the power simplifies to

$$ {}^{\mathcal{F}}P = -mg\mathbf{E}_y\cdot{}^{\mathcal{F}}v\mathbf{e}_t \tag{3–437} $$

Using the expression for ${}^{\mathcal{F}}v$ from Eq. (3–84) and the expression for $\mathbf{e}_t$ from Eq. (3–87), we obtain

$$ {}^{\mathcal{F}}P = -mg\mathbf{E}_y\cdot\left[\dot{r}\sqrt{1+\left(\frac{r}{R}\right)^2}\right]\frac{\mathbf{E}_x+\frac{r}{R}\mathbf{E}_y}{\sqrt{1+\left(\frac{r}{R}\right)^2}} = -\frac{mgr\dot{r}}{R} \tag{3–438} $$

Next, the rate of change of the kinetic energy in reference frame $\mathcal{F}$ is given as

$$ \frac{d}{dt}\left({}^{\mathcal{F}}T\right) = m\dot{r}\left[\ddot{r}\left(1+\frac{r^2}{R^2}\right)+\frac{r\dot{r}^2}{R^2}\right] \tag{3–439} $$

Then, applying the work-energy theorem by setting $d({}^{\mathcal{F}}T)/dt$ from Eq. (3–439) equal to ${}^{\mathcal{F}}P$ from Eq. (3–438), we obtain

$$ m\dot{r}\left[\ddot{r}\left(1+\frac{r^2}{R^2}\right)+\frac{r\dot{r}^2}{R^2}\right] = -\frac{mgr\dot{r}}{R} \tag{3–440} $$

Rearranging Eq. (3–440) gives

$$\ddot{r}\left(1+\frac{r^2}{R^2}\right)+\frac{r\dot{r}^2}{R^2}=-\frac{gr}{R} \tag{3–441}$$

Simplifying further gives

$$\ddot{r}\left(1+\frac{r^2}{R^2}\right)+\frac{r}{R}\left(\frac{\dot{r}^2}{R}+g\right)=0 \tag{3–442}$$

Multiplying Eq. (3–442) through by R^2, we obtain

$$(r^2+R^2)\ddot{r}+r(\dot{r}^2+Rg)=0 \tag{3–443}$$

which is identical to the result obtained in Eq. (3–105) on page 167.

(b) Differential Equation Using Alternate Form of Work-Energy Theorem

In order to apply the alternate form of the work-energy theorem as given in Eq. (3–332), we need to compute the total energy in reference frame $\mathcal{F}$ and the power produced by all of the *nonconservative* forces in reference frame $\mathcal{F}$. First, the kinetic energy in reference frame $\mathcal{F}$ is given from Eq. (3–433) of part (a). Next, it is observed that the only two forces act on the particle are the force of gravity, $m\mathbf{g}$, and the reaction force of the track on the particle, **N**. First, the gravity is a constant force. Consequently, $m\mathbf{g}$ is conservative and its potential energy is given by Eq. (3–290) as

$${}^{\mathcal{F}}U_g=-m\mathbf{g}\cdot\mathbf{r} \tag{3–444}$$

Using the fact that $m\mathbf{g}=-mg\mathbf{E}_y$, the potential energy of the force of gravity is given as

$${}^{\mathcal{F}}U_g=-(-mg\mathbf{E}_y)\cdot\left(r\mathbf{E}_x+\frac{r^2}{2R}\mathbf{E}_y\right)=mg\frac{r^2}{2R} \tag{3–445}$$

Because $m\mathbf{g}$ is the only conservative force acting on the particle, the total potential energy of the system is

$${}^{\mathcal{F}}U={}^{\mathcal{F}}U_g=mg\frac{r^2}{2R} \tag{3–446}$$

The *total energy* of the system in reference frame $\mathcal{F}$ is then given as

$${}^{\mathcal{F}}E={}^{\mathcal{F}}T+{}^{\mathcal{F}}U=\tfrac{1}{2}m\dot{r}^2\left[1+\frac{r^2}{R^2}\right]+mg\frac{r^2}{2R} \tag{3–447}$$

Noting that the term $\mathbf{F}^{nc}\cdot{}^{\mathcal{F}}\mathbf{v}$ is accounted for by the force **N**, we have

$$\mathbf{F}^{nc}\cdot{}^{\mathcal{F}}\mathbf{v}=\mathbf{N}\cdot{}^{\mathcal{F}}\mathbf{v} \tag{3–448}$$

But from part (a) we know that $\mathbf{N}\cdot{}^{\mathcal{F}}\mathbf{v}=0$ (because **N** lies in the direction of $\mathbf{e}_n$ and ${}^{\mathcal{F}}\mathbf{v}$ lies in the direction of $\mathbf{e}_t$). Then, applying Eq. (3–332), we obtain

$$\frac{d}{dt}\left({}^{\mathcal{F}}E\right)=\mathbf{F}^{nc}\cdot{}^{\mathcal{F}}\mathbf{v}=0 \tag{3–449}$$

Computing the rate of change of ${}^{\mathcal{F}}E$ in Eq. (3-447), we obtain

$$\frac{d}{dt}\left({}^{\mathcal{F}}E\right) = m\dot{r}\ddot{r}\left(1 + \frac{r^2}{R^2}\right) + m\dot{r}^2\frac{r\dot{r}}{R^2} + mg\frac{r\dot{r}}{R} = 0 \tag{3-450}$$

Simplifying Eq. (3-450) by dropping m and multiplying by R^2, we obtain

$$\dot{r}\ddot{r}\left(r^2 + R^2\right) + \dot{r}r\left(\dot{r}^2 + Rg\right) = 0 \tag{3-451}$$

Finally, dropping the common factor of $\dot{r}$ (noting that $\dot{r} \neq 0$ as a function of time), we obtain

$$\ddot{r}\left(r^2 + R^2\right) + r\left(\dot{r}^2 + Rg\right) = 0 \tag{3-452}$$

which is identical to the results obtained both in part (a) and in Eq. (3-105) on page 167.

■

Example 3–10

A particle P of mass m moves in the plane under the gravitational attraction of a particle of mass M as shown in Fig. 3-31. Knowing that mass M is located at the inertially fixed point O, determine (a) a system of two differential equations describing the motion of the particle, (b) the angular momentum of the particle relative to point O, and (c) the total energy of the system. In addition, show that (d) the angular momentum of the particle is conserved in the $\mathbf{E}_z$-direction and (e) the energy of the system is conserved.

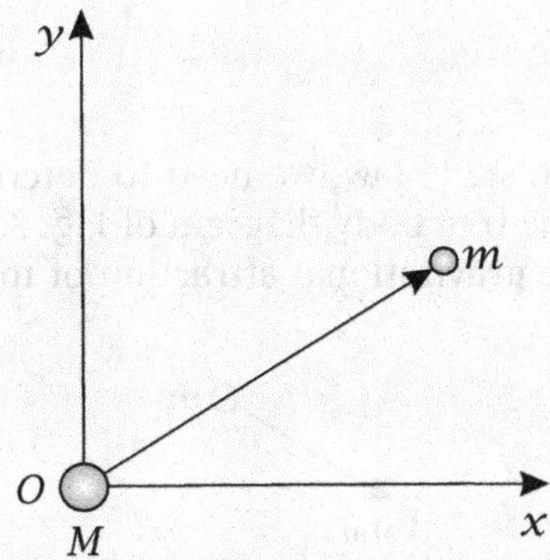

Figure 3–31 Particle of mass m moving in two dimensions under the influence of a fixed mass M.

Solution to Example 3–10

Kinematics

For this problem it is convenient to use a fixed reference frame $\mathcal{F}$ to observe the motion. Corresponding to reference frame $\mathcal{F}$, the following Cartesian coordinate system fixed in $\mathcal{F}$ is chosen to describe the motion of the particle:

$$\begin{array}{ccc} & \text{Origin at } O & \\ \mathbf{E}_x & = & \text{Along } Ox \\ \mathbf{E}_y & = & \text{Along } Oy \\ \mathbf{E}_z & = & \mathbf{E}_x \times \mathbf{E}_y \end{array}$$

The position of the particle is then given in terms of the basis $\{\mathbf{E}_x, \mathbf{E}_y, \mathbf{E}_z\}$ as

$$\mathbf{r} = x\mathbf{E}_x + y\mathbf{E}_y \tag{3–453}$$

Noting that $\mathcal{F}$ is an inertial reference frame, we have trivially that

$$^{\mathcal{F}}\boldsymbol{\omega}^{\mathcal{F}} = \mathbf{0} \tag{3–454}$$

Then, computing the rate of change of $\mathbf{r}$ in reference frame $\mathcal{F}$, we obtain the velocity of P as

$$^{\mathcal{F}}\mathbf{v} = \frac{^{\mathcal{F}}d\mathbf{r}}{dt} = \dot{x}\mathbf{E}_x + \dot{y}\mathbf{E}_y \tag{3–455}$$

Computing the rate of change of $^{\mathcal{F}}\mathbf{v}$ in reference frame $\mathcal{F}$ using $^{\mathcal{F}}\mathbf{v}$ from Eq. (3–455), we obtain the acceleration of point P in reference frame $\mathcal{F}$ as

$$^{\mathcal{F}}\mathbf{a} = \frac{^{\mathcal{F}}d}{dt}\left(^{\mathcal{F}}\mathbf{v}\right) = \ddot{x}\mathbf{E}_x + \ddot{y}\mathbf{E}_y \tag{3–456}$$

Kinetics

Next, in order to apply Newton's 2nd law, we need to determine the resultant force $\mathbf{F}$ acting on the particle. From the free body diagram of Fig. 3–32 it is seen that the only force acting on P is due to the gravitational attraction of mass M located at the fixed point O.

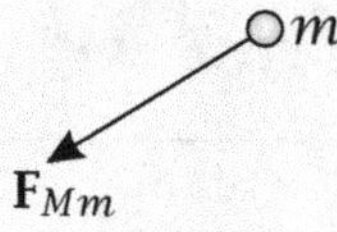

Figure 3–32 Free body diagram of mass m for Example 3–10.

Now we know that the force of gravitational attraction of M on m lies in the direction from P to O. Denoting the direction from O to P by $\mathbf{e}_r$ and applying Newton's law of gravitation, the resultant force on P is given as

$$\mathbf{F} = \mathbf{F}_{Mm} = -\frac{GmM}{r^2}\mathbf{e}_r \tag{3–457}$$

where $r = \|\mathbf{r}\|$ is the distance from O to P. In terms of the Cartesian coordinate system chosen for this problem, we have

$$r = \|\mathbf{r}\| = \sqrt{x^2 + y^2} \tag{3–458}$$

Furthermore, using the expression for **r** from Eq. (3–453), we have

$$\mathbf{e}_r = \frac{\mathbf{r}}{r} = \frac{x\mathbf{E}_x + y\mathbf{E}_y}{\sqrt{x^2 + y^2}} \tag{3–459}$$

Substituting r and $\mathbf{e}_r$ from Eqs. (3–458) and (3–459), respectively, into Eq. (3–457), the resultant force on m is given as

$$\mathbf{F} = -\frac{GmM}{x^2 + y^2}\frac{x\mathbf{E}_x + y\mathbf{E}_y}{\sqrt{x^2 + y^2}} = -\frac{GmM}{(x^2 + y^2)^{3/2}}\left(x\mathbf{E}_x + y\mathbf{E}_y\right) \tag{3–460}$$

(a) System of Two Differential Equations

Applying Newton's 2nd law to the particle by setting **F** from Eq. (3–460) equal to $m^{\mathcal{F}}\mathbf{a}$ using ${}^{\mathcal{F}}\mathbf{a}$ from Eq. (3–456), we obtain

$$m\left(\ddot{x}\mathbf{E}_x + \ddot{y}\mathbf{E}_y\right) = -\frac{GmM}{(x^2 + y^2)^{3/2}}\left(x\mathbf{E}_x + y\mathbf{E}_y\right) \tag{3–461}$$

Setting the $\mathbf{E}_x$ and $\mathbf{E}_y$ components equal yields the following two scalar equations:

$$m\ddot{x} = -\frac{GmM}{(x^2 + y^2)^{3/2}}x \tag{3–462}$$

$$m\ddot{y} = -\frac{GmM}{(x^2 + y^2)^{3/2}}y \tag{3–463}$$

Rearranging Eqs. (3–462) and (3–463), we obtain the system of two differential equations for mass m as

$$m\ddot{x} + \frac{GmM}{(x^2 + y^2)^{3/2}}x = 0 \tag{3–464}$$

$$m\ddot{y} + \frac{GmM}{(x^2 + y^2)^{3/2}}y = 0 \tag{3–465}$$

(b) Angular Momentum of Particle Relative to Point O in Reference Frame $\mathcal{F}$

The angular momentum of point P relative to point O in reference frame $\mathcal{F}$ is obtained using Eq. (3–195) as

$${}^{\mathcal{F}}\mathbf{H}_O = (\mathbf{r} - \mathbf{r}_O) \times m({}^{\mathcal{F}}\mathbf{v} - {}^{\mathcal{F}}\mathbf{v}_O) \tag{3–466}$$

Since point O is fixed in reference frame $\mathcal{F}$, we have

$${}^{\mathcal{F}}\mathbf{v}_O = \mathbf{0} \tag{3–467}$$

Then, substituting **r** from Eq. (3–453) and ${}^{\mathcal{F}}\mathbf{v}$ from Eq. (3–455) into (3–466), we obtain

$${}^{\mathcal{F}}\mathbf{H}_O = \left(x\mathbf{E}_x + y\mathbf{E}_y\right) \times m\left(\dot{x}\mathbf{E}_x + \dot{y}\mathbf{E}_y\right) \tag{3–468}$$

Taking the vector products in Eq. (3–468) gives

$${}^{\mathcal{F}}\mathbf{H}_O = m(x\dot{y} - y\dot{x})\mathbf{E}_z \tag{3–469}$$

(c) Total Energy of System in Reference Frame $\mathcal{F}$

The total energy of the system in reference frame $\mathcal{F}$ is obtained using Eq. (3–331) as

$$^{\mathcal{F}}E = {}^{\mathcal{F}}T + {}^{\mathcal{F}}U \tag{3-470}$$

First, using the expression for $^{\mathcal{F}}\mathbf{v}$ from Eq. (3–455), the kinetic energy of point P in reference frame $\mathcal{F}$ is obtained using Eq. (3–271) on page 189 as

$$^{\mathcal{F}}T = \tfrac{1}{2}m{}^{\mathcal{F}}\mathbf{v}\cdot{}^{\mathcal{F}}\mathbf{v} = \tfrac{1}{2}m\left(\dot{x}^2+\dot{y}^2\right) \tag{3-471}$$

Next, for this problem, the only force acting on the particle is the conservative force of gravitational attraction of M on m. Using the expression for the potential energy due to gravitational attraction from Eq. (3–310), we obtain the potential energy in reference frame $\mathcal{F}$ as

$$^{\mathcal{F}}U = {}^{\mathcal{F}}U_g = -\frac{GmM}{r} = -\frac{GmM}{\sqrt{x^2+y^2}} \tag{3-472}$$

where $r = \sqrt{x^2+y^2}$ from Eq. (3–458) has been used in Eq. (3–472). Then, using the kinetic energy in reference frame $\mathcal{F}$ from Eq. (3–471) and the potential energy in reference frame $\mathcal{F}$ from Eq. (3–472), the total energy of the system in reference frame $\mathcal{F}$ is obtained from Eq. (3–470) as

$$^{\mathcal{F}}E = \tfrac{1}{2}m\left(\dot{x}^2+\dot{y}^2\right) - \frac{GmM}{\sqrt{x^2+y^2}} \tag{3-473}$$

(d) Conservation of Angular Momentum of Particle in $\mathbf{E}_z$-Direction

In accordance with Eq. (3–215), in order to show that the angular momentum about point O in the $\mathbf{E}_z$-direction is conserved, we must show that

$$\left(\mathbf{M}_O - (\mathbf{r}-\mathbf{r}_O)\times m{}^{\mathcal{F}}\mathbf{a}_O\right)\cdot\mathbf{E}_z + {}^{\mathcal{F}}\mathbf{H}_O\cdot\frac{^{\mathcal{F}}d\mathbf{E}_z}{dt} = 0 \tag{3-474}$$

Now, since O is fixed in the inertial reference frame $\mathcal{F}$, it follows that $^{\mathcal{F}}\mathbf{a}_O = \mathbf{0}$. Furthermore, since $\mathbf{E}_z$ is fixed in reference frame $\mathcal{F}$, we have $^{\mathcal{F}}d\mathbf{E}_z/dt = \mathbf{0}$. Therefore, we must show that

$$\mathbf{M}_O\cdot\mathbf{E}_z = 0 \tag{3-475}$$

Because $\mathbf{r}_O = \mathbf{0}$, we have the moment of the resultant force on P relative to point O as

$$\mathbf{M}_O = \mathbf{r}\times\mathbf{F} \tag{3-476}$$

Using $\mathbf{r}$ from Eq. (3–453) and $\mathbf{F}$ from Eq. (3–460), we obtain $\mathbf{M}_O$ as

$$\mathbf{M}_O = \left(x\mathbf{E}_x+y\mathbf{E}_y\right)\times\left[-\frac{GmM}{(x^2+y^2)^{3/2}}\left(x\mathbf{E}_x+y\mathbf{E}_y\right)\right] \tag{3-477}$$

Computing the vector products in Eq. (3–477), we obtain

$$\mathbf{M}_O = -\frac{GmM}{(x^2+y^2)^{3/2}}xy\mathbf{E}_z + \frac{GmM}{(x^2+y^2)^{3/2}}xy\mathbf{E}_z \equiv \mathbf{0} \tag{3-478}$$

Then, because $\mathbf{M}_O = \mathbf{0}$, we have trivially that

$$\mathbf{M}_O \cdot \mathbf{E}_z \equiv 0 \tag{3-479}$$

which implies that the angular momentum of the particle in the $\mathbf{E}_z$-direction is conserved.

(e) Conservation of Energy

In order to show that energy is conserved, we need to show that the rate of change of ${}^{\mathcal{F}}E$ is zero. Differentiating the total energy from Eq. (3–473), we have

$$\frac{d}{dt}\left({}^{\mathcal{F}}E\right) = m\dot{x}\ddot{x} + m\dot{y}\ddot{y} + \frac{1}{2}\frac{GmM}{(x^2+y^2)^{3/2}}(2x\dot{x} + 2y\dot{y}) \tag{3-480}$$

Rearranging Eq. (3–480) gives

$$\frac{d}{dt}\left({}^{\mathcal{F}}E\right) = \dot{x}\left(m\ddot{x} + \frac{GmM}{(x^2+y^2)^{3/2}}x\right) + \dot{y}\left(m\ddot{y} + \frac{GmM}{(x^2+y^2)^{3/2}}y\right) \tag{3-481}$$

It is seen from Eqs. (3–464) and (3–465) that the two terms in parentheses in Eq. (3–481) are *zero*. Consequently, we have

$$\frac{d}{dt}\left({}^{\mathcal{F}}E\right) = \dot{x}\,(0) + \dot{y}\,(0) \equiv 0 \tag{3-482}$$

which implies that the total energy of the particle in reference frame $\mathcal{F}$ is conserved.

■

Summary of Chapter 3

This chapter was devoted to developing the framework for analyzing the kinetics of a particle. As a precursor to stating the laws of kinetics of a particle, models were derived for several common types of forces that arise in dynamics. In particular, friction forces, spring forces, and gravitational forces were described quantitatively. Two types of friction forces were modeled: Coulomb friction and viscous friction. It was shown that the force of friction depends on the relative velocity, $\mathbf{v}_{\text{rel}}$, between the particle and the point on the surface with which the particle is in contact. More specifically, it was shown that $\mathbf{v}_{\text{rel}}$ is *independent* of the reference frame. Then the model for the force of dynamic Coulomb friction was stated as

$$\mathbf{F}_f = -\mu_d \|\mathbf{N}\| \frac{\mathbf{v}_{\text{rel}}}{\|\mathbf{v}_{\text{rel}}\|} \tag{3-8}$$

where μ_d is the coefficient of dynamic friction and $\mathbf{N}$ is the normal force (or reaction force) applied by the surface on the particle. Next, the model for the force of viscous friction was stated as

$$\mathbf{F}_f = -c\mathbf{v}_{\text{rel}} \tag{3-9}$$

where c is the coefficient of viscous friction. Next, models were developed for the force exerted by a linear spring and a curvilinear spring. The model derived for the force exerted by a linear spring was

$$\mathbf{F}_s = \mathbf{F}_s = -K(\ell - \ell_0)\mathbf{u}_s \tag{3-19}$$

where K is the spring constant, ℓ_0 is the unstretched length of the spring, $\ell = \|\mathbf{r} - \mathbf{r}_Q\|$ is the stretched length of the spring, $\mathbf{u}_s = (\mathbf{r} - \mathbf{r}_Q)/\ell$ is the unit vector in the direction from the attachment point of the spring to the particle, and $\mathbf{r}$ and $\mathbf{r}_Q$ are the position of the particle and the attachment point of the spring, respectively. The model derived for the force exerted by a curvilinear spring (i.e., a spring whose shape conforms to a curve C in $\mathbb{E}^3$) was

$$\mathbf{F}_s = -K(\ell - \ell_0)\mathbf{e}_t \tag{3-20}$$

where $\mathbf{e}_t$ is the tangent vector to the curve at the end of the spring where the particle is located. The last type of force that was modeled was that of gravitational attraction between a mass M and a mass m. In particular, it was shown that the force of gravitational attraction of a particle of mass M on a particle of mass m is

$$\mathbf{F}_g = -\frac{GmM}{r^3}\mathbf{r} \tag{3-25}$$

where $\mathbf{r}$ is the position of m relative to M, $r = \|\mathbf{r}\|$ is the distance between M and m, and G is the *constant of gravitation*. Equation (3-25) is called *Newton's law of gravitation*. A special case of gravitational forces was then discussed where a particle of mass m is near the surface of a large spherical body of mass M. In particular, it was shown that, when a particle lies a small distance from the large spherical body, the acceleration due to gravity is given as

$$\mathbf{g} = -\frac{GM}{R^2}\mathbf{e}_r = -g\mathbf{e}_r \tag{3-36}$$

where R is the radius of the spherical body, $g = GM/R^2$, and $\mathbf{e}_r$ is the unit vector in the local vertical direction (i.e., the direction from the center of the sphere to the particle).

The next topics covered in this chapter were *Newton's laws* for a particle. In particular, Newton's 2nd law was stated as

$$\mathbf{F} = m^{\mathcal{N}}\mathbf{a} \tag{3-41}$$

where $\mathbf{F}$ is the resultant force acting on a particle of mass m and ${}^{\mathcal{N}}\mathbf{a}$ is the acceleration of the particle as viewed by an observer in an inertial reference frame $\mathcal{N}$. Several examples were then solved using Newton's laws.

The next topics covered in this chapter were linear momentum and linear impulse for a particle. The linear momentum of a particle in an inertial reference frame $\mathcal{N}$ was defined as

$${}^{\mathcal{N}}\mathbf{G} = m^{\mathcal{N}}\mathbf{v} \tag{3-174}$$

It was then shown that the resultant force applied to a particle is related to the rate of change of its linear momentum in an inertial reference frame $\mathcal{N}$ as

$$\mathbf{F} = \frac{{}^{\mathcal{N}}d}{dt}\left({}^{\mathcal{N}}\mathbf{G}\right) \tag{3-176}$$

Moreover, it was shown that Eq. (3-176) is an alternate form of Newton's 2nd law for a particle. Next, the linear impulse of a force $\mathbf{F}$ on a time interval $t \in [t_1, t_2]$ was defined as

$$\hat{\mathbf{F}} = \int_{t_1}^{t_2} \mathbf{F} dt \tag{3-177}$$

Using the definition of linear impulse as given in Eq. (3-177), the *principle of linear impulse and momentum for a particle* in an inertial reference frame $\mathcal{N}$ was derived as

$$\hat{\mathbf{F}} = {}^{\mathcal{N}}\mathbf{G}(t_2) - {}^{\mathcal{N}}\mathbf{G}(t_1) \tag{3-179}$$

The next topics covered in this chapter were moment of a force, angular momentum, and angular impulse for a particle. The moment of a force $\mathbf{F}$ applied to a particle P relative to an arbitrary reference point Q was defined as

$$\mathbf{M}_Q = (\mathbf{r} - \mathbf{r}_Q) \times \mathbf{F} \tag{3-190}$$

The angular momentum of a particle P of mass m in an inertial reference frame $\mathcal{N}$ relative to an arbitrary reference point Q was defined as

$${}^{\mathcal{N}}\mathbf{H}_Q = (\mathbf{r} - \mathbf{r}_Q) \times m({}^{\mathcal{N}}\mathbf{v} - {}^{\mathcal{N}}\mathbf{v}_Q) \tag{3-195}$$

It was then shown that the rate of change of angular momentum of a particle in an inertial reference frame $\mathcal{N}$ relative to an arbitrary point Q is related to the moment of all forces applied to the particle relative to point Q as

$$\frac{{}^{\mathcal{N}}d}{dt}\left({}^{\mathcal{N}}\mathbf{H}_Q\right) = \mathbf{M}_Q - (\mathbf{r} - \mathbf{r}_Q) \times m^{\mathcal{N}}\mathbf{a}_Q \tag{3-203}$$

where the quantity $-(\mathbf{r} - \mathbf{r}_Q) \times m^{\mathcal{N}}\mathbf{a}_Q$ was defined as the inertial moment of point Q relative to the particle. For the special case of motion relative to a point O fixed in an inertial reference frame $\mathcal{N}$, it was shown that

$$\frac{{}^{\mathcal{N}}d}{dt}\left({}^{\mathcal{N}}\mathbf{H}_O\right) = \mathbf{M}_O \tag{3-205}$$

Next, the angular impulse of the moment $\mathbf{M}_Q$ was defined as

$$\hat{\mathbf{M}}_Q = \int_{t_1}^{t_2} \mathbf{M}_Q dt \tag{3-206}$$

Using the definition of angular impulse, the *principle of angular impulse and angular momentum for a particle* was derived as

$$\hat{\mathbf{M}}_Q - \int_{t_1}^{t_2} \left[(\mathbf{r} - \mathbf{r}_Q) \times m^{\mathcal{N}}\mathbf{a}_Q \right] dt = {}^{\mathcal{N}}\mathbf{H}_Q(t_2) - {}^{\mathcal{N}}\mathbf{H}_Q(t_1) \tag{3-210}$$

Finally, for motion relative to a point O fixed in an inertial reference frame $\mathcal{N}$, it was shown that the principle of angular impulse and angular momentum simplifies to

$$\hat{\mathbf{M}}_O = {}^{\mathcal{N}}\mathbf{H}_O(t_2) - {}^{\mathcal{N}}\mathbf{H}_O(t_1) \tag{3-211}$$

The last topics covered in this Chapter were the power, work, and energy for a particle. First, the power of a force $\mathbf{F}$ acting on a particle P of mass m in an inertial reference frame $\mathcal{N}$ was defined as

$${}^{\mathcal{N}}P = \mathbf{F} \cdot {}^{\mathcal{N}}\mathbf{v} \tag{3-267}$$

Using the definition of the power of a force, the work of a force in an inertial reference frame $\mathcal{N}$ over a time interval $t \in [t_1, t_2]$ was defined as

$${}^{\mathcal{N}}W_{12} = \int_{t_1}^{t_2} {}^{\mathcal{N}}P dt = \int_{t_1}^{t_2} \mathbf{F} \cdot {}^{\mathcal{N}}\mathbf{v} dt \tag{3-268}$$

Next, the kinetic energy of a particle in an inertial reference frame $\mathcal{N}$ was defined as

$${}^{\mathcal{N}}T = \tfrac{1}{2} m {}^{\mathcal{N}}\mathbf{v} \cdot {}^{\mathcal{N}}\mathbf{v} \tag{3-271}$$

The work-energy theorem for a particle was then derived as

$$\frac{d}{dt}\left({}^{\mathcal{N}}T\right) = \mathbf{F} \cdot {}^{\mathcal{N}}\mathbf{v} = {}^{\mathcal{N}}P \tag{3-275}$$

Then, considering motion on a time interval $t \in [t_1, t_2]$, the principle of work and energy for a particle was derived as

$${}^{\mathcal{N}}T_2 - {}^{\mathcal{N}}T_1 = {}^{\mathcal{N}}W_{12} \tag{3-279}$$

Next, a special class of forces, called *conservative forces*, was defined. It was stated that a force is conservative if it is independent of the trajectory taken in moving a particle from an initial position $\mathbf{r}(t_1) = \mathbf{r}_1$ to a final position $\mathbf{r}(t_2) = \mathbf{r}_2$. Furthermore, it was shown for a conservative force that there exists a scalar function ${}^{\mathcal{N}}U = {}^{\mathcal{N}}U(\mathbf{r})$ such that

$$\mathbf{F}_c = -\nabla_{\mathbf{r}} {}^{\mathcal{N}}U \tag{3-281}$$

The function ${}^{\mathcal{N}}U$ is called a potential energy of the force $\mathbf{F}_c$ in an inertial reference frame $\mathcal{N}$ and $\nabla_{\mathbf{r}} {}^{\mathcal{N}}U$ is the gradient of ${}^{\mathcal{N}}U$ with respect to $\mathbf{r}$. Then, the *total energy* of a particle in an inertial reference frame was defined as

$${}^{\mathcal{N}}E = {}^{\mathcal{N}}T + {}^{\mathcal{N}}U \tag{3-331}$$

where ${}^{\mathcal{N}}T$ and ${}^{\mathcal{N}}U$ are the kinetic energy and potential energy, respectively, in the inertial reference frame $\mathcal{N}$. Using the total energy, an *alternate form* of the work-energy theorem for a particle was derived as

$$\frac{d}{dt}\left({}^{\mathcal{N}}E\right) = \mathbf{F}^{nc} \cdot {}^{\mathcal{N}}\mathbf{v} = {}^{\mathcal{N}}p^{nc} \tag{3-332}$$

where $\mathbf{F}^{nc}$ is the resultant nonconservative force acting on the particle and ${}^{\mathcal{N}}p^{nc}$ is the power of the resultant nonconservative force $\mathbf{F}^{nc}$. It was shown that, for the special case where the nonconservative forces do no work, the energy of the system is a *conserved*, i.e., if $\mathbf{F}^{nc} \cdot \mathbf{v} = 0$, then ${}^{\mathcal{N}}E$ = constant. Furthermore, considering motion over a time interval $t \in [t_1, t_2]$, an alternate form of the principle of work and energy for particle was derived as

$${}^{\mathcal{N}}E_2 - {}^{\mathcal{N}}E_1 = {}^{\mathcal{N}}W_{12}^{nc} \tag{3-339}$$

where ${}^{\mathcal{N}}W_{12}^{nc}$ is the work done by all *nonconservative* forces in reference frame $\mathcal{N}$ on the time interval $t \in [t_1, t_2]$. As with the alternate form of the work-energy theorem, if the work done by the nonconservative forces is zero, then ${}^{\mathcal{N}}E$ = constant.

Problems for Chapter 3

3–1 A particle of mass m moves in the vertical plane along a track in the form of a circle as shown in Fig. P3-1. The equation for the track is

$$r = r_0 \cos\theta$$

Knowing that gravity acts downward and assuming the initial conditions $\theta(t = 0) = 0$ and $\dot{\theta}(t = 0) = \dot{\theta}_0$, determine (a) the differential equation of motion for the particle and (b) the force exerted by the track on the particle as a function of θ.

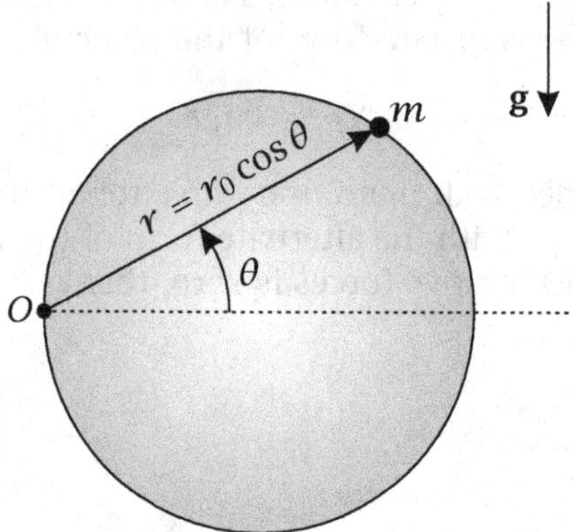

Figure P3-1

3–2 A collar of mass m slides without friction along a rigid massless rod as shown in Fig. P3-2. The collar is attached to a linear spring with spring constant K and unstretched length L. Assuming no gravity, determine the differential equation of motion for the collar.

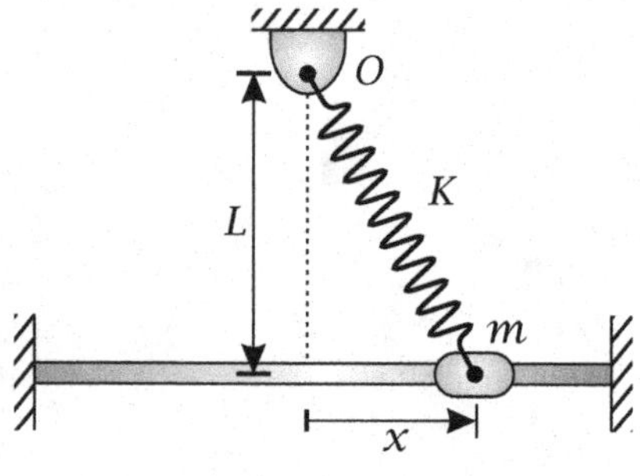

Figure P3-2

3–3 A bead of mass m slides along a fixed circular helix of radius R and constant helical inclination angle ϕ as shown in Fig. P3-3. The equation for the helix is given in cylindrical coordinates as

$$z = R\theta\tan\phi$$

Knowing that gravity acts vertically downward, determine the differential equation of motion for the bead in terms of the angle θ using (a) Newton's 2nd law and (b) the work-energy theorem for a particle. In addition, assuming the initial conditions $\theta(t = 0) = \theta_0$ and $\dot{\theta}(t = 0) = \dot{\theta}_0$, determine (c) the displacement attained by the bead when it reaches its maximum height on the helix.

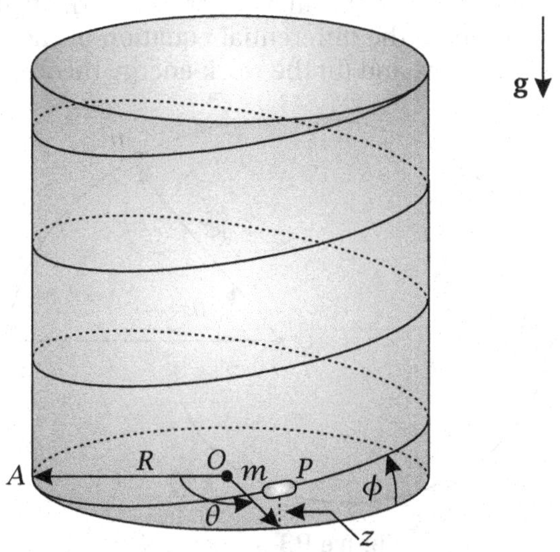

Figure P3-3

3–4 A particle of mass m slides inside a circular slot of radius r cut out of a massless disk of radius $R > r$ as shown in Fig. P3-4. The particle is attached to a curvilinear spring with spring constant K and unstressed angle θ_0, where the angle θ describes the position of the particle relative to the attachment point, A, of the spring. Knowing that the disk rotates with constant angular velocity Ω about an axis through its center at O and assuming no gravity, determine the differential equation of motion for the particle in terms of the angle θ.

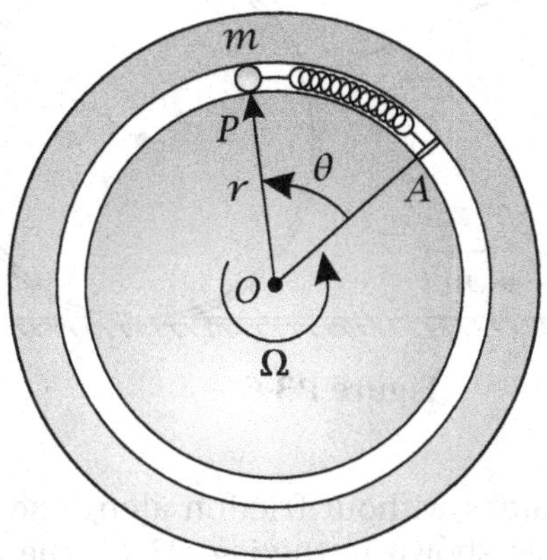

Figure P3-4

3–5 A collar of mass m is constrained to move along a frictionless track in the form of a logarithmic spiral as shown in Fig. P3-5. The equation for the spiral is given as

$$r = r_0 e^{-a\theta}$$

where r_0 and a are constants and θ is the angle as shown in the figure. Assuming that gravity acts downward, determine the differential equation of motion in terms of the angle θ using (a) Newton's 2nd law and (b) the work-energy theorem for a particle.

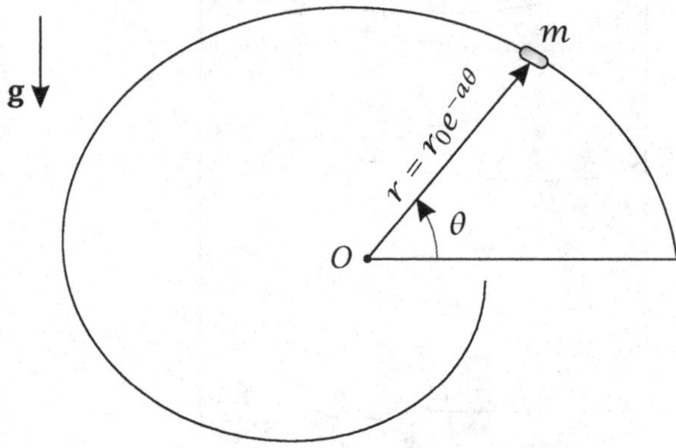

Figure P3-5

3–6 A block of mass m is initially at rest atop a frictionless horizontal surface when it is struck by a horizontal impulse $\hat{\mathbf{P}}$ as shown in Fig. P3-6. After the impulse is applied, the block slides along the surface until it reaches a frictionless circular track of radius r. Knowing that gravity acts downward, determine (a) the velocity of the block immediately after the application of the impulse $\hat{\mathbf{P}}$ and (b) the magnitude of the impulse $\hat{\mathbf{P}}$ for which the block is moving with zero velocity as it reaches the top of the track.

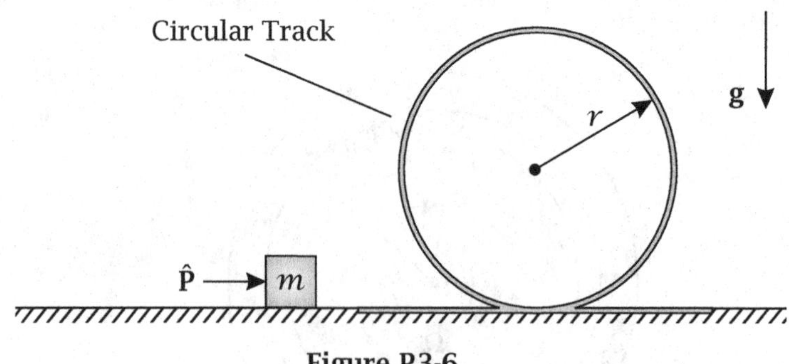

Figure P3-6

3–7 A particle of mass m slides without friction along the inner surface of a fixed cone of semi-vertex angle β as shown in the Fig. P3-7. The equation for the cone is given in cylindrical coordinates as

$$z = r \cot \beta$$

Knowing that the basis $\{\mathbf{E}_x, \mathbf{E}_y, \mathbf{E}_z\}$ is fixed to the cone; θ is the angle between the $\mathbf{E}_x$-direction and the direction OQ, where Q is the projection of the particle into the

$\{\mathbf{E}_x, \mathbf{E}_y\}$-plane; and gravity acts vertically downward, determine a system of two differential equations in terms of r and θ that describes the motion of the particle.

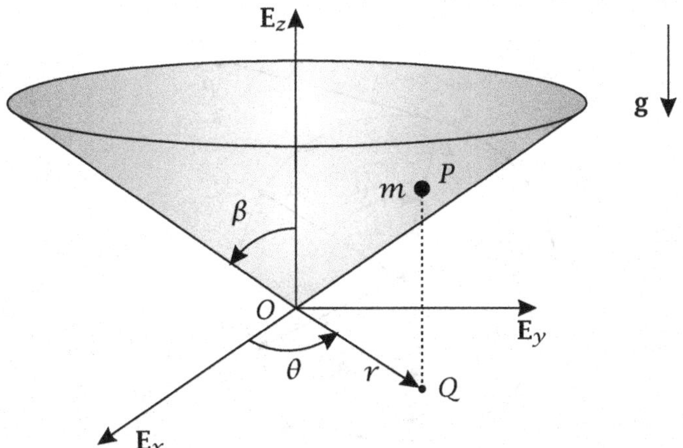

Figure P3-7

3–8 A particle of mass m is connected to an inelastic cord of length ℓ as shown in Fig. P3-8. The cord is connected at its other end to a fixed support at point O. The cord is initially slack and the particle is located a distance ℓ_0 directly above the support when the particle is struck by a horizontal impulse $\hat{\mathbf{P}}$. After being struck by the impulse, the particle travels to the right until the cord becomes taut. Assuming no gravity, determine the following quantities at the instant the cord becomes taut: (a) the velocity of the particle and (b) the impulse applied by the cord on the particle.

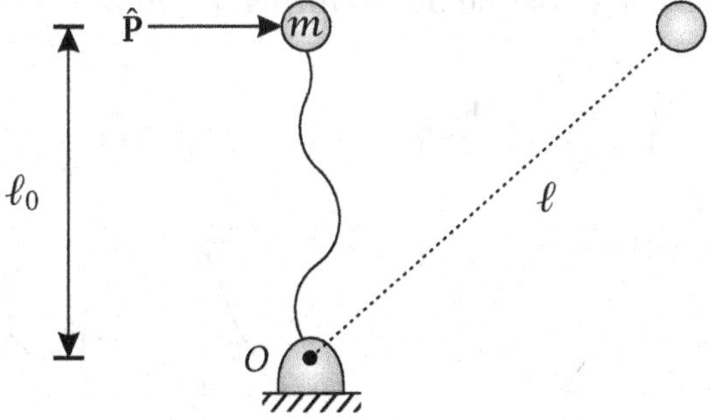

Figure P3-8

3–9 A particle of mass m is attached to one end of a flexible but inextensible massless rope as shown in Fig. P3-9. The rope is wrapped around a cylinder of radius R, where the cylinder rotates with constant angular velocity Ω relative to the ground. The rope unravels from the cylinder in such a manner that it never becomes slack. Furthermore, point A is *fixed to the cylinder* and corresponds to a configuration where no portion of the rope is exposed while point B is the *instantaneous* point of contact of the exposed portion of the rope with the cylinder. Knowing that the exposed portion of the rope is tangent to the cylinder at every instant of the motion; θ is the angle between points A

and B; and assuming the initial conditions $\theta(t=0)=0$, $\dot{\theta}(t=0)=\Omega$ (where $\Omega=\|\mathbf{\Omega}\|$), determine (a) the differential equation for the particle in terms of the variable θ, and (b) the tension in the rope as a function of time.

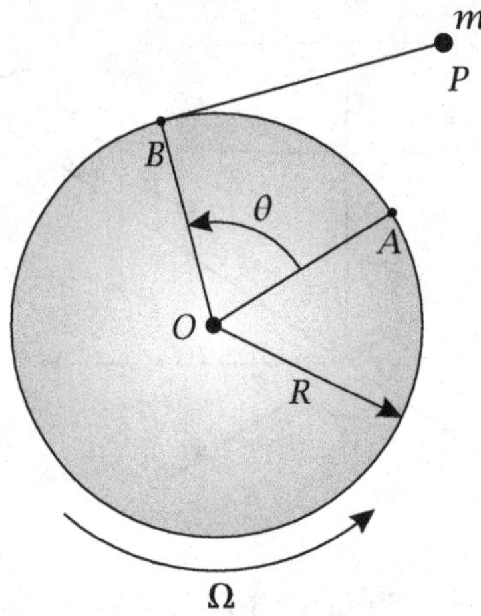

Figure P3-9

3–10 A particle of mass m moves under the influence of gravity in the vertical plane along a track as shown in Fig. P3-10. The equation for the track is given in Cartesian coordinates as

$$y = -\ln\cos x$$

where $-\pi/2 < x < \pi/2$. Using the horizontal component of position, x, as the variable to describe the motion, determine the differential equation of motion for the particle using (a) Newton's 2nd law and (b) one of the forms of the work-energy theorem for a particle.

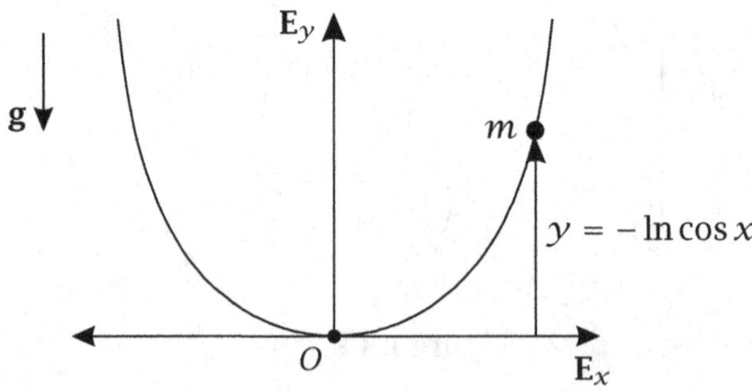

Figure P3-10

3–11 A particle of mass m moves in the horizontal plane as shown in Fig. P3-11. The particle is attached to a linear spring with spring constant K and unstretched length ℓ while the spring is attached at its other end to the fixed point O. Assuming no gravity, (a) determine a system of two differential equations of motion for the particle in terms of the variables r and θ, (b) show that the total energy of the system is conserved, and (c) show that the angular momentum relative to point O is conserved.

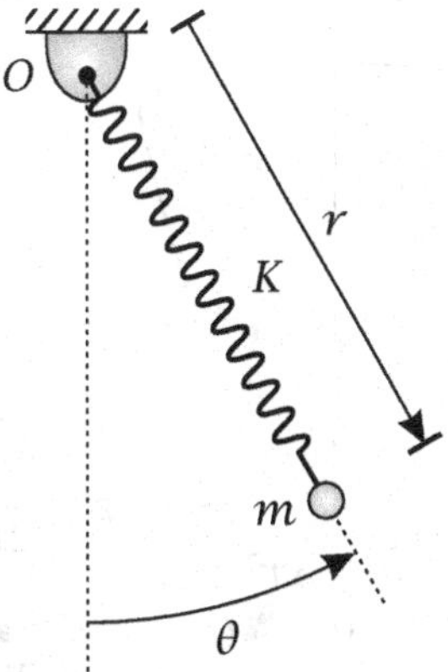

Figure P3-11

3–12 A particle of mass m is attached to a linear spring with spring constant K and unstretched length r_0 as shown in Fig. P3-12. The spring is attached at its other end at point P to the free end of a rigid massless arm of length l. The arm is hinged at its other end and rotates in a circular path at a constant angular rate ω. Knowing that the angle θ is measured from the downward direction and assuming no friction, determine a system of two differential equations of motion for the particle in terms of r and θ.

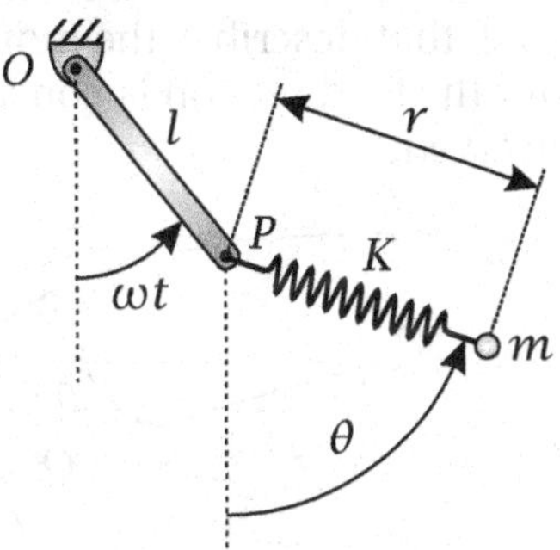

Figure P3-12

3–13 A particle of mass m slides without friction along a surface in form of a paraboloid as shown in Fig. P3-13. The equation for the paraboloid is

$$z = \frac{r^2}{2R}$$

where z is the height of the particle above the horizontal plane; r is the distance from O to Q, where Q is the projection of P onto the horizontal plane; and R is a constant. Knowing that θ is the angle formed by the direction OQ with the x-axis and that gravity acts downward, determine a system of two differential equations for the motion of the particle.

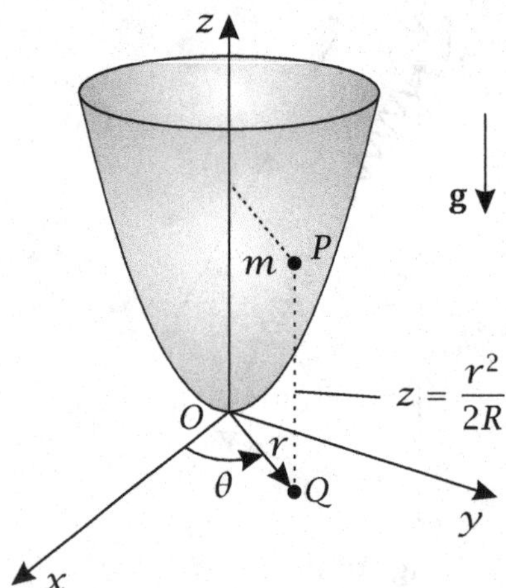

Figure P3-13

3–14 A particle of mass m slides without friction along the inside of a fixed hemispherical bowl of radius R as shown in Fig. P3-14. Furthermore, the angle θ is measured from the $\mathbf{E}_x$-direction to the direction OQ, where point Q lies on the rim of the bowl while the angle ϕ is measured from the OQ-direction to the position of the particle. Knowing that gravity acts downward, (a) determine a system of two differential equations in terms of the angles θ and ϕ that describes the motion of the particle, (b) show that the angular momentum about the $\mathbf{E}_z$-direction is conserved, and (c) show that the total energy of the system is conserved.

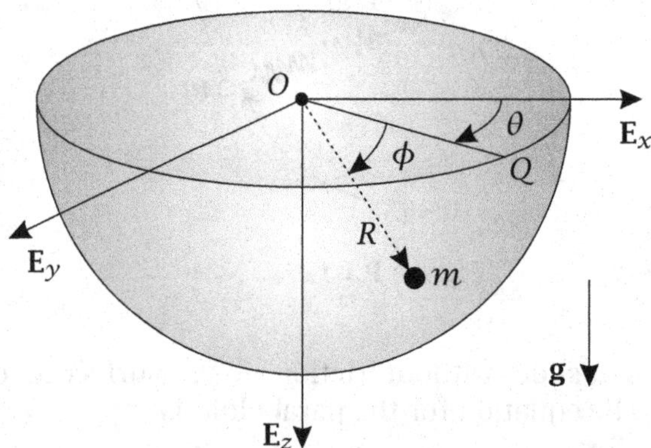

Figure P3-14

3–15 A massless disk of radius R rolls without slip at a constant angular rate ω along a horizontal surface as shown in Fig. P3-15. A particle of mass m slides without friction in a slot cut along a radial direction of the disk. Attached to the particle is a linear spring with spring constant K, unstretched length r_0, and attachment point O, where O is located at the center of the disk. Knowing that gravity acts downward, determine the differential equation of motion for the particle in terms of the variable r.

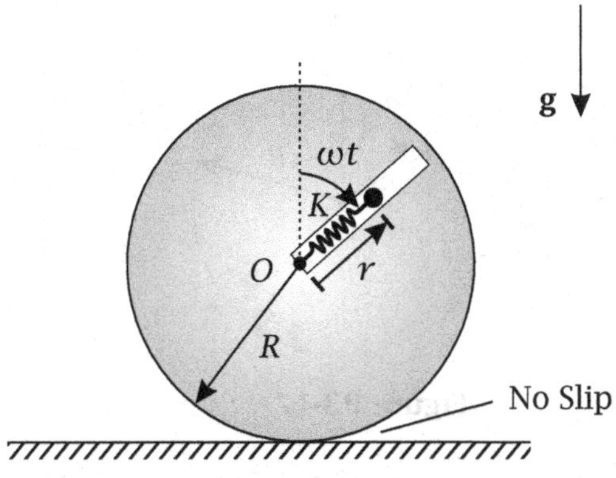

Figure P3-15

3–16 A particle of mass m moves under the influence of gravity in the vertical plane along a fixed curve of the form of a hyperbola $y = a/x$ (where a is a constant and $x > 0$) as shown in Fig. P3-16. Using the initial condition $x(t = 0) = x_0$, determine the differential equation of motion in terms of the variable x using (a) Newton's 2nd law and (b) the work-energy theorem for a particle.

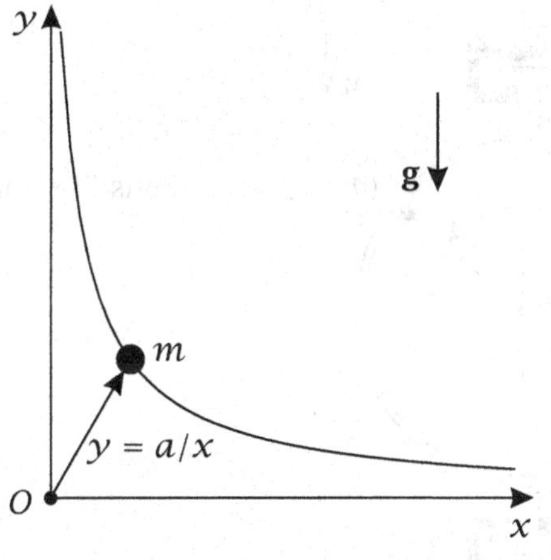

Figure P3-16

3–17 A particle of mass m is attached to an inextensible massless rope of length l as shown in Fig. P3-17. The rope is attached at its other end to point A, located at the top of a fixed cylinder of radius R. As the particle moves, the rope wraps itself around the cylinder and never becomes slack. Knowing that θ is the angle measured from the vertical to the point of tangency of the exposed portion of the rope with the cylinder and that gravity acts downward, determine the differential equation of motion for the particle in terms of the angle θ. You may assume in your solution that the angle θ is always positive.

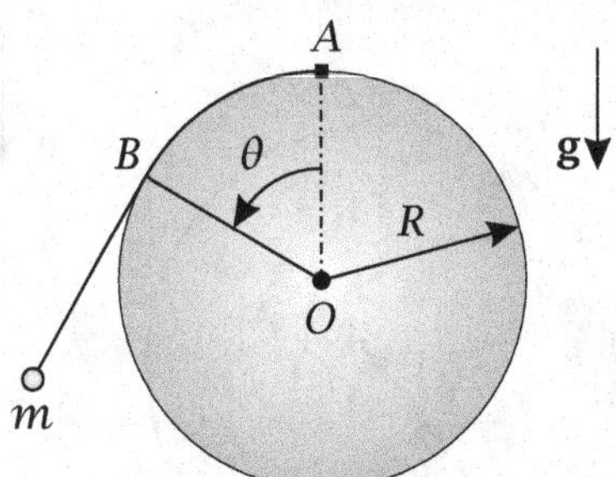

Figure P3-17

3–18 A collar of mass m slides along a massless circular annulus of radius r as shown in Fig. P3-18. The annulus rotates with constant angular velocity $\mathbf{\Omega}$ (where $\Omega = \|\mathbf{\Omega}\|$) about the vertical direction. Furthermore, a viscous friction force with a coefficient of friction c acts at the point of contact of the collar with the annulus. Knowing that θ is the angle measured from the vertical and that gravity acts downward, determine the differential equation of motion for the collar.

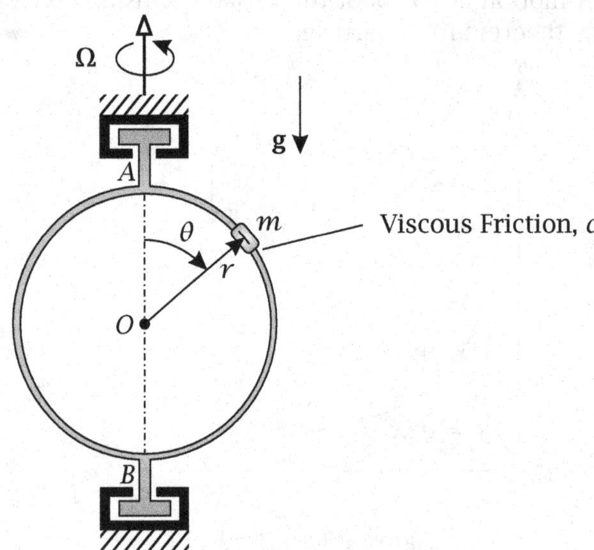

Figure P3-18

3–19 A collar of mass m slides without friction along a circular track of radius R as shown in Fig. P3-19. Attached to the collar is a linear spring with spring constant K and unstretched length zero. The spring is attached at the fixed point A located a distance $2R$ from the center of the circle. Assuming no gravity and the initial conditions $\theta(t = 0) = \theta_0$ and $\dot{\theta}(t = 0) = \dot{\theta}_0$, determine (a) the differential equation of motion for the collar in terms the angle θ and (b) the reaction force exerted by the track on the collar as a function of the angle θ.

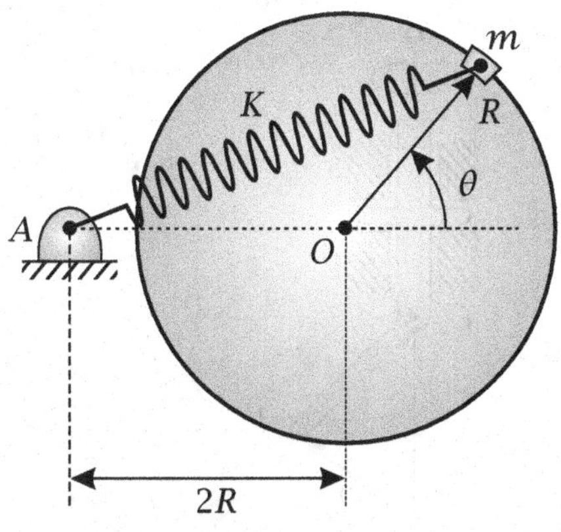

Figure P3-19

3–20 A particle of mass m slides without friction along a fixed horizontal table as shown in Fig. P3-20. The particle is attached to an inextensible rope. The rope itself is threaded through a tiny hole in the table at point O such that the portion of the rope that hangs below the table remains vertical. Knowing that a constant vertical force $\mathbf{F}$ is applied to the rope, that the rope remains taut, and that gravity acts vertically downward, (a) determine a system of two differential equations in terms of r and θ that describes the motion of the particle, (b) show that the angular momentum of the particle relative to point O is conserved, and (c) show that the total energy of the system is conserved.

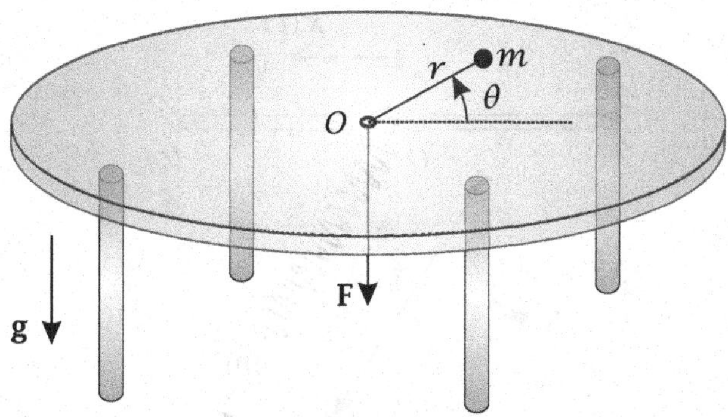

Figure P3-20

3–21 A particle mass m is suspended from one end of a rigid massless arm. The arm is hinged at one of its ends at point O, where point O lies on a vertical shaft that rotates with constant angular velocity $\boldsymbol{\Omega}$ (where $\Omega = \|\boldsymbol{\Omega}\|$) about the vertical direction as shown in Fig. P3-21. Knowing that the arm is free to rotate in the plane of the vertical and horizontal shafts, that θ is the angle between the arm and the horizontal shaft, and that gravity acts downward, determine the differential equation of motion for the particle in terms of the angle θ.

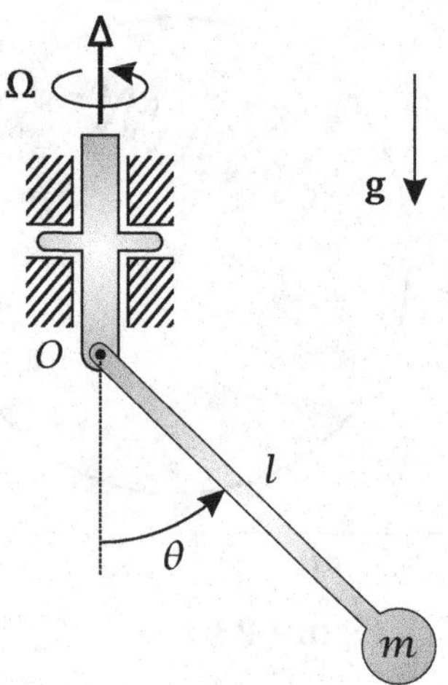

Figure P3-21

3–22 A particle of mass m is attached to a linear spring with spring constant K and unstretched length r_0 as shown in Fig. P3-22. The spring is attached at its other end to a massless collar where the collar slides along a frictionless horizontal track with a *known* displacement $x(t)$. Knowing that gravity acts downward, determine a system of two differential equations in terms of the variables r and θ that describes the motion of the particle.

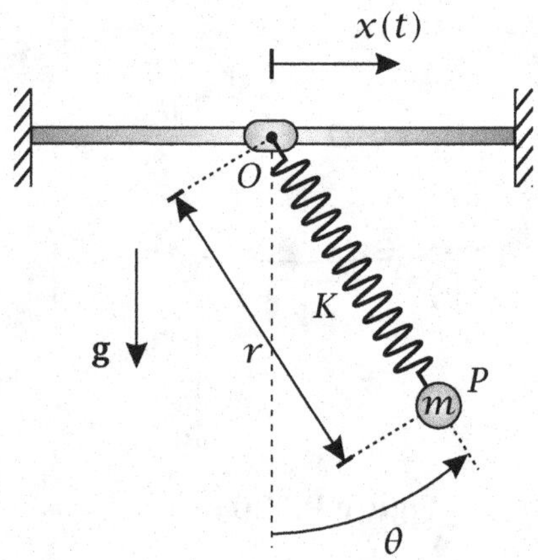

Figure P3-22

3–23 A collar of mass m slides with friction along a rod that is welded rigidly at a constant angle β with the vertical to a shaft as shown in Fig. P3-23. The shaft rotates about the vertical with constant angular velocity $\boldsymbol{\Omega}$ (where $\Omega = \|\boldsymbol{\Omega}\|$). Knowing that r is the radial distance from the point of the weld to the collar, that the friction is viscous with viscous friction coefficient c, and that gravity acts vertically downward, determine the differential equation of motion for the collar in terms of r.

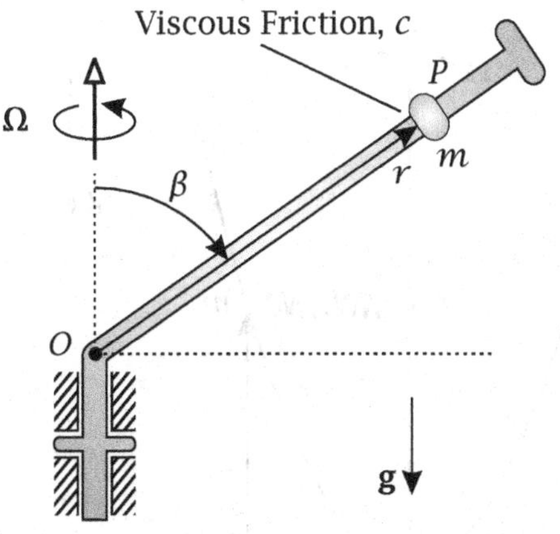

Figure P3-23

3–24 A particle of mass m slides without friction along a track in the form of a spiral as shown in Fig. P3-24. The equation for the spiral is

$$r = a\theta$$

where a is a constant and θ is the angle measured from the horizontal. Assuming no gravity, determine the differential equation of motion for the particle in terms of the angle θ using (a) Newton's 2nd law and (b) the work-energy theorem for a particle.

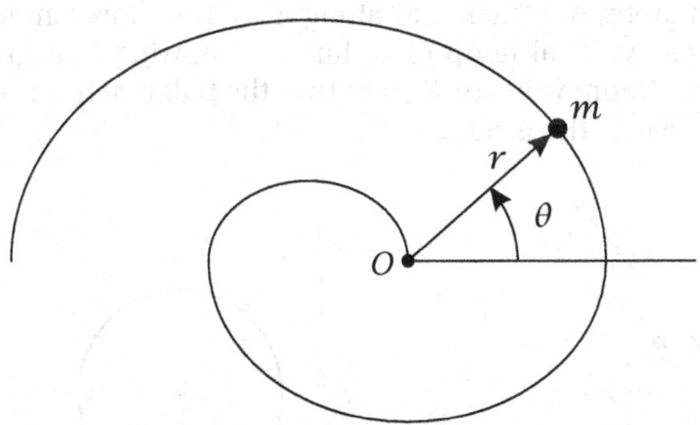

Figure P3-24

3–25 A particle of mass m slides without friction along a track in the form of a parabola as shown in Fig. P3-25. The equation for the parabola is

$$y = \frac{r^2}{2a}$$

where a is a constant, r is the distance from point O to point P, point P is the projection of the particle onto the horizontal direction, and y is the vertical distance. Furthermore, the particle is attached to a linear spring with spring constant K and unstretched length x_0. The spring is always aligned horizontally such that its attachment point is free to slide along a vertical shaft through the center of the parabola. Knowing that the parabola rotates with constant angular velocity Ω (where $\Omega = \|\mathbf{\Omega}\|$) about the

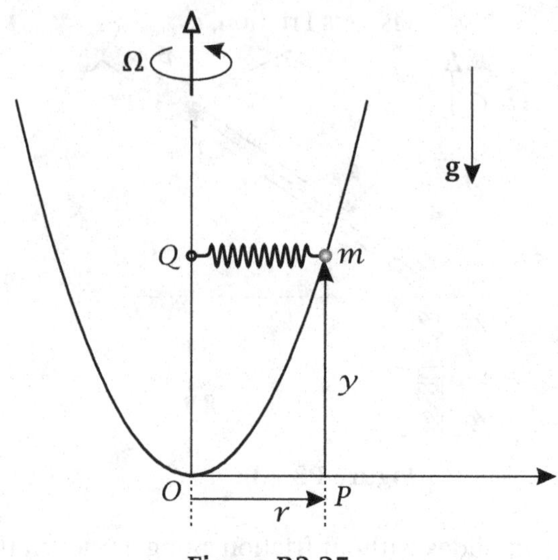

Figure P3-25

vertical direction and that gravity acts vertically downward, determine the differential equation of motion for the particle in terms of the variable r.

3–26 A ball bearing of mass m is released from rest at the top of a frictionless circular track of radius R as shown in Fig. P3-26. After being released, the ball bearing slides without friction along the track and along a friction horizontal surface when it encounters a frictionless vertical hoop of radius r. Knowing that gravity acts downward, determine the minimum value of R such that the ball reaches the top of the hoop with zero velocity relative to the ground.

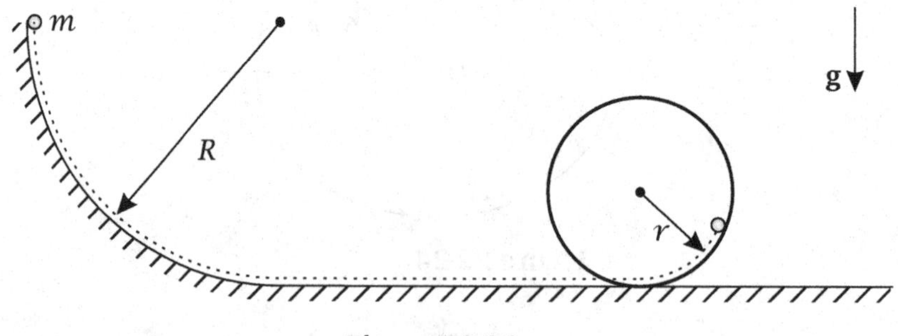

Figure P3-26

3–27 A particle of mass m slides without friction along the surface of a semicircular wedge as shown in Fig. P3-27. The wedge translates horizontally with a known displacement $x(t)$. Knowing that the radius of the semi-circle along which the particle slides is R, that θ describes the position of the particle relative to the vertically downward direction, and that gravity acts vertically downward, determine the differential equation for the particle in terms of the angle θ. Assume in your answers that the particle is constrained to stay on the semi-circular surface of the wedge.

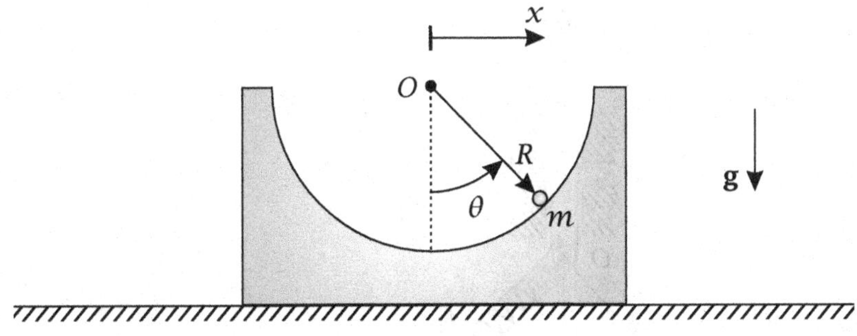

Figure P3-27

3–28 A particle of mass m slides without friction inside a straight slot cut out of a rigid massless disk as shown in Fig. P3-28. The disk rotates in the vertical plane with constant angular velocity $\mathbf{\Omega}$ (where $\Omega = \|\mathbf{\Omega}\|$) about an axis through point O, where O lies along the circumference of the disk. Furthermore, the slot is cut orthogonal to the diameter of the disk that contains point O. Attached to the particle is a linear spring with spring constant K and unstretched length ℓ_0. Knowing that gravity acts downward and that the variable x describes the position of the particle relative to the midpoint of the slot, determine the differential equation of motion for the particle in terms of the variable x.

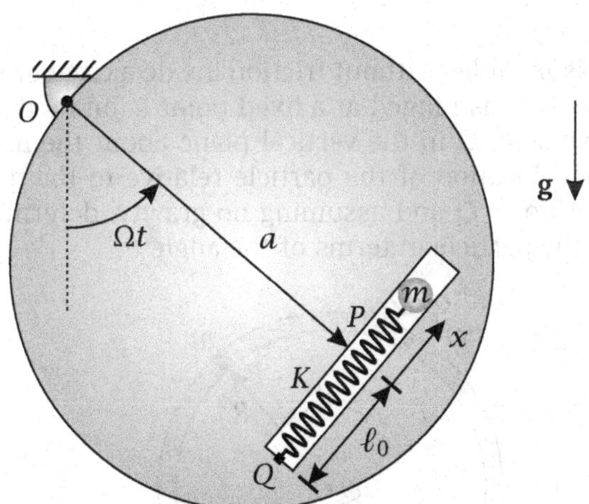

Figure P3-28

3–29 A collar of mass m slides without friction along a rigid massless arm. The arm is hinged at one of its ends at point O where point O lies on a vertical shaft that rotates with constant angular velocity $\mathbf{\Omega}$ (where $\Omega = \|\mathbf{\Omega}\|$) about the vertical direction as shown in Fig. P3-29. Furthermore, the collar is attached to a linear spring with spring constant K and unstretched length r_0 while the spring is attached to the hinge at point O. Knowing that the arm is free to rotate in the plane of the vertical and horizontal shafts, that θ is the angle between the arm and the horizontal shaft, and that gravity acts downward, determine a system of two differential equations of motion that describes the motion of the collar in terms of the variables r and θ.

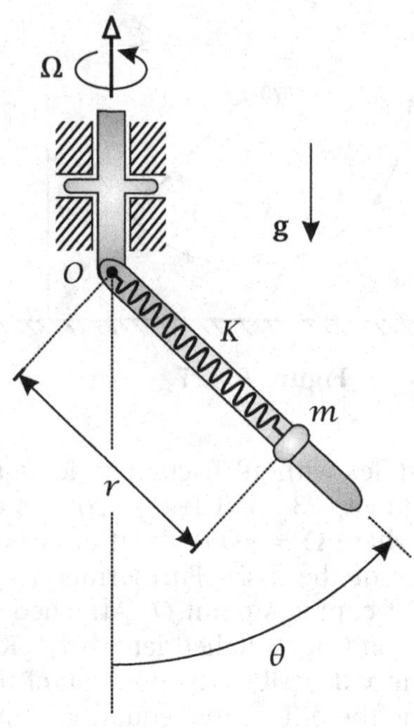

Figure P3-29

3–30 A particle of mass m slides without friction inside a circular tube of radius R as shown in Fig. P3-30. The tube is hinged at a fixed point O on its diameter and rotates with constant angular velocity Ω in the vertical plane about the hinge. Knowing that the angle θ describes the location of the particle relative to the direction from O to the center of the tube at point Q and assuming no gravity, determine the differential equation of motion for the particle in terms of the angle θ.

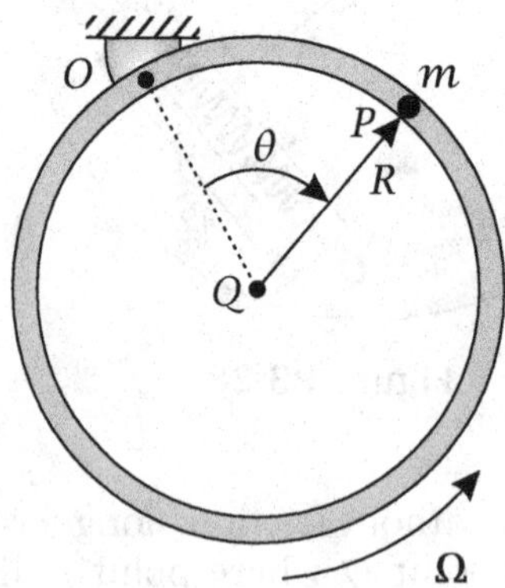

Figure P3-30

Chapter 4

Kinetics of a System of Particles

The most incomprehensible thing about the world is that it is at all comprehensible.

- Albert Einstein (1879–1955)
German and American Physicist

In Chapter 3 we discussed the important principles and methods used in the formulation, solution, and analysis of the motion of a single particle. In this chapter we extend the results of particle kinetics to systems consisting of two or more particles.

The first topic covered in this chapter is the center of mass of a system of particles. Using the definition of the center of mass, the linear momentum of a system of particles is defined. Then, using the definition of linear momentum, the velocity and acceleration of the center of mass of the system are defined.

The second topic covered in this chapter is the angular momentum of a system of particles. In particular, expressions for the angular momentum are derived relative to an arbitrary point, an inertially fixed point, and the center of mass of the system. Then, relationships between these three different forms of angular momentum are derived.

The third and fourth topics covered in this chapter are Newton's 2nd law and the rate of change of angular momentum for a system of particles. In particular, it is shown that the center of mass of the system satisfies Newton's 2nd law. Furthermore, the key results relating the rate of change of angular momentum for a system of particles to moment applied to the system are derived.

The fifth topic covered in this chapter is impulse and momentum for a system of particles. First, using Newton's 2nd law for a system of particles, the principle of linear impulse and linear momentum for a system of particles is derived. Then, using the results from the rate of change of the angular momentum for a system of particles, the principle of angular impulse and angular momentum for a system of particles is derived.

The sixth topic covered in this chapter is work and energy for a system of particles. First, the kinetic energy for a system of particles is defined. Then, an alternate expression for the kinetic energy, called *Koenig's decomposition*, is derived. Using the kinetic energy, the work-energy theorem for a system of particles is derived. Then, using the definition of a conservative force as defined in Chapter 3, an alternate form of the work-energy theorem for a system of particles is derived.

The seventh and final topic in this chapter is the collision of particles. In particular, a simplified model for the collision between two particles is derived in terms of an ad hoc parameter called the coefficient of restitution. Using the coefficient of restitution, the post-impact velocities of two colliding particles are derived in terms of the pre-impact velocities.

4.1 Center of Mass and Linear Momentum of a System of Particles

4.1.1 Center of Mass of a System of Particles

Consider a system that consists of n particles $P_1, \ldots, P_n$ of mass $m_1, \ldots, m_n$, respectively, as shown in Fig. 4–1. Furthermore, let $\mathcal{N}$ be an inertial reference frame and let $\mathbf{r}_i$ $(i = 1, \ldots, n)$ be the position of particle i in reference frame $\mathcal{N}$. Then the *position of the center of mass* of the system is defined as

$$\bar{\mathbf{r}} = \frac{\sum_{i=1}^{n} m_i \mathbf{r}_i}{\sum_{i=1}^{n} m_i} = \frac{\sum_{i=1}^{n} m_i \mathbf{r}_i}{m} \tag{4–1}$$

where the quantity

$$m = \sum_{i=1}^{n} m_i \tag{4–2}$$

is the *total mass* of the system. It is noted that m is a constant for a system of particles.

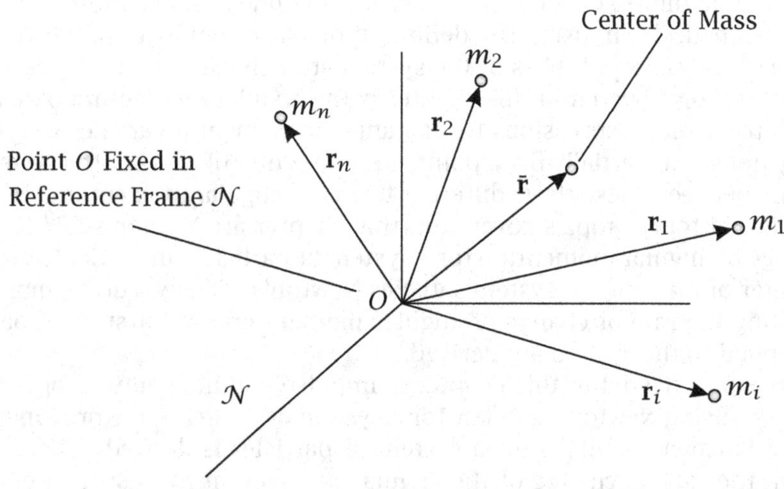

Figure 4–1 A system of n particles moving in an inertial reference frame $\mathcal{N}$.

4.1.2 Linear Momentum of a System of Particles

The linear momentum of a system of particles is defined as

$$
{}^{\mathcal{N}}\mathbf{G} = \sum_{i=1}^{n} m_i {}^{\mathcal{N}}\mathbf{v}_i \tag{4-3}
$$

where ${}^{\mathcal{N}}\mathbf{v}_i$ is the velocity of particle i $(i = 1, \ldots, n)$ in the inertial reference frame $\mathcal{N}$. Now we note that

$$
{}^{\mathcal{N}}\mathbf{v}_i = \frac{{}^{\mathcal{N}}d\mathbf{r}_i}{dt} \quad (i = 1, \ldots, n) \tag{4-4}
$$

Substituting the result of Eq. (4–4) into Eq. (4–3), we obtain

$$
{}^{\mathcal{N}}\mathbf{G} = \sum_{i=1}^{n} m_i \frac{{}^{\mathcal{N}}d\mathbf{r}_i}{dt} \tag{4-5}
$$

Now, since the summation is being computed over the system of particles, the rate of change is independent of the summation. Therefore, the order of summation and integration in Eq. (4–5) can be interchanged to give

$$
{}^{\mathcal{N}}\mathbf{G} = \frac{{}^{\mathcal{N}}d}{dt} \sum_{i=1}^{n} m_i \mathbf{r}_i \tag{4-6}
$$

Then, using Eq. (4–1), we have

$$
\sum_{i=1}^{n} m_i \mathbf{r}_i = m\bar{\mathbf{r}} \tag{4-7}
$$

Consequently, Eq. (4–6) can be rewritten as

$$
{}^{\mathcal{N}}\mathbf{G} = \frac{{}^{\mathcal{N}}d}{dt}(m\bar{\mathbf{r}}) = m\frac{{}^{\mathcal{N}}d\bar{\mathbf{r}}}{dt} = m{}^{\mathcal{N}}\bar{\mathbf{v}} \tag{4-8}
$$

The quantity ${}^{\mathcal{N}}\bar{\mathbf{v}}$ is called the *velocity of the center of mass* of the system in reference frame $\mathcal{N}$. Using Eqs. (4–3) and (4–8), the velocity of the center of mass of a system of particles can be written as

$$
{}^{\mathcal{N}}\bar{\mathbf{v}} = \frac{\sum_{i=1}^{n} m_i {}^{\mathcal{N}}\mathbf{v}_i}{m} \tag{4-9}
$$

Next, let

$$
{}^{\mathcal{N}}\mathbf{a}_i = \frac{{}^{\mathcal{N}}d}{dt}\left({}^{\mathcal{N}}\mathbf{v}_i\right) \quad (i = 1, \ldots, n) \tag{4-10}
$$

be the acceleration of particle i in the inertial reference frame $\mathcal{N}$. Then the *acceleration of the center of mass* of the system in reference frame $\mathcal{N}$, denoted ${}^{\mathcal{N}}\bar{\mathbf{a}}$, is defined as

$$
{}^{\mathcal{N}}\bar{\mathbf{a}} = \frac{\sum_{i=1}^{n} m_i {}^{\mathcal{N}}\mathbf{a}_i}{m} \tag{4-11}
$$

Using Eq. (4–10), the acceleration of the center of mass can be written as

$$
{}^{\mathcal{N}}\bar{\mathbf{a}} = \frac{1}{m}\sum_{i=1}^{n} m_i \frac{{}^{\mathcal{N}}d}{dt}\left({}^{\mathcal{N}}\mathbf{v}_i\right) \tag{4-12}
$$

Observing again that the rate of change is independent of the summation, Eq. (4–12) can be written as

$$ {}^{\mathcal{N}}\bar{\mathbf{a}} = \frac{1}{m}\frac{{}^{\mathcal{N}}d}{dt}\sum_{i=1}^{n} m_i {}^{\mathcal{N}}\mathbf{v}_i \tag{4–13} $$

Finally, using the definition of the velocity of the center of mass of the system as given in Eq. (4–9), the acceleration of the center of mass of the system in reference frame $\mathcal{N}$ can be written as

$$ {}^{\mathcal{N}}\bar{\mathbf{a}} = \frac{{}^{\mathcal{N}}d}{dt}\left({}^{\mathcal{N}}\bar{\mathbf{v}}\right) \tag{4–14} $$

4.2 Angular Momentum of a System of Particles

The angular momentum of the system of n particles relative to an arbitrary point Q in an inertial reference frame $\mathcal{N}$ is defined as[1]

$$ {}^{\mathcal{N}}\mathbf{H}_Q = \sum_{i=1}^{n}(\mathbf{r}_i - \mathbf{r}_Q)\times m_i({}^{\mathcal{N}}\mathbf{v}_i - {}^{\mathcal{N}}\mathbf{v}_Q) \tag{4–15} $$

Furthermore, the angular momentum relative to a point O fixed in the inertial reference frame $\mathcal{N}$ is defined as

$$ {}^{\mathcal{N}}\mathbf{H}_O = \sum_{i=1}^{n}(\mathbf{r}_i - \mathbf{r}_O)\times m_i {}^{\mathcal{N}}\mathbf{v}_i \tag{4–16} $$

where we note that, because point O is fixed in $\mathcal{N}$, the quantity ${}^{\mathcal{N}}\mathbf{v}_O = \mathbf{0}$. Finally, the angular momentum relative to the center of mass of the system is defined as

$$ {}^{\mathcal{N}}\bar{\mathbf{H}} = \sum_{i=1}^{n}(\mathbf{r}_i - \bar{\mathbf{r}})\times m_i({}^{\mathcal{N}}\mathbf{v}_i - {}^{\mathcal{N}}\bar{\mathbf{v}}) \tag{4–17} $$

A relationship between ${}^{\mathcal{N}}\mathbf{H}_Q$ and ${}^{\mathcal{N}}\mathbf{H}_O$ can be obtained as follows. Subtracting Eq. (4–16) from (4–15), we have

$$ {}^{\mathcal{N}}\mathbf{H}_Q - {}^{\mathcal{N}}\mathbf{H}_O = \sum_{i=1}^{n}(\mathbf{r}_i - \mathbf{r}_Q)\times m_i({}^{\mathcal{N}}\mathbf{v}_i - {}^{\mathcal{N}}\mathbf{v}_Q) - \sum_{i=1}^{n}(\mathbf{r}_i - \mathbf{r}_O)\times m_i {}^{\mathcal{N}}\mathbf{v}_i \tag{4–18} $$

Expanding Eq. (4–18), we obtain

$$ \begin{aligned} {}^{\mathcal{N}}\mathbf{H}_Q - {}^{\mathcal{N}}\mathbf{H}_O = & \sum_{i=1}^{n}\mathbf{r}_i\times m_i {}^{\mathcal{N}}\mathbf{v}_i - \sum_{i=1}^{n}\mathbf{r}_i\times m_i {}^{\mathcal{N}}\mathbf{v}_Q - \sum_{i=1}^{n}\mathbf{r}_Q\times m_i {}^{\mathcal{N}}\mathbf{v}_i \\ & + \sum_{i=1}^{n}\mathbf{r}_Q\times m_i {}^{\mathcal{N}}\mathbf{v}_Q - \sum_{i=1}^{n}\mathbf{r}_i\times m_i {}^{\mathcal{N}}\mathbf{v}_i + \sum_{i=1}^{n}\mathbf{r}_O\times m_i {}^{\mathcal{N}}\mathbf{v}_i \end{aligned} \tag{4–19} $$

[1]It is noted again that the angular momentum is defined using the *relative velocity* ${}^{\mathcal{N}}\mathbf{v}_i - {}^{\mathcal{N}}\mathbf{v}_Q$ and is consistent with the definition used in Greenwood (1977; 1988).

Observing that the first and fifth summations in Eq. (4–19) cancel, we obtain

$$\begin{aligned} {}^{\mathcal{N}}\mathbf{H}_Q - {}^{\mathcal{N}}\mathbf{H}_O = &- \sum_{i=1}^{n} \mathbf{r}_i \times m_i {}^{\mathcal{N}}\mathbf{v}_Q - \sum_{i=1}^{n} \mathbf{r}_Q \times m_i {}^{\mathcal{N}}\mathbf{v}_i \\ &+ \sum_{i=1}^{n} \mathbf{r}_Q \times m_i {}^{\mathcal{N}}\mathbf{v}_Q + \sum_{i=1}^{n} \mathbf{r}_O \times m_i {}^{\mathcal{N}}\mathbf{v}_i \end{aligned} \tag{4–20}$$

Next, because $\mathbf{r}_Q$, $\mathbf{r}_O$, and ${}^{\mathcal{N}}\mathbf{v}_Q$ are independent of the summation, we have

$$\begin{aligned} \textstyle\sum_{i=1}^{n} \mathbf{r}_i \times m_i {}^{\mathcal{N}}\mathbf{v}_Q &= \bar{\mathbf{r}} \times m {}^{\mathcal{N}}\mathbf{v}_Q \\ \textstyle\sum_{i=1}^{n} \mathbf{r}_Q \times m_i {}^{\mathcal{N}}\mathbf{v}_i &= \mathbf{r}_Q \times m {}^{\mathcal{N}}\bar{\mathbf{v}} \\ \textstyle\sum_{i=1}^{n} \mathbf{r}_Q \times m_i {}^{\mathcal{N}}\mathbf{v}_Q &= \mathbf{r}_Q \times m {}^{\mathcal{N}}\mathbf{v}_Q \\ \textstyle\sum_{i=1}^{n} \mathbf{r}_O \times m_i {}^{\mathcal{N}}\mathbf{v}_i &= \mathbf{r}_O \times m {}^{\mathcal{N}}\bar{\mathbf{v}} \end{aligned} \tag{4–21}$$

Substituting the results of Eq. (4–21) into (4–20), we obtain

$${}^{\mathcal{N}}\mathbf{H}_Q - {}^{\mathcal{N}}\mathbf{H}_O = -\bar{\mathbf{r}} \times m {}^{\mathcal{N}}\mathbf{v}_Q - \mathbf{r}_Q \times m {}^{\mathcal{N}}\bar{\mathbf{v}} + \mathbf{r}_Q \times m {}^{\mathcal{N}}\mathbf{v}_Q + \mathbf{r}_O \times m {}^{\mathcal{N}}\bar{\mathbf{v}} \tag{4–22}$$

Rewriting Eq. (4–22), we obtain

$$\boxed{{}^{\mathcal{N}}\mathbf{H}_Q = {}^{\mathcal{N}}\mathbf{H}_O - (\bar{\mathbf{r}} - \mathbf{r}_Q) \times m {}^{\mathcal{N}}\mathbf{v}_Q - (\mathbf{r}_Q - \mathbf{r}_O) \times m {}^{\mathcal{N}}\bar{\mathbf{v}}} \tag{4–23}$$

Equation (4–23) relates the angular momentum of a system of particles relative to a point O fixed in an inertial reference frame to the angular momentum relative to an arbitrary reference point.

Next, a relationship between ${}^{\mathcal{N}}\mathbf{H}_O$ and ${}^{\mathcal{N}}\bar{\mathbf{H}}$ can be obtained using the result of Eq. (4–23). First, setting the reference point in Eq. (4–23) to the center of mass, we have

$${}^{\mathcal{N}}\bar{\mathbf{H}} = {}^{\mathcal{N}}\mathbf{H}_O - (\bar{\mathbf{r}} - \bar{\mathbf{r}}) \times m {}^{\mathcal{N}}\bar{\mathbf{v}} - (\bar{\mathbf{r}} - \mathbf{r}_O) \times m {}^{\mathcal{N}}\bar{\mathbf{v}} \tag{4–24}$$

Observing that the second term in Eq. (4–24) is zero, we have

$$\boxed{{}^{\mathcal{N}}\bar{\mathbf{H}} = {}^{\mathcal{N}}\mathbf{H}_O - (\bar{\mathbf{r}} - \mathbf{r}_O) \times m {}^{\mathcal{N}}\bar{\mathbf{v}}} \tag{4–25}$$

Equation (4–25) relates the angular momentum of a system of particles relative to a point O fixed in an inertial reference frame to the angular momentum of the system relative to the center of mass of the system.

Finally, a relationship between ${}^{\mathcal{N}}\mathbf{H}_Q$ and ${}^{\mathcal{N}}\bar{\mathbf{H}}$ can be obtained by using the results of Eqs. (4–23) and (4–25). First, solving Eq. (4–23) for ${}^{\mathcal{N}}\mathbf{H}_O$ gives

$${}^{\mathcal{N}}\mathbf{H}_O = {}^{\mathcal{N}}\mathbf{H}_Q + (\bar{\mathbf{r}} - \mathbf{r}_Q) \times m {}^{\mathcal{N}}\mathbf{v}_Q + (\mathbf{r}_Q - \mathbf{r}_O) \times m {}^{\mathcal{N}}\bar{\mathbf{v}} \tag{4–26}$$

Then, substituting the result of Eq. (4–26) into (4–25), we obtain

$${}^{\mathcal{N}}\bar{\mathbf{H}} = {}^{\mathcal{N}}\mathbf{H}_Q + (\bar{\mathbf{r}} - \mathbf{r}_Q) \times m {}^{\mathcal{N}}\mathbf{v}_Q + (\mathbf{r}_Q - \mathbf{r}_O) \times m {}^{\mathcal{N}}\bar{\mathbf{v}} - (\bar{\mathbf{r}} - \mathbf{r}_O) \times m {}^{\mathcal{N}}\bar{\mathbf{v}} \tag{4–27}$$

Rearranging and simplifying Eq. (4-27), we obtain

$$\boxed{{}^{\mathcal{N}}\mathbf{H}_Q = {}^{\mathcal{N}}\bar{\mathbf{H}} + (\mathbf{r}_Q - \bar{\mathbf{r}}) \times m({}^{\mathcal{N}}\mathbf{v}_Q - {}^{\mathcal{N}}\bar{\mathbf{v}})} \tag{4-28}$$

Equation (4-28) relates the angular momentum of a system of particles relative to an arbitrary point Q to the angular momentum relative to the center of mass of the system.

4.3 Newton's 2nd Law for a System of Particles

Suppose now that we let $\mathbf{R}_i$ be the resultant force acting on particle i in a system of n particles. The resultant force acting on particle i can be decomposed into two parts: (1) a force that is *external* to the system and (2) a force due to all of the other particles in the system. For convenience, let

$$\left.\begin{array}{lcl} \mathbf{F}_i & = & \text{Force exerted on particle } i \text{ external to system} \\ \mathbf{f}_{ij} & = & \text{Force exerted on particle } i \text{ by particle } j \;\; (j = 1,\ldots,n) \end{array}\right\} \;\; (i = 1,\ldots,n)$$

Now from Newton's 3rd law we know that the force exerted by particle i on particle j is equal and opposite the force exerted by particle j on particle i. Consequently,

$$\mathbf{f}_{ji} = -\mathbf{f}_{ij} \quad (i,j = 1,\ldots,n) \tag{4-29}$$

Furthermore, because a particle cannot exert a force on itself, we have

$$\mathbf{f}_{ii} = 0 \quad (i = 1,\ldots,n) \tag{4-30}$$

In addition, it is assumed that the force exerted by particle j on particle i lies along the line between particle i and particle j.[2] Consequently, the resultant force acting on particle i, denoted $\mathbf{R}_i$, is given as

$$\mathbf{R}_i = \mathbf{F}_i + \sum_{j=1}^{n} \mathbf{f}_{ij} \; (i = 1,2,\ldots,n) \tag{4-31}$$

Then, applying Newton's 2nd law to particle i, we have

$$\mathbf{R}_i = m_i {}^{\mathcal{N}}\mathbf{a}_i \quad (i = 1,2,\ldots,n) \tag{4-32}$$

Substituting Eq. (4-31) into (4-32), we obtain

$$\mathbf{F}_i + \sum_{j=1}^{n} \mathbf{f}_{ij} = m_i {}^{\mathcal{N}}\mathbf{a}_i \quad (i = 1,2,\ldots,n) \tag{4-33}$$

Next, summing forces over all particles in the system, we obtain

$$\sum_{i=1}^{n}\left[\mathbf{F}_i + \sum_{j=1}^{n} \mathbf{f}_{ij}\right] = \sum_{i=1}^{n} \mathbf{F}_i + \sum_{i=1}^{n}\sum_{j=1}^{n} \mathbf{f}_{ij} = \sum_{i=1}^{n} m_i {}^{\mathcal{N}}\mathbf{a}_i \tag{4-34}$$

[2]The assumption that $\mathbf{f}_{ij}$ lies along the line between particles i and j is commonly referred to as the *strong form of Newton's* 3rd *law* (Thornton and Marion, 2004).

Using Eq. (4–29), we have

$$\sum_{i=1}^{n}\sum_{j=1}^{n}\mathbf{f}_{ij}=\mathbf{0} \tag{4-35}$$

Equation (4–34) then reduces to

$$\sum_{i=1}^{n}\mathbf{F}_i=\sum_{i=1}^{n}m_i{}^{\mathcal{N}}\mathbf{a}_i \tag{4-36}$$

Suppose now that we define the total *external* force acting on the system as

$$\mathbf{F}=\sum_{i=1}^{n}\mathbf{F}_i \tag{4-37}$$

Then, using the definition of the acceleration of the center of mass from Eq. (4–11), the expression in Eq. (4–36) becomes

$$\boxed{\mathbf{F}=m^{\mathcal{N}}\bar{\mathbf{a}}} \tag{4-38}$$

Equation (4–38) states that the resultant external force acting on a system of particles is equal to the product of the mass of the system and the acceleration of the center of mass of the system.

Example 4–1

A wedge of mass m_1 and inclination angle β slides along a fixed horizontal surface while a block of mass m_2 slides along the angled surface of the wedge as shown in Fig. 4-2. Knowing that x_1 describes the displacement of the wedge relative to the surface, x_2 describes the displacement of the block relative to the wedge, all surfaces are frictionless and gravity acts downward, determine a system of two differential equations that describes the motion of the wedge and the block.

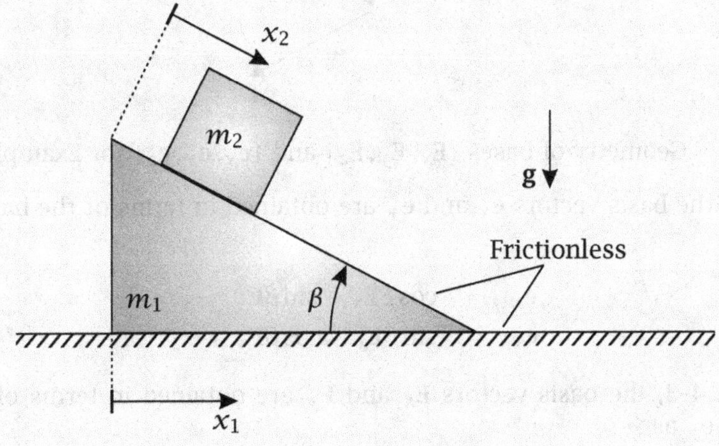

Figure 4–2 Block sliding on wedge.

Solution to Example 4–1

Kinematics

Let $\mathcal{F}$ be a fixed reference frame. Then, choose the following coordinate system fixed in reference frame $\mathcal{F}$:

$$
\begin{array}{ccc}
 & \text{Origin at wedge at } t = 0 & \\
\mathbf{E}_x & = & \text{To the right} \\
\mathbf{E}_z & = & \text{Out of page} \\
\mathbf{E}_y & = & \mathbf{E}_x \times \mathbf{E}_x
\end{array}
$$

Next, let $\mathcal{A}$ be a reference frame fixed to the wedge. Then, choose the following coordinate system fixed in reference frame $\mathcal{A}$:

$$
\begin{array}{ccc}
 & \text{Origin at block at } t = 0 & \\
\mathbf{e}_x & = & \text{Down incline of wedge} \\
\mathbf{e}_z & = & \text{Out of page} \\
\mathbf{e}_y & = & \mathbf{e}_z \times \mathbf{e}_x
\end{array}
$$

The geometry of the bases $\{\mathbf{E}_x, \mathbf{E}_y, \mathbf{E}_z\}$ and $\{\mathbf{e}_x, \mathbf{e}_y, \mathbf{e}_z\}$ is shown in Fig. 4-3.

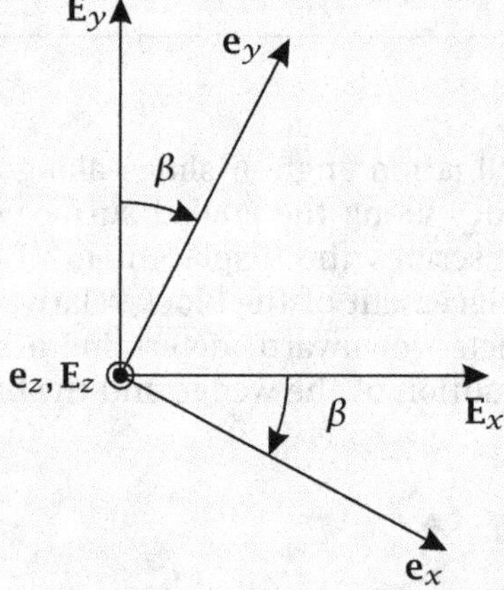

Figure 4-3 Geometry of bases $\{\mathbf{E}_x, \mathbf{E}_y, \mathbf{E}_z\}$ and $\{\mathbf{e}_x, \mathbf{e}_y, \mathbf{e}_z\}$ for Example 4-1.

Using Fig. 4-3, the basis vectors $\mathbf{e}_x$ and $\mathbf{e}_y$ are obtained in terms of the basis vectors $\mathbf{E}_x$ and $\mathbf{E}_y$ as

$$\mathbf{e}_x = \cos\beta\mathbf{E}_x - \sin\beta\mathbf{E}_y \tag{4-39}$$
$$\mathbf{e}_y = \sin\beta\mathbf{E}_x + \cos\beta\mathbf{E}_y \tag{4-40}$$

Also using Fig. 4-3, the basis vectors $\mathbf{E}_x$ and $\mathbf{E}_y$ are obtained in terms of the basis vectors $\mathbf{e}_x$ and $\mathbf{e}_y$ as

$$\mathbf{E}_x = \cos\beta\mathbf{e}_x + \sin\beta\mathbf{e}_y \tag{4-41}$$
$$\mathbf{E}_y = -\sin\beta\mathbf{e}_x + \cos\beta\mathbf{e}_y \tag{4-42}$$

Kinematics of Wedge

In terms of the basis $\{\mathbf{e}_x, \mathbf{e}_y, \mathbf{E}_z\}$, the position of the wedge is given as

$$\mathbf{r}_1 = x_1\mathbf{E}_x \tag{4-43}$$

Then, the velocity of the wedge in reference frame $\mathcal{F}$ is given as

$${}^{\mathcal{F}}\mathbf{v}_1 = \frac{{}^{\mathcal{F}}d\mathbf{r}_1}{dt} = \dot{x}_1\mathbf{E}_x \tag{4-44}$$

Finally, the acceleration of the wedge in reference frame $\mathcal{F}$ is obtained as

$${}^{\mathcal{F}}\mathbf{a}_1 = \frac{{}^{\mathcal{F}}d}{dt}\left({}^{\mathcal{F}}\mathbf{v}_1\right) = \ddot{x}_1\mathbf{E}_x \tag{4-45}$$

Kinematics of Block

The position of the block is given as

$$\mathbf{r}_2 = \mathbf{r}_1 + \mathbf{r}_{2/1} \tag{4-46}$$

where $\mathbf{r}_1$ and $\mathbf{r}_{2/1}$ are the position of the wedge and the position of the block relative to the wedge, respectively. Now, because the block slides along the wedge, we have

$$\mathbf{r}_{2/1} = x_2\mathbf{e}_x \tag{4-47}$$

Substituting the expression for $\mathbf{r}_1$ from Eq. (4-43) and the expression for $\mathbf{r}_{2/1}$ from Eq. (4-47) into Eq. (4-46), we obtain the position of the block as

$$\mathbf{r}_2 = x_1\mathbf{E}_x + x_2\mathbf{e}_x \tag{4-48}$$

Differentiating the position of the block as given in Eq. (4-48) in reference frame $\mathcal{F}$ (observing that the bases $\{\mathbf{E}_x, \mathbf{E}_y, \mathbf{E}_z\}$ and $\{\mathbf{e}_x, \mathbf{e}_y, \mathbf{e}_z\}$ do not rotate), we obtain the velocity of the block in reference frame $\mathcal{F}$ as

$${}^{\mathcal{F}}\mathbf{v}_2 = \frac{{}^{\mathcal{F}}d\mathbf{r}_2}{dt} = \dot{x}_1\mathbf{E}_x + \dot{x}_2\mathbf{e}_x \tag{4-49}$$

Furthermore, differentiating ${}^{\mathcal{F}}\mathbf{v}_2$ as given in Eq. (4-49) in reference frame $\mathcal{F}$, we obtain the acceleration of the block in reference frame $\mathcal{F}$ as

$${}^{\mathcal{F}}\mathbf{a}_2 = \frac{{}^{\mathcal{F}}d}{dt}\left({}^{\mathcal{F}}\mathbf{v}_2\right) = \ddot{x}_1\mathbf{E}_x + \ddot{x}_2\mathbf{e}_x \tag{4-50}$$

Kinematics of Center of Mass of System

The acceleration of the center of mass of the wedge-block system is given as

$${}^{\mathcal{F}}\bar{\mathbf{a}} = \frac{m_1{}^{\mathcal{F}}\mathbf{a}_1 + m_2{}^{\mathcal{F}}\mathbf{a}_2}{m_1 + m_2} \tag{4-51}$$

Substituting the expressions for ${}^{\mathcal{F}}\mathbf{a}_1$ and ${}^{\mathcal{F}}\mathbf{a}_2$ from Eqs. (4-45) and (4-50), respectively, the acceleration of the center of mass in reference frame $\mathcal{F}$ is obtained as

$${}^{\mathcal{F}}\bar{\mathbf{a}} = \frac{m_1\ddot{x}_1\mathbf{E}_x + m_2(\ddot{x}_1\mathbf{E}_x + \ddot{x}_2\mathbf{e}_x)}{m_1 + m_2} = \ddot{x}_1\mathbf{E}_x + \frac{m_2}{m_1 + m_2}\ddot{x}_2\mathbf{e}_x \tag{4-52}$$

Kinetics

For this problem it is convenient to apply Newton's 2nd law to the following systems: (1) the block and (2) the block and wedge. We now perform each of these steps.

Kinetics of Block

Applying Newton's 2nd law to the block, we have

$$\mathbf{F}_2 = m_2 {}^{\mathcal{F}}\mathbf{a}_2 \tag{4-53}$$

where $\mathbf{F}_2$ is the resultant force acting on the block. The free body diagram of the block is shown in Fig. 4-4.

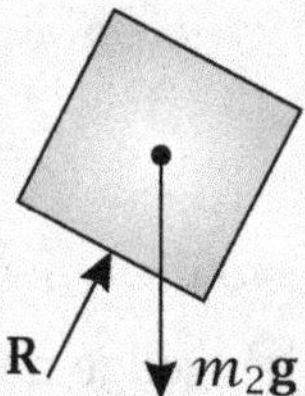

Figure 4-4 Free body diagram of block for Example 4-1.

Using Fig. 4-4, it is seen that the following forces act on the block:

$$\begin{aligned} \mathbf{R} &= \text{Reaction force of wedge on block} \\ m_2\mathbf{g} &= \text{Force of gravity} \end{aligned}$$

Consequently, the resultant force acting on the block is given as

$$\mathbf{F}_2 = \mathbf{R} + m_2\mathbf{g} \tag{4-54}$$

Now from the geometry of the problem we have

$$\mathbf{R} = R\mathbf{e}_y \tag{4-55}$$

$$m_2\mathbf{g} = -m_2 g\mathbf{E}_y \tag{4-56}$$

where we note that the force $\mathbf{R}$ acts in the direction orthogonal to the side of the block that is in contact with the wedge. Substituting the results of Eqs. (4-55) and (4-56) into Eq. (4-54), we obtain the resultant force acting on the block as

$$\mathbf{F}_2 = R\mathbf{e}_y - m_2 g\mathbf{E}_y \tag{4-57}$$

Then, using the expression for $\mathbf{E}_y$ in terms of $\mathbf{e}_x$ and $\mathbf{e}_y$ as given in (4-42), the resultant force on the wedge in Eq. (4-57) can be written in terms of the basis $\{\mathbf{e}_x, \mathbf{e}_y, \mathbf{e}_z\}$ as

$$\mathbf{F}_2 = R\mathbf{e}_y - m_2 g(-\sin\beta\mathbf{e}_x + \cos\beta\mathbf{e}_y) = m_2 g\sin\beta\mathbf{e}_x + (R - m_2 g\cos\beta)\mathbf{e}_y \tag{4-58}$$

Setting $\mathbf{F}_2$ from Eq. (4-58) equal to $m_2 {}^{\mathcal{F}}\mathbf{a}_2$ using ${}^{\mathcal{F}}\mathbf{a}_2$ from Eq. (4-50), we have

$$m_2 g\sin\beta\mathbf{e}_x + (R - m_2 g\cos\beta)\mathbf{e}_y = m_2(\ddot{x}_1\mathbf{E}_x + \ddot{x}_2\mathbf{e}_x) \tag{4-59}$$

Next, substituting the expression for $\mathbf{E}_x$ in terms of $\mathbf{e}_x$ and $\mathbf{e}_y$ from Eq. (4–41) into (4–59), we obtain

$$m_2 g \sin\beta \mathbf{e}_x + (R - m_2 g\cos\beta)\mathbf{e}_y = m_2\left[\ddot{x}_1(\cos\beta\mathbf{e}_x + \sin\beta\mathbf{e}_y) + \ddot{x}_2\mathbf{e}_x\right] \quad (4\text{–}60)$$

Rearranging the right-hand side of Eq. (4–60), we obtain

$$-m_2 g \sin\beta \mathbf{e}_x + (R - m_2 g\cos\beta)\mathbf{e}_y = (m_2\ddot{x}_1\cos\beta + m_2\ddot{x}_2)\mathbf{e}_x + m_2\ddot{x}_1\sin\beta\mathbf{e}_y \quad (4\text{–}61)$$

Equating components in Eq. (4–61), we obtain the following two scalar equations:

$$\begin{aligned} m_2 g\sin\beta &= m_2\ddot{x}_1\cos\beta + m_2\ddot{x}_2 && (4\text{–}62)\\ R - m_2 g\cos\beta &= m_2\ddot{x}_1\sin\beta && (4\text{–}63)\end{aligned}$$

Kinetics of System Consisting of Wedge and Block

Applying Newton's 2nd to the system consisting of the wedge and block, we have

$$\mathbf{F} = m\,{}^{\mathcal{F}}\bar{\mathbf{a}} \quad (4\text{–}64)$$

where $\mathbf{F}$ is the resultant force acting on the wedge-block system, $m = m_1 + m_2$ is the mass of the system, and ${}^{\mathcal{F}}\bar{\mathbf{a}}$ is the acceleration of the center of mass of the system. The free body diagram of the wedge-block system is shown in Fig. 4-5.

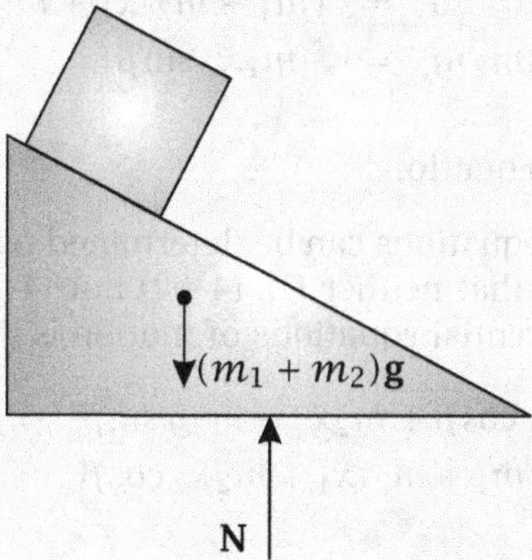

Figure 4–5 Free body diagram of wedge-block system for Example 4–1.

Using Fig. 4-5, it is seen that the following forces act on the wedge-block system:

$$\begin{aligned} \mathbf{N} &= \text{Reaction force of ground}\\ (m_1 + m_2)\mathbf{g} &= \text{Force of gravity}\end{aligned}$$

Now, from the geometry of the problem we have

$$\begin{aligned} \mathbf{N} &= N\mathbf{E}_y && (4\text{–}65)\\ (m_1 + m_2)g &= -(m_1 + m_2)g\mathbf{E}_y && (4\text{–}66)\end{aligned}$$

where we note that the force of gravity acts at the *center of mass* of the system. Using Eqs. (4-65) and (4-66), the resultant force acting on the wedge-block system is then given as

$$\mathbf{F} = \mathbf{N} + (m_1 + m_2)\mathbf{g} \tag{4-67}$$

Then, using the expression for $\mathbf{N}$ from Eq. (4-65) and the expression for $(m_1 + m_2)\mathbf{g}$ from Eq. (4-66), we obtain $\mathbf{F}$ as

$$\mathbf{F} = N\mathbf{E}_y - (m_1 + m_2)g\mathbf{E}_y = [N - (m_1 + m_2)g]\,\mathbf{E}_y \tag{4-68}$$

Next, substituting the expression for $\mathbf{F}$ and the expression for ${}^{\mathcal{F}}\bar{\mathbf{a}}$ from Eq. (4-52) into (4-64), we obtain

$$\begin{aligned} [N - (m_1 + m_2)g]\,\mathbf{E}_y &= (m_1 + m_2)\left[\ddot{x}_1\mathbf{E}_x + \frac{m_2}{m_1 + m_2}\ddot{x}_2\mathbf{e}_x\right] \\ &= (m_1 + m_2)\ddot{x}_1\mathbf{E}_x + m_2\ddot{x}_2\mathbf{e}_x \end{aligned} \tag{4-69}$$

Then, substituting the expression for $\mathbf{e}_x$ in terms of $\mathbf{E}_x$ and $\mathbf{E}_y$ from Eq. (4-39), we can rewrite Eq. (4-69) as

$$\begin{aligned} [N - (m_1 + m_2)g]\,\mathbf{E}_y &= (m_1 + m_2)\ddot{x}_1\mathbf{E}_x + m_2\ddot{x}_2(\cos\beta\mathbf{E}_x - \sin\beta\mathbf{E}_y) \\ &= [(m_1 + m_2)\ddot{x}_1 + m_2\ddot{x}_2\cos\beta]\,\mathbf{E}_x - m_2\ddot{x}_2\sin\beta\mathbf{E}_y \end{aligned} \tag{4-70}$$

Equating components in Eq. (4-70) yields the following two scalar equations:

$$0 = (m_1 + m_2)\ddot{x}_1 + m_2\ddot{x}_2\cos\beta \tag{4-71}$$

$$N - (m_1 + m_2)g = -m_2\ddot{x}_2\sin\beta \tag{4-72}$$

System of Two Differential Equations

A system of two differential equations can be determined using the Eqs. (4-62) and (4-71). In particular, observing that neither Eq. (4-62) nor (4-71) has unknown reaction forces, a system of two differential equations of motion is given as

$$m_2\ddot{x}_1\cos\beta + m_2\ddot{x}_2 - m_2 g\sin\beta = 0 \tag{4-73}$$

$$(m_1 + m_2)\ddot{x}_1 + m_2\ddot{x}_2\cos\beta = 0 \tag{4-74}$$

■

4.4 Moment of a System of Forces Acting on a System of Particles

Consider a system of particles $P_1, \ldots, P_n$ of mass $m_1, \ldots, m_n$, respectively. Furthermore, let $\mathbf{F}_1, \ldots, \mathbf{F}_n$ be the resultant external forces acting, respectively, on $P_1, \ldots, P_n$. Finally, let $\mathcal{N}$ be an inertial reference frame and let Q be an arbitrary point. Then the moment due to all external forces relative to point Q is defined as

$$\mathbf{M}_Q = \sum_{i=1}^{n}(\mathbf{r}_i - \mathbf{r}_Q) \times \mathbf{F}_i \tag{4-75}$$

Next, the moment due to all external forces relative to a point O fixed in the inertial reference frame $\mathcal{N}$ is defined as

$$\mathbf{M}_O = \sum_{i=1}^{n} (\mathbf{r}_i - \mathbf{r}_O) \times \mathbf{F}_i \tag{4-76}$$

Finally, the moment due to all external forces relative to the center of mass of the system is defined as

$$\bar{\mathbf{M}} = \sum_{i=1}^{n} (\mathbf{r}_i - \bar{\mathbf{r}}) \times \mathbf{F}_i \tag{4-77}$$

4.5 Rate of Change of Angular Momentum for a System of Particles

Suppose that we compute the rate of change of ${}^{\mathcal{N}}\mathbf{H}_Q$ in Eq. (4-15). We then obtain

$$\begin{aligned}\frac{{}^{\mathcal{N}}d}{dt}\left({}^{\mathcal{N}}\mathbf{H}_Q\right) = {} & \sum_{i=1}^{n} ({}^{\mathcal{N}}\mathbf{v}_i - {}^{\mathcal{N}}\mathbf{v}_Q) \times m_i({}^{\mathcal{N}}\mathbf{v}_i - {}^{\mathcal{N}}\mathbf{v}_Q) \\ & + \sum_{i=1}^{n} (\mathbf{r}_i - \mathbf{r}_Q) \times m_i({}^{\mathcal{N}}\mathbf{a}_i - {}^{\mathcal{N}}\mathbf{a}_Q)\end{aligned} \tag{4-78}$$

We note that

$$\sum_{i=1}^{n} ({}^{\mathcal{N}}\mathbf{v}_i - {}^{\mathcal{N}}\mathbf{v}_Q) \times m_i({}^{\mathcal{N}}\mathbf{v}_i - {}^{\mathcal{N}}\mathbf{v}_Q) = \mathbf{0}$$

Consequently, Eq. (4-78) reduces to

$$\frac{{}^{\mathcal{N}}d}{dt}\left({}^{\mathcal{N}}\mathbf{H}_Q\right) = \sum_{i=1}^{n} (\mathbf{r}_i - \mathbf{r}_Q) \times m_i({}^{\mathcal{N}}\mathbf{a}_i - {}^{\mathcal{N}}\mathbf{a}_Q) \tag{4-79}$$

Equation (4-79) can be rewritten as

$$\frac{{}^{\mathcal{N}}d}{dt}\left({}^{\mathcal{N}}\mathbf{H}_Q\right) = \sum_{i=1}^{n} (\mathbf{r}_i - \mathbf{r}_Q) \times m_i{}^{\mathcal{N}}\mathbf{a}_i - \sum_{i=1}^{n} (\mathbf{r}_i - \mathbf{r}_Q) \times m_i{}^{\mathcal{N}}\mathbf{a}_Q \tag{4-80}$$

Separating the second summation in Eq. (4-80), we obtain

$$\frac{{}^{\mathcal{N}}d}{dt}\left({}^{\mathcal{N}}\mathbf{H}_Q\right) = \sum_{i=1}^{n} (\mathbf{r}_i - \mathbf{r}_Q) \times m_i{}^{\mathcal{N}}\mathbf{a}_i - \sum_{i=1}^{n} \mathbf{r}_i \times m_i{}^{\mathcal{N}}\mathbf{a}_Q + \sum_{i=1}^{n} \mathbf{r}_Q \times m_i{}^{\mathcal{N}}\mathbf{a}_Q \tag{4-81}$$

Then, applying the results of Eqs. (4-2), (4-1), and (4-11) and observing that $\mathbf{r}_Q$ and ${}^{\mathcal{N}}\mathbf{a}_Q$ are independent of the summation, we have

$$\frac{{}^{\mathcal{N}}d}{dt}\left({}^{\mathcal{N}}\mathbf{H}_Q\right) = \sum_{i=1}^{n} (\mathbf{r}_i - \mathbf{r}_Q) \times m_i{}^{\mathcal{N}}\mathbf{a}_i - (\bar{\mathbf{r}} - \mathbf{r}_Q) \times m{}^{\mathcal{N}}\mathbf{a}_Q \tag{4-82}$$

Substituting Eq. (4-33) into (4-82) gives

$$\frac{{}^{\mathcal{N}}d}{dt}\left({}^{\mathcal{N}}\mathbf{H}_Q\right) = \sum_{i=1}^{n} (\mathbf{r}_i - \mathbf{r}_Q) \times \left[\mathbf{F}_i + \sum_{j=1}^{n} \mathbf{f}_{ij}\right] - (\bar{\mathbf{r}} - \mathbf{r}_Q) \times m{}^{\mathcal{N}}\mathbf{a}_Q \tag{4-83}$$

Separating the first summation of Eq. (4–83), we obtain

$$\frac{^{\mathcal{N}}d}{dt}\left(^{\mathcal{N}}\mathbf{H}_Q\right) = \sum_{i=1}^{n}(\mathbf{r}_i - \mathbf{r}_Q)\times\mathbf{F}_i + \sum_{i=1}^{n}\sum_{j=1}^{n}(\mathbf{r}_i - \mathbf{r}_Q)\times\mathbf{f}_{ij} - (\bar{\mathbf{r}} - \mathbf{r}_Q)\times m\,^{\mathcal{N}}\mathbf{a}_Q \tag{4-84}$$

Then, from the assumption that the forces $\mathbf{f}_{ij}$ $(i,j = 1,2,\ldots,n)$ satisfy the strong form of Newton's 3rd law, we have

$$\sum_{i=1}^{n}\sum_{j=1}^{n}(\mathbf{r}_i - \mathbf{r}_Q)\times\mathbf{f}_{ij} = \mathbf{0} \tag{4-85}$$

Consequently, Eq. (4–84) reduces to

$$\frac{^{\mathcal{N}}d}{dt}\left(^{\mathcal{N}}\mathbf{H}_Q\right) = \sum_{i=1}^{n}(\mathbf{r}_i - \mathbf{r}_Q)\times\mathbf{F}_i - (\bar{\mathbf{r}} - \mathbf{r}_Q)\times m\,^{\mathcal{N}}\mathbf{a}_Q \tag{4-86}$$

Then, using the definition of the moment of a force for a system of particles relative to an arbitrary point Q as given in Eq. (4–75), Eq. (4–86) becomes

$$\boxed{\mathbf{M}_Q - (\bar{\mathbf{r}} - \mathbf{r}_Q)\times m\,^{\mathcal{N}}\mathbf{a}_Q = \frac{^{\mathcal{N}}d}{dt}\left(^{\mathcal{N}}\mathbf{H}_Q\right)} \tag{4-87}$$

Equation (4–87) relates the rate of change of the angular momentum of a system of particles relative to an arbitrary reference point Q to the resultant moment due to all external forces applied relative to Q. Similar to Eq. (3–204) on page 180 for a single particle, the quantity

$$-(\bar{\mathbf{r}} - \mathbf{r}_Q)\times m\,^{\mathcal{N}}\mathbf{a}_Q \tag{4-88}$$

is called the *inertial moment* of the reference point Q relative to the center of mass of the system.

Now consider motion relative to a reference point O, where O is fixed in an inertial reference frame $\mathcal{N}$. In this case the acceleration of point O is zero, i.e., $^{\mathcal{N}}\mathbf{a}_Q = \mathbf{0}$, which implies that $-(\bar{\mathbf{r}} - \mathbf{r}_Q)\times m\,^{\mathcal{N}}\mathbf{a}_Q = \mathbf{0}$. Then, substituting the expression for $\mathbf{M}_O$ given in Eq. (4–76) for the moment due to all external forces relative to a point fixed in an inertial reference frame, Eq. (4–87) becomes

$$\boxed{\mathbf{M}_O = \frac{^{\mathcal{N}}d}{dt}\left(^{\mathcal{N}}\mathbf{H}_O\right)} \tag{4-89}$$

Next, consider motion relative to the center of mass of the system. Because the reference point is the center of mass, we have $\bar{\mathbf{r}} - \mathbf{r}_Q = \mathbf{0}$, which again implies that $-(\bar{\mathbf{r}} - \bar{\mathbf{r}})\times m\,^{\mathcal{N}}\mathbf{a}_Q = \mathbf{0}$. Then, substituting the expression for $\bar{\mathbf{M}}$ given in Eq. (4–77) for the moment due to all external forces relative to the center of mass of the system, Eq. (4–87) simplifies to

$$\boxed{\bar{\mathbf{M}} = \frac{^{\mathcal{N}}d}{dt}\left(^{\mathcal{N}}\bar{\mathbf{H}}\right)} \tag{4-90}$$

It is seen from the results of Eqs. (4–89) and (4–90) that, for the *special cases* where the reference point is either fixed in an inertial reference frame $\mathcal{N}$ or is the center of mass of the system, the sum of the moments due to all external forces relative to the reference point is equal to the rate of change of angular momentum relative to the reference point. Observing that the results of Eqs. (4–89) and (4–90) have the same mathematical form, when the reference point is either fixed in an inertial reference frame or is the center of mass of the system, we have

$$\mathbf{M} = \frac{{}^{\mathcal{N}}d}{dt}\left({}^{\mathcal{N}}\mathbf{H}\right) \tag{4–91}$$

It is important to emphasize that, while Eq. (4–87) is a general result that can *always* be applied, Eqs. (4–89) and (4–90) can only be applied, respectively, in the special cases where the reference point is either fixed in an inertial reference frame or is the center of mass. In particular, in either of these special cases, the inertial moment as give in Eq. (4–88) is *zero*. It is noted that the choice of the reference point for computing angular momentum is highly problem-dependent.

Example 4–2

A collar of mass m_1 is constrained to slide along a frictionless horizontal track as shown in Fig. 4-6. The collar is attached to one end of a rigid massless arm of length l while a particle of mass m_2 is attached to the other end of the arm. Knowing that the angle θ is measured from the downward direction and that gravity acts downward, determine (a) a system of two differential equations that describes the motion of the collar-particle system; (b) an alternate system of differential equations via algebraic manipulation of the system obtained in part (a); and (c) one of the differential equations via a balance of angular momentum relative to the collar.

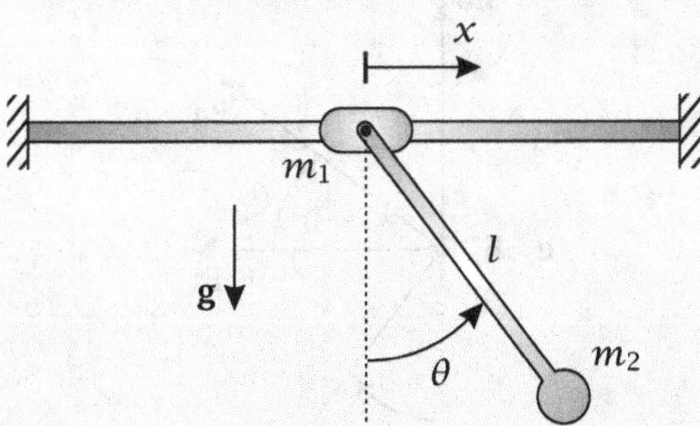

Figure 4–6 Particle on rigid massless arm attached to sliding collar.

Solution to Example 4–2

Preliminaries

It is important to recognize for this problem that, because the system consists of two particles (i.e., the collar and the particle), the analysis can be performed on the following three systems: (1) the collar, (2) the particle, and (3) the collar and the particle. However, because the sum of system (1) and system (2) is equal to system (3), the three systems are *not* independent. Moreover, analyzing system (1) and system (2) separately is equivalent to analyzing system (3). Consequently, in order to obtain a sufficient number of independent results, it is necessary to analyze either systems (1) and (3) *or* systems (2) and (3). While it is possible to analyze either combination of systems, in solving this problem we will analyze systems (1) and (3).

Kinematics

First, let $\mathcal{F}$ be a reference frame fixed to the track. Then, choose the following coordinate system fixed in reference frame $\mathcal{F}$:

$$\begin{array}{ccc} & \text{Origin at } Q & \\ & \text{when } x = 0 & \\ \mathbf{E}_x & = & \text{To the right} \\ \mathbf{E}_z & = & \text{Out of page} \\ \mathbf{E}_y & = & \mathbf{E}_z \times \mathbf{E}_x \end{array}$$

Next, let $\mathcal{A}$ be a reference frame fixed to the arm. Then, choose the following coordinate system fixed in reference frame $\mathcal{A}$:

$$\begin{array}{ccc} & \text{Origin at } Q & \\ \mathbf{e}_r & = & \text{Along } QP \\ \mathbf{e}_z & = & \text{Out of page} \\ \mathbf{e}_\theta & = & \mathbf{e}_z \times \mathbf{e}_r \end{array}$$

The geometry of the bases $\{\mathbf{E}_x, \mathbf{E}_y, \mathbf{E}_z\}$ and $\{\mathbf{e}_r, \mathbf{e}_\theta, \mathbf{e}_z\}$ is shown in Fig. 4-7.

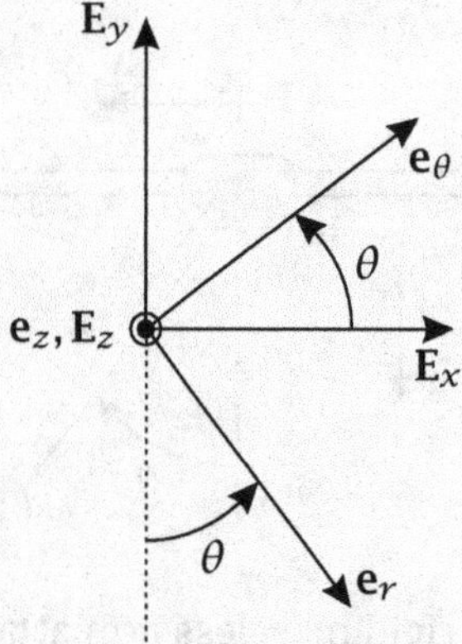

Figure 4-7 Geometry of bases $\{\mathbf{E}_x, \mathbf{E}_y, \mathbf{E}_z\}$ and $\{\mathbf{e}_r, \mathbf{e}_\theta, \mathbf{e}_z\}$ for Example 4-2.

Using Fig. 4-7, we have

$$\mathbf{e}_r = \sin\theta\,\mathbf{E}_x - \cos\theta\,\mathbf{E}_y \tag{4-92}$$

$$\mathbf{e}_\theta = \cos\theta\,\mathbf{E}_x + \sin\theta\,\mathbf{E}_y \tag{4-93}$$

Now, consistent with the discussion at the beginning of this problem, we establish the kinematics relevant to the system consisting of the collar and the system consisting of the collar and the particle.

Kinematics of Collar

The position of the collar is given as

$$\mathbf{r}_1 = x\mathbf{E}_x \tag{4-94}$$

Computing the rate of change of $\mathbf{r}_1$ in reference frame $\mathcal{F}$, we obtain the velocity of the collar in reference frame $\mathcal{F}$ as

$${}^{\mathcal{F}}\mathbf{v}_1 = \frac{{}^{\mathcal{F}}d}{dt}(\mathbf{r}_1) = \dot{x}\mathbf{E}_x \tag{4-95}$$

Finally, computing the rate of change of ${}^{\mathcal{F}}\mathbf{v}_1$ in reference frame $\mathcal{F}$, we obtain the acceleration of the collar in reference frame $\mathcal{F}$ as

$${}^{\mathcal{F}}\mathbf{a}_1 = \frac{{}^{\mathcal{F}}d}{dt}\left({}^{\mathcal{F}}\mathbf{v}_1\right) = \ddot{x}\mathbf{E}_x \tag{4-96}$$

Kinematics of Particle

The kinematics of the collar-particle system are governed by the motion of the center of mass of the system. Consequently, in order to determine the kinematics of the center of mass of the collar-particle system, it is first necessary to determine the position, velocity, and The position of the particle is given as

$$\mathbf{r}_2 = \mathbf{r}_1 + \mathbf{r}_{2/1} \tag{4-97}$$

Now we have

$$\mathbf{r}_{2/1} = l\mathbf{e}_r \tag{4-98}$$

Then, adding Eqs. (4-98) and (4-97), we obtain the position of the particle as

$$\mathbf{r}_2 = x\mathbf{E}_x + l\mathbf{e}_r \tag{4-99}$$

Next, the angular velocity of reference frame $\mathcal{A}$ in reference frame $\mathcal{F}$ is given as

$${}^{\mathcal{F}}\boldsymbol{\omega}^{\mathcal{A}} = \dot{\theta}\mathbf{e}_z \tag{4-100}$$

Consequently, the velocity of the particle in reference frame $\mathcal{F}$ is obtained as

$${}^{\mathcal{F}}\mathbf{v}_2 = \frac{{}^{\mathcal{F}}d}{dt}(\mathbf{r}_2) = \frac{{}^{\mathcal{F}}d}{dt}(\mathbf{r}_1) + \frac{{}^{\mathcal{F}}d}{dt}(\mathbf{r}_{2/1}) = {}^{\mathcal{F}}\mathbf{v}_1 + {}^{\mathcal{F}}\mathbf{v}_{2/1} \tag{4-101}$$

We already have ${}^{\mathcal{F}}\mathbf{v}_1$ from Eq. (4-95). Applying the rate of change transport theorem to $\mathbf{r}_{2/1}$ between reference frames $\mathcal{A}$ and $\mathcal{F}$ gives

$${}^{\mathcal{F}}\mathbf{v}_{2/1} = \frac{{}^{\mathcal{F}}d}{dt}(\mathbf{r}_{2/1}) = \frac{{}^{\mathcal{A}}d}{dt}(\mathbf{r}_{2/1}) + {}^{\mathcal{F}}\boldsymbol{\omega}^{\mathcal{A}} \times \mathbf{r}_{2/1} \tag{4-102}$$

Now we have

$$\frac{{}^{\mathcal{A}}d}{dt}(\mathbf{r}_{2/1}) = \mathbf{0} \tag{4-103}$$

$${}^{\mathcal{F}}\boldsymbol{\omega}^{\mathcal{A}} \times \mathbf{r}_{2/1} = \dot{\theta}\mathbf{e}_z \times l\mathbf{e}_r = l\dot{\theta}\mathbf{e}_\theta \tag{4-104}$$

Adding Eqs. (4–103) and (4–104), we obtain the velocity of the particle relative to the collar in reference frame $\mathcal{F}$ as

$$^{\mathcal{F}}\mathbf{v}_{2/1} = l\dot{\theta}\mathbf{e}_\theta \tag{4–105}$$

Then, substituting the results of Eqs. (4–95) and (4–105) into Eq. (4–101), we obtain the velocity of the particle in reference frame $\mathcal{F}$ as

$$^{\mathcal{F}}\mathbf{v}_2 = \dot{x}\mathbf{E}_x + l\dot{\theta}\mathbf{e}_\theta \tag{4–106}$$

Computing the rate of change of $^{\mathcal{F}}\mathbf{v}_2$ in reference frame $\mathcal{F}$ using the general expression for $^{\mathcal{F}}\mathbf{v}_2$ as given in Eq. (4–101), the acceleration of the particle in reference frame $\mathcal{F}$ is given as

$$^{\mathcal{F}}\mathbf{a}_2 = \frac{^{\mathcal{F}}d}{dt}\left(^{\mathcal{F}}\mathbf{v}_2\right) = \frac{^{\mathcal{F}}d}{dt}\left(^{\mathcal{F}}\mathbf{v}_1\right) + \frac{^{\mathcal{F}}d}{dt}\left(^{\mathcal{F}}\mathbf{v}_{2/1}\right) = {^{\mathcal{F}}\mathbf{a}_1} + {^{\mathcal{F}}\mathbf{a}_{2/1}} \tag{4–107}$$

Now we already have $^{\mathcal{F}}\mathbf{a}_1$ from Eq. (4–96). Applying the rate of change transport theorem between reference frames $\mathcal{A}$ and $\mathcal{F}$, we obtain $^{\mathcal{F}}\mathbf{a}_{2/1}$ as

$$^{\mathcal{F}}\mathbf{a}_{2/1} = \frac{^{\mathcal{F}}d}{dt}\left(^{\mathcal{F}}\mathbf{v}_{2/1}\right) = \frac{^{\mathcal{A}}d}{dt}\left(^{\mathcal{F}}\mathbf{v}_{2/1}\right) + {^{\mathcal{F}}\boldsymbol{\omega}^{\mathcal{A}}} \times {^{\mathcal{F}}\mathbf{v}_{2/1}} \tag{4–108}$$

Now we have

$$\frac{^{\mathcal{A}}d}{dt}\left(^{\mathcal{F}}\mathbf{v}_{2/1}\right) = l\ddot{\theta}\mathbf{e}_\theta \tag{4–109}$$

$$^{\mathcal{F}}\boldsymbol{\omega}^{\mathcal{A}} \times {^{\mathcal{F}}\mathbf{v}_{2/1}} = \dot{\theta}\mathbf{e}_z \times l\dot{\theta}\mathbf{e}_\theta = -l\dot{\theta}^2\mathbf{e}_r \tag{4–110}$$

Adding Eqs. (4–109) and (4–110), we obtain

$$^{\mathcal{F}}\mathbf{a}_{2/1} = -l\dot{\theta}^2\mathbf{e}_r + l\ddot{\theta}\mathbf{e}_\theta \tag{4–111}$$

Finally, adding Eqs. (4–111) and (4–96), we obtain the acceleration of the particle in reference frame $\mathcal{F}$ as

$$^{\mathcal{F}}\mathbf{a}_2 = \ddot{x}\mathbf{E}_x - l\dot{\theta}^2\mathbf{e}_r + l\ddot{\theta}\mathbf{e}_\theta \tag{4–112}$$

Kinematics of Center of Mass of Collar-Particle System

The position of the center of mass of the collar-particle system is given as

$$\bar{\mathbf{r}} = \frac{m_1\mathbf{r}_1 + m_2\mathbf{r}_2}{m_1 + m_2} \tag{4–113}$$

Substituting the expressions for $\mathbf{r}_1$ and $\mathbf{r}_2$ from Eqs. (4–94) and (4–97), respectively, into (4–113), we obtain $\bar{\mathbf{r}}$ as

$$\bar{\mathbf{r}} = \frac{m_1 x\mathbf{E}_x + m_2(x\mathbf{E}_x + l\mathbf{e}_r)}{m_1 + m_2} = x\mathbf{E}_x + \frac{m_2}{m_1 + m_2}l\mathbf{e}_r \tag{4–114}$$

Next, the velocity of the center of mass of the collar-particle system in reference frame $\mathcal{F}$ is given as

$$^{\mathcal{F}}\bar{\mathbf{v}} = \frac{m_1{^{\mathcal{F}}\mathbf{v}_1} + m_2{^{\mathcal{F}}\mathbf{v}_2}}{m_1 + m_2} \tag{4–115}$$

Substituting the expression for ${}^{\mathcal{F}}\mathbf{v}_1$ from Eq. (4–95) and the expression for ${}^{\mathcal{F}}\mathbf{v}_2$ from Eq. (4–106) into (4–115), we obtain ${}^{\mathcal{F}}\bar{\mathbf{v}}$ as

$$ {}^{\mathcal{F}}\bar{\mathbf{v}} = \frac{m_1\dot{x}\mathbf{E}_x + m_2(\dot{x}\mathbf{E}_x + l\dot{\theta}\mathbf{e}_\theta)}{m_1+m_2} = \dot{x}\mathbf{E}_x + \frac{m_2}{m_1+m_2} l\dot{\theta}\mathbf{e}_\theta \tag{4–116} $$

Finally, the acceleration of the center of mass of the collar-particle system in reference frame $\mathcal{F}$ is given as

$$ {}^{\mathcal{F}}\bar{\mathbf{a}} = \frac{m_1{}^{\mathcal{F}}\mathbf{a}_1 + m_2{}^{\mathcal{F}}\mathbf{a}_2}{m_1+m_2} \tag{4–117} $$

Substituting the expressions for ${}^{\mathcal{F}}\mathbf{a}_1$ and ${}^{\mathcal{F}}\mathbf{a}_2$ from Eqs. (4–96) and (4–112), respectively, into (4–117), we obtain

$$ {}^{\mathcal{F}}\bar{\mathbf{a}} = \frac{m_1\ddot{x}\mathbf{E}_x + m_2(\ddot{x}\mathbf{E}_x - l\dot{\theta}^2\mathbf{e}_r + l\ddot{\theta}\mathbf{e}_\theta)}{m_1+m_2} = \ddot{x}\mathbf{E}_x + \frac{m_2}{m_1+m_2}(-l\dot{\theta}^2\mathbf{e}_r + l\ddot{\theta}\mathbf{e}_\theta) \tag{4–118} $$

Kinetics

As discussed earlier, this problem will be solved by analyzing the following two systems: (1) the collar and (2) the collar and the particle. The kinetic relationships for each of these two systems is now established.

Kinetics of Collar

The free body diagram of the collar is shown in Fig. 4–8.

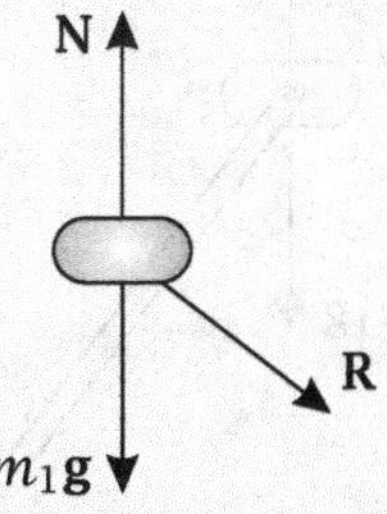

Figure 4–8 Free body diagram of collar for Example 4–2.

Using Fig. 4–8, it is seen that the following forces act on the collar:

$\mathbf{N}$ = Reaction force of track on collar
$\mathbf{R}$ = Tension force in arm due to particle
$m_1\mathbf{g}$ = Force of gravity

The forces acting on the collar are given in terms of the bases $\{\mathbf{E}_x, \mathbf{E}_y, \mathbf{E}_z\}$ and $\{\mathbf{e}_r, \mathbf{e}_\theta, \mathbf{e}_z\}$ as

$$ \mathbf{N} = N\mathbf{E}_y \tag{4–119} $$
$$ \mathbf{R} = R\mathbf{e}_r \tag{4–120} $$
$$ m_1\mathbf{g} = -m_1 g\mathbf{E}_y \tag{4–121} $$

It is noted in Eq. (4–120) that, from the strong form of Newton's 3rd law, the force $\mathbf{R}$ must lie along the line of action connecting the collar and the particle. The resultant force acting on the collar is then given as

$$\mathbf{F}_1 = \mathbf{N} + \mathbf{R} + m_1\mathbf{g} = N\mathbf{E}_y + R\mathbf{e}_r - m_1 g\mathbf{E}_y \tag{4–122}$$

Then, substituting the expression for $\mathbf{e}_r$ from Eq. (4–92) into (4–122), we have

$$\begin{aligned} \mathbf{F}_1 &= N\mathbf{E}_y + R(\sin\theta\mathbf{E}_x - \cos\theta\mathbf{E}_y) - m_1 g\mathbf{E}_y \\ &= R\sin\theta\mathbf{E}_x + (N - R\cos\theta - m_1 g)\mathbf{E}_y \end{aligned} \tag{4–123}$$

Applying Newton's 2nd law to the collar by setting $\mathbf{F}_1$ in Eq. (4–123) equal to $m_1{}^{\mathcal{F}}\mathbf{a}_1$ using the expression for ${}^{\mathcal{F}}\mathbf{a}_1$ from Eq. (4–96), we obtain

$$R\sin\theta\mathbf{E}_x + (N - R\cos\theta - m_1 g)\mathbf{E}_y = m_1\ddot{x}\mathbf{E}_x \tag{4–124}$$

Equation (4–124) yields the following two scalar equations:

$$R\sin\theta = m_1\ddot{x} \tag{4–125}$$

$$N - R\cos\theta - m_1 g = 0 \tag{4–126}$$

Kinetics of Collar-Particle System

The free body diagram of the collar-particle system is shown in Fig. 4–9.

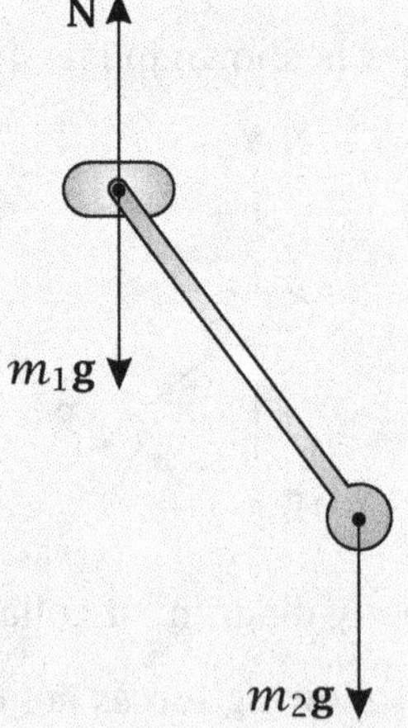

Figure 4–9 Free body diagram of collar-particle system for Example 4–2.

Using Fig. 4–9, it is seen that the following forces act on the collar-particle system:

$$\begin{aligned} \mathbf{N} &= \text{Reaction force of track on collar} \\ (m_1 + m_2)\mathbf{g} &= \text{Force of gravity} \end{aligned}$$

We already have $\mathbf{N}$ from Eq. (4–119). Furthermore, the force of gravity acting on the collar-particle system is given as

$$(m_1 + m_2)\mathbf{g} = -(m_1 + m_2)g\mathbf{E}_y \tag{4–127}$$

Consequently, the resultant force acting on the collar is given as

$$\mathbf{F} = \mathbf{N} + (m_1 + m_2)\mathbf{g} = N\mathbf{E}_y - (m_1 + m_2)g\mathbf{E}_y = [N - (m_1 + m_2)g]\,\mathbf{E}_y \tag{4-128}$$

Applying Newton's 2nd law to the collar-particle system by setting $\mathbf{F}$ in Eq. (4-128) equal to $(m_1 + m_2)^{\mathcal{F}}\bar{\mathbf{a}}$ using the expression for ${}^{\mathcal{F}}\bar{\mathbf{a}}$ from Eq. (4-118), we obtain

$$[N - (m_1 + m_2)g]\,\mathbf{E}_y = (m_1 + m_2)\left[\ddot{x}\mathbf{E}_x + \frac{m_2}{m_1 + m_2}(-l\dot{\theta}^2\mathbf{e}_r + l\ddot{\theta}\mathbf{e}_\theta)\right] \tag{4-129}$$

Equation (4-129) can be rewritten as

$$[N - (m_1 + m_2)g]\,\mathbf{E}_y = (m_1 + m_2)\ddot{x}\mathbf{E}_x - m_2 l\dot{\theta}^2\mathbf{e}_r + m_2 l\ddot{\theta}\mathbf{e}_\theta \tag{4-130}$$

Then, substituting the expressions for $\mathbf{e}_r$ and $\mathbf{e}_\theta$ from Eqs. (4-92) and (4-93), respectively, into Eq. (4-130), we obtain

$$\begin{aligned}[N - (m_1 + m_2)g]\,\mathbf{E}_y = {} & (m_1 + m_2)\ddot{x}\mathbf{E}_x - m_2 l\dot{\theta}^2(\sin\theta\,\mathbf{E}_x - \cos\theta\,\mathbf{E}_y) \\ & + m_2 l\ddot{\theta}(\cos\theta\,\mathbf{E}_x + \sin\theta\,\mathbf{E}_y)\end{aligned} \tag{4-131}$$

Rearranging Eq. (4-131), we obtain

$$\begin{aligned}[N - (m_1 + m_2)g]\,\mathbf{E}_y = {} & \left[(m_1 + m_2)\ddot{x} + m_2 l\ddot{\theta}\cos\theta - m_2 l\dot{\theta}^2\sin\theta\right]\mathbf{E}_x \\ & + \left[m_2 l\ddot{\theta}\sin\theta + m_2 l\dot{\theta}^2\cos\theta\right]\mathbf{E}_y\end{aligned} \tag{4-132}$$

Equation (4-132) yields the following two scalar equations:

$$0 = (m_1 + m_2)\ddot{x} + m_2 l\ddot{\theta}\cos\theta - m_2 l\dot{\theta}^2\sin\theta \tag{4-133}$$

$$N - (m_1 + m_2)g = m_2 l\ddot{\theta}\sin\theta + m_2 l\dot{\theta}^2\cos\theta \tag{4-134}$$

(a) System of Two Differential Equations

Using the results of Eqs. (4-125), (4-126), (4-133), and (4-134), a system of two differential equations is now determined. Because Eq. (4-133) has no unknown forces, it is the first differential equation. The second differential equation is obtained as follows. Multiplying Eqs. (4-125) and (4-126) by $\cos\theta$ and $\sin\theta$, respectively, we have

$$R\sin\theta\cos\theta = m_1\ddot{x}\cos\theta \tag{4-135}$$

$$N\sin\theta - R\cos\theta\sin\theta - m_1 g\sin\theta = 0 \tag{4-136}$$

Adding Eqs. (4-135) and (4-136), we obtain

$$N\sin\theta - m_1 g\sin\theta = m_1\ddot{x}\cos\theta \tag{4-137}$$

Next, multiplying Eq. (4-134) by $\sin\theta$ gives

$$N\sin\theta - (m_1 + m_2)g\sin\theta = m_2 l\ddot{\theta}\sin^2\theta + m_2 l\dot{\theta}^2\cos\theta\sin\theta \tag{4-138}$$

Then, subtracting Eq. (4-138) from (4-137), we obtain

$$m_2 g\sin\theta = m_1\ddot{x}\cos\theta - m_2 l\ddot{\theta}\sin^2\theta - m_2 l\dot{\theta}^2\cos\theta\sin\theta \tag{4-139}$$

Rearranging Eq. (4–139), we obtain the second differential equation of motion as

$$m_1\ddot{x}\cos\theta - m_2 l\ddot{\theta}\sin^2\theta - m_2 l\dot{\theta}^2\cos\theta\sin\theta - m_2 g\sin\theta = 0 \qquad (4\text{–}140)$$

A system of two differential equations that describes the motion of the collar-particle system is then given as

$$\begin{aligned} (m_1+m_2)\ddot{x} + m_2 l\ddot{\theta}\cos\theta - m_2 l\dot{\theta}^2\sin\theta &= 0 \qquad (4\text{–}141) \\ m_1\ddot{x}\cos\theta - m_2 l\ddot{\theta}\sin^2\theta - m_2 l\dot{\theta}^2\cos\theta\sin\theta - m_2 g\sin\theta &= 0 \qquad (4\text{–}142) \end{aligned}$$

(b) Alternate System of Differential Equations

While Eqs. (4–141) and (4–142) are independent, a slightly simpler set of equations can be derived by manipulating Eqs. (4–141) and (4–142) as follows. Multiplying Eq. (4–141) by $\cos\theta$, we have

$$\begin{aligned} (m_1+m_2)\ddot{x}\cos\theta + m_2 l\ddot{\theta}\cos^2\theta - m_2 l\dot{\theta}^2\sin\theta\cos\theta &= 0 \qquad (4\text{–}143) \\ m_1\ddot{x}\cos\theta - m_2 l\ddot{\theta}\sin^2\theta - m_2 l\dot{\theta}^2\cos\theta\sin\theta - m_2 g\sin\theta &= 0 \qquad (4\text{–}144) \end{aligned}$$

Then, subtracting Eq. (4–144) from (4–143), we obtain

$$m_2\ddot{x}\cos\theta + m_2 l\ddot{\theta}(\cos^2\theta + \sin^2\theta) + m_2 g\sin\theta = 0 \qquad (4\text{–}145)$$

Simplifying Eq. (4–145), we obtain

$$m_2\ddot{x}\cos\theta + m_2 l\ddot{\theta} + m_2 g\sin\theta = 0 \qquad (4\text{–}146)$$

Now, because the two differential equations in Eqs. (4–141) and (4–142) are independent and we have obtained Eqs. (4–141) and (4–146) via a nonsingular transformation, the two differential equations in Eqs. (4–141) and (4–146) are also independent. Consequently, an alternate system of two differential equations describing the motion of the collar-particle system is given as

$$\begin{aligned} (m_1+m_2)\ddot{x} + m_2 l\ddot{\theta}\cos\theta - m_2 l\dot{\theta}^2\sin\theta &= 0 \qquad (4\text{–}147) \\ m_2\ddot{x}\cos\theta + m_2 l\ddot{\theta} + m_2 g\sin\theta &= 0 \qquad (4\text{–}148) \end{aligned}$$

(c) Determination of One Differential Equation via Balance of Angular Momentum Relative to Collar

It was seen above that the two differential equations of motion were obtained purely by using Newton's 2nd law. However, it is useful to note that the one of the differential equations can also be obtained by performing a balance of angular momentum of the collar-particle system relative to the collar. Observing that the collar is *not* an inertially fixed point, it is necessary to apply Eq. (4–87) as

$$\mathbf{M}_1 - (\bar{\mathbf{r}} - \mathbf{r}_1) \times (m_1+m_2)\,{}^{\mathcal{F}}\mathbf{a}_1 = \frac{{}^{\mathcal{F}}d}{dt}\left({}^{\mathcal{F}}\mathbf{H}_1\right) \qquad (4\text{–}149)$$

where $\mathbf{M}_1$ is the resultant moment applied to the collar-particle system relative to the collar while ${}^{\mathcal{F}}d({}^{\mathcal{F}}\mathbf{H}_1)/dt$ is the rate of change of angular momentum of the collar-particle system relative to the collar. First, the rate of change of angular momentum of the collar-particle system relative to the collar is given from Eq. (4–79) as

$$\frac{^{\mathcal{F}}d}{dt}\left(^{\mathcal{F}}\mathbf{H}_1\right) = (\mathbf{r}_1 - \mathbf{r}_1) \times m_1(^{\mathcal{F}}\mathbf{a}_1 - {^{\mathcal{F}}}\mathbf{a}_1) + (\mathbf{r}_2 - \mathbf{r}_1) \times m_2(^{\mathcal{F}}\mathbf{a}_2 - {^{\mathcal{F}}}\mathbf{a}_1) \quad (4\text{–}150)$$

Observing that the first term in Eq. (4–150) is zero, we have

$$\frac{^{\mathcal{F}}d}{dt}\left(^{\mathcal{F}}\mathbf{H}_1\right) = (\mathbf{r}_2 - \mathbf{r}_1) \times m_2(^{\mathcal{F}}\mathbf{a}_2 - {^{\mathcal{F}}}\mathbf{a}_1) \quad (4\text{–}151)$$

Using the expressions for $\mathbf{r}_1$ and $\mathbf{r}_2$ from Eqs. (4–94) and (4–97), respectively, we have

$$\mathbf{r}_2 - \mathbf{r}_1 = l\mathbf{e}_r \quad (4\text{–}152)$$

Next, using the expressions for $^{\mathcal{F}}\mathbf{a}_2$ and $^{\mathcal{F}}\mathbf{a}_1$ from Eqs. (4–96) and (4–112), respectively, we have

$$^{\mathcal{F}}\mathbf{a}_2 - {^{\mathcal{F}}}\mathbf{a}_1 = -l\dot{\theta}^2\mathbf{e}_r + l\ddot{\theta}\mathbf{e}_\theta \quad (4\text{–}153)$$

Consequently, we obtain

$$\frac{^{\mathcal{F}}d}{dt}\left(^{\mathcal{F}}\mathbf{H}_1\right) = l\mathbf{e}_r \times m_2(-l\dot{\theta}^2\mathbf{e}_r + l\ddot{\theta}\mathbf{e}_\theta) = m_2 l^2\ddot{\theta}\mathbf{e}_z \quad (4\text{–}154)$$

Next, observing from the free body diagram of the collar-particle system in Fig. 4-9 that the forces $\mathbf{N}$ and $m_1\mathbf{g}$ pass through the collar, the resultant moment relative to the collar is due entirely to $m_2\mathbf{g}$ and is given as

$$\begin{aligned}\mathbf{M}_1 &= (\mathbf{r}_2 - \mathbf{r}_1) \times m_2\mathbf{g} = l\mathbf{e}_r \times (-m_2 g\mathbf{E}_y) \\ &= -m_2 g l(\sin\theta\mathbf{E}_x - \cos\theta\mathbf{E}_y) \times \mathbf{E}_y = -m_2 g l\sin\theta\,\mathbf{e}_z\end{aligned} \quad (4\text{–}155)$$

Finally, using the expressions for $\bar{\mathbf{r}}$ and $^{\mathcal{F}}\mathbf{a}_1$ from Eqs. (4–114) and (4–96), respectively, the inertial moment due to the acceleration of the collar is given as

$$\begin{aligned}-(\bar{\mathbf{r}} - \mathbf{r}_1) \times (m_1 + m_2)^{\mathcal{F}}\mathbf{a}_1 &= -\frac{m_2}{m_1 + m_2} l\mathbf{e}_r \times (m_1 + m_2) \times (\ddot{x}\mathbf{E}_x) \\ &= -m_2 l\ddot{x}\mathbf{e}_r \times \mathbf{E}_x = -m_2 l\ddot{x}(\sin\theta\mathbf{E}_x - \cos\theta\mathbf{E}_y) \times \mathbf{E}_x \\ &= -m_2 l\ddot{x}\cos\theta\,\mathbf{e}_z\end{aligned} \quad (4\text{–}156)$$

where we have used the expression for $\mathbf{e}_r$ from Eq. (4–92) in (4–156). Substituting the results of Eqs. (4–154), (4–155), and (4–156) into Eq. (4–149), we obtain

$$m_2 l^2\ddot{\theta}\mathbf{e}_z = -m_2 g l\sin\theta\,\mathbf{e}_z - m_2 l\ddot{x}\cos\theta\,\mathbf{e}_z \quad (4\text{–}157)$$

Rearranging and dropping the dependence on $\mathbf{e}_z$ in Eq. (4–157), we obtain

$$m_2 l\ddot{x}\cos\theta + m_2 l^2\ddot{\theta} + m_2 g l\sin\theta = 0 \quad (4\text{–}158)$$

Finally, dropping the common factor of l from Eq. (4–158), the second differential equation is obtained as

$$m_2\ddot{x}\cos\theta + m_2 l\ddot{\theta} + m_2 g\sin\theta = 0 \quad (4\text{–}159)$$

It is seen that Eq. (4–159) is identical to Eq. (4–148).

■

Example 4–3

A dumbbell consists of two particles, each of mass m, connected by a rigid massless rod of length l as shown in Fig. 4–10. The upper end of the dumbbell, located at point A, slides without friction along a vertical wall while the lower end of the dumbbell, located at point B, slides without friction along a horizontal floor. Knowing that θ is the angle between the vertical wall and the dumbbell, that gravity acts vertically downward, and assuming the initial conditions $\theta(0) = 0$ and $\dot{\theta}(0) = 0$, determine (a) the differential equation of motion while the dumbbell maintains contact with the wall and the floor and (b) the angle θ at which the dumbbell loses contact with the vertical wall.

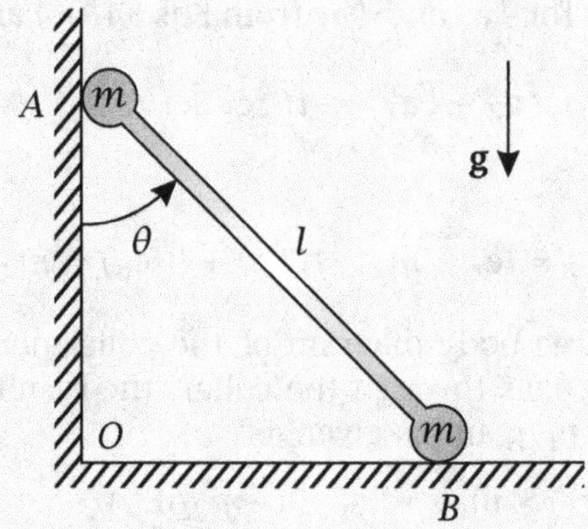

Figure 4–10 Dumbbell sliding on wall and floor.

Solution to Example 4–3

Kinematics

Kinematics of Each Particle

For this problem, it is convenient to describe the motion using a reference frame $\mathcal{F}$ that is fixed to the ground. Corresponding to reference frame $\mathcal{F}$, we choose the following coordinate system:

$$
\begin{array}{ccc}
 & \text{Origin at } O & \\
\mathbf{E}_x & = & \text{Along } OB \\
\mathbf{E}_y & = & \text{Along } OA \\
\mathbf{E}_z & = & \mathbf{E}_x \times \mathbf{E}_y
\end{array}
$$

Then the positions of the particles A and B are given as

$$\mathbf{r}_A = y\mathbf{E}_y \tag{4–160}$$

$$\mathbf{r}_B = x\mathbf{E}_x \tag{4–161}$$

Now from the geometry we note that

$$
\begin{aligned}
y &= l\cos\theta \\
x &= l\sin\theta
\end{aligned}
\tag{4–162}
$$

Therefore, the positions are given as

$$\mathbf{r}_A = l\cos\theta\mathbf{E}_y \tag{4-163}$$

$$\mathbf{r}_B = l\sin\theta\mathbf{E}_x \tag{4-164}$$

Noting that reference frame $\mathcal{F}$ is fixed, the velocities of the particles in reference frame $\mathcal{F}$ are obtained as

$${}^{\mathcal{F}}\mathbf{v}_A = \frac{{}^{\mathcal{F}}d}{dt}(\mathbf{r}_A) = -l\dot{\theta}\sin\theta\mathbf{E}_y \tag{4-165}$$

$${}^{\mathcal{F}}\mathbf{v}_B = \frac{{}^{\mathcal{F}}d}{dt}(\mathbf{r}_B) = l\dot{\theta}\cos\theta\mathbf{E}_x \tag{4-166}$$

Computing the rate of change of ${}^{\mathcal{F}}\mathbf{v}_A$ and ${}^{\mathcal{F}}\mathbf{v}_B$ in Eqs. (4–165) and (4–166), respectively, we obtain the accelerations of the particles in reference frame $\mathcal{F}$ as

$${}^{\mathcal{F}}\mathbf{a}_A = \frac{{}^{\mathcal{F}}d}{dt}\left({}^{\mathcal{F}}\mathbf{v}_A\right) = -l(\ddot{\theta}\sin\theta + \dot{\theta}^2\cos\theta)\mathbf{E}_y \tag{4-167}$$

$${}^{\mathcal{F}}\mathbf{a}_B = \frac{{}^{\mathcal{F}}d}{dt}\left({}^{\mathcal{F}}\mathbf{v}_B\right) = l(\ddot{\theta}\cos\theta - \dot{\theta}^2\sin\theta)\mathbf{E}_x \tag{4-168}$$

Kinematics of Center of Mass of Dumbbell

The position of the center of mass of the rod is given as

$$\bar{\mathbf{r}} = \frac{m_A\mathbf{r}_A + m_B\mathbf{r}_B}{m_A + m_B} \tag{4-169}$$

Substituting the expression for $\mathbf{r}_A$ and $\mathbf{r}_B$ from Eqs. (4–163) and (4–164), respectively, and noting that $m_A = m_B = m$, we obtain $\bar{\mathbf{r}}$ as

$$\bar{\mathbf{r}} = \frac{ml\cos\theta\mathbf{E}_y + ml\sin\theta\mathbf{E}_x}{2m} = \frac{l}{2}\sin\theta\mathbf{E}_x + \frac{l}{2}\cos\theta\mathbf{E}_y \tag{4-170}$$

Furthermore, the velocity of the center of mass is given as

$${}^{\mathcal{F}}\bar{\mathbf{v}} = \frac{m_A{}^{\mathcal{F}}\mathbf{v}_A + m_B{}^{\mathcal{F}}\mathbf{v}_A}{m_A + m_B} = \frac{l}{2}\dot{\theta}\cos\theta\mathbf{E}_x - \frac{l}{2}\dot{\theta}\sin\theta\mathbf{E}_y \tag{4-171}$$

Then the acceleration of the center of mass is given as

$${}^{\mathcal{F}}\bar{\mathbf{a}} = \frac{m_A{}^{\mathcal{F}}\mathbf{a}_A + m_B{}^{\mathcal{F}}\mathbf{a}_A}{m_A + m_B} = \frac{l}{2}(\ddot{\theta}\cos\theta - \dot{\theta}^2\sin\theta)\mathbf{E}_x - \frac{l}{2}(\ddot{\theta}\sin\theta + \dot{\theta}^2\cos\theta)\mathbf{E}_y \tag{4-172}$$

Kinematics Relative to Center of Mass of Dumbbell

In order to solve this problem we will also need to perform a balance of angular momentum. Given the geometry of this problem, it is convenient to choose the center of mass of the system as the reference point. Correspondingly, the angular momentum of the system relative to the center of mass of the rod in reference frame $\mathcal{F}$ is given as

$${}^{\mathcal{F}}\bar{\mathbf{H}} = (\mathbf{r}_A - \bar{\mathbf{r}}) \times m_A({}^{\mathcal{F}}\mathbf{v}_A - {}^{\mathcal{F}}\bar{\mathbf{v}}) + (\mathbf{r}_B - \bar{\mathbf{r}}) \times m_B({}^{\mathcal{F}}\mathbf{v}_B - {}^{\mathcal{F}}\bar{\mathbf{v}}) \tag{4-173}$$

Using the expressions for $\mathbf{r}_A$ and $\mathbf{r}_B$ from Eqs. (4–163) and (4–164), respectively, and the expression for $\bar{\mathbf{r}}$ from Eq. (4–170), we have

$$\begin{aligned} \mathbf{r}_A - \bar{\mathbf{r}} &= -\frac{l}{2}\sin\theta\,\mathbf{E}_x + \frac{l}{2}\cos\theta\,\mathbf{E}_y \\ \mathbf{r}_B - \bar{\mathbf{r}} &= \frac{l}{2}\sin\theta\,\mathbf{E}_x - \frac{l}{2}\cos\theta\,\mathbf{E}_y \end{aligned} \tag{4–174}$$

Furthermore, using the expressions for ${}^{\mathcal{F}}\mathbf{v}_A$, ${}^{\mathcal{F}}\mathbf{v}_B$, and ${}^{\mathcal{F}}\bar{\mathbf{v}}$ from Eqs. (4–165), (4–166), and (4–171), respectively, we obtain

$$\begin{aligned} {}^{\mathcal{F}}\mathbf{v}_A - {}^{\mathcal{F}}\bar{\mathbf{v}} &= -\frac{l}{2}\dot{\theta}\cos\theta\,\mathbf{E}_x - \frac{l}{2}\dot{\theta}\sin\theta\,\mathbf{E}_y \\ {}^{\mathcal{F}}\mathbf{v}_B - {}^{\mathcal{F}}\bar{\mathbf{v}} &= \frac{l}{2}\dot{\theta}\cos\theta\,\mathbf{E}_x + \frac{l}{2}\dot{\theta}\sin\theta\,\mathbf{E}_y \end{aligned} \tag{4–175}$$

Finally, substituting the results of Eqs. (4–174) and (4–175) into (4–173) and using the fact that $m_A = m_B = m$, we obtain ${}^{\mathcal{F}}\bar{\mathbf{H}}$ as

$$\begin{aligned} {}^{\mathcal{F}}\bar{\mathbf{H}} = & \left(-\frac{l}{2}\sin\theta\,\mathbf{E}_x + \frac{l}{2}\cos\theta\,\mathbf{E}_y\right) \times m\left(-\frac{l}{2}\dot{\theta}\cos\theta\,\mathbf{E}_x - \frac{l}{2}\dot{\theta}\sin\theta\,\mathbf{E}_y\right) \\ & + \left(\frac{l}{2}\sin\theta\,\mathbf{E}_x - \frac{l}{2}\cos\theta\,\mathbf{E}_y\right) \times m\left(\frac{l}{2}\dot{\theta}\cos\theta\,\mathbf{E}_x + \frac{l}{2}\dot{\theta}\sin\theta\,\mathbf{E}_y\right) \end{aligned} \tag{4–176}$$

Simplifying Eq. (4–176) gives

$${}^{\mathcal{F}}\bar{\mathbf{H}} = \frac{ml^2}{2}\dot{\theta}\mathbf{E}_z \tag{4–177}$$

Computing the rate of change of ${}^{\mathcal{F}}\bar{\mathbf{H}}$ in reference frame $\mathcal{F}$, we obtain

$$\frac{{}^{\mathcal{F}}d}{dt}\left({}^{\mathcal{F}}\bar{\mathbf{H}}\right) = \frac{ml^2}{2}\ddot{\theta}\mathbf{E}_z \tag{4–178}$$

Kinetics

It is convenient to solve this problem by applying the following two balance laws for a system of particles: (1) Newton's 2nd law to the center of mass of the dumbbell and (2) a balance of angular momentum relative to the center of mass of the dumbbell. Each of these balance laws is now applied.

Application of Newton's 2nd Law to Center of Mass of Dumbbell

The free body diagram of the system is shown in Fig. 4–11.

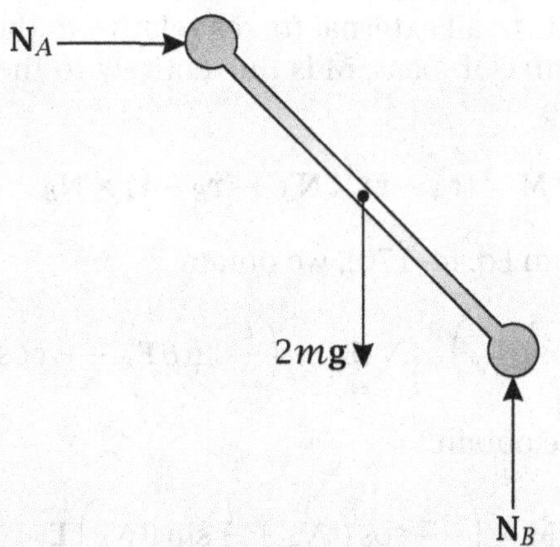

Figure 4–11 Free body diagram for Example 4–3.

Using Fig. 4–11, it is seen that three forces act on the dumbbell, namely, the two reaction forces $\mathbf{N}_A$ and $\mathbf{N}_B$ and the force of gravity $2m\mathbf{g}$ (we note that the force of gravity is multiplied by 2 because the mass of the dumbbell is $2m$). Using the fact that $\mathbf{N}_A$ acts horizontally, $\mathbf{N}_B$ acts vertically, and $2m\mathbf{g}$ acts vertically downward, we have

$$\mathbf{N}_A = N_A\mathbf{E}_x \tag{4–179}$$

$$\mathbf{N}_B = N_B\mathbf{E}_y \tag{4–180}$$

$$2m\mathbf{g} = -2mg\mathbf{E}_y \tag{4–181}$$

Then, applying Newton's 2nd law using the acceleration of the center of mass obtained in Eq. (4–172), we have

$$\mathbf{F} = \mathbf{N}_A + \mathbf{N}_B + 2m\mathbf{g} = 2m\,{}^{\mathcal{F}}\bar{\mathbf{a}} \tag{4–182}$$

Substituting $\mathbf{N}_A$, $\mathbf{N}_B$, and $2m\mathbf{g}$ from Eqs. (4–179)-(4–181), respectively, and the expression for ${}^{\mathcal{F}}\bar{\mathbf{a}}$ from Eq. (4–172), we obtain

$$N_A\mathbf{E}_x + N_B\mathbf{E}_y - 2mg\mathbf{E}_y = 2m\left(\frac{l}{2}(\ddot{\theta}\cos\theta - \dot{\theta}^2\sin\theta)\mathbf{E}_x - \frac{l}{2}(\ddot{\theta}\sin\theta + \dot{\theta}^2\cos\theta)\mathbf{E}_y\right) \tag{4–183}$$

Simplifying Eq. (4–183), we have

$$N_A\mathbf{E}_x + (N_B - 2mg)\mathbf{E}_y = ml(\ddot{\theta}\cos\theta - \dot{\theta}^2\sin\theta)\mathbf{E}_x - ml(\ddot{\theta}\sin\theta + \dot{\theta}^2\cos\theta)\mathbf{E}_y \tag{4–184}$$

Equating components in Eq. (4–184), we obtain the following two scalar equations:

$$N_A = ml(\ddot{\theta}\cos\theta - \dot{\theta}^2\sin\theta) \tag{4–185}$$

$$N_B - 2mg = -ml(\ddot{\theta}\sin\theta + \dot{\theta}^2\cos\theta) \tag{4–186}$$

Balance of Angular Momentum Relative to Center of Mass of Dumbbell

Applying a balance of angular momentum relative to the center of mass of the dumbbell, we have

$$\bar{\mathbf{M}} = \frac{{}^{\mathcal{F}}d}{dt}\left({}^{\mathcal{F}}\bar{\mathbf{H}}\right)$$

where $\bar{\mathbf{M}}$ is the moment due to all external forces relative to the center of mass. Noting that gravity acts at the center of mass, $\bar{\mathbf{M}}$ is due entirely to the reaction forces $\mathbf{N}_A$ and $\mathbf{N}_B$. Consequently, we have

$$\bar{\mathbf{M}} = (\mathbf{r}_A - \bar{\mathbf{r}}) \times \mathbf{N}_A + (\mathbf{r}_B - \bar{\mathbf{r}}) \times \mathbf{N}_B \tag{4-187}$$

Using $\mathbf{r}_A - \bar{\mathbf{r}}$ and $\mathbf{r}_B - \bar{\mathbf{r}}$ from Eq. (4–170), we obtain

$$\bar{\mathbf{M}} = \left(-\frac{l}{2}\sin\theta\mathbf{E}_x + \frac{l}{2}\cos\theta\mathbf{E}_y\right) \times N_A\mathbf{E}_x + \left(\frac{l}{2}\sin\theta\mathbf{E}_x - \frac{l}{2}\cos\theta\mathbf{E}_y\right) \times N_B\mathbf{E}_y \tag{4-188}$$

Simplifying Eq. (4–188), we obtain

$$\bar{\mathbf{M}} = \left(-\frac{l}{2}\cos\theta N_A + \frac{l}{2}\sin\theta N_B\right)\mathbf{E}_z \tag{4-189}$$

Setting $\bar{\mathbf{M}}$ from Eq. (4–188) equal to ${}^{\mathcal{F}}d({}^{\mathcal{F}}\bar{\mathbf{H}})/dt$ from Eq. (4–178) gives

$$\left(-\frac{l}{2}\cos\theta N_A + \frac{l}{2}\sin\theta N_B\right)\mathbf{E}_z = \frac{ml^2}{2}\ddot{\theta}\mathbf{E}_z \tag{4-190}$$

Dropping the dependence on $\mathbf{E}_z$ in Eq. (4–190), we obtain the scalar equation

$$-\frac{l}{2}\cos\theta N_A + \frac{l}{2}\sin\theta N_B = \frac{ml^2}{2}\ddot{\theta} \tag{4-191}$$

Dropping the common factor of $l/2$, Eq. (4–191) simplifies to

$$-N_A\cos\theta + N_B\sin\theta = ml\ddot{\theta} \tag{4-192}$$

(a) Differential Equation While Dumbbell Maintains Contact with Vertical Wall

Equations (4–185), (4–186), and (4–192) can now be used together to obtain the differential equation of motion. Multiplying Eqs. (4–185) and (4–186) by $\cos\theta$ and $\sin\theta$, respectively, and subtracting the result, we obtain

$$\begin{aligned} N_A\cos\theta - N_B\sin\theta + 2mg\sin\theta &= ml(\ddot{\theta}\cos\theta - \dot{\theta}^2\sin\theta)\cos\theta \\ &\quad + ml(\ddot{\theta}\sin\theta + \dot{\theta}^2\cos\theta)\sin\theta \end{aligned} \tag{4-193}$$

Equation (4–193) simplifies to

$$N_A\cos\theta - N_B\sin\theta = -2mg\sin\theta + ml\ddot{\theta} \tag{4-194}$$

Then, adding Eqs. (4–194) and (4–192), we obtain

$$2ml\ddot{\theta} - 2mg\sin\theta = 0 \tag{4-195}$$

Simplifying Eq. (4–195), we obtain the differential equation of motion as

$$\ddot{\theta} - \frac{g}{l}\sin\theta = 0 \tag{4-196}$$

(b) Angle θ at Which Dumbbell Loses Contact with Vertical Wall

We note that the rod will lose contact with the vertical wall when $N_A = 0$, which, from Eq. (4–185), implies that

$$\ddot{\theta}\cos\theta - \dot{\theta}^2\sin\theta = 0 \tag{4–197}$$

Now, because we want to solve for the angle θ at which contact is lost, in order to make Eq. (4–197) an algebraic equation, we need to obtain expressions for $\dot{\theta}^2$ and $\ddot{\theta}$ as functions of θ. We can get $\ddot{\theta}$ as a function of θ simply by rearranging Eq. (4–196) as

$$\ddot{\theta} = \frac{g}{l}\sin\theta \tag{4–198}$$

We note that $\ddot{\theta} = \dot{\theta}d\dot{\theta}/d\theta$. Therefore, Eq. (4–198) can be rewritten as

$$\dot{\theta}\frac{d\dot{\theta}}{d\theta} = \frac{g}{l}\sin\theta \tag{4–199}$$

Separating variables in Eq. (4–199) gives

$$\dot{\theta}d\dot{\theta} = \frac{g}{l}\sin\theta\, d\theta \tag{4–200}$$

Integrating both sides of Eq. (4–200) gives

$$\int_{\dot{\theta}_0}^{\dot{\theta}} \eta d\eta = \int_{\theta_0}^{\theta} \frac{g}{l}\sin\nu d\nu \tag{4–201}$$

where η and ν are dummy variables of integration. From Eq. (4–201) we have

$$\left.\frac{\eta^2}{2}\right|_{\dot{\theta}_0}^{\dot{\theta}} = -\frac{g}{l}\cos\nu\Big|_{\theta_0}^{\theta} \tag{4–202}$$

Equation (4–202) gives

$$\dot{\theta}^2 - \dot{\theta}_0^2 = -\frac{2g}{l}(\cos\theta - \cos\theta_0) \tag{4–203}$$

Rearranging Eq. (4–203), we obtain

$$\dot{\theta}^2 = \dot{\theta}_0^2 + \frac{2g}{l}(\cos\theta_0 - \cos\theta) \tag{4–204}$$

Substituting $\dot{\theta}^2$ and $\ddot{\theta}$ from Eq. (4–204) and Eq. (4–198), respectively, into Eq. (4–197) gives

$$\frac{g}{l}\sin\theta\cos\theta - \left(\dot{\theta}_0^2 + \frac{2g}{l}(\cos\theta_0 - \cos\theta)\right)\sin\theta = 0 \tag{4–205}$$

Then, using the given initial conditions $\theta(t = 0) = 0$ and $\dot{\theta}(t = 0) = 0$, Eq. (4–205) simplifies to

$$\frac{g}{l}\sin\theta\cos\theta - \frac{2g}{l}(1 - \cos\theta)\sin\theta = 0 \tag{4–206}$$

Dropping g and l from Eq. (4–206) and simplifying, we obtain

$$\sin\theta\cos\theta - 2(1 - \cos\theta)\sin\theta = 0 \tag{4–207}$$

Equation (4–207) can be written as

$$\sin\theta\,(3\cos\theta - 2) = 0 \tag{4–208}$$

which implies that *either* $\sin\theta = 0$ *or* $3\cos\theta - 2 = 0$. However, we note that the case $\sin\theta = 0$ implies that $\theta = 0$, which corresponds to the instant when the dumbbell is vertical and, therefore, has not yet begun its motion. Therefore, the dumbbell will lose contact with the vertical wall when

$$3\cos\theta - 2 = 0 \tag{4–209}$$

Solving Eq. (4–209) for θ, we obtain

$$\theta = \cos^{-1}(2/3) \approx 48.1897 \text{ deg} \tag{4–210}$$

We note that any similarity between Eqs. (4–210) and (3–173) on page 176 from Example 3–4 is purely coincidental.

■

4.6 Impulse and Momentum for a System of Particles

4.6.1 Linear Impulse and Linear Momentum for a System of Particles

Consider again Eq. (4–38) from Section 4.3, i.e.,

$$\mathbf{F} = m^{\mathcal{N}}\bar{\mathbf{a}} \tag{4–211}$$

Suppose now that we consider motion over a time interval $t \in [t_1, t_2]$. Then, integrating Eq. (4–211) from t_1 to t_2 gives

$$\int_{t_1}^{t_2} \mathbf{F}dt = \int_{t_1}^{t_2} m^{\mathcal{N}}\bar{\mathbf{a}}dt = m({}^{\mathcal{N}}\bar{\mathbf{v}}(t_2) - {}^{\mathcal{N}}\bar{\mathbf{v}}(t_1)) = m^{\mathcal{N}}\bar{\mathbf{v}}_2 - m^{\mathcal{N}}\bar{\mathbf{v}}_1 \tag{4–212}$$

Now from Eq. (4–9) we have

$${}^{\mathcal{N}}\mathbf{G} = m^{\mathcal{N}}\bar{\mathbf{v}} = \sum_{i=1}^{n} m_i{}^{\mathcal{N}}\mathbf{v}_i = \sum_{i=1}^{n} {}^{\mathcal{N}}\mathbf{G}_i \tag{4–213}$$

where

$${}^{\mathcal{N}}\mathbf{G}_i = m_i{}^{\mathcal{N}}\mathbf{v}_i \;(i = 1, \ldots, n) \tag{4–214}$$

are the linear momenta of each particle in the system. In other words, the linear momentum of a system of particles is the sum of the linear momenta of each particle in the system. Furthermore, as we saw in Chapter 3, the quantity

$$\int_{t_1}^{t_2} \mathbf{F}dt$$

is the linear impulse of the total external force $\mathbf{F}$ applied to the system on the time interval $t \in [t_1, t_2]$. Defining

$$\hat{\mathbf{F}} \equiv \int_{t_1}^{t_2} \mathbf{F}dt \tag{4-215}$$

we have

$$\boxed{\hat{\mathbf{F}} = {}^{\mathcal{N}}\mathbf{G}(t_2) - {}^{\mathcal{N}}\mathbf{G}(t_1)} \tag{4-216}$$

Equation (4–216) is called the *principle of linear impulse and linear momentum* for a system of particles.

4.6.2 Angular Impulse and Angular Momentum for a System of Particles

Consider again the result of Eq. (4–87) from Section 4.5, i.e.,

$$\mathbf{M}_Q - (\bar{\mathbf{r}} - \mathbf{r}_Q) \times m^{\mathcal{N}}\mathbf{a}_Q = \frac{{}^{\mathcal{N}}d}{dt}\left({}^{\mathcal{N}}\mathbf{H}_Q\right) \tag{4-217}$$

Suppose now that we consider motion over a time interval $t \in [t_1, t_2]$. Then, integrating Eq. (4–217), we obtain

$$\int_{t_1}^{t_2} \left[\mathbf{M}_Q - (\bar{\mathbf{r}} - \mathbf{r}_Q) \times m^{\mathcal{N}}\mathbf{a}_Q\right] dt = \int_{t_1}^{t_2} \frac{{}^{\mathcal{N}}d}{dt}\left({}^{\mathcal{N}}\mathbf{H}_Q\right) dt \tag{4-218}$$

We observe that the first term in Eq. (4–218) is the angular impulse applied to the system relative to point Q, i.e.,

$$\hat{\mathbf{M}}_Q = \int_{t_1}^{t_2} \mathbf{M}_Q dt \tag{4-219}$$

Furthermore, the right-hand side of Eq. (4–218) is the difference between the angular momentum relative to Q at t_2 and t_1, i.e.,

$${}^{\mathcal{N}}\mathbf{H}_Q(t_2) - {}^{\mathcal{N}}\mathbf{H}_Q(t_1) = \int_{t_1}^{t_2} \frac{{}^{\mathcal{N}}d}{dt}\left({}^{\mathcal{N}}\mathbf{H}_Q\right) dt \tag{4-220}$$

Equation (4–218) then becomes

$$\boxed{\hat{\mathbf{M}}_Q - \int_{t_1}^{t_2} (\bar{\mathbf{r}} - \mathbf{r}_Q) \times m^{\mathcal{N}}\mathbf{a}_Q dt = {}^{\mathcal{N}}\mathbf{H}_Q(t_2) - {}^{\mathcal{N}}\mathbf{H}_Q(t_1)} \tag{4-221}$$

which is the *principle of angular impulse and angular momentum for a system of particles relative to an arbitrary reference point Q*. It is noted that the quantity

$$-\int_{t_1}^{t_2} (\bar{\mathbf{r}} - \mathbf{r}_Q) \times m^{\mathcal{N}}\mathbf{a}_Q dt \tag{4-222}$$

is called the *inertial angular impulse* of the reference point Q relative to the center of mass on the time interval $t \in [t_1, t_2]$.

Now suppose that we choose a point O fixed in an inertial reference frame $\mathcal{N}$ as the reference point. Because the acceleration of point O in $\mathcal{N}$ is zero, i.e., ${}^{\mathcal{N}}\mathbf{a}_O = \mathbf{0}$, the inertial moment is zero and hence the inertial angular impulse in Eq. (4-222) is zero. Equation (4-221) then reduces to

$$\boxed{\hat{\mathbf{M}}_O = {}^{\mathcal{N}}\mathbf{H}_O(t_2) - {}^{\mathcal{N}}\mathbf{H}_O(t_1)} \tag{4-223}$$

and is called the *principle of angular impulse and angular momentum for a system of particles relative to a point O fixed in an inertial reference frame* $\mathcal{N}$. Next, suppose that we choose the center of mass of the system as the reference point. Then, again, the inertial moment is zero and hence the inertial angular impulse in Eq. (4-222) is zero. Furthermore, using the definition of angular momentum for the center of mass as given in Eq. (4-17), Eq. (4-221) becomes

$$\boxed{\hat{\bar{\mathbf{M}}} = {}^{\mathcal{N}}\bar{\mathbf{H}}(t_2) - {}^{\mathcal{N}}\bar{\mathbf{H}}(t_1)} \tag{4-224}$$

where $\hat{\bar{\mathbf{M}}}$ is obtained from Eq. (4-77) and is the angular impulse due to all external forces relative to the center of mass of the system, i.e.,

$$\hat{\bar{\mathbf{M}}} = \int_{t_1}^{t_2} \bar{\mathbf{M}} dt \tag{4-225}$$

Example 4-4

A particle of mass m_1 is moving with velocity $\mathbf{v}_0$ at an angle β below the horizontal when it strikes a collar of mass m_2 that is constrained to slide along a horizontal track as shown in Fig. 4-12. Knowing that the collar is initially at rest and slides without friction, and assuming that the particle sticks to the collar immediately after impact, determine: (a) the velocity of the collar immediately after impact and (b) the impulse exerted by the collar on the particle during the impact.

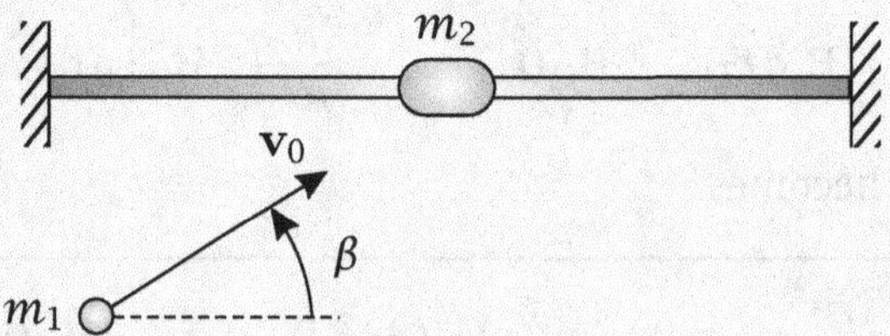

Figure 4-12 Particle of mass m_1 striking collar of mass m_2.

Solution to Example 4-4

Kinematics

Since this problem involves a translational impact between more than one body, it is necessary to apply the principle of linear impulse and momentum for a system of particles. Let $\mathcal{F}$ be a fixed reference frame. Then, choose the following coordinate system fixed in reference frame $\mathcal{F}$:

Origin at Collar when $t = 0$		
$\mathbf{E}_x$	=	To the right
$\mathbf{E}_z$	=	Into page
$\mathbf{E}_y$	=	$\mathbf{E}_z \times \mathbf{E}_x$ (=down)

The geometry of the basis $\{\mathbf{E}_x, \mathbf{E}_y, \mathbf{E}_z\}$ is shown in Fig. 4-13.

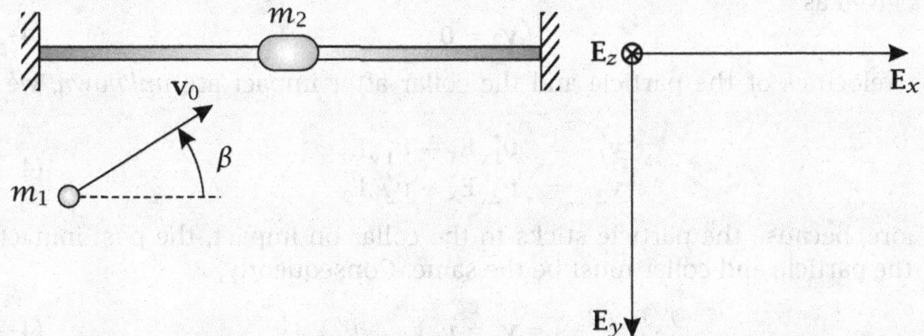

Figure 4-13 Geometry of basis $\{\mathbf{E}_x, \mathbf{E}_y, \mathbf{E}_z\}$ for Example 4-4.

The velocities of the particle and block are then given, respectively, in reference frame $\mathcal{F}$ as

$$^{\mathcal{F}}\mathbf{v}_1 = v_{1x}\mathbf{E}_x + v_{1y}\mathbf{E}_y \quad (4\text{-}226)$$

$$^{\mathcal{F}}\mathbf{v}_2 = v_{2x}\mathbf{E}_x + v_{2y}\mathbf{E}_y \quad (4\text{-}227)$$

Kinetics

This problem will be solved by applying the principle of linear impulse and momentum. In particular, linear impulse and momentum will be applied to the following systems: (1) the particle and collar and (2) the collar.

Application of Linear Impulse and Linear Momentum to Particle-Collar System

The principle of linear impulse and momentum for the system consisting of the particle and collar is applied using Eq. (4-216), i.e.,

$$\hat{\mathbf{F}}_{12} = {}^{\mathcal{F}}\mathbf{G}' - {}^{\mathcal{F}}\mathbf{G} \quad (4\text{-}228)$$

where $\hat{\mathbf{F}}_{12}$ is the linear impulse applied to the system during the impact, and ${}^{\mathcal{F}}\mathbf{G}$ and ${}^{\mathcal{F}}\mathbf{G}'$ are the linear momenta of the system before and after impact, respectively. Consequently, we need to determine the linear momentum of the center of mass of the system before and after impact and the external impulse applied to the system. It is important to understand that obtaining expressions for the linear momentum of the particle and collar before and after impact has to do with the *kinematics* of the problem while determining the external impulse applied to the system has to do with the *kinetics* of the problem. The linear momentum of the center of mass *before* impact is found as

$$^{\mathcal{F}}\mathbf{G} = (m_1 + m_2)^{\mathcal{F}}\bar{\mathbf{v}} = m_1{}^{\mathcal{F}}\mathbf{v}_1 + m_2{}^{\mathcal{F}}\mathbf{v}_2 \qquad (4\text{-}229)$$

Using the chosen coordinate system $\{\mathbf{E}_x, \mathbf{E}_y, \mathbf{E}_z\}$, the velocity of the particle before impact is given in terms of the angle β as

$$^{\mathcal{F}}\mathbf{v}_1 = \mathbf{v}_0 = v_0 \cos\beta\mathbf{E}_x - v_0 \sin\beta\mathbf{E}_y \qquad (4\text{-}230)$$

where $v_0 = \|\mathbf{v}_0\|$. Next, because the collar is initially motionless, its velocity before impact is given as

$$^{\mathcal{F}}\mathbf{v}_2 = \mathbf{0} \qquad (4\text{-}231)$$

Since the velocities of the particle and the collar after impact are *unknown*, we can write

$$\begin{aligned} ^{\mathcal{F}}\mathbf{v}_1' &= v_{1x}'\mathbf{E}_x + v_{1y}'\mathbf{E}_y \\ ^{\mathcal{F}}\mathbf{v}_2' &= v_{2x}'\mathbf{E}_x + v_{2y}'\mathbf{E}_y \end{aligned} \qquad (4\text{-}232)$$

Furthermore, because the particle sticks to the collar on impact, the post-impact velocity of the particle and collar must be the same. Consequently,

$$^{\mathcal{F}}\mathbf{v}_1' = {}^{\mathcal{F}}\mathbf{v}_2' = \mathbf{v}' = v_x'\mathbf{E}_x + v_y'\mathbf{E}_y \qquad (4\text{-}233)$$

Also, because the system must move in the $\mathbf{E}_x$-direction after impact, we have

$$v_y' = 0 \qquad (4\text{-}234)$$

Consequently, the post-impact velocity of the particle and collar can be written as

$$^{\mathcal{F}}\mathbf{v}' = v_x'\mathbf{E}_x \qquad (4\text{-}235)$$

where v_x' has yet to be determined.

The external impulse applied to the particle-collar system is determined by examining the free body diagram of the system during impact as shown in Fig. 4-14.

Figure 4-14 Free body diagram of system consisting of particle and collar during impact for Example 4-4.

It can be seen that the only impulse that is external *to the particle-collar system* is the reaction impulse, $\hat{\mathbf{R}}$, of the track. Because the collar slides without friction, this reaction impulse of the track on the particle-collar system can be written as

$$\hat{\mathbf{R}} = \hat{R}\mathbf{E}_y \qquad (4\text{-}236)$$

Consequently, the total external impulse applied to the system, denoted $\hat{\mathbf{F}}_{12}$, is given as

$$\hat{\mathbf{F}}_{12} = \hat{\mathbf{R}} = \hat{R}\mathbf{E}_y \qquad (4\text{-}237)$$

Substituting ${}^{\mathcal{F}}\mathbf{v}_1$ from Eq. (4-230), ${}^{\mathcal{F}}\mathbf{v}_2$ from Eq. (4-231), ${}^{\mathcal{F}}\mathbf{v}'$ from Eq. (4-235), and $\hat{\mathbf{F}}_{12}$ from Eq. (4-237) into (4-216), we obtain

$$\hat{R}\mathbf{E}_y = m_1 v_x' \mathbf{E}_x + m_2 v_x' \mathbf{E}_x - m_1(v_0 \cos\beta \mathbf{E}_x - v_0 \sin\beta \mathbf{E}_y) \tag{4-238}$$

Rearranging Eq. (4-238), we obtain

$$\hat{R}\mathbf{E}_y = ((m_1 + m_2)v_x' - m_1 v_0 \cos\beta)\mathbf{E}_x + m_1 v_0 \sin\beta \mathbf{E}_y \tag{4-239}$$

Equating components in Eq. (4-239) results in the following two scalar equations:

$$\hat{R} = m_1 v_0 \sin\beta \tag{4-240}$$

$$0 = (m_1 + m_2)v_x' - m_1 v_0 \cos\beta \tag{4-241}$$

It is noted that $\hat{R}$ is unknown in Eq. (4-240) while v_x' is unknown in Eq. (4-241). These two unknowns are solved for at the end of this example.

Application of Linear Impulse and Linear Momentum to Particle

Because the particle is a system consisting of only a single body, the principle of linear impulse and momentum is applied to the particle using Eq. (3-179) on page 177, i.e.,

$$\hat{\mathbf{F}}_1 = {}^{\mathcal{F}}\mathbf{G}_1' - {}^{\mathcal{F}}\mathbf{G}_1 \tag{4-242}$$

where $\hat{\mathbf{F}}_1$ is the linear impulse applied to the particle during the impact, and ${}^{\mathcal{F}}\mathbf{G}_1$ and ${}^{\mathcal{F}}\mathbf{G}_1'$ are the linear momenta of the particle before and after impact, respectively. In this case we need to determine the linear momentum of only the particle before and after impact and the external impulse applied to the particle. The linear momentum of the particle before impact is found using Eq. (4-230) as

$${}^{\mathcal{F}}\mathbf{G}_1 = m_1 {}^{\mathcal{F}}\mathbf{v}_1 = m_1(v_0 \cos\beta \mathbf{E}_x - v_0 \sin\beta \mathbf{E}_y) \tag{4-243}$$

Furthermore, using Eq. (4-235), the linear momentum of the particle after impact is given as

$${}^{\mathcal{F}}\mathbf{G}_1' = m_1 {}^{\mathcal{F}}\mathbf{v}_1' = m_1 v_x' \mathbf{E}_x \tag{4-244}$$

The external impulse applied to the particle is determined by examining the free body diagram of the particle during impact as shown in Fig. 4-15.

Figure 4-15 Free body diagram of particle during impact for Example 4-4.

It can be seen from Fig. 4-15 that the only external impulse applied to the particle during impact is the impulse applied by the collar, denoted $\hat{\mathbf{P}}$. Because the direction of $\hat{\mathbf{P}}$ is not known at this point, we have

$$\hat{\mathbf{P}} = \hat{P}_x \mathbf{E}_x + \hat{P}_y \mathbf{E}_y \tag{4-245}$$

The total external impulse applied to the particle, denoted $\hat{\mathbf{F}}_1$, is then given as

$$\hat{\mathbf{F}}_1 = \hat{\mathbf{P}} = \hat{P}_x \mathbf{E}_x + \hat{P}_y \mathbf{E}_y \tag{4-246}$$

Then, applying Eq. (3-179) to the particle, we obtain

$$\hat{P}_x \mathbf{E}_x + \hat{P}_y \mathbf{E}_y = m_1 v'_x \mathbf{E}_x - m_1 (v_0 \cos\beta \mathbf{E}_x - v_0 \sin\beta \mathbf{E}_y) \tag{4-247}$$

Equation (4-247) can be rewritten as

$$\hat{P}_x \mathbf{E}_x + \hat{P}_y \mathbf{E}_y = (m_1 v'_x - m_1 v_0 \cos\beta)\mathbf{E}_x + m_1 v_0 \sin\beta \mathbf{E}_y \tag{4-248}$$

Equating components in Eq. (4-248) results in the following two scalar equations:

$$\hat{P}_x = m_1 v'_x - m_1 v_0 \cos\beta \tag{4-249}$$
$$\hat{P}_y = m_1 v_0 \sin\beta \tag{4-250}$$

Now that the principle of linear impulse and linear momentum has been applied to the two aforementioned systems, we can proceed to determine the solutions to the questions asked in parts (a), (b), and (c).

(a) Velocity of Collar the Instant After Impact

The velocity of the collar the instant after impact is obtained by solving Eq. (4-241) for v'_x as

$$v'_x = \frac{m_1}{m_1 + m_2} v_0 \cos\beta \tag{4-251}$$

The velocity of the collar the instant after impact is then given as

$$\mathbf{v}' = \frac{m_1}{m_1 + m_2} v_0 \cos\beta \mathbf{E}_x \tag{4-252}$$

(b) Impulse Exerted by Collar on Particle During Impact

As stated earlier, the impulse exerted by the collar on the particle is $\hat{\mathbf{P}}$. In order to determine $\hat{\mathbf{P}}$, we use the results obtained in Eqs. (4-249) and (4-250) in conjunction with the result from part (a). First, substituting v'_x from Eq. (4-251) into (4-249), we obtain $\hat{P}_x$ as

$$\hat{P}_x = m_1 \frac{m_1}{m_1 + m_2} v_0 \cos\beta - m_1 v_0 \cos\beta = -\frac{m_1 m_2}{m_1 + m_2} v_0 \cos\beta \tag{4-253}$$

Next, $\hat{P}_y$ is obtained directly from Eq. (4-250) as

$$\hat{P}_y = m_1 v_0 \sin\beta \tag{4-254}$$

Consequently, using Eqs. (4-253) and (4-254), the impulse $\hat{\mathbf{P}}$ applied by the collar on the particle during impact is given as

$$\hat{\mathbf{P}} = \hat{P}_x \mathbf{E}_x + \hat{P}_y \mathbf{E}_y = -\frac{m_1 m_2}{m_1 + m_2} v_0 \cos\beta \mathbf{E}_x + m_1 v_0 \sin\beta \mathbf{E}_y \tag{4-255}$$

Note that the impulse $\hat{\mathbf{P}}$ does not lie in the direction of the pre-impact velocity of the particle, i.e., $\hat{\mathbf{P}}$ does not lie in the direction of $\mathbf{v}_0$.

■

Example 4–5

A dumbbell consists of two particles of mass m_1 and m_2 connected by a rigid massless rod as shown in Fig. 4-16. The dumbbell is initially motionless at an angle θ with the horizontal when mass m_1 is struck by a horizontal impulse $\hat{\mathbf{P}}$. Determine (a) the angular velocity of the dumbbell immediately after impact and (b) the velocity of each particle immediately after impact.

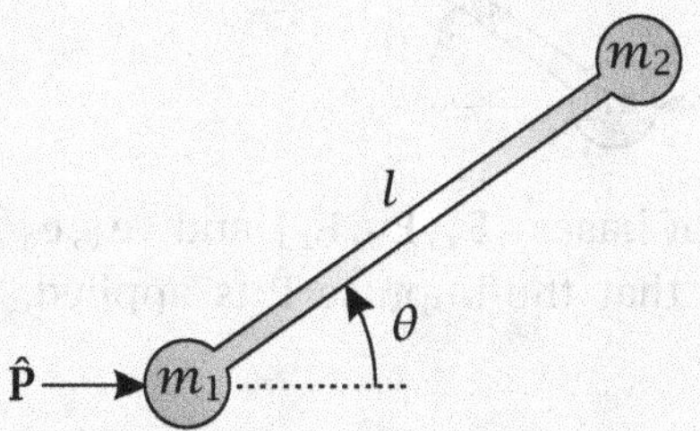

Figure 4-16 Impulse striking dumbbell.

Solution to Example 4–5

The objective of this problem is to demonstrate the proper use of the principle of linear impulse and linear momentum and the principle of angular impulse and angular momentum for a system of particles. Furthermore, this problem is an excellent example of how a proper choice of reference frames can increase the tractability of a problem.

Kinematics

First, let $\mathcal{F}$ be a fixed reference frame. Then, choose the following coordinate system fixed in reference frame $\mathcal{F}$:

$$
\begin{array}{ccc}
 & \text{Origin at } m_1 \text{ at } t = 0^- & \\
\mathbf{E}_y & = & \text{Along } m_1 m_2 \\
\mathbf{E}_z & = & \text{Out of page} \\
\mathbf{E}_x & = & \mathbf{E}_y \times \mathbf{E}_z
\end{array}
$$

Next, let $\mathcal{R}$ be a reference frame that is fixed to the dumbbell. Then, choose the following coordinate system fixed in reference frame $\mathcal{R}$:

$$
\begin{array}{ccc}
 & \text{Origin at } m_1 & \\
\mathbf{e}_y & = & \text{Along } m_1 m_2 \\
\mathbf{e}_z & = & \text{Out of page } (= \mathbf{E}_z) \\
\mathbf{e}_x & = & \mathbf{e}_y \times \mathbf{e}_z
\end{array}
$$

The geometry of the bases $\{\mathbf{E}_x, \mathbf{E}_y, \mathbf{E}_z\}$ and $\{\mathbf{e}_x, \mathbf{e}_y, \mathbf{e}_z\}$ is shown in Fig. 4-17. It is noted that, at the instant that the impulse $\hat{\mathbf{P}}$ is applied, the bases $\{\mathbf{E}_x, \mathbf{E}_y, \mathbf{E}_z\}$ and $\{\mathbf{e}_x, \mathbf{e}_y, \mathbf{e}_z\}$ are aligned.

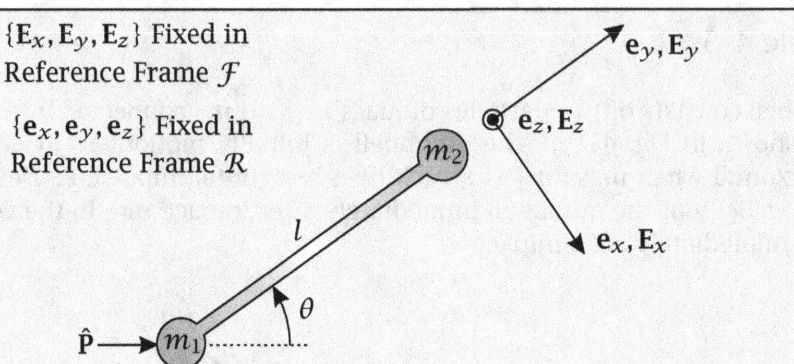

Figure 4–17 Geometry of bases $\{\mathbf{E}_x, \mathbf{E}_y, \mathbf{E}_z\}$ and $\{\mathbf{e}_x, \mathbf{e}_y, \mathbf{e}_z\}$ for Example 4–5. It is noted that, at the instant that the impulse $\hat{\mathbf{P}}$ is applied, the bases $\{\mathbf{E}_x, \mathbf{E}_y, \mathbf{E}_z\}$ and $\{\mathbf{e}_x, \mathbf{e}_y, \mathbf{e}_z\}$ are aligned.

Kinematics of Each Particle

Since the motion of the dumbbell is planar, the angular velocity of the dumbbell in reference frame $\mathcal{F}$ is given as

$$^{\mathcal{F}}\boldsymbol{\omega}^{\mathcal{R}} = \omega \mathbf{e}_z \tag{4–256}$$

In terms of the basis $\{\mathbf{E}_x, \mathbf{E}_y, \mathbf{E}_z\}$, the position of particle m_1 can be written as

$$\mathbf{r}_1 = x_1 \mathbf{E}_x + y_1 \mathbf{E}_y \tag{4–257}$$

Computing the rate of change of $\mathbf{r}_1$ in reference frame $\mathcal{F}$, we obtain the velocity of particle m_1 as

$$^{\mathcal{F}}\mathbf{v}_1 = \frac{^{\mathcal{F}}d\mathbf{r}_1}{dt} = \dot{x}_1 \mathbf{E}_x + \dot{y}_1 \mathbf{E}_y = v_{1x}\mathbf{E}_x + v_{1y}\mathbf{E}_y \tag{4–258}$$

Furthermore, the position of particle m_2 is given as

$$\mathbf{r}_2 = \mathbf{r}_1 + \mathbf{r}_{2/1} \tag{4–259}$$

Now, because the dumbbell is rigid and has length l, the position of particle m_2 relative to m_1 is given as

$$\mathbf{r}_{2/1} = l\mathbf{e}_y \tag{4–260}$$

Using the results of Eqs. (4–257) and (4–260), we obtain the position of particle m_2 as

$$\mathbf{r}_2 = x_1 \mathbf{E}_x + y_1 \mathbf{E}_y + l\mathbf{e}_y \tag{4–261}$$

Computing the rate of change of $\mathbf{r}_2$ in reference frame $\mathcal{F}$, we have

$$^{\mathcal{F}}\mathbf{v}_2 = \frac{^{\mathcal{F}}d\mathbf{r}_2}{dt} = \frac{^{\mathcal{F}}d\mathbf{r}_1}{dt} + \frac{^{\mathcal{F}}d\mathbf{r}_{2/1}}{dt} = {}^{\mathcal{F}}\mathbf{v}_1 + {}^{\mathcal{F}}\mathbf{v}_{2/1} \tag{4–262}$$

The quantity $^{\mathcal{F}}\mathbf{v}_1$ has already been computed in Eq. (4–258). Computing the rate of change of $\mathbf{r}_{2/1}$ in reference frame $\mathcal{F}$, we obtain

$$^{\mathcal{F}}\mathbf{v}_{2/1} = \frac{^{\mathcal{F}}d}{dt}(\mathbf{r}_{2/1}) = \frac{^{\mathcal{R}}d}{dt}(\mathbf{r}_{2/1}) + {}^{\mathcal{F}}\boldsymbol{\omega}^{\mathcal{R}} \times \mathbf{r}_{2/1} \tag{4–263}$$

Now we have

$$\frac{{}^{\mathcal{R}}d}{dt}(\mathbf{r}_{2/1}) = \mathbf{0} \tag{4-264}$$

$${}^{\mathcal{F}}\boldsymbol{\omega}^{\mathcal{R}} \times \mathbf{r}_{2/1} = \omega\mathbf{e}_z \times l\mathbf{e}_y = -l\omega\mathbf{e}_x \tag{4-265}$$

Adding Eqs. (4-264) and (4-265), we obtain ${}^{\mathcal{F}}\mathbf{v}_{2/1}$ as

$${}^{\mathcal{F}}\mathbf{v}_{2/1} = -l\omega\mathbf{e}_x \tag{4-266}$$

Then, adding Eqs. (4-258) and (4-266), the velocity of particle m_2 in reference frame $\mathcal{F}$ is given as

$${}^{\mathcal{F}}\mathbf{v}_2 = {}^{\mathcal{F}}\mathbf{v}_1 + {}^{\mathcal{F}}\mathbf{v}_{2/1} = v_{1x}\mathbf{E}_x + v_{1y}\mathbf{E}_y - l\omega\mathbf{e}_x \tag{4-267}$$

Kinematics of Center of Mass of Dumbbell

The position of the center of mass of the dumbbell is given as

$$\bar{\mathbf{r}} = \frac{m_1\mathbf{r}_1 + m_2\mathbf{r}_2}{m_1 + m_2} \tag{4-268}$$

Substituting the expressions for $\mathbf{r}_1$ and $\mathbf{r}_2$ from Eqs. (4-257) and (4-261), respectively, we obtain $\bar{\mathbf{r}}$ as

$$\begin{aligned}\bar{\mathbf{r}} &= \frac{m_1(x_1\mathbf{E}_x + y_1\mathbf{E}_y) + m_2(x_1\mathbf{E}_x + y_1\mathbf{E}_y + l\mathbf{e}_y)}{m_1 + m_2} \\ &= x_1\mathbf{E}_x + y_1\mathbf{E}_y + \frac{m_2}{m_1 + m_2}l\mathbf{e}_y\end{aligned} \tag{4-269}$$

The velocity of the center of mass of the dumbbell in reference frame $\mathcal{F}$ is then given as

$${}^{\mathcal{F}}\bar{\mathbf{v}} = \frac{m_1{}^{\mathcal{F}}\mathbf{v}_1 + m_2{}^{\mathcal{F}}\mathbf{v}_2}{m_1 + m_2} \tag{4-270}$$

Substituting the expressions for ${}^{\mathcal{F}}\mathbf{v}_1$ and ${}^{\mathcal{F}}\mathbf{v}_2$ from Eqs. (4-258) and (4-266), respectively, we obtain the velocity of the center of mass of the dumbbell in reference frame $\mathcal{F}$ as

$$\begin{aligned}{}^{\mathcal{F}}\bar{\mathbf{v}} &= \frac{m_1(v_{1x}\mathbf{E}_x + v_{1y}\mathbf{E}_y) + m_2(v_{1x}\mathbf{E}_x + v_{1y}\mathbf{E}_y - l\omega\mathbf{e}_y)}{m_1 + m_2} \\ &= v_{1x}\mathbf{E}_x + v_{1y}\mathbf{E}_y - \frac{m_2}{m_1 + m_2}l\omega\mathbf{e}_x\end{aligned} \tag{4-271}$$

Kinematics Relative to Center of Mass of Dumbbell

For this problem the important kinematic quantities relative to the center of mass of the dumbbell are the position and velocity of each particle and the angular momentum. Using the expressions for $\mathbf{r}_1$, $\mathbf{r}_2$, and $\bar{\mathbf{r}}$ from Eqs. (4-257), (4-261), and (4-269), respectively, we have

$$\mathbf{r}_1 - \bar{\mathbf{r}} = x_1\mathbf{E}_x + y_1\mathbf{E}_y - \left(x_1\mathbf{E}_x + y_1\mathbf{E}_y + \frac{m_2}{m_1 + m_2}l\mathbf{e}_y\right)$$

$$= -\frac{m_2}{m_1+m_2} l\mathbf{e}_y \tag{4-272}$$

$$\begin{aligned}\mathbf{r}_2 - \bar{\mathbf{r}} &= x_1\mathbf{E}_x + y_1\mathbf{E}_y + l\mathbf{e}_y - \left(x_1\mathbf{E}_x + y_1\mathbf{E}_y + \frac{m_2}{m_1+m_2} l\mathbf{e}_y\right) \\ &= \frac{m_1}{m_1+m_2} l\mathbf{e}_y \end{aligned} \tag{4-273}$$

Furthermore, using the expressions for ${}^{\mathcal{F}}\mathbf{v}_1$, ${}^{\mathcal{F}}\mathbf{v}_2$, and ${}^{\mathcal{F}}\bar{\mathbf{v}}$, from Eqs. (4-258), (4-267), and (4-271), respectively, the velocities of each particle relative to the center of mass of the dumbbell are given as

$$\begin{aligned}{}^{\mathcal{F}}\mathbf{v}_1 - {}^{\mathcal{F}}\bar{\mathbf{v}} &= v_{1x}\mathbf{E}_x + v_{1y}\mathbf{E}_y - \left(v_{1x}\mathbf{E}_x + v_{1y}\mathbf{E}_y - \frac{m_2}{m_1+m_2} l\omega\mathbf{e}_x\right) \\ &= \frac{m_2}{m_1+m_2} l\omega\mathbf{e}_x \end{aligned} \tag{4-274}$$

$$\begin{aligned}{}^{\mathcal{F}}\mathbf{v}_2 - {}^{\mathcal{F}}\bar{\mathbf{v}} &= v_{1x}\mathbf{E}_x + v_{1y}\mathbf{E}_y - l\omega\mathbf{e}_x - \left(v_{1x}\mathbf{E}_x + v_{1y}\mathbf{E}_y - \frac{m_2}{m_1+m_2} l\omega\mathbf{e}_x\right) \\ &= -\frac{m_1}{m_1+m_2} l\omega\mathbf{e}_x \end{aligned} \tag{4-275}$$

Finally, the angular momentum of the dumbbell relative to the center of mass of the dumbbell in reference frame $\mathcal{F}$ is given as

$${}^{\mathcal{F}}\bar{\mathbf{H}} = (\mathbf{r}_1 - \bar{\mathbf{r}}) \times m_1({}^{\mathcal{F}}\mathbf{v}_1 - {}^{\mathcal{F}}\bar{\mathbf{v}}) + (\mathbf{r}_2 - \bar{\mathbf{r}}) \times m_2({}^{\mathcal{F}}\mathbf{v}_2 - {}^{\mathcal{F}}\bar{\mathbf{v}}) \tag{4-276}$$

Using the expressions for $\mathbf{r}_1 - \bar{\mathbf{r}}$, $\mathbf{r}_2 - \bar{\mathbf{r}}$, ${}^{\mathcal{F}}\mathbf{v}_1 - {}^{\mathcal{F}}\bar{\mathbf{v}}$, and ${}^{\mathcal{F}}\mathbf{v}_2 - {}^{\mathcal{F}}\bar{\mathbf{v}}$, from Eqs. (4-272), (4-273), (4-274), and (4-275), respectively, we obtain ${}^{\mathcal{F}}\bar{\mathbf{H}}$ as

$${}^{\mathcal{F}}\bar{\mathbf{H}} = -\frac{m_2}{m_1+m_2} l\mathbf{e}_y \times m_1 \frac{m_2}{m_1+m_2} l\omega\mathbf{e}_x + \frac{m_1}{m_1+m_2} l\mathbf{e}_y \times \left(-\frac{m_1}{m_1+m_2} l^2\omega\mathbf{e}_x\right) \tag{4-277}$$

Simplifying Eq. (4-277) gives

$${}^{\mathcal{F}}\bar{\mathbf{H}} = \frac{m_1 m_2^2 + m_2 m_1^2}{(m_1+m_2)^2} l^2\omega\mathbf{e}_z = \frac{m_1 m_2}{m_1+m_2} l^2\omega\mathbf{e}_z \tag{4-278}$$

Kinetics

For this problem it is sufficient to consider the *entire* system consisting of both particles and the dumbbell. The free body diagram of the system is shown in Fig. 4-18.

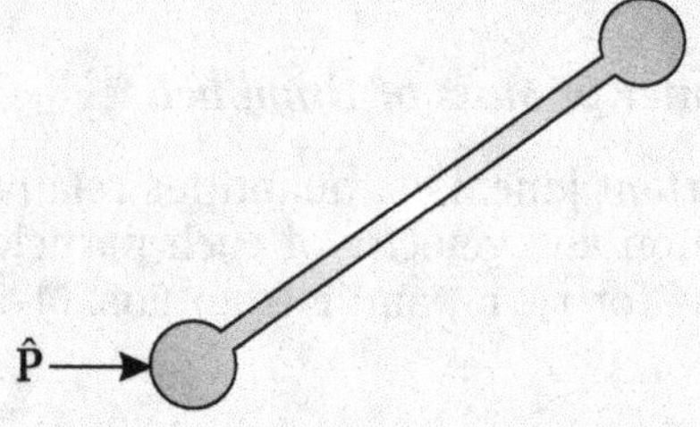

Figure 4-18 Free body diagram of entire system for Example 4-5.

It is seen that the only external impulse acting on the system is $\hat{\mathbf{P}}$. Resolving the impulse $\hat{\mathbf{P}}$ in the basis $\{\mathbf{E}_x, \mathbf{E}_y, \mathbf{E}_z\}$, we obtain

$$\hat{\mathbf{P}} = \hat{P}\sin\theta\,\mathbf{E}_x + \hat{P}\cos\theta\,\mathbf{E}_y \tag{4-279}$$

Now for this problem we need to apply both the principle of linear impulse and momentum and the principle of angular impulse and momentum.

Application of Linear Impulse and Linear Momentum to Dumbbell

Applying the principle of linear impulse and linear momentum to the entire system in reference frame $\mathcal{F}$, we have

$$\hat{\mathbf{F}} = {}^{\mathcal{F}}\mathbf{G}' - {}^{\mathcal{F}}\mathbf{G} \tag{4-280}$$

where ${}^{\mathcal{F}}\mathbf{G}$ and ${}^{\mathcal{F}}\mathbf{G}'$ are the linear momenta of the system the instants before and after the application of $\hat{\mathbf{P}}$, respectively. Now it is seen that $\hat{\mathbf{P}}$ is the only impulse acting on the system. Using the expression for $\hat{\mathbf{P}}$ from Eq. (4-279), we have

$$\hat{\mathbf{F}} = \hat{\mathbf{P}} = \hat{P}\sin\theta\,\mathbf{E}_x + \hat{P}\cos\theta\,\mathbf{E}_y \tag{4-281}$$

Furthermore, because the dumbbell is at rest before $\hat{\mathbf{P}}$ is applied, we have

$${}^{\mathcal{F}}\mathbf{G} = \mathbf{0} \tag{4-282}$$

Finally, the linear momentum of the system immediately after the impulse $\hat{\mathbf{P}}$ is applied is given as

$${}^{\mathcal{F}}\mathbf{G}' = m\,{}^{\mathcal{F}}\bar{\mathbf{v}}' = (m_1 + m_2)\,{}^{\mathcal{F}}\bar{\mathbf{v}}' \tag{4-283}$$

where ${}^{\mathcal{F}}\bar{\mathbf{v}}'$ is the velocity of the center of mass in reference frame $\mathcal{F}$ immediately after $\hat{\mathbf{P}}$ is applied. Then, using the expression for ${}^{\mathcal{F}}\bar{\mathbf{v}}'$ from Eq. (4-271), we obtain ${}^{\mathcal{F}}\mathbf{G}'$ as

$${}^{\mathcal{F}}\mathbf{G}' = (m_1 + m_2)\left(v_{1x}'\mathbf{E}_x + v_{1y}'\mathbf{E}_y - \frac{m_2}{m_1 + m_2}l\omega'\mathbf{e}_x\right) \tag{4-284}$$

Rewriting Eq. (4-284), we obtain

$${}^{\mathcal{F}}\mathbf{G}' = (m_1 + m_2)v_{1x}'\mathbf{E}_x + (m_1 + m_2)v_{1y}'\mathbf{E}_y - m_2 l\omega'\mathbf{e}_x \tag{4-285}$$

Because $\mathbf{e}_x$ and $\mathbf{E}_x$ are aligned at the instant the impulse $\hat{\mathbf{P}}$ is applied, Eq. (4-285) simplifies to

$${}^{\mathcal{F}}\mathbf{G}' = [(m_1 + m_2)v_{1x}' - m_2 l\omega']\,\mathbf{E}_x + (m_1 + m_2)v_{1y}'\mathbf{E}_y \tag{4-286}$$

Substituting ${}^{\mathcal{F}}\mathbf{G}'$ from Eq. (4-286) and $\hat{\mathbf{F}}$ from Eq. (4-281) into (4-280), we obtain

$$\hat{P}\sin\theta\,\mathbf{E}_x + \hat{P}\cos\theta\,\mathbf{E}_y = [(m_1 + m_2)v_{1x}' - m_2 l\omega']\,\mathbf{E}_x + (m_1 + m_2)v_{1y}'\mathbf{E}_y \tag{4-287}$$

Equating components in Eq. (4-287), we obtain the following two scalar equations:

$$\hat{P}\sin\theta = (m_1 + m_2)v_{1x}' - m_2 l\omega' \tag{4-288}$$

$$\hat{P}\cos\theta = (m_1 + m_2)v_{1y}' \tag{4-289}$$

Application of Angular Impulse and Angular Momentum to Dumbbell

For this problem it is convenient to apply angular impulse and angular momentum in reference frame $\mathcal{F}$ relative to the center of mass of the dumbbell. Consequently, we have

$$\hat{\bar{\mathbf{M}}} = {}^{\mathcal{F}}\bar{\mathbf{H}}' - {}^{\mathcal{F}}\bar{\mathbf{H}} \tag{4-290}$$

Because the only impulse applied to the dumbbell is $\hat{\mathbf{P}}$ and $\hat{\mathbf{P}}$ is applied at the location of particle m_1, the angular impulse applied to the dumbbell relative to the center of mass of the dumbbell is given as

$$\hat{\bar{\mathbf{M}}} = (\mathbf{r}_1 - \bar{\mathbf{r}}) \times \hat{\mathbf{P}} \tag{4-291}$$

Using the expressions for $\mathbf{r}_1 - \bar{\mathbf{r}}$ and $\hat{\mathbf{P}}$ from Eqs. (4-272) and (4-279), respectively, we obtain $\hat{\bar{\mathbf{M}}}$ as

$$\hat{\bar{\mathbf{M}}} = -\frac{m_2}{m_1 + m_2} l\mathbf{e}_y \times (\hat{P}\sin\theta\mathbf{E}_x + \hat{P}\cos\theta\mathbf{E}_y) \tag{4-292}$$

However, because $\mathbf{e}_y$ and $\mathbf{E}_y$ are aligned at the instant that $\hat{\mathbf{P}}$ is applied, the angular impulse applied relative to the center of mass of the dumbbell simplifies to

$$\begin{aligned}\hat{\bar{\mathbf{M}}} &= -\frac{m_2}{m_1 + m_2} l\mathbf{E}_y \times (\hat{P}\sin\theta\mathbf{E}_x + \hat{P}\cos\theta\mathbf{E}_y) \\ &= \frac{m_2}{m_1 + m_2} l\hat{P}\sin\theta\mathbf{E}_z = \frac{m_2}{m_1 + m_2} l\hat{P}\sin\theta\mathbf{e}_z\end{aligned} \tag{4-293}$$

Next, because the dumbbell is at rest the instant before the impulse $\hat{\mathbf{P}}$ is applied, we have

$${}^{\mathcal{F}}\bar{\mathbf{H}} = \mathbf{0} \tag{4-294}$$

Furthermore, using the expression for ${}^{\mathcal{F}}\bar{\mathbf{H}}$ from Eq. (4-278), the angular momentum of the dumbbell relative to the center of mass of the dumbbell in reference frame $\mathcal{F}$ the instant after the impulse $\hat{\mathbf{P}}$ is applied is given as

$${}^{\mathcal{F}}\bar{\mathbf{H}}' = \frac{m_1 m_2}{m_1 + m_2} l^2 \omega' \mathbf{e}_z \tag{4-295}$$

Setting the results of Eqs. (4-293) and (4-295) equal, we obtain

$$\frac{m_2}{m_1 + m_2} l\hat{P}\sin\theta\mathbf{e}_z = \frac{m_1 m_2}{m_1 + m_2} l^2 \omega' \mathbf{e}_z \tag{4-296}$$

Simplifying Eq. (4-296) gives

$$\omega' = \frac{\hat{P}\sin\theta}{m_1 l} \tag{4-297}$$

(a) Determination of Angular Velocity of Dumbbell the Instant After $\hat{\mathbf{P}}$ Is Applied

The angular velocity of the dumbbell the instant after $\hat{\mathbf{P}}$ is applied is obtained directly from Eq. (4-297) and is given as

$$\left({}^{\mathcal{F}}\boldsymbol{\omega}^{\mathcal{R}}\right)' = \omega'\mathbf{e}_z = \frac{\hat{P}\sin\theta}{m_1 l}\mathbf{e}_z \tag{4-298}$$

(b) Determination of Velocity of Each Particle the Instant After $\hat{\mathbf{P}}$ Is Applied

The velocities of each particle in reference frame $\mathcal{F}$ the instant after the impulse $\hat{\mathbf{P}}$ is applied are obtained using Eqs. (4-288), (4-289), and (4-297. Substituting the result of Eq. (4-297) into (4-288), we obtain

$$\hat{P}\sin\theta = (m_1 + m_2)v'_{1x} - m_2 l \frac{\hat{P}\sin\theta}{m_1 l} \tag{4-299}$$

Rearranging Eq. (4-299), we obtain

$$(m_1 + m_2)v'_{1x} = \left(1 + \frac{m_2}{m_1}\right)\hat{P}\sin\theta \tag{4-300}$$

Equation (4-300) can be rewritten as

$$(m_1 + m_2)v'_{1x} = \frac{m_1 + m_2}{m_1}\hat{P}\sin\theta \tag{4-301}$$

Solving Eq. (4-301) for v'_{1x}, we obtain

$$v'_{1x} = \frac{\hat{P}\sin\theta}{m_1} \tag{4-302}$$

Next, solving Eq. (4-289) for v'_{1y} gives

$$v'_{1y} = \frac{\hat{P}\cos\theta}{m_1 + m_2} \tag{4-303}$$

Substituting the results of Eqs. (4-302) and (4-303) into Eq. (4-258), we obtain the velocity of particle m_1 in reference frame $\mathcal{F}$ the instant after the impulse $\hat{\mathbf{P}}$ is applied as

$$^{\mathcal{F}}\mathbf{v}'_1 = \frac{\hat{P}\sin\theta}{m_1}\mathbf{E}_x + \frac{\hat{P}\cos\theta}{m_1 + m_2}\mathbf{E}_y \tag{4-304}$$

Then, substituting the results of Eqs. (4-297), (4-302), and (4-303) into Eq. (4-267), we obtain the velocity of particle m_2 in reference frame $\mathcal{F}$ the instant after the impulse $\hat{\mathbf{P}}$ is applied as

$$^{\mathcal{F}}\mathbf{v}'_2 = \frac{\hat{P}\sin\theta}{m_1}\mathbf{E}_x + \frac{\hat{P}\cos\theta}{m_1 + m_2}\mathbf{E}_y - \frac{\hat{P}\sin\theta}{m_1}\mathbf{e}_x \tag{4-305}$$

Finally, because the directions $\mathbf{e}_x$ and $\mathbf{E}_x$ are aligned at the instant that the impulse $\hat{\mathbf{P}}$ is applied, Eq. (4-305) simplifies to

$$^{\mathcal{F}}\mathbf{v}'_2 = \frac{\hat{P}\cos\theta}{m_1 + m_2}\mathbf{E}_y \tag{4-306}$$

■

4.7 Work and Energy for a System of Particles

4.7.1 Kinetic Energy for a System of Particles

The kinetic energy for a system of particles in an inertial reference frame $\mathcal{N}$ is defined as

$$^{\mathcal{N}}T = \frac{1}{2}\sum_{i=1}^{n} m_i {}^{\mathcal{N}}\mathbf{v}_i \cdot {}^{\mathcal{N}}\mathbf{v}_i \tag{4-307}$$

An alternate expression for the kinetic energy is obtained as follows. First, we have

$$^{\mathcal{N}}\mathbf{v}_i = {}^{\mathcal{N}}\mathbf{v}_i - {}^{\mathcal{N}}\bar{\mathbf{v}} + {}^{\mathcal{N}}\bar{\mathbf{v}} \tag{4-308}$$

where it is noted in Eq. (4–308) that the velocity of the center of mass of the system, ${}^{\mathcal{N}}\bar{\mathbf{v}}$, has merely been added and subtracted. Substituting Eq. (4–308) into (4–307), we obtain

$$^{\mathcal{N}}T = \frac{1}{2}\sum_{i=1}^{n} m_i \left({}^{\mathcal{N}}\mathbf{v}_i - {}^{\mathcal{N}}\bar{\mathbf{v}} + {}^{\mathcal{N}}\bar{\mathbf{v}}\right) \cdot \left({}^{\mathcal{N}}\mathbf{v}_i - {}^{\mathcal{N}}\bar{\mathbf{v}} + {}^{\mathcal{N}}\bar{\mathbf{v}}\right) \tag{4-309}$$

Expanding Eq. (4–309) gives

$$\begin{aligned} ^{\mathcal{N}}T = \frac{1}{2}\Bigg[&\sum_{i=1}^{n} m_i \left({}^{\mathcal{N}}\mathbf{v}_i - {}^{\mathcal{N}}\bar{\mathbf{v}}\right) \cdot \left({}^{\mathcal{N}}\mathbf{v}_i - {}^{\mathcal{N}}\bar{\mathbf{v}}\right) + \sum_{i=1}^{n} m_i \left({}^{\mathcal{N}}\mathbf{v}_i - {}^{\mathcal{N}}\bar{\mathbf{v}}\right) \cdot {}^{\mathcal{N}}\bar{\mathbf{v}} \\ &+ \sum_{i=1}^{n} m_i \left({}^{\mathcal{N}}\mathbf{v}_i - {}^{\mathcal{N}}\bar{\mathbf{v}}\right) \cdot {}^{\mathcal{N}}\bar{\mathbf{v}} + \sum_{i=1}^{n} m_i {}^{\mathcal{N}}\bar{\mathbf{v}} \cdot {}^{\mathcal{N}}\bar{\mathbf{v}}\Bigg] \end{aligned} \tag{4-310}$$

Combining the second and third terms in Eq. (4–310), we obtain

$$\begin{aligned} ^{\mathcal{N}}T = \frac{1}{2}\Bigg[&\sum_{i=1}^{n} m_i \left({}^{\mathcal{N}}\mathbf{v}_i - {}^{\mathcal{N}}\bar{\mathbf{v}}\right) \cdot \left({}^{\mathcal{N}}\mathbf{v}_i - {}^{\mathcal{N}}\bar{\mathbf{v}}\right) \\ &+2\sum_{i=1}^{n} m_i \left({}^{\mathcal{N}}\mathbf{v}_i - {}^{\mathcal{N}}\bar{\mathbf{v}}\right) \cdot {}^{\mathcal{N}}\bar{\mathbf{v}} + \sum_{i=1}^{n} m_i {}^{\mathcal{N}}\bar{\mathbf{v}} \cdot {}^{\mathcal{N}}\bar{\mathbf{v}}\Bigg] \end{aligned} \tag{4-311}$$

Because ${}^{\mathcal{N}}\bar{\mathbf{v}}$ is independent of the summation, we have

$$\sum_{i=1}^{n} m_i {}^{\mathcal{N}}\bar{\mathbf{v}} \cdot {}^{\mathcal{N}}\bar{\mathbf{v}} = m {}^{\mathcal{N}}\bar{\mathbf{v}} \cdot {}^{\mathcal{N}}\bar{\mathbf{v}} \tag{4-312}$$

Furthermore,

$$\sum_{i=1}^{n} m_i \left({}^{\mathcal{N}}\mathbf{v}_i - {}^{\mathcal{N}}\bar{\mathbf{v}}\right) = \sum_{i=1}^{n} m_i {}^{\mathcal{N}}\mathbf{v}_i - \sum_{i=1}^{n} m_i {}^{\mathcal{N}}\bar{\mathbf{v}} = m {}^{\mathcal{N}}\bar{\mathbf{v}} - m {}^{\mathcal{N}}\bar{\mathbf{v}} = \mathbf{0} \tag{4-313}$$

The kinetic energy of Eq. (4–311) then simplifies to

$$\boxed{^{\mathcal{N}}T = \frac{1}{2} m {}^{\mathcal{N}}\bar{\mathbf{v}} \cdot {}^{\mathcal{N}}\bar{\mathbf{v}} + \frac{1}{2}\sum_{i=1}^{n} m_i \left({}^{\mathcal{N}}\mathbf{v}_i - {}^{\mathcal{N}}\bar{\mathbf{v}}\right) \cdot \left({}^{\mathcal{N}}\mathbf{v}_i - {}^{\mathcal{N}}\bar{\mathbf{v}}\right)} \tag{4-314}$$

For compactness we can write

$$^{\mathcal{N}}T = {}^{\mathcal{N}}T^{(1)} + {}^{\mathcal{N}}T^{(2)} \tag{4-315}$$

where

$$^{\mathcal{N}}T^{(1)} = \tfrac{1}{2}m\,{}^{\mathcal{N}}\bar{\mathbf{v}} \cdot {}^{\mathcal{N}}\bar{\mathbf{v}} \tag{4-316}$$

$$^{\mathcal{N}}T^{(2)} = \tfrac{1}{2}\sum_{i=1}^{n} m_i \left({}^{\mathcal{N}}\mathbf{v}_i - {}^{\mathcal{N}}\bar{\mathbf{v}}\right) \cdot \left({}^{\mathcal{N}}\mathbf{v}_i - {}^{\mathcal{N}}\bar{\mathbf{v}}\right) \tag{4-317}$$

Equation (4-314) is called *Koenig's decomposition* and states that the kinetic energy of a system of particles, $^{\mathcal{N}}T$, is equal to the sum of the kinetic energy of the center of mass of the system, $^{\mathcal{N}}T^{(1)}$, and the kinetic energy relative to the center of mass of the system, $^{\mathcal{N}}T^{(2)}$.

4.7.2 Work-Energy Theorem for a System of Particles

Computing the rate of change of $^{\mathcal{N}}T^{(1)}$ as given in Eq. (4-316), we have

$$\frac{d}{dt}\left({}^{\mathcal{N}}T^{(1)}\right) = \frac{1}{2}m\left[{}^{\mathcal{N}}\bar{\mathbf{a}} \cdot {}^{\mathcal{N}}\bar{\mathbf{v}} + {}^{\mathcal{N}}\bar{\mathbf{v}} \cdot {}^{\mathcal{N}}\bar{\mathbf{a}}\right] \tag{4-318}$$

Using the fact that the scalar product is commutative, Eq. (4-318) simplifies to

$$\frac{d}{dt}\left({}^{\mathcal{N}}T^{(1)}\right) = m\,{}^{\mathcal{N}}\bar{\mathbf{a}} \cdot {}^{\mathcal{N}}\bar{\mathbf{v}} \tag{4-319}$$

Then, substituting the result of Eq. (4-38) into (4-319), we have

$$\boxed{\frac{d}{dt}\left({}^{\mathcal{N}}T^{(1)}\right) = \mathbf{F} \cdot {}^{\mathcal{N}}\bar{\mathbf{v}} = {}^{\mathcal{N}}P^{(1)}} \tag{4-320}$$

Equation (4-320) states that the power of all *external* forces acting on a system of particles is equal to the rate of change of the kinetic energy of the center of mass of the system. Integrating Eq. (4-320) from time $t = t_1$ to $t = t_2$, we have

$$\int_{t_1}^{t_2} \frac{d}{dt}\left({}^{\mathcal{N}}T^{(1)}\right) dt = \int_{t_1}^{t_2} \mathbf{F} \cdot {}^{\mathcal{N}}\bar{\mathbf{v}}dt \tag{4-321}$$

Now it can be seen that the left-hand side of Eq. (4-321) is equal to $^{\mathcal{N}}T_2^{(1)} - {}^{\mathcal{N}}T_1^{(1)}$. Furthermore, it is seen that

$$\int_{t_1}^{t_2} \mathbf{F} \cdot {}^{\mathcal{N}}\bar{\mathbf{v}}dt$$

is the work done by all of the external forces on the time interval $t \in [t_1, t_2]$. Consequently, Eq. (4-321) can be written as

$$\boxed{{}^{\mathcal{N}}T^{(1)}(t_2) - {}^{\mathcal{N}}T^{(1)}(t_1) = {}^{\mathcal{N}}W_{12}^{(1)}} \tag{4-322}$$

Equation (4-322) states that the work done by all external forces acting on a system in moving the center of mass of the system over a time interval $t \in [t_1, t_2]$ is equal to the change in the kinetic energy associated with the center of mass of the system.

Next, computing the rate of change of the *total* kinetic energy ${}^{\mathcal{N}}T$, we have

$$\frac{d}{dt}\left({}^{\mathcal{N}}T\right) = \frac{1}{2}m\left[{}^{\mathcal{N}}\bar{\mathbf{a}}\cdot{}^{\mathcal{N}}\bar{\mathbf{v}} + {}^{\mathcal{N}}\bar{\mathbf{v}}\cdot{}^{\mathcal{N}}\bar{\mathbf{a}}\right] + \frac{1}{2}\sum_{i=1}^{n} m_i\left[\left({}^{\mathcal{N}}\mathbf{a}_i - {}^{\mathcal{N}}\bar{\mathbf{a}}\right)\cdot\left({}^{\mathcal{N}}\mathbf{v}_i - {}^{\mathcal{N}}\bar{\mathbf{v}}\right) + \left({}^{\mathcal{N}}\mathbf{v}_i - {}^{\mathcal{N}}\bar{\mathbf{v}}\right)\cdot\left({}^{\mathcal{N}}\mathbf{a}_i - {}^{\mathcal{N}}\bar{\mathbf{a}}\right)\right] \tag{4-323}$$

Using the fact that the scalar product is commutative, Eq. (4–323) simplifies to

$$\frac{d}{dt}\left({}^{\mathcal{N}}T\right) = m{}^{\mathcal{N}}\bar{\mathbf{a}}\cdot{}^{\mathcal{N}}\bar{\mathbf{v}} + \sum_{i=1}^{n} m_i\left({}^{\mathcal{N}}\mathbf{a}_i - {}^{\mathcal{N}}\bar{\mathbf{a}}\right)\cdot\left({}^{\mathcal{N}}\mathbf{v}_i - {}^{\mathcal{N}}\bar{\mathbf{v}}\right) \tag{4-324}$$

Equation (4–324) can be rewritten as

$$\frac{d}{dt}\left({}^{\mathcal{N}}T\right) = m{}^{\mathcal{N}}\bar{\mathbf{a}}\cdot{}^{\mathcal{N}}\bar{\mathbf{v}} + \sum_{i=1}^{n} m_i{}^{\mathcal{N}}\mathbf{a}_i\cdot\left({}^{\mathcal{N}}\mathbf{v}_i - {}^{\mathcal{N}}\bar{\mathbf{v}}\right) - \sum_{i=1}^{n} m_i{}^{\mathcal{N}}\bar{\mathbf{a}}\cdot\left({}^{\mathcal{N}}\mathbf{v}_i - {}^{\mathcal{N}}\bar{\mathbf{v}}\right) \tag{4-325}$$

Substituting the result of Eq. (4–32) into (4–325) and noting that ${}^{\mathcal{N}}\bar{\mathbf{a}}$ is independent of the summation, we obtain

$$\frac{d}{dt}\left({}^{\mathcal{N}}T\right) = \mathbf{F}\cdot{}^{\mathcal{N}}\bar{\mathbf{v}} + \sum_{i=1}^{n} \mathbf{R}_i\cdot\left({}^{\mathcal{N}}\mathbf{v}_i - {}^{\mathcal{N}}\bar{\mathbf{v}}\right) - {}^{\mathcal{N}}\bar{\mathbf{a}}\cdot\sum_{i=1}^{n} m_i\left({}^{\mathcal{N}}\mathbf{v}_i - {}^{\mathcal{N}}\bar{\mathbf{v}}\right) \tag{4-326}$$

From the definition of the center of mass, we have

$$\sum_{i=1}^{n} m_i\left({}^{\mathcal{N}}\mathbf{v}_i - {}^{\mathcal{N}}\bar{\mathbf{v}}\right) = \mathbf{0} \tag{4-327}$$

Consequently, Eq. (4–326) simplifies to

$$\frac{d}{dt}\left({}^{\mathcal{N}}T\right) = \mathbf{F}\cdot{}^{\mathcal{N}}\bar{\mathbf{v}} + \sum_{i=1}^{n} \mathbf{R}_i\cdot\left({}^{\mathcal{N}}\mathbf{v}_i - {}^{\mathcal{N}}\bar{\mathbf{v}}\right) \tag{4-328}$$

Furthermore, we have

$$\sum_{i=1}^{n} \mathbf{R}_i\cdot{}^{\mathcal{N}}\bar{\mathbf{v}} = {}^{\mathcal{N}}\bar{\mathbf{v}}\cdot\sum_{i=1}^{n} \mathbf{R}_i = {}^{\mathcal{N}}\bar{\mathbf{v}}\cdot\sum_{i=1}^{n}\left[\mathbf{F}_i + \sum_{j=1}^{n}\mathbf{f}_{ij}\right] = {}^{\mathcal{N}}\bar{\mathbf{v}}\cdot\sum_{i=1}^{n}\mathbf{F}_i = \mathbf{F}\cdot{}^{\mathcal{N}}\bar{\mathbf{v}} \tag{4-329}$$

Therefore, Eq. (4–328) reduces to

$$\boxed{\frac{d}{dt}\left({}^{\mathcal{N}}T\right) = \sum_{i=1}^{n} \mathbf{R}_i\cdot{}^{\mathcal{N}}\mathbf{v}_i} \tag{4-330}$$

Equation (4–330) is the *work-energy theorem* for a system of particles and states that the rate of change of kinetic energy of a system of particles is equal to the power produced by *all* internal and external forces acting on the system.

4.7.3 Alternate Form of Work-Energy Theorem for a System of Particles

Suppose now that the external force applied to each particle in the system is decomposed as

$$\mathbf{F}_i = \mathbf{F}_i^c + \mathbf{F}_i^{nc} \quad (i = 1, \dots, n) \tag{4-331}$$

where

$$\begin{aligned} \mathbf{F}_i^c &= \text{External conservative force applied to particle } i \\ \mathbf{F}_i^{nc} &= \text{External nonconservative force applied to particle } i \end{aligned}$$

Then the resultant force acting on particle i can be written as

$$\mathbf{R}_i = \mathbf{F}_i^c + \mathbf{F}_i^{nc} + \sum_{j=1}^{n} \mathbf{f}_{ij} \quad (i = 1, \dots, n) \tag{4-332}$$

Substituting the result of Eq. (4-332) into the work-energy theorem of Eq. (4-330), we obtain

$$\frac{d}{dt}\left({}^{\mathcal{N}}T\right) = \sum_{i=1}^{n} \left[\mathbf{F}_i^c + \mathbf{F}_i^{nc} + \sum_{j=1}^{n} \mathbf{f}_{ij}\right] \cdot {}^{\mathcal{N}}\mathbf{v}_i \tag{4-333}$$

Expanding Eq. (4-333) gives

$$\frac{d}{dt}\left({}^{\mathcal{N}}T\right) = \sum_{i=1}^{n} \mathbf{F}_i^c \cdot {}^{\mathcal{N}}\mathbf{v}_i + \sum_{i=1}^{n} \left[\mathbf{F}_i^{nc} + \sum_{j=1}^{n} \mathbf{f}_{ij}\right] \cdot {}^{\mathcal{N}}\mathbf{v}_i \tag{4-334}$$

Now, because the forces $\mathbf{F}_i^c$, $(i = 1, \dots, n)$ are conservative, we know that there exist potential energy functions ${}^{\mathcal{N}}U_i$, $(i = 1, \dots, n)$ such that

$$\mathbf{F}_i^c \cdot {}^{\mathcal{N}}\mathbf{v}_i = -\frac{d}{dt}\left({}^{\mathcal{N}}U_i\right) \quad (i = 1, \dots, n) \tag{4-335}$$

Using the result of Eq. (4-335) in (4-334), we have

$$\frac{d}{dt}\left({}^{\mathcal{N}}T\right) = -\sum_{i=1}^{n} \frac{d}{dt}\left({}^{\mathcal{N}}U_i\right) + \sum_{i=1}^{n} \left[\mathbf{F}_i^{nc} + \sum_{i=1}^{n}\sum_{j=1}^{n} \mathbf{f}_{ij}\right] \cdot {}^{\mathcal{N}}\mathbf{v}_i \tag{4-336}$$

Suppose now that we let the resultant *nonconservative* force acting on particle i, $(i = 1, \dots, n)$ be defined as

$$\mathbf{R}_i^{nc} = \mathbf{F}_i^{nc} + \sum_{j=1}^{n} \mathbf{f}_{ij} \tag{4-337}$$

Equation (4-336) can then be written as

$$\frac{d}{dt}\left({}^{\mathcal{N}}T\right) = -\sum_{i=1}^{n} \frac{d}{dt}\left({}^{\mathcal{N}}U_i\right) + \sum_{i=1}^{n} \mathbf{R}_i^{nc} \cdot {}^{\mathcal{N}}\mathbf{v}_i \tag{4-338}$$

where the quantity

$$\sum_{i=1}^{n} \mathbf{R}_i^{nc} \cdot {}^{\mathcal{N}}\mathbf{v}_i \tag{4-339}$$

is the power produced by all of the *nonconservative* forces acting on the system. Now let

$$ {}^{\mathcal{N}}E = {}^{\mathcal{N}}T + \sum_{i=1}^{n} {}^{\mathcal{N}}U_i \tag{4-340} $$

The quantity ${}^{\mathcal{N}}E$ is the *total energy* for a system of particles. In terms of the total energy, we have

$$ \boxed{\frac{d}{dt}\left({}^{\mathcal{N}}E\right) = \sum_{i=1}^{n} \mathbf{R}_i^{nc} \cdot {}^{\mathcal{N}}\mathbf{v}_i} \tag{4-341} $$

Equation (4-341) is the *alternate form of the work-energy theorem* for a system of particles and states that the rate of change of total energy of a system of particles is equal to the power produced by all of the nonconservative forces acting on the system.

Example 4-6

Recall Example 4-3 on page 260 of a dumbbell consisting of two particles, each of mass m, connected by a rigid massless rod of length l as shown again in Fig. 4-19. Determine the differential equation of motion for the dumbbell in terms of the angle θ during the phase where the dumbbell maintains contact with the wall and the floor using (a) the work-energy theorem for a system of particles and (b) the alternate form of the work-energy theorem for a system of particles.

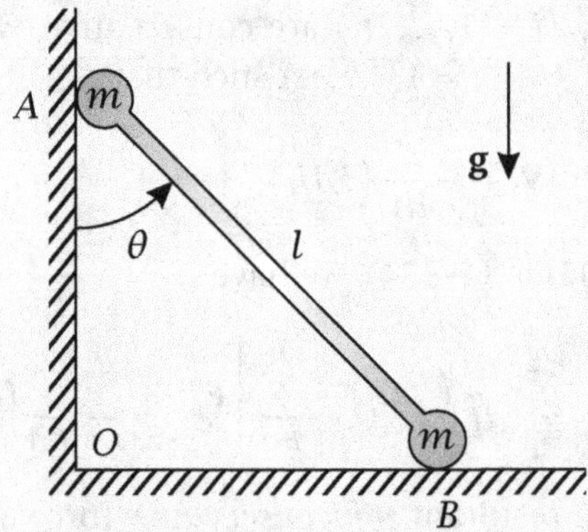

Figure 4-19 Dumbbell sliding on wall and ground.

Solution to Example 4-6

(a) Differential Equation Using Work-Energy Theorem

Recall from Example 4-3 on page 260 that we chose an inertial reference frame $\mathcal{F}$ that is fixed to the wall and the ground. The work-energy theorem for the system of two particles on the dumbbell is given in reference frame $\mathcal{F}$ as

$$ \frac{d}{dt}\left({}^{\mathcal{F}}T\right) = \mathbf{R}_A \cdot {}^{\mathcal{F}}\mathbf{v}_A + \mathbf{R}_B \cdot {}^{\mathcal{F}}\mathbf{v}_B \tag{4-342} $$

Consequently, in order to apply Eq. (4–342), we need to compute the kinetic energy, the rate of change of kinetic energy, and the power produced by all forces acting on the system. The kinetic energy of the system in reference frame $\mathcal{F}$ is computed as follows. First, from Eqs. (4–165) and (4–166) we have

$$
\begin{aligned}
{}^{\mathcal{F}}\mathbf{v}_A &= -l\dot{\theta}\sin\theta\,\mathbf{E}_y && \text{(4–343)}\\
{}^{\mathcal{F}}\mathbf{v}_B &= l\dot{\theta}\cos\theta\,\mathbf{E}_x && \text{(4–344)}
\end{aligned}
$$

The kinetic energy of the system in reference frame $\mathcal{F}$ is given as

$$
\begin{aligned}
{}^{\mathcal{F}}T &= \tfrac{1}{2}m\,{}^{\mathcal{F}}\mathbf{v}_A\cdot{}^{\mathcal{F}}\mathbf{v}_A + \tfrac{1}{2}m\,{}^{\mathcal{F}}\mathbf{v}_B\cdot{}^{\mathcal{F}}\mathbf{v}_B\\
&= \tfrac{1}{2}ml^2\dot{\theta}^2\sin^2\theta + \tfrac{1}{2}ml^2\dot{\theta}^2\cos^2\theta && \text{(4–345)}\\
&= \tfrac{1}{2}ml^2\dot{\theta}^2
\end{aligned}
$$

Computing the rate of change of ${}^{\mathcal{F}}T$ gives

$$
\frac{d}{dt}\left({}^{\mathcal{F}}T\right) = ml^2\dot{\theta}\ddot{\theta} \qquad \text{(4–346)}
$$

Next, the power produced by all of the forces is computed as follows. Using the free body diagram of particle A as given in Fig. 4–20, it is seen that

$$
\begin{aligned}
\mathbf{N}_A &= \text{Reaction force of wall on particle } A\\
\mathbf{f}_{AB} &= \text{Force of particle } B \text{ on particle } A\\
m\mathbf{g} &= \text{Force of gravity}
\end{aligned}
$$

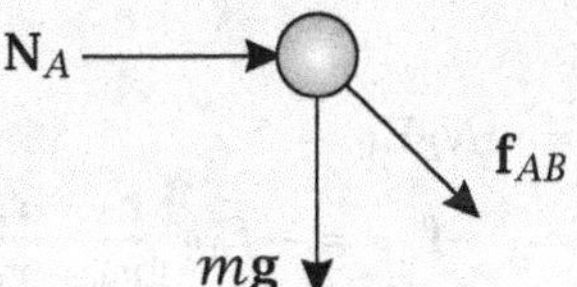

Figure 4–20 Free body diagram of particle A for Example 4–6.

Now we have

$$
\begin{aligned}
\mathbf{N}_A &= N_A\mathbf{E}_x && \text{(4–347)}\\
\mathbf{f}_{AB} &= f_{AB}\frac{\mathbf{r}_A-\mathbf{r}_B}{\|\mathbf{r}_A-\mathbf{r}_B\|} && \text{(4–348)}\\
m\mathbf{g} &= -mg\mathbf{E}_y && \text{(4–349)}
\end{aligned}
$$

where $(\mathbf{r}_A-\mathbf{r}_B)/\|\mathbf{r}_A-\mathbf{r}_B\|$ is the unit vector in the direction from particle B to particle A. The resultant force acting on particle A is then given as

$$
\mathbf{R}_A = \mathbf{N}_A + \mathbf{f}_{AB} + m\mathbf{g} \qquad \text{(4–350)}
$$

Then, using Eqs. (4–163) and (4–164), we have

$$
\begin{aligned}
\frac{\mathbf{r}_A-\mathbf{r}_B}{\|\mathbf{r}_A-\mathbf{r}_B\|} &= \frac{l\cos\theta\,\mathbf{E}_y - l\sin\theta\,\mathbf{E}_x}{\|l\cos\theta\,\mathbf{E}_y - l\sin\theta\,\mathbf{E}_x\|}\\
&= \frac{l\cos\theta\,\mathbf{E}_y - l\sin\theta\,\mathbf{E}_x}{l} && \text{(4–351)}\\
&= -\sin\theta\,\mathbf{E}_x + \cos\theta\,\mathbf{E}_y
\end{aligned}
$$

The resultant force acting on particle A is then given as

$$\mathbf{R}_A = N_A\mathbf{E}_x + f_{AB}(-\sin\theta\,\mathbf{E}_x + \cos\theta\,\mathbf{E}_y) - mg\mathbf{E}_y \tag{4-352}$$

Simplifying Eq. (4-352), we obtain

$$\mathbf{R}_A = (N_A\mathbf{E}_x - f_{AB}\sin\theta)\mathbf{E}_x + (f_{AB}\cos\theta - mg)\mathbf{E}_y \tag{4-353}$$

The free body diagram of particle B is shown in Fig. 4-21, where

$$\begin{aligned} \mathbf{N}_B &= \text{Reaction force of ground on particle } B \\ \mathbf{f}_{BA} &= \text{Force of particle } A \text{ on particle } B \\ m\mathbf{g} &= \text{Force of gravity} \end{aligned}$$

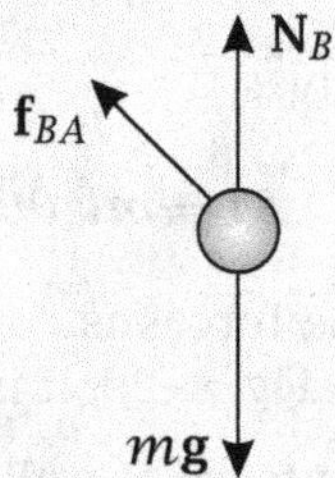

Figure 4-21 Free body diagram of particle B for Example 4-6.

Now we have

$$\begin{aligned} \mathbf{N}_B &= N_B\mathbf{E}_y && (4\text{-}354)\\ \mathbf{f}_{BA} &= -\mathbf{f}_{AB} = -f_{AB}\frac{\mathbf{r}_A - \mathbf{r}_B}{\|\mathbf{r}_A - \mathbf{r}_B\|} && (4\text{-}355)\\ m\mathbf{g} &= -mg\mathbf{E}_y && (4\text{-}356) \end{aligned}$$

The resultant force acting on particle B is then given as

$$\mathbf{R}_B = \mathbf{N}_A + \mathbf{f}_{AB} + m\mathbf{g} \tag{4-357}$$

where we note that, from Newton's 3rd law, $\mathbf{f}_{BA}$ is equal and opposite $\mathbf{f}_{AB}$. The resultant force acting on particle B is then given as

$$\mathbf{R}_B = N_B\mathbf{E}_y - f_{AB}(\cos\theta\,\mathbf{E}_y - \sin\theta\,\mathbf{E}_x) - mg\mathbf{E}_y \tag{4-358}$$

Simplifying Eq. (4-358), we obtain

$$\mathbf{R}_B = f_{AB}\sin\theta\,\mathbf{E}_x + (N_B - f_{AB}\cos\theta - mg)\mathbf{E}_y \tag{4-359}$$

The power produced by all forces acting on the system is then given as

$$\begin{aligned} \mathbf{R}_A \cdot {}^{\mathcal{F}}\mathbf{v}_A + \mathbf{R}_B \cdot {}^{\mathcal{F}}\mathbf{v}_B = {} & \left[(N_A - f_{AB}\sin\theta)\mathbf{E}_x + (f_{AB}\cos\theta - mg)\mathbf{E}_y\right]\cdot\left[-l\dot\theta\sin\theta\,\mathbf{E}_y\right] \\ & + \left[f_{AB}\sin\theta\,\mathbf{E}_x + (N_B - f_{AB}\cos\theta - mg)\mathbf{E}_y\right]\cdot\left[l\dot\theta\cos\theta\,\mathbf{E}_x\right] \end{aligned} \tag{4-360}$$

Simplifying Eq. (4–360), we have

$$\begin{aligned}\mathbf{R}_A \cdot {}^{\mathcal{F}}\mathbf{v}_A + \mathbf{R}_B \cdot {}^{\mathcal{F}}\mathbf{v}_B &= -f_{AB} l\dot{\theta}\cos\theta\,\sin\theta + mgl\dot{\theta}\sin\theta + f_{AB} l\dot{\theta}\sin\theta\,\cos\theta \\ &= mgl\dot{\theta}\sin\theta\end{aligned} \tag{4-361}$$

Setting the rate of change of kinetic energy as given in Eq. (4–346) equal to the power produced by all forces as given in Eq. (4–361), we obtain

$$ml^2\dot{\theta}\ddot{\theta} = mgl\dot{\theta}\sin\theta \tag{4-362}$$

Rearranging Eq. (4–362), we obtain

$$\dot{\theta}(ml^2\ddot{\theta} - mgl\sin\theta) = 0 \tag{4-363}$$

Since $\dot{\theta} \neq 0$ as a function of time (otherwise the dumbbell would not move), we have

$$ml^2\ddot{\theta} - mgl\sin\theta = 0 \tag{4-364}$$

Simplifying Eq. (4–364), the differential equation of motion is obtained as

$$\ddot{\theta} - \frac{g}{l}\sin\theta = 0 \tag{4-365}$$

It is observed that the result of Eq. (4–365) is identical to that obtained in Eq. (4–196) on page 264 of Example 4–3.

(b) Differential Equation Using Alternate Form of Work-Energy Theorem

For this problem the alternate form of the work-energy theorem is given in reference frame $\mathcal{F}$ as

$$\frac{d}{dt}\left({}^{\mathcal{F}}E\right) = \mathbf{R}_A^{nc} \cdot {}^{\mathcal{F}}\mathbf{v}_A + \mathbf{R}_B^{nc} \cdot {}^{\mathcal{F}}\mathbf{v}_B \tag{4-366}$$

where $\mathbf{R}_A^{nc}$ and $\mathbf{R}_B^{nc}$ are the resultant nonconservative forces acting on particles A and B, respectively. First, the total energy is given as

$${}^{\mathcal{F}}E = {}^{\mathcal{F}}T + {}^{\mathcal{F}}U \tag{4-367}$$

We already have the kinetic energy from Eq. (4–345). Next, because the only conservative force acting on the dumbbell is that of gravity, the potential energy in reference frame $\mathcal{F}$ is given as

$${}^{\mathcal{F}}U = {}^{\mathcal{F}}U_g = -2m\mathbf{g} \cdot \bar{\mathbf{r}} \tag{4-368}$$

Recalling the position of the center of mass of the dumbbell and the force due to gravity from Eqs. (4–170) and (4–181) on pages 261 and 263, respectively, we have

$${}^{\mathcal{F}}U = -(-2mg\mathbf{E}_y) \cdot \left(\tfrac{l}{2}\sin\theta\,\mathbf{E}_x + \tfrac{l}{2}\cos\theta\,\mathbf{E}_y\right) = mgl\cos\theta \tag{4-369}$$

Then, adding Eqs. (4–345) and (4–369), the total energy is given as

$${}^{\mathcal{F}}E = \tfrac{1}{2}ml^2\dot{\theta}^2 + mgl\cos\theta \tag{4-370}$$

Computing the rate of change of the total energy, we obtain

$$\frac{d}{dt}\left({}^{\mathcal{F}}E\right) = ml^2\dot\theta\ddot\theta - mgl\dot\theta\sin\theta \tag{4-371}$$

Now the nonconservative force acting on particle A is given as

$$\mathbf{R}_A^{nc} = \mathbf{N}_A + \mathbf{f}_{AB} \tag{4-372}$$

where $\mathbf{N}_A$ is the force exerted by the vertical wall and $\mathbf{f}_{AB}$ is the force exerted by particle B. Using the expressions for $\mathbf{N}_A$ and $\mathbf{f}_{AB}$ from Eqs. (4-347) and (4-348), respectively, we obtain

$$\mathbf{R}_A^{nc} = N_A\mathbf{E}_x + f_{AB}\frac{\mathbf{r}_A - \mathbf{r}_B}{\|\mathbf{r}_A - \mathbf{r}_B\|} \tag{4-373}$$

Using the expression for $(\mathbf{r}_A - \mathbf{r}_B)/\|\mathbf{r}_A - \mathbf{r}_B\|$ from Eq. (4-351), we have

$$\mathbf{R}_A^{nc} = N_A\mathbf{E}_x + f_{AB}(-\sin\theta\,\mathbf{E}_x + \cos\theta\,\mathbf{E}_y) = (N_A - f_{AB}\sin\theta\,)\mathbf{E}_x + f_{AB}\cos\theta\,\mathbf{E}_y \tag{4-374}$$

Similarly, the nonconservative force acting on particle B is given as

$$\mathbf{R}_B^{nc} = \mathbf{N}_B + \mathbf{f}_{BA} \tag{4-375}$$

where $\mathbf{N}_B$ is the force exerted by the vertical wall and $\mathbf{f}_{BA}$ is the force exerted by particle A. Using the expressions for $\mathbf{N}_B$ and $\mathbf{f}_{BA}$ from Eqs. (4-354) and (4-355), respectively, we obtain

$$\mathbf{R}_B^{nc} = N_B\mathbf{E}_y - f_{AB}\frac{\mathbf{r}_A - \mathbf{r}_B}{\|\mathbf{r}_A - \mathbf{r}_B\|} \tag{4-376}$$

Again, using the expression for $(\mathbf{r}_A - \mathbf{r}_B)/\|\mathbf{r}_A - \mathbf{r}_B\|$ from Eq. (4-351), we have

$$\mathbf{R}_B^{nc} = N_B\mathbf{E}_y - f_{AB}(-\sin\theta\,\mathbf{E}_x + \cos\theta\,\mathbf{E}_y) = f_{AB}\sin\theta\, + (N_B - f_{AB}\cos\theta\,)\mathbf{E}_y \tag{4-377}$$

Then, using the expressions for ${}^{\mathcal{F}}\mathbf{v}_A$ and ${}^{\mathcal{F}}\mathbf{v}_B$ from Eqs. (4-343) and (4-344), the power produced by all nonconservative forces is given as

$$\begin{aligned}\mathbf{R}_A^{nc}\cdot{}^{\mathcal{F}}\mathbf{v}_A + \mathbf{R}_B^{nc}\cdot{}^{\mathcal{F}}\mathbf{v}_B &= \left[(N_A - f_{AB}\sin\theta\,)\mathbf{E}_x + f_{AB}\cos\theta\,\mathbf{E}_y\right]\cdot\left[-l\dot\theta\sin\theta\,\mathbf{E}_y\right] \\ &+ \left[f_{AB}\sin\theta\,\mathbf{E}_x + (N_B - f_{AB}\cos\theta\,)\mathbf{E}_y\right]\cdot\left[l\dot\theta\cos\theta\,\mathbf{E}_x\right] \\ &= -f_{AB}l\dot\theta\cos\theta\sin\theta + f_{AB}l\dot\theta\sin\theta\cos\theta = 0\end{aligned} \tag{4-378}$$

It is noted that Eq. (4-371) implies that the total energy in reference frame $\mathcal{F}$ is conserved. Setting the rate of change of total energy in Eq. (4-371) equal to the power produced by all nonconservative forces in Eq. (4-378, we obtain

$$ml^2\dot\theta\ddot\theta - mgl\dot\theta\sin\theta = 0 \tag{4-379}$$

Noting that $\dot\theta$ cannot be zero as a function of time (otherwise the dumbbell would not move), we have

$$ml^2\ddot\theta - mgl\sin\theta = 0 \tag{4-380}$$

Simplifying Eq. (4-380), we obtain the differential equation of motion as

$$\ddot\theta - \frac{g}{l}\sin\theta = 0 \tag{4-381}$$

It is observed that the result of Eq. (4-381) is identical to that obtained in Eq. (4-365) using the work-energy theorem and is also identical to the result obtained in Eq. (4-196) on page 264 of Example 4-3.

■

4.8 Collision of Particles

A *collision* is defined as the impact between two objects that occurs over a very short duration of time and where the forces exerted by each object on the other are extremely large.[3] It can be seen that a real collision can be difficult to model because two real physical objects are generally oddly shaped, are generally composed of non-homogeneous materials, and occupy a nonzero amount of space. Consequently, the deformations that real physical objects undergo during a collision are highly nontrivial. However, by making an appropriate set of assumptions and simplifications, it is possible to arrive at a representative model to express the post-collision velocities of two bodies in terms of the pre-collision velocities of the bodies. In particular, a commonly used collision model is obtained by making the following assumptions: (1) the collision occurs over a very short time interval, and, thus, the positions of the bodies do not change during the collision; (2) the colliding bodies are either particles (i.e., the bodies occupy no physical space) or are homogeneous spheres (so that the direction along which the collision takes place passes through the geometric centers of the bodies), and, thus, the impact is a *central impact*;[4] (3) The surfaces of the objects are "smooth," and, thus there is no friction between the particles during the collision; (4) The kinetics associated with the deformations of the bodies are lumped into an ad hoc scalar parameter called the *coefficient of restitution.* Using these assumptions, a model for the collision between two particles is now developed. It is noted that the development of this section is functionally similar to that found in many other dynamics books, including Beer and Johnston (1997), Hibbeler (2001), Greenwood (1988), and O'Reilly (2001).

Let P_1 and P_2 be particles of mass m_1 and m_2, respectively, moving in an inertial reference frame $\mathcal{N}$. Furthermore, assume that P_1 and P_2 are following independent trajectories in reference frame $\mathcal{N}$ when the paths of the particles coincide, thereby resulting in a collision. The geometry of the collision is shown in Fig. 4-22.

It is evident that the collision will change the velocity of each particle in a manner different from the change that would have occurred had the particles *not* collided. In order to determine the effect that the collision has on the velocity of each particle, we focus on the interval of time over which the collision occurs. In particular, during the collision it is assumed that there exists a well-defined tangent plane, $\mathcal{T}_P$, to the surface of each particle at the point of contact, P. The impact is assumed to occur in a direction $\mathbf{n}$ that is normal to $\mathcal{T}_P$. The direction $\mathbf{n}$ is referred to as the *direction of impact* or the *line of impact* (Beer and Johnston, 1997; Bedford and Fowler, 2005). Also, in order to describe motion in the tangent plane $\mathcal{T}_P$,we define two unit vectors $\mathbf{u}$ and $\mathbf{w}$ whose directions lie in $\mathcal{T}_P$ during the collision (Greenwood, 1988; O'Reilly, 2001). It is assumed that the vectors $\mathbf{n}$, $\mathbf{u}$, and $\mathbf{w}$ are constant during the collision and that $\mathbf{u}$ and $\mathbf{w}$ are chosen such that $\{\mathbf{u}, \mathbf{w}, \mathbf{n}\}$ forms a right-handed system (Greenwood, 1988; O'Reilly, 2001).

[3]The terms "collision" and "impact" are, generally speaking, used interchangeably.

[4]In many dynamics books (e.g., Beer and Johnston (1997) and Bedford and Fowler (2005)), the derivation of collision of particles is divided into two parts, commonly referred to as *direct central impact* and *oblique central impact.* However, direct central impact is a special case of oblique central impact and need not be considered separately. In the development presented here, we discuss the general case of oblique central impact and use the general term *central impact.* Finally, it is noted that all of the results presented here can be applied to the special case of direct central impact.

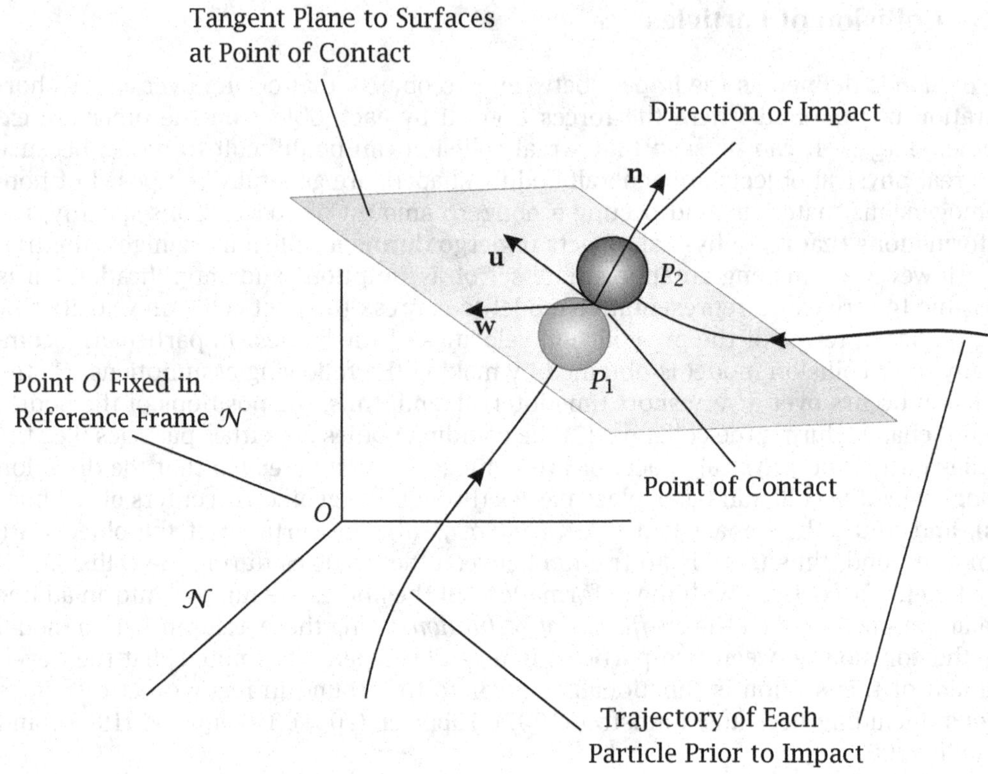

Figure 4–22 Collision of two particles moving in an inertial reference frame $\mathcal{N}$.

4.8.1 Collision Model

The physical process of the collision can be decomposed into two phases (Beer and Johnston, 1997; Bedford and Fowler, 2005): compression and restitution. The geometry of the compression and restitution phase is shown qualitatively in Fig. 4–22. The compression phase begins at $t = t_0$, when the particles first make contact, and ends at a time $t = t_1$, when the particles have attained their *maximum* deformation. The restitution phase begins at the instant of time that the compression phase ends, i.e., the restitution phase begins at time t_1 and ends at the instant of time t_2, when the particles first separate. The velocities of each particle *in the inertial reference frame* $\mathcal{N}$ that correspond to the times t_0, t_1, and t_2 are given as follows:

$$
\begin{array}{rcl}
{}^{\mathcal{N}}\mathbf{v}_1 & = & \text{Velocity of particle } P_1 \text{ at beginning of compression} \\
{}^{\mathcal{N}}\mathbf{v}_2 & = & \text{Velocity of particle } P_2 \text{ at beginning of compression} \\
{}^{\mathcal{N}}\tilde{\mathbf{v}}_1 & = & \text{Velocity of particle } P_1 \text{ at end of compression} \\
 & = & \text{Velocity of particle } P_1 \text{ at beginning of restitution} \\
{}^{\mathcal{N}}\tilde{\mathbf{v}}_2 & = & \text{Velocity of particle } P_2 \text{ at end of compression} \\
 & = & \text{Velocity of particle } P_2 \text{ at beginning of restitution} \\
{}^{\mathcal{N}}\mathbf{v}_1' & = & \text{Velocity of particle } P_1 \text{ at end of restitution} \\
{}^{\mathcal{N}}\mathbf{v}_2' & = & \text{Velocity of particle } P_2 \text{ at end of restitution}
\end{array}
\tag{4-382}
$$

In terms of the basis $\{\mathbf{u}, \mathbf{w}, \mathbf{n}\}$, the velocity of each particle as given in Eq. (4-382) can be expressed as

$$\begin{aligned} {}^{\mathcal{N}}\mathbf{v}_1 &= v_{1u}\mathbf{u} + v_{1w}\mathbf{w} + v_{1n}\mathbf{n} \\ {}^{\mathcal{N}}\mathbf{v}_2 &= v_{2u}\mathbf{u} + v_{2w}\mathbf{w} + v_{2n}\mathbf{n} \\ {}^{\mathcal{N}}\tilde{\mathbf{v}}_1 &= \tilde{v}_{1u}\mathbf{u} + \tilde{v}_{1w}\mathbf{w} + \tilde{v}_{1n}\mathbf{n} \\ {}^{\mathcal{N}}\tilde{\mathbf{v}}_2 &= \tilde{v}_{2u}\mathbf{u} + \tilde{v}_{2w}\mathbf{w} + \tilde{v}_{2n}\mathbf{n} \\ {}^{\mathcal{N}}\mathbf{v}_1' &= v_{1u}'\mathbf{u} + v_{1w}'\mathbf{w} + v_{1n}'\mathbf{n} \\ {}^{\mathcal{N}}\mathbf{v}_2' &= v_{2u}'\mathbf{u} + v_{2w}'\mathbf{w} + v_{2n}'\mathbf{n} \end{aligned} \tag{4-383}$$

During the collision, each particle exerts an impulse on the other particle. Suppose that we let

$\mathbf{C}_1$ = Force exerted by particle P_2 on particle P_1 during compression
$\mathbf{R}_1$ = Force exerted by particle P_2 on particle P_1 during restitution
$\mathbf{C}_2$ = Force exerted by particle P_1 on particle P_2 during compression
$\mathbf{R}_2$ = Force exerted By particle P_1 on particle P_2 during restitution

Then the impulses due to the collision are given as

$$\begin{aligned} \hat{\mathbf{C}}_1 &= \int_{t_0}^{t_1} \mathbf{C}_1 dt \\ \hat{\mathbf{C}}_2 &= \int_{t_0}^{t_1} \mathbf{C}_2 dt \\ \hat{\mathbf{R}}_1 &= \int_{t_1}^{t_2} \mathbf{R}_1 dt \\ \hat{\mathbf{R}}_2 &= \int_{t_1}^{t_2} \mathbf{R}_2 dt \end{aligned} \tag{4-384}$$

where

$\hat{\mathbf{C}}_1$ = Impulse exerted by particle P_2 on particle P_1 during compression
$\hat{\mathbf{R}}_1$ = Impulse exerted by particle P_2 on particle P_1 during restitution
$\hat{\mathbf{C}}_2$ = Impulse exerted by particle P_1 on particle P_2 during compression
$\hat{\mathbf{R}}_2$ = Impulse exerted by particle P_1 on particle P_2 during restitution

The compression impulses $\hat{\mathbf{C}}_1$ and $\hat{\mathbf{C}}_2$ are shown in Fig. 4-23 while the restitution impulses $\hat{\mathbf{R}}_1$ and $\hat{\mathbf{R}}_2$ are shown in Fig. 4-24.
Applying Newton's 3rd law, it is seen that the impulse applied by particle 2 on particle 1 must be equal and opposite the impulse applied by particle 1 on particle 2, i.e.,

$$\begin{aligned} \hat{\mathbf{C}}_1 &= -\hat{\mathbf{C}}_2 \\ \hat{\mathbf{R}}_1 &= -\hat{\mathbf{R}}_2 \end{aligned} \tag{4-385}$$

Now, because the particles are assumed to be smooth (i.e., the surfaces are frictionless), the impulses during the collision must lie in the direction of $\mathbf{n}$. Consequently, we have

$$\begin{aligned} \hat{\mathbf{C}}_1 &= \hat{C}_1\mathbf{n} \\ \hat{\mathbf{C}}_2 &= \hat{C}_2\mathbf{n} \\ \hat{\mathbf{R}}_1 &= \hat{R}_1\mathbf{n} \\ \hat{\mathbf{R}}_2 &= \hat{R}_2\mathbf{n} \end{aligned} \tag{4-386}$$

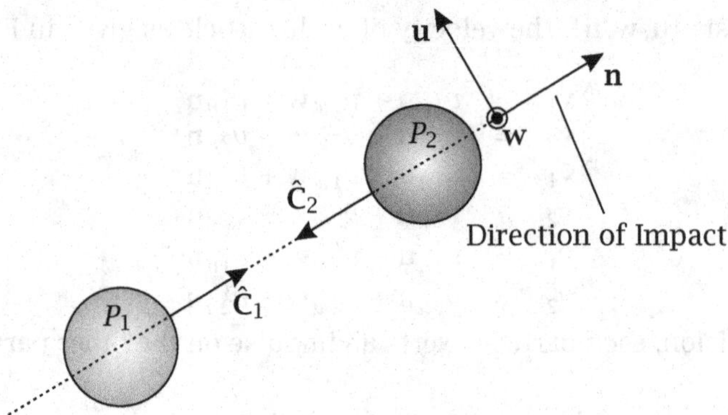

Figure 4–23 Free body diagram during compression phase of the collision between two particles.

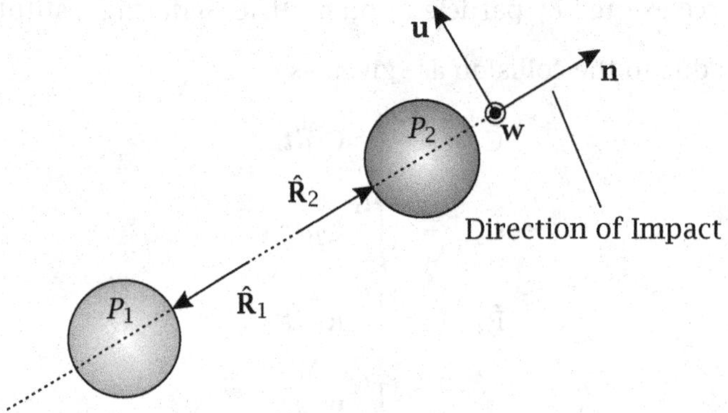

Figure 4–24 Free body diagram during restitution phase of the collision between two particles.

Furthermore, it is assumed that, during the collision of the two particles, *all other impulses* are negligible in comparison to the $\hat{\mathbf{C}}_1$, $\hat{\mathbf{C}}_2$, $\hat{\mathbf{R}}_1$, and $\hat{\mathbf{R}}_2$. Using the impulses given in Eq. (4–386), we define the scalar parameter e called the *coefficient of restitution* as

$$e = \frac{\hat{\mathbf{R}}_1 \cdot \mathbf{n}}{\hat{\mathbf{C}}_1 \cdot \mathbf{n}} = \frac{\hat{\mathbf{R}}_2 \cdot \mathbf{n}}{\hat{\mathbf{C}}_2 \cdot \mathbf{n}} \tag{4–387}$$

Finally, using Eq. (4–387), we can solve for $\hat{\mathbf{R}}_1 \cdot \mathbf{n}$ and $\hat{\mathbf{R}}_2 \cdot \mathbf{n}$ in terms of $\hat{\mathbf{C}}_1 \cdot \mathbf{n}$ and $\hat{\mathbf{C}}_2 \cdot \mathbf{n}$ as

$$\begin{aligned} \hat{\mathbf{R}}_1 \cdot \mathbf{n} &= e\hat{\mathbf{C}}_1 \cdot \mathbf{n} \\ \hat{\mathbf{R}}_2 \cdot \mathbf{n} &= e\hat{\mathbf{C}}_2 \cdot \mathbf{n} \end{aligned} \tag{4–388}$$

Application of Linear Impulse and Linear Momentum During Compression

Applying the principle of linear impulse and linear momentum of Eq. (3–179) from page 177 to each particle during *compression*, we have

$$\begin{aligned} \hat{\mathbf{C}}_1 &= {}^{\mathcal{N}}\tilde{\mathbf{G}}_1 - {}^{\mathcal{N}}\mathbf{G}_1 \\ \hat{\mathbf{C}}_2 &= {}^{\mathcal{N}}\tilde{\mathbf{G}}_2 - {}^{\mathcal{N}}\mathbf{G}_2 \end{aligned} \tag{4–389}$$

where $\hat{\mathbf{C}}_1$ and $\hat{\mathbf{C}}_2$ are the impulses applied to P_1 and P_2, respectively, during compression as shown in Fig. 4-23. Furthermore, the linear momenta of P_1 and P_2 at the beginning of compression are given as

$$\begin{aligned} {}^{\mathcal{N}}\mathbf{G}_1 &= m_1 {}^{\mathcal{N}}\mathbf{v}_1 \\ {}^{\mathcal{N}}\mathbf{G}_2 &= m_2 {}^{\mathcal{N}}\mathbf{v}_2 \end{aligned} \tag{4-390}$$

Similarly, the linear momenta of P_1 and P_2 at the end of compression are given as

$$\begin{aligned} {}^{\mathcal{N}}\tilde{\mathbf{G}}_1 &= m_1 {}^{\mathcal{N}}\tilde{\mathbf{v}}_1 \\ {}^{\mathcal{N}}\tilde{\mathbf{G}}_2 &= m_2 {}^{\mathcal{N}}\tilde{\mathbf{v}}_2 \end{aligned} \tag{4-391}$$

Substituting the results of Eqs. (4-390) and (4-391) into Eq. (4-389), we obtain

$$\begin{aligned} \hat{\mathbf{C}}_1 &= m_1 {}^{\mathcal{N}}\tilde{\mathbf{v}}_1 - m_1 {}^{\mathcal{N}}\mathbf{v}_1 \\ \hat{\mathbf{C}}_2 &= m_2 {}^{\mathcal{N}}\tilde{\mathbf{v}}_2 - m_2 {}^{\mathcal{N}}\mathbf{v}_2 \end{aligned} \tag{4-392}$$

Taking the scalar products in Eq. (4-392) with $\mathbf{n}$ gives

$$\begin{aligned} \hat{\mathbf{C}}_1 \cdot \mathbf{n} &= m_1 {}^{\mathcal{N}}\tilde{\mathbf{v}}_1 \cdot \mathbf{n} - m_1 {}^{\mathcal{N}}\mathbf{v}_1 \cdot \mathbf{n} \\ \hat{\mathbf{C}}_2 \cdot \mathbf{n} &= m_2 {}^{\mathcal{N}}\tilde{\mathbf{v}}_2 \cdot \mathbf{n} - m_2 {}^{\mathcal{N}}\mathbf{v}_2 \cdot \mathbf{n} \end{aligned} \tag{4-393}$$

Then, using the expressions for the velocities of the particles in terms of the coordinate system $\{\mathbf{u}, \mathbf{w}, \mathbf{n}\}$ as given in Eq. (4-383) and the expressions for the impulses as given in Eq. (4-393), Eq. (4-393) can be written as

$$\begin{aligned} \hat{\mathbf{C}}_1 \cdot \mathbf{n} &= m_1 \tilde{v}_{1n} - m_1 v_{1n} \\ \hat{\mathbf{C}}_2 \cdot \mathbf{n} &= m_2 \tilde{v}_{2n} - m_2 v_{2n} \end{aligned} \tag{4-394}$$

Application of Linear Impulse and Linear Momentum During Restitution

Applying the principle of linear impulse and linear momentum of Eq. (3-179) from page 177 to each particle during *restitution*, we have

$$\begin{aligned} \hat{\mathbf{R}}_1 &= {}^{\mathcal{N}}\mathbf{G}_1' - {}^{\mathcal{N}}\tilde{\mathbf{G}}_1 \\ \hat{\mathbf{R}}_2 &= {}^{\mathcal{N}}\mathbf{G}_2' - {}^{\mathcal{N}}\tilde{\mathbf{G}}_2 \end{aligned} \tag{4-395}$$

where $\hat{\mathbf{R}}_1$ and $\hat{\mathbf{R}}_2$ are the impulses applied to P_1 and P_2, respectively, during restitution, as shown in Fig. 4-24. Because the beginning of restitution corresponds to the end of compression, the linear momentum of each particle at the beginning of restitution is given by ${}^{\mathcal{N}}\tilde{\mathbf{G}}_1$ and ${}^{\mathcal{N}}\tilde{\mathbf{G}}_2$, respectively, in Eq. (4-391). The linear momenta of each particle at the *end* of restitution are given as

$$\begin{aligned} {}^{\mathcal{N}}\mathbf{G}_1' &= m_1 {}^{\mathcal{N}}\mathbf{v}_1' \\ {}^{\mathcal{N}}\mathbf{G}_2' &= m_2 {}^{\mathcal{N}}\mathbf{v}_2' \end{aligned} \tag{4-396}$$

Substituting the results of Eqs. (4-391) and (4-396) into Eq. (4-395), we obtain

$$\begin{aligned} \hat{\mathbf{R}}_1 &= m_1 {}^{\mathcal{N}}\mathbf{v}_1' - m_1 {}^{\mathcal{N}}\tilde{\mathbf{v}}_1 \\ \hat{\mathbf{R}}_2 &= m_2 {}^{\mathcal{N}}\mathbf{v}_2' - m_2 {}^{\mathcal{N}}\tilde{\mathbf{v}}_2 \end{aligned} \tag{4-397}$$

Taking the scalar products in Eq. (4–397) with $\mathbf{n}$, we obtain

$$\begin{aligned} \hat{\mathbf{R}}_1 \cdot \mathbf{n} &= m_1 {}^{\mathcal{N}}\mathbf{v}_1' \cdot \mathbf{n} - m_1 {}^{\mathcal{N}}\tilde{\mathbf{v}}_1 \cdot \mathbf{n} \\ \hat{\mathbf{R}}_2 \cdot \mathbf{n} &= m_2 {}^{\mathcal{N}}\mathbf{v}_2' \cdot \mathbf{n} - m_2 {}^{\mathcal{N}}\tilde{\mathbf{v}}_2 \cdot \mathbf{n} \end{aligned} \tag{4–398}$$

Then, using the expressions for the velocities of the particles in terms of the coordinate system $\{\mathbf{u}, \mathbf{w}, \mathbf{n}\}$ as given in Eq. (4–383), Eq. (4–398) can be written as

$$\begin{aligned} \hat{\mathbf{R}}_1 \cdot \mathbf{n} &= m_1 v_{1n}' - m_1 \tilde{v}_{1n} \\ \hat{\mathbf{R}}_2 \cdot \mathbf{n} &= m_2 v_{2n}' - m_2 \tilde{v}_{2n} \end{aligned} \tag{4–399}$$

Now, because at the end of compression (equivalently, at the beginning of restitution) the component of velocity of each particle in the $\mathbf{n}$-direction must be the same, we have

$${}^{\mathcal{N}}\tilde{\mathbf{v}}_1 \cdot \mathbf{n} = {}^{\mathcal{N}}\tilde{\mathbf{v}}_2 \cdot \mathbf{n} \equiv {}^{\mathcal{N}}\tilde{\mathbf{v}} \cdot \mathbf{n} \tag{4–400}$$

Using the expression for ${}^{\mathcal{N}}\tilde{\mathbf{v}}$ from Eq. (4–383), Eq. (4–400) can be written as

$${}^{\mathcal{N}}\tilde{\mathbf{v}}_1 \cdot \mathbf{n} = {}^{\mathcal{N}}\tilde{\mathbf{v}}_2 \cdot \mathbf{n} = \tilde{v}_{1n} = \tilde{v}_{2n} \equiv \tilde{v}_n \tag{4–401}$$

Equations (4–388), (4–394), (4–399), and (4–401) can then be used to obtain the following two expressions for $\tilde{v}_n$:

$$\tilde{v}_n = \frac{v_{1n}' + e v_{1n}}{1+e} = \frac{v_{2n}' + e v_{2n}}{1+e} \tag{4–402}$$

Solving Eq. (4–402) for the coefficient of restitution, we obtain

$$\boxed{e = \frac{v_{2n}' - v_{1n}'}{v_{1n} - v_{2n}}} \tag{4–403}$$

Using vector notation, the coefficient of restitution can be written as (O'Reilly, 2001)

$$\boxed{e = \frac{{}^{\mathcal{N}}\mathbf{v}_2' \cdot \mathbf{n} - {}^{\mathcal{N}}\mathbf{v}_1' \cdot \mathbf{n}}{{}^{\mathcal{N}}\mathbf{v}_1 \cdot \mathbf{n} - {}^{\mathcal{N}}\mathbf{v}_2 \cdot \mathbf{n}}} \tag{4–404}$$

It is seen that the coefficient of restitution is the ratio between the component of the restitution impulse in the direction of impact over the compression impulse in the direction of impact. Essentially, e is a measure of how much size and shape of the colliding bodies is restored during the collision. In general, e depends on various factors, including the material properties of the colliding bodies and the relative velocity of the particles on impact. Furthermore, because the restitution impulse can neither be less than zero nor can exceed the compression impulse, e can take on values only between zero and unity. A value of $e = 1$ corresponds to a *perfectly elastic* impact while a value of $e = 0$ corresponds to a perfectly *inelastic* (also known as a perfectly *plastic*) impact. It is noted that a perfectly elastic impact is one where no energy is lost during the collision. In general, the value of e will be somewhere between zero and one, i.e., the collision will be only *partially* restitutive.

Application of Linear impulse and Linear momentum During Entire Collision

Applying the principle of linear impulse and linear momentum of Eq. (3–179) from page 177 to each particle during the entire collision (i.e., during both compression and restitution), we have

$$\begin{aligned}\hat{\mathbf{F}}_1 &= {}^{\mathcal{N}}\mathbf{G}_1' - {}^{\mathcal{N}}\mathbf{G}_1 \\ \hat{\mathbf{F}}_2 &= {}^{\mathcal{N}}\mathbf{G}_2' - {}^{\mathcal{N}}\mathbf{G}_2\end{aligned} \tag{4–405}$$

where $\hat{\mathbf{F}}_1$ and $\hat{\mathbf{F}}_2$ are the impulses applied to P_1 and P_2, respectively, during the entire collision. Now it is seen that the impulses applied to P_1 and P_2, respectively, during the entire collision are the *sum* of the impulses during compression and restitution, i.e.,

$$\begin{aligned}\hat{\mathbf{F}}_1 &= \hat{\mathbf{C}}_1 + \hat{\mathbf{R}}_1 \\ \hat{\mathbf{F}}_2 &= \hat{\mathbf{C}}_2 + \hat{\mathbf{R}}_2\end{aligned} \tag{4–406}$$

Substituting the result of Eq. (4–406) together with the pre-collision and post-collision linear momenta as given, respectively, in Eqs. (4–390) and (4–396) into Eq. (4–405), we obtain

$$\begin{aligned}\hat{\mathbf{C}}_1 + \hat{\mathbf{R}}_1 &= m_1{}^{\mathcal{N}}\mathbf{v}_1' - m_1{}^{\mathcal{N}}\mathbf{v}_1 \\ \hat{\mathbf{C}}_2 + \hat{\mathbf{R}}_2 &= m_2{}^{\mathcal{N}}\mathbf{v}_2' - m_2{}^{\mathcal{N}}\mathbf{v}_2\end{aligned} \tag{4–407}$$

Then, taking the scalar product of Eq. (4–407) in the **u**-direction, we obtain

$$\begin{aligned}\hat{\mathbf{C}}_1 \cdot \mathbf{u} + \hat{\mathbf{R}}_1 \cdot \mathbf{u} &= m_1{}^{\mathcal{N}}\mathbf{v}_1' \cdot \mathbf{u} - m_1{}^{\mathcal{N}}\mathbf{v}_1 \cdot \mathbf{u} \\ \hat{\mathbf{C}}_2 \cdot \mathbf{u} + \hat{\mathbf{R}}_2 \cdot \mathbf{u} &= m_2{}^{\mathcal{N}}\mathbf{v}_2' \cdot \mathbf{u} - m_2{}^{\mathcal{N}}\mathbf{v}_2 \cdot \mathbf{u}\end{aligned} \tag{4–408}$$

Now because both the compression and restitution impulses lie in the **n**-direction, we have

$$\begin{aligned}\hat{\mathbf{C}}_1 \cdot \mathbf{u} + \hat{\mathbf{R}}_1 \cdot \mathbf{u} &= 0 \\ \hat{\mathbf{C}}_2 \cdot \mathbf{u} + \hat{\mathbf{R}}_2 \cdot \mathbf{u} &= 0\end{aligned} \tag{4–409}$$

Consequently, Eq. (4–408) simplifies to

$$\begin{aligned}m_1{}^{\mathcal{N}}\mathbf{v}_1' \cdot \mathbf{u} - m_1{}^{\mathcal{N}}\mathbf{v}_1 \cdot \mathbf{u} &= 0 \\ m_2{}^{\mathcal{N}}\mathbf{v}_2' \cdot \mathbf{u} - m_2{}^{\mathcal{N}}\mathbf{v}_2 \cdot \mathbf{u} &= 0\end{aligned} \tag{4–410}$$

Rearranging Eq. (4–410), we obtain

$$\begin{aligned}{}^{\mathcal{N}}\mathbf{v}_1' \cdot \mathbf{u} &= {}^{\mathcal{N}}\mathbf{v}_1 \cdot \mathbf{u} \\ {}^{\mathcal{N}}\mathbf{v}_2' \cdot \mathbf{u} &= {}^{\mathcal{N}}\mathbf{v}_2 \cdot \mathbf{u}\end{aligned} \tag{4–411}$$

Then, substituting the expressions for the velocities from Eq. (4–383) into (4–411), we obtain

$$\begin{aligned}v_{1u}' &= v_{1u} \\ v_{2u}' &= v_{2u}\end{aligned} \tag{4–412}$$

Similarly, taking the scalar product of Eq. (4–407) in the **w**-direction, we obtain

$$\begin{aligned}\hat{\mathbf{C}}_1 \cdot \mathbf{w} + \hat{\mathbf{R}}_1 \cdot \mathbf{w} &= m_1{}^{\mathcal{N}}\mathbf{v}_1' \cdot \mathbf{w} - m_1{}^{\mathcal{N}}\mathbf{v}_1 \cdot \mathbf{w} \\ \hat{\mathbf{C}}_2 \cdot \mathbf{w} + \hat{\mathbf{R}}_2 \cdot \mathbf{w} &= m_2{}^{\mathcal{N}}\mathbf{v}_2' \cdot \mathbf{w} - m_2{}^{\mathcal{N}}\mathbf{v}_2 \cdot \mathbf{w}\end{aligned} \tag{4–413}$$

Again, because both the compression and restitution impulses lie in the $\mathbf{n}$-direction, we have

$$\begin{aligned}\hat{\mathbf{C}}_1\cdot\mathbf{w}+\hat{\mathbf{R}}_1\cdot\mathbf{w} &= 0\\ \hat{\mathbf{C}}_2\cdot\mathbf{w}+\hat{\mathbf{R}}_2\cdot\mathbf{w} &= 0\end{aligned}\tag{4-414}$$

Consequently, Eq. (4–413) simplifies to

$$\begin{aligned}m_1{}^{\mathcal{N}}\mathbf{v}_1'\cdot\mathbf{w}-m_1{}^{\mathcal{N}}\mathbf{v}_1\cdot\mathbf{w} &= 0\\ m_2{}^{\mathcal{N}}\mathbf{v}_2'\cdot\mathbf{w}-m_2{}^{\mathcal{N}}\mathbf{v}_2\cdot\mathbf{w} &= 0\end{aligned}\tag{4-415}$$

Rearranging Eq. (4–415), we obtain

$$\begin{aligned}{}^{\mathcal{N}}\mathbf{v}_1'\cdot\mathbf{w} &= {}^{\mathcal{N}}\mathbf{v}_1\cdot\mathbf{w}\\ {}^{\mathcal{N}}\mathbf{v}_2'\cdot\mathbf{w} &= {}^{\mathcal{N}}\mathbf{v}_2\cdot\mathbf{w}\end{aligned}\tag{4-416}$$

Then, substituting the expressions for the velocities from Eq. (4–383) into (4–416), we obtain

$$\begin{aligned}v_{1w}' &= v_{1w}\\ v_{2w}' &= v_{2w}\end{aligned}\tag{4-417}$$

Finally, taking the scalar product of Eq. (4–407) in the $\mathbf{n}$-direction, we obtain

$$\begin{aligned}\hat{\mathbf{C}}_1\cdot\mathbf{n}+\hat{\mathbf{R}}_1\cdot\mathbf{n} &= m_1{}^{\mathcal{N}}\mathbf{v}_1'\cdot\mathbf{n}-m_1{}^{\mathcal{N}}\mathbf{v}_1\cdot\mathbf{n}\\ \hat{\mathbf{C}}_2\cdot\mathbf{n}+\hat{\mathbf{R}}_2\cdot\mathbf{n} &= m_2{}^{\mathcal{N}}\mathbf{v}_2'\cdot\mathbf{n}-m_2{}^{\mathcal{N}}\mathbf{v}_2\cdot\mathbf{n}\end{aligned}\tag{4-418}$$

Then, since both the compression and restitution impulses lie in the $\mathbf{n}$-direction, we have

$$\begin{aligned}\hat{\mathbf{C}}_1\cdot\mathbf{n}+\hat{\mathbf{R}}_1\cdot\mathbf{n} &= \hat{C}_1+\hat{R}_1\\ \hat{\mathbf{C}}_2\cdot\mathbf{n}+\hat{\mathbf{R}}_2\cdot\mathbf{n} &= \hat{C}_2+\hat{R}_2\end{aligned}\tag{4-419}$$

Furthermore, substituting Eq. (4–388) into (4–419), we obtain

$$\begin{aligned}\hat{C}_1+\hat{R}_1 &= (1+e)\hat{C}_1\\ \hat{C}_2+\hat{R}_2 &= (1+e)\hat{C}_2\end{aligned}\tag{4-420}$$

Then, substituting the result of Eq. (4–420) into (4–418) and using the fact that $\hat{C}_2 = -\hat{C}_1$, we obtain

$$\begin{aligned}(1+e)\hat{C}_1 &= m_1{}^{\mathcal{N}}\mathbf{v}_1'\cdot\mathbf{n}-m_1{}^{\mathcal{N}}\mathbf{v}_1\cdot\mathbf{n}\\ -(1+e)\hat{C}_1 &= m_2{}^{\mathcal{N}}\mathbf{v}_2'\cdot\mathbf{n}-m_2{}^{\mathcal{N}}\mathbf{v}_2\cdot\mathbf{n}\end{aligned}\tag{4-421}$$

Adding the expressions in Eq. (4–421) gives

$$m_1{}^{\mathcal{N}}\mathbf{v}_1'\cdot\mathbf{n}+m_2{}^{\mathcal{N}}\mathbf{v}_2'\cdot\mathbf{n}-\left(m_1{}^{\mathcal{N}}\mathbf{v}_1\cdot\mathbf{n}+m_2{}^{\mathcal{N}}\mathbf{v}_2\cdot\mathbf{n}\right)=0\tag{4-422}$$

Rearranging Eq. (4–422), we obtain

$$m_1{}^{\mathcal{N}}\mathbf{v}_1'\cdot\mathbf{n}+m_2{}^{\mathcal{N}}\mathbf{v}_2'\cdot\mathbf{n}=m_1{}^{\mathcal{N}}\mathbf{v}_1\cdot\mathbf{n}+m_2{}^{\mathcal{N}}\mathbf{v}_2\cdot\mathbf{n}\tag{4-423}$$

Then, substituting the expressions for the velocities from Eq. (4–383) into (4–423), we obtain

$$m_1v_{1n}'+m_2v_{2n}'=m_1v_{1n}+m_2v_{2n}\tag{4-424}$$

Equation (4–424) states that the component of the linear momentum of the *system* in the $\mathbf{n}$-direction is conserved during the collision.

Solving for Post-Collision Velocities

The results of Eqs. (4–403), (4–412), (4–417), and (4–424) form a system of six equations in six unknowns that can be used to solve for the post-collision velocities ${}^{\mathcal{N}}\mathbf{v}_1'$ and ${}^{\mathcal{N}}\mathbf{v}_2'$ in terms of the coordinate system $\{\mathbf{u}, \mathbf{w}, \mathbf{n}\}$. First, we recall from Eqs. (4–412) and (4–417) that the components of the velocities in the **u**-direction and **w**-direction are the same before and after the collision. Therefore, the only remaining quantities that must be determined are the components of the velocities of each particle in the **n**-direction after the collision, v_{1n}' and v_{2n}'. Multiplying both sides of Eq. (4–403) by $v_{1n} - v_{2n}$, we have

$$v_{2n}' - v_{1n}' = e(v_{1n} - v_{2n}) \tag{4–425}$$

Rearranging Eq. (4–424) and adjoining Eq. (4–425) gives the following two equations:

$$\begin{aligned} m_1 v_{1n}' + m_2 v_{2n}' &= m_1 v_{1n} + m_2 v_{2n} && (4–426) \\ v_{2n}' - v_{1n}' &= e(v_{1n} - v_{2n}) && (4–427) \end{aligned}$$

Multiplying Eq. (4–427) by m_2 and subtracting from Eq. (4–426) gives

$$(m_1 + m_2) v_{1n}' = (m_1 - e m_2) v_{1n} + m_2(1+e) v_{2n} \tag{4–428}$$

Solving Eq. (4–428) for v_{1n}' gives

$$v_{1n}' = \frac{m_1 - e m_2}{m_1 + m_2} v_{1n} + \frac{m_2(1+e)}{m_1 + m_2} v_{2n} \tag{4–429}$$

Similarly, multiplying Eq. (4–427) by m_1 and adding Eq. (4–426) gives

$$(m_1 + m_2) v_{2n}' = m_1(1+e) v_{1n} + (m_2 - e m_1) v_{2n} \tag{4–430}$$

Solving Eq. (4–430) for v_{2n}' gives

$$v_{2n}' = \frac{m_1(1+e)}{m_1 + m_2} v_{1n} + \frac{m_2 - e m_1}{m_1 + m_2} v_{2n} \tag{4–431}$$

Consequently, the post-collision velocities of the two particles are given in terms of the $\{\mathbf{u}, \mathbf{w}, \mathbf{n}\}$ coordinate system as

$$\boxed{\begin{aligned} {}^{\mathcal{N}}\mathbf{v}_1' &= v_{1u}\mathbf{u} + v_{1w}\mathbf{w} + \left(\frac{m_1 - e m_2}{m_1 + m_2} v_{1n} + \frac{m_2(1+e)}{m_1 + m_2} v_{2n}\right)\mathbf{n} \\ {}^{\mathcal{N}}\mathbf{v}_2' &= v_{2u}\mathbf{u} + v_{2w}\mathbf{w} + \left(\frac{m_1(1+e)}{m_1 + m_2} v_{1n} + \frac{m_2 - e m_1}{m_1 + m_2} v_{2n}\right)\mathbf{n} \end{aligned}} \tag{4–432}$$

Alternatively, the post-collision velocities can be expressed elegantly using vector notation as (O'Reilly, 2001)

$$\boxed{\begin{aligned} {}^{\mathcal{N}}\mathbf{v}_1' &= ({}^{\mathcal{N}}\mathbf{v}_1 \cdot \mathbf{u})\,\mathbf{u} + ({}^{\mathcal{N}}\mathbf{v}_1 \cdot \mathbf{w})\,\mathbf{w} \\ &+ \left(\frac{m_1 - e m_2}{m_1 + m_2} {}^{\mathcal{N}}\mathbf{v}_1 \cdot \mathbf{n} + \frac{m_2(1+e)}{m_1 + m_2} {}^{\mathcal{N}}\mathbf{v}_2 \cdot \mathbf{n}\right)\mathbf{n} \\ {}^{\mathcal{N}}\mathbf{v}_2' &= ({}^{\mathcal{N}}\mathbf{v}_2 \cdot \mathbf{u})\,\mathbf{u} + ({}^{\mathcal{N}}\mathbf{v}_2 \cdot \mathbf{w})\,\mathbf{w} \\ &+ \left(\frac{m_1(1+e)}{m_1 + m_2} {}^{\mathcal{N}}\mathbf{v}_1 \cdot \mathbf{n} + \frac{m_2 - e m_1}{m_1 + m_2} {}^{\mathcal{N}}\mathbf{v}_2 \cdot \mathbf{n}\right)\mathbf{n} \end{aligned}} \tag{4–433}$$

4.8.2 Further Discussion About Collision Model

The first key assumption made in the collision model of Section 4.8.1 is that, during a collision, all forces *other* than those that arise due to the collision are considered *small* in comparison to the forces of the collision. As an example, consider the collision of two automobiles. During an automobile collision, the forces of impact are so great that, over the short time interval that the collision occurs, any other forces acting on the vehicles (such as those of friction and gravity) have little effect on the outcome of the collision. In fact, the forces of collision are, for all practical purposes, infinite in comparison to any other force. Consequently, in solving a collision problem, it is only necessary to model the forces due to the collision.

The second key assumption made in the collision model of Section 4.8.1 is that the complexities of the deformations that are induced during the collision are subsumed into the coefficient of restitution. In applying the collision model for the first time, a student may be confused because this simplified model is counter to one's intuition that the kinetics during a collision are extremely complex and that this complexity needs to be modeled. Instead, this complexity has not been modeled and has been replaced by a drastic simplification. Consequently, a student has to take it on faith that a collision model based on the coefficient of restitution model is a good approximation to reality.

Example 4–7

A particle of mass m_2 initially hangs motionless from an inelastic cord when it is struck by another particle of mass m_1 moving with velocity $\mathbf{v}_0$ at an angle θ below the horizontal as shown in Fig. 4–25. Assuming a perfectly elastic collision and that $m_1 = m_2 = m$, determine (a) the velocity of each particle the instant after impact and (b) the impulse exerted by the string on the system during the collision.

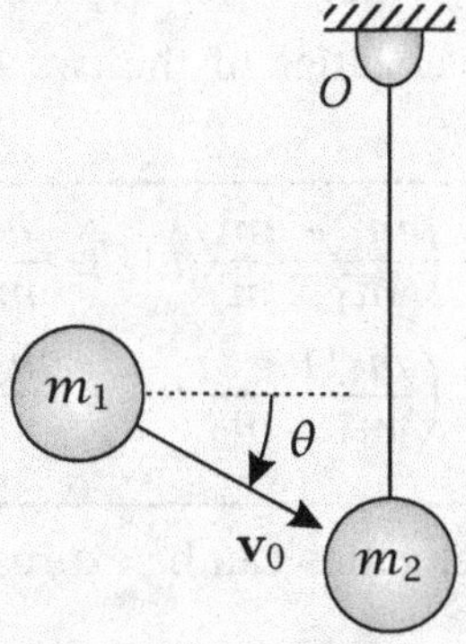

Figure 4–25 Sphere colliding with sphere on inelastic cord.

Solution to Example 4–7

Kinematics

Let $\mathcal{F}$ be a fixed reference frame. Then choose the following coordinate system fixed in reference frame $\mathcal{F}$:

$$\begin{array}{ccc} & \text{Origin at } O & \\ \mathbf{E}_x & = & \text{Along } Om_2 \text{ at } t=0^- \\ \mathbf{E}_z & = & \text{Out of page} \\ \mathbf{E}_y & = & \mathbf{E}_z \times \mathbf{E}_x \end{array}$$

where $t = 0^-$ is the instant before impact. Because $\{\mathbf{E}_x, \mathbf{E}_y, \mathbf{E}_z\}$ is a fixed basis, the velocity of each particle can be written as

$$^{\mathcal{F}}\mathbf{v}_1 = v_{1x}\mathbf{E}_x + v_{1y}\mathbf{E}_y \tag{4–434}$$

$$^{\mathcal{F}}\mathbf{v}_2 = v_{2x}\mathbf{E}_x + v_{2y}\mathbf{E}_y \tag{4–435}$$

It is noted that, because we will be using the principle of linear impulse and linear momentum to solve this problem, it is not necessary to compute the acceleration of each particle.

Kinetics

The next step is to apply linear impulse and linear momentum to the particles. For this problem, it is convenient to apply linear impulse and linear momentum to the following two systems: (1) particle m_1 and (2) particles m_1 and m_2. Each of these kinetic analyses is now performed.

Application of Linear Impulse and Linear Momentum to Particle m_1

Applying linear impulse and linear momentum to particle A, we have

$$\hat{\mathbf{F}}_1 = {}^{\mathcal{F}}\mathbf{G}_1' - {}^{\mathcal{F}}\mathbf{G}_1 \tag{4–436}$$

Examining the free body diagram of particle m_1 as shown in Fig. 4-26, it is seen that the only impulse acting on m_1 during the collision is due to particle m_2.

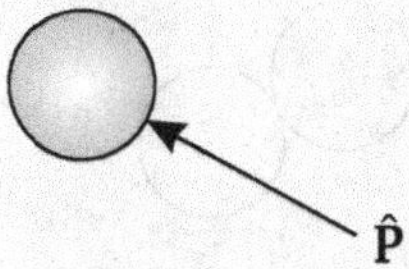

Figure 4–26 Free body diagram of particle m_1 for Example 4–25.

We denote the impulse applied by m_2 on m_1 by $\hat{\mathbf{P}}$, and observe that $\hat{\mathbf{P}}$ must lie along the line of $\mathbf{v}_0$, i.e.,

$$\hat{\mathbf{P}} = \hat{P}\sin\theta\,\mathbf{E}_x + \hat{P}\cos\theta\,\mathbf{E}_y \tag{4–437}$$

Next the linear momentum of particle m_1 before and after the collision are given, respectively, as

$$ {}^{\mathcal{F}}\mathbf{G}_1 = m\mathbf{v}_0 = mv_0 \sin\theta \mathbf{E}_x + mv_0 \cos\theta \mathbf{E}_y \tag{4-438}$$

$$ {}^{\mathcal{F}}\mathbf{G}_1' = mv_{1x}'\mathbf{E}_x + mv_{1y}'\mathbf{E}_y \tag{4-439}$$

Then, substituting the results of Eqs. (4-437), (4-438), and (4-439) into Eq. (4-436), we obtain

$$ \hat{P}\sin\theta \mathbf{E}_x + \hat{P}\cos\theta \mathbf{E}_y = mv_{1x}'\mathbf{E}_x + mv_{1y}'\mathbf{E}_y - (mv_0 \sin\theta \mathbf{E}_x + mv_0 \cos\theta \mathbf{E}_y) \tag{4-440}$$

Simplifying Eq. (4-440), we obtain

$$ \hat{P}\sin\theta \mathbf{E}_x + \hat{P}\cos\theta \mathbf{E}_y = (mv_{1x}' - mv_0 \sin\theta)\mathbf{E}_x + (mv_{1y}' - mv_0 \cos\theta)\mathbf{E}_y \tag{4-441}$$

Equating components in Eq. (4-441), we obtain the following two scalar equations:

$$ \hat{P}\sin\theta = mv_{1x}' - mv_0 \sin\theta \tag{4-442}$$

$$ \hat{P}\cos\theta = mv_{1y}' - mv_0 \cos\theta \tag{4-443}$$

Dividing Eqs. (4-442) and (4-443) by m, we obtain

$$ v_{1x}' - v_0 \sin\theta = \frac{\hat{P}}{m}\sin\theta \tag{4-444}$$

$$ v_{1y}' - v_0 \cos\theta = \frac{\hat{P}}{m}\cos\theta \tag{4-445}$$

Application of Linear Impulse and Linear Momentum to Particles m_1 and m_2

Applying linear impulse and linear momentum to the system consisting of particles m_1 and m_2, we have

$$ \hat{\mathbf{F}} = {}^{\mathcal{F}}\mathbf{G}' - {}^{\mathcal{F}}\mathbf{G} \tag{4-446}$$

Examining the free body diagram of the system consisting of both particles as shown in Fig. 4-27, it is seen that the only impulse acting on the system during the collision is due to string.

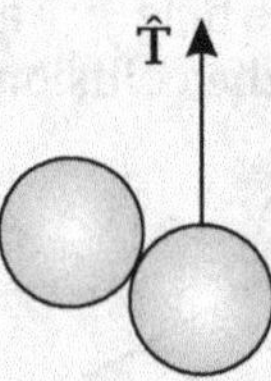

Figure 4-27 Free body diagram of particles m_1 and m_2 for Example 4-25.

Denoting the impulse applied by the string on the system as $\hat{\mathbf{T}}$ and observing that $\hat{\mathbf{T}}$ must lie in the direction of $\mathbf{E}_x$-direction, we have

$$ \hat{\mathbf{T}} = \hat{T}\mathbf{E}_x \tag{4-447}$$

Next, the linear momenta of the system before and after the collision are given, respectively, as

$$^{\mathcal{F}}\mathbf{G} = m\mathbf{v}_0 = mv_0 \sin\theta \mathbf{E}_x + mv_0 \cos\theta \mathbf{E}_y \tag{4-448}$$

$$^{\mathcal{F}}\mathbf{G}' = mv'_{1x}\mathbf{E}_x + mv'_{1y}\mathbf{E}_y + mv'_{2x}\mathbf{E}_x + mv'_{2y}\mathbf{E}_y \tag{4-449}$$

Substituting the results of Eqs. (4-447), (4-448), and (4-449) into Eq. (4-446), we obtain

$$\hat{T}\mathbf{E}_x = mv'_{1x}\mathbf{E}_x + mv'_{1y}\mathbf{E}_y + mv'_{2x}\mathbf{E}_x + mv'_{2y}\mathbf{E}_y - (mv_0 \sin\theta \mathbf{E}_x + mv_0 \cos\theta \mathbf{E}_y) \tag{4-450}$$

Simplifying Eq. (4-450), we have

$$\hat{T}\mathbf{E}_x = (mv'_{1x} + mv'_{2x} - mv_0 \sin\theta)\mathbf{E}_x + (mv'_{1y} + mv'_{2y} - mv_0 \cos\theta)\mathbf{E}_y \tag{4-451}$$

Equating components in Eq. (4-451), we obtain the following two scalar equations:

$$\hat{T} = mv'_{1x} + mv'_{2x} - mv_0 \sin\theta \tag{4-452}$$

$$0 = mv'_{1y} + mv'_{2y} - mv_0 \cos\theta \tag{4-453}$$

Dividing Eqs. (4-452) and (4-453) by m, we obtain

$$v'_{1x} + v'_{2x} - v_0 \sin\theta = \frac{\hat{T}}{m} \tag{4-454}$$

$$v'_{1y} + v'_{2y} - v_0 \cos\theta = 0 \tag{4-455}$$

Kinematic Constraints During Impact

It can be seen that Eqs. (4-444), (4-445), (4-454), and (4-455) are a system of four equations in the six unknowns v'_{1x}, v'_{1y}, v'_{2x}, v'_{2y}, $\hat{P}$, and $\hat{T}$. Consequently, two more independent equations are required in order to determine the solution. These two equations are obtained from the kinematic constraints that act on the particles during impact. First, because the cord is inelastic, the component of velocity of mass m_2 the instant after impact must be zero, i.e.,

$$v'_{2x} = 0 \tag{4-456}$$

Second, because the collision is perfectly elastic, the coefficient of restitution, e, between the two particles must be unity, i.e., $e = 1$. Applying the coefficient of restitution condition of Eq. (4-404) on page 294, we have

$$e = \frac{^{\mathcal{F}}\mathbf{v}'_2 \cdot \mathbf{n} - {}^{\mathcal{F}}\mathbf{v}'_1 \cdot \mathbf{n}}{^{\mathcal{F}}\mathbf{v}_1 \cdot \mathbf{n} - {}^{\mathcal{F}}\mathbf{v}_2 \cdot \mathbf{n}} = 1 \tag{4-457}$$

where $\mathbf{n}$ is the direction of impact. Now, because in this problem the impact occurs in the direction of $\mathbf{v}_0$, we have

$$\mathbf{n} = \frac{\mathbf{v}_0}{\|\mathbf{v}_0\|} = \frac{v_0 \sin\theta \mathbf{E}_x + v_0 \cos\theta \mathbf{E}_y}{v_0} = \sin\theta \mathbf{E}_x + \cos\theta \mathbf{E}_y \tag{4-458}$$

Substituting the expression for $\mathbf{n}$ from Eq. (4-458) into Eq. (4-457) along with the expressions for the pre-impact and post-impact velocities of the particles, we obtain

$$\frac{v'_{2x} \sin\theta + v'_{2y} \cos\theta - (v'_{1x} \sin\theta + v'_{1y} \cos\theta)}{v_0} = 1 \tag{4-459}$$

Rearranging Eq. (4–459) gives

$$v'_{2x}\sin\theta + v'_{2y}\cos\theta - v'_{1x}\sin\theta - v'_{1y}\cos\theta - v_0 = 0 \tag{4–460}$$

Determination of Post-Impact Velocities and Impulse Exerted by Cord

The post-impact velocity of each particle and the impulse exerted by the cord on the system can now be determined using the results of Eqs. (4–444), (4–445), (4–454), (4–455), (4–456), and (4–460). First, multiplying Eq. (4–444) by $\sin\theta$ and multiplying Eq. (4–445) by $\cos\theta$, we obtain

$$\frac{\hat{P}}{m}\sin^2\theta = v'_{1x}\sin\theta - v_0\sin^2\theta \tag{4–461}$$

$$\frac{\hat{P}}{m}\cos^2\theta = v'_{1y}\cos\theta - v_0\cos^2\theta \tag{4–462}$$

Adding Eqs. (4–461) and (4–462), we obtain

$$\frac{\hat{P}}{m} = v'_{1x}\sin\theta + v'_{1y}\cos\theta - v_0 \tag{4–463}$$

Rearranging Eq. (4–463), we obtain

$$v'_{1x}\sin\theta + v'_{1y}\cos\theta = \frac{\hat{P}}{m} + v_0 \tag{4–464}$$

Then, substituting the result of Eq. (4–456) into (4–460), we have

$$v'_{2y}\cos\theta - v'_{1x}\sin\theta - v'_{1y}\cos\theta - v_0 = 0 \tag{4–465}$$

Next, substituting the result of Eq. (4–464) into (4–465), we obtain

$$v'_{2y}\cos\theta - \left(\frac{\hat{P}}{m} + v_0\right) - v_0 = 0 \tag{4–466}$$

Simplifying Eq. (4–466), we obtain

$$v'_{2y}\cos\theta = \frac{\hat{P}}{m} + 2v_0 \tag{4–467}$$

Then, multiplying Eq. (4–455) by $\cos\theta$ and solving for $v'_{2y}\cos\theta$ gives

$$v'_{2y}\cos\theta = -v'_{1y}\cos\theta + v_0\cos^2\theta \tag{4–468}$$

Furthermore, solving Eq. (4–455) for $v'_{1y}\cos\theta$ gives

$$v'_{1y}\cos\theta = \frac{\hat{P}}{m}\cos^2\theta + v_0\cos^2\theta \tag{4–469}$$

Substituting the result of Eq. (4–469) into (4–468) gives

$$v'_{2y}\cos\theta = -\left(\frac{\hat{P}}{m}\cos^2\theta + v_0\cos^2\theta\right) + v_0\cos^2\theta = -\frac{\hat{P}}{m}\cos^2\theta \tag{4–470}$$

Then, setting the results of Eqs. (4–467) and (4–470) equal, we have

$$\frac{\hat{P}}{m} + 2v_0 = -\frac{\hat{P}}{m}\cos^2\theta \tag{4–471}$$

Solving Eq. (4–471) for $\hat{P}$ gives

$$\hat{P} = -\frac{2mv_0}{1+\cos^2\theta} \tag{4–472}$$

Then, substituting the result of Eq. (4–472) into (4–444) gives

$$-\frac{2v_0}{1+\cos^2\theta}\sin\theta = v'_{1x} - v_0\sin\theta \tag{4–473}$$

Solving Eq. (4–473) for v'_{1x} gives

$$v'_{1x} = -\frac{2v_0}{1+\cos^2\theta}\sin\theta + v_0\sin\theta = v_0\left(1-\frac{2}{1+\cos^2\theta}\right)\sin\theta \tag{4–474}$$

Simplifying Eq. (4–474), we obtain v'_{1x} as

$$v'_{1x} = -v_0\frac{\sin^2\theta}{1+\cos^2\theta}\sin\theta \tag{4–475}$$

Similarly, substituting the result of Eq. (4–472) into (4–445) gives

$$-\frac{2v_0\cos\theta}{1+\cos^2\theta} = v'_{1y} - v_0\cos\theta \tag{4–476}$$

Solving Eq. (4–476) for v'_{1y} gives

$$v'_{1y} = v_0\left(1-\frac{2}{1+\cos^2\theta}\right)\cos\theta = -v_0\frac{\sin^2\theta}{1+\cos^2\theta}\cos\theta \tag{4–477}$$

Then, substituting the result of Eqs. (4–477) into (4–455), we obtain

$$v_0\left(1-\frac{2}{1+\cos^2\theta}\right)\cos\theta + v'_{2y} - v_0\cos\theta = 0 \tag{4–478}$$

Solving Eq. (4–478) for v'_{2y}, we obtain

$$v'_{2y} = v_0\cos\theta - v_0\left(1-\frac{2}{1+\cos^2\theta}\right)\cos\theta = \frac{2v_0\cos\theta}{1+\cos^2\theta} \tag{4–479}$$

Substituting the results of Eqs. (4–474) and (4–456) into Eq. (4–454), we have

$$\frac{\hat{T}}{m} = v_0\left(1-\frac{2}{1+\cos^2\theta}\right)\sin\theta - v_0\sin\theta = -\frac{2v_0\sin\theta}{1+\cos^2\theta} \tag{4–480}$$

Solving Eq. (4–480) for $\hat{T}$, we obtain

$$\hat{T} = -\frac{2mv_0\sin\theta}{1+\cos^2\theta} \tag{4–481}$$

Then, substituting the result of Eq. (4–481) into (4–447), the impulse exerted by the cord on the system during the impact is given as

$$\hat{\mathbf{T}} = -\frac{2mv_0 \sin\theta}{1+\cos^2\theta}\mathbf{E}_x \tag{4–482}$$

Finally, using the results of Eqs. (4–475), (4–477), and (4–479), the post-impact velocities of particles m_1 and m_2 are given, respectively, as

$$\begin{aligned} {}^{\mathcal{F}}\mathbf{v}_1' &= -v_0\frac{\sin^2\theta}{1+\cos^2\theta}\sin\theta\,\mathbf{E}_x + -v_0\frac{\sin^2\theta}{1+\cos^2\theta}\cos\theta\,\mathbf{E}_y && (4–483) \\ {}^{\mathcal{F}}\mathbf{v}_2' &= \frac{2v_0\cos\theta}{1+\cos^2\theta}\mathbf{E}_y && (4–484) \end{aligned}$$

■

Summary of Chapter 4

This chapter was devoted to developing the framework for solving and analyzing problems involving a system of particles. The first topics covered in this Chapter were the center of mass, linear momentum, and angular momentum of a rigid body. The position of the center of mass for a system of particles was defined as

$$\bar{\mathbf{r}} = \frac{\sum_{i=1}^{n} m_i \mathbf{r}_i}{\sum_{i=1}^{n} m_i} = \frac{\sum_{i=1}^{n} m_i \mathbf{r}_i}{m} \tag{4-1}$$

where $\mathbf{r}_i$ $(i = 1, \ldots, n)$ is the position of particle i measured relative to a point O fixed in an inertial reference frame $\mathcal{N}$. The linear momentum for a system of particles was then defined as

$${}^{\mathcal{N}}\mathbf{G} = \sum_{i=1}^{n} m_i {}^{\mathcal{N}}\mathbf{v}_i \tag{4-3}$$

where ${}^{\mathcal{N}}\mathbf{v}_i$ $(i = 1, \ldots, n)$ is the velocity of particle i in the inertial reference frame $\mathcal{N}$. It was then shown that the linear momentum for a system of particles can be written as

$${}^{\mathcal{N}}\mathbf{G} = \frac{{}^{\mathcal{N}}d}{dt}(m\bar{\mathbf{r}}) = m\frac{{}^{\mathcal{N}}d\bar{\mathbf{r}}}{dt} = m{}^{\mathcal{N}}\bar{\mathbf{v}} \tag{4-8}$$

where ${}^{\mathcal{N}}\bar{\mathbf{v}}$ is the velocity of the center of mass of the system in the inertial reference frame $\mathcal{N}$. It was then shown that the velocity of the center of mass in reference frame $\mathcal{N}$ can be written as

$${}^{\mathcal{N}}\bar{\mathbf{v}} = \frac{\sum_{i=1}^{n} m_i {}^{\mathcal{N}}\mathbf{v}_i}{m} \tag{4-9}$$

Finally, the acceleration of the center of mass of the system in the inertial reference frame $\mathcal{N}$ was defined as

$${}^{\mathcal{N}}\bar{\mathbf{a}} = \frac{\sum_{i=1}^{n} m_i {}^{\mathcal{N}}\mathbf{a}_i}{m} \tag{4-11}$$

The next topic covered in this chapter was the angular momentum for a system of particles. First, the angular momentum of a system of particles in an inertial reference frame $\mathcal{N}$ relative to an arbitrary reference point Q was defined as

$${}^{\mathcal{N}}\mathbf{H}_Q = \sum_{i=1}^{n}(\mathbf{r}_i - \mathbf{r}_Q) \times m_i({}^{\mathcal{N}}\mathbf{v}_i - {}^{\mathcal{N}}\mathbf{v}_Q) \tag{4-15}$$

Furthermore, the angular momentum relative to a point O fixed in the inertial reference frame $\mathcal{N}$, denoted ${}^{\mathcal{N}}\mathbf{H}_O$, was defined as

$${}^{\mathcal{N}}\mathbf{H}_O = \sum_{i=1}^{n}(\mathbf{r}_i - \mathbf{r}_O) \times m_i {}^{\mathcal{N}}\mathbf{v}_i \tag{4-16}$$

Finally, the angular momentum relative to the center of mass of the system, denoted ${}^{\mathcal{N}}\bar{\mathbf{H}}$, was defined as

$${}^{\mathcal{N}}\bar{\mathbf{H}} = \sum_{i=1}^{n}(\mathbf{r}_i - \bar{\mathbf{r}}) \times m_i({}^{\mathcal{N}}\mathbf{v}_i - {}^{\mathcal{N}}\bar{\mathbf{v}}) \tag{4-17}$$

Using the three definitions of angular momentum, it was shown that ${}^{\mathcal{N}}\mathbf{H}_Q$ and ${}^{\mathcal{N}}\mathbf{H}_O$ are related as

$${}^{\mathcal{N}}\mathbf{H}_Q = {}^{\mathcal{N}}\mathbf{H}_O - (\bar{\mathbf{r}} - \mathbf{r}_Q) \times m{}^{\mathcal{N}}\mathbf{v}_Q - (\mathbf{r}_Q - \mathbf{r}_O) \times m{}^{\mathcal{N}}\bar{\mathbf{v}} \tag{4-23}$$

Next, it was shown that ${}^{\mathcal{N}}\bar{\mathbf{H}}$ and ${}^{\mathcal{N}}\mathbf{H}_O$ are related as

$${}^{\mathcal{N}}\bar{\mathbf{H}} = {}^{\mathcal{N}}\mathbf{H}_O - (\bar{\mathbf{r}} - \mathbf{r}_O) \times m{}^{\mathcal{N}}\bar{\mathbf{v}} \tag{4-25}$$

Finally, it was shown that ${}^{\mathcal{N}}\mathbf{H}_Q$ and ${}^{\mathcal{N}}\bar{\mathbf{H}}$ are related as

$${}^{\mathcal{N}}\mathbf{H}_Q = {}^{\mathcal{N}}\bar{\mathbf{H}} + (\bar{\mathbf{r}} - \mathbf{r}_Q) \times m({}^{\mathcal{N}}\bar{\mathbf{v}} - {}^{\mathcal{N}}\mathbf{v}_Q) \tag{4-28}$$

The next topic covered in this chapter was Newton's 2nd law for a system of particles. In particular, it was shown that

$$\mathbf{R}_i = m_i{}^{\mathcal{N}}\mathbf{a}_i \quad (i = 1, 2, \ldots, n) \tag{4-32}$$

where ${}^{\mathcal{N}}\mathbf{a}_i$ is the acceleration of particle i in an inertial reference frame $\mathcal{N}$ and $\mathbf{R}_i$ is the resultant force acting on particle i. Furthermore, it was shown that the resultant force on particle i is given as

$$\mathbf{R}_i = \mathbf{F}_i + \sum_{j=1}^{n} \mathbf{f}_{ij} \quad (i = 1, 2, \ldots, n) \tag{4-33}$$

where $\mathbf{F}_i$ is the resultant *external* force applied to particle i and $\sum_{j=1}^{n} \mathbf{f}_{ij}$ is the resultant force on particle i applied by every other particle in the system. It was assumed that the forces of interaction between particles satisfy the *strong form of Newton's* 3rd *law*, i.e., that $\mathbf{f}_{ij} = -\mathbf{f}_{ji}$ and $\mathbf{f}_{ij}$ lies along the line between particle i and particle j. Using the properties of the forces of interaction, it was shown that the motion of the center of mass of the system satisfies Newton's 2nd law, i.e.,

$$\mathbf{F} = m{}^{\mathcal{N}}\bar{\mathbf{a}} \tag{4-38}$$

where $\mathbf{F}$ is the resultant external force applied to the system and ${}^{\mathcal{N}}\bar{\mathbf{a}}$ is the acceleration of the center of mass of the system in an inertial reference frame $\mathcal{N}$.

The next topic covered in this chapter was the moment applied by a system of forces to a system of particles. First, the moment due to a system of forces $\mathbf{F}_1, \ldots, \mathbf{F}_n$ relative to an arbitrary reference point Q, denoted $\mathbf{M}_Q$, was defined as

$$\mathbf{M}_Q = \sum_{i=1}^{n} (\mathbf{r}_i - \mathbf{r}_Q) \times \mathbf{F}_i \tag{4-75}$$

Next, the moment due to all external forces relative to a point O fixed in the inertial reference frame $\mathcal{N}$, denoted $\mathbf{M}_O$, was defined as

$$\mathbf{M}_O = \sum_{i=1}^{n} (\mathbf{r}_i - \mathbf{r}_O) \times \mathbf{F}_i \tag{4-76}$$

Finally, the moment due to all external forces relative to the center of mass of the system, denoted $\bar{\mathbf{M}}$, was defined as

$$\bar{\mathbf{M}} = \sum_{i=1}^{n} (\mathbf{r}_i - \bar{\mathbf{r}}) \times \mathbf{F}_i \tag{4-77}$$

The next topic covered in this chapter was the rate of change of angular momentum for a system of particles. First, the rate of change of ${}^{\mathcal{N}}\mathbf{H}_Q$ in the inertial reference frame $\mathcal{N}$ was derived to be

$$\mathbf{M}_Q - (\bar{\mathbf{r}} - \mathbf{r}_Q) \times m{}^{\mathcal{N}}\mathbf{a}_Q = \frac{{}^{\mathcal{N}}d}{dt}\left({}^{\mathcal{N}}\mathbf{H}_Q\right) \tag{4-87}$$

where $\mathbf{M}_Q$ is the moment due to all external forces relative to point Q as given in Eq. (4-75). Furthermore, the quantity

$$-(\bar{\mathbf{r}} - \mathbf{r}_Q) \times m{}^{\mathcal{N}}\mathbf{a}_Q \tag{4-88}$$

was called the *inertial moment* of the reference point Q relative to the center of mass of the system. Then, for the case of a reference point O fixed in an inertial reference frame $\mathcal{N}$, it was shown that

$$\mathbf{M}_O = \frac{{}^{\mathcal{N}}d}{dt}\left({}^{\mathcal{N}}\mathbf{H}_O\right) \tag{4-89}$$

Similarly, for the case where the reference point is located at the center of mass of the system, it was shown that

$$\bar{\mathbf{M}} = \frac{{}^{\mathcal{N}}d}{dt}\left({}^{\mathcal{N}}\bar{\mathbf{H}}\right) \tag{4-90}$$

Because Eqs. (4-89) and (4-90) have the same mathematical form, when the reference point is either fixed in an inertial reference frame or is the center of mass of the system, we have

$$\mathbf{M} = \frac{{}^{\mathcal{N}}d}{dt}\left({}^{\mathcal{N}}\mathbf{H}\right) \tag{4-91}$$

Finally, it was emphasized that the relationship of Eq. (4-91) is *not* valid for any points other than a point fixed in an inertial reference frame or the center of mass. In the case of an arbitrary reference point, it is necessary to use the result of Eq. (4-87).

The next topic covered in this chapter was impulse and momentum for a system of particles. First, the principle of linear impulse and linear momentum for a system of particles was derived as

$$\hat{\mathbf{F}} = {}^{\mathcal{N}}\mathbf{G}(t_2) - {}^{\mathcal{N}}\mathbf{G}(t_1) \tag{4-216}$$

Next, the principle of angular impulse and angular momentum relative to an arbitrary reference point Q was derived as

$$\hat{\mathbf{M}}_Q - \int_{t_1}^{t_2} (\bar{\mathbf{r}} - \mathbf{r}_Q) \times m{}^{\mathcal{N}}\mathbf{a}_Q dt = {}^{\mathcal{N}}\mathbf{H}_Q(t_2) - {}^{\mathcal{N}}\mathbf{H}_Q(t_1) \tag{4-221}$$

Then, for the case of a reference point O fixed in an inertial reference frame $\mathcal{N}$, it was shown that

$$\hat{\mathbf{M}}_O = {}^{\mathcal{N}}\mathbf{H}_O(t_2) - {}^{\mathcal{N}}\mathbf{H}_O(t_1) \tag{4-223}$$

Finally, for the case where the reference point is the center of mass of the system, it was shown that

$$\hat{\bar{\mathbf{M}}} = {}^{\mathcal{N}}\bar{\mathbf{H}}(t_2) - {}^{\mathcal{N}}\bar{\mathbf{H}}(t_1) \tag{4-224}$$

The next topics covered in this chapter were work and energy for a system of particles. First, kinetic energy for a system of particles in an inertial reference frame $\mathcal{N}$ was defined as

$${}^{\mathcal{N}}T = \frac{1}{2}\sum_{i=1}^{n} m_i {}^{\mathcal{N}}\mathbf{v}_i \cdot {}^{\mathcal{N}}\mathbf{v}_i \tag{4-307}$$

An alternate expression for the kinetic energy for a system of particles was also obtained as

$$^{\mathcal{N}}T = \frac{1}{2}m^{\mathcal{N}}\bar{\mathbf{v}} \cdot {}^{\mathcal{N}}\bar{\mathbf{v}} + \frac{1}{2}\sum_{i=1}^{n} m_i \left({}^{\mathcal{N}}\mathbf{v}_i - {}^{\mathcal{N}}\bar{\mathbf{v}}\right) \cdot \left({}^{\mathcal{N}}\mathbf{v}_i - {}^{\mathcal{N}}\bar{\mathbf{v}}\right) \tag{4-314}$$

The kinetic energy was then written in the more compact form

$$^{\mathcal{N}}T = {}^{\mathcal{N}}T^{(1)} + {}^{\mathcal{N}}T^{(2)} \tag{4-315}$$

where

$$^{\mathcal{N}}T^{(1)} = \frac{1}{2}m^{\mathcal{N}}\bar{\mathbf{v}} \cdot {}^{\mathcal{N}}\bar{\mathbf{v}} \tag{4-316}$$

$$^{\mathcal{N}}T^{(2)} = \frac{1}{2}\sum_{i=1}^{n} m_i \left({}^{\mathcal{N}}\mathbf{v}_i - {}^{\mathcal{N}}\bar{\mathbf{v}}\right) \cdot \left({}^{\mathcal{N}}\mathbf{v}_i - {}^{\mathcal{N}}\bar{\mathbf{v}}\right) \tag{4-317}$$

Using Eq. (4-316), it was shown that

$$\frac{d}{dt}\left({}^{\mathcal{N}}T^{(1)}\right) = \mathbf{F} \cdot {}^{\mathcal{N}}\bar{\mathbf{v}} = {}^{\mathcal{N}}P^{(1)} \tag{4-320}$$

which states that the power of all *external* forces acting on a system of particles is equal to the rate of change of the kinetic energy due to the motion of the center of mass of the system. Integrating Eq. (4-320) from time $t = t_1$ to $t = t_2$, it was shown that

$$^{\mathcal{N}}T^{(1)}(t_2) - {}^{\mathcal{N}}T^{(1)}(t_1) = {}^{\mathcal{N}}W_{12}^{(1)} \tag{4-322}$$

Next, it was shown that the rate of change of the kinetic energy of a system of particles satisfies

$$\frac{d}{dt}\left({}^{\mathcal{N}}T\right) = \sum_{i=1}^{n} \mathbf{R}_i \cdot {}^{\mathcal{N}}\mathbf{v}_i \tag{4-330}$$

which is the *work-energy theorem* for a system of particles and states that the rate of change of kinetic energy of a system of particles is equal to the power produced by *all* internal and external forces acting on the system. Finally, it was shown that

$$\frac{d}{dt}\left({}^{\mathcal{N}}E\right) = \sum_{i=1}^{n} \mathbf{R}_i^{nc} \cdot {}^{\mathcal{N}}\mathbf{v}_i \tag{4-341}$$

which is the *alternate form of the work-energy theorem* for a system of particles and states that the rate of change of total energy of a system of particles is equal to the power produced by all of the nonconservative forces acting on the system.

The last topic covered in this Chapter was the collision of two particles. In particular, a simplified collision model was derived that used an ad hoc parameter called the *coefficient of restitution*, e. In terms of e, it was shown that the pre-collision and post-collision velocities of two particles undergoing an impact are related as

$$e = \frac{{}^{\mathcal{N}}\mathbf{v}_2' \cdot \mathbf{n} - {}^{\mathcal{N}}\mathbf{v}_1' \cdot \mathbf{n}}{{}^{\mathcal{N}}\mathbf{v}_1 \cdot \mathbf{n} - {}^{\mathcal{N}}\mathbf{v}_2 \cdot \mathbf{n}} \tag{4-404}$$

where $\mathbf{n}$ is the direction of the impact (also called the *line of impact*) and the notation $(\cdot)'$ denotes a post-collision quantity. Furthermore, it was shown that the post-collision velocities of the particles are given as

$$\begin{aligned}
{}^{\mathcal{N}}\mathbf{v}_1' &= ({}^{\mathcal{N}}\mathbf{v}_1 \cdot \mathbf{u})\,\mathbf{u} + ({}^{\mathcal{N}}\mathbf{v}_1 \cdot \mathbf{w})\,\mathbf{w} \\
&+ \left(\frac{m_1 - em_2}{m_1 + m_2}{}^{\mathcal{N}}\mathbf{v}_1 \cdot \mathbf{n} + \frac{m_2(1+e)}{m_1 + m_2}{}^{\mathcal{N}}\mathbf{v}_2 \cdot \mathbf{n}\right)\mathbf{n} \\
{}^{\mathcal{N}}\mathbf{v}_2' &= ({}^{\mathcal{N}}\mathbf{v}_2 \cdot \mathbf{u})\,\mathbf{u} + ({}^{\mathcal{N}}\mathbf{v}_2 \cdot \mathbf{w})\,\mathbf{w} \\
&+ \left(\frac{m_1(1+e)}{m_1 + m_2}{}^{\mathcal{N}}\mathbf{v}_1 \cdot \mathbf{n} + \frac{m_2 - em_1}{m_1 + m_2}{}^{\mathcal{N}}\mathbf{v}_2 \cdot \mathbf{n}\right)\mathbf{n}
\end{aligned} \tag{4-433}$$

where $\mathbf{u}$ and $\mathbf{w}$ lie in the plane tangent to the surface of each particle at the point of contact during the collision. Finally, it was discussed that the collision model is highly simplified in that all of the kinetics associated with the impact are lumped into the coefficient of restitution.

Problems for Chapter 4

4–1 A particle of mass m is connected to a block of mass M via a rigid massless rod of length l as shown in Fig. P4-1. The rod is free to pivot about a hinge attached to the block at point O. Furthermore, the block rolls without friction along a horizontal surface. Knowing that a horizontal force $\mathbf{F}$ is applied to the block and that gravity acts downward, determine a system of two differential equations describing the motion of the block and the particle.

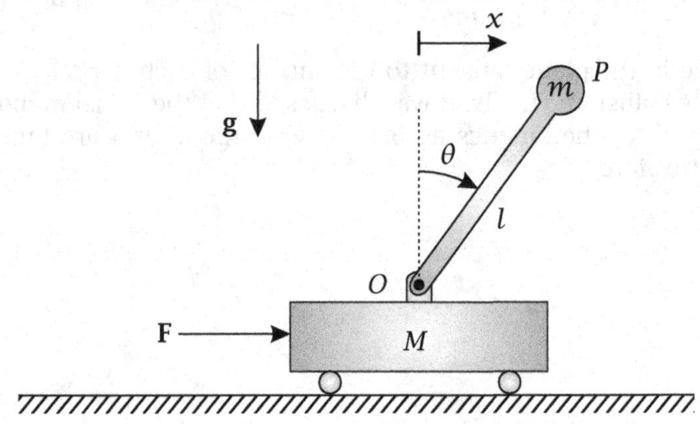

Figure P4-1

4–2 A block of mass M is initially at rest atop a horizontal surface when it is struck by a ball bearing of mass m as shown in Fig. P4-2. The velocity of the particle as it strikes the block is $\mathbf{v}_0$ at an angle θ below the horizontal. Assuming a perfectly inelastic impact, that the block is constrained to remain on the horizontal surface, and no gravity, determine the velocity of the block and the particle immediately after impact.

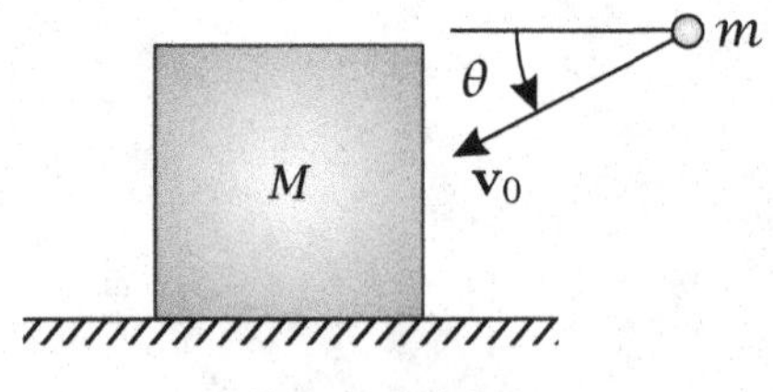

Figure P4-2

4–3 A block of mass m is dropped from a height h above a plate of mass M as shown in Fig. P4-3. The plate is supported by three linear springs, each with spring constant K, and is initially in static equilibrium. Assuming that the compression of the springs due to the weight of the plate is negligible, that the impact is perfectly inelastic, that the block strikes the vertical center of the plate, and that gravity acts downward, determine (a) the velocity of the block and plate immediately after impact and (b) the maximum compression, x_{max}, attained by the springs after impact.

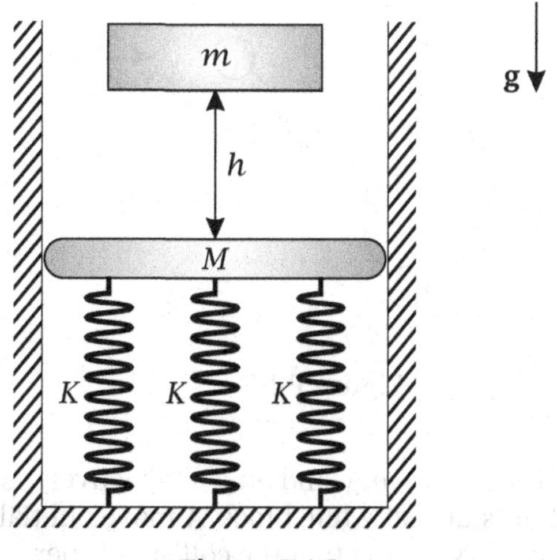

Figure P4-3

4–4 A system consists of two blocks of mass m_1 and m_2 connected to three springs with spring constants k_1, k_2, and k_3 as shown in Fig. P4-4. The blocks are constrained to move horizontally and the surface along which they slide is frictionless. Furthermore, the positions of m_1 and m_2 are denoted x_1 and x_2, respectively. Knowing that the springs are unstretched when x_1 and x_2 are *simultaneously* zero, determine a system of two differential equations of motion for the blocks.

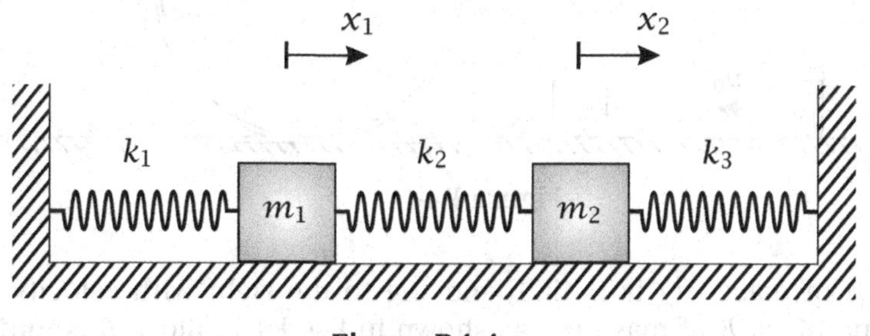

Figure P4-4

4–5 A system consists of three particles A, B, and C, each with mass m, that can slide on a frictionless horizontal surface as shown in Fig. P4-5. Furthermore, particles A and B are connected via an inelastic cord of length ℓ. Particles A and B are initially at rest when particle C, moving horizontally and leftward with velocity $\mathbf{v}_0$, strikes particle B. After impact, particles B and C continue to move to the left until the cord becomes taut. Knowing that the impact is perfectly inelastic, determine (a) the velocity of particles B and C immediately after impact and (b) the velocity of each particle at the instant that the cord becomes taut.

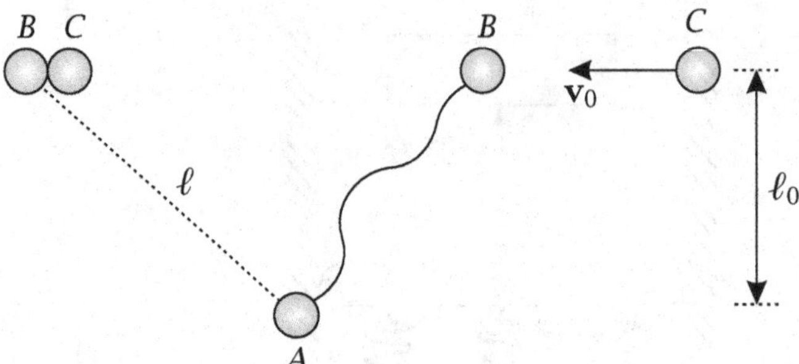

Figure P4-5

4–6 Two blocks A and B of mass m_A and m_B, respectively, slide on a frictionless horizontal surface. Block B is at rest while block A moves initially with velocity $\mathbf{v}_0$ to the right as shown in Fig. P4-6. Knowing that the collision is perfectly elastic, determine (a) the velocity of each block immediately after block A strikes block B and (b) the value of the initial speed $v_0 = \|\mathbf{v}_0\|$ of block A such that block B is moving with zero velocity as it reaches the top of the circular track. Assume in your answers that $m_A \le em_B$.

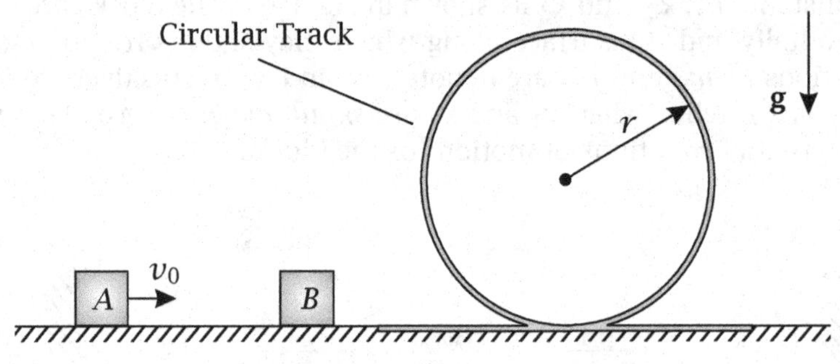

Figure P4-6

4–7 A block A of mass m_A is initially moving to the right with velocity $\mathbf{v}_0$ when it strikes a second block B of mass m_B as shown in Fig. P4-7. Block B is initially at rest and is connected to a linear spring with spring constant K and unstressed length l. Knowing that the spring is initially unstressed and assuming a perfectly elastic impact, determine (a) the velocity of each block immediately after impact and (b) the maximum displacement of block B after the collision. Repeat (a) and (b) for the case of a perfectly inelastic impact. Assume in your answers that $m_A \le em_B$.

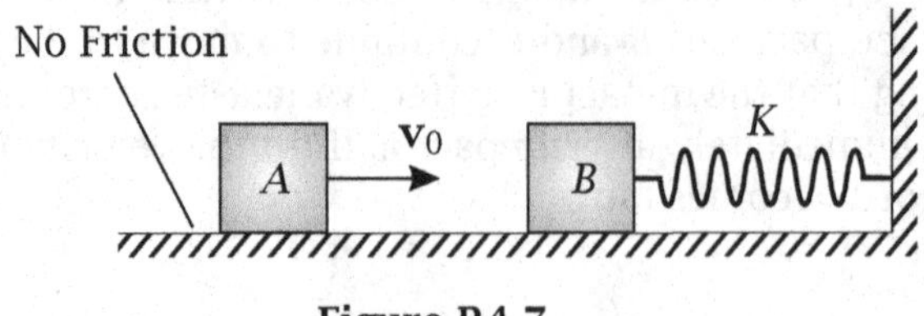

Figure P4-7

4–8 A particle of mass m_A slides without friction along a fixed vertical rigid rod. The particle is attached via a rigid massless arm to a particle of mass m_B, where m_B

slides without friction along a fixed horizontal rigid rod. Assuming that θ is the angle between the horizontal rod and the arm and that gravity acts downward, determine the differential equation of motion for the system in terms of the angle θ.

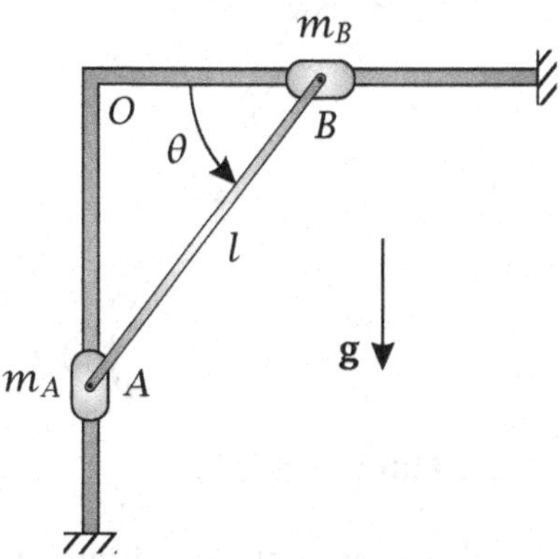

Figure P4-8

4–9 Two particles of mass M and m, with positions $\mathbf{r}_1$ and $\mathbf{r}_2$, respectively, move in three dimensions as shown in Fig. P4-9. Knowing that the only force acting on each particle is due to the gravitational attraction of the other body and that the gravitational attraction obeys the universal law of gravitation, determine (a) the differential equation describing the motion of mass m, (b) the differential equation describing the motion of mass M, and (c) the differential equation describing the motion of mass m relative to mass M.

Assume now that $m << M$ and $\mu = GM$. Using this approximation and the result of part (c) determine (d) the differential equation of mass m relative to mass M. In addition, show that (e) the energy of the system in (d) is conserved and that (f) the angular momentum of the system in (d) relative to M is conserved.

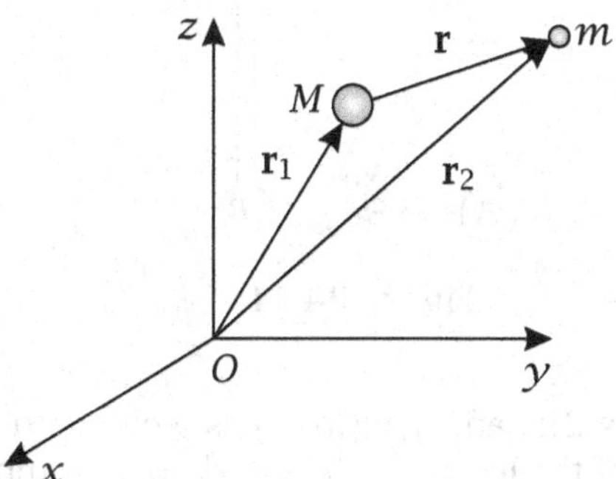

Figure P4-9

4–10 A particle of mass m_2 is suspended from a rigid massless rod as shown in Fig. P4-10. The rod is attached at its other end to a collar of mass m_1. The collar, in turn, slides without friction along a horizontal track. Furthermore, the collar is attached to a linear spring with spring constant K and unstretched length ℓ_0. Assuming that gravity acts downward, determine a system of two differential equations in terms of the variables x and θ that describes the motion of the collar and the particle.

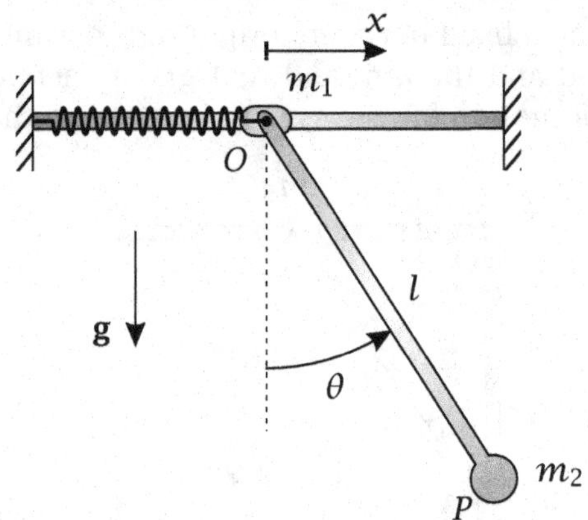

Figure P4-10

4–11 Particle A of mass $m/2$ is moving with an initial velocity $\mathbf{v}_0$ to the right when it strikes particle B, also of mass $m/2$, as shown in Fig. P4-11. Particle B is connected via a rigid massless rod to a cart C of mass m. The cart rolls without friction along a fixed horizontal surface. Knowing that the collision is perfectly inelastic, determine (a) the velocity of each particle and the velocity of the cart immediately after impact and (b) the maximum angle θ attained by the rod after the collision.

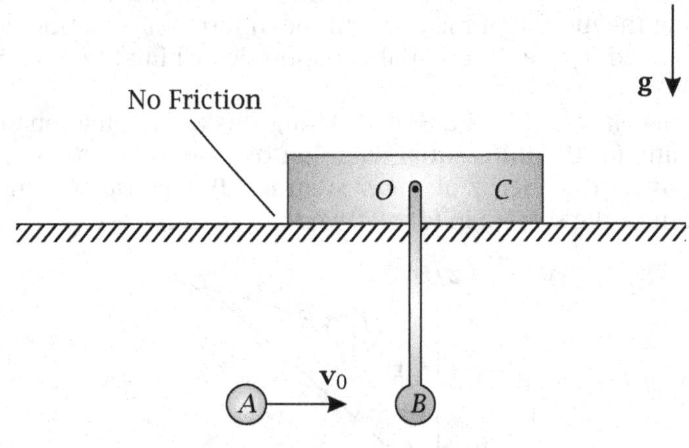

Figure P4-11

4–12 A dumbbell of mass $2m$ and length $2a$ is welded to a rigid massless rod of length l to form the shape of the letter "T". The T-shaped dumbbell is hinged at the free end of the height of the "T" to the fixed point O as shown in Fig. P4-12. Knowing that θ is the angle formed by the height of the "T" with the vertically downward direction and that gravity acts downward, determine the differential equation of motion of the dumbbell in terms of the angle θ using (a) a balance of angular momentum (b) one of the forms of the work-energy theorem.

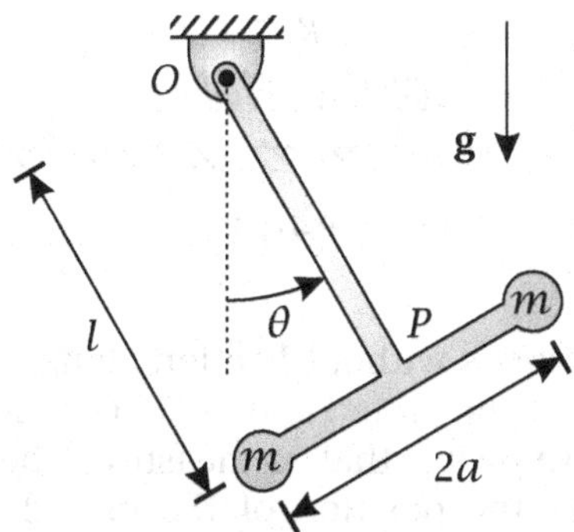

Figure P4-12

4–13 A system consists of two small blocks of mass m_1 and m_2. The blocks slide without friction inside a circular slot of radius R cut from a disk of radius R as shown in Fig. P4-13. Furthermore, a linear spring with spring constant K and unstretched length ℓ_0 is attached between the two blocks. Knowing that the angle θ_1 describes the position of m_1 while the angle θ_2 describes the position m_2, that the spring cannot bend, and that gravity acts downward, determine a system of two differential equations of motion for the blocks.

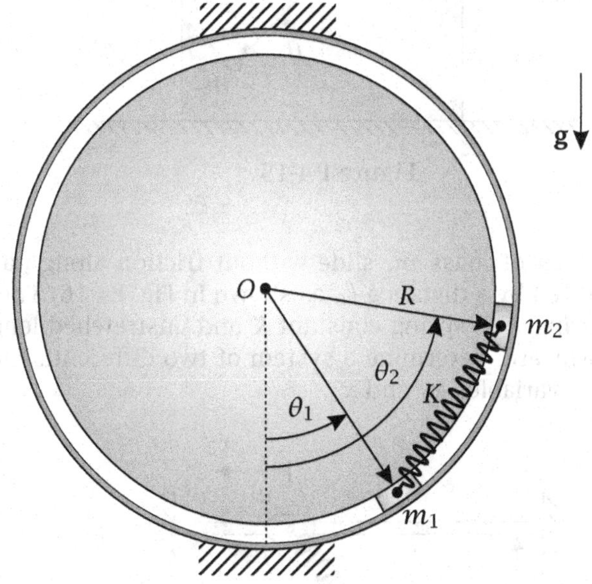

Figure P4-13

4–14 A system consists of two blocks A and B of masses m_A and m_B, respectively, that slide along a frictionless surface as shown in Fig. P4-14. The blocks are connected by a linear spring with spring constant K and unstretched length ℓ. Assuming that the blocks are initially at rest and the spring is initially unstretched when block A is struck by a horizontal impulse $\hat{\mathbf{P}}$, determine the velocity of each block immediately after the application of the impulse.

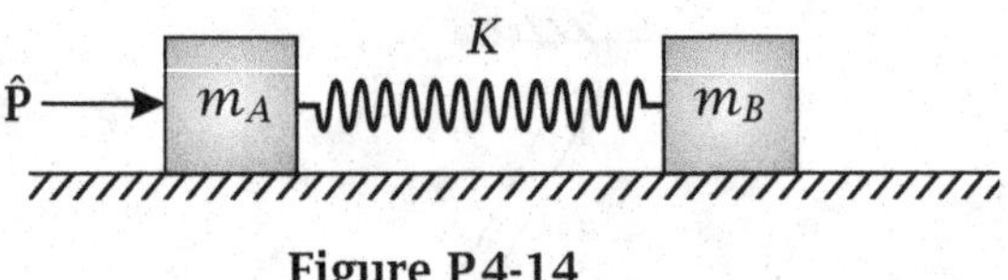

Figure P4-14

4–15 A particle of mass m_1 slides without friction along a circular track of radius R cut from a block of mass m_2 as shown in Fig. P4-15. The block slides without friction along a horizontal surface. Knowing that x measures the horizontal displacement of the block, that θ describes the position of the particle relative to the vertically downward direction, and that gravity acts vertically downward, determine a system of two differential equations in terms of x and θ that describes the motion of the system.

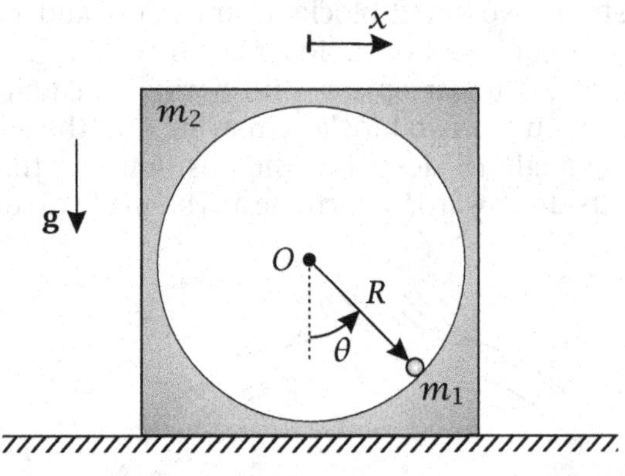

Figure P4-15

4–16 Two collars, each of mass m, slide without friction along parallel horizontal tracks that are separated by a distance ℓ_0 as shown in Fig. P4-16. The collars are connected by a linear spring with spring constant K and unstretched length ℓ_0. Knowing that gravity acts downward, determine a system of two differential equations for the collars in terms of the variables x_1 and x_2.

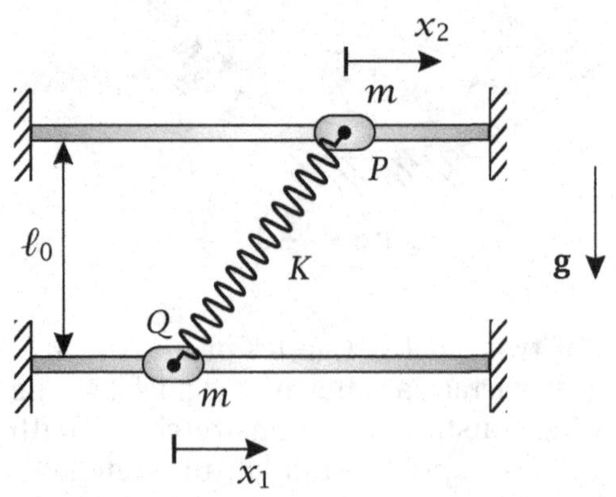

Figure P4-16

4–17 A dumbbell consists of two particles A and B, each of mass m, connected by a rigid massless rod of length $2l$. Each end of the dumbbell slides without friction along a fixed circular track of radius R as shown in Fig. P4-17. Knowing that θ is the angle from the vertical to the center of the rod and that gravity acts downward, determine the differential equation of motion for the dumbbell.

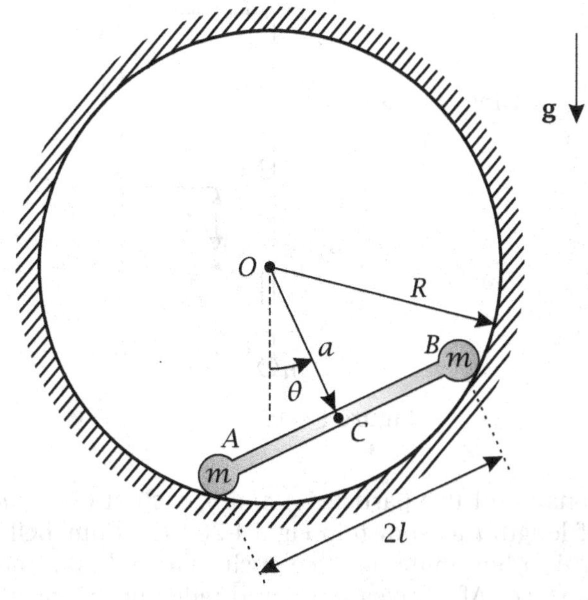

Figure P4-17

4–18 A wedge of mass M and wedge angle β is initially at rest atop a horizontal surface when it is struck by a particle of mass m as shown in Fig. P4-18. The velocity of the particle as it strikes the wedge is $\mathbf{v}_0$ in a direction orthogonal to the surface of the wedge. Knowing that the coefficient of restitution between the particle and the wedge is e and that the wedge is constrained to remain on the horizontal surface after impact, determine the velocity of the wedge and the particle immediately after impact.

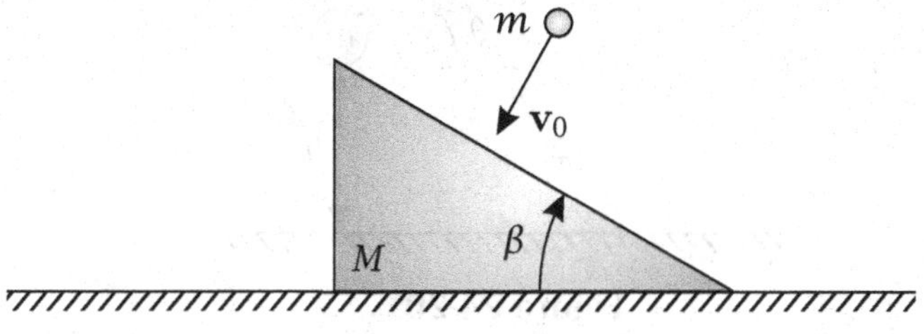

Figure P4-18

4–19 Two massless arms OA and AB, each of length l, are connected in tandem as shown in Fig. P4-19. Arm OA is hinged at one end to the fixed point O and at its other end to point A that is located at one end of rod AB. Three particles, each of mass m, are attached to the rods as follows. The first particle is attached to rod OA at point A, the second particle is attached to rod AB at point C located an unknown distance

x from point A, and the third particle is attached to rod AB at point B. Assuming that the system initially hangs motionless, determine the location x on rod AB at which the second mass should be attached such that the two rods maintain their relative alignment immediately after a transverse impulse $\hat{\mathbf{P}}$ is applied at point C.

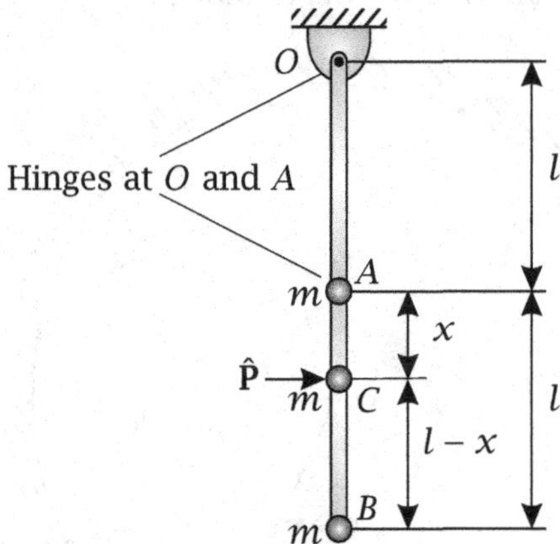

Figure P4-19

4–20 A dumbbell consists of two particles A and B, each of mass m, connected by a rigid massless rod of length l as shown in Fig. P4-20. The dumbbell is released from rest at an angle θ with a horizontal surface such that its lower end at point B is a height h above the surface. After release, the rod falls until point B hits the surface. Knowing that the coefficient of restitution between particle B and the surface is e (where $0 \le e \le 1$), determine (a) the velocity of the center of mass of the rod and (b) the angular velocity of the rod immediately after impact.

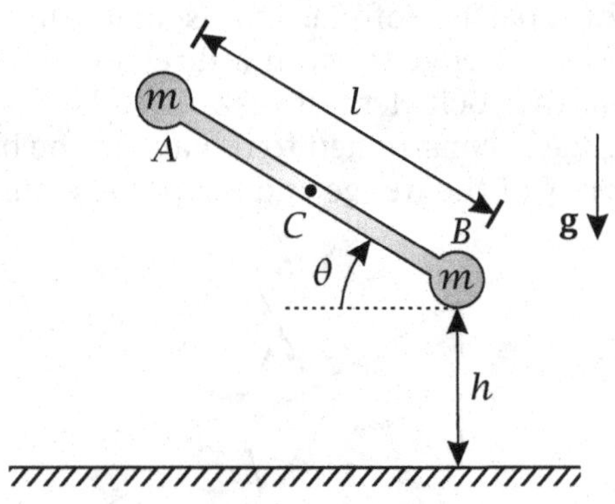

Figure P4-20

4–21 A system consists of two blocks of mass m_1 and m_2. The blocks slide without friction inside a circular slot of radius R cut from a disk of radius R as shown in Fig. P4-21. Furthermore, a curvilinear spring with spring constant K and unstretched length ℓ_0 is attached between the two blocks. Knowing that the angle θ_1 describes the position of m_1 while the angle θ_2 describes the position m_2, and that gravity acts downward, determine a system of two differential equations of motion for the blocks.

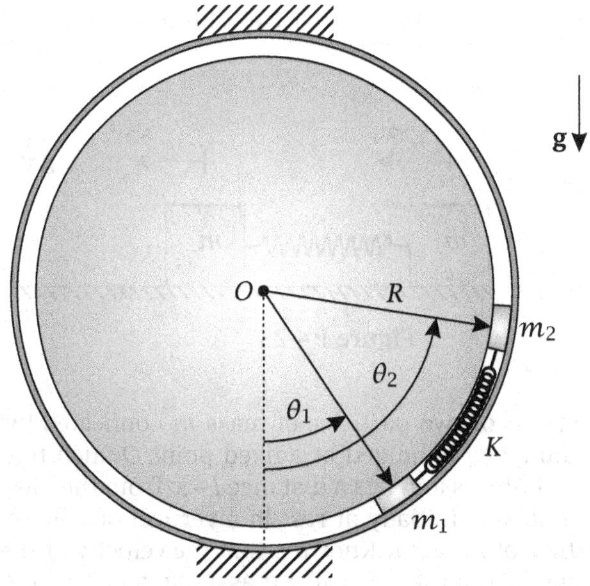

Figure P4-21

4–22 A collar of mass m_1 slides without friction along a fixed vertical rigid rod while a second collar of mass m_2 slides without friction along a rigid horizontal rod. The collars are connected via a linear spring with spring constant K and unstretched length ℓ_0. Assuming that the spring does not bend and that gravity acts downward, determine a system of two differential equations of motion for the system in terms of x_1 and x_2.

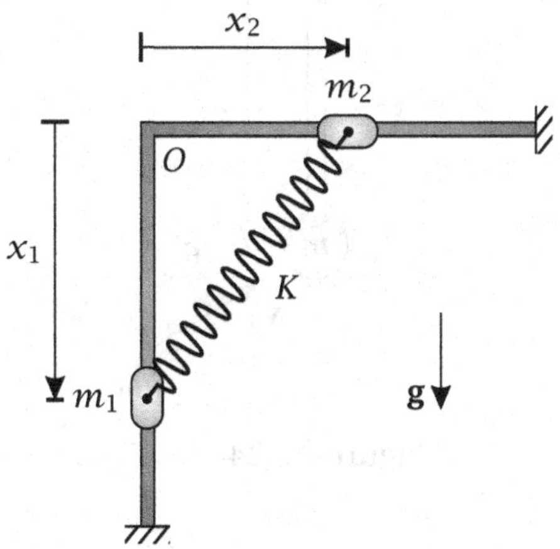

Figure P4-22

4–23 Two blocks of mass m_1 and m_2 slide without friction along a horizontal surface as shown in Fig. P4-23. The blocks are connected to one another by a linear spring of spring constant K and unstretched length ℓ_0. Assuming that the blocks are constrained to stay on the surface, determine a system of two differential equations for the blocks in terms of the displacements x_1 and x_2.

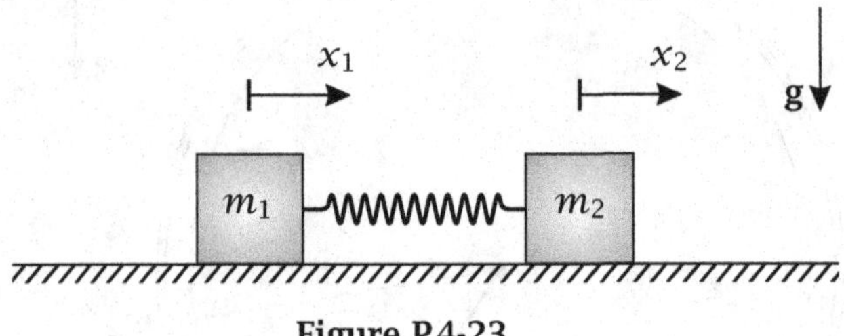

Figure P4-23

4–24 A dumbbell consists of two particles of mass m connected by a rigid massless rod of length l. The dumbbell is hinged at a fixed point O such that the hinge lies a distance x from the upper mass and lies a distance $l-x$ from the lower mass as shown in Fig. P4-24. The dumbbell is initially at rest in a vertical orientation when its lower end is struck by a particle of mass m. Knowing that the velocity of the particle prior to impact is $\mathbf{v}_0$ at an angle θ from the vertical and assuming a perfectly inelastic impact, determine (a) the angular velocity of the dumbbell the instant after the impact and (b) the maximum angular deflection attained by the dumbbell after impact.

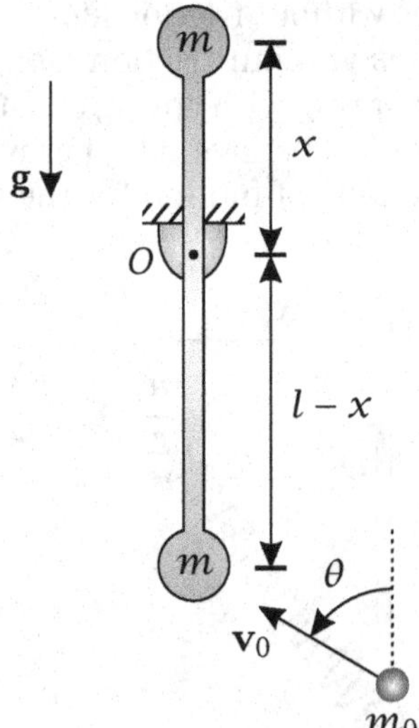

Figure P4-24

Chapter 5

Kinetics of Rigid Bodies

If we all worked on the assumption that what is accepted as true is really true, there would be little hope of advance.

- Orville Wright (1871–1948)
U.S. Inventor Who, with His Brother Wilbur Wright, Achieved the First Powered, Sustained, and Controlled Airplane Flight.

Until now we have been concerned with the kinetics of particles, i.e., the kinetics of objects that have nonzero finite mass but do not occupy any physical space. Furthermore, in Section 2.15 of Chapter 2 we studied the *kinematics* of motion of a rigid body. In this chapter we turn our attention to the *kinetics* of rigid bodies. To this end, the objectives of this chapter are threefold: (1) to describe quantitatively the forces and moments that act on a rigid body; (2) to determine the motion that results from the application of these forces and moments using postulated laws of physics; and (3) to analyze the motion.

The key difference between a particle and rigid body is that a particle can undergo only translational motion whereas a rigid body can undergo *both* translational and rotational motion. In general, for motion in $\mathbb{E}^3$ it is necessary to specify three variables for the translational motion and to specify another three variables for the rotational motion of the rigid body. Consequently, a particle moving in $\mathbb{E}^3$ is said to have *three degrees of freedom* (one for each variable required to specify the translational motion of the particle) while a rigid body moving in $\mathbb{E}^3$ is said to have *six degrees of freedom* (three degrees of freedom for the translational motion of an arbitrary point on the body and another three degrees of freedom for the rotational motion). Consequently, two balance laws, one for translation and another for rotation, are required to specify completely the motion of a rigid body.

This chapter is organized as follows. The first section establishes the mass properties of a rigid body. In particular, the center of mass, linear momentum, and angular momentum of a rigid body are defined. The angular momentum of a rigid body is then expressed conveniently in terms of the moment of inertia tensor for a rigid body. Using the definition of the moment of inertia tensor, the parallel-axis theorem for a rigid body is derived. In particular, the parallel-axis theorem provides a convenient way to relate the angular momentum of a rigid body about an arbitrary reference point to the angular momentum of the rigid body about the center of mass of the rigid body.

Using the mass properties given in the first section, the second section establishes the laws of kinetics of a rigid body. First, the two different types of allowable actions that can be applied to a rigid body, namely forces and pure torques, are described. Then, using the definitions of linear momentum and angular momentum of a rigid body, the fundamental postulates for the kinetics of a rigid body, called *Euler's laws*, are stated. Using the postulated laws of kinetics of a rigid body, two alternate forms of Euler's 2nd law are derived. Next, the kinetic energy of a rigid body is defined from which the work-energy theorem for a rigid body is derived. Finally, using Euler's 1st law and three forms of Euler's 2nd law, the concepts of linear impulse and linear momentum and angular impulse and angular momentum for a rigid body are derived. Similar to the approach used in previous chapters, the understanding of rigid body kinetics is facilitated through the systematic application of the concepts to example problems.

5.1 Center of Mass and Linear Momentum of a Rigid Body

5.1.1 Center of Mass of a Rigid Body

Let $\mathcal{R}$ be a rigid body (as defined in Chapter 2) and let $\mathbf{r}$ be the position of a point P on $\mathcal{R}$. Finally, let $\mathcal{N}$ be an inertial reference frame. Then the *center of mass* of the rigid body $\mathcal{R}$ is defined as

$$\bar{\mathbf{r}} = \frac{\int_{\mathcal{R}} \mathbf{r}dm}{\int_{\mathcal{R}} dm} \tag{5-1}$$

where the notation

$$\int_{\mathcal{R}} dm \tag{5-2}$$

means that the integral is being taken over all points in $\mathbb{E}^3$ occupied by $\mathcal{R}$ and dm is a differential mass element. Denoting the volume of the rigid body by V, the differential mass element dm is given as

$$dm = \rho dV \tag{5-3}$$

where ρ is mass density per unit volume of $\mathcal{R}$ and dV is a differential volume element. Consequently, it is seen that the integral of Eq. (5-2) represents a volume integral over all points on $\mathcal{R}$.[1] Furthermore, it is seen that the mass of the rigid body is given as

$$m = \int_{\mathcal{R}} dm = \int_{\mathcal{R}} \rho dV \tag{5-4}$$

Now because a rigid body cannot change in mass, we note that m is a constant. The center of mass of a rigid body can then be written as

$$\bar{\mathbf{r}} = \frac{\int_{\mathcal{R}} \mathbf{r}dm}{m} \tag{5-5}$$

[1] From this point forth the notation of Eq. (5-2) will always be used to denote a volume integral.

5.1.2 Linear Momentum of a Rigid Body

Let ${}^{\mathcal{N}}\mathbf{v} = {}^{\mathcal{N}}d\mathbf{r}/dt$ be the velocity of a point P on a rigid body $\mathcal{R}$ as viewed by an observer in an inertial reference frame $\mathcal{N}$. Then the *linear momentum* of $\mathcal{R}$ in reference frame $\mathcal{N}$ is defined as

$$ {}^{\mathcal{N}}\mathbf{G} = \int_{\mathcal{R}} {}^{\mathcal{N}}\mathbf{v}dm \tag{5-6} $$

Equation (5-6) can also be written as

$$ {}^{\mathcal{N}}\mathbf{G} = \int_{\mathcal{R}} \frac{{}^{\mathcal{N}}d\mathbf{r}}{dt}dm \tag{5-7} $$

Because the integral is being computed over the body, the rate of change is independent of the integral. Therefore, the order of integration and differentiation in Eq. (5-7) can be interchanged to give

$$ {}^{\mathcal{N}}\mathbf{G} = \frac{{}^{\mathcal{N}}d}{dt}\int_{\mathcal{R}} \mathbf{r}dm \tag{5-8} $$

Substituting Eq. (5-5) into (5-8), we obtain

$$ {}^{\mathcal{N}}\mathbf{G} = \frac{{}^{\mathcal{N}}d}{dt}(m\bar{\mathbf{r}}) = m\,\frac{{}^{\mathcal{N}}d\bar{\mathbf{r}}}{dt} = m\,{}^{\mathcal{N}}\bar{\mathbf{v}} \tag{5-9} $$

The quantity ${}^{\mathcal{N}}\bar{\mathbf{v}} = {}^{\mathcal{N}}d\bar{\mathbf{r}}/dt$ is the *velocity of the center of mass* of the rigid body in reference frame $\mathcal{N}$. It is noted that ${}^{\mathcal{N}}\bar{\mathbf{v}}$ can also be written as

$$ {}^{\mathcal{N}}\bar{\mathbf{v}} = \frac{\int_{\mathcal{R}} {}^{\mathcal{N}}\mathbf{v}dm}{m} \tag{5-10} $$

Finally, the *acceleration of the center of mass* of a rigid body as viewed by an observer in an inertial reference frame $\mathcal{N}$, denoted ${}^{\mathcal{N}}\bar{\mathbf{a}}$, is defined as

$$ {}^{\mathcal{N}}\bar{\mathbf{a}} = \frac{\int_{\mathcal{R}} {}^{\mathcal{N}}\mathbf{a}dm}{m} \tag{5-11} $$

It is noted that the acceleration of the center of mass is also given as

$$ {}^{\mathcal{N}}\bar{\mathbf{a}} = \frac{{}^{\mathcal{N}}d}{dt}\left({}^{\mathcal{N}}\bar{\mathbf{v}}\right) \tag{5-12} $$

5.2 Angular Momentum of a Rigid Body

The angular momentum of a rigid body $\mathcal{R}$ relative to an arbitrary reference point Q in an inertial reference frame $\mathcal{N}$ is defined as[2]

$$ {}^{\mathcal{N}}\mathbf{H}_Q = \int_{\mathcal{R}} (\mathbf{r} - \mathbf{r}_Q) \times ({}^{\mathcal{N}}\mathbf{v} - {}^{\mathcal{N}}\mathbf{v}_Q)dm \tag{5-13} $$

[2]It is noted again that the angular momentum is defined using the *relative velocity* ${}^{\mathcal{N}}\mathbf{v} - {}^{\mathcal{N}}\mathbf{v}_Q$ and is consistent with the definition used in Greenwood (1977; 1988).

where $\mathbf{r}_Q$ and ${}^{\mathcal{N}}\mathbf{v}_Q$ are the position and velocity of point Q, respectively. The angular momentum of a rigid body relative to a point O fixed in an inertial reference frame $\mathcal{N}$ is defined as

$$ {}^{\mathcal{N}}\mathbf{H}_O = \int_{\mathcal{R}} (\mathbf{r} - \mathbf{r}_O) \times {}^{\mathcal{N}}\mathbf{v} dm \tag{5-14} $$

The angular momentum of a rigid body relative to the center of mass of the rigid body in an inertial reference frame $\mathcal{N}$ is defined as

$$ {}^{\mathcal{N}}\bar{\mathbf{H}} = \int_{\mathcal{R}} (\mathbf{r} - \bar{\mathbf{r}}) \times ({}^{\mathcal{N}}\mathbf{v} - {}^{\mathcal{N}}\bar{\mathbf{v}}) dm \tag{5-15} $$

We can obtain a relationship between ${}^{\mathcal{N}}\mathbf{H}_Q$ and ${}^{\mathcal{N}}\mathbf{H}_O$ by subtracting Eq. (5–14) from (5–13). This gives

$$ {}^{\mathcal{N}}\mathbf{H}_Q - {}^{\mathcal{N}}\mathbf{H}_O = \int_{\mathcal{R}} (\mathbf{r} - \mathbf{r}_Q) \times ({}^{\mathcal{N}}\mathbf{v} - {}^{\mathcal{N}}\mathbf{v}_Q) dm - \int_{\mathcal{R}} (\mathbf{r} - \mathbf{r}_O) \times {}^{\mathcal{N}}\mathbf{v} dm \tag{5-16} $$

Expanding Eq. (5–16), we have

$$ \begin{aligned} {}^{\mathcal{N}}\mathbf{H}_Q - {}^{\mathcal{N}}\mathbf{H}_O = & \int_{\mathcal{R}} \left(\mathbf{r} \times {}^{\mathcal{N}}\mathbf{v} - \mathbf{r} \times {}^{\mathcal{N}}\mathbf{v}_Q - \mathbf{r}_Q \times {}^{\mathcal{N}}\mathbf{v} + \mathbf{r}_Q \times {}^{\mathcal{N}}\mathbf{v}_Q\right) dm \\ & + \int_{\mathcal{R}} \left(-\mathbf{r} \times {}^{\mathcal{N}}\mathbf{v} + \mathbf{r}_O \times {}^{\mathcal{N}}\mathbf{v}\right) dm \end{aligned} \tag{5-17} $$

Observing that the first and fifth terms in the integral cancel, Eq. (5–17) simplifies to

$$ {}^{\mathcal{N}}\mathbf{H}_Q - {}^{\mathcal{N}}\mathbf{H}_O = \int_{\mathcal{R}} \left(-\mathbf{r} \times {}^{\mathcal{N}}\mathbf{v}_Q - \mathbf{r}_Q \times {}^{\mathcal{N}}\mathbf{v} + \mathbf{r}_Q \times {}^{\mathcal{N}}\mathbf{v}_Q + \mathbf{r}_O \times {}^{\mathcal{N}}\mathbf{v}\right) dm \tag{5-18} $$

Now, because $\mathbf{r}_O$, $\mathbf{r}_Q$ and ${}^{\mathcal{N}}\mathbf{v}_Q$ are independent of the integral, we have

$$ \begin{aligned} \int_{\mathcal{R}} \mathbf{r} \times {}^{\mathcal{N}}\mathbf{v}_Q dm &= \bar{\mathbf{r}} \times m {}^{\mathcal{N}}\mathbf{v}_Q \\ \int_{\mathcal{R}} \mathbf{r}_Q \times {}^{\mathcal{N}}\mathbf{v} dm &= \mathbf{r}_Q \times m {}^{\mathcal{N}}\bar{\mathbf{v}} \\ \int_{\mathcal{R}} \mathbf{r}_Q \times {}^{\mathcal{N}}\mathbf{v}_Q dm &= \mathbf{r}_Q \times m {}^{\mathcal{N}}\mathbf{v}_Q \\ \int_{\mathcal{R}} \mathbf{r}_O \times {}^{\mathcal{N}}\mathbf{v} dm &= \mathbf{r}_O \times m {}^{\mathcal{N}}\bar{\mathbf{v}} \end{aligned} \tag{5-19} $$

Equation (5–18) can then be written as

$$ {}^{\mathcal{N}}\mathbf{H}_Q - {}^{\mathcal{N}}\mathbf{H}_O = -\bar{\mathbf{r}} \times m {}^{\mathcal{N}}\mathbf{v}_Q - \mathbf{r}_Q \times m {}^{\mathcal{N}}\bar{\mathbf{v}} + \mathbf{r}_Q \times m {}^{\mathcal{N}}\mathbf{v}_Q + \mathbf{r}_O \times m {}^{\mathcal{N}}\bar{\mathbf{v}} \tag{5-20} $$

Equation (5–20) simplifies to

$$ \boxed{{}^{\mathcal{N}}\mathbf{H}_Q = {}^{\mathcal{N}}\mathbf{H}_O - (\bar{\mathbf{r}} - \mathbf{r}_Q) \times m {}^{\mathcal{N}}\mathbf{v}_Q - (\mathbf{r}_Q - \mathbf{r}_O) \times m {}^{\mathcal{N}}\bar{\mathbf{v}}} \tag{5-21} $$

Next, using the result of Eq. (5–21), a relationship between ${}^{\mathcal{N}}\mathbf{H}_O$ and ${}^{\mathcal{N}}\bar{\mathbf{H}}$ is obtained by setting the reference point in Eq. (5–21) equal to the center of mass of the rigid body as

$$ \boxed{{}^{\mathcal{N}}\bar{\mathbf{H}} = {}^{\mathcal{N}}\mathbf{H}_O - (\bar{\mathbf{r}} - \mathbf{r}_O) \times m {}^{\mathcal{N}}\bar{\mathbf{v}}} \tag{5-22} $$

Finally, by substituting the result of Eq. (5–22) into (5–21), we can obtain a relationship between ${}^{\mathcal{N}}\mathbf{H}_Q$ and ${}^{\mathcal{N}}\bar{\mathbf{H}}$ as

$$
{}^{\mathcal{N}}\mathbf{H}_Q = {}^{\mathcal{N}}\bar{\mathbf{H}} + (\bar{\mathbf{r}} - \mathbf{r}_O) \times m{}^{\mathcal{N}}\bar{\mathbf{v}} - (\bar{\mathbf{r}} - \mathbf{r}_Q) \times m{}^{\mathcal{N}}\mathbf{v}_Q - (\mathbf{r}_Q - \mathbf{r}_O) \times m{}^{\mathcal{N}}\bar{\mathbf{v}} \tag{5–23}
$$

Equation (5–23) then simplifies to

$$
\boxed{{}^{\mathcal{N}}\mathbf{H}_Q = {}^{\mathcal{N}}\bar{\mathbf{H}} + (\mathbf{r}_Q - \bar{\mathbf{r}}) \times m({}^{\mathcal{N}}\mathbf{v}_Q - {}^{\mathcal{N}}\bar{\mathbf{v}})} \tag{5–24}
$$

5.3 Moment of Inertia Tensor of a Rigid Body

Quite often in rigid body dynamics it is useful to compute the angular momentum relative to a point that is fixed in the rigid body. In such cases an elegant expression for the angular momentum of the rigid body can be derived in terms of the *moment of inertia tensor*. In this section we derive the moment of inertia tensor of a rigid body and describe how to compute the angular momentum of the rigid body in terms of the moment of inertia tensor.

5.3.1 Moment of Inertia Tensor Relative to a Body-Fixed Point

Let P be a point fixed in a rigid body $\mathcal{R}$ and let ${}^{\mathcal{N}}\mathbf{v}$ be the velocity of P in an inertial reference frame $\mathcal{N}$. Furthermore, let Q be an arbitrary point *fixed in* $\mathcal{R}$ and let ${}^{\mathcal{N}}\mathbf{v}_Q$ be the velocity of point Q in the inertial reference frame $\mathcal{N}$. Finally, let ${}^{\mathcal{N}}\boldsymbol{\omega}^{\mathcal{R}}$ be the angular velocity of $\mathcal{R}$ in the inertial reference frame. Then, because P and Q are both points that are fixed in $\mathcal{R}$, the velocity P relative to Q in reference frame $\mathcal{N}$ is given from Eq. (2–516) on page 106 as

$$
{}^{\mathcal{N}}\mathbf{v} - {}^{\mathcal{N}}\mathbf{v}_Q = {}^{\mathcal{N}}\boldsymbol{\omega}^{\mathcal{R}} \times (\mathbf{r} - \mathbf{r}_Q) \tag{5–25}
$$

Substituting Eq. (5–25) into the expression for ${}^{\mathcal{N}}\mathbf{H}_Q$ from Eq. (5–13), we obtain

$$
{}^{\mathcal{N}}\mathbf{H}_Q = \int_{\mathcal{R}} (\mathbf{r} - \mathbf{r}_Q) \times \left[{}^{\mathcal{N}}\boldsymbol{\omega}^{\mathcal{R}} \times (\mathbf{r} - \mathbf{r}_Q)\right] dm \tag{5–26}
$$

Now suppose that we let

$$
\boldsymbol{\rho}_Q = \mathbf{r} - \mathbf{r}_Q \tag{5–27}
$$

Equation (5–26) can then be written in terms of $\boldsymbol{\rho}_Q$ as

$$
{}^{\mathcal{N}}\mathbf{H}_Q = \int_{\mathcal{R}} \boldsymbol{\rho}_Q \times \left({}^{\mathcal{N}}\boldsymbol{\omega}^{\mathcal{R}} \times \boldsymbol{\rho}_Q\right) dm \tag{5–28}
$$

Then, using the vector triple product as given in Eq. (1–29) on page 10, we have

$$
\boldsymbol{\rho}_Q \times \left({}^{\mathcal{N}}\boldsymbol{\omega}^{\mathcal{R}} \times \boldsymbol{\rho}_Q\right) = \left(\boldsymbol{\rho}_Q \cdot \boldsymbol{\rho}_Q\right){}^{\mathcal{N}}\boldsymbol{\omega}^{\mathcal{R}} - \left(\boldsymbol{\rho}_Q \cdot {}^{\mathcal{N}}\boldsymbol{\omega}^{\mathcal{R}}\right)\boldsymbol{\rho}_Q \tag{5–29}
$$

Using the result of Eq. (5–29), Eq. (5–28) becomes

$$
{}^{\mathcal{N}}\mathbf{H}_Q = \int_{\mathcal{R}} \left[\left(\boldsymbol{\rho}_Q \cdot \boldsymbol{\rho}_Q\right){}^{\mathcal{N}}\boldsymbol{\omega}^{\mathcal{R}} - \left(\boldsymbol{\rho}_Q \cdot {}^{\mathcal{N}}\boldsymbol{\omega}^{\mathcal{R}}\right)\boldsymbol{\rho}_Q\right] dm \tag{5–30}
$$

Using the definition of the product between a tensor and a vector from Eq. (1–41) on page 11, the second term in the integrand of Eq. (5–30) can be written as

$$\left(\boldsymbol{\rho}_Q \cdot {}^{\mathcal{N}}\boldsymbol{\omega}^{\mathcal{R}}\right)\boldsymbol{\rho}_Q = \boldsymbol{\rho}_Q\left(\boldsymbol{\rho}_Q \cdot {}^{\mathcal{N}}\boldsymbol{\omega}^{\mathcal{R}}\right) = \left(\boldsymbol{\rho}_Q \otimes \boldsymbol{\rho}_Q\right) \cdot {}^{\mathcal{N}}\boldsymbol{\omega}^{\mathcal{R}} \tag{5–31}$$

Substituting the result of Eq. (5–31) into (5–30), we obtain

$${}^{\mathcal{N}}\mathbf{H}_Q = \int_{\mathcal{R}} \left[\left(\boldsymbol{\rho}_Q \cdot \boldsymbol{\rho}_Q\right){}^{\mathcal{N}}\boldsymbol{\omega}^{\mathcal{R}} - \left(\boldsymbol{\rho}_Q \otimes \boldsymbol{\rho}_Q\right) \cdot {}^{\mathcal{N}}\boldsymbol{\omega}^{\mathcal{R}}\right] dm \tag{5–32}$$

Furthermore, recalling the identity tensor $\mathbf{U}$ from Eq. (1–34), we have

$${}^{\mathcal{N}}\boldsymbol{\omega}^{\mathcal{R}} = \mathbf{U} \cdot {}^{\mathcal{N}}\boldsymbol{\omega}^{\mathcal{R}} \tag{5–33}$$

Substituting the result of Eq. (5–33), we obtain

$${}^{\mathcal{N}}\mathbf{H}_Q = \int_{\mathcal{R}} \left[\left(\boldsymbol{\rho}_Q \cdot \boldsymbol{\rho}_Q\right)\mathbf{U} \cdot {}^{\mathcal{N}}\boldsymbol{\omega}^{\mathcal{R}} - \left(\boldsymbol{\rho}_Q \otimes \boldsymbol{\rho}_Q\right) \cdot {}^{\mathcal{N}}\boldsymbol{\omega}^{\mathcal{R}}\right] dm \tag{5–34}$$

Now we note that ${}^{\mathcal{N}}\boldsymbol{\omega}^{\mathcal{R}}$ is a property of the rigid body and, hence, is independent of the integral. Therefore, Eq. (5–34) can be written as

$${}^{\mathcal{N}}\mathbf{H}_Q = \left\{\int_{\mathcal{R}} \left[\left(\boldsymbol{\rho}_Q \cdot \boldsymbol{\rho}_Q\right)\mathbf{U} - \left(\boldsymbol{\rho}_Q \otimes \boldsymbol{\rho}_Q\right)\right] dm\right\} \cdot {}^{\mathcal{N}}\boldsymbol{\omega}^{\mathcal{R}} \tag{5–35}$$

The quantity

$$\mathbf{I}_Q^{\mathcal{R}} = \int_{\mathcal{R}} \left[\left(\boldsymbol{\rho}_Q \cdot \boldsymbol{\rho}_Q\right)\mathbf{U} - \left(\boldsymbol{\rho}_Q \otimes \boldsymbol{\rho}_Q\right)\right] dm \tag{5–36}$$

is called the *moment of inertia tensor of the rigid body $\mathcal{R}$ relative to the body-fixed point Q*. Now, we know that the identity tensor is symmetric, i.e., $\mathbf{U} = \mathbf{U}^T$. Furthermore, we have

$$\left(\boldsymbol{\rho}_Q \otimes \boldsymbol{\rho}_Q\right)^T = \boldsymbol{\rho}_Q \otimes \boldsymbol{\rho}_Q \tag{5–37}$$

Consequently, the moment of inertia tensor, $\mathbf{I}_Q^{\mathcal{R}}$, is itself symmetric, i.e.,

$$\left[\mathbf{I}_Q^{\mathcal{R}}\right]^T = \mathbf{I}_Q^{\mathcal{R}} \tag{5–38}$$

The angular momentum of the rigid body $\mathcal{R}$ relative to a point Q (where we remind the reader that Q is fixed in $\mathcal{R}$) in the inertial reference frame $\mathcal{N}$ can then be written compactly as

$$\boxed{{}^{\mathcal{N}}\mathbf{H}_Q = \mathbf{I}_Q^{\mathcal{R}} \cdot {}^{\mathcal{N}}\boldsymbol{\omega}^{\mathcal{R}}} \tag{5–39}$$

5.3.2 Moment of Inertia Tensor Relative to Center of Mass

Consider again the form of the angular momentum of Eq. (5–30). However, now let the reference point be the center of mass of the rigid body. We then have

$$\mathbf{r}_Q = \bar{\mathbf{r}} \Longrightarrow \boldsymbol{\rho}_Q = \mathbf{r} - \bar{\mathbf{r}} \equiv \boldsymbol{\rho} \tag{5–40}$$

Now, because the center of mass is fixed in the rigid body, the angular momentum of the rigid body relative to the center of mass can be written as

$${}^{\mathcal{N}}\bar{\mathbf{H}} = \int_{\mathcal{R}} \left[(\boldsymbol{\rho} \cdot \boldsymbol{\rho})\,{}^{\mathcal{N}}\boldsymbol{\omega}^{\mathcal{R}} - \left(\boldsymbol{\rho} \cdot {}^{\mathcal{N}}\boldsymbol{\omega}^{\mathcal{R}}\right)\boldsymbol{\rho}\right] dm \tag{5–41}$$

It can be seen that Eq. (5–41) has the exact same mathematical form as Eq. (5–30). Consequently, Eq. (5–41) can be rewritten in a form similar to Eq. (5–35) as

$$^{\mathcal{N}}\bar{\mathbf{H}} = \left\{ \int_{\mathcal{R}} [(\boldsymbol{\rho} \cdot \boldsymbol{\rho})\mathbf{U} - (\boldsymbol{\rho} \otimes \boldsymbol{\rho})]\, dm \right\} \cdot {}^{\mathcal{N}}\boldsymbol{\omega}^{\mathcal{R}} \tag{5–42}$$

The quantity

$$\bar{\mathbf{I}}^{\mathcal{R}} = \int_{\mathcal{R}} [(\boldsymbol{\rho} \cdot \boldsymbol{\rho})\mathbf{U} - (\boldsymbol{\rho} \otimes \boldsymbol{\rho})]\, dm \tag{5–43}$$

is called the *moment of inertia tensor of the rigid body $\mathcal{R}$ relative to the center of mass of the rigid body*. Using $\bar{\mathbf{I}}^{\mathcal{R}}$, the quantity $^{\mathcal{N}}\bar{\mathbf{H}}$ can be written as

$$\boxed{^{\mathcal{N}}\bar{\mathbf{H}} = \bar{\mathbf{I}}^{\mathcal{R}} \cdot {}^{\mathcal{N}}\boldsymbol{\omega}^{\mathcal{R}}} \tag{5–44}$$

We note that both $^{\mathcal{N}}\mathbf{H}_Q$ in Eq. (5–39) and $^{\mathcal{N}}\bar{\mathbf{H}}$ in Eq. (5–44) are independent of the choice of coordinate system.

5.3.3 Moments and Products of Inertia

It is seen that, consistent with the definition of a general tensor (see Chapter 1), the moment of inertia tensor is a coordinate-free quantity. Consequently, using the moment of inertia tensor, it is possible to arrive at a coordinate-free expression for the angular momentum of a rigid body. However, in practice it is necessary to choose a coordinate system in which to express the moment of inertia tensor. The choice of a coordinate system in which to express the moment of inertia tensor leads to a set of scalar quantities called the *moments of inertia* and *products of inertia*. In this subsection expressions for the moments and products of inertia of a rigid body are derived in terms of a body-fixed coordinate system.

Consider a coordinate system E^{Q} with origin Q and basis $\mathbf{E} = \{\mathbf{e}_1, \mathbf{e}_2, \mathbf{e}_3\}$ fixed in a rigid body $\mathcal{R}$. Then the moment of inertia tensor relative to the body-fixed point Q, $\mathbf{I}_Q^{\mathcal{R}}$, can be represented in the basis $\mathbf{E}$ as

$$\mathbf{I}_Q^{\mathcal{R}} = \sum_{i=1}^{3} \sum_{j=1}^{3} I_{ij}^{Q} \mathbf{e}_i \otimes \mathbf{e}_j \tag{5–45}$$

Now, because we know from Eq. (5–38) that $\mathbf{I}_Q^{\mathcal{R}}$ is symmetric, we have

$$I_{ij}^{Q} = I_{ji}^{Q} \qquad (i, j = 1, 2, 3) \tag{5–46}$$

The quantities I_{11}^{Q}, I_{22}^{Q}, and I_{33}^{Q} are called the *moments of inertia of $\mathcal{R}$* relative to the body-fixed point Q in the basis $\mathbf{E}$ while the quantities I_{12}^{Q}, I_{13}^{Q}, and I_{23}^{Q} are called the *products of inertia of $\mathcal{R}$* relative to the body-fixed point Q in the basis $\mathbf{E}$.

Suppose now that the quantity $\boldsymbol{\rho}_Q$ in Eq. (5–27) is expressed in terms of the basis $\mathbf{E}$ as

$$\boldsymbol{\rho}_Q = \rho_1 \mathbf{e}_1 + \rho_2 \mathbf{e}_2 + \rho_3 \mathbf{e}_3 \tag{5–47}$$

Then expressions for the moments and products of inertia in the basis $\{\mathbf{e}_1, \mathbf{e}_2, \mathbf{e}_3\}$ can be determined as follows. First, using the expression for $\boldsymbol{\rho}_Q$ in Eq. (5–47), we have

$$\boldsymbol{\rho}_Q \cdot \boldsymbol{\rho}_Q = \rho_1^2 + \rho_2^2 + \rho_3^2 \tag{5–48}$$

$$\boldsymbol{\rho}_Q \otimes \boldsymbol{\rho}_Q = \sum_{i=1}^{3}\sum_{j=1}^{3} \rho_i\rho_j \mathbf{e}_i \otimes \mathbf{e}_j \tag{5–49}$$

Next, recalling the identity tensor from Eq. (1–51) on page 12, we have

$$\mathbf{U} = \sum_{i=1}^{3} \mathbf{e}_i \otimes \mathbf{e}_i \tag{5–50}$$

Using Eqs. (5–48) and (5–50), the quantity $\left(\boldsymbol{\rho}_Q \cdot \boldsymbol{\rho}_Q\right)\mathbf{U}$ can be written as

$$\left(\boldsymbol{\rho}_Q \cdot \boldsymbol{\rho}_Q\right)\mathbf{U} = (\rho_1^2 + \rho_2^2 + \rho_3^2)\sum_{i=1}^{3} \mathbf{e}_i \otimes \mathbf{e}_i \tag{5–51}$$

We can then write the quantity $\left(\boldsymbol{\rho}_Q \cdot \boldsymbol{\rho}_Q\right)\mathbf{U} - \left(\boldsymbol{\rho}_Q \otimes \boldsymbol{\rho}_Q\right)$ in Eq. (5–36) as

$$\left(\boldsymbol{\rho}_Q \cdot \boldsymbol{\rho}_Q\right)\mathbf{U} - \left(\boldsymbol{\rho}_Q \otimes \boldsymbol{\rho}_Q\right) = \left[(\rho_1^2 + \rho_2^2 + \rho_3^2)\sum_{i=1}^{3} \mathbf{e}_i \otimes \mathbf{e}_i\right] - \sum_{i=1}^{3}\sum_{j=1}^{3} \rho_i\rho_j \mathbf{e}_i \otimes \mathbf{e}_j \tag{5–52}$$

Expanding Eq. (5–52), we obtain

$$\begin{aligned}\left(\boldsymbol{\rho}_Q \cdot \boldsymbol{\rho}_Q\right)\mathbf{U} - \left(\boldsymbol{\rho}_Q \otimes \boldsymbol{\rho}_Q\right) = {} & (\rho_2^2 + \rho_3^2)\mathbf{e}_1 \otimes \mathbf{e}_1 - \rho_1\rho_2\mathbf{e}_1 \otimes \mathbf{e}_2 - \rho_1\rho_3\mathbf{e}_1 \otimes \mathbf{e}_3 \\ & - \rho_2\rho_1\mathbf{e}_2 \otimes \mathbf{e}_1 + (\rho_1^2 + \rho_3^2)\mathbf{e}_2 \otimes \mathbf{e}_2 - \rho_2\rho_3\mathbf{e}_2 \otimes \mathbf{e}_3 \\ & - \rho_3\rho_1\mathbf{e}_3 \otimes \mathbf{e}_1 - \rho_3\rho_2\mathbf{e}_3 \otimes \mathbf{e}_2 + (\rho_1^2 + \rho_2^2)\mathbf{e}_3 \otimes \mathbf{e}_3\end{aligned} \tag{5–53}$$

Now, since the tensor products $\mathbf{e}_i \otimes \mathbf{e}_j$ $(i, j = 1, 2, 3)$ are mutually orthogonal, we know from Eq. (5–36) that the integral of every term in Eq. (5–53) must be equal to the corresponding term in Eq. (5–45). Consequently, the moments of inertia I_{11}^Q, I_{22}^Q, and I_{33}^Q in the basis $\{\mathbf{e}_1, \mathbf{e}_2, \mathbf{e}_3\}$ are given as

$$\begin{aligned} I_{11}^Q &= \int_{\mathcal{R}} (\rho_2^2 + \rho_3^2)dm \\ I_{22}^Q &= \int_{\mathcal{R}} (\rho_1^2 + \rho_3^2)dm \\ I_{33}^Q &= \int_{\mathcal{R}} (\rho_1^2 + \rho_2^2)dm \end{aligned} \tag{5–54}$$

Similarly, the products of inertia I_{12}^Q, I_{13}^Q, and I_{23}^Q in the basis $\{\mathbf{e}_1, \mathbf{e}_2, \mathbf{e}_3\}$ are given as

$$\begin{aligned} I_{12}^Q = I_{21}^Q &= -\int_{\mathcal{R}} \rho_1\rho_2 dm \\ I_{13}^Q = I_{31}^Q &= -\int_{\mathcal{R}} \rho_1\rho_3 dm \\ I_{23}^Q = I_{32}^Q &= -\int_{\mathcal{R}} \rho_2\rho_3 dm \end{aligned} \tag{5–55}$$

We note that, when the reference point is the center of mass, the moments of inertia in the basis $\{\mathbf{e}_1, \mathbf{e}_2, \mathbf{e}_3\}$ are given as

$$
\begin{aligned}
\bar{I}_{11} &= \int_{\mathcal{R}} (\rho_2^2 + \rho_3^2) dm \\
\bar{I}_{22} &= \int_{\mathcal{R}} (\rho_1^2 + \rho_3^2) dm \\
\bar{I}_{33} &= \int_{\mathcal{R}} (\rho_1^2 + \rho_2^2) dm
\end{aligned}
\tag{5-56}
$$

Similarly, the products of inertia relative to the center of mass in the basis $\{\mathbf{e}_1, \mathbf{e}_2, \mathbf{e}_3\}$ are given as

$$
\begin{aligned}
\bar{I}_{12} = \bar{I}_{21} &= -\int_{\mathcal{R}} \rho_1 \rho_2 dm \\
\bar{I}_{13} = \bar{I}_{31} &= -\int_{\mathcal{R}} \rho_1 \rho_3 dm \\
\bar{I}_{23} = \bar{I}_{32} &= -\int_{\mathcal{R}} \rho_2 \rho_3 dm
\end{aligned}
\tag{5-57}
$$

It is noted in Eqs. (5-56) and (5-57) that the quantities ρ_1, ρ_2, and ρ_3 are the components of the position $\boldsymbol{\rho} = \mathbf{r} - \bar{\mathbf{r}}$ in the body-fixed basis $\{\mathbf{e}_1, \mathbf{e}_2, \mathbf{e}_3\}$.

It is important to re-emphasize that the preceding derivation assumes that the moment of inertia tensor is represented in a body-fixed coordinate system. Using this assumption, it is known that the moments of and products of inertia relative to a point fixed in a rigid body are *constant*, i.e., the quantities I_{ij}^Q $(i, j = 1, 2, 3)$ as given in Eqs. (5-54) and (5-55) and the quantities $\bar{I}_{ij}$ $(i, j = 1, 2, 3)$ as given in Eqs. (5-56) and (5-57) are all constants. However, in the general case where the moment of inertia tensor is represented in terms of a *non-body-fixed* coordinate system, the moments and products of inertia will, in general, *not* be constant.[3] Finally, in terms of the body-fixed moments and products of inertia, the angular momentum in an inertial reference frame $\mathcal{N}$ of a rigid body relative to a point Q fixed in the body is given as

$$
\begin{aligned}
{}^{\mathcal{N}}\mathbf{H}_Q &= \mathbf{I}^{\mathcal{R}} \cdot {}^{\mathcal{N}}\boldsymbol{\omega}^{\mathcal{R}} \\
&= (I_{11}\omega_1 + I_{12}\omega_2 + I_{13}\omega_3)\mathbf{e}_1 \\
&+ (I_{12}\omega_1 + I_{22}\omega_2 + I_{23}\omega_3)\mathbf{e}_2 \\
&+ (I_{13}\omega_1 + I_{23}\omega_2 + I_{33}\omega_3)\mathbf{e}_3
\end{aligned}
\tag{5-58}
$$

where we recall that ${}^{\mathcal{N}}\boldsymbol{\omega}^{\mathcal{R}} = \omega_1\mathbf{e}_1 + \omega_2\mathbf{e}_2 + \omega_3\mathbf{e}_3$ is the angular velocity of the rigid body in the inertial reference frame $\mathcal{N}$. Equation (5-58) can be written in matrix form as

$$
\left\{ {}^{\mathcal{N}}\mathbf{H}_Q \right\}_{\mathbf{E}} = \left\{ \begin{array}{ccc} I_{11} & I_{12} & I_{13} \\ I_{12} & I_{22} & I_{23} \\ I_{13} & I_{23} & I_{33} \end{array} \right\} \left\{ \begin{array}{c} \omega_1 \\ \omega_2 \\ \omega_3 \end{array} \right\}
\tag{5-59}
$$

where $\mathbf{E} = \{\mathbf{e}_1, \mathbf{e}_2, \mathbf{e}_3\}$. It is noted once again that the quantity $\left\{ {}^{\mathcal{N}}\mathbf{H}_Q \right\}_{\mathbf{E}}$ is not the angular momentum relative to the body-fixed point Q, but is the *representation* of the angular momentum relative to the body-fixed point Q in the basis $\mathbf{e}$.

[3]Practically speaking, the moments and products of inertia are generally not expressed in terms of a non-body-fixed coordinate system.

5.3.4 Parallel-Axis Theorem (Huygens-Steiner Theorem)

In general, the most convenient reference point about which to compute angular momentum of a rigid body is problem-dependent. For example, for some problems it may be most convenient to choose the center of mass as the reference point while for other problems it may be most convenient to choose a reference point that is *not* the center of mass. More specifically, for some problems it is often convenient to compute the angular momentum relative to a body-fixed point Q that is not the center of mass. Consequently, it is useful to develop a way to relate the angular momentum of a rigid body relative to the center of mass to the angular momentum of a rigid body relative to another body-fixed reference point.

Now, it is seen that the angular momentum of a rigid body relative to a body-fixed point Q is a function of the moment of inertia tensor relative to point Q. Consequently, a relationship between the angular momentum of a rigid body relative to the center of mass and the angular momentum relative of a rigid body relative to a body-fixed point Q can be obtained by relating the moment of inertia tensors between these two points. The relationship between the moments and products of inertia relative to the center of mass and the moments and products of inertia relative to a body-fixed point Q is called the *parallel-axis theorem* or the *Huygens-Steiner theorem* (Papastavridis, 2002) and is now derived.

Suppose that it is desired to determine the relationship between the moment of inertia tensor of a rigid body $\mathcal{R}$ relative to the center of mass of $\mathcal{R}$ and the moment of inertia tensor relative to an alternate body-fixed point Q. Recall from Eq. (5–24) the relationship between the angular momentum of a rigid body $\mathcal{R}$ relative to the center of mass of $\mathcal{R}$ and the angular momentum of $\mathcal{R}$ relative to an arbitrary reference point Q as

$$ {}^{\mathcal{N}}\mathbf{H}_Q = {}^{\mathcal{N}}\bar{\mathbf{H}} + (\mathbf{r}_Q - \bar{\mathbf{r}}) \times m({}^{\mathcal{N}}\mathbf{v}_Q - {}^{\mathcal{N}}\bar{\mathbf{v}}) \tag{5–60} $$

Now let

$$ \boldsymbol{\pi} = \mathbf{r}_Q - \bar{\mathbf{r}} \tag{5–61} $$

where $\boldsymbol{\pi}$ is the position of the point Q relative to the center of mass of $\mathcal{R}$. Now, because Q is a body-fixed point, we have from Eq. (2–517) on page 106 that

$$ {}^{\mathcal{N}}\mathbf{v}_Q - {}^{\mathcal{N}}\bar{\mathbf{v}} = {}^{\mathcal{N}}\boldsymbol{\omega}^{\mathcal{R}} \times \boldsymbol{\pi} \tag{5–62} $$

Equation (5–60) can then be written as

$$ {}^{\mathcal{N}}\mathbf{H}_Q = {}^{\mathcal{N}}\bar{\mathbf{H}} + \boldsymbol{\pi} \times m({}^{\mathcal{N}}\boldsymbol{\omega}^{\mathcal{R}} \times \boldsymbol{\pi}) \tag{5–63} $$

Next, recall from Eqs. (5–39) and (5–44) that ${}^{\mathcal{N}}\mathbf{H}_Q$ and ${}^{\mathcal{N}}\bar{\mathbf{H}}$ can be written in terms of their respective moment of inertia tensors as

$$ \begin{aligned} {}^{\mathcal{N}}\mathbf{H}_Q &= \mathbf{I}_Q^{\mathcal{R}} \cdot {}^{\mathcal{N}}\boldsymbol{\omega}^{\mathcal{R}} \\ {}^{\mathcal{N}}\bar{\mathbf{H}} &= \bar{\mathbf{I}}^{\mathcal{R}} \cdot {}^{\mathcal{N}}\boldsymbol{\omega}^{\mathcal{R}} \end{aligned} \tag{5–64} $$

Substituting the results of Eq. (5–64) into (5–63), we obtain

$$ \mathbf{I}_Q^{\mathcal{R}} \cdot {}^{\mathcal{N}}\boldsymbol{\omega}^{\mathcal{R}} = \bar{\mathbf{I}}^{\mathcal{R}} \cdot {}^{\mathcal{N}}\boldsymbol{\omega}^{\mathcal{R}} + \boldsymbol{\pi} \times m({}^{\mathcal{N}}\boldsymbol{\omega}^{\mathcal{R}} \times \boldsymbol{\pi}) \tag{5–65} $$

Now we have from the scalar triple product that

$$ \boldsymbol{\pi} \times ({}^{\mathcal{N}}\boldsymbol{\omega}^{\mathcal{R}} \times \boldsymbol{\pi}) = (\boldsymbol{\pi} \cdot \boldsymbol{\pi}){}^{\mathcal{N}}\boldsymbol{\omega}^{\mathcal{R}} - (\boldsymbol{\pi} \cdot {}^{\mathcal{N}}\boldsymbol{\omega}^{\mathcal{R}})\boldsymbol{\pi} \tag{5–66} $$

Also, using the property of a tensor from Eq. (1-41), we see that

$$(\boldsymbol{\pi} \cdot {}^{\mathcal{N}}\boldsymbol{\omega}^{\mathcal{R}})\boldsymbol{\pi} = (\boldsymbol{\pi} \otimes \boldsymbol{\pi})^{\mathcal{N}}\boldsymbol{\omega}^{\mathcal{R}} \tag{5-67}$$

Substituting the result of Eq. (5-67) into (5-66), we obtain

$$\boldsymbol{\pi} \times ({}^{\mathcal{N}}\boldsymbol{\omega}^{\mathcal{R}} \times \boldsymbol{\pi}) = (\boldsymbol{\pi} \cdot \boldsymbol{\pi})^{\mathcal{N}}\boldsymbol{\omega}^{\mathcal{R}} - (\boldsymbol{\pi} \otimes \boldsymbol{\pi}) \cdot {}^{\mathcal{N}}\boldsymbol{\omega}^{\mathcal{R}} \tag{5-68}$$

Finally, recalling the identity tensor U, we can write Eq. (5-68) as

$$\boldsymbol{\pi} \times ({}^{\mathcal{N}}\boldsymbol{\omega}^{\mathcal{R}} \times \boldsymbol{\pi}) = (\boldsymbol{\pi} \cdot \boldsymbol{\pi})\mathrm{U} \cdot {}^{\mathcal{N}}\boldsymbol{\omega}^{\mathcal{R}} - (\boldsymbol{\pi} \otimes \boldsymbol{\pi}) \cdot {}^{\mathcal{N}}\boldsymbol{\omega}^{\mathcal{R}} \tag{5-69}$$

Factoring out ${}^{\mathcal{N}}\boldsymbol{\omega}^{\mathcal{R}}$ from Eq. (5-69) gives

$$\boldsymbol{\pi} \times ({}^{\mathcal{N}}\boldsymbol{\omega}^{\mathcal{R}} \times \boldsymbol{\pi}) = [(\boldsymbol{\pi} \cdot \boldsymbol{\pi})\mathrm{U} - (\boldsymbol{\pi} \otimes \boldsymbol{\pi})] \cdot {}^{\mathcal{N}}\boldsymbol{\omega}^{\mathcal{R}} \tag{5-70}$$

Then, substituting the result of Eq. (5-70) into (5-65), we obtain

$$\mathbf{I}_Q^{\mathcal{R}} \cdot {}^{\mathcal{N}}\boldsymbol{\omega}^{\mathcal{R}} = \bar{\mathbf{I}}^{\mathcal{R}} \cdot {}^{\mathcal{N}}\boldsymbol{\omega}^{\mathcal{R}} + m[(\boldsymbol{\pi} \cdot \boldsymbol{\pi})\mathrm{U} - (\boldsymbol{\pi} \otimes \boldsymbol{\pi})] \cdot {}^{\mathcal{N}}\boldsymbol{\omega}^{\mathcal{R}} \tag{5-71}$$

Because Eq. (5-71) must hold for *all* values of ${}^{\mathcal{N}}\boldsymbol{\omega}^{\mathcal{R}}$, we can drop the dependence on ${}^{\mathcal{N}}\boldsymbol{\omega}^{\mathcal{R}}$ to obtain

$$\boxed{\mathbf{I}_Q^{\mathcal{R}} = \bar{\mathbf{I}}^{\mathcal{R}} + m[(\boldsymbol{\pi} \cdot \boldsymbol{\pi})\mathrm{U} - (\boldsymbol{\pi} \otimes \boldsymbol{\pi})]} \tag{5-72}$$

Equation (5-72) is a relationship between the moment of inertia tensor of a rigid body $\mathcal{R}$ relative to a body-fixed point Q and the moment of inertia tensor of $\mathcal{R}$ relative to the center of mass.

Now consider two coordinate systems, denoted $\bar{E}$ and E^Q, each fixed in a rigid body $\mathcal{R}$. Furthermore, suppose that $\bar{E}$ and E^Q share the same basis $\mathbf{E} = \{\mathbf{e}_1, \mathbf{e}_2, \mathbf{e}_3\}$ but have different origins. In particular, suppose that the origin of $\bar{E}$ is the center of mass of $\mathcal{R}$ while the origin of E^Q is an arbitrary body-fixed point Q. It can be seen that $\bar{E}$ and E^Q are *parallel* coordinate systems because they differ only in their origins. A schematic of the parallel coordinate systems $\bar{E}$ and E^Q is shown in Fig. 5-1.

In terms of the body-fixed basis $\mathbf{E}$, the moment of inertia tensors of $\mathbf{I}_Q^{\mathcal{R}}$ and $\bar{\mathbf{I}}^{\mathcal{R}}$ can be written, respectively, as

$$\begin{aligned} \mathbf{I}_Q^{\mathcal{R}} &= \textstyle\sum_{i=1}^3 \sum_{j=1}^3 I_{ij}^Q \mathbf{e}_i \otimes \mathbf{e}_j \\ \bar{\mathbf{I}}^{\mathcal{R}} &= \textstyle\sum_{i=1}^3 \sum_{j=1}^3 \bar{I}_{ij} \mathbf{e}_i \otimes \mathbf{e}_j \end{aligned} \tag{5-73}$$

Then, substituting the expressions from Eq. (5-73) into (5-72), we have

$$\sum_{i=1}^3 \sum_{j=1}^3 I_{ij}^Q \mathbf{e}_i \otimes \mathbf{e}_j = \sum_{i=1}^3 \sum_{j=1}^3 \bar{I}_{ij} \mathbf{e}_i \otimes \mathbf{e}_j + m[(\boldsymbol{\pi} \cdot \boldsymbol{\pi})\mathrm{U} - (\boldsymbol{\pi} \otimes \boldsymbol{\pi})] \tag{5-74}$$

Now it is seen that the quantity $(\boldsymbol{\pi} \cdot \boldsymbol{\pi})\mathrm{U} - (\boldsymbol{\pi} \otimes \boldsymbol{\pi})$ is identical in form to the quantity $\left(\boldsymbol{\rho}_Q \cdot \boldsymbol{\rho}_Q\right)\mathrm{U} - \left(\boldsymbol{\rho}_Q \otimes \boldsymbol{\rho}_Q\right)$ from Eq. (5-53). Consequently, we have

$$\begin{aligned} (\boldsymbol{\pi} \cdot \boldsymbol{\pi})\mathrm{U} - (\boldsymbol{\pi} \otimes \boldsymbol{\pi}) = {} & (\pi_2^2 + \pi_3^2)\mathbf{e}_1 \otimes \mathbf{e}_1 - \pi_1\pi_2\mathbf{e}_1 \otimes \mathbf{e}_2 - \pi_1\pi_3\mathbf{e}_1 \otimes \mathbf{e}_3 \\ & - \pi_2\pi_1\mathbf{e}_2 \otimes \mathbf{e}_1 + (\pi_1^2 + \pi_3^2)\mathbf{e}_2 \otimes \mathbf{e}_2 - \pi_2\pi_3\mathbf{e}_2 \otimes \mathbf{e}_3 \\ & - \pi_3\pi_1\mathbf{e}_3 \otimes \mathbf{e}_1 - \pi_3\pi_2\mathbf{e}_3 \otimes \mathbf{e}_2 + (\pi_1^2 + \pi_2^2)\mathbf{e}_3 \otimes \mathbf{e}_3 \end{aligned} \tag{5-75}$$

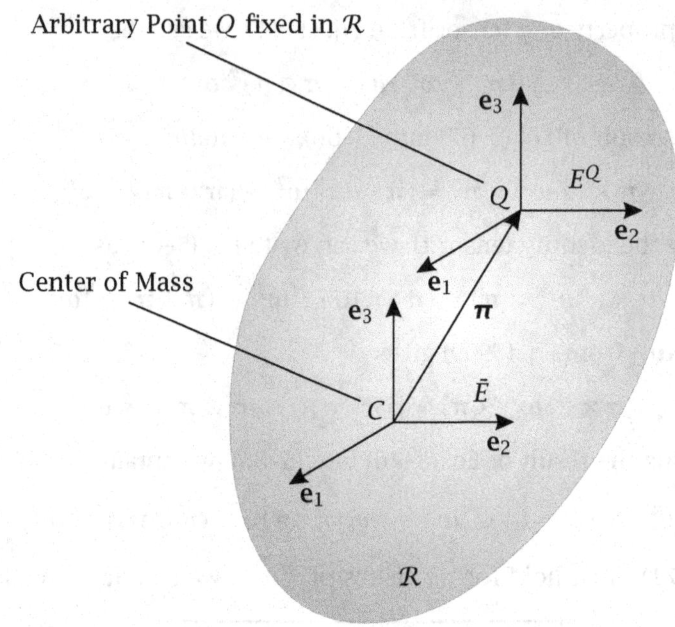

Figure 5–1 Coordinate system $\bar{E}$ consisting of an origin located at the center of mass of the rigid body $\mathcal{R}$ with body-fixed basis $\{\mathbf{e}_1, \mathbf{e}_2, \mathbf{e}_3\}$ alongside a parallel coordinate system E^Q consisting of an origin located at a point Q Fixed in $\mathcal{R}$ and body-fixed basis $\{\mathbf{e}_1, \mathbf{e}_2, \mathbf{e}_3\}$.

Substituting the result of Eq. (5–75) into Eq. (5–74) and combining terms, we obtain

$$\begin{aligned}
&\textstyle\sum_{i=1}^{3}\sum_{j=1}^{3} I_{ij}^{Q}\mathbf{e}_i \otimes \mathbf{e}_j \\
&= \left(\bar{I}_{11} + m(\pi_2^2 + \pi_3^2)\right)\mathbf{e}_1 \otimes \mathbf{e}_1 + (\bar{I}_{12} - m\pi_1\pi_2)\,\mathbf{e}_1 \otimes \mathbf{e}_2 + (\bar{I}_{13} - m\pi_1\pi_3)\,\mathbf{e}_1 \otimes \mathbf{e}_3 \\
&+ (\bar{I}_{21} - m\pi_2\pi_1)\,\mathbf{e}_2 \otimes \mathbf{e}_1 + \left(\bar{I}_{22} + m(\pi_1^2 + \pi_3^2)\right)\mathbf{e}_2 \otimes \mathbf{e}_2 + (\bar{I}_{23} - m\pi_2\pi_3)\,\mathbf{e}_2 \otimes \mathbf{e}_3 \\
&+ (\bar{I}_{31} - m\pi_3\pi_1)\,\mathbf{e}_3 \otimes \mathbf{e}_1 + (\bar{I}_{32} - m\pi_3\pi_2)\,\mathbf{e}_3 \otimes \mathbf{e}_2 + \left(\bar{I}_{33} + m(\pi_1^2 + \pi_2^2)\right)\mathbf{e}_3 \otimes \mathbf{e}_3
\end{aligned} \tag{5–76}$$

In matrix form, Eq. (5–76) is given as

$$\left\{\begin{array}{ccc} I_{11}^{Q} & I_{12}^{Q} & I_{13}^{Q} \\ I_{21}^{Q} & I_{22}^{Q} & I_{23}^{Q} \\ I_{31}^{Q} & I_{32}^{Q} & I_{33}^{Q} \end{array}\right\} = \left\{\begin{array}{ccc} \bar{I}_{11} + m(\pi_2^2 + \pi_3^2) & \bar{I}_{12} - m\pi_1\pi_2 & \bar{I}_{13} - m\pi_1\pi_3 \\ \bar{I}_{21} - m\pi_2\pi_1 & \bar{I}_{22} + m(\pi_1^2 + \pi_3^2) & \bar{I}_{23} - m\pi_2\pi_3 \\ \bar{I}_{31} - m\pi_3\pi_1 & \bar{I}_{32} - m\pi_3\pi_2 & \bar{I}_{33} + m(\pi_1^2 + \pi_2^2) \end{array}\right\} \tag{5–77}$$

Equating matrix elements in Eq. (5–77) and using the fact that the moment of inertia tensor is symmetric, we obtain

$$
\boxed{
\begin{array}{lcl}
I_{11}^Q & = & \bar{I}_{11} + m(\pi_2^2 + \pi_3^2) \\
I_{22}^Q & = & \bar{I}_{22} + m(\pi_1^2 + \pi_3^2) \\
I_{33}^Q & = & \bar{I}_{33} + m(\pi_1^2 + \pi_2^2) \\
I_{12}^Q = I_{21}^Q & = & \bar{I}_{12} - m\pi_1\pi_2 \\
I_{13}^Q = I_{31}^Q & = & \bar{I}_{13} - m\pi_1\pi_3 \\
I_{23}^Q = I_{32}^Q & = & \bar{I}_{23} - m\pi_2\pi_3
\end{array}
}
\tag{5-78}
$$

It can be seen from Eq. (5–78) that the moments and products of inertia in coordinate system E^Q are different from the moments and products of inertia in coordinate system $\bar{E}$. Equation (5–78) is called the *parallel-axis theorem* or the *Huygens-Steiner Theorem* (Papastavridis, 2002; Huygens, 1673) and is stated in words as follows:

The moment of inertia I_{ii}^Q $(i = 1,2,3)$ of a rigid body $\mathcal{R}$ in a body-fixed coordinate system E^Q about the ith axis $\mathbf{e}_i^Q$ $(i = 1,2,3)$ that passes through a body-fixed point Q is equal to the sum of the moment of inertia of the rigid body in a body-fixed coordinate system $\bar{E}$ about the ith axis $\bar{\mathbf{e}}_i$ $(i = 1,2,3)$ that is parallel to $\mathbf{e}_i^Q$ $(i = 1,2,3)$ and passes through the center of mass of the body and the product of the mass of the rigid body with the square of the distance between the two axes $\mathbf{e}^Q$ and $\bar{\mathbf{e}}$. The product of inertia I_{ij}^Q $(i \neq j,\ i,j = 1,2,3)$ of a rigid body $\mathcal{R}$ in a body-fixed coordinate system E^Q between the ith and j^{th} axes $\mathbf{e}_i^Q$ and $\mathbf{e}_j^Q$ $(i \neq j,\ i,j = 1,2,3)$ that passes through a body-fixed point Q is equal to the difference between the product of inertia of the rigid body in a body-fixed coordinate system $\bar{E}$ between the ith and j^{th} axes $\bar{\mathbf{e}}_i$ and $\bar{\mathbf{e}}_j$ $(i \neq j,\ i,j = 1,2,3)$ that passes through the center of mass of the body and the product of the mass of the rigid body with the product of the components of the position of Q relative to the center of mass in the directions of $\bar{\mathbf{e}}_i$ and $\bar{\mathbf{e}}_j$.

It is emphasized that, in order to apply the result of Eq. (5–78), it is only necessary to have knowledge of the moments and products of inertia relative to the center of mass in the body-fixed coordinate system and the position of the point Q relative to the center of mass.

5.3.5 Limitation of the Parallel-Axis Theorem

It is seen from Eq. (5–78) that the parallel-axis theorem can be used to compute products of inertia relative to a body-fixed point Q given the moments and products of inertia relative to the center of mass. Conversely, the parallel-axis theorem can be used to compute the moments and products of inertia relative to the center of mass given the moments and products of inertia relative to the body-fixed point Q. This second result arises from a "backward" application of Eq. (5–78) as

$$\boxed{\begin{array}{lcc} \bar{I}_{11} & = & I_{11}^Q - m(\pi_2^2 + \pi_3^2) \\ \bar{I}_{22} & = & I_{22}^Q - m(\pi_1^2 + \pi_3^2) \\ \bar{I}_{33} & = & I_{33}^Q - m(\pi_1^2 + \pi_2^2) \\ \bar{I}_{12} = \bar{I}_{21} & = & I_{12}^Q + m\pi_1\pi_2 \\ \bar{I}_{13} = \bar{I}_{31} & = & I_{13}^Q + m\pi_1\pi_3 \\ \bar{I}_{23} = \bar{I}_{32} & = & I_{23}^Q + m\pi_2\pi_3 \end{array}} \tag{5–79}$$

Now, while the parallel-axis theorem provides a way to compute moments and products of inertia between the center of mass and an arbitrary body-fixed point, it *cannot* be used to compute the moments and products of inertia between two arbitrary body-fixed points. In the case where it is desired to translate the moments and products of inertia between two arbitrary body-fixed points P and Q, it is necessary to apply the parallel-axis theorem twice. In the first application, the moments and products of inertia relative to the center of mass are obtained via translation from point P to the center of mass using Eq. (5–79). In the second application, the moments and products of inertia relative to point Q are obtained via translation from the center of mass using Eq. (5–78).

5.3.6 Limitation of Moment of Inertia Form of Angular Momentum

It is important to understand that the angular momentum of a rigid body can only be computed in terms of the moment of inertia tensor when the reference point is fixed in the rigid body. This limitation arises from the fact that Eq. (2–516) was used to derive the moment of inertia tensor, i.e., in deriving the moment of inertia tensor we used the fact that the velocity between two points on a rigid body $\mathcal{R}$ are related as

$${}^{\mathcal{N}}\mathbf{v}_P - {}^{\mathcal{N}}\mathbf{v}_Q = {}^{\mathcal{N}}\boldsymbol{\omega}^{\mathcal{R}} \times (\mathbf{r}_P - \mathbf{r}_Q) \tag{5–80}$$

Because Eq. (2–516) is valid only for two points fixed in a rigid body, for the case where it is desired to compute the angular momentum relative to a point that is *not* fixed in the body, Eq. (2–516) no longer holds. Consequently, for an arbitrary point it is necessary to compute the angular momentum using either the integral of Eq. (5–13) or, if given the angular momentum relative to either an inertially fixed point or the center of mass of the rigid body, to apply either Eq. (5–21) or (5–24).

5.4 Principal-Axis Coordinate Systems

5.4.1 Rotation About a Principal Axis

Suppose now that we consider the case of a rigid body $\mathcal{R}$ rotating about a direction such that the angular momentum of $\mathcal{R}$ relative to a body-fixed point Q in an inertial reference frame $\mathcal{N}$, ${}^{\mathcal{N}}\mathbf{H}_Q$, is *parallel* to the angular velocity of $\mathcal{R}$ in $\mathcal{N}$, ${}^{\mathcal{N}}\boldsymbol{\omega}^{\mathcal{R}}$. Then we have

$${}^{\mathcal{N}}\mathbf{H}_Q = I\,{}^{\mathcal{N}}\boldsymbol{\omega}^{\mathcal{R}} \tag{5–81}$$

where $I \in \mathbb{R}$. Then, because Q is a body-fixed point, ${}^{\mathcal{N}}\mathbf{H}_Q$ can be written in terms of the moment of inertia tensor relative to point Q as

$$ {}^{\mathcal{N}}\mathbf{H}_Q = \mathbf{I}_Q^{\mathcal{R}} \cdot {}^{\mathcal{N}}\boldsymbol{\omega}^{\mathcal{R}} \tag{5-82} $$

Setting Eqs. (5-81) and (5-82) equal, we obtain

$$ \mathbf{I}_Q^{\mathcal{R}} \cdot {}^{\mathcal{N}}\boldsymbol{\omega}^{\mathcal{R}} = I\,{}^{\mathcal{N}}\boldsymbol{\omega}^{\mathcal{R}} \tag{5-83} $$

It can be seen that Eq. (5-83) is identical in form to Eq. (1-131). Consequently, the condition that ${}^{\mathcal{N}}\mathbf{H}_Q$ is parallel to ${}^{\mathcal{N}}\boldsymbol{\omega}^{\mathcal{R}}$ results in an *eigenvalue problem*. Moreover, any vector that satisfies Eq. (5-83) is an *eigenvector* of the moment of inertia tensor $\mathbf{I}_Q^{\mathcal{R}}$. In general, the eigenvectors of the moment of inertia tensor $\mathbf{I}_Q^{\mathcal{R}}$ are said to be *principal-axis* directions. In other words, the principal axes of the inertia tensor $\mathbf{I}_Q^{\mathcal{R}}$ are the eigenvectors of the tensor $\mathbf{I}_Q^{\mathcal{R}}$. Now, because the moment of inertia tensor $\mathbf{I}_Q^{\mathcal{R}}$ is real and symmetric (i.e., $\mathbf{I}_Q^{\mathcal{R}} = \left[\mathbf{I}_Q^{\mathcal{R}}\right]^T$), we know that the eigenvalues of $\mathbf{I}_Q^{\mathcal{R}}$ are real. It is also known that a moment of inertia tensor is positive semi-definite. Consequently, the eigenvalues of $\mathbf{I}_Q^{\mathcal{R}}$ must be nonnegative. Finally, it is known that the eigenvectors of $\mathbf{I}_Q^{\mathcal{R}}$ are mutually orthogonal.

5.4.2 Determination of a Principal-Axis Basis

Suppose that we choose to express the moment of inertia tensor, $\mathbf{I}_Q^{\mathcal{R}}$, and the angular velocity of $\mathcal{R}$, ${}^{\mathcal{N}}\boldsymbol{\omega}^{\mathcal{R}}$, in terms of a body-fixed basis $\mathbf{E} = \{\mathbf{e}_1, \mathbf{e}_2, \mathbf{e}_3\}$. We then can write

$$ \mathbf{I}_Q^{\mathcal{R}} = \sum_{i=1}^{3}\sum_{j=1}^{3} I_{ij}^Q \mathbf{e}_i \otimes \mathbf{e}_j \tag{5-84} $$

and

$$ {}^{\mathcal{N}}\boldsymbol{\omega}^{\mathcal{R}} = \sum_{k=1}^{3} \omega_k \mathbf{e}_k \tag{5-85} $$

Substituting the results of Eqs. (5-84) and (5-85) into Eq. (5-83) gives

$$ \sum_{i=1}^{3}\sum_{j=1}^{3} I_{ij}^Q \mathbf{e}_i \otimes \mathbf{e}_j \cdot \sum_{k=1}^{3} \omega_k \mathbf{e}_k = I \sum_{k=1}^{3} \omega_k \mathbf{e}_k \tag{5-86} $$

Then, using the results of Section 1.4.3 on page 15, Eq. (5-86) can be written in matrix form as

$$ \left\{ \begin{array}{ccc} I_{11}^Q & I_{12}^Q & I_{13}^Q \\ I_{12}^Q & I_{22}^Q & I_{23}^Q \\ I_{13}^Q & I_{23}^Q & I_{33}^Q \end{array} \right\} \left\{ \begin{array}{c} \omega_1 \\ \omega_2 \\ \omega_3 \end{array} \right\} = I \left\{ \begin{array}{c} \omega_1 \\ \omega_2 \\ \omega_3 \end{array} \right\} \tag{5-87} $$

Rearranging Eq. (5-87), we obtain

$$ \left\{ \begin{array}{ccc} I_{11}^Q - I & I_{12}^Q & I_{13}^Q \\ I_{12}^Q & I_{22}^Q - I & I_{23}^Q \\ I_{13}^Q & I_{23}^Q & I_{33}^Q - I \end{array} \right\} \left\{ \begin{array}{c} \omega_1 \\ \omega_2 \\ \omega_3 \end{array} \right\} = \left\{ \begin{array}{c} 0 \\ 0 \\ 0 \end{array} \right\} \tag{5-88} $$

Now we know that any value of I that satisfies Eq. (5–88) is such that

$$\det\left\{\begin{array}{ccc} I_{11}^{Q}-I & I_{12}^{Q} & I_{13}^{Q} \\ I_{12}^{Q} & I_{22}^{Q}-I & I_{23}^{Q} \\ I_{13}^{Q} & I_{23}^{Q} & I_{33}^{Q}-I \end{array}\right\}=0 \tag{5-89}$$

Equation (5–89) results in a cubic polynomial in the variable I and is called the *characteristic equation.* Suppose further that we let

$$\begin{aligned} \left\{\mathbf{I}_{Q}^{\mathcal{R}}\right\}_{\mathrm{E}} &= \left\{\begin{array}{ccc} I_{11}^{Q} & I_{12}^{Q} & I_{13}^{Q} \\ I_{12}^{Q} & I_{22}^{Q} & I_{23}^{Q} \\ I_{13}^{Q} & I_{23}^{Q} & I_{33}^{Q} \end{array}\right\} & \quad (5\text{-}90) \\ \left\{{}^{\mathcal{N}}\boldsymbol{\omega}^{\mathcal{R}}\right\}_{\mathrm{E}} &= \left\{\begin{array}{c} \omega_1 \\ \omega_2 \\ \omega_3 \end{array}\right\} & \quad (5\text{-}91) \end{aligned}$$

Equation (5–87) can then be written compactly as

$$\left\{\mathbf{I}_{Q}^{\mathcal{R}}\right\}_{\mathrm{E}}\left\{{}^{\mathcal{N}}\boldsymbol{\omega}^{\mathcal{R}}\right\}_{\mathrm{E}}=I\left\{{}^{\mathcal{N}}\boldsymbol{\omega}^{\mathcal{R}}\right\}_{\mathrm{E}} \tag{5-92}$$

Equation (5–92) is a matrix representation of the eigenvalue problem of Eq. (5–83) in the body-fixed basis $\mathbf{E}=\{\mathbf{e}_1,\mathbf{e}_2,\mathbf{e}_3\}$. Alternatively, we can write

$$\left[\left\{\mathbf{I}_{Q}^{\mathcal{R}}\right\}_{\mathrm{E}}-I\{\mathrm{U}\}\right]\left\{{}^{\mathcal{N}}\boldsymbol{\omega}^{\mathcal{R}}\right\}_{\mathrm{E}}=\mathbf{0} \tag{5-93}$$

where $\{\mathrm{U}\}$ is the identity matrix.

Now, suppose we let $\mathbf{e}_1^p$, $\mathbf{e}_2^p$, and $\mathbf{e}_3^p$ be the eigenvectors of the tensor $\mathbf{I}_{Q}^{\mathcal{R}}$. Then, using the result of Eq. (1–151) on page 24, the *eigenvector matrix* of the moment of inertia matrix $\left\{\mathbf{I}_{Q}^{\mathcal{R}}\right\}_{\mathrm{E}}$ can be composed column-wise as

$$\mathbf{P}=\left\{\ \left\{\mathbf{e}_1^p\right\}_{\mathrm{E}}\ \ \left\{\mathbf{e}_2^p\right\}_{\mathrm{E}}\ \ \left\{\mathbf{e}_3^p\right\}_{\mathrm{E}}\ \right\} \tag{5-94}$$

where $\left\{\mathbf{e}_1^p\right\}_{\mathrm{E}}$, $\left\{\mathbf{e}_2^p\right\}_{\mathrm{E}}$, and $\left\{\mathbf{e}_3^p\right\}_{\mathrm{E}}$ are the column-vector *representations* of the eigenvectors in the basis $\mathbf{E}$. Now by definition we know that the eigenvectors satisfy the conditions

$$\begin{aligned} \mathbf{I}_{Q}^{\mathcal{R}}\cdot\mathbf{e}_1^p &= I_1^Q\mathbf{e}_1^p \\ \mathbf{I}_{Q}^{\mathcal{R}}\cdot\mathbf{e}_2^p &= I_2^Q\mathbf{e}_2^p \\ \mathbf{I}_{Q}^{\mathcal{R}}\cdot\mathbf{e}_3^p &= I_3^Q\mathbf{e}_3^p \end{aligned} \tag{5-95}$$

where I_1^Q, I_2^Q, and I_3^Q are the eigenvalues of the tensor $\mathbf{I}_{Q}^{\mathcal{R}}$. In terms of the eigenvector matrix $\mathbf{P}$ and the matrix representation of $\mathbf{I}_{Q}^{\mathcal{R}}$ in the basis $\mathbf{E}$, the results of Eq. (5–95) can be combined into the single matrix equation

$$\left\{\mathbf{I}_{Q}^{\mathcal{R}}\right\}_{\mathrm{E}}\mathbf{P}=\mathbf{P}\left\{\mathbf{I}_{Q}^{\mathcal{R}}\right\}_{\mathrm{E}^p} \tag{5-96}$$

where $\left\{\mathbf{I}_Q^{\mathcal{R}}\right\}_{\mathbf{E}^p}$ is the matrix representation of the tensor $\mathbf{I}_Q^{\mathcal{R}}$ in the basis $\mathbf{E}^p$. Now, because $\mathbf{P}$ diagonalizes $\left\{\mathbf{I}_Q^{\mathcal{R}}\right\}_{\mathbf{E}}$, we have

$$\left\{\mathbf{I}_Q^{\mathcal{R}}\right\}_{\mathbf{E}^p} = \left\{\begin{array}{ccc} I_1^Q & 0 & 0 \\ 0 & I_2^Q & 0 \\ 0 & 0 & I_3^Q \end{array}\right\} \tag{5-97}$$

and the quantities I_1^Q, I_2^Q, and I_3^Q are the eigenvalues of the matrix $\left\{\mathbf{I}_Q^{\mathcal{R}}\right\}_{\mathbf{E}}$. Furthermore, since the eigenvectors of a real symmetric matrix are mutually orthogonal, the eigenvector matrix $\mathbf{P}$ is an *orthogonal* matrix, i.e., $\mathbf{P}^{-1} = \mathbf{P}^T$. Therefore, from Eq. (5-96) we have

$$\mathbf{P}^T \left\{\mathbf{I}_Q^{\mathcal{R}}\right\}_{\mathbf{E}} \mathbf{P} = \left\{\mathbf{I}_Q^{\mathcal{R}}\right\}_{\mathbf{E}^p} \tag{5-98}$$

Then, comparing Eq. (5-98) and (1-130) on page 22, we see that the matrix $\mathbf{P}$ is the transformation from the basis $\mathbf{E}^p$ to the basis $\mathbf{E}$, i.e.,

$$\mathbf{P} \equiv \mathbf{P}_{\mathbf{E}^p}^{\mathbf{E}} \tag{5-99}$$

Using the result of Eq. (5-99), Eq. (5-98) can be written as

$$\mathbf{P}_{\mathbf{E}}^{\mathbf{E}^p} \left\{\mathbf{I}_Q^{\mathcal{R}}\right\}_{\mathbf{E}} \mathbf{P}_{\mathbf{E}^p}^{\mathbf{E}} = \left\{\mathbf{I}_Q^{\mathcal{R}}\right\}_{\mathbf{E}^p} = \left\{\begin{array}{ccc} I_1^Q & 0 & 0 \\ 0 & I_2^Q & 0 \\ 0 & 0 & I_3^Q \end{array}\right\} \tag{5-100}$$

A basis $\mathbf{E}^p$ that results in the moment of inertia matrix having the form of Eq. (5-97) is called a *principal-axis basis* and the corresponding coordinate system is called a *principal-axis coordinate system.* Furthermore, the eigenvalues of the matrix $\left\{\mathbf{I}_Q^{\mathcal{R}}\right\}_{\mathbf{E}}$ are called the *principal-axis moments of inertia* of the rigid body $\mathcal{R}$. It can be seen that, when the moment of inertia matrix is expressed in a principal-axis basis, the products of inertia are *zero.* Consequently, in terms of a principal-axis (i.e., eigenvector) basis, the moment of inertia tensor is given as

$$\mathbf{I}_Q^{\mathcal{R}} = \sum_{i=1}^{3} I_i^Q \mathbf{e}_i^p \otimes \mathbf{e}_i^p \tag{5-101}$$

Example 5-1

Consider a uniform circular cylinder of mass m, radius R, and height h as shown in Fig. 5-2. Determine the principal-axis moments of inertia for the following cases: (a) relative to the center of mass of the cylinder, (b) the limiting cases from part (a) where $R = 0$ and $h = 0$, (c) relative to the point P located a distance $h/2$ in the positive z-direction from the center of mass of the cylinder, and (d) relative to the point Q located a distance R along an axis that lies in the plane of the center of mass of the cylinder at an angle of $\pi/4$ from the x-direction.

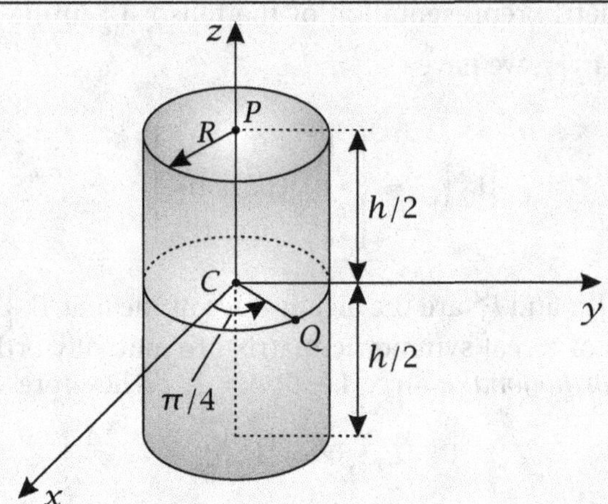

Figure 5–2 Uniform circular cylinder of mass m, radius R, and height h.

Solution to Example 5–1

(a) Principal-Axis Moments of Inertia Relative to Center of Mass of Cylinder

It is seen from the geometry of the cylinder that the body-fixed principal-axis coordinate system that passes through the center of mass is aligned such that the x-axis and y-axis are orthogonal to the direction along the height of the cylinder while the z-axis is in the direction along the height of the cylinder. Furthermore, by symmetry it is seen that the principal moments of inertia in the x-direction and y-direction are equal, i.e., $\bar{I}_{xx} = \bar{I}_{yy}$. First, using Eq. (5–56), the moment of inertia along the x-direction is given as

$$\bar{I}_{xx} = \int_{\mathcal{R}} (\rho_y^2 + \rho_z^2) dm \tag{5-102}$$

Now we know that an elemental unit of mass, dm, can be written as

$$dm = \rho dV \tag{5-103}$$

where ρ is the mass density per unit volume and dV is an elemental unit of volume. Furthermore, for a uniform circular cylinder we have

$$V = \pi R^2 h \tag{5-104}$$

Consequently, the density of cylinder is constant and is given as

$$\rho = \frac{m}{V} = \frac{m}{\pi R^2 h} \tag{5-105}$$

Next, we can write the volume element in terms of cylindrical coordinates as

$$dV = r dr d\theta dz \tag{5-106}$$

Therefore,

$$dm = \rho dV = \frac{m}{\pi R^2 h} r dr d\theta dz \tag{5-107}$$

where

$$\begin{array}{rcccl} 0 & \le & r & \le & R \\ 0 & \le & \theta & \le & 2\pi \\ -h/2 & \le & z & \le & h/2 \end{array} \tag{5-108}$$

Therefore, we have $\bar{I}_{xx}$ as

$$\bar{I}_{xx} = \int_{-h/2}^{h/2} \int_0^{2\pi} \int_0^R (\rho_y^2 + \rho_z^2) \frac{m}{\pi R^2 h} r dr d\theta dz \tag{5-109}$$

Now we know that

$$\begin{array}{rcl} \rho_y & = & y \\ \rho_z & = & z \end{array} \tag{5-110}$$

Furthermore, using the transformation from cylindrical to Cartesian coordinates, we have

$$\begin{array}{rcl} x & = & r\cos\theta \\ y & = & r\sin\theta \end{array} \tag{5-111}$$

Consequently, Eq. (5-109) can be written as

$$\bar{I}_{xx} = \int_{-h/2}^{h/2} \int_0^{2\pi} \int_0^R (r^2 \sin^2\theta + z^2) \frac{m}{\pi R^2 h} r dr d\theta dz \tag{5-112}$$

Noting that the quantity $m/(\pi R^2 h)$ is constant, Eq. (5-112) can be written as

$$\bar{I}_{xx} = \frac{m}{\pi R^2 h} \left[\int_{-h/2}^{h/2} \int_0^{2\pi} \int_0^R r^3 \sin^2\theta dr d\theta dz + \int_{-h/2}^{h/2} \int_0^{2\pi} \int_0^R z^2 r dr d\theta dz \right] \tag{5-113}$$

Equation (5-113) can then be rewritten as

$$\bar{I}_{xx} = \frac{m}{\pi R^2 h} \left[\int_{-h/2}^{h/2} dz \int_0^{2\pi} \sin^2\theta d\theta \int_0^R r^3 dr + \int_{-h/2}^{h/2} z^2 dz \int_0^{2\pi} d\theta \int_0^R r dr \right] \tag{5-114}$$

Now we have

$$\begin{array}{rclcrcl} \int_{-h/2}^{h/2} dz & = & h & , & \int_{-h/2}^{h/2} z^2 dz & = & \frac{h^3}{12} \\ \int_0^{2\pi} d\theta & = & 2\pi & , & \int_0^{2\pi} \sin^2\theta d\theta & = & \pi \\ \int_0^R r dr & = & \frac{R^2}{2} & , & \int_0^R r^3 dr & = & \frac{R^4}{4} \end{array} \tag{5-115}$$

Using Eq. (5-115), Eq. (5-114) simplifies to

$$\bar{I}_{xx} = \frac{m}{\pi R^2 h} \left[\frac{\pi h R^4}{4} + \frac{\pi h^3 R^2}{12} \right] \tag{5-116}$$

Equation (5-116) simplifies further to

$$\bar{I}_{xx} = \frac{mR^2}{4} + \frac{mh^2}{12} \tag{5-117}$$

Rewriting Eq. (5-117), we obtain $\bar{I}_{xx}$ as

$$\bar{I}_{xx} = \frac{m(3R^2 + h^2)}{12} \tag{5-118}$$

Using the fact that $\bar{I}_{yy} = \bar{I}_{xx}$, we have

$$\bar{I}_{yy} = \frac{m(3R^2 + h^2)}{12} \tag{5-119}$$

Next, the moment of inertia relative to the center of mass of the cylinder about the z-direction is given as

$$\bar{I}_{zz} = \int_{\mathcal{R}} (\rho_x^2 + \rho_y^2) dm \tag{5-120}$$

Now we have

$$\begin{aligned} \rho_x &= x \\ \rho_y &= y \end{aligned} \tag{5-121}$$

Furthermore, using Eq. (5-110), we have that

$$\rho_x^2 + \rho_y^2 = x^2 + y^2 = (r\cos\theta)^2 + (r\sin\theta)^2 = r^2 \tag{5-122}$$

Equation (5-120) can then be written as

$$\bar{I}_{zz} = \int_{\mathcal{R}} r^2 dm \tag{5-123}$$

Using dm from Eq. (5-107) and the limits of integration from Eq. (5-108), we have

$$\bar{I}_{zz} = \int_{-h/2}^{h/2} \int_0^{2\pi} \int_0^R \frac{m}{\pi R^2 h} r^3 dr d\theta dz \tag{5-124}$$

Rearranging Eq. (5-124), we obtain

$$\bar{I}_{zz} = \frac{m}{\pi R^2 h} \int_{-h/2}^{h/2} dz \int_0^{2\pi} d\theta \int_0^R r^3 dr \tag{5-125}$$

Then, using the results of Eq. (5-115), we have that

$$\bar{I}_{zz} = \frac{m}{\pi R^2 h} h(2\pi) \frac{R^4}{4} \tag{5-126}$$

Equation (5-126) simplifies to

$$\bar{I}_{zz} = \frac{1}{2} m R^2 \tag{5-127}$$

The moments of inertia $\bar{I}_{xx}$, $\bar{I}_{yy}$ and $\bar{I}_{zz}$ are then given as

$$\bar{I}_{xx} = \frac{m(3R^2 + h^2)}{12} \tag{5-128}$$

$$\bar{I}_{yy} = \frac{m(3R^2 + h^2)}{12} \tag{5-129}$$

$$\bar{I}_{zz} = \frac{mR^2}{2} \tag{5-130}$$

(b) Principal-Axis Moments of Inertia for $R = 0$ and $h = 0$

Using the results of Eqs. (5–118), (5–119, and (5–127), for the case where $R = 0$ we have

$$\bar{I}_{xx} = \frac{mh^2}{12} \tag{5–131}$$

$$\bar{I}_{yy} = \frac{mh^2}{12} \tag{5–132}$$

$$\bar{I}_{zz} = 0 \tag{5–133}$$

It is noted that the moments of inertia of the cylinder for case where $R = 0$ correspond to those for a *uniform slender rod of length $l = h$*. Next, using the results of Eqs. (5–118), (5–119, and (5–127), for the case where $h = 0$ we have

$$\bar{I}_{xx} = \frac{mR^2}{4} \tag{5–134}$$

$$\bar{I}_{yy} = \frac{mR^2}{4} \tag{5–135}$$

$$\bar{I}_{zz} = \frac{mR^2}{2} \tag{5–136}$$

It is noted that the moments of inertia of the cylinder for the case where $h = 0$ correspond to those for a *thin circular disk of radius R*.

(c) Principal-Axis Moments of Inertia Relative to Point P

The moments of inertia relative to the point P that lies a distance $h/2$ from the center of mass of the cylinder in the z-direction are obtained using the parallel-axis theorem. In particular, we have

$$I_{xx}^P = \bar{I}_{xx} + m(\pi_y^2 + \pi_z^2) \tag{5–137}$$

$$I_{yy}^P = \bar{I}_{yy} + m(\pi_x^2 + \pi_z^2) \tag{5–138}$$

$$I_{zz}^P = \bar{I}_{zz} + m(\pi_x^2 + \pi_y^2) \tag{5–139}$$

where I_{xx}^P, I_{yy}^P, and I_{zz}^P are the moments of inertia in the coordinate system parallel to (x, y, z) and passing through point P. Because for this case the center of mass is translated in only the z-direction, we have

$$\pi_x = 0 \quad ; \quad \pi_y = 0 \quad ; \quad \pi_z = \frac{h}{2} \tag{5–140}$$

Substituting the results of Eq. (5–140) into Eqs. (5–137)-(5–139), we obtain I_{xx}, I_{yy}, and I_{zz} as

$$I_{xx}^P = \bar{I}_{xx} + m\left(\frac{h}{2}\right)^2 \tag{5–141}$$

$$I_{yy}^P = \bar{I}_{yy} + m\left(\frac{h}{2}\right)^2 \tag{5–142}$$

$$I_{zz}^P = \bar{I}_{zz} \tag{5–143}$$

Using the expressions for $\bar{I}_{xx}$, $\bar{I}_{yy}$, and $\bar{I}_{zz}$ from Eqs. (5-118), (5-119), and (5-127), respectively, Eqs. (5-141)-(5-143) become

$$\begin{aligned} I_{xx}^P &= \frac{m(3R^2+h^2)}{12} + m\left(\frac{h}{2}\right)^2 && (5\text{-}144) \\ I_{yy}^P &= \frac{m(3R^2+h^2)}{12} + m\left(\frac{h}{2}\right)^2 && (5\text{-}145) \\ I_{zz}^P &= \frac{mR^2}{2} && (5\text{-}146) \end{aligned}$$

Simplifying the expressions in Eqs. (5-144)-(5-146), we obtain the moments of inertia translated a distance $h/2$ in the positive z-direction as

$$\begin{aligned} I_{xx}^P &= \frac{mR^2}{4} + \frac{mh^2}{3} && (5\text{-}147) \\ I_{yy}^P &= \frac{mR^2}{4} + \frac{mh^2}{3} && (5\text{-}148) \\ I_{zz}^P &= \frac{mR^2}{2} && (5\text{-}149) \end{aligned}$$

(d) Principal-Axis Moments of Inertia Relative to Point Q

The moments of inertia relative to the point Q that lies a distance R from the center of mass along an axis that lies along an angle $\pi/4$ from the x-direction are obtained using the parallel-axis theorem. In particular, we have

$$\begin{aligned} I_{xx}^Q &= \bar{I}_{xx} + m(\pi_y^2 + \pi_z^2) \\ I_{yy}^Q &= \bar{I}_{yy} + m(\pi_x^2 + \pi_z^2) \\ I_{zz}^Q &= \bar{I}_{zz} + m(\pi_x^2 + \pi_y^2) \end{aligned} \qquad (5\text{-}150)$$

where I_{xx}^Q, I_{yy}^Q, and I_{zz}^Q are the moments of inertia in the coordinate system parallel to (x, y, z) and passing through point Q. Now it is seen from Fig. 5-2 that the point Q is obtained by translating the center of mass in *both* the x-direction *and* the y-direction. Furthermore, given that point Q lies on the circle of radius R in the plane of the center of mass of the cylinder at an angle of $\pi/4$ from the x-direction, we have

$$\pi_x = R\cos\frac{\pi}{4} = R\frac{\sqrt{2}}{2} \quad ; \quad \pi_y = R\sin\frac{\pi}{4} = R\frac{\sqrt{2}}{2} \quad ; \quad \pi_z = 0 \qquad (5\text{-}151)$$

Substituting the results of Eq. (5-151) into (5-150), we obtain I_{xx}^Q, I_{yy}^Q, and I_{zz}^Q as

$$\begin{aligned} I_{xx}^Q &= \bar{I}_{xx} + m\left(R\frac{\sqrt{2}}{2}\right)^2 = \bar{I}_{xx} + \frac{mR^2}{2} \\ I_{yy}^Q &= \bar{I}_{yy} + m\left(R\frac{\sqrt{2}}{2}\right)^2 = \bar{I}_{yy} + \frac{mR^2}{2} \\ I_{zz}^Q &= \bar{I}_{zz} + m\left[\left(R\frac{\sqrt{2}}{2}\right)^2 + \left(R\frac{\sqrt{2}}{2}\right)^2\right] = \bar{I}_{zz} + mR^2 \end{aligned} \qquad (5\text{-}152)$$

Using the expressions for $\bar{I}_{xx}$, $\bar{I}_{yy}$, and $\bar{I}_{zz}$ from Eqs. (5-118), (5-119), and (5-127), respectively, Eq. (5-152) becomes

$$\begin{aligned} I_{xx}^{Q} &= \frac{m(3R^2+h^2)}{12} + \frac{mR^2}{2} \\ I_{yy}^{Q} &= \frac{m(3R^2+h^2)}{12} + \frac{mR^2}{2} \\ I_{zz}^{Q} &= \frac{mR^2}{2} + mR^2 \end{aligned} \tag{5-153}$$

Simplifying the expressions in Eq. (5-153), we obtain the moments of inertia translated a distance $h/2$ in the positive z-direction as

$$\begin{aligned} I_{xx}^{Q} &= \frac{3mR^2}{4} + \frac{mh^2}{12} \\ I_{yy}^{Q} &= \frac{3mR^2}{4} + \frac{mh^2}{12} \\ I_{zz}^{Q} &= \frac{3mR^2}{2} \end{aligned} \tag{5-154}$$

■

Example 5-2

Given a uniform sphere of mass m and radius R as shown in Fig. 5-3, determine the principal-axis moments of inertia for the following two cases: (a) relative to the center of mass of the sphere and (b) relative to the point Q, where Q is located a distance R in the positive z-direction from the center of mass.

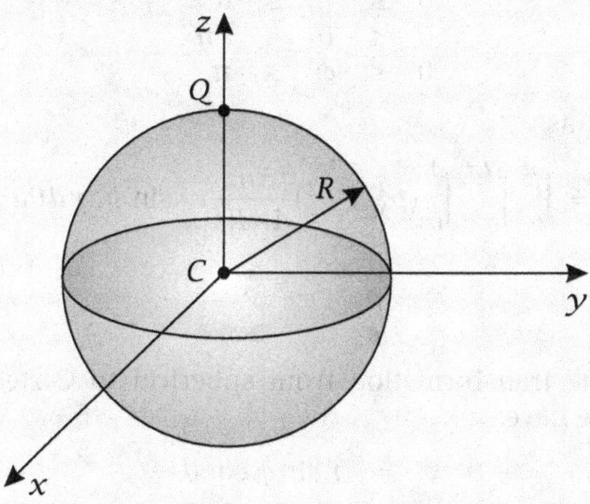

Figure 5-3 Uniform sphere of mass m and radius R.

Solution to Example 5–2

(a) Principal-Axis Moments of Inertia Relative to Center of Mass of Sphere

It is seen from the geometry of the sphere that the body-fixed principal-axis coordinate system that passes through the center of mass is aligned such that all three axes lie along a diameter of the sphere. Furthermore, due to the symmetry of the sphere, it is seen that $\bar{I}_{xx} = \bar{I}_{yy} = \bar{I}_{zz}$. Consequently, it is sufficient to compute one of the moments of inertia. Using Eq. (5–56), the moment of inertia along the x-direction is given as

$$\bar{I}_{xx} = \int_{\mathcal{R}} (\rho_y^2 + \rho_z^2) dm \tag{5–155}$$

We know that an elemental unit of mass, dm, can be written as

$$dm = \rho dV \tag{5–156}$$

where ρ is the mass density per unit volume and dV is an elemental unit of volume. Furthermore, for a uniform sphere we have

$$V = \frac{4}{3}\pi R^3 \tag{5–157}$$

Consequently, the density of cylinder is constant and is given as

$$\rho = \frac{m}{V} = \frac{m}{\frac{4}{3}\pi R^3} = \frac{3m}{4\pi R^3} \tag{5–158}$$

Next, we can write the volume element in terms of spherical coordinates (Thomas and Finney, 2002) as

$$dV = r^2 \sin\phi dr d\theta d\phi \tag{5–159}$$

Therefore,

$$dm = \rho dV = \frac{3m}{4\pi R^3} r^2 \sin\phi dr d\theta d\phi \tag{5–160}$$

where

$$\begin{array}{ccccc} 0 & \leq & r & \leq & R \\ 0 & \leq & \theta & \leq & 2\pi \\ 0 & \leq & \phi & \leq & \pi \end{array} \tag{5–161}$$

Therefore, we have $\bar{I}_{xx}$ as

$$\bar{I}_{xx} = \int_0^{\pi}\int_0^{2\pi}\int_0^{R} (\rho_y^2 + \rho_z^2) \frac{3m}{4\pi R^3} r^2 \sin\phi dr d\theta d\phi \tag{5–162}$$

Now we know that

$$\begin{array}{rcl} \rho_y & = & y \\ \rho_z & = & z \end{array} \tag{5–163}$$

Furthermore, using the transformation from spherical to Cartesian coordinates as given in Eq. (2–241), we have

$$\begin{array}{rcl} x & = & r \sin\phi \cos\theta \\ y & = & r \sin\phi \sin\theta \\ z & = & r \cos\phi \end{array} \tag{5–164}$$

We have from Eq. (5-164)

$$\rho_y^2 + \rho_z^2 = y^2 + z^2 = r^2 \sin^2\phi \sin^2\theta + r^2\cos^2\phi = r^2(\sin^2\phi\sin^2\theta + \cos^2\phi) \quad (5\text{-}165)$$

Then, using the fact that $\sin^2\phi + \cos^2\phi = 1$, Eq. (5-165) can be rewritten as

$$\rho_y^2 + \rho_z^2 = r^2(\sin^2\theta + \cos^2\phi\cos^2\theta) \quad (5\text{-}166)$$

Equation (5-162) then becomes

$$\bar{I}_{xx} = \int_0^\pi \int_0^{2\pi} \int_0^R r^2(\sin^2\theta + \cos^2\phi\cos^2\theta)\frac{3m}{4\pi R^3} r^2 \sin\phi dr d\theta d\phi \quad (5\text{-}167)$$

Noting that the quantity $3m/(4\pi R^3)$ is a constant, Eq. (5-167) can be rearranged to give

$$\bar{I}_{xx} = \frac{3m}{4\pi R^3}\int_0^\pi \int_0^{2\pi} \int_0^R r^2(\sin^2\theta + \cos^2\phi\cos^2\theta) r^2 \sin\phi dr d\theta d\phi \quad (5\text{-}168)$$

Decomposing Eq. (5-168), we obtain

$$\begin{aligned}\bar{I}_{xx} = {} & \frac{3m}{4\pi R^3}\int_0^\pi \int_0^{2\pi} \int_0^R r^4 \sin^2\theta \sin\phi dr d\theta d\phi \\ & + \frac{3m}{4\pi R^3}\int_0^\pi \int_0^{2\pi} \int_0^R r^4 \cos^2\phi \sin\phi \cos^2\theta dr d\theta d\phi\end{aligned} \quad (5\text{-}169)$$

Decoupling the integrals in Eq. (5-169) gives

$$\begin{aligned}\bar{I}_{xx} = {} & \frac{3m}{4\pi R^3}\int_0^\pi \sin\phi d\phi \int_0^{2\pi} \sin^2\theta d\theta \int_0^R r^4 dr \\ & + \frac{3m}{4\pi R^3}\int_0^\pi \cos^2\phi \sin\phi d\phi \int_0^{2\pi} \cos^2\theta d\theta \int_0^R r^4 dr\end{aligned} \quad (5\text{-}170)$$

Now we have the following:

$$\begin{aligned}\int_0^\pi \sin\phi d\phi &= 2 \\ \int_0^\pi \cos^2\phi\sin\phi d\phi &= \frac{2}{3} \\ \int_0^{2\pi} \sin^2\theta d\theta &= \pi \\ \int_0^{2\pi} \cos^2\theta d\theta &= \pi \\ \int_0^R r^4 dr &= \frac{R^5}{5}\end{aligned} \quad (5\text{-}171)$$

Substituting the results of Eq. (5-171) into (5-170), we obtain

$$\bar{I}_{xx} = \frac{3m}{4\pi R^3}\left[2\pi\frac{R^5}{5} + \frac{2}{3}\pi\frac{R^5}{5}\right] \quad (5\text{-}172)$$

Simplifying Eq. (5–172) gives

$$\bar{I}_{xx} = \frac{3mR^2}{10} + \frac{mR^2}{10} = \frac{4mR^2}{10} = \frac{2mR^2}{5} \tag{5–173}$$

Finally, due to the symmetry of the sphere, we have

$$\bar{I}_{yy} = \bar{I}_{zz} = \bar{I}_{xx} = \frac{2mR^2}{5} \tag{5–174}$$

(b) Principal-Axis Moments of Inertia Relative to Point Q

The moments of inertia relative to the point Q that lies a distance R from the center of mass along the z-direction are obtained using the parallel-axis theorem. In particular, we have

$$\begin{aligned} I_{xx}^Q &= \bar{I}_{xx} + m(\pi_y^2 + \pi_z^2) \\ I_{yy}^Q &= \bar{I}_{yy} + m(\pi_x^2 + \pi_z^2) \\ I_{zz}^Q &= \bar{I}_{zz} + m(\pi_x^2 + \pi_y^2) \end{aligned} \tag{5–175}$$

where I_{xx}^Q, I_{yy}^Q, and I_{zz}^Q are the moments of inertia in the coordinate system parallel to (x, y, z) and passing through point Q. Now it is seen from Fig. 5-3 that the point Q is obtained by translating the center of mass in only the z-direction. Furthermore, given that point Q lies on the circle of radius R and in the plane of the center of mass of the sphere, we have

$$\begin{aligned} \pi_x &= 0 \\ \pi_y &= 0 \\ \pi_z &= R \end{aligned} \tag{5–176}$$

Substituting the results of Eq. (5–176) into (5–175), we obtain I_{xx}^Q, I_{yy}^Q, and I_{zz}^Q as

$$\begin{aligned} I_{xx}^Q &= \bar{I}_{xx} + mR^2 \\ I_{yy}^Q &= \bar{I}_{yy} + mR^2 \\ I_{zz}^Q &= \bar{I}_{zz} \end{aligned} \tag{5–177}$$

Using the expressions for $\bar{I}_{xx}$, $\bar{I}_{yy}$, and $\bar{I}_{zz}$ from Eqs. (5–173) and (5–174), respectively, Eq. (5–177) becomes

$$\begin{aligned} I_{xx}^Q &= \frac{2mR^2}{5} + mR^2 \\ I_{yy}^Q &= \frac{2mR^2}{5} + mR^2 \\ I_{zz}^Q &= \frac{2mR^2}{5} \end{aligned} \tag{5–178}$$

Simplifying the expressions in Eq. (5–178), we obtain the moments of inertia translated a distance R in the positive z-direction as

$$\begin{aligned} I_{xx}^{Q} &= \frac{7mR^2}{5} \\ I_{yy}^{Q} &= \frac{7mR^2}{5} \\ I_{zz}^{Q} &= \frac{2mR^2}{5} \end{aligned} \qquad (5\text{–}179)$$

■

Example 5–3

Consider a rigid body $\mathcal{R}$ whose moment of inertia tensor relative to the center of mass of $\mathcal{R}$ is given as

$$\begin{aligned} \bar{\mathbf{I}}^{\mathcal{R}} = {} & 14\mathbf{e}_1 \otimes \mathbf{e}_1 - 4\mathbf{e}_1 \otimes \mathbf{e}_2 - 2\mathbf{e}_1 \otimes \mathbf{e}_3 \\ & - 4\mathbf{e}_2 \otimes \mathbf{e}_1 + 14\mathbf{e}_2 \otimes \mathbf{e}_2 - 2\mathbf{e}_2 \otimes \mathbf{e}_3 \\ & - 2\mathbf{e}_3 \otimes \mathbf{e}_1 - 2\mathbf{e}_3 \otimes \mathbf{e}_2 + 12\mathbf{e}_3 \otimes \mathbf{e}_3 \end{aligned}$$

where $\{\mathbf{e}_1, \mathbf{e}_2, \mathbf{e}_3\}$ is a basis fixed in $\mathcal{R}$. Determine (a) the principal moments of inertia, (b) a principal-axis basis $\mathbf{E}^p = \{\mathbf{e}_1^p, \mathbf{e}_2^p, \mathbf{e}_3^p\}$ of $\mathcal{R}$ in terms of the body-fixed basis $\mathbf{E} = \{\mathbf{e}_1, \mathbf{e}_2, \mathbf{e}_3\}$, and (c) the moment of inertia tensor in terms of the principal-axis basis $\mathbf{E}^p = \{\mathbf{e}_1^p, \mathbf{e}_2^p, \mathbf{e}_3^p\}$.

Solution to Example 5–3

First, let $\mathbf{E} = \{\mathbf{e}_1, \mathbf{e}_2, \mathbf{e}_3\}$. Then, the matrix representation of $\bar{\mathbf{I}}^{\mathcal{R}}$ in the basis $\mathbf{E}$ is given as

$$\left\{\bar{\mathbf{I}}^{\mathcal{R}}\right\}_{\mathbf{E}} = \left\{\begin{array}{rrr} 14 & -4 & -2 \\ -4 & 14 & -2 \\ -2 & -2 & 12 \end{array}\right\} \qquad (5\text{–}180)$$

Furthermore, from Eq. (5–89) the eigenvalues of the matrix $\left\{\bar{\mathbf{I}}^{\mathcal{R}}\right\}_{\mathbf{E}}$ are obtained by solving for the roots of the equation

$$\det\left\{\begin{array}{ccc} 14-I & -4 & -2 \\ -4 & 14-I & -2 \\ -2 & -2 & 12-I \end{array}\right\} = 0 \qquad (5\text{–}181)$$

Expanding the determinant in Eq. (5–181) and setting the result equal to zero, we obtain the following cubic polynomial in I:

$$I^3 - 40I^2 + 508I - 2016 = 0 \qquad (5\text{–}182)$$

Computing the roots of the cubic polynomial in Eq. (5–182), we have

$$\begin{aligned}\bar{I}_1 &= 8\\ \bar{I}_2 &= 14\\ \bar{I}_3 &= 18\end{aligned} \tag{5-183}$$

The values $\bar{I}_1$, $\bar{I}_2$, and $\bar{I}_3$ are the *principal-axis moments of inertia* of the rigid body $\mathcal{R}$.

Suppose now that we let $\mathbf{e}_1^p$, $\mathbf{e}_2^p$, and $\mathbf{e}_3^p$ be the eigenvectors of the tensor $\bar{\mathbf{I}}^{\mathcal{R}}$. Then, after some algebra, the column-vector representations of the eigenvectors in the basis $\mathbf{E} = \{\mathbf{e}_1, \mathbf{e}_2, \mathbf{e}_3\}$ are given, respectively, as

$$\left\{\mathbf{e}_1^p\right\}_{\mathbf{E}} = \left\{\begin{array}{c}\frac{1}{\sqrt{3}}\\ \frac{1}{\sqrt{3}}\\ \frac{1}{\sqrt{3}}\end{array}\right\} \tag{5-184}$$

$$\left\{\mathbf{e}_2^p\right\}_{\mathbf{E}} = \left\{\begin{array}{c}-\frac{1}{\sqrt{6}}\\ -\frac{1}{\sqrt{6}}\\ \frac{2}{\sqrt{6}}\end{array}\right\} \tag{5-185}$$

$$\left\{\mathbf{e}_3^p\right\}_{\mathbf{E}} = \left\{\begin{array}{c}-\frac{1}{\sqrt{2}}\\ \frac{1}{\sqrt{2}}\\ 0\end{array}\right\} \tag{5-186}$$

The eigenvector matrix $\mathbf{P}_{\mathbf{E}^p}^{\mathbf{E}}$ is then given as

$$\mathbf{P}_{\mathbf{E}^p}^{\mathbf{E}} = \left\{\begin{array}{ccc}\frac{1}{\sqrt{3}} & -\frac{1}{\sqrt{6}} & -\frac{1}{\sqrt{2}}\\ \frac{1}{\sqrt{3}} & -\frac{1}{\sqrt{6}} & \frac{1}{\sqrt{2}}\\ \frac{1}{\sqrt{3}} & \frac{2}{\sqrt{6}} & 0\end{array}\right\} \tag{5-187}$$

Correspondingly, the matrix $\mathbf{P}_{\mathbf{E}}^{\mathbf{E}^p}$ (which, by definition, is the transpose of the matrix $\mathbf{P}_{\mathbf{E}^p}^{\mathbf{E}}$ because $\mathbf{P}_{\mathbf{E}^p}^{\mathbf{E}}$ is an orthogonal matrix) is given as

$$\mathbf{P}_{\mathbf{E}}^{\mathbf{E}^p} = \left\{\begin{array}{ccc}\frac{1}{\sqrt{3}} & \frac{1}{\sqrt{3}} & \frac{1}{\sqrt{3}}\\ -\frac{1}{\sqrt{6}} & -\frac{1}{\sqrt{6}} & \frac{2}{\sqrt{6}}\\ -\frac{1}{\sqrt{2}} & \frac{1}{\sqrt{2}} & 0\end{array}\right\} \tag{5-188}$$

Then, computing the quantity $\mathbf{P}_{\mathbf{E}}^{\mathbf{E}^p}\left\{\mathbf{I}_Q^{\mathcal{R}}\right\}_{\mathbf{E}}\mathbf{P}_{\mathbf{E}^p}^{\mathbf{E}}$, we obtain

$$\mathbf{P}_{\mathbf{E}}^{\mathbf{E}^p}\left\{\mathbf{I}_Q^{\mathcal{R}}\right\}_{\mathbf{E}}\mathbf{P}_{\mathbf{E}^p}^{\mathbf{E}} = \left\{\begin{array}{ccc}8 & 0 & 0\\ 0 & 14 & 0\\ 0 & 0 & 18\end{array}\right\} \tag{5-189}$$

Furthermore, the eigenvectors of the inertia tensor $\bar{\mathbf{I}}^{\mathcal{R}}$ are given as

$$\mathbf{e}_1^p = \frac{1}{\sqrt{3}}\mathbf{e}_1 + \frac{1}{\sqrt{3}}\mathbf{e}_2 + \frac{1}{\sqrt{3}}\mathbf{e}_3 \tag{5-190}$$

$$\mathbf{e}_2^p = -\frac{1}{\sqrt{6}}\mathbf{e}_1 - \frac{1}{\sqrt{6}}\mathbf{e}_2 + \frac{2}{\sqrt{6}}\mathbf{e}_3 \tag{5-191}$$

$$\mathbf{e}_3^p = -\frac{1}{\sqrt{2}}\mathbf{e}_1 + \frac{1}{\sqrt{2}}\mathbf{e}_2 \tag{5-192}$$

It is noted that the eigenvectors $\mathbf{e}_1^p$, $\mathbf{e}_2^p$, and $\mathbf{e}_3^p$ form a principal-axis basis. The moment of inertia tensor $\bar{\mathbf{I}}^{\mathcal{R}}$ can then be represented in the principal-axis basis $\mathbf{E}^p = \{\mathbf{e}_1^p, \mathbf{e}_2^p, \mathbf{e}_3^p\}$ as

$$\bar{\mathbf{I}}^{\mathcal{R}} = 8\mathbf{e}_1^p \otimes \mathbf{e}_1^p + 14\mathbf{e}_2^p \otimes \mathbf{e}_2^p + 18\mathbf{e}_3^p \otimes \mathbf{e}_3^p \tag{5-193}$$

■

5.4.3 Common Usage of Principal-Axis Coordinate Systems

While in principal it is possible to start with a moment of inertia tensor in a non-principal-axis coordinate system and transform to principal-axis coordinates, in this book we will restrict our attention to rigid bodies whose principal-axis moments of inertia (and corresponding principal-axes themselves) are already known. Appendix A provides a table of principal-axis moments of inertia relative to the center of mass of many common rigid bodies. These shapes will form the basis for many of the examples considered throughout the remainder of this chapter.

As a short aside, it is noted that for most applications a principal-axis coordinate system will necessarily be fixed in a rigid body. However, in some (relatively simple) situations it is possible to find a principal-axis coordinate system that is *not* fixed in a rigid body.[4] In terms of a principal-axis basis $\{\mathbf{e}_1^p, \mathbf{e}_2^p, \mathbf{e}_3^p\}$, the angular momentum of the rigid body in an inertial reference frame $\mathcal{N}$ relative to a point Q fixed in the body is given as

$${}^{\mathcal{N}}\mathbf{H}_Q = \mathbf{I}^{\mathcal{R}} \cdot {}^{\mathcal{N}}\boldsymbol{\omega}^{\mathcal{R}} = I_1^Q \omega_1 \mathbf{e}_1^p + I_2^Q \omega_2 \mathbf{e}_2^p + I_3^Q \omega_3 \mathbf{e}_3^p \tag{5-194}$$

where we recall that ${}^{\mathcal{N}}\boldsymbol{\omega}^{\mathcal{R}} = \omega_1 \mathbf{e}_1^p + \omega_2 \mathbf{e}_2^p + \omega_3 \mathbf{e}_3^p$ is the angular velocity of the rigid body in the inertial reference frame $\mathcal{N}$.

5.5 Actions on a Rigid Body

Because a particle can undergo only translational motion, the only fundamental action that can be applied to a particle is that of a *force*; any moments arise from the application of a force about a point. However, because a rigid body can undergo *both* translational and rotational motion, *two* fundamental actions can be applied to a rigid body. The first fundamental action that can be applied to a rigid body is the same as that of a particle, namely, a force. However, in addition to forces, a rigid body can also be subjected to a so-called *pure torque* that affects only the rotational motion of the body. We now describe the effects that a force and a torque have on a rigid body.

---

[4]A homogeneous circular disk is an example of a rigid body for which a principal-axis coordinate system exists that is not fixed in the body. In this case, because the disk is circular, a principal-axis coordinate system can be constructed by choosing the origin to be the center of mass of the disk and choosing the basis such that one basis vector is orthogonal to the plane of the disk while the other two basis vectors lie in the plane of the disk. In particular, the two basis vectors that lie in the plane of the disk need not be fixed to the disk.

5.5.1 Force Applied to a Rigid Body

It is evident that the action of a force on a particle must occur at the location of the particle because a particle is a single point. However, unlike a particle, a rigid body is a collection of points that occupies a nonzero volume of $\mathbb{E}^3$. Consequently, the motion of a rigid body due to the application of a force depends on the particular point on the rigid body to which the force is applied. Therefore, when describing the action of a force on a rigid body, it is necessary to specify both the force and the *location* or *position* in $\mathbb{E}^3$ where the force acts. In particular, as we will see, the point of application of a force on a rigid body affects both the translational and rotational motion of the body.

5.5.2 Pure Torque Applied to a Rigid Body

Unlike particles, whose motions are only affected by forces, the motion of a rigid body is affected by forces and so-called *pure torques.* A pure torque is a moment applied to a body that *does not* arise from a unique force. A pure torque applied to a rigid body has an important property, which we will now derive.

Consider a pure torque $\boldsymbol{\tau}_P$ acting about a point P on a rigid body $\mathcal{R}$ as shown in Fig. 5-4. It is noted that, because the body $\mathcal{R}$ is rigid, the application of $\boldsymbol{\tau}_P$ does not

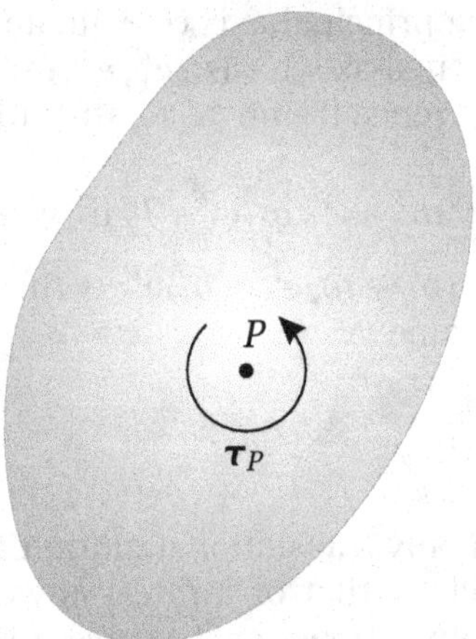

Figure 5-4 Pure torque $\boldsymbol{\tau}_P$ acting on a rigid body relative to point P.

change the relative position of any of the points on the body.[5] Suppose now that we replace $\boldsymbol{\tau}_P$ with a pair of forces $\mathbf{F}$ and $-\mathbf{F}$, where $\mathbf{F}$ and $-\mathbf{F}$ are applied at points A and B, respectively, as shown in Fig. 5-5.
The points A and B are such that

$$\begin{aligned} \boldsymbol{\rho} &= \mathbf{r}_{A/P} = \text{Position of } A \text{ relative to } P \\ -\boldsymbol{\rho} &= \mathbf{r}_{B/P} = \text{Position of } B \text{ relative to } P \end{aligned} \tag{5-195}$$

The pair of forces $(\mathbf{F}, -\mathbf{F})$, together with the relative positions $(\boldsymbol{\rho}, -\boldsymbol{\rho})$, form a so-called *couple* and is denoted $(\boldsymbol{\rho}, \mathbf{F})$. It is seen that, because the $(\mathbf{F}, -\mathbf{F})$ are equal and opposite,

[5]It is extremely important to note that, for an elastic (nonrigid) body, the application of a pure torque *does* change the relative location of points on the body. Consequently, the result derived here is not valid for elastic bodies.

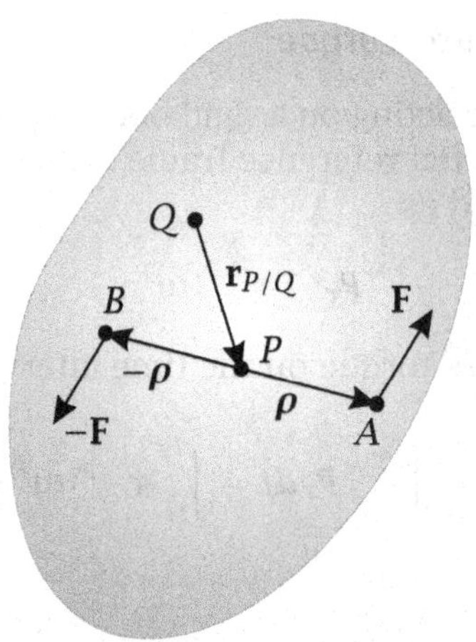

Figure 5–5 Couple $(\boldsymbol{\rho}, \mathbf{F})$ that replaces a pure torque $\boldsymbol{\tau}_P$ acting a rigid body relative to point P.

their sum is zero. In terms of the couple $(\boldsymbol{\rho}, \mathbf{F})$, the pure torque $\boldsymbol{\tau}$ can be written as

$$\boldsymbol{\tau}_P = \boldsymbol{\rho} \times \mathbf{F} + (-\boldsymbol{\rho}) \times (-\mathbf{F}) = 2\boldsymbol{\rho} \times \mathbf{F} \tag{5–196}$$

It can be seen that no unique couple $(\boldsymbol{\rho}, \mathbf{F})$ exists to replace a pure torque because the relative position $\boldsymbol{\rho}$ and the force $\mathbf{F}$ can be scaled in a reciprocal manner by an arbitrary nonzero scalar a without changing the value of $\boldsymbol{\tau}$, i.e.,

$$a\boldsymbol{\rho} \times (\mathbf{F}/a) + (-a\boldsymbol{\rho}) \times (-\mathbf{F}/a) = 2\boldsymbol{\rho} \times \mathbf{F} = \boldsymbol{\tau}_P \tag{5–197}$$

Suppose now that an arbitrary point Q (different from point P) is chosen on the rigid body as shown in Fig. 5-5. Also, let $\mathbf{r}_{P/Q}$ be the position of point P relative to Q and let $\mathbf{r}_{A/Q}$ and $\mathbf{r}_{B/Q}$ be the positions of points A and B relative to point Q, respectively. Then, using the result of Eq. (5–195), we have

$$\begin{aligned} \mathbf{r}_{A/Q} &= \mathbf{r}_{P/Q} + \mathbf{r}_{A/P} = \mathbf{r}_{P/Q} + \boldsymbol{\rho} \\ \mathbf{r}_{B/Q} &= \mathbf{r}_{P/Q} + \mathbf{r}_{B/P} = \mathbf{r}_{P/Q} - \boldsymbol{\rho} \end{aligned} \tag{5–198}$$

The moment about point Q due to the couple $(\boldsymbol{\rho}, \mathbf{F})$, denoted $\boldsymbol{\tau}_Q$, is then given as

$$\boldsymbol{\tau}_Q = \mathbf{r}_{A/Q} \times \mathbf{F} + \mathbf{r}_{B/Q} \times (-\mathbf{F}) \tag{5–199}$$

Substituting Eq. (5–198) into (5–199), we have

$$\boldsymbol{\tau}_Q = (\mathbf{r}_{P/Q} + \boldsymbol{\rho}) \times \mathbf{F} + (\mathbf{r}_{P/Q} - \boldsymbol{\rho}) \times (-\mathbf{F}) \tag{5–200}$$

Equation (5–200) can be expanded to give

$$\boldsymbol{\tau}_Q = \mathbf{r}_{P/Q} \times \mathbf{F} + \boldsymbol{\rho} \times \mathbf{F} - \mathbf{r}_{P/Q} \times \mathbf{F} + \boldsymbol{\rho} \times \mathbf{F} = 2\boldsymbol{\rho} \times \mathbf{F} \equiv \boldsymbol{\tau}_P \tag{5–201}$$

It is seen from Eq. (5–201) that the moment $\boldsymbol{\tau}_Q$ about the point Q and the moment $\boldsymbol{\tau}_P$ about the point P due to the couple $(\boldsymbol{\rho}, \mathbf{F})$ are identical. In other words, a pure torque can be translated between any two points on a rigid body without changing its effect on the body. Consequently, a pure torque applied to a rigid body is a free vector that may be translated without change between any two points on a rigid body.

5.5.3 Power and Work of a Pure Torque

Suppose that $\boldsymbol{\tau}$ is a pure torque acting on a rigid body $\mathcal{R}$. Furthermore, let ${}^{\mathcal{N}}\boldsymbol{\omega}^{\mathcal{R}}$ be the angular velocity of $\mathcal{R}$ in an inertial reference frame $\mathcal{N}$. Then the *power* produced by $\boldsymbol{\tau}$ in reference frame $\mathcal{N}$ is defined as

$$ {}^{\mathcal{N}}P_{\tau} = \boldsymbol{\tau} \cdot {}^{\mathcal{N}}\boldsymbol{\omega}^{\mathcal{R}} \tag{5-202} $$

Furthermore, the work of a pure torque on the time interval from t_1 to t_2 in reference frame $\mathcal{N}$ is given as

$$ {}^{\mathcal{N}}W_{\tau} = \int_{t_1}^{t_2} {}^{\mathcal{N}}P_{\tau} dt = \int_{t_1}^{t_2} \boldsymbol{\tau} \cdot {}^{\mathcal{N}}\boldsymbol{\omega}^{\mathcal{R}} dt \tag{5-203} $$

5.5.4 Conservative Torques

As with a force applied to a particle or a rigid body, the work done by a pure torque over a time interval $t \in [t_1, t_2]$ in rotating a rigid body between two orientations will depend on both the trajectory taken by the particle on $[t_1, t_2]$ and the orientations of the rigid body at the endpoints t_1 and t_2. Suppose now that we let $\boldsymbol{\Theta} \in \mathbb{R}^3$ be a column-vector whose elements describe the orientation of a rigid body $\mathcal{R}$ in an inertial reference frame $\mathcal{N}$, i.e., $\boldsymbol{\Theta}$ is given as

$$ \boldsymbol{\Theta} = \left\{ \begin{array}{c} \theta_1 \\ \theta_2 \\ \theta_3 \end{array} \right\} \tag{5-204} $$

Then a torque $\boldsymbol{\tau}^c$ is said to be conservative if the work done in rotating the rigid body from an initial orientation $\boldsymbol{\Theta}_1$ at $t = t_1$ to a final orientation $\boldsymbol{\Theta}_2$ at $t = t_2$ is *independent* of the trajectory taken to rotate the rigid body from $\boldsymbol{\Theta}(t_1) = \boldsymbol{\Theta}_1$ to $\boldsymbol{\Theta}(t_2) = \boldsymbol{\Theta}_2$. In other words, if $[{}^{\mathcal{N}}\boldsymbol{\omega}^{\mathcal{R}}]^{(1)}, \ldots, [{}^{\mathcal{N}}\boldsymbol{\omega}^{\mathcal{R}}]^{(n)}$ are the angular velocities in an inertial reference frame $\mathcal{N}$ that correspond to an arbitrary set of n trajectories that start at $\boldsymbol{\Theta}_1$ and end at $\boldsymbol{\Theta}_2$, then the pure torque $\boldsymbol{\tau}^c$ is conservative if, for all trajectories $[{}^{\mathcal{N}}\boldsymbol{\omega}^{\mathcal{R}}]^{(1)}, \ldots, [{}^{\mathcal{N}}\boldsymbol{\omega}^{\mathcal{R}}]^{(n)}$, $(i = 1, 2, \ldots, n)$, we have

$$ {}^{\mathcal{N}}W_{\tau 12}^{c} = \int_{t_1}^{t_2} \boldsymbol{\tau}^c \cdot \left[{}^{\mathcal{N}}\boldsymbol{\omega}^{\mathcal{R}}\right]^{(1)} dt = \cdots = \int_{t_1}^{t_2} \boldsymbol{\tau}^c \cdot \left[{}^{\mathcal{N}}\boldsymbol{\omega}^{\mathcal{R}}\right]^{(n)} dt \tag{5-205} $$

A consequence of the fact that a conservative torque is independent of the trajectory is that there exists a scalar function ${}^{\mathcal{N}}V = {}^{\mathcal{N}}V(\boldsymbol{\Theta})$ such that

$$ \boxed{\boldsymbol{\tau}^c = -\nabla_{\boldsymbol{\Theta}} {}^{\mathcal{N}}V = -\frac{\partial}{\partial \boldsymbol{\Theta}} \left({}^{\mathcal{N}}V\right)} \tag{5-206} $$

The power of a conservative torque can then be written as

$$ \boxed{{}^{\mathcal{N}}P_{\tau}^{c} = \boldsymbol{\tau}^c \cdot {}^{\mathcal{N}}\boldsymbol{\omega}^{\mathcal{R}} = -\nabla_{\boldsymbol{\Theta}} {}^{\mathcal{N}}V \cdot {}^{\mathcal{N}}\boldsymbol{\omega}^{\mathcal{R}}} \tag{5-207} $$

Now, as was seen in Chapter 2, the orientation of a rigid body cannot be obtained via direct integration of the angular velocity. As a result, an alternate parameterization of the orientation of the rigid body was developed using *Euler angles*. In particular, in terms of Type I Euler angles, it was shown in Eq. (2-600) on page 120 that the angular velocity of a rigid body $\mathcal{R}$ in an arbitrary reference frame $\mathcal{A}$ can be written in terms of the *Euler basis* (O'Reilly and Srinivasa, 2002) as

$$^{\mathcal{A}}\boldsymbol{\omega}^{\mathcal{R}} = \dot{\psi}\mathbf{k}_1 + \dot{\theta}\mathbf{k}_2 + \dot{\phi}\mathbf{k}_3 \tag{5-208}$$

where ψ, θ, and ϕ are the set of Type I Euler angles and $\{\mathbf{k}_1, \mathbf{k}_2, \mathbf{k}_3\}$ is the corresponding Euler basis. Using the expression for $^{\mathcal{A}}\boldsymbol{\omega}^{\mathcal{R}}$ in Eq. (5-208) and the definition of the power of a conservative torque from Eq. (5-207), we can write

$$\boldsymbol{\tau}^c \cdot {}^{\mathcal{N}}\boldsymbol{\omega}^{\mathcal{R}} = -\nabla_{\Theta}{}^{\mathcal{N}}V \cdot \left(\dot{\psi}\mathbf{k}_1 + \dot{\theta}\mathbf{k}_2 + \dot{\phi}\mathbf{k}_3\right) \tag{5-209}$$

Furthermore, we can obtain a second expression for $\boldsymbol{\tau}^c \cdot {}^{\mathcal{N}}\boldsymbol{\omega}^{\mathcal{R}}$ by noting that the power of $\boldsymbol{\tau}^c$ can be obtained by computing the rate of change of $^{\mathcal{N}}V$ as

$$^{\mathcal{N}}P_{\tau} = -\frac{d}{dt}\left(^{\mathcal{N}}V\right) = -\frac{\partial}{\partial\psi}\left(^{\mathcal{N}}V\right)\dot{\psi} - \frac{\partial}{\partial\theta}\left(^{\mathcal{N}}V\right)\dot{\theta} - \frac{\partial}{\partial\phi}\left(^{\mathcal{N}}V\right)\dot{\phi} \tag{5-210}$$

Now, in order for the expressions in Eqs. (5-209) and (5-210) to be equal, it is necessary to determine a basis to express $\nabla_{\Theta}{}^{\mathcal{N}}V$ such that the scalar product of $\nabla_{\Theta}{}^{\mathcal{N}}V$ with $^{\mathcal{N}}\boldsymbol{\omega}^{\mathcal{R}}$ equals the expression in Eq. (5-210). In other words, we seek a basis $\{\mathbf{u}_1, \mathbf{u}_2, \mathbf{u}_3\}$ such that

$$\nabla_{\Theta}{}^{\mathcal{N}}V = \frac{\partial}{\partial\psi}\left(^{\mathcal{N}}V\right)\mathbf{u}_1 + \frac{\partial}{\partial\theta}\left(^{\mathcal{N}}V\right)\mathbf{u}_2 + \frac{\partial}{\partial\phi}\left(^{\mathcal{N}}V\right)\mathbf{u}_3 \tag{5-211}$$

and

$$\nabla_{\Theta}{}^{\mathcal{N}}V \cdot {}^{\mathcal{N}}\boldsymbol{\omega}^{\mathcal{R}} = \frac{\partial}{\partial\psi}\left(^{\mathcal{N}}V\right)\dot{\psi} + \frac{\partial}{\partial\theta}\left(^{\mathcal{N}}V\right)\dot{\theta} + \frac{\partial}{\partial\phi}\left(^{\mathcal{N}}V\right)\dot{\phi} \tag{5-212}$$

Recall now the *dual Euler basis* (O'Reilly and Srinivasa, 2002; O'Reilly, 2004) $\{\mathbf{k}_1^*, \mathbf{k}_2^*, \mathbf{k}_3^*\}$ which, from Eq. (2-601) on page 120, has the property that

$$\mathbf{k}_i^* \cdot \mathbf{k}_j = \begin{cases} 1 & , \quad i = j \\ 0 & , \quad i \neq j \end{cases} \qquad (i, j = 1, 2, 3) \tag{5-213}$$

Using the property of Eq. (5-213), it is seen that $\nabla_{\Theta}{}^{\mathcal{N}}V$ can be written in terms of the dual Euler basis as

$$\nabla_{\Theta}{}^{\mathcal{N}}V = \frac{\partial}{\partial\psi}\left(^{\mathcal{N}}V\right)\mathbf{k}_1^* + \frac{\partial}{\partial\theta}\left(^{\mathcal{N}}V\right)\mathbf{k}_2^* + \frac{\partial}{\partial\phi}\left(^{\mathcal{N}}V\right)\mathbf{k}_3^* \tag{5-214}$$

Equation (5-214) is an expression for the gradient of the potential energy of a conservative pure torque in terms of Type I Euler angles expressed in the Type I dual Euler basis. Using Eq. (5-214), the conservative torque is given in terms of Type I Euler angles and the corresponding dual Euler basis as

$$\boldsymbol{\tau}^c = -\frac{\partial}{\partial\psi}\left(^{\mathcal{N}}V\right)\mathbf{k}_1^* - \frac{\partial}{\partial\theta}\left(^{\mathcal{N}}V\right)\mathbf{k}_2^* - \frac{\partial}{\partial\phi}\left(^{\mathcal{N}}V\right)\mathbf{k}_3^* \tag{5-215}$$

5.6 Moment Transport Theorem for a Rigid Body

Let $\mathbf{F}_1, \ldots, \mathbf{F}_n$ be the forces that act on a rigid body $\mathcal{R}$. Furthermore, let $\mathbf{r}_1, \ldots, \mathbf{r}_n$ be the positions of the forces $\mathbf{F}_1, \ldots, \mathbf{F}_n$, respectively. Finally, let $\boldsymbol{\tau}$ be the resultant *pure torque* acting on $\mathcal{R}$. Then the resultant moment applied to $\mathcal{R}$ relative to an arbitrary point Q is defined as

$$\mathbf{M}_Q = \sum_{i=1}^{n} (\mathbf{r}_i - \mathbf{r}_Q) \times \mathbf{F}_i + \boldsymbol{\tau} \tag{5-216}$$

Furthermore, the resultant moment applied to $\mathcal{R}$ relative to a point O fixed in an inertial reference frame is defined as

$$\mathbf{M}_O = \sum_{i=1}^{n} (\mathbf{r}_i - \mathbf{r}_O) \times \mathbf{F}_i + \boldsymbol{\tau} \tag{5-217}$$

Lastly, the resultant moment applied to $\mathcal{R}$ relative to the center of mass of $\mathcal{R}$ is defined as

$$\bar{\mathbf{M}} = \sum_{i=1}^{n} (\mathbf{r}_i - \bar{\mathbf{r}}) \times \mathbf{F}_i + \boldsymbol{\tau} \tag{5-218}$$

It is seen that the pure torque in Eqs. (5-216), (5-217), and (5-218) is the same due to the fact that a pure torque $\boldsymbol{\tau}$ is a free vector and thus can be translated to any point on a rigid body without change.

Suppose now that we choose a second arbitrary point Q'. Then, from Eq. (5-216), the moment due to all forces and all pure torques relative to Q' is given as

$$\mathbf{M}_{Q'} = \sum_{i=1}^{n} (\mathbf{r}_i - \mathbf{r}_{Q'}) \times \mathbf{F}_i + \boldsymbol{\tau} \tag{5-219}$$

Subtracting Eq. (5-216) from (5-219), we have

$$\mathbf{M}_{Q'} - \mathbf{M}_Q = \sum_{i=1}^{n} (\mathbf{r}_Q - \mathbf{r}_{Q'}) \times \mathbf{F}_i \tag{5-220}$$

Observing that $\mathbf{r}_Q$ and $\mathbf{r}_{Q'}$ are independent of the summation, Eq. (5-220) can be written as

$$\mathbf{M}_{Q'} - \mathbf{M}_Q = (\mathbf{r}_Q - \mathbf{r}_{Q'}) \times \sum_{i=1}^{n} \mathbf{F}_i \tag{5-221}$$

Now we recall that

$$\mathbf{F} = \sum_{i=1}^{n} \mathbf{F}_i \tag{5-222}$$

Therefore, Eq. (5-221) becomes

$$\mathbf{M}_{Q'} - \mathbf{M}_Q = (\mathbf{r}_Q - \mathbf{r}_{Q'}) \times \mathbf{F} \tag{5-223}$$

Rearranging Eq. (5-223), we obtain

$$\boxed{\mathbf{M}_{Q'} = \mathbf{M}_Q + (\mathbf{r}_Q - \mathbf{r}_{Q'}) \times \mathbf{F}} \tag{5-224}$$

Equation (5-224) is called the *moment transport theorem* for a rigid body and can be used to determine the moment applied to a rigid body relative to a point Q' when the moment applied relative to a point Q is already known.

5.7 Euler's Laws for a Rigid Body

Because a rigid body has both translational and rotational motion, two independent balance laws are required to fully specify the motion of the body. In particular, one balance law is needed to describe the translational motion and another law is needed to describe the rotational motion of the rigid body. These two balance laws for a rigid body are called *Euler's laws.* We now state Euler's laws.

5.7.1 Euler's 1st Law:

The resultant force applied to a rigid body $\mathcal{R}$ is equal to the product of the mass of the rigid body and the acceleration of the center of mass of the rigid body in an inertial reference frame $\mathcal{N}$, i.e.,

$$\boxed{\mathbf{F} = m\,{}^{\mathcal{N}}\bar{\mathbf{a}}} \tag{5-225}$$

5.7.2 Euler's 2nd Law:

The resultant moment applied to a rigid body $\mathcal{R}$ relative to a point O fixed point in an inertial reference frame $\mathcal{N}$ is equal to the rate of change of angular momentum of the rigid body relative to point O in reference frame $\mathcal{N}$, i.e.,

$$\boxed{\mathbf{M}_O = \frac{{}^{\mathcal{N}}d}{dt}\left({}^{\mathcal{N}}\mathbf{H}_O\right)} \tag{5-226}$$

5.7.3 Alternate Forms of Euler's 2nd Law

Because of its simplicity, it is most desirable to apply Euler's 2nd law about an inertially fixed point. However, for many problems it is difficult to find a convenient inertially fixed point about which to apply Euler's 2nd law. For such situations, it is important to have ways to apply Euler's 2nd law relative to a point that moves relative to any inertial reference frame. In this section we develop two alternate forms of Euler's 2nd law. The first is applicable relative to an arbitrary reference point while the second is applicable relative to the center of mass of the rigid body.

Euler's 2nd Law Relative to an Arbitrary Reference Point

Recalling Eq. (5-21), the relationship between the angular momentum of a rigid body in an inertial reference frame $\mathcal{N}$ relative to an arbitrary reference point Q, ${}^{\mathcal{N}}\mathbf{H}_Q$, and the angular of a rigid body in an inertial reference frame $\mathcal{N}$ relative to a point O fixed in $\mathcal{N}$, ${}^{\mathcal{N}}\mathbf{H}_O$, is given as

$${}^{\mathcal{N}}\mathbf{H}_Q = {}^{\mathcal{N}}\mathbf{H}_O + (\bar{\mathbf{r}} - \mathbf{r}_Q) \times m({}^{\mathcal{N}}\bar{\mathbf{v}} - {}^{\mathcal{N}}\mathbf{v}_Q) - (\bar{\mathbf{r}} - \mathbf{r}_O) \times m\,{}^{\mathcal{N}}\bar{\mathbf{v}} \tag{5-227}$$

Computing the rate of change of ${}^{\mathcal{N}}\mathbf{H}_Q$ in Eq. (5-227) in reference frame $\mathcal{N}$, we have

$$\begin{aligned}\frac{{}^{\mathcal{N}}d}{dt}\left({}^{\mathcal{N}}\mathbf{H}_Q\right) = {} & \frac{{}^{\mathcal{N}}d}{dt}\left({}^{\mathcal{N}}\mathbf{H}_O\right) + ({}^{\mathcal{N}}\bar{\mathbf{v}} - {}^{\mathcal{N}}\mathbf{v}_Q) \times m({}^{\mathcal{N}}\bar{\mathbf{v}} - {}^{\mathcal{N}}\mathbf{v}_Q) \\ & + (\bar{\mathbf{r}} - \mathbf{r}_Q) \times m({}^{\mathcal{N}}\bar{\mathbf{a}} - {}^{\mathcal{N}}\mathbf{a}_Q) - {}^{\mathcal{N}}\bar{\mathbf{v}} \times m\,{}^{\mathcal{N}}\bar{\mathbf{v}} - (\bar{\mathbf{r}} - \mathbf{r}_O) \times m\,{}^{\mathcal{N}}\bar{\mathbf{a}}\end{aligned} \tag{5-228}$$

Now we observe that the second and fourth terms in Eq. (5-228) are zero. Therefore, Eq. (5-228) simplifies to

$$\frac{{}^{\mathcal{N}}d}{dt}\left({}^{\mathcal{N}}\mathbf{H}_Q\right) = \frac{{}^{\mathcal{N}}d}{dt}\left({}^{\mathcal{N}}\mathbf{H}_O\right) + (\bar{\mathbf{r}} - \mathbf{r}_Q) \times m({}^{\mathcal{N}}\bar{\mathbf{a}} - {}^{\mathcal{N}}\mathbf{a}_Q) - (\bar{\mathbf{r}} - \mathbf{r}_O) \times m{}^{\mathcal{N}}\bar{\mathbf{a}} \tag{5-229}$$

Then, applying Euler's 2nd law from Eq. (5-226), the expression in Eq. (5-229) becomes

$$\frac{{}^{\mathcal{N}}d}{dt}\left({}^{\mathcal{N}}\mathbf{H}_Q\right) = \mathbf{M}_O + (\bar{\mathbf{r}} - \mathbf{r}_Q) \times m({}^{\mathcal{N}}\bar{\mathbf{a}} - {}^{\mathcal{N}}\mathbf{a}_Q) - (\bar{\mathbf{r}} - \mathbf{r}_O) \times m{}^{\mathcal{N}}\bar{\mathbf{a}} \tag{5-230}$$

Furthermore, applying the moment transport theorem of Eq. (5-224), we have

$$\mathbf{M}_O = \mathbf{M}_Q + (\mathbf{r}_Q - \mathbf{r}_O) \times \mathbf{F} \tag{5-231}$$

Substituting the result of Eq. (5-231) into (5-230) gives

$$\frac{{}^{\mathcal{N}}d}{dt}\left({}^{\mathcal{N}}\mathbf{H}_Q\right) = \mathbf{M}_Q + (\mathbf{r}_Q - \mathbf{r}_O) \times \mathbf{F} + (\bar{\mathbf{r}} - \mathbf{r}_Q) \times m({}^{\mathcal{N}}\bar{\mathbf{a}} - {}^{\mathcal{N}}\mathbf{a}_Q) - (\bar{\mathbf{r}} - \mathbf{r}_O) \times m{}^{\mathcal{N}}\bar{\mathbf{a}} \tag{5-232}$$

Then, applying Euler's 1st law from Eq. (5-225), the expression in Eq. (5-232) can be written as

$$\frac{{}^{\mathcal{N}}d}{dt}\left({}^{\mathcal{N}}\mathbf{H}_Q\right) = \mathbf{M}_Q + (\mathbf{r}_Q - \mathbf{r}_O) \times m{}^{\mathcal{N}}\bar{\mathbf{a}} + (\bar{\mathbf{r}} - \mathbf{r}_Q) \times m({}^{\mathcal{N}}\bar{\mathbf{a}} - {}^{\mathcal{N}}\mathbf{a}_Q) - (\bar{\mathbf{r}} - \mathbf{r}_O) \times m{}^{\mathcal{N}}\bar{\mathbf{a}} \tag{5-233}$$

Equation (5-233) simplifies to

$$\mathbf{M}_Q - (\bar{\mathbf{r}} - \mathbf{r}_Q) \times m{}^{\mathcal{N}}\mathbf{a}_Q = \frac{{}^{\mathcal{N}}d}{dt}\left({}^{\mathcal{N}}\mathbf{H}_Q\right) \tag{5-234}$$

It is observed that, similar to the result for a system of particles, the quantity

$$-(\bar{\mathbf{r}} - \mathbf{r}_Q) \times m{}^{\mathcal{N}}\mathbf{a}_Q$$

is the *inertial moment* of the reference point Q relative to the center of mass of the system. From Eq. (5-234), we obtain the following statement of Euler's 2nd law relative to an arbitrary reference point:

The sum of the resultant moment applied to a rigid body relative to an arbitrary reference point Q and the inertial moment due to the acceleration of the reference point Q relative to the center of mass of a rigid body is equal to the rate of change of angular momentum of the rigid body relative to the reference point Q, i.e.,

$$\boxed{\mathbf{M}_Q - (\bar{\mathbf{r}} - \mathbf{r}_Q) \times m{}^{\mathcal{N}}\mathbf{a}_Q = \frac{{}^{\mathcal{N}}d}{dt}\left({}^{\mathcal{N}}\mathbf{H}_Q\right)} \tag{5-235}$$

Euler's 2nd Law Relative to the Center of Mass

Suppose that we choose the reference point in Eq. (5-234) to be the center of mass of the rigid body. Then $\mathbf{r}_Q = \bar{\mathbf{r}}$, which implies that $(\bar{\mathbf{r}} - \mathbf{r}_Q) \times m{}^{\mathcal{N}}\mathbf{a}_Q = \mathbf{0}$. Furthermore, $\mathbf{M}_Q = \bar{\mathbf{M}}$, and Eq. (5-234) simplifies to

$$\bar{\mathbf{M}} = \frac{{}^{\mathcal{N}}d}{dt}\left({}^{\mathcal{N}}\bar{\mathbf{H}}\right) \tag{5-236}$$

From Eq. (5-236), we obtain the following statement of Euler's 2nd law relative to the center of mass of a rigid body:

The resultant moment applied to a rigid body about its center of mass is equal to the rate of change of angular momentum of the rigid body relative to its center of mass, i.e.,

$$\boxed{\bar{\mathbf{M}} = \frac{{}^{\mathcal{N}}d}{dt}\left({}^{\mathcal{N}}\bar{\mathbf{H}}\right)} \tag{5-237}$$

5.7.4 Important Results About Euler's 2nd Law

In the previous section, two important results were obtained regarding Euler's 2nd law for a rigid body. First, it is noted that the relationship

$$\mathbf{M} = \frac{{}^{\mathcal{N}}d}{dt}\left({}^{\mathcal{N}}\mathbf{H}\right) \tag{5-238}$$

is valid *only* when the reference point is fixed in an inertial reference frame or is the center of mass of a rigid body. However, if an *arbitrary* reference point (i.e., a reference point that is *neither* a point fixed in an inertial reference frame *nor* the center of mass of the rigid body), then the rate of change of angular momentum is *not* equal to the resultant moment. Instead, in the latter case of an arbitrary reference point Q, we have

$$\mathbf{M}_Q - (\bar{\mathbf{r}} - \mathbf{r}_Q) \times m\,{}^{\mathcal{N}}\mathbf{a}_Q = \frac{{}^{\mathcal{N}}d}{dt}\left({}^{\mathcal{N}}\mathbf{H}_Q\right) \tag{5-239}$$

where $-(\bar{\mathbf{r}} - \mathbf{r}_Q) \times m\,{}^{\mathcal{N}}\mathbf{a}_Q$ is the inertial moment of the reference point Q relative to the center of mass of the system. It is important to re-emphasize that Eq. (5-239) must be applied when neither a fixed point nor the center of mass of the system is convenient for performing the analysis.

Example 5-4

A uniform circular disk of mass m and radius r rolls along a fixed plane inclined at a constant inclination angle β with the horizontal as shown in Fig. 5-6. Knowing that gravity acts downward, determine the differential equation of motion for the disk in terms of the angle θ using the following reference points: (a) the center of mass of the disk; (b) the instantaneous point of contact fixed to the disk; and (c) the instantaneous point of contact that moves parallel to the center of mass in a straight line down the incline.

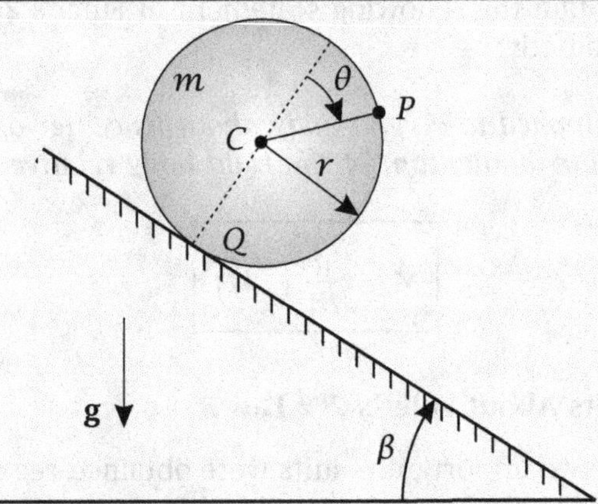

Figure 5–6 Uniform circular disk rolling on plane inclined at a constant inclination angle β with horizontal.

Solution to Example 5–4

Preliminaries

For this problem the differential equation of motion can be obtained by applying Euler's laws choosing as the reference point either the center of mass of the disk or the point of contact of the disk with the incline. We will now solve this problem using each of these reference points.

Kinematics

First, let $\mathcal{F}$ be a reference frame fixed to the incline. Then choose the following coordinate system fixed in $\mathcal{F}$:

$$
\begin{array}{ccc}
 & \text{Origin at } C & \\
 & \text{when } \theta = 0 & \\
\mathbf{E}_x & = & \text{Along incline} \\
\mathbf{E}_z & = & \text{Into page} \\
\mathbf{E}_y & = & \mathbf{E}_z \times \mathbf{E}_x
\end{array}
$$

Next, let $\mathcal{D}$ be a reference frame fixed to the disk. Noting that point P is fixed to the disk, the following coordinate system fixed in $\mathcal{D}$ is used:

$$
\begin{array}{ccc}
 & \text{Origin at } C & \\
\mathbf{e}_x & = & \text{Along } CP \\
\mathbf{e}_z & = & \mathbf{E}_z \\
\mathbf{e}_y & = & \mathbf{e}_z \times \mathbf{e}_x
\end{array}
$$

The bases $\{\mathbf{E}_x, \mathbf{E}_y, \mathbf{E}_z\}$ and $\{\mathbf{e}_x, \mathbf{e}_y, \mathbf{e}_z\}$ are shown in Fig. 5–7.

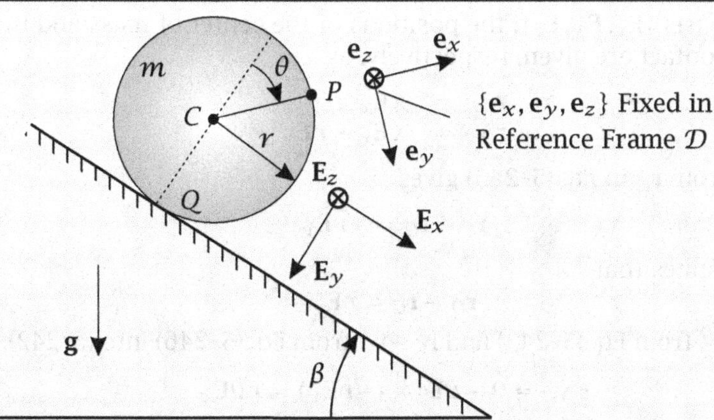

Figure 5-7 Basis $\{\mathbf{E}_x, \mathbf{E}_y, \mathbf{E}_z\}$ fixed in reference frame $\mathcal{F}$ for Example 5-4.

Now, while the kinematics of this problem are the same as those of Example 2-11, for completeness we re-derive the key kinematic results. In particular, in order to apply Euler's laws relative to the center of mass and relative to the instantaneous point of contact, we need to compute the following quantities: (1) the acceleration of the center of mass and (2) the acceleration of the instantaneous point of contact.

Acceleration of Center of Mass

First, let $\mathcal{D}$ and $\mathcal{S}$ be the references frame of the disk and the incline, respectively. Then, because the incline is a fixed surface, the velocity of the instantaneous point of contact on the incline with the disk is zero, i.e.,

$$^{\mathcal{F}}\mathbf{v}_Q^{\mathcal{S}} = \mathbf{0} \tag{5-240}$$

Furthermore, because the disk rolls without slip along the incline, we have from the rolling condition of Eq. (2-531) on page 109 that

$$^{\mathcal{F}}\mathbf{v}_Q^{\mathcal{D}} = {}^{\mathcal{F}}\mathbf{v}_Q^{\mathcal{S}} = \mathbf{0} \tag{5-241}$$

It is emphasized in Eqs. (5-240) and (5-241) that $^{\mathcal{F}}\mathbf{v}_Q^{\mathcal{D}}$ is the velocity in reference frame $\mathcal{F}$ of the instantaneous point of contact *on the disk* while $^{\mathcal{F}}\mathbf{v}_Q^{\mathcal{S}}$ is the velocity of the instantaneous point of contact in reference frame $\mathcal{F}$ *on the incline.* Applying Eq. (2-517) on page 106, the velocity of the center of mass of the disk is then given as

$$^{\mathcal{F}}\mathbf{v}_C = {}^{\mathcal{F}}\mathbf{v}_Q^{\mathcal{D}} + {}^{\mathcal{F}}\boldsymbol{\omega}^{\mathcal{D}} \times (\mathbf{r}_C - \mathbf{r}_Q) \tag{5-242}$$

Now the angular velocity of the disk in the inertial reference frame $\mathcal{F}$ is given as

$$^{\mathcal{F}}\boldsymbol{\omega}^{\mathcal{D}} = \dot{\theta}\mathbf{E}_z \tag{5-243}$$

Furthermore, computing the rate of change of $^{\mathcal{F}}\boldsymbol{\omega}^{\mathcal{D}}$ in reference frame $\mathcal{F}$, the angular acceleration of the disk in reference frame $\mathcal{F}$ is obtained as

$$^{\mathcal{F}}\boldsymbol{\alpha}^{\mathcal{D}} = \frac{^{\mathcal{F}}d}{dt}\left({}^{\mathcal{F}}\boldsymbol{\omega}^{\mathcal{D}}\right) = \ddot{\theta}\mathbf{E}_z \tag{5-244}$$

In terms of the basis $\{\mathbf{E}_x, \mathbf{E}_y, \mathbf{E}_z\}$, the positions of the center of mass and the instantaneous point of contact are given, respectively, as

$$\begin{aligned} \mathbf{r}_C &= x\mathbf{E}_x \\ \mathbf{r}_Q &= x\mathbf{E}_x + r\mathbf{E}_y \end{aligned} \tag{5-245}$$

Subtracting $\mathbf{r}_Q$ from $\mathbf{r}_C$ in Eq. (5-245) gives

$$\mathbf{r}_C - \mathbf{r}_Q = -r\mathbf{E}_y \tag{5-246}$$

which further implies that

$$\mathbf{r}_Q - \mathbf{r}_C = r\mathbf{E}_y \tag{5-247}$$

Substituting ${}^{\mathcal{F}}\boldsymbol{\omega}^{\mathcal{D}}$ from Eq. (5-243) and $\mathbf{r}_C - \mathbf{r}_Q$ from Eq. (5-246) into (5-242), we obtain

$${}^{\mathcal{F}}\mathbf{v}_C = \mathbf{0} + \dot{\theta}\mathbf{E}_z \times (-r\mathbf{E}_y) = r\dot{\theta}\mathbf{E}_x \tag{5-248}$$

Next, differentiating $\mathbf{r}_C$ from Eq. (5-245) gives

$${}^{\mathcal{F}}\mathbf{v}_C = \frac{{}^{\mathcal{F}}d}{dt}(\mathbf{r}_C) = \dot{x}\mathbf{E}_x \tag{5-249}$$

Setting the two expressions for ${}^{\mathcal{F}}\mathbf{v}_C$ from Eq. (5-248) and (5-249) equal, we obtain

$$\dot{x} = r\dot{\theta} \tag{5-250}$$

which implies that

$${}^{\mathcal{F}}\mathbf{v}_C = r\dot{\theta}\mathbf{E}_x \tag{5-251}$$

Furthermore, integrating Eq. (5-250) using the fact that the disk is in the reference configuration when $x(t = 0) = 0$ and $\theta(t = 0) = 0$, we obtain

$$x = r\theta \tag{5-252}$$

Equation (5-252) implies that

$$\mathbf{r}_C = r\theta\mathbf{E}_x \tag{5-253}$$

$$\mathbf{r}_Q = r\theta\mathbf{E}_x + r\mathbf{E}_y \tag{5-254}$$

The acceleration of point C is then obtained by computing the rate of change of ${}^{\mathcal{F}}\mathbf{v}_C$ from Eq. (5-251) in reference frame $\mathcal{F}$ as

$${}^{\mathcal{F}}\mathbf{a}_C = \frac{{}^{\mathcal{F}}d}{dt}\left({}^{\mathcal{F}}\mathbf{v}_C\right) = r\ddot{\theta}\mathbf{E}_x \tag{5-255}$$

Acceleration of Instantaneous Point of Contact on Disk

The acceleration of the instantaneous point of contact fixed in the disk is obtained from Eq. (2-524) as

$${}^{\mathcal{F}}\mathbf{a}_Q^{\mathcal{D}} = {}^{\mathcal{F}}\mathbf{a}_C + {}^{\mathcal{F}}\boldsymbol{\alpha}^{\mathcal{D}} \times (\mathbf{r}_Q - \mathbf{r}_C) + {}^{\mathcal{F}}\boldsymbol{\omega}^{\mathcal{D}} \times \left[{}^{\mathcal{F}}\boldsymbol{\omega}^{\mathcal{D}} \times (\mathbf{r}_Q - \mathbf{r}_C)\right] \tag{5-256}$$

Substituting the expressions for ${}^{\mathcal{F}}\boldsymbol{\omega}^{\mathcal{D}}$, ${}^{\mathcal{F}}\boldsymbol{\alpha}^{\mathcal{D}}$, ${}^{\mathcal{F}}\mathbf{a}_C$, and $\mathbf{r}_Q - \mathbf{r}_C$ from Eq. (5-243), (5-244), (5-255), and (5-247), respectively, we obtain

$${}^{\mathcal{F}}\mathbf{a}_Q^{\mathcal{D}} = r\ddot{\theta}\mathbf{E}_x + \ddot{\theta}\mathbf{E}_z \times r\mathbf{E}_y + \dot{\theta}\mathbf{E}_z \times \left[\dot{\theta}\mathbf{E}_z \times r\mathbf{E}_y\right] \tag{5-257}$$

Computing the vector products in Eq. (5-257) gives

$${}^{\mathcal{F}}\mathbf{a}_Q^{\mathcal{D}} = r\ddot{\theta}\mathbf{E}_x - r\ddot{\theta}\mathbf{E}_x - r\dot{\theta}^2\mathbf{E}_y = -r\dot{\theta}^2\mathbf{E}_y \tag{5-258}$$

Kinetics

The free body diagram of the disk is shown in Fig. 5-8.

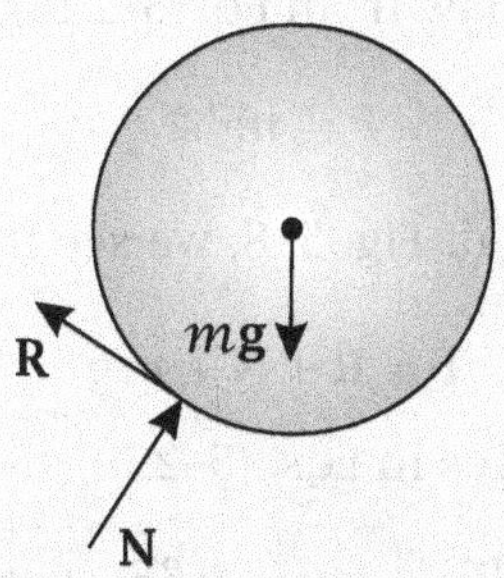

Figure 5-8 Free body diagram of disk for Example 5-4.

Using Fig. 5-8, we have

$\mathbf{N}$	=	Normal force of incline on disk
$\mathbf{R}$	=	Rolling force
$m\mathbf{g}$	=	Force of gravity

Now, from the geometry we have

$$\mathbf{N} = N\mathbf{E}_y \tag{5-259}$$

$$\mathbf{R} = R\mathbf{E}_x \tag{5-260}$$

$$m\mathbf{g} = mg\mathbf{u}_v \tag{5-261}$$

where $\mathbf{u}_v$ is the vertically downward direction. Using Fig. 5-9, the vector $\mathbf{u}_v$ is obtained as

$$\mathbf{u}_v = \sin\beta\mathbf{E}_x + \cos\beta\mathbf{E}_y \tag{5-262}$$

The force of gravity then becomes

$$m\mathbf{g} = mg\sin\beta\mathbf{E}_x + mg\cos\beta\mathbf{E}_y \tag{5-263}$$

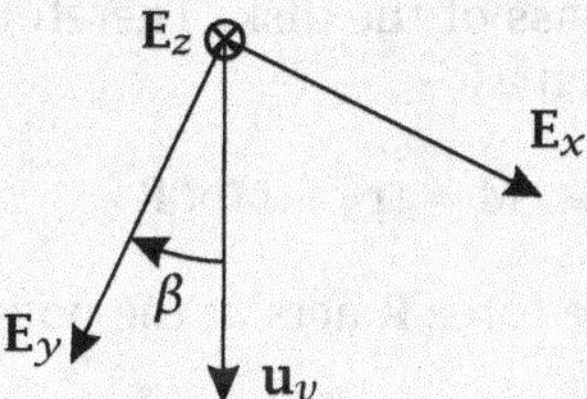

Figure 5-9 Unit vector in vertically downward direction in terms of basis $\{\mathbf{E}_x, \mathbf{E}_y, \mathbf{E}_z\}$ for Example 5-4.

As stated at the start of this problem, it is convenient to determine the differential equation of motion by applying Euler's laws relative to one of the following two reference points: (1) the center of mass of the disk or (2) the point of contact of the disk with the incline. We now proceed to apply Euler's laws using each of these reference points.

(a) Differential Equation Using Center of Mass of Disk as Reference Point

Using the center of mass of the disk as a reference point, we need to apply *both* of Euler's laws. Recalling Euler's 1st law from Eq. (5-225), we have

$$\mathbf{F} = m^{\mathcal{F}}\bar{\mathbf{a}} \tag{5-264}$$

Examining the free body diagram of Fig. 5-8, we see that the resultant force acting on the disk is given as

$$\mathbf{F} = \mathbf{R} + \mathbf{N} + m\mathbf{g} \tag{5-265}$$

Using the expressions for the forces in Eqs. (5-259)-(5-261), we have

$$\mathbf{F} = R\mathbf{E}_x + N\mathbf{E}_y + mg\sin\beta\mathbf{E}_x + mg\cos\beta\mathbf{E}_y \tag{5-266}$$

Then, noting that ${}^{\mathcal{F}}\bar{\mathbf{a}} = {}^{\mathcal{F}}\mathbf{a}_C$ and using the expression ${}^{\mathcal{F}}\mathbf{a}_C$ from Eq. (5-255), we can equate $\mathbf{F}$ and $m^{\mathcal{F}}\bar{\mathbf{a}}$ to give

$$R\mathbf{E}_x + N\mathbf{E}_y + mg\sin\beta\mathbf{E}_x + mg\cos\beta\mathbf{E}_y = mr\ddot{\theta}\mathbf{E}_x \tag{5-267}$$

Rearranging Eq. (5-267), we obtain

$$(R + mg\sin\beta)\mathbf{E}_x + (N + mg\cos\beta)\mathbf{E}_y = mr\ddot{\theta}\mathbf{E}_x \tag{5-268}$$

Equation (5-268) yields the following two scalar equations:

$$\begin{aligned} mr\ddot{\theta} &= R + mg\sin\beta && (5\text{-}269) \\ 0 &= N + mg\cos\beta && (5\text{-}270) \end{aligned}$$

Recalling Euler's 2nd about the center of mass of a rigid body from Eq. (5-237), we have

$$\bar{\mathbf{M}} = \frac{{}^{\mathcal{F}}d}{dt}\left({}^{\mathcal{F}}\bar{\mathbf{H}}\right) \tag{5-271}$$

Examining the free body diagram of Fig. 5-8, we see that the forces $m\mathbf{g}$ and $\mathbf{N}$ pass through the center of mass of the disk. Consequently, only the rolling force $\mathbf{R}$ produces a moment about the center of mass of the disk. Therefore, the moment relative to the center of mass of the disk is given as

$$\bar{\mathbf{M}} = (\mathbf{r}_R - \bar{\mathbf{r}}) \times \mathbf{R} \tag{5-272}$$

Now we observe that because the force $\mathbf{R}$ acts at the point of contact, we have

$$\mathbf{r}_R = \mathbf{r}_Q = r\theta\mathbf{E}_x + r\mathbf{E}_y \tag{5-273}$$

where we have used the expression for $\mathbf{r}_Q$ from Eq. (5-254). We then have

$$\mathbf{r}_R - \bar{\mathbf{r}} = \mathbf{r}_R - \mathbf{r}_C = r\theta\mathbf{E}_x + r\mathbf{E}_y - r\theta\mathbf{E}_x = r\mathbf{E}_y \tag{5-274}$$

Substituting $\mathbf{r}_R$ from Eq. (5-274) and using the expression for $\mathbf{R}$ from Eq. (5-260), we obtain

$$\bar{\mathbf{M}} = r\mathbf{E}_y \times R\mathbf{E}_x = -rR\mathbf{E}_z = -rR\mathbf{e}_z \tag{5-275}$$

Also, the angular momentum about the center of mass is given as

$$^{\mathcal{F}}\bar{\mathbf{H}} = \bar{\mathbf{I}}^{\mathcal{D}} \cdot {}^{\mathcal{F}}\boldsymbol{\omega}^{\mathcal{D}} \tag{5-276}$$

Because the $\{\mathbf{e}_x, \mathbf{e}_y, \mathbf{e}_z\}$ is a principal-axis basis, we have

$$\bar{\mathbf{I}}^{\mathcal{D}} = \bar{I}_{xx}\mathbf{e}_x \otimes \mathbf{e}_x + \bar{I}_{yy}\mathbf{e}_y \otimes \mathbf{e}_y + \bar{I}_{zz}\mathbf{e}_z \otimes \mathbf{e}_z \tag{5-277}$$

Substituting $\bar{\mathbf{I}}^{\mathcal{D}}$ from Eq. (5-277) and ${}^{\mathcal{F}}\boldsymbol{\omega}^{\mathcal{D}}$ from Eq. (5-243), we obtain ${}^{\mathcal{F}}\bar{\mathbf{H}}$ as

$$^{\mathcal{F}}\bar{\mathbf{H}} = \bar{I}_{zz}\mathbf{e}_z \otimes \mathbf{e}_z \cdot \dot{\theta}\mathbf{e}_z = \bar{I}_{zz}\dot{\theta}\mathbf{e}_z \tag{5-278}$$

Noting that the moment of inertia of a disk about the center of mass along the direction $\mathbf{E}_z$ is $I_{zz} = mr^2/2$, we have

$$^{\mathcal{F}}\bar{\mathbf{H}} = \frac{mr^2}{2}\dot{\theta}\mathbf{e}_z \tag{5-279}$$

Computing the rate of change of ${}^{\mathcal{F}}\bar{\mathbf{H}}$ in reference frame $\mathcal{F}$, we obtain

$$\frac{^{\mathcal{F}}d}{dt}\left({}^{\mathcal{F}}\bar{\mathbf{H}}\right) = \frac{mr^2}{2}\ddot{\theta}\mathbf{e}_z \tag{5-280}$$

Setting $\bar{\mathbf{M}}$ from Eq. (5-275) equal to ${}^{\mathcal{F}}d({}^{\mathcal{F}}\bar{\mathbf{H}})/dt$ from Eq. (5-280), we obtain

$$-rR = \frac{mr^2}{2}\ddot{\theta} \tag{5-281}$$

which yields the scalar equation

$$R = -\frac{mr}{2}\ddot{\theta} \tag{5-282}$$

The results of Eqs. (5-269) and (5-282) can now be used together to obtain the differential equation of motion. Substituting R from Eq. (5-282) into Eq. (5-269) gives

$$mr\ddot{\theta} = -\frac{mr}{2}\ddot{\theta} + mg\sin\beta \tag{5-283}$$

Equation (5-283) can be rearranged to give

$$\frac{3mr}{2}\ddot{\theta} = mg\sin\beta \tag{5-284}$$

Simplifying Eq. (5-284), we obtain the differential equation of motion as

$$\ddot{\theta} = \frac{2g\sin\beta}{3r} \tag{5-285}$$

(b) Differential Equation Using Instantaneous Point of Contact Fixed in Disk as Reference Point

A second method to obtain the differential equation of motion is to use the instantaneous point of contact P between the disk and the incline and *fixed in the disk* as the reference. Now it is seen from Eq. (5-241) that the velocity of the instantaneous point of contact fixed in the disk is zero. However, from Eq. (5-258) it is seen that the acceleration of the instantaneous point of contact is *not* zero (see Example 2-11 for this same result). Consequently, while it may be counterintuitive, the instantaneous point of contact of the disk with the incline is *not* a fixed point. Therefore, in order to use point P as the reference, it is necessary to apply the form of Euler's 2nd law for an arbitrary reference point as given in Eq. (5-234) on page 356, i.e.,

$$\mathbf{M}_Q - (\mathbf{r}_C - \mathbf{r}_Q) \times m^{\mathcal{F}}\mathbf{a}_Q^{\mathcal{D}} = \frac{^{\mathcal{F}}d}{dt}\left(^{\mathcal{F}}\mathbf{H}_Q\right) \tag{5-286}$$

Using $\mathbf{r}_C - \mathbf{r}_Q$ and $^{\mathcal{F}}\mathbf{a}_Q$ from Eqs. (5-246) and (5-258), respectively, the inertial moment due to the acceleration of point P is given as

$$-(\mathbf{r}_C - \mathbf{r}_Q) \times m^{\mathcal{F}}\mathbf{a}_Q^{\mathcal{D}} = r\mathbf{E}_y \times m(-r\dot{\theta}^2\mathbf{E}_y) = \mathbf{0} \tag{5-287}$$

It is seen from Eq. (5-287) that, while P is not a fixed point, the inertial moment due to the acceleration of point P is *zero.* This last result arises from the fact that $^{\mathcal{F}}\mathbf{a}_Q^{\mathcal{D}}$ is in the same direction as $\mathbf{r}_C - \mathbf{r}_Q$.[6] Equation (5-286) then simplifies to

$$\mathbf{M}_Q = \frac{^{\mathcal{F}}d}{dt}\left(^{\mathcal{F}}\mathbf{H}_Q\right) \tag{5-288}$$

In other words, for a uniform disk that rolls without slip along a flat surface, the instantaneous point of contact *appears* to be a fixed point[7] and it *happens* to be the case that $\mathbf{M}_Q$ is equal to $^{\mathcal{F}}d(^{\mathcal{F}}\mathbf{H}_Q)/dt$.

Next, using the free body diagram of Fig. 5-8, it is seen that only the force of gravity produces a moment relative to point P. Consequently, the moment relative to point P is given as

$$\mathbf{M}_Q = (\mathbf{r}_C - \mathbf{r}_Q) \times m\mathbf{g} \tag{5-289}$$

Substituting $\mathbf{r}_C - \mathbf{r}_Q$ from Eq. (5-246) and $m\mathbf{g}$ from Eq. (5-263) into Eq. (5-289), we obtain

$$\mathbf{M}_Q = -r\mathbf{E}_y \times (mg\sin\beta\mathbf{E}_x + mg\cos\beta\mathbf{E}_y) = mgr\sin\beta\mathbf{E}_z \tag{5-290}$$

Furthermore, from Eq. (5-24) on page 325, we have

$$^{\mathcal{F}}\mathbf{H}_Q = {}^{\mathcal{F}}\mathbf{H}_C + (\mathbf{r}_Q - \mathbf{r}_C) \times m(^{\mathcal{F}}\mathbf{v}_Q^{\mathcal{D}} - {}^{\mathcal{F}}\mathbf{v}_C) \tag{5-291}$$

[6] It is noted that, in the general case of a rigid body rolling on a surface, the acceleration of the point of contact fixed in the disk, $^{\mathcal{F}}\mathbf{a}_Q^{\mathcal{D}}$, would *not* lie in the same direction as the position of the center of mass relative to the point of contact, $\mathbf{r}_C - \mathbf{r}_Q$. In this case the inertial moment of the center of mass relative to the point of contact, $-(\mathbf{r}_C - \mathbf{r}_Q) \times m^{\mathcal{F}}\mathbf{a}_Q^{\mathcal{D}}$, would *not* be zero.

[7] For the case of a uniform disk rolling on a flat surface, the point of contact P appears to be a fixed point because it is the *instantaneous center of rotation* for the rigid body. While the concept of an instantaneous center of rotation can be used on certain simple problems, it is restricted to two-dimensional motion. Moreover, even in two dimensions, great care must be taken when using an instantaneous center of rotation as a reference point. Because of the potential for confusion, instantaneous centers of rotation are not used as the basis for solving problems in this book. See Beer and Johnston (1997), Bedford and Fowler (2005), or Greenwood (1988) for discussions about instantaneous centers of rotation.

Now, because the disk rolls without slip we have ${}^{\mathcal{F}}\mathbf{v}_Q^{\mathcal{D}} = \mathbf{0}$. Consequently,

$$ {}^{\mathcal{F}}\mathbf{v}_Q^{\mathcal{D}} - {}^{\mathcal{F}}\mathbf{v}_C = -{}^{\mathcal{F}}\mathbf{v}_C = -r\dot{\theta}\mathbf{E}_x \tag{5-292} $$

Substituting ${}^{\mathcal{F}}\bar{\mathbf{H}}$, $\mathbf{r}_Q - \mathbf{r}_C$, and ${}^{\mathcal{F}}\mathbf{v}_Q^{\mathcal{D}} - {}^{\mathcal{F}}\mathbf{v}_C$ from Eqs. (5-279), (5-247), and (5-292), respectively, into Eq. (5-291), we obtain

$$ {}^{\mathcal{F}}\mathbf{H}_Q = \frac{mr^2}{2}\dot{\theta}\mathbf{e}_z + r\mathbf{E}_y \times m(-r\dot{\theta}\mathbf{E}_x) = \frac{3mr^2}{2}\dot{\theta}\mathbf{E}_z \tag{5-293} $$

Computing the rate of change of ${}^{\mathcal{F}}\mathbf{H}_Q$ in reference frame $\mathcal{F}$, we obtain

$$ \frac{{}^{\mathcal{F}}d}{dt}\left({}^{\mathcal{F}}\mathbf{H}_Q\right) = \frac{3mr^2}{2}\ddot{\theta}\mathbf{E}_z \tag{5-294} $$

Then, setting $\mathbf{M}_Q$ from Eq. (5-290) equal to ${}^{\mathcal{F}}d({}^{\mathcal{F}}\mathbf{H}_Q)/dt$ from Eq. (5-294), we have

$$ mgr\sin\beta\mathbf{E}_z = \frac{3mr^2}{2}\ddot{\theta}\mathbf{E}_z \tag{5-295} $$

Dropping $\mathbf{E}_z$ from both sides of Eq. (5-295) and rearranging, we obtain

$$ \frac{3mr^2}{2}\ddot{\theta} = mgr\sin\beta \tag{5-296} $$

Simplifying Eq. (5-296) yields the differential equation of motion as

$$ \ddot{\theta} = \frac{2g\sin\beta}{3r} \tag{5-297} $$

It is seen that the differential equation of Eq. (5-297) obtained using the second method is identical to that of Eq. (5-285) obtained using the first method.

(c) Differential Equation Using Point of Contact Moving in Straight Line Down Incline as Reference Point

A third method to obtain the differential equation of motion is to use the instantaneous point of contact P between the disk and the incline and *moving in a straight line along the incline* as the reference point. It is important to note in this case that the reference point is *not* fixed to the disk, but rather is the projection of the center of mass onto the plane of the incline. Now in this case it is seen that point P moves with the same velocity as the center of mass. Using the expression for the velocity of the center of mass in reference frame $\mathcal{F}$ from Eq. (5-251), we have

$$ {}^{\mathcal{F}}\mathbf{v}_P = {}^{\mathcal{F}}\mathbf{v}_C = r\dot{\theta}\mathbf{E}_x \tag{5-298} $$

Consequently, the acceleration of point P is given as

$$ {}^{\mathcal{F}}\mathbf{a}_Q = \frac{{}^{\mathcal{F}}d}{dt}\left({}^{\mathcal{F}}\mathbf{v}_Q\right) = r\ddot{\theta}\mathbf{E}_x \tag{5-299} $$

Again, since ${}^{\mathcal{F}}\mathbf{a}_Q \neq \mathbf{0}$, it is seen that point P is *not* an inertially fixed point. Therefore, it is necessary to apply the general form of Euler's 2nd law as given in Eq. (5-234) on page 356, i.e.,

$$\mathbf{M}_Q - (\mathbf{r}_C - \mathbf{r}_Q) \times m^{\mathcal{F}}\mathbf{a}_Q = \frac{{}^{\mathcal{F}}d}{dt}\left({}^{\mathcal{F}}\mathbf{H}_Q\right) \tag{5-300}$$

Now we already have $\mathbf{M}_Q$ from Eq. (5-290). Next, using the expressions for $\mathbf{r}_C - \mathbf{r}_Q$ and ${}^{\mathcal{F}}\mathbf{a}_Q$ from Eqs. (5-246) and (5-299), the inertial moment due to the acceleration of point P relative to the center of mass of the disk is given as

$$-(\mathbf{r}_C - \mathbf{r}_Q) \times m^{\mathcal{F}}\mathbf{a}_Q = -(-r\mathbf{E}_y) \times mr\ddot{\theta}\mathbf{E}_x = -mr^2\ddot{\theta}\mathbf{E}_z \tag{5-301}$$

Furthermore, the angular momentum relative to point P is obtained from Eq. (5-24) on page 325 as

$${}^{\mathcal{F}}\mathbf{H}_Q = {}^{\mathcal{F}}\mathbf{H}_C + (\mathbf{r}_Q - \mathbf{r}_C) \times m({}^{\mathcal{F}}\mathbf{v}_Q - {}^{\mathcal{F}}\mathbf{v}_C) \tag{5-302}$$

Now, from Eq. (5-298), we have

$${}^{\mathcal{F}}\mathbf{v}_Q - {}^{\mathcal{F}}\mathbf{v}_C = \mathbf{0} \tag{5-303}$$

The second term in Eq. (5-302) is then zero and Eq. (5-302) reduces to

$${}^{\mathcal{F}}\mathbf{H}_Q = {}^{\mathcal{F}}\mathbf{H}_C = \frac{mr^2}{2}\dot{\theta}\mathbf{E}_z \tag{5-304}$$

where we have used the expression for ${}^{\mathcal{F}}\mathbf{H}_C$ from Eq. (5-279). Computing the rate of change of ${}^{\mathcal{F}}\mathbf{H}_Q$ in Eq. (5-304), we have

$$\frac{{}^{\mathcal{F}}d}{dt}\left({}^{\mathcal{F}}\mathbf{H}_Q\right) = \frac{mr^2}{2}\ddot{\theta}\mathbf{E}_z \tag{5-305}$$

Finally, substituting the results of Eqs. (5-290), (5-301), and (5-305) into Eq. (5-300), we obtain

$$mgr\sin\beta\mathbf{E}_z - mr^2\ddot{\theta}\mathbf{E}_z = \frac{mr^2}{2}\ddot{\theta}\mathbf{E}_z \tag{5-306}$$

Dropping the common factor of $\mathbf{E}_z$ from both sides of Eq. (5-306) gives

$$mgr\sin\beta - mr^2\ddot{\theta} = \frac{mr^2}{2}\ddot{\theta} \tag{5-307}$$

Rearranging and simplifying Eq. (5-307), we obtain the differential equation of motion as

$$\ddot{\theta} = \frac{2g\sin\beta}{3r} \tag{5-308}$$

It is seen that the differential equation of Eq. (5-308) obtained using this third method is identical to the results obtained in Eqs. (5-285) and (5-297) using the first and second methods, respectively.

■

Example 5–5

A homogeneous disk of mass m and radius r rolls in the vertical plane without slip along an inextensible cord as shown in Fig. 5–10. The center of the disk is located at point C while points A and B lie along the ends of the horizontal diameter of the disk. The cord is attached at one end to a fixed point O and is attached at its other end to a linear spring with spring constant K and damper with damping constant c. The damper exerts a force that is in the direction opposite the velocity at point B and is proportional to the cube of the magnitude of the velocity of point B, i.e., the force of the damper has the form

$$\mathbf{F}_d = -c\|\mathbf{v}_B\|^3 \frac{\mathbf{v}_B}{\|\mathbf{v}_B\|}$$

Assuming that the cord never becomes slack, that the spring is unstretched with length L when the angle θ is zero, and that gravity acts downward, determine the differential equation of motion for the disk in terms of the angle θ.

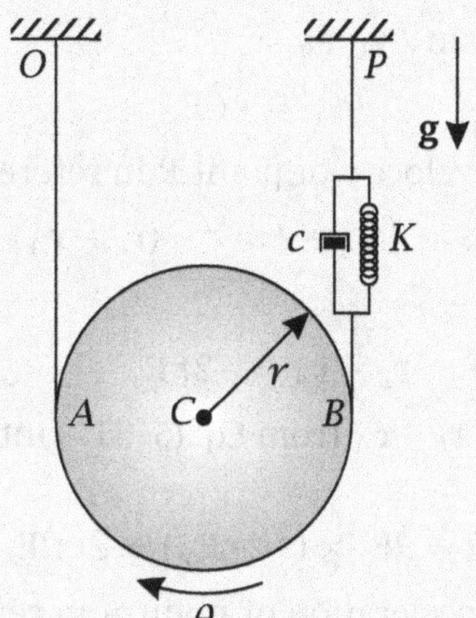

Figure 5–10 Disk rolling on cord attached to linear spring and nonlinear damper.

Solution to Example 5–5

Kinematics

For this problem, it is convenient to choose a fixed inertial reference frame $\mathcal{F}$. Corresponding to reference frame $\mathcal{F}$, the following coordinate system is chosen to express the motion of the disk:

	Origin at C when $\theta = 0$	
$\mathbf{E}_x$	=	Down
$\mathbf{E}_z$	=	Into page
$\mathbf{E}_y$	=	$\mathbf{E}_z \times \mathbf{E}_x$

We know that the disk rolls without slip on the cord. Since O is a fixed point, we know that the velocity of *every* point on the cord between O and A must be zero in reference frame $\mathcal{F}$. In particular, we know that

$$^{\mathcal{F}}\mathbf{v}_A = \mathbf{0} \tag{5-309}$$

Also, the angular velocity of the disk in reference frame $\mathcal{F}$ is given as

$$^{\mathcal{F}}\boldsymbol{\omega}^{\mathcal{R}} = \dot{\theta}\mathbf{E}_z \tag{5-310}$$

Then, from Eq. (2-517) on page 106 we have

$$^{\mathcal{F}}\mathbf{v}_C = {}^{\mathcal{F}}\mathbf{v}_A + {}^{\mathcal{F}}\boldsymbol{\omega}^{\mathcal{R}} \times (\mathbf{r}_C - \mathbf{r}_A) \tag{5-311}$$

Furthermore, from the geometry we have

$$\mathbf{r}_C - \mathbf{r}_A = -r\mathbf{E}_y \tag{5-312}$$

which implies that

$$\mathbf{r}_A - \mathbf{r}_C = r\mathbf{E}_y \tag{5-313}$$

Substituting $\mathbf{r}_C - \mathbf{r}_A$ from Eq. (5-312) into (5-311), we obtain

$$^{\mathcal{F}}\mathbf{v}_C = {}^{\mathcal{F}}\mathbf{v}_A + \dot{\theta}\mathbf{E}_z \times (-r\mathbf{E}_y) = {}^{\mathcal{F}}\mathbf{v}_A + r\dot{\theta}\mathbf{E}_x \tag{5-314}$$

Then, because $^{\mathcal{F}}\mathbf{v}_A = \mathbf{0}$, we obtain $^{\mathcal{F}}\mathbf{v}_C$ as

$$^{\mathcal{F}}\mathbf{v}_C = r\dot{\theta}\mathbf{E}_x \tag{5-315}$$

Next, using Eq. (2-517), the velocity of point B in reference frame $\mathcal{F}$ is given as

$$^{\mathcal{F}}\mathbf{v}_B = {}^{\mathcal{F}}\mathbf{v}_A + {}^{\mathcal{F}}\boldsymbol{\omega}^{\mathcal{R}} \times (\mathbf{r}_B - \mathbf{r}_A) \tag{5-316}$$

Now from the geometry we have

$$\mathbf{r}_B - \mathbf{r}_A = -2r\mathbf{E}_y \tag{5-317}$$

Substituting the expression for $\mathbf{r}_B - \mathbf{r}_A$ from Eq. (5-317) into (5-316) and using the fact that $^{\mathcal{F}}\mathbf{v}_A = \mathbf{0}$, we obtain $^{\mathcal{F}}\mathbf{v}_B$ as

$$^{\mathcal{F}}\mathbf{v}_B = \dot{\theta}\mathbf{E}_z \times (-2r\mathbf{E}_y) = 2r\dot{\theta}\mathbf{E}_x \tag{5-318}$$

Finally, using Eq. (2-524), the acceleration of point A in reference frame $\mathcal{F}$ is given as

$$^{\mathcal{F}}\mathbf{a}_A = {}^{\mathcal{F}}\mathbf{a}_C + {}^{\mathcal{F}}\boldsymbol{\alpha}^{\mathcal{R}} \times (\mathbf{r}_A - \mathbf{r}_C) + {}^{\mathcal{F}}\boldsymbol{\omega}^{\mathcal{R}} \times \left[{}^{\mathcal{F}}\boldsymbol{\omega}^{\mathcal{R}} \times (\mathbf{r}_A - \mathbf{r}_C)\right] \tag{5-319}$$

Using the expression for $^{\mathcal{F}}\mathbf{v}_C$ from Eq. (5-315), the acceleration of point C in reference frame $\mathcal{F}$ is given as

$$^{\mathcal{F}}\mathbf{a}_C = \frac{^{\mathcal{F}}d}{dt}\left({}^{\mathcal{F}}\mathbf{v}_C\right) = r\ddot{\theta}\mathbf{E}_x \tag{5-320}$$

Next, using the angular velocity of the disk from Eq. (5-310), we obtain the angular acceleration of the disk as

$$^{\mathcal{F}}\boldsymbol{\alpha}^{\mathcal{R}} = \frac{^{\mathcal{F}}d}{dt}\left({}^{\mathcal{F}}\boldsymbol{\omega}^{\mathcal{R}}\right) = \ddot{\theta}\mathbf{E}_z \tag{5-321}$$

Then, substituting the results of Eqs. (5-310), (5-313), (5-320), and (5-321) into (5-319), we obtain

$$^{\mathcal{F}}\mathbf{a}_A = r\ddot{\theta}\mathbf{E}_x + \ddot{\theta}\mathbf{E}_z \times r\mathbf{E}_y + \dot{\theta}\mathbf{E}_z \times (\dot{\theta}\mathbf{E}_z \times r\mathbf{E}_y) \tag{5-322}$$

Taking the vector products Eq. (5-322), we obtain

$$^{\mathcal{F}}\mathbf{a}_A = r\ddot{\theta}\mathbf{E}_x - r\ddot{\theta}\mathbf{E}_x - r\dot{\theta}^2\mathbf{E}_y = -r\dot{\theta}^2\mathbf{E}_y \tag{5-323}$$

It is seen that, while the velocity of point A is zero, its acceleration is *not* zero. Consequently, point A is *not* inertially fixed.

Kinetics

We have a few options in deciding on a reference point about which to apply Euler's laws to the disk. The first option for a reference point is the center of mass of the disk. While this is a perfectly reasonable choice, using the center of mass will require that we apply both of Euler's laws. Another option is to use point A. Because the motion is two-dimensional, choosing point A as the reference will only require that we apply Euler's 2nd law. Because the latter choice of point A will require only one balance law, we choose point A as the reference for the kinetic analysis.

Now, in applying Euler's 2nd law to the disk about point A, it is important to remember that point A is *not* fixed because its acceleration is not zero (see Eq. (5-323)). Therefore, it is necessary to apply the form of Euler's 2nd law given in Eq. (5-234) relative to an arbitrary reference point. Substituting the appropriate values for this example into Eq. (5-234), we have

$$\mathbf{M}_A - (\mathbf{r}_C - \mathbf{r}_A) \times m^{\mathcal{F}}\mathbf{a}_A = \frac{^{\mathcal{F}}d}{dt}\left(^{\mathcal{F}}\mathbf{H}_A\right) \tag{5-324}$$

The free body diagram of the disk is shown in Fig. 5-11.

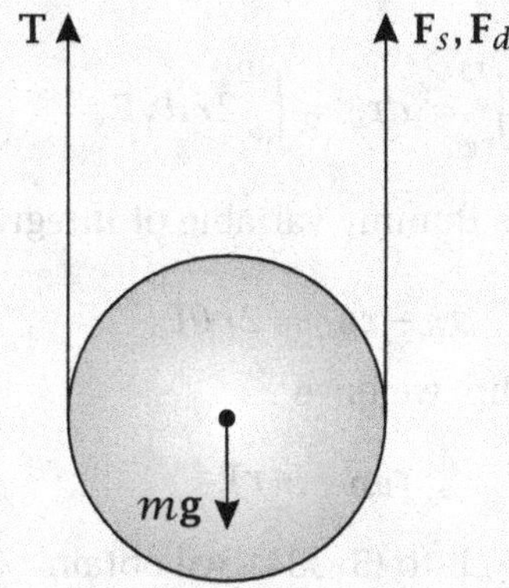

Figure 5-11 Free body diagram for Example 5-5.

Using Fig. 5-11, the forces acting the disk are given as

$$\begin{aligned} \mathbf{T} &= \text{Force of cord on disk} \\ \mathbf{F}_s &= \text{Spring force} \\ \mathbf{F}_d &= \text{Force of damper} \\ m\mathbf{g} &= \text{Force of gravity} \end{aligned}$$

Now we have

$$\mathbf{T} = T\mathbf{E}_x \tag{5-325}$$

$$\mathbf{F}_s = -K(\ell - \ell_0)\mathbf{u}_s \tag{5-326}$$

$$\mathbf{F}_d = -c\|\mathbf{v}_B\|^3 \frac{\mathbf{v}_B}{\|\mathbf{v}_B\|} \tag{5-327}$$

$$m\mathbf{g} = mg\mathbf{E}_x \tag{5-328}$$

Because the force **T** passes through point A, the moment applied to the disk relative to point A is due entirely to the forces $\mathbf{F}_s$, $\mathbf{F}_d$, and $m\mathbf{g}$, i.e.,

$$\mathbf{M}_A = (\mathbf{r}_s - \mathbf{r}_A) \times \mathbf{F}_s + (\mathbf{r}_d - \mathbf{r}_A) \times \mathbf{F}_d + (\mathbf{r}_g - \mathbf{r}_A) \times m\mathbf{g} \tag{5-329}$$

where $\mathbf{r}_s$, $\mathbf{r}_d$, and $\mathbf{r}_g$ are the positions at which the forces $\mathbf{F}_s$, $\mathbf{F}_d$, and $m\mathbf{g}$ act, respectively. From the geometry we have

$$\begin{aligned} \mathbf{r}_s &= \mathbf{r}_B \\ \mathbf{r}_d &= \mathbf{r}_B \\ \mathbf{r}_g &= \mathbf{r}_C \end{aligned} \tag{5-330}$$

Now, in order to obtain $\mathbf{r}_B$, we need to integrate ${}^{\mathcal{F}}\mathbf{v}_B$. First, recall ${}^{\mathcal{F}}\mathbf{v}_B$ from Eq. (5-318) as

$${}^{\mathcal{F}}\mathbf{v}_B = 2r\dot{\theta}\mathbf{E}_x \equiv \frac{{}^{\mathcal{F}}d\mathbf{r}_B}{dt} \tag{5-331}$$

Separating variables in Eq. (5-331), we obtain

$${}^{\mathcal{F}}d\mathbf{r}_B = 2r\dot{\theta}\mathbf{E}_x dt = 2rd\theta\mathbf{E}_x \tag{5-332}$$

Integrating Eq. (5-332) gives

$$\int_{\mathbf{r}_{B,0}}^{\mathbf{r}_B} {}^{\mathcal{F}}d\mathbf{r}_B = \int_0^\theta 2rd\nu\mathbf{E}_x \tag{5-333}$$

where $\mathbf{r}_{B,0} = \mathbf{r}_B(\theta = 0)$ and ν is a dummy variable of integration. From Eq. (5-333), we obtain

$$\mathbf{r}_B - \mathbf{r}_{B,0} = 2r\theta\mathbf{E}_x \tag{5-334}$$

Because the disk rolls without slip, we have

$$\mathbf{r}_{B,0} = -r\mathbf{E}_y \tag{5-335}$$

Substituting $\mathbf{r}_{B,0}$ from Eq. (5-335) into (5-334), we obtain

$$\mathbf{r}_B + r\mathbf{E}_y = 2r\theta\mathbf{E}_x \tag{5-336}$$

which implies that

$$\mathbf{r}_B = -r\mathbf{E}_y + 2r\theta\mathbf{E}_x \tag{5-337}$$

Similarly, from Eq. (5-315) we have

$${}^{\mathcal{F}}\mathbf{v}_C = \frac{{}^{\mathcal{F}}d}{dt}(\mathbf{r}_C) = r\dot{\theta}\mathbf{E}_x \tag{5-338}$$

Separating variables, in Eq. (5-338), we obtain

$${}^{\mathcal{F}}d\mathbf{r}_C = r\dot{\theta}\mathbf{E}_x dt = rd\theta\mathbf{E}_x \tag{5-339}$$

Integrating Eq. (5-339) gives

$$\int_{\mathbf{r}_{C,0}}^{\mathbf{r}_C} {}^{\mathcal{F}}d\mathbf{r}_C = \int_0^\theta rd\nu\mathbf{E}_x \tag{5-340}$$

where $\mathbf{r}_{C,0} = \mathbf{r}_C(\theta = 0)$ and ν is a dummy variable of integration. From Eq. (5-340), we obtain

$$\mathbf{r}_C - \mathbf{r}_{C,0} = r\theta\mathbf{E}_x \tag{5-341}$$

Now, because the disk rolls without slip, $\mathbf{r}_{C,0} = \mathbf{0}$. Therefore,

$$\mathbf{r}_C = r\theta\mathbf{E}_x \tag{5-342}$$

Consequently, the locations of the forces $\mathbf{F}_s$, $\mathbf{F}_d$, and $m\mathbf{g}$ are given, respectively, as

$$\begin{aligned} \mathbf{r}_s &= 2r\theta\mathbf{E}_x - r\mathbf{E}_y \\ \mathbf{r}_d &= 2r\theta\mathbf{E}_x - r\mathbf{E}_y \\ \mathbf{r}_g &= r\theta\mathbf{E}_x \end{aligned} \tag{5-343}$$

Next, from the geometry of the problem, the position of point A is given as

$$\mathbf{r}_A = \mathbf{r}_C + r\mathbf{E}_y = r\theta\mathbf{E}_x + r\mathbf{E}_y \tag{5-344}$$

Subtracting $\mathbf{r}_A$ in Eq. (5-344) from $\mathbf{r}_s$, $\mathbf{r}_d$, and $\mathbf{r}_g$, respectively, in Eq. (5-343), we obtain

$$\begin{aligned} \mathbf{r}_s - \mathbf{r}_A &= r\theta\mathbf{E}_x - 2r\mathbf{E}_y \\ \mathbf{r}_d - \mathbf{r}_A &= r\theta\mathbf{E}_x - 2r\mathbf{E}_y \\ \mathbf{r}_g - \mathbf{r}_A &= -r\mathbf{E}_y \end{aligned} \tag{5-345}$$

Now the position of point P is given in terms of the basis $\{\mathbf{E}_x, \mathbf{E}_y, \mathbf{E}_z\}$ as

$$\mathbf{r}_P = -L\mathbf{E}_x - r\mathbf{E}_y \tag{5-346}$$

Substituting $\mathbf{r}_B$ from Eq. (5-337) and $\mathbf{r}_P$ from Eq. (5-346), we obtain

$$\mathbf{r}_B - \mathbf{r}_P = 2r\theta\mathbf{E}_x - r\mathbf{E}_y - (-L\mathbf{E}_x - r\mathbf{E}_y) = (2r\theta + L)\mathbf{E}_x \tag{5-347}$$

The length of the spring is then given as

$$\ell = \|\mathbf{r}_B - \mathbf{r}_P\| = \|(2r\theta + L)\mathbf{E}_x\| = 2r\theta + L \tag{5-348}$$

Furthermore, the unit vector in the direction along the spring is given as

$$\mathbf{u}_s = \frac{\mathbf{r}_B - \mathbf{r}_P}{\|\mathbf{r}_B - \mathbf{r}_P\|} = \frac{(2r\theta + L)\mathbf{E}_x}{2r\theta + L} = \mathbf{E}_x \tag{5-349}$$

Then, substituting the results of Eqs. (5-348) and (5-349) into (5-326) and using the fact that $\ell_0 = L$, we obtain the force in the linear spring as

$$\mathbf{F}_s = -K[(2r\theta + L) - L]\mathbf{E}_x = -2Kr\theta\mathbf{E}_x \tag{5-350}$$

Furthermore, using the expression for ${}^{\mathcal{F}}\mathbf{v}_B$ from Eq. (5-318), the force of the damper is obtained as

$$\begin{aligned} \mathbf{F}_d &= -c\|2r\dot{\theta}\mathbf{E}_x\|^3 \frac{2r\dot{\theta}\mathbf{E}_x}{\|2r\dot{\theta}\mathbf{E}_x\|} \\ &= -c(2r\dot{\theta})^3\mathbf{E}_x \\ &= -8cr^3\dot{\theta}^3\mathbf{E}_x \end{aligned} \tag{5-351}$$

Then, using the values of $\mathbf{r}_s - \mathbf{r}_A$, $\mathbf{r}_d - \mathbf{r}_A$, and $\mathbf{r}_g - \mathbf{r}_A$ from Eq. (5-345), the moment applied about the point A is obtained as

$$\begin{aligned}\mathbf{M}_A &= (r\theta\mathbf{E}_x - 2r\mathbf{E}_y) \times [-2Kr\theta\mathbf{E}_x] \\ &+ (r\theta\mathbf{E}_x - 2r\mathbf{E}_y) \times \left[-8cr^3\dot{\theta}^3\mathbf{E}_x\right] - r\mathbf{E}_y \times mg\mathbf{E}_x\end{aligned} \tag{5-352}$$

Noting that $\mathbf{E}_x \times \mathbf{E}_x = \mathbf{0}$ and $\mathbf{E}_y \times \mathbf{E}_x = -\mathbf{E}_z$, Eq. (5-352) simplifies to

$$\mathbf{M}_A = \left[-4Kr^2\theta - 16cr^4\dot{\theta}^3 + mgr\right]\mathbf{E}_z \tag{5-353}$$

Next, looking at Eq. (5-324), it is seen that it is necessary to compute the inertial moment, i.e., we need to compute

$$-(\mathbf{r}_C - \mathbf{r}_A) \times m\,{}^{\mathcal{F}}\mathbf{a}_A \tag{5-354}$$

Substituting $\mathbf{r}_C - \mathbf{r}_A$ and ${}^{\mathcal{F}}\mathbf{a}_A$ from Eqs. (5-312) and (5-323), respectively, we have

$$-(\mathbf{r}_C - \mathbf{r}_A) \times m\,{}^{\mathcal{F}}\mathbf{a}_A = -(-r\mathbf{E}_y) \times m(-r\dot{\theta}^2\mathbf{E}_y) = \mathbf{0} \tag{5-355}$$

It is interesting to note that, although point A is not a fixed point, for this particular problem, the inertial moment is *zero*. As a result, for the *particular* case of a uniform disk rolling without slip along a flat surface, we can apply the fixed point form of Euler's 2nd law as given in Eq. (5-226).

Lastly, it is necessary to compute the rate of change of angular momentum of the disk about point A. Using Eq. (5-24), we have the angular momentum about point A as

$${}^{\mathcal{F}}\mathbf{H}_A = {}^{\mathcal{F}}\mathbf{H}_C + (\mathbf{r}_A - \mathbf{r}_C) \times m({}^{\mathcal{F}}\mathbf{v}_A - {}^{\mathcal{F}}\mathbf{v}_C) \tag{5-356}$$

We have ${}^{\mathcal{F}}\mathbf{H}_C$ as

$${}^{\mathcal{F}}\mathbf{H}_C = \bar{\mathbf{I}}^{\mathcal{R}} \cdot {}^{\mathcal{F}}\boldsymbol{\omega}^{\mathcal{R}} = \mathbf{I}_C^{\mathcal{R}} \cdot {}^{\mathcal{F}}\boldsymbol{\omega}^{\mathcal{R}} \tag{5-357}$$

Because $\{\mathbf{E}_x, \mathbf{E}_y, \mathbf{E}_z\}$ is a principal-axis basis, the moment of inertia tensor $\mathbf{I}_C^{\mathcal{R}}$ is given as

$$\mathbf{I}_C^{\mathcal{R}} = \bar{I}_{xx}\mathbf{E}_x \otimes \mathbf{E}_x + \bar{I}_{yy}\mathbf{E}_y \otimes \mathbf{E}_y + \bar{I}_{zz}\mathbf{E}_z \otimes \mathbf{E}_z \tag{5-358}$$

Then, substituting $\mathbf{I}_C^{\mathcal{R}}$ from Eq. (5-358) and ${}^{\mathcal{F}}\boldsymbol{\omega}^{\mathcal{R}}$ from Eq. (5-310), we obtain ${}^{\mathcal{F}}\mathbf{H}_C$ as

$${}^{\mathcal{F}}\mathbf{H}_C = \bar{I}_{zz}\dot{\theta}\mathbf{E}_z \tag{5-359}$$

Now we note that $\bar{I}_{zz}$ is given as

$$\bar{I}_{zz} = \frac{mr^2}{2} \tag{5-360}$$

Consequently, we obtain ${}^{\mathcal{F}}\mathbf{H}_C$ as

$${}^{\mathcal{F}}\mathbf{H}_C = \frac{mr^2}{2}\dot{\theta}\mathbf{E}_z \tag{5-361}$$

Then, using Eqs. (5-310), (5-313), and (5-315), and noting that ${}^{\mathcal{F}}\mathbf{v}_A = \mathbf{0}$, the second term of Eq. (5-356) is obtained as

$$(\mathbf{r}_A - \mathbf{r}_C) \times m({}^{\mathcal{F}}\mathbf{v}_A - {}^{\mathcal{F}}\mathbf{v}_C) = r\mathbf{E}_y \times m(-r\dot{\theta}\mathbf{E}_x) = mr^2\dot{\theta}\mathbf{E}_z \tag{5-362}$$

Adding the results of Eqs. (5-361) and (5-362), we obtain ${}^{\mathcal{F}}\mathbf{H}_A$ as

$$ {}^{\mathcal{F}}\mathbf{H}_A = \frac{mr^2}{2}\dot{\theta}\mathbf{E}_z + mr^2\dot{\theta}\mathbf{E}_z = \frac{3}{2}mr^2\dot{\theta}\mathbf{E}_z \tag{5-363} $$

Computing the rate of change of ${}^{\mathcal{F}}\mathbf{H}_A$ in reference frame $\mathcal{F}$, we obtain

$$ \frac{{}^{\mathcal{F}}d}{dt}\left({}^{\mathcal{F}}\mathbf{H}_A\right) = \frac{3}{2}mr^2\ddot{\theta}\mathbf{E}_z \tag{5-364} $$

Then, because the inertial moment as given in Eq. (5-355) is zero, we can set $\mathbf{M}_A$ from Eq. (5-353) equal to ${}^{\mathcal{F}}d({}^{\mathcal{F}}\mathbf{H}_A)/dt$ from Eq. (5-364) to give

$$ (-4Kr^2\theta\mathbf{E}_x - 16cr^4\dot{\theta}^3 + mgr)\mathbf{E}_z = \tfrac{3}{2}mr^2\ddot{\theta}\mathbf{E}_z \tag{5-365} $$

Rearranging Eq. (5-365) and dropping the dependence on $\mathbf{E}_z$, we obtain the differential equation of motion as

$$ \tfrac{3}{2}mr^2\ddot{\theta} + 16cr^4\dot{\theta}^3 + 4Kr^2\theta - mgr = 0 \tag{5-366} $$

■

Example 5-6

A uniform sphere of mass m and radius r rolls without slip along the surface of a fixed sphere, also of radius r, as shown in Fig. 5-12. The angle ϕ describes the rotation of the moving sphere from the vertical direction while the angle θ describes the location of the center of mass of the sphere relative to the vertical. Knowing that gravity acts vertically downward, determine the differential equation of motion for the sphere in terms of the angle θ.

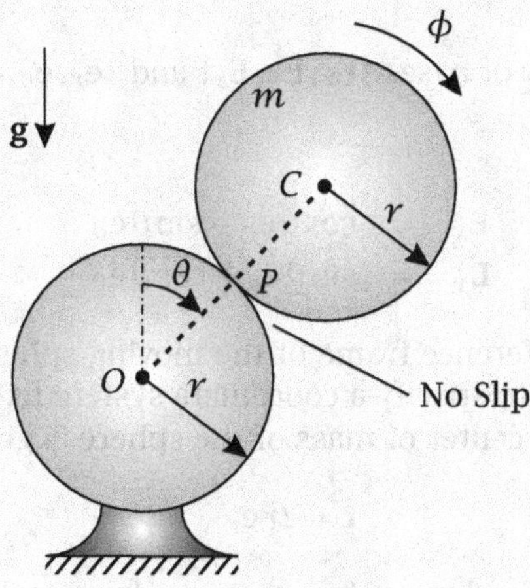

Figure 5-12 Sphere rolling on fixed spherical surface.

Solution to Example 5–6

Kinematics

First, let $\mathcal{F}$ be a reference frame that is fixed to the ground. Then, choose the following coordinate system fixed in reference frame $\mathcal{F}$:

$$\begin{array}{lcl} & \text{Origin at } O & \\ \mathbf{E}_x & = & \text{Along } OP \text{ when } \theta = 0 \\ \mathbf{E}_z & = & \text{Into page} \\ \mathbf{E}_y & = & \mathbf{E}_z \times \mathbf{E}_x \end{array}$$

Next, let $\mathcal{A}$ be a reference frame fixed to the direction of OC. Then, choose the following coordinate system fixed in reference frame $\mathcal{A}$:

$$\begin{array}{lcl} & \text{Origin at } O & \\ \mathbf{e}_r & = & \text{Along } OP \\ \mathbf{e}_z & = & \text{Into page} \\ \mathbf{e}_\theta & = & \mathbf{E}_z \times \mathbf{e}_r \end{array}$$

The geometry of the bases $\{\mathbf{E}_x, \mathbf{E}_y, \mathbf{E}_z\}$ and $\{\mathbf{e}_r, \mathbf{e}_\theta, \mathbf{e}_z\}$ is shown in Fig. 5–13.

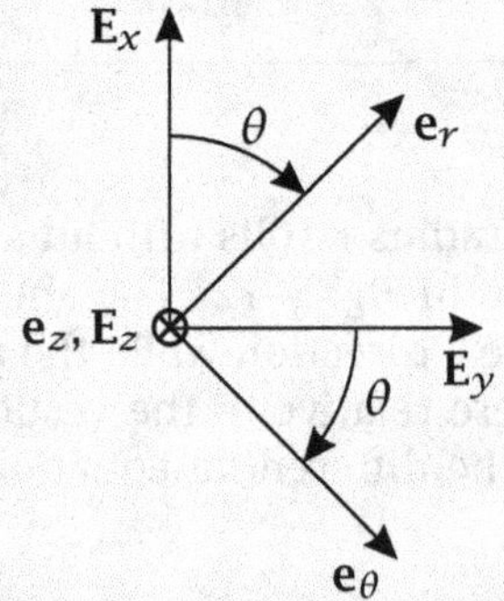

Figure 5–13 Geometry of bases $\{\mathbf{E}_x, \mathbf{E}_y, \mathbf{E}_z\}$ and $\{\mathbf{e}_r, \mathbf{e}_\theta, \mathbf{e}_z\}$ for Example 5–6.

Using Fig. 5–13, we have

$$\mathbf{E}_x = \cos\theta\,\mathbf{e}_r - \sin\theta\,\mathbf{e}_\theta \qquad (5\text{–}367)$$

$$\mathbf{E}_y = \sin\theta\,\mathbf{e}_r + \cos\theta\,\mathbf{e}_\theta \qquad (5\text{–}368)$$

Finally, let $\mathcal{R}$ denote the reference frame of the moving sphere. We note that for this problem it is not necessary to specify a coordinate system fixed in $\mathcal{R}$.

Now, the position of the center of mass of the sphere is given as

$$\bar{\mathbf{r}} = 2r\mathbf{e}_r \qquad (5\text{–}369)$$

Also, the angular velocity of reference frame $\mathcal{A}$ in reference frame $\mathcal{F}$ is given as

$${}^{\mathcal{F}}\boldsymbol{\omega}^{\mathcal{A}} = \dot{\theta}\mathbf{e}_z \qquad (5\text{–}370)$$

Furthermore, because the angle ϕ is measured relative to the vertical direction, the angular velocity of reference frame $\mathcal{R}$ in reference frame $\mathcal{F}$ is given as

$$^{\mathcal{F}}\boldsymbol{\omega}^{\mathcal{R}} = \dot{\phi}\mathbf{e}_z \tag{5-371}$$

Applying the rate of change transport theorem of Eq. (2-128) on page 47 between reference frames $\mathcal{A}$ and $\mathcal{F}$, the velocity of the center of mass of the sphere is

$$^{\mathcal{F}}\bar{\mathbf{v}} = \frac{^{\mathcal{F}}d\bar{\mathbf{r}}}{dt} = \frac{^{\mathcal{A}}d\bar{\mathbf{r}}}{dt} + {}^{\mathcal{F}}\boldsymbol{\omega}^{\mathcal{A}} \times \bar{\mathbf{r}} \tag{5-372}$$

Now we have

$$\frac{^{\mathcal{A}}d\bar{\mathbf{r}}}{dt} = \mathbf{0} \tag{5-373}$$

$$^{\mathcal{F}}\boldsymbol{\omega}^{\mathcal{A}} \times \bar{\mathbf{r}} = \dot{\theta}\mathbf{e}_z \times (2r\mathbf{e}_r) = 2r\dot{\theta}\mathbf{e}_\theta \tag{5-374}$$

Adding Eqs. (5-373) and (5-374), we obtain

$$^{\mathcal{F}}\bar{\mathbf{v}} = 2r\dot{\theta}\mathbf{e}_\theta \tag{5-375}$$

Applying the rate of change transport theorem of Eq. (2-128) on page 47 to $^{\mathcal{F}}\bar{\mathbf{v}}$ between reference frames $\mathcal{A}$ and $\mathcal{F}$, the acceleration of the center of mass of the sphere is given as

$$^{\mathcal{F}}\bar{\mathbf{a}} = \frac{^{\mathcal{F}}d}{dt}\left(^{\mathcal{F}}\bar{\mathbf{v}}\right) = \frac{^{\mathcal{A}}d}{dt}\left(^{\mathcal{F}}\bar{\mathbf{v}}\right) + {}^{\mathcal{F}}\boldsymbol{\omega}^{\mathcal{A}} \times {}^{\mathcal{F}}\bar{\mathbf{v}} \tag{5-376}$$

Now we have

$$\frac{^{\mathcal{A}}d}{dt}\left(^{\mathcal{F}}\bar{\mathbf{v}}\right) = 2r\ddot{\theta}\mathbf{e}_\theta \tag{5-377}$$

$$^{\mathcal{F}}\boldsymbol{\omega}^{\mathcal{A}} \times {}^{\mathcal{F}}\bar{\mathbf{v}} = \dot{\theta}\mathbf{e}_z \times (2r\dot{\theta}\mathbf{e}_\theta) = -2r\dot{\theta}^2\mathbf{e}_r \tag{5-378}$$

Adding Eqs. (5-377) and (5-378), we obtain the acceleration of the center of mass in reference frame $\mathcal{F}$ as

$$^{\mathcal{F}}\bar{\mathbf{a}} = -2r\dot{\theta}^2\mathbf{e}_r + 2r\ddot{\theta}\mathbf{e}_\theta \tag{5-379}$$

A second expression for $^{\mathcal{F}}\bar{\mathbf{v}}$ is obtained as follows. Observing that points C and P are fixed in the sphere, from Eq. (2-517) we have

$$^{\mathcal{F}}\bar{\mathbf{v}} = {}^{\mathcal{F}}\mathbf{v}_P^{\mathcal{R}} + {}^{\mathcal{F}}\boldsymbol{\omega}^{\mathcal{R}} \times (\bar{\mathbf{r}} - \mathbf{r}_P) \tag{5-380}$$

where

$$\bar{\mathbf{r}} - \mathbf{r}_P = r\mathbf{e}_r \tag{5-381}$$

Then, applying the rolling condition of Eq. (2-531) on page 109, we have

$$^{\mathcal{F}}\mathbf{v}_P^{\mathcal{R}} = \mathbf{0} \tag{5-382}$$

Consequently,

$$^{\mathcal{F}}\bar{\mathbf{v}} = \dot{\phi}\mathbf{e}_z \times r\mathbf{e}_r = r\dot{\phi}\mathbf{e}_\theta \tag{5-383}$$

Setting the results of Eqs. (5-375) and (5-383) equal, we obtain

$$2r\dot{\theta} = r\dot{\phi} \tag{5-384}$$

Equation (5-384) implies that

$$\dot{\phi} = 2\dot{\theta} \tag{5-385}$$

Therefore, the angular velocity of the upper sphere in reference frame $\mathcal{F}$ is given as

$$^{\mathcal{F}}\boldsymbol{\omega}^{\mathcal{R}} = 2\dot{\theta}\mathbf{e}_z \tag{5-386}$$

Finally, computing the rate of change of $\dot{\phi}$ in Eq. (5-385), we obtain

$$\ddot{\phi} = 2\ddot{\theta} \tag{5-387}$$

Kinetics

This problem will be solved by applying both of Euler's laws using the center of mass of the sphere as the reference point. The free body diagram of the sphere is shown in Fig. 5-14.

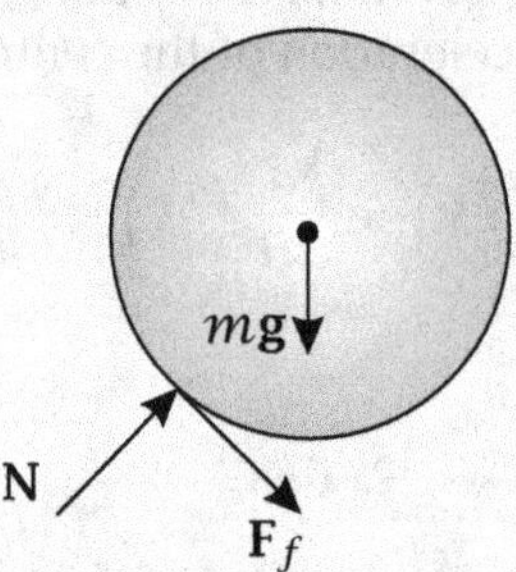

Figure 5-14 Free body diagram for Example 5-6.

Using Fig. 5-14, we see that the following forces act on the sphere:

$$\begin{array}{lcl} \mathbf{N} & = & \text{Normal force} \\ \mathbf{F}_f & = & \text{Rolling force} \\ m\mathbf{g} & = & \text{Force of gravity} \end{array}$$

In terms of the basis $\{\mathbf{e}_r, \mathbf{e}_\theta, \mathbf{e}_z\}$, these forces are resolved as

$$\mathbf{N} = N\mathbf{e}_r \tag{5-388}$$

$$\mathbf{F}_f = F_f\mathbf{e}_\theta \tag{5-389}$$

$$m\mathbf{g} = -mg\mathbf{E}_x \tag{5-390}$$

Using the expression for $\mathbf{E}_x$ from Eq. (5-367), the force of gravity can be written as

$$m\mathbf{g} = -mg(\cos\theta\,\mathbf{e}_r - \sin\theta\,\mathbf{e}_\theta) = -mg\cos\theta\,\mathbf{e}_r + mg\sin\theta\,\mathbf{e}_\theta \tag{5-391}$$

Application of Euler's* 1st *Law to Center of Mass of Sphere

Using Eqs. (5-388), (5-389), and (5-391), the resultant force acting the sphere is given as

$$\begin{aligned}\mathbf{F} &= N\mathbf{e}_r + F_f\mathbf{e}_\theta - mg\cos\theta\,\mathbf{e}_r + mg\sin\theta\,\mathbf{e}_\theta \\ &= (N - mg\cos\theta)\mathbf{e}_r + (F_f + mg\sin\theta)\mathbf{e}_\theta\end{aligned} \tag{5-392}$$

Applying Euler's 1st law by setting $\mathbf{F}$ in Eq. (5-392) equal to $m^{\mathcal{F}}\bar{\mathbf{a}}$ using ${}^{\mathcal{F}}\bar{\mathbf{a}}$ from Eq. (5-379), we obtain

$$(N - mg\cos\theta)\mathbf{e}_r + (F_f + mg\sin\theta)\mathbf{e}_\theta = -2mr\dot{\theta}^2\mathbf{e}_r + 2mr\ddot{\theta}\mathbf{e}_\theta \tag{5-393}$$

Equation (5-393) yields the following two scalar equations:

$$N - mg\cos\theta = -2mr\dot{\theta}^2 \tag{5-394}$$

$$F_f + mg\sin\theta = 2mr\ddot{\theta} \tag{5-395}$$

Application of Euler's* 2nd *Law Relative to Center of Mass of Sphere

It is seen from Fig. 5-14 that the forces $\mathbf{N}$ and mg pass through the center of mass of the sphere while the force $\mathbf{F}_f$ acts at point P. Consequently, the moment applied to the sphere relative to the center of mass of the sphere is given as

$$\bar{\mathbf{M}} = (\mathbf{r}_P - \bar{\mathbf{r}}) \times \mathbf{F}_f \tag{5-396}$$

Now from Eq. (5-381) we have

$$\mathbf{r}_P - \bar{\mathbf{r}} = -r\mathbf{e}_r \tag{5-397}$$

Therefore,

$$\bar{\mathbf{M}} = -r\mathbf{e}_r \times F_f\mathbf{e}_\theta = -rF_f\mathbf{e}_z \tag{5-398}$$

Next, the angular momentum of the sphere relative to the center of mass of the sphere is given as

$${}^{\mathcal{F}}\bar{\mathbf{H}} = \bar{\mathbf{I}}^{\mathcal{R}} \cdot {}^{\mathcal{F}}\boldsymbol{\omega}^{\mathcal{R}} \tag{5-399}$$

Then, since $\{\mathbf{e}_r, \mathbf{e}_\theta, \mathbf{e}_z\}$ is a principal-axis basis, the moment of inertia tensor $\bar{\mathbf{I}}^{\mathcal{R}}$ is given as

$$\bar{\mathbf{I}}^{\mathcal{R}} = \bar{I}_{rr}\mathbf{e}_r \otimes \mathbf{e}_r + \bar{I}_{\theta\theta}\mathbf{e}_\theta \otimes \mathbf{e}_r + \bar{I}_{zz}\mathbf{e}_z \otimes \mathbf{e}_z \tag{5-400}$$

Substituting $\bar{\mathbf{I}}^{\mathcal{R}}$ from Eq. (5-400) and ${}^{\mathcal{F}}\boldsymbol{\omega}^{\mathcal{R}}$ from Eq. (5-386), we obtain

$${}^{\mathcal{F}}\bar{\mathbf{H}} = \bar{I}_{zz}2\dot{\theta}\mathbf{e}_z \tag{5-401}$$

Using the fact that $\bar{I}_{zz} = 2mr^2/5$, we have

$${}^{\mathcal{F}}\bar{\mathbf{H}} = \frac{2}{5}mr^2(2\dot{\theta}\mathbf{e}_z) = \frac{4}{5}mr^2\dot{\theta}\mathbf{e}_z \tag{5-402}$$

Computing the rate of change of ${}^{\mathcal{F}}\bar{\mathbf{H}}$ in reference frame $\mathcal{F}$ gives

$$\frac{{}^{\mathcal{F}}d}{dt}\left({}^{\mathcal{F}}\bar{\mathbf{H}}\right) = \frac{4}{5}mr^2\ddot{\theta}\mathbf{e}_z \tag{5-403}$$

Applying Euler's 2nd law relative to the center of mass of the sphere by setting ${}^{\mathcal{F}}d({}^{\mathcal{F}}\bar{\mathbf{H}})/dt$ in Eq. (5-403) equal to $\bar{\mathbf{M}}$ in Eq. (5-398), we obtain

$$-rF_f = \frac{4}{5}mr^2\ddot{\theta} \tag{5-404}$$

Solving Eq. (5-404) for F_f gives

$$F_f = -\frac{4}{5}mr\ddot{\theta} \tag{5-405}$$

Differential Equation of Motion

Substituting Eq. (5-405) into (5-395), we obtain

$$-\tfrac{4}{5}mr^2\ddot{\theta} + mg\sin\theta = 2mr\ddot{\theta} \tag{5-406}$$

Simplifying Eq. (5-406), we obtain the differential equation of motion for the sphere as

$$\ddot{\theta} - \frac{5g}{14r}\sin\theta = 0 \tag{5-407}$$

■

Example 5–7

A uniform slender rod of mass m and length l is hinged at one of its ends to a point O, where point O is located at one end of a massless assembly of a vertical shaft as shown in Fig. 5-15. The shaft assembly rotates relative to the ground with constant angular velocity Ω. Assuming that θ describes the angle formed by the shaft with the vertically downward direction and that gravity acts vertically downward, determine the differential equation of motion for the rod in terms of the angle θ.

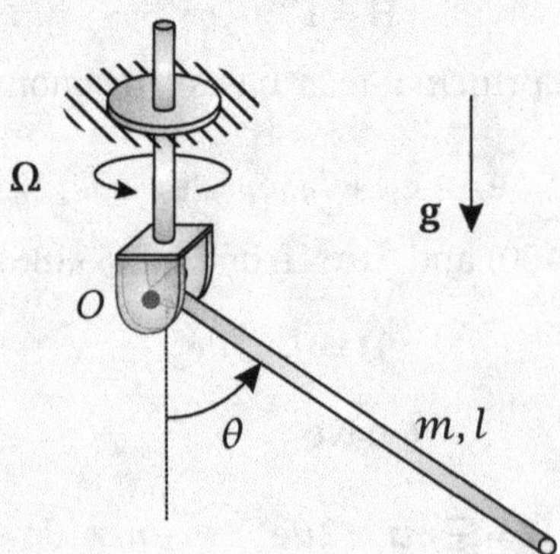

Figure 5–15 Rod rotating on vertical shaft assembly.

Solution to Example 5-7

Kinematics

Let $\mathcal{F}$ be a fixed reference frame. Then, choose the following coordinate system fixed in reference frame $\mathcal{F}$:

$$\begin{array}{ccl} & \text{Origin at } O & \\ \mathbf{E}_y & = & \text{Along } \boldsymbol{\Omega} \\ \mathbf{E}_z & = & \text{Out of page at } t = 0 \\ \mathbf{E}_x & = & \mathbf{E}_y \times \mathbf{E}_z \end{array}$$

Next, let $\mathcal{A}$ be a reference frame fixed to the vertical shaft. Then, choose the following coordinate system fixed in reference frame $\mathcal{A}$:

$$\begin{array}{ccl} & \text{Origin at } O & \\ \mathbf{u}_y & = & \text{Along } \boldsymbol{\Omega} \\ \mathbf{u}_z & = & \text{Orthogonal to plane of} \\ & & \text{shaft and rod} \\ \mathbf{u}_x & = & \mathbf{u}_y \times \mathbf{u}_z \end{array}$$

Finally, let $\mathcal{R}$ be a reference frame fixed to the rod. Then, choose the following coordinate system fixed in reference frame $\mathcal{R}$:

$$\begin{array}{ccl} & \text{Origin at } O & \\ \mathbf{e}_r & = & \text{Along Rod} \\ \mathbf{e}_z & = & \mathbf{u}_z \\ \mathbf{e}_\theta & = & \mathbf{e}_z \times \mathbf{e}_r \end{array}$$

The geometry of the bases $\{\mathbf{u}_x, \mathbf{u}_y, \mathbf{u}_z\}$ and $\{\mathbf{e}_r, \mathbf{e}_\theta, \mathbf{e}_z\}$ is shown in Figs. 5-16 and 5-17.

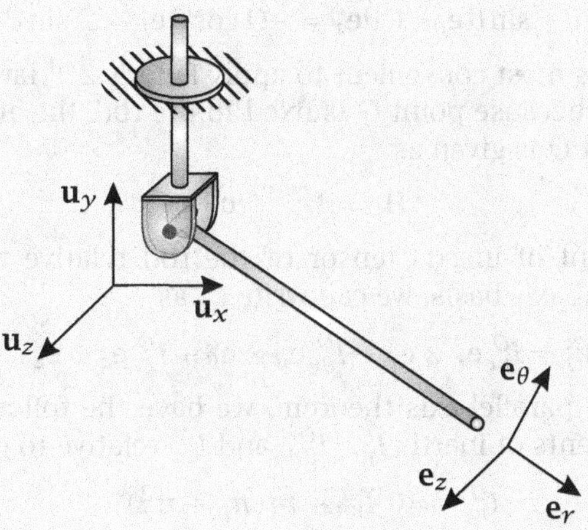

Figure 5-16 Three-dimensional view of bases $\{\mathbf{u}_x, \mathbf{u}_y, \mathbf{u}_z\}$ and $\{\mathbf{e}_r, \mathbf{e}_\theta, \mathbf{e}_z\}$ for Example 5-7.

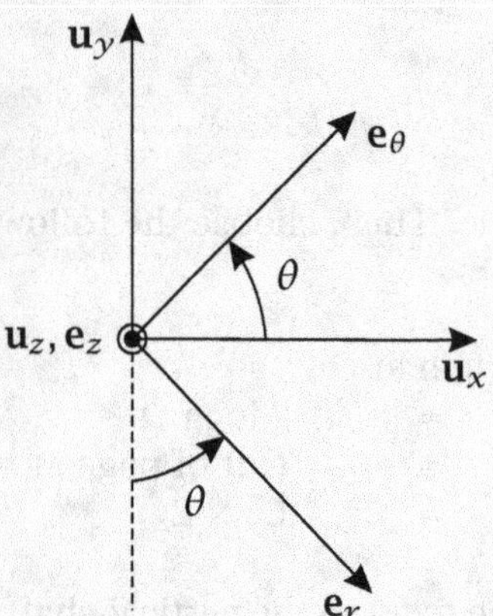

Figure 5-17 Two-dimensional view of bases $\{\mathbf{u}_x, \mathbf{u}_y, \mathbf{u}_z\}$ and $\{\mathbf{e}_r, \mathbf{e}_\theta, \mathbf{e}_z\}$ for Example 5-7.

Using Fig. 5-17, the relationship between the bases $\{\mathbf{u}_x, \mathbf{u}_y, \mathbf{u}_z\}$ and $\{\mathbf{e}_r, \mathbf{e}_\theta, \mathbf{e}_z\}$ is given as

$$\begin{aligned} \mathbf{u}_x &= \sin\theta\,\mathbf{e}_r + \cos\theta\,\mathbf{e}_\theta \\ \mathbf{u}_y &= -\cos\theta\,\mathbf{e}_r + \sin\theta\,\mathbf{e}_\theta \end{aligned} \tag{5-408}$$

Using the reference frames chosen above, the angular velocity of $\mathcal{A}$ in $\mathcal{F}$ and the angular velocity of $\mathcal{R}$ in $\mathcal{A}$, are given, respectively, as

$$ {}^{\mathcal{F}}\boldsymbol{\omega}^{\mathcal{A}} = \boldsymbol{\Omega} = \Omega\mathbf{u}_y \tag{5-409}$$

$$ {}^{\mathcal{A}}\boldsymbol{\omega}^{\mathcal{R}} = \dot{\theta}\mathbf{e}_z \tag{5-410}$$

Then, applying the angular velocity addition theorem using the results of Eqs. (5-409) and (5-410), we obtain the angular velocity of reference frame $\mathcal{R}$ in reference frame $\mathcal{F}$ as

$$ {}^{\mathcal{F}}\boldsymbol{\omega}^{\mathcal{R}} = {}^{\mathcal{F}}\boldsymbol{\omega}^{\mathcal{A}} + {}^{\mathcal{A}}\boldsymbol{\omega}^{\mathcal{R}} = \Omega\mathbf{u}_y + \dot{\theta}\mathbf{e}_z \tag{5-411}$$

Substituting the expression for $\mathbf{e}_y$ from Eq. (5-408) into (5-411), we obtain

$$ {}^{\mathcal{F}}\boldsymbol{\omega}^{\mathcal{R}} = \Omega(-\cos\theta\,\mathbf{e}_r + \sin\theta\,\mathbf{e}_\theta) + \dot{\theta}\mathbf{e}_z = -\Omega\cos\theta\,\mathbf{e}_r + \Omega\sin\theta\,\mathbf{e}_\theta + \dot{\theta}\mathbf{e}_z \tag{5-412}$$

For this problem it is most convenient to apply Euler's 2nd law relative to the fixed hinge at point O. First, because point O is fixed to the rod, the angular momentum of the rod relative to point O is given as

$$ {}^{\mathcal{F}}\mathbf{H}_O = \mathbf{I}_O^{\mathcal{R}} \cdot {}^{\mathcal{F}}\boldsymbol{\omega}^{\mathcal{R}} \tag{5-413}$$

where $\mathbf{I}_O^{\mathcal{R}}$ is the moment of inertia tensor of the rod relative to point O. Because $\{\mathbf{e}_r, \mathbf{e}_\theta, \mathbf{e}_z\}$ is a principal-axis basis, we can write $\mathbf{I}_O^{\mathcal{R}}$ as

$$ \mathbf{I}_O^{\mathcal{R}} = I_{rr}^O\mathbf{e}_r \otimes \mathbf{e}_r + I_{\theta\theta}^O\mathbf{e}_\theta \otimes \mathbf{e}_\theta + I_{zz}^O\mathbf{e}_z \otimes \mathbf{e}_z \tag{5-414}$$

Furthermore, using the parallel-axis theorem, we have the following expressions for the principal-axis moments of inertia I_{rr}^O, $I_{\theta\theta}^O$, and I_{zz}^O relative to point O:

$$ I_{rr}^O = \bar{I}_{rr} + m(\pi_\theta^2 + \pi_z^2) \tag{5-415}$$

$$ I_{\theta\theta}^O = \bar{I}_{\theta\theta} + m(\pi_r^2 + \pi_z^2) \tag{5-416}$$

$$ I_{zz}^O = \bar{I}_{zz} + m(\pi_r^2 + \pi_\theta^2) \tag{5-417}$$

where $\bar{I}_{rr}$, $\bar{I}_{\theta\theta}$, and $\bar{I}_{zz}$ are the principal-axis moments of inertia of the rod relative to the center of mass along the directions $\mathbf{e}_r$, $\mathbf{e}_\theta$, and $\mathbf{e}_z$, respectively, and π_r, π_θ, and π_z are the distances in the $\mathbf{e}_r$, $\mathbf{e}_\theta$, and $\mathbf{e}_z$ directions, respectively, by which the principal-axis system relative to the center of mass of the rod is translated to get to point O. Now we know for a uniform slender rod that

$$\bar{I}_{rr} = 0 \tag{5-418}$$

$$\bar{I}_{\theta\theta} = \frac{ml^2}{12} \tag{5-419}$$

$$\bar{I}_{zz} = \frac{ml^2}{12} \tag{5-420}$$

Furthermore, in this case it is seen that only a translation along the $\mathbf{e}_r$-direction by a distance $l/2$ is made in order to get from the center of mass to point O. Consequently, we have

$$\pi_r = \frac{l}{2} \tag{5-421}$$

$$\pi_\theta = 0 \tag{5-422}$$

$$\pi_z = 0 \tag{5-423}$$

Substituting the results of Eqs. (5-418)-(5-420) and Eqs. (5-421)-(5-423) appropriately into Eqs. (5-415)-(5-417), we obtain

$$I_{rr}^O = 0 \tag{5-424}$$

$$I_{\theta\theta}^O = \frac{ml^2}{12} + m\left(\frac{l}{2}\right)^2 = \frac{ml^2}{3} \tag{5-425}$$

$$I_{zz}^O = \frac{ml^2}{12} + m\left(\frac{l}{2}\right)^2 = \frac{ml^2}{3} \tag{5-426}$$

The moment of inertia tensor of Eq. (5-414) is then given as

$$\mathbf{I}_O^{\mathcal{R}} = \frac{ml^2}{3}\mathbf{e}_\theta \otimes \mathbf{e}_\theta + \frac{ml^2}{3}\mathbf{e}_z \otimes \mathbf{e}_z \tag{5-427}$$

Using the result of Eq. (5-427) together with the angular velocity ${}^{\mathcal{F}}\boldsymbol{\omega}^{\mathcal{R}}$ from Eq. (5-412), we obtain the angular momentum of the rod relative to point O as

$${}^{\mathcal{F}}\mathbf{H}_O = \left(\frac{ml^2}{3}\mathbf{e}_\theta \otimes \mathbf{e}_\theta + \frac{ml^2}{3}\mathbf{e}_z \otimes \mathbf{e}_z\right) \cdot \left(-\Omega\cos\theta\,\mathbf{e}_r + \Omega\sin\theta\,\mathbf{e}_\theta + \dot{\theta}\mathbf{e}_z\right) \tag{5-428}$$

Simplifying Eq. (5-428), we obtain

$${}^{\mathcal{F}}\mathbf{H}_O = \frac{ml^2}{3}\Omega\sin\theta\,\mathbf{e}_\theta + \frac{ml^2}{3}\dot{\theta}\mathbf{e}_z \tag{5-429}$$

Using the expression for ${}^{\mathcal{F}}\mathbf{H}_O$ as given in Eq. (5-429), we need to obtain an expression for ${}^{\mathcal{F}}d({}^{\mathcal{F}}\mathbf{H}_O)/dt$. We have from the transport theorem that

$$\frac{{}^{\mathcal{F}}d}{dt}\left({}^{\mathcal{F}}\mathbf{H}_O\right) = \frac{{}^{\mathcal{R}}d}{dt}\left({}^{\mathcal{F}}\mathbf{H}_O\right) + {}^{\mathcal{F}}\boldsymbol{\omega}^{\mathcal{R}} \times {}^{\mathcal{F}}\mathbf{H}_O \tag{5-430}$$

where

$$\begin{aligned}
\frac{{}^{\mathcal{R}}d}{dt}\left({}^{\mathcal{F}}\mathbf{H}_O\right) &= \frac{ml^2}{3}\Omega\dot{\theta}\cos\theta\,\mathbf{e}_\theta + \frac{ml^2}{3}\ddot{\theta}\mathbf{e}_z \qquad (5\text{-}431)\\
{}^{\mathcal{F}}\boldsymbol{\omega}^{\mathcal{R}} \times {}^{\mathcal{F}}\mathbf{H}_O &= \left(-\Omega\cos\theta\,\mathbf{e}_r + \Omega\sin\theta\,\mathbf{e}_\theta + \dot{\theta}\mathbf{e}_z\right) \times \left(\frac{ml^2}{3}\Omega\sin\theta\,\mathbf{e}_\theta + \frac{ml^2}{3}\dot{\theta}\mathbf{e}_z\right)\\
&= -\frac{ml^2}{3}\Omega^2\cos\theta\,\sin\theta\,\mathbf{e}_z + \frac{ml^2}{3}\Omega\cos\theta\,\dot{\theta}\mathbf{e}_\theta\\
&+ \frac{ml^2}{3}\Omega\dot{\theta}\sin\theta\,\mathbf{e}_r - \frac{ml^2}{3}\Omega\dot{\theta}\sin\theta\,\mathbf{e}_r\\
&= \frac{ml^2}{3}\Omega\dot{\theta}\cos\theta\,\mathbf{e}_\theta - \frac{ml^2}{3}\Omega^2\cos\theta\,\sin\theta\,\mathbf{e}_z \qquad (5\text{-}432)
\end{aligned}$$

Then, adding Eqs. (5-431) and (5-432), we obtain

$$\frac{{}^{\mathcal{F}}d}{dt}\left({}^{\mathcal{F}}\mathbf{H}_O\right) = \frac{2ml^2}{3}\Omega\dot{\theta}\cos\theta\,\mathbf{e}_\theta + \left(\frac{ml^2}{3}\ddot{\theta} - \frac{ml^2}{3}\Omega^2\cos\theta\,\sin\theta\right)\mathbf{e}_z \qquad (5\text{-}433)$$

Kinetics

The free body diagram of the rod is shown in Fig. 5-18, from which it is seen that the following forces act on the rod:

$\mathbf{R}$ = Reaction force of hinge on rod
$m\mathbf{g}$ = Force of gravity
$\boldsymbol{\tau}$ = Pure torque due to prescribed rotation of shaft

It is emphasized that the pure torque applied by the shaft on the rod arises from the fact that the shaft is constrained to rotate with a *known* angular velocity. Furthermore, since this pure torque must be applied in the $\mathbf{e}_y$-direction (since the shaft rotates about the $\mathbf{e}_y$-direction) and $\mathbf{e}_y$ lies in the plane contained by $\mathbf{e}_r$ and $\mathbf{e}_\theta$, the torque applied by the shaft on the rod must have the form

$$\boldsymbol{\tau} = \boldsymbol{\tau}_r + \boldsymbol{\tau}_\theta = \tau_r\mathbf{e}_r + \tau_\theta\mathbf{e}_\theta \qquad (5\text{-}434)$$

where $\boldsymbol{\tau}_r$ and $\boldsymbol{\tau}_\theta$ are the components of the torque $\boldsymbol{\tau}$ in the directions of $\mathbf{e}_r$ and $\mathbf{e}_\theta$, respectively.

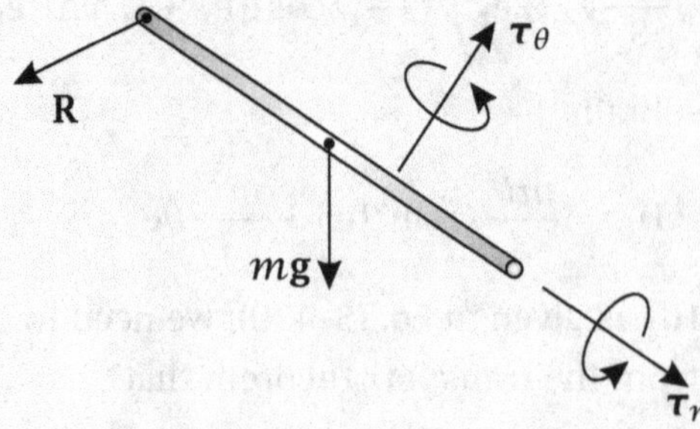

Figure 5-18 Free body diagram of rod for Example 5-7.

Now, from the geometry of the problem we have

$$\begin{aligned} \mathbf{R} &= R_r\mathbf{e}_r + R_\theta\mathbf{e}_\theta + R_z\mathbf{e}_z \\ m\mathbf{g} &= -mg\mathbf{u}_y = mg\cos\theta\,\mathbf{e}_r - mg\sin\theta\,\mathbf{e}_\theta \\ \boldsymbol{\tau}_r &= \tau_r\mathbf{e}_r \\ \boldsymbol{\tau}_\theta &= \tau_\theta\mathbf{e}_\theta \end{aligned} \tag{5-435}$$

Then, since $\mathbf{R}$ passes through point O, the moment due to all forces and torques relative to point O is given as

$$\mathbf{M}_O = (\bar{\mathbf{r}} - \mathbf{r}_O) \times m\mathbf{g} + \boldsymbol{\tau}_s + \boldsymbol{\tau}_r + \boldsymbol{\tau}_\theta \tag{5-436}$$

Now we note that

$$\bar{\mathbf{r}} - \mathbf{r}_O = \frac{l}{2}\mathbf{e}_r \tag{5-437}$$

Consequently, the moment relative to point O is given as

$$\mathbf{M}_O = \frac{l}{2}\mathbf{e}_r \times (mg\cos\theta\,\mathbf{e}_r - mg\sin\theta\,\mathbf{e}_\theta) + \tau_r\mathbf{e}_r + \tau_\theta\mathbf{e}_\theta \tag{5-438}$$

Expanding the cross products in Eq. (5–438), we obtain

$$\mathbf{M}_O = \tau_r\mathbf{e}_r + \tau_\theta\mathbf{e}_\theta - \frac{mgl}{2}\sin\theta\,\mathbf{e}_z \tag{5-439}$$

Then, setting $\mathbf{M}_O$ equal to ${}^{\mathcal{F}}d({}^{\mathcal{F}}\mathbf{H}_O)/dt$ using the expression for ${}^{\mathcal{F}}d({}^{\mathcal{F}}\mathbf{H}_O)/dt$ from Eq. (5–433), we obtain

$$\begin{aligned} \tau_r\mathbf{e}_r + \tau_\theta\mathbf{e}_\theta - \frac{mgl}{2}\sin\theta\,\mathbf{e}_z = {} & \frac{2ml^2}{3}\Omega\dot{\theta}\cos\theta\,\mathbf{e}_\theta \\ & + \left(\frac{ml^2}{3}\ddot{\theta} - \frac{ml^2}{3}\Omega^2\cos\theta\sin\theta\right)\mathbf{e}_z \end{aligned} \tag{5-440}$$

Equation (5–440) yields the following three scalar equations:

$$\tau_r = 0 \tag{5-441}$$

$$\tau_\theta = \frac{2ml^2}{3}\Omega\dot{\theta}\cos\theta \tag{5-442}$$

$$-\frac{mgl}{2}\sin\theta = \frac{ml^2}{3}\ddot{\theta} - \frac{ml^2}{3}\Omega^2\cos\theta\sin\theta \tag{5-443}$$

Upon close examination of Eq. (5–441), it is seen that the component of the pure torque due to the shaft in the $\mathbf{e}_r$-direction is *zero* whereas the component of pure torque in the $\mathbf{e}_\theta$-direction is not zero. These results are absolutely consistent with the fact that the moment of inertia of the rod along the $\mathbf{e}_r$-direction is zero (i.e., the shaft cannot exert a torque along a direction where the moment of inertia is zero) while the moment of inertia of the rod along the $\mathbf{e}_\theta$-direction is nonzero. Next, examining Eqs. (5–442) and (5–443), it is seen that Eq. (5–442) is indeterminate because the reaction torque in the $\mathbf{e}_\theta$-direction, τ_θ, is unknown. However, Eq. (5–443) contains no unknown forces or torques. Consequently, the differential equation of motion is obtained from Eq. (5-443). First, Eq. (5-443) can be rearranged to give

$$\frac{ml^2}{3}\ddot{\theta} - \frac{ml^2}{3}\Omega^2 \cos\theta \sin\theta = -\frac{mgl}{2}\sin\theta \tag{5-444}$$

Then, simplifying Eq. (5-444) by collecting all terms to one side and dividing through by $ml^2/3$, we obtain the differential equation of motion for the rod as

$$\ddot{\theta} + \frac{3g}{2l}\sin\theta - \Omega^2 \cos\theta \sin\theta = 0 \tag{5-445}$$

■

Example 5-8

An imbalanced wheel rolls without slip along a surface inclined at a constant angle β with the horizontal as shown in Fig. 5-19. The wheel is modeled as a massless circular disk of radius R with a particle of mass m lodged a distance r from the center of the wheel. Knowing that θ describes the location of the particle relative to the direction from Q to O (where Q is the point of contact point of the wheel with the incline and O is the center of the wheel) and that gravity acts downward, determine the differential equation for the wheel.

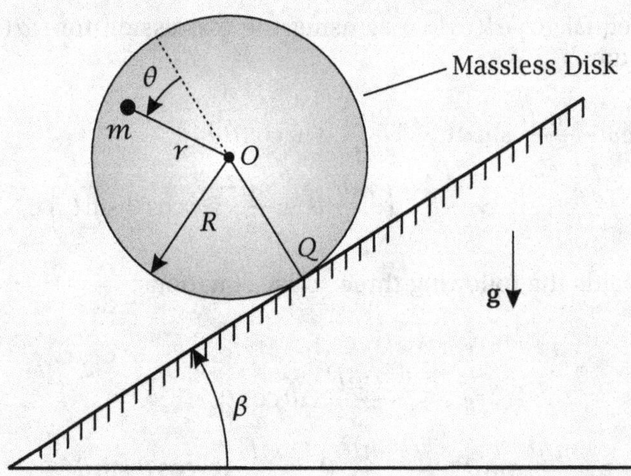

Figure 5-19 Imbalanced wheel rolling along surface inclined at angle β with the horizontal.

Solution to Example 5-8

Kinematics

First, let $\mathcal{F}$ be a fixed reference frame. Then, choose the following coordinate system fixed in reference frame $\mathcal{F}$:

Origin at center of disk when $\theta = 0$		
$\mathbf{E}_x$	=	Down incline
$\mathbf{E}_z$	=	Out of page
$\mathbf{E}_y$	=	$\mathbf{E}_z \times \mathbf{E}_x$

Next, let $\mathcal{R}$ be a reference frame fixed to the wheel. Then, choose the following coordinate system fixed in reference frame $\mathcal{R}$:

Origin at center of wheel		
$\mathbf{e}_r$	=	Along Om
$\mathbf{e}_z$	=	$\mathbf{E}_z$ = Out of page
$\mathbf{e}_\theta$	=	$\mathbf{E}_z \times \mathbf{e}_r$

The geometry of the bases $\{\mathbf{E}_x, \mathbf{E}_y, \mathbf{E}_z\}$ and $\{\mathbf{e}_r, \mathbf{e}_\theta, \mathbf{e}_z\}$ is shown in Fig. 5-20. Also shown in Fig. 5-20 is the unit vector in the vertically downward direction, $\mathbf{u}_v$.

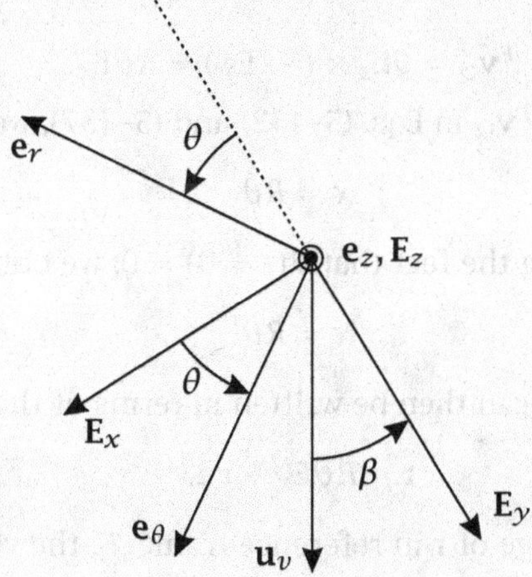

Figure 5-20 Geometry of bases $\{\mathbf{E}_x, \mathbf{E}_y, \mathbf{E}_z\}$ and $\{\mathbf{e}_r, \mathbf{e}_\theta, \mathbf{e}_z\}$ for Example 5-8.

Using Fig. 5-20, we have the following relationship between the bases $\{\mathbf{E}_x, \mathbf{E}_y, \mathbf{E}_z\}$ and $\{\mathbf{e}_r, \mathbf{e}_\theta, \mathbf{e}_z\}$:

$$\mathbf{e}_r = \sin\theta\,\mathbf{E}_x - \cos\theta\,\mathbf{E}_y \tag{5-446}$$

$$\mathbf{e}_\theta = \cos\theta\,\mathbf{E}_x + \sin\theta\,\mathbf{E}_y \tag{5-447}$$

$$\mathbf{u}_v = \sin\theta\,\mathbf{E}_x + \cos\theta\,\mathbf{E}_y \tag{5-448}$$

The position of the particle can then be written in terms of the bases $\{\mathbf{E}_x, \mathbf{E}_y, \mathbf{E}_z\}$ and $\{\mathbf{e}_r, \mathbf{e}_\theta, \mathbf{e}_z\}$ as

$$\mathbf{r} = \mathbf{r}_O + \mathbf{r}_{m/O} = x\mathbf{E}_x + r\mathbf{e}_r \tag{5-449}$$

where

$$\mathbf{r}_O = x\mathbf{E}_x \tag{5-450}$$

$$\mathbf{r}_{m/O} = r\mathbf{e}_r \tag{5-451}$$

We have from Eq. (5-450) that

$${}^{\mathcal{F}}\mathbf{v}_O = \dot{x}\mathbf{E}_x \tag{5-452}$$

Now, because the wheel rolls without slip along a fixed surface, we have

$${}^{\mathcal{F}}\mathbf{v}_Q = \mathbf{0} \tag{5-453}$$

Furthermore, because θ is measured from a direction fixed in the inertial reference frame $\mathcal{F}$, the angular velocity of the wheel in reference frame $\mathcal{F}$ is given as

$${}^{\mathcal{F}}\boldsymbol{\omega}^{\mathcal{R}} = \dot{\theta}\mathbf{e}_z = \dot{\theta}\mathbf{E}_z \tag{5-454}$$

Then, applying Eq. (2-517) on page 106, the velocity of point O in reference frame $\mathcal{F}$ is obtained as

$${}^{\mathcal{F}}\mathbf{v}_O = {}^{\mathcal{F}}\mathbf{v}_Q + {}^{\mathcal{F}}\boldsymbol{\omega}^{\mathcal{R}} \times (\mathbf{r}_O - \mathbf{r}_Q) = {}^{\mathcal{F}}\boldsymbol{\omega}^{\mathcal{R}} \times (\mathbf{r}_O - \mathbf{r}_Q) \tag{5-455}$$

Now we have

$$\mathbf{r}_O - \mathbf{r}_Q = -R\mathbf{E}_y \tag{5-456}$$

Substituting the expressions for ${}^{\mathcal{F}}\boldsymbol{\omega}^{\mathcal{R}}$ and $\mathbf{r}_O - \mathbf{r}_Q$ from Eqs. (5-454) and (5-456, respectively, we have

$${}^{\mathcal{F}}\mathbf{v}_O = \dot{\theta}\mathbf{E}_z \times (-r\mathbf{E}_y) = R\dot{\theta}\mathbf{E}_x \tag{5-457}$$

Setting the expressions for ${}^{\mathcal{F}}\mathbf{v}_O$ in Eqs. (5-452) and (5-457), we obtain

$$\dot{x} = R\dot{\theta} \tag{5-458}$$

Integrating Eq. (5-458) using the fact that $\theta(t = 0) = 0$, we obtain

$$x = R\theta \tag{5-459}$$

The position of the particle can then be written in terms of the angle θ as

$$\mathbf{r} = R\theta\mathbf{E}_x + r\mathbf{e}_r \tag{5-460}$$

Computing the rate of change of $\mathbf{r}$ in reference frame $\mathcal{F}$, the velocity of the particle in reference frame $\mathcal{F}$ is given as

$${}^{\mathcal{F}}\mathbf{v} = \frac{{}^{\mathcal{F}}d\mathbf{r}}{dt} = \frac{{}^{\mathcal{F}}d}{dt}(\mathbf{r}_O) + \frac{{}^{\mathcal{F}}d}{dt}(\mathbf{r}_{m/O}) = {}^{\mathcal{F}}\mathbf{v}_O + {}^{\mathcal{F}}\mathbf{v}_{m/O} \tag{5-461}$$

Because the basis $\{\mathbf{E}_x, \mathbf{E}_y, \mathbf{E}_z\}$ is fixed in reference frame $\mathcal{F}$, we have

$${}^{\mathcal{F}}\mathbf{v}_O = \frac{{}^{\mathcal{F}}d\mathbf{r}_O}{dt} = r\dot{\theta}\mathbf{E}_x \tag{5-462}$$

Furthermore, because $\mathbf{r}_{m/O}$ is expressed in the basis $\{\mathbf{e}_r, \mathbf{e}_\theta, \mathbf{e}_z\}$, we obtain

$${}^{\mathcal{F}}\mathbf{v}_{m/O} = \frac{{}^{\mathcal{F}}d}{dt}(\mathbf{r}_{m/O}) = \frac{{}^{\mathcal{R}}d}{dt}(\mathbf{r}_{m/O}) + {}^{\mathcal{F}}\boldsymbol{\omega}^{\mathcal{R}} \times \mathbf{r}_{m/O} \tag{5-463}$$

Now we have

$$\begin{aligned} \frac{{}^{\mathcal{R}}d}{dt}\left(\mathbf{r}_{m/O}\right) &= \mathbf{0} & (5\text{-}464) \\ {}^{\mathcal{F}}\boldsymbol{\omega}^{\mathcal{R}} \times \mathbf{r}_{m/O} &= \dot{\theta}\mathbf{e}_z \times r\mathbf{e}_r = r\dot{\theta}\mathbf{e}_\theta & (5\text{-}465) \end{aligned}$$

Adding the expressions in Eqs. (5-464) and (5-465), we obtain

$${}^{\mathcal{F}}\mathbf{v}_{m/O} = r\dot{\theta}\mathbf{e}_\theta \qquad (5\text{-}466)$$

Then, adding Eqs. (5-462) and (5-466), the velocity of the particle in reference frame $\mathcal{F}$ is given as

$${}^{\mathcal{F}}\mathbf{v} = R\dot{\theta}\mathbf{E}_x + r\dot{\theta}\mathbf{e}_\theta \qquad (5\text{-}467)$$

Next, the acceleration of the particle in reference frame $\mathcal{F}$ is given from the transport theorem of Eq. (2-128) on page 47 as

$${}^{\mathcal{F}}\mathbf{a} = \frac{{}^{\mathcal{F}}d}{dt}\left({}^{\mathcal{F}}\mathbf{v}\right) = \frac{{}^{\mathcal{F}}d}{dt}\left({}^{\mathcal{F}}\mathbf{v}_O\right) + \frac{{}^{\mathcal{F}}d}{dt}\left({}^{\mathcal{F}}\mathbf{v}_{m/O}\right) = {}^{\mathcal{F}}\mathbf{a}_O + {}^{\mathcal{F}}\mathbf{a}_{m/O} \qquad (5\text{-}468)$$

First, we have

$${}^{\mathcal{F}}\mathbf{a}_O = \frac{{}^{\mathcal{F}}d}{dt}\left({}^{\mathcal{F}}\mathbf{v}_O\right) = R\ddot{\theta}\mathbf{E}_x \qquad (5\text{-}469)$$

Next, applying the rate of change transport theorem to ${}^{\mathcal{F}}\mathbf{v}_{m/O}$ between reference frames $\mathcal{R}$ and $\mathcal{F}$, we obtain ${}^{\mathcal{F}}\mathbf{a}_{m/O}$ as

$${}^{\mathcal{F}}\mathbf{a}_{m/O} = \frac{{}^{\mathcal{F}}d}{dt}\left({}^{\mathcal{F}}\mathbf{v}_{m/O}\right) = \frac{{}^{\mathcal{R}}d}{dt}\left({}^{\mathcal{F}}\mathbf{v}_{m/O}\right) + {}^{\mathcal{F}}\boldsymbol{\omega}^{\mathcal{R}} \times {}^{\mathcal{F}}\mathbf{v}_{m/O} \qquad (5\text{-}470)$$

Now the two terms on the right-hand side of Eq. (5-470) are given as

$$\begin{aligned} \frac{{}^{\mathcal{R}}d}{dt}\left({}^{\mathcal{F}}\mathbf{v}_{m/O}\right) &= r\ddot{\theta}\mathbf{e}_\theta & (5\text{-}471) \\ {}^{\mathcal{F}}\boldsymbol{\omega}^{\mathcal{R}} \times {}^{\mathcal{F}}\mathbf{v}_{m/O} &= \dot{\theta}\mathbf{e}_z \times r\dot{\theta}\mathbf{e}_\theta = -r\dot{\theta}^2\mathbf{e}_r & (5\text{-}472) \end{aligned}$$

Adding the expressions in Eqs. (5-471) and (5-472), we obtain

$${}^{\mathcal{F}}\mathbf{a}_{m/O} = -r\dot{\theta}^2\mathbf{e}_r + r\ddot{\theta}\mathbf{e}_\theta \qquad (5\text{-}473)$$

Then, adding Eqs. (5-469) and (5-473), we obtain the acceleration of the particle in reference frame $\mathcal{F}$ as

$${}^{\mathcal{F}}\mathbf{a} = R\ddot{\theta}\mathbf{E}_x - r\dot{\theta}^2\mathbf{e}_r + r\ddot{\theta}\mathbf{e}_\theta \qquad (5\text{-}474)$$

Kinetics

Because the wheel rolls without slip, it is convenient to use the point of contact, Q, as the reference point. Now, as discussed in Section 2.15.4 of Chapter 2, while the velocity of the point of contact in reference frame $\mathcal{F}$ is zero, i.e., ${}^{\mathcal{F}}\mathbf{v}_Q = \mathbf{0}$, the acceleration of the point of contact in reference frame $\mathcal{F}$ is *not* zero. Therefore, we must apply the general form of Euler's 2nd law when using the contact point as the reference, i.e.,

$$\mathbf{M}_Q - (\bar{\mathbf{r}} - \mathbf{r}_Q) \times m\,{}^{\mathcal{F}}\mathbf{a}_Q = \frac{{}^{\mathcal{F}}d}{dt}\left({}^{\mathcal{F}}\mathbf{H}_Q\right) \qquad (5\text{-}475)$$

The free body diagram of the wheel is shown in Fig. 5-21.

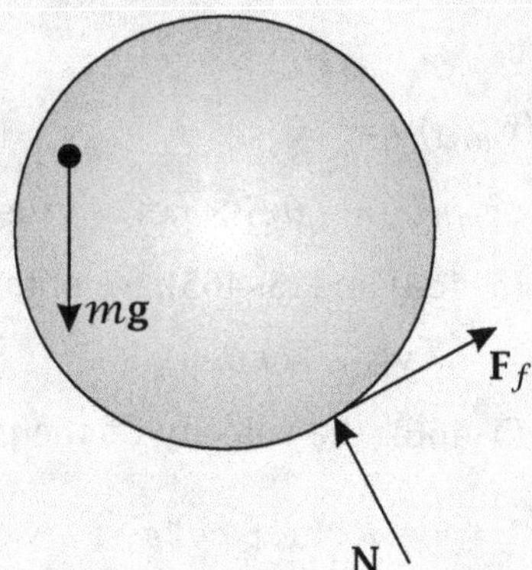

Figure 5–21 Free body diagram for Example 5–8.

Using Fig. 5–21, it is seen that the following forces act on the wheel:

$$\begin{aligned} \mathbf{F}_f &= \text{Force of friction} \\ \mathbf{N} &= \text{Reaction force of incline} \\ m\mathbf{g} &= \text{Force of gravity} \end{aligned} \tag{5-476}$$

Now, because the forces $\mathbf{F}_f$ and $\mathbf{N}$ pass through point Q, the moment relative to point Q is due entirely to the force of gravity, i.e.,

$$\mathbf{M}_Q = (\mathbf{r} - \mathbf{r}_Q) \times m\mathbf{g} \tag{5-477}$$

It is noted in Eq. (5–477) that, because the entire mass of the wheel is concentrated at a single point, the force of gravity acts at the location of the particle itself. Furthermore, from the definition of the basis $\{\mathbf{E}_x, \mathbf{E}_y, \mathbf{E}_z\}$, we have

$$\mathbf{r}_Q = R\theta\mathbf{E}_x + R\mathbf{E}_y \tag{5-478}$$

Then, using the result of Eq. (5–460), we obtain

$$\mathbf{r} - \mathbf{r}_Q = r\mathbf{e}_r - R\mathbf{E}_y \tag{5-479}$$

Next, the force of gravity is given as

$$m\mathbf{g} = mg\mathbf{u}_v \tag{5-480}$$

where $\mathbf{u}_v$ is the vertically downward direction, as shown in Fig. 5–20. Using the expression for $\mathbf{u}_v$ from Eq. (5–448), we have

$$m\mathbf{g} = mg\sin\beta\mathbf{E}_x + mg\cos\beta\mathbf{E}_y \tag{5-481}$$

Therefore, the resultant moment applied to the wheel relative to point Q is given as

$$\mathbf{M}_Q = (\mathbf{r} - \mathbf{r}_Q) \times m\mathbf{g} = (r\mathbf{e}_r - R\mathbf{E}_y) \times (mg\sin\beta\mathbf{E}_x + mg\cos\beta\mathbf{E}_y) \tag{5-482}$$

Expanding Eq. (5–482), we obtain

$$\mathbf{M}_Q = mgr\sin\beta\mathbf{e}_r \times \mathbf{E}_x + mgr\cos\beta\mathbf{e}_r \times \mathbf{E}_y + mgR\sin\beta\mathbf{E}_z \tag{5-483}$$

Using the expression for $\mathbf{e}_r$ in Eq. (5–446), we have

$$\begin{aligned}\mathbf{e}_r \times \mathbf{E}_x &= (\sin\theta\,\mathbf{E}_x - \cos\theta\,\mathbf{E}_y)\times\mathbf{E}_x = \cos\theta\,\mathbf{E}_z = \cos\theta\,\mathbf{e}_z && (5\text{–}484)\\ \mathbf{e}_r \times \mathbf{E}_y &= (\sin\theta\,\mathbf{E}_x - \cos\theta\,\mathbf{E}_y)\times\mathbf{E}_y = \sin\theta\,\mathbf{E}_z = \sin\theta\,\mathbf{e}_z && (5\text{–}485)\end{aligned}$$

Substituting the results of Eqs. (5–484) and (5–485), respectively, into Eq. (5–483), we have

$$\mathbf{M}_Q = mgr\sin\beta\cos\theta\,\mathbf{e}_z + mgr\cos\beta\sin\theta\,\mathbf{e}_z + mgR\sin\beta\mathbf{E}_z \tag{5–486}$$

Simplifying Eq. (5–486) and noting that $\mathbf{E}_z = \mathbf{e}_z$, we obtain

$$\mathbf{M}_Q = mg\left[r(\sin\beta\cos\theta + \cos\beta\sin\theta) + R\sin\beta\right]\mathbf{e}_z \tag{5–487}$$

Now from trigonometry we have

$$\sin\beta\cos\theta + \cos\beta\sin\theta = \sin(\theta+\beta) \tag{5–488}$$

Consequently, the moment applied to the disk relative to point Q is given as

$$\mathbf{M}_Q = mg\left[r\sin(\theta+\beta) + R\sin\beta\right]\mathbf{e}_z \tag{5–489}$$

Next, the acceleration of the point of contact is given as

$${}^{\mathcal{F}}\mathbf{a}_Q = \mathbf{a}_O + {}^{\mathcal{F}}\boldsymbol{\alpha}^{\mathcal{R}} \times (\mathbf{r}_Q - \mathbf{r}_O) + {}^{\mathcal{F}}\boldsymbol{\omega}^{\mathcal{R}} \times \left[{}^{\mathcal{F}}\boldsymbol{\omega}^{\mathcal{R}} \times (\mathbf{r}_Q - \mathbf{r}_O)\right] \tag{5–490}$$

where we note that

$${}^{\mathcal{F}}\boldsymbol{\alpha}^{\mathcal{R}} = \frac{{}^{\mathcal{F}}d}{dt}\left({}^{\mathcal{F}}\boldsymbol{\omega}^{\mathcal{R}}\right) = \ddot{\theta}\mathbf{e}_z = \ddot{\theta}\mathbf{E}_z \tag{5–491}$$

Also, noting that $\mathbf{r}_Q - \mathbf{r}_O = R\mathbf{E}_y$, we obtain

$${}^{\mathcal{F}}\mathbf{a}_Q = R\ddot{\theta}\mathbf{E}_x + \ddot{\theta}\mathbf{E}_z \times (R\mathbf{E}_y) + \dot{\theta}\mathbf{E}_z \times \left[\dot{\theta}\mathbf{E}_z \times (R\mathbf{E}_y)\right] = -R\dot{\theta}^2\mathbf{E}_y \tag{5–492}$$

Therefore, the inertial moment is given as

$$\begin{aligned}-(\bar{\mathbf{r}} - \mathbf{r}_Q) \times m{}^{\mathcal{F}}\mathbf{a}_Q &= -(\mathbf{r} - \mathbf{r}_Q) \times m{}^{\mathcal{F}}\mathbf{a}_Q = -(r\mathbf{e}_r - R\mathbf{E}_y) \times m(-R\dot{\theta}^2\mathbf{E}_y)\\ &= mRr\dot{\theta}^2\mathbf{e}_r \times \mathbf{E}_y = mRr\dot{\theta}^2\sin\theta\,\mathbf{e}_z\end{aligned} \tag{5–493}$$

It is noted in Eq. (5–493) that we have used the expression for $\mathbf{e}_r \times \mathbf{E}_y$ from Eq. (5–485) and have also used the fact that the mass of the system is concentrated at a single point.

Next, the angular momentum of the disk relative to point Q in reference frame $\mathcal{F}$ can be obtained from Eq. (5–24) as

$${}^{\mathcal{F}}\mathbf{H}_Q = {}^{\mathcal{F}}\bar{\mathbf{H}} + (\mathbf{r}_Q - \bar{\mathbf{r}}) \times m({}^{\mathcal{F}}\mathbf{v}_Q - {}^{\mathcal{F}}\bar{\mathbf{v}}) \tag{5–494}$$

Because the mass is concentrated at a single point, the angular momentum relative to the center of mass of the system is zero, i.e.,

$${}^{\mathcal{F}}\bar{\mathbf{H}} = \mathbf{0} \tag{5–495}$$

Consequently, we obtain ${}^{\mathcal{F}}\mathbf{H}_Q$ as

$$ {}^{\mathcal{F}}\mathbf{H}_Q = (\mathbf{r}_Q - \bar{\mathbf{r}}) \times m({}^{\mathcal{F}}\mathbf{v}_Q - {}^{\mathcal{F}}\bar{\mathbf{v}}) \tag{5-496} $$

Furthermore, observing that $\bar{\mathbf{r}} = \mathbf{r}$ and ${}^{\mathcal{F}}\bar{\mathbf{v}} = {}^{\mathcal{F}}\mathbf{v}$, where $\mathbf{r}$ is the position of the particle and ${}^{\mathcal{F}}\bar{\mathbf{v}} = {}^{\mathcal{F}}\mathbf{v}$ is the velocity of the particle in reference frame $\mathcal{F}$, Eq. (5-496) can be written as

$$ {}^{\mathcal{F}}\mathbf{H}_Q = (\mathbf{r}_Q - \mathbf{r}) \times m({}^{\mathcal{F}}\mathbf{v}_Q - {}^{\mathcal{F}}\mathbf{v}) \tag{5-497} $$

Computing the rate of change of ${}^{\mathcal{F}}\mathbf{H}_Q$ from Eq. (5-497) in reference frame $\mathcal{F}$, we obtain

$$ \frac{{}^{\mathcal{F}}d}{dt}\left({}^{\mathcal{F}}\mathbf{H}_Q\right) = (\mathbf{r}_Q - \mathbf{r}) \times m({}^{\mathcal{F}}\mathbf{a}_Q - {}^{\mathcal{F}}\mathbf{a}) \tag{5-498} $$

Then, using $\mathbf{r} - \mathbf{r}_Q$, ${}^{\mathcal{F}}\mathbf{a}$, and ${}^{\mathcal{F}}\mathbf{a}_Q$ from Eqs. (5-479), (5-474), and (5-492), respectively, we obtain ${}^{\mathcal{F}}d\left({}^{\mathcal{F}}\mathbf{H}_Q\right)/dt$ as

$$ \begin{aligned} \frac{{}^{\mathcal{F}}d}{dt}\left({}^{\mathcal{F}}\mathbf{H}_Q\right) &= (-r\mathbf{e}_r + R\mathbf{E}_y) \times m\left[-R\dot{\theta}^2\mathbf{E}_y - (R\ddot{\theta}\mathbf{E}_x - r\dot{\theta}^2\mathbf{e}_r + r\ddot{\theta}\mathbf{e}_\theta)\right] \\ &= (-r\mathbf{e}_r + R\mathbf{E}_y) \times m(-R\dot{\theta}^2\mathbf{E}_y - R\ddot{\theta}\mathbf{E}_x + r\dot{\theta}^2\mathbf{e}_r - r\ddot{\theta}\mathbf{e}_\theta) \end{aligned} \tag{5-499} $$

Expanding Eq. (5-499) gives

$$ \begin{aligned} \frac{{}^{\mathcal{F}}d}{dt}\left({}^{\mathcal{F}}\mathbf{H}_Q\right) &= mRr\dot{\theta}^2\mathbf{e}_r \times \mathbf{E}_y + mRr\ddot{\theta}\mathbf{e}_r \times \mathbf{E}_x + mr^2\ddot{\theta}\mathbf{E}_z \\ &\quad + mR^2\ddot{\theta}\mathbf{E}_z + mRr\dot{\theta}^2\mathbf{E}_y \times \mathbf{e}_r - mRr\ddot{\theta}\mathbf{E}_y \times \mathbf{e}_\theta \end{aligned} \tag{5-500} $$

Now we already have expressions for $\mathbf{e}_r \times \mathbf{E}_x$ and $\mathbf{e}_r \times \mathbf{E}_y$ from Eqs. (5-484) and (5-485), respectively. In addition, we have

$$ \mathbf{E}_y \times \mathbf{e}_\theta = \mathbf{E}_y \times (\cos\theta\,\mathbf{E}_x + \sin\theta\,\mathbf{E}_y) = -\cos\theta\,\mathbf{E}_z \tag{5-501} $$

Substituting the results of Eqs. (5-484), (5-485), and (5-501) into Eq. (5-500), we obtain

$$ \begin{aligned} \frac{{}^{\mathcal{F}}d}{dt}\left({}^{\mathcal{F}}\mathbf{H}_Q\right) &= mRr\dot{\theta}^2\sin\theta\,\mathbf{E}_z + mRr\ddot{\theta}\cos\theta\,\mathbf{E}_z + mr^2\ddot{\theta}\mathbf{E}_z \\ &\quad + mR^2\ddot{\theta}\mathbf{E}_z - mRr\dot{\theta}^2\sin\theta\,\mathbf{E}_z + mRr\ddot{\theta}\cos\theta\,\mathbf{E}_z \end{aligned} \tag{5-502} $$

Equation (5-502) simplifies to

$$ \frac{{}^{\mathcal{F}}d}{dt}\left({}^{\mathcal{F}}\mathbf{H}_Q\right) = m(r^2 + R^2 + 2rR\cos\theta)\ddot{\theta}\mathbf{E}_z = m(r^2 + R^2 + 2rR\cos\theta)\ddot{\theta}\mathbf{e}_z \tag{5-503} $$

Finally, substituting $\mathbf{M}_Q$, $-(\bar{\mathbf{r}}-\mathbf{r}_Q)\times m{}^{\mathcal{F}}\mathbf{a}_Q$, and ${}^{\mathcal{F}}d\left({}^{\mathcal{F}}\mathbf{H}_Q\right)/dt$ from Eq. (5-489), (5-493), and (5-503), respectively, into (5-475) we obtain

$$ mg\left[r\sin(\theta+\beta) + R\sin\beta\right]\mathbf{e}_z + mRr\dot{\theta}^2\sin\theta\,\mathbf{e}_z = m(r^2 + R^2 + 2rR\cos\theta)\ddot{\theta}\mathbf{e}_z \tag{5-504} $$

Rearranging and dropping the common component of $\mathbf{e}_z$, we obtain the differential equation of motion as

$$ m(r^2 + R^2 + 2rR\cos\theta)\ddot{\theta} - mrR\dot{\theta}^2\sin\theta - mgr\sin(\theta+\beta) - mgR\sin\beta = 0 \tag{5-505} $$

■

5.8 Systems of Rigid Bodies

Problems involving a system of rigid bodies are solved in a manner similar to problems involving a system of particles, i.e., in a problem involving multiple rigid bodies the system divided into subsystems of the entire system and the motion of each relevant subsystem are analyzed . However, unlike a particle, which has only one fundamental law of kinetics, a rigid body has *two* fundamental laws of kinetics. Consequently, when solving a problem involving a system of particles, it may be possible to obtain a sufficient number of equations by dividing the problem into subsystems each consisting of only one body (as compared with a system of particles where it is necessary that at least one of the subsystems must involve multiple particles). In the next two examples we consider systems involving a system of rigid bodies and show the key approach required to determine the differential equations of motion.

Example 5–9

A homogeneous disk of mass m and radius r rolls without slip on a uniform slender rod of mass m and length l as shown in Fig. 5-22. The rod is hinged at the fixed point O. Furthermore, attached to the rod is a torsional spring with spring constant K_t and uncoiled angle zero. Also, attached to the center of the disk is a linear spring with spring constant K_l and unstretched length x_0. The spring is attached at its other end to point A, where A lies a distance r from point O along a massless extension of the rod. Furthermore, the massless extension is orthogonal to the rod. Knowing that gravity acts downward, determine a system of two differential equations in terms of x and θ that describes the motion of the rod and disk.

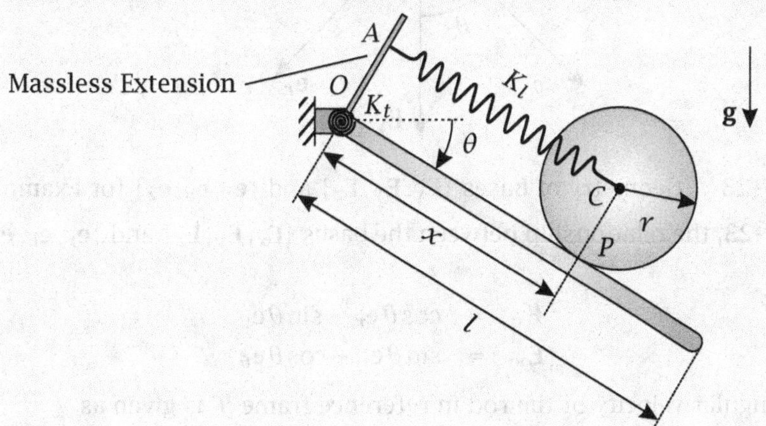

Figure 5–22 Disk rolling on rotating rod.

Solution to Example 5–9

Preliminaries

This problem will be solved by applying the following balance laws: Euler's 1st to the disk; Euler's 2nd law to the disk relative to the center of mass of the disk; and Euler's

2nd law to the rod relative to the hinge point. To this end, it the following kinematic quantities are required: the acceleration of the center of mass of the disk; the angular velocity of the disk; and the angular velocity of the rod. Each of these quantities is now determined.

Kinematics

First, let $\mathcal{F}$ be a fixed reference frame. Then, choose the following coordinate system fixed in $\mathcal{F}$:

$$
\begin{array}{ccl}
 & \text{Origin at } O & \\
\mathbf{E}_x & = & \text{Along rod when } \theta = 0 \\
\mathbf{E}_z & = & \text{Into page} \\
\mathbf{E}_y & = & \mathbf{E}_z \times \mathbf{E}_x
\end{array}
$$

Next, let $\mathcal{R}$ be a reference frame fixed to the rod. Then, choose the following coordinate system fixed in reference frame $\mathcal{R}$:

$$
\begin{array}{ccl}
 & \text{Origin at } O & \\
\mathbf{e}_r & = & \text{Along rod} \\
\mathbf{e}_z & = & \text{Into page} \\
\mathbf{e}_\theta & = & \mathbf{E}_z \times \mathbf{e}_r
\end{array}
$$

The geometry of the bases $\{\mathbf{E}_x, \mathbf{E}_y, \mathbf{E}_z\}$ and $\{\mathbf{e}_r, \mathbf{e}_\theta, \mathbf{e}_z\}$ is given in Fig. 5-23.

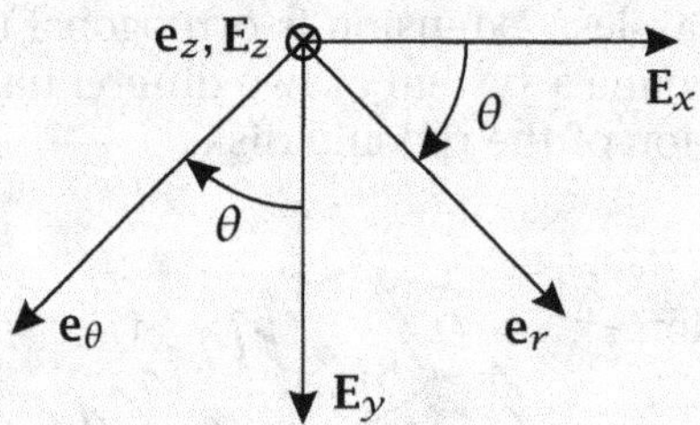

Figure 5-23 Geometry of bases $\{\mathbf{E}_x, \mathbf{E}_y, \mathbf{E}_z\}$ and $\{\mathbf{e}_r, \mathbf{e}_\theta, \mathbf{e}_z\}$ for Example 5-9.

Using Fig. 5-23, the relationship between the bases $\{\mathbf{E}_x, \mathbf{E}_y, \mathbf{E}_z\}$ and $\{\mathbf{e}_r, \mathbf{e}_\theta, \mathbf{e}_z\}$ is given as

$$\mathbf{E}_x = \cos\theta\,\mathbf{e}_r - \sin\theta\,\mathbf{e}_\theta \tag{5-506}$$

$$\mathbf{E}_y = \sin\theta\,\mathbf{e}_r + \cos\theta\,\mathbf{e}_\theta \tag{5-507}$$

Then, the angular velocity of the rod in reference frame $\mathcal{F}$ is given as

$$^{\mathcal{F}}\boldsymbol{\omega}^{\mathcal{R}} = \dot{\theta}\mathbf{e}_z \tag{5-508}$$

Next, the position of the center of mass of the disk is given as

$$\mathbf{r}_C = x\mathbf{e}_r - r\mathbf{e}_\theta \tag{5-509}$$

The velocity of the center of mass of the disk in reference frame $\mathcal{F}$ is then obtained by applying the rate of change transport theorem between reference frames $\mathcal{R}$ and $\mathcal{F}$ as

$$^{\mathcal{F}}\mathbf{v}_C = \frac{^{\mathcal{F}}d\mathbf{r}_C}{dt} = \frac{^{\mathcal{R}}d\mathbf{r}_C}{dt} + {}^{\mathcal{F}}\boldsymbol{\omega}^{\mathcal{R}} \times \mathbf{r}_C \tag{5-510}$$

Now we have

$$\begin{aligned} \frac{{}^{\mathcal{R}}d\mathbf{r}_C}{dt} &= \dot{x}\mathbf{e}_r & (5\text{-}511) \\ {}^{\mathcal{F}}\boldsymbol{\omega}^{\mathcal{R}} \times \mathbf{r}_C &= \dot{\theta}\mathbf{e}_z \times (x\mathbf{e}_r - r\mathbf{e}_\theta) = r\dot{\theta}\mathbf{e}_r + x\dot{\theta}\mathbf{e}_\theta & (5\text{-}512) \end{aligned}$$

Adding Eqs. (5-511) and (5-512), the velocity of the center of mass of the disk in reference frame $\mathcal{F}$ is obtained as

$$ {}^{\mathcal{F}}\mathbf{v}_C = (\dot{x} + r\dot{\theta})\mathbf{e}_r + x\dot{\theta}\mathbf{e}_\theta \qquad (5\text{-}513)$$

Next, the acceleration of the center of mass of the disk in reference frame $\mathcal{F}$ is obtained by applying the rate of change transport theorem between reference frames $\mathcal{R}$ and $\mathcal{F}$ to ${}^{\mathcal{F}}\mathbf{v}_C$ as

$$ {}^{\mathcal{F}}\mathbf{a}_C = \frac{{}^{\mathcal{F}}d}{dt}\left({}^{\mathcal{F}}\mathbf{v}_C\right) = \frac{{}^{\mathcal{R}}d}{dt}\left({}^{\mathcal{F}}\mathbf{v}_C\right) + {}^{\mathcal{F}}\boldsymbol{\omega}^{\mathcal{R}} \times {}^{\mathcal{F}}\mathbf{v}_C \qquad (5\text{-}514)$$

Now we have

$$\begin{aligned} \frac{{}^{\mathcal{R}}d}{dt}\left({}^{\mathcal{F}}\mathbf{v}_C\right) &= (\ddot{x} + r\ddot{\theta})\mathbf{e}_r + (\dot{x}\dot{\theta} + x\ddot{\theta})\mathbf{e}_\theta & (5\text{-}515) \\ {}^{\mathcal{F}}\boldsymbol{\omega}^{\mathcal{R}} \times {}^{\mathcal{F}}\mathbf{v}_C &= \dot{\theta}\mathbf{e}_z \times \left[(\dot{x} + r\dot{\theta})\mathbf{e}_r + x\dot{\theta}\mathbf{e}_\theta\right] \\ &= -x\dot{\theta}^2\mathbf{e}_r + \dot{\theta}(\dot{x} + r\dot{\theta})\mathbf{e}_\theta & (5\text{-}516) \end{aligned}$$

Adding Eqs. (5-515) and (5-516) gives

$$ {}^{\mathcal{F}}\mathbf{a}_C = (\ddot{x} + r\ddot{\theta})\mathbf{e}_r + (\dot{x}\dot{\theta} + x\ddot{\theta})\mathbf{e}_\theta - x\dot{\theta}^2\mathbf{e}_r + \dot{\theta}(\dot{x} + r\dot{\theta})\mathbf{e}_\theta \qquad (5\text{-}517)$$

Simplifying Eq. (5-517), we obtain the acceleration of the center of mass of the disk in reference frame $\mathcal{F}$ as

$$ {}^{\mathcal{F}}\mathbf{a}_C = (\ddot{x} + r\ddot{\theta} - x\dot{\theta}^2)\mathbf{e}_r + (2\dot{x}\dot{\theta} + x\ddot{\theta} + r\dot{\theta}^2)\mathbf{e}_\theta \qquad (5\text{-}518)$$

Next, we need to determine the angular velocity of the disk in reference frame $\mathcal{F}$. Suppose we denote the reference frame of the disk by $\mathcal{D}$. Then, since the motion is planar, we have

$$ {}^{\mathcal{F}}\boldsymbol{\omega}^{\mathcal{D}} = \omega\mathbf{e}_z \qquad (5\text{-}519)$$

Furthermore, considering point P to be fixed to the disk, we have from Eq. (2-517) on page 106 that

$$ {}^{\mathcal{F}}\mathbf{v}_C = {}^{\mathcal{F}}\mathbf{v}_P + {}^{\mathcal{F}}\boldsymbol{\omega}^{\mathcal{D}} \times (\mathbf{r}_C - \mathbf{r}_P) \qquad (5\text{-}520)$$

Furthermore, from the geometry we have

$$\begin{aligned} \mathbf{r}_C &= x\mathbf{e}_r - r\mathbf{e}_\theta & (5\text{-}521) \\ \mathbf{r}_P &= x\mathbf{e}_r & (5\text{-}522) \end{aligned}$$

Therefore,

$$ \mathbf{r}_C - \mathbf{r}_P = x\mathbf{e}_r - r\mathbf{e}_\theta - x\mathbf{e}_r = -r\mathbf{e}_\theta \qquad (5\text{-}523)$$

Substituting ${}^{\mathcal{F}}\boldsymbol{\omega}^{\mathcal{D}}$ and $\mathbf{r}_C - \mathbf{r}_P$ from Eqs. (5-519) and (5-523), respectively, into Eq. (5-520), we obtain

$$ {}^{\mathcal{F}}\mathbf{v}_C = {}^{\mathcal{F}}\mathbf{v}_P + \omega\mathbf{e}_z \times (-r\mathbf{e}_\theta) = {}^{\mathcal{F}}\mathbf{v}_P + r\omega\mathbf{e}_r \qquad (5\text{-}524)$$

Next, considering point P to be fixed to the rod, we have from Eq. (2-517) on page 106 that

$$^{\mathcal{F}}\mathbf{v}_P = {}^{\mathcal{F}}\mathbf{v}_O + {}^{\mathcal{F}}\boldsymbol{\omega}^{\mathcal{R}} \times (\mathbf{r}_P - \mathbf{r}_O) = \dot{\theta}\mathbf{E}_z \times x\mathbf{e}_r = x\dot{\theta}\mathbf{e}_\theta \tag{5-525}$$

Noting that ${}^{\mathcal{F}}\mathbf{v}_O = \mathbf{0}$, we obtain

$$^{\mathcal{F}}\mathbf{v}_P = {}^{\mathcal{F}}\boldsymbol{\omega}^{\mathcal{R}} \times (\mathbf{r}_P - \mathbf{r}_O) = \dot{\theta}\mathbf{E}_z \times x\mathbf{e}_r = x\dot{\theta}\mathbf{e}_\theta \tag{5-526}$$

Then, subtracting Eq. (5-526) from Eq. (5-513), we obtain ${}^{\mathcal{F}}\mathbf{v}_C - {}^{\mathcal{F}}\mathbf{v}_P$ as

$$^{\mathcal{F}}\mathbf{v}_C - {}^{\mathcal{F}}\mathbf{v}_P = (\dot{x} + r\dot{\theta})\mathbf{e}_r + x\dot{\theta}\mathbf{e}_\theta - x\dot{\theta}\mathbf{e}_\theta = (\dot{x} + r\dot{\theta})\mathbf{e}_r \tag{5-527}$$

Rearranging Eq. (5-527), we obtain

$$^{\mathcal{F}}\mathbf{v}_C = {}^{\mathcal{F}}\mathbf{v}_P + (\dot{x} + r\dot{\theta})\mathbf{e}_r + x\dot{\theta}\mathbf{e}_\theta - x\dot{\theta}\mathbf{e}_\theta = {}^{\mathcal{F}}\mathbf{v}_P + (\dot{x} + r\dot{\theta})\mathbf{e}_r \tag{5-528}$$

Setting the results of Eqs. (5-524) and (5-527) equal, we obtain

$$^{\mathcal{F}}\mathbf{v}_P + r\omega\mathbf{e}_r = {}^{\mathcal{F}}\mathbf{v}_P + (\dot{x} + r\dot{\theta})\mathbf{e}_r \tag{5-529}$$

From Eq. (5-529) we obtain

$$r\omega = \dot{x} + r\dot{\theta} \tag{5-530}$$

Solving Eq. (5-530) for ω yields

$$\omega = \dot{\theta} + \frac{\dot{x}}{r} \tag{5-531}$$

The angular velocity of the disk in reference frame $\mathcal{F}$ is then given as

$$^{\mathcal{F}}\boldsymbol{\omega}^{\mathcal{D}} = \left(\dot{\theta} + \frac{\dot{x}}{r}\right)\mathbf{e}_z \tag{5-532}$$

Kinetics

As stated above, this problem will be solved by applying the following balance laws: Euler's 1st law to the disk; Euler's 2nd law to the disk relative to the center of mass of the disk; and Euler's 2nd law to the rod relative to the hinge point.

Application of Euler's 1st Law to Disk

Applying Euler's 1st law to the center of mass of the disk, we have

$$\mathbf{F} = m\,{}^{\mathcal{F}}\bar{\mathbf{a}} = m\,{}^{\mathcal{F}}\mathbf{a}_C \tag{5-533}$$

The free body diagram of the disk is shown in Fig. 5-24.

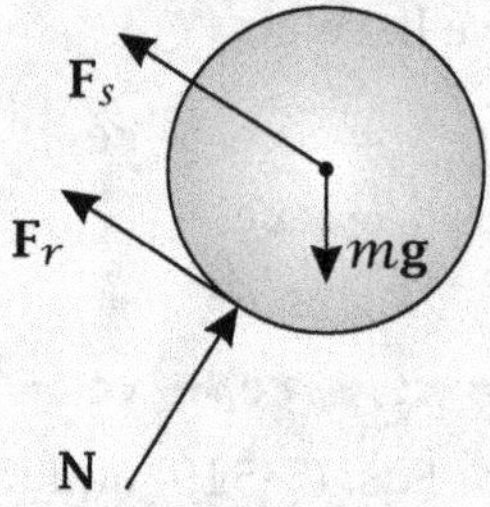

Figure 5-24 Free body diagram of disk for Example 5-9.

Using Fig. 5-24, we have

$$\begin{array}{lcl} \mathbf{F}_R & = & \text{Rolling force of rod on disk} \\ \mathbf{N} & = & \text{Normal force of rod on disk} \\ m\mathbf{g} & = & \text{Force of gravity} \\ \mathbf{F}_s & = & \text{Force of linear spring} \end{array}$$

Now from the geometry we have

$$\mathbf{F}_R = F_R\mathbf{e}_\theta \tag{5-534}$$

$$\mathbf{N} = N\mathbf{e}_\theta \tag{5-535}$$

$$m\mathbf{g} = mg\mathbf{E}_y = mg\sin\theta\,\mathbf{e}_r + mg\cos\theta\,\mathbf{e}_\theta \tag{5-536}$$

$$\mathbf{F}_s = -K_l(x - x_0)\mathbf{e}_r \tag{5-537}$$

where we have used in Eq. (5-536) the expression for $\mathbf{E}_y$ from Eq. (5-507). Consequently,

$$\begin{aligned} \mathbf{F} &= F_R\mathbf{e}_r + N\mathbf{e}_\theta + mg\sin\theta\,\mathbf{e}_r + mg\cos\theta\,\mathbf{e}_\theta - K_l(x - x_0)\mathbf{e}_r \\ &= [F_R + mg\sin\theta - K_l(x - x_0)]\,\mathbf{e}_r + [mg\cos\theta + N]\,\mathbf{e}_\theta \end{aligned} \tag{5-538}$$

Then, setting $\mathbf{F}$ equal to $m^{\mathcal{F}}\mathbf{a}_C$ using the expression for ${}^{\mathcal{F}}\mathbf{a}_C$ from Eq. (5-518), we obtain

$$\begin{aligned} [F_R + mg\sin\theta - K_l(x - x_0)]\,\mathbf{e}_r + [mg\cos\theta + N]\,\mathbf{e}_\theta \\ = m(\ddot{x} + r\ddot{\theta} - x\dot{\theta}^2)\mathbf{e}_r \\ + m(2\dot{x}\dot{\theta} + x\ddot{\theta} + r\dot{\theta}^2)\mathbf{e}_\theta \end{aligned} \tag{5-539}$$

Equating components, we obtain the following two scalar equations:

$$F_R + mg\sin\theta - K_l(x - x_0) = m(\ddot{x} + r\ddot{\theta} - x\dot{\theta}^2) \tag{5-540}$$

$$mg\cos\theta + N = m(2\dot{x}\dot{\theta} + x\ddot{\theta} + r\dot{\theta}^2) \tag{5-541}$$

Application of Euler's 2nd Law to Disk Relative to Center of Mass of Disk

Applying Euler's law relative to the center of mass of the disk, we have

$$\mathbf{M}_C = \frac{{}^{\mathcal{F}}d}{dt}\left({}^{\mathcal{F}}\mathbf{H}_C\right) \tag{5-542}$$

Again it is seen from Fig. 5-24 that the forces $m\mathbf{g}$, $\mathbf{F}_s$, and $\mathbf{N}$ pass through point C. Consequently, the moment relative to the center of mass of the disk is given as

$$\mathbf{M}_C = (\mathbf{r}_P - \mathbf{r}_C) \times \mathbf{F}_R \tag{5-543}$$

Using the fact that $\mathbf{r}_P - \mathbf{r}_C = r\mathbf{e}_\theta$, we have

$$\mathbf{M}_C = r\mathbf{e}_\theta \times F_R\mathbf{e}_r = -rF_R\mathbf{E}_z \tag{5-544}$$

Now the angular momentum of the disk relative to its center of mass is given as

$${}^{\mathcal{F}}\mathbf{H}_C = \bar{\mathbf{I}}^{\mathcal{D}} \cdot {}^{\mathcal{F}}\boldsymbol{\omega}^{\mathcal{D}} \tag{5-545}$$

Expressing the moment of inertia tensor $\bar{\mathbf{I}}^{\mathcal{D}}$ in terms of the basis $\{\mathbf{e}_r, \mathbf{e}_\theta, \mathbf{e}_z\}$, we obtain

$$\bar{\mathbf{I}}^{\mathcal{D}} = \bar{I}_{rr}\mathbf{e}_r \otimes \mathbf{e}_r + \bar{I}_{\theta\theta}\mathbf{e}_\theta \otimes \mathbf{e}_\theta + \bar{I}_{zz}\mathbf{e}_z \otimes \mathbf{e}_z \tag{5-546}$$

Furthermore, using ${}^{\mathcal{F}}\boldsymbol{\omega}^{\mathcal{D}}$ from Eq. (5-532), we have

$$^{\mathcal{F}}\mathbf{H}_C = \bar{I}_{zz}\left(\dot{\theta} + \frac{\dot{x}}{r}\right)\mathbf{e}_z \tag{5-547}$$

Now the moment of inertia for a uniform disk relative to its center of mass is given as

$$\bar{I}_{zz} = \frac{mr^2}{2} \tag{5-548}$$

Therefore, the angular momentum of the disk in reference frame $\mathcal{F}$ is given as

$$^{\mathcal{F}}\mathbf{H}_C = \frac{mr^2}{2}\left(\dot{\theta} + \frac{\dot{x}}{r}\right)\mathbf{e}_z \tag{5-549}$$

Computing the rate of change of ${}^{\mathcal{F}}\mathbf{H}_C$ in reference frame $\mathcal{F}$, we obtain

$$\frac{^{\mathcal{F}}d}{dt}\left({}^{\mathcal{F}}\mathbf{H}_C\right) = \frac{mr^2}{2}\left(\ddot{\theta} + \frac{\ddot{x}}{r}\right)\mathbf{E}_z \tag{5-550}$$

Setting $\mathbf{M}_C$ from Eq. (5-544) equal to ${}^{\mathcal{F}}d({}^{\mathcal{F}}\mathbf{H}_C)/dt$ from Eq. (5-550), we obtain

$$-rF_R = \frac{mr^2}{2}\left(\ddot{\theta} + \frac{\ddot{x}}{r}\right) \tag{5-551}$$

Solving for F_R, we obtain

$$F_R = -\frac{mr}{2}\left(\ddot{\theta} + \frac{\ddot{x}}{r}\right) \tag{5-552}$$

Application of Euler's 2nd Law to Rod Relative to Hinge Point

Applying Euler's 2nd law to the rod relative to the hinge point, observing that the hinge point is inertially fixed, we have

$$\mathbf{M}_O = \frac{^{\mathcal{F}}d}{dt}\left({}^{\mathcal{F}}\mathbf{H}_O\right) \tag{5-553}$$

The free body diagram of the rod is shown in Fig. 5-25.

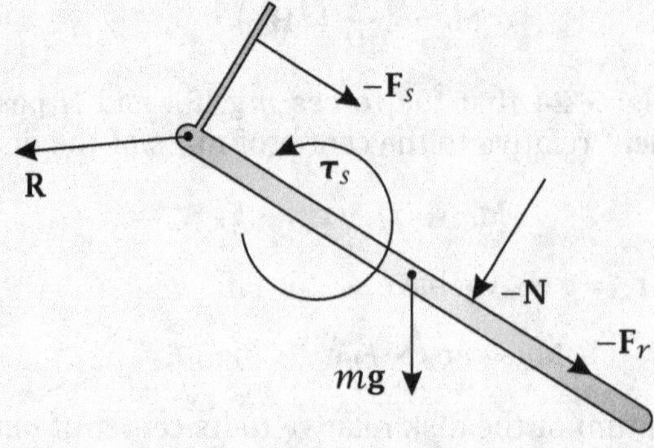

Figure 5-25 Free body diagram of rod for Example 5-9.

Using Fig. 5-25, we have

$$
\begin{aligned}
\mathbf{R} &= \text{Reaction force of hinge on rod} \\
-\mathbf{N} &= \text{Reaction force of disk on rod} \\
-\mathbf{F}_R &= \text{Rolling force of disk on rod} \\
m\mathbf{g} &= \text{Force of gravity} \\
-\mathbf{F}_s &= \text{Force of linear spring} \\
\boldsymbol{\tau}_s &= \text{Pure torque due to torsional spring}
\end{aligned}
$$

First, the reaction force of the hinge on the rod is given in terms of the basis $\{\mathbf{e}_r, \mathbf{e}_\theta, \mathbf{e}_z\}$ as

$$\mathbf{R} = R_r\mathbf{e}_r + R_\theta\mathbf{e}_\theta \tag{5-554}$$

Furthermore, using the expressions for $\mathbf{F}_R$, $\mathbf{N}$, $m\mathbf{g}$, and $\mathbf{F}_s$ from Eqs. (5-534)-(5-537) we have, respectively,

$$
\begin{aligned}
-\mathbf{N} &= -N\mathbf{e}_\theta & (5\text{-}555)\\
-\mathbf{F}_R &= -F_R\mathbf{e}_r & (5\text{-}556)\\
m\mathbf{g} &= mg\mathbf{E}_y = mg\sin\theta\,\mathbf{e}_r + \cos\theta\,\mathbf{e}_\theta & (5\text{-}557)\\
-\mathbf{F}_s &= K_l(x - x_0)\mathbf{e}_r & (5\text{-}558)
\end{aligned}
$$

where in Eq. (5-557) we have used the expression for $\mathbf{E}_y$ from Eq. (5-507). Finally, the pure torque due to the torsional spring is given as

$$\boldsymbol{\tau}_s = -K_\tau\theta\mathbf{e}_z \tag{5-559}$$

The moment applied to the rod relative to point O is then given as

$$
\begin{aligned}
\mathbf{M}_O = \mathbf{r}_O \times \mathbf{R} + (\bar{\mathbf{r}} - \mathbf{r}_O) \times m\mathbf{g} + (\mathbf{r}_P - \mathbf{r}_O) \times (-\mathbf{N} - \mathbf{F}_R)\\
+ (\mathbf{r}_A - \mathbf{r}_O) \times (-\mathbf{F}_s) + \boldsymbol{\tau}_s
\end{aligned}
\tag{5-560}
$$

Observing that $\bar{\mathbf{r}} - \mathbf{r}_O = (l/2)\mathbf{e}_r$, $\mathbf{r}_A - \mathbf{r}_O = -r\mathbf{e}_\theta$, and $\mathbf{r}_P - \mathbf{r}_O = x\mathbf{e}_r$, Eq. (5-560) becomes

$$
\begin{aligned}
\mathbf{M}_O &= \tfrac{l}{2}\mathbf{e}_r \times [mg\sin\theta\,\mathbf{e}_r + mg\cos\theta\,\mathbf{e}_\theta] + x\mathbf{e}_r \times (-N\mathbf{e}_\theta - F_R\mathbf{e}_r)\\
&\quad + (-r\mathbf{e}_\theta) \times (K_l(x - x_0)\mathbf{e}_r) - K_\tau\theta\mathbf{e}_z\\
&= \left(\frac{mgl}{2}\cos\theta - xN + rK_l(x - x_0) - K_\tau\theta\right)\mathbf{e}_z
\end{aligned}
\tag{5-561}
$$

Next, the angular momentum of the rod relative to point O in reference frame $\mathcal{F}$ is given as

$${}^{\mathcal{F}}\mathbf{H}_O = \mathbf{I}^{\mathcal{R}} \cdot {}^{\mathcal{F}}\boldsymbol{\omega}^{\mathcal{R}} \tag{5-562}$$

Now since $\{\mathbf{e}_r, \mathbf{e}_\theta, \mathbf{e}_z\}$ is a principle-axis basis for the rod, we have

$$\mathbf{I}^{\mathcal{R}} = I_{rr}^O\mathbf{e}_r \otimes \mathbf{e}_r + I_{\theta\theta}^O\mathbf{e}_\theta \otimes \mathbf{e}_\theta + I_{zz}^O\mathbf{e}_z \otimes \mathbf{e}_z \tag{5-563}$$

Then, using the expression for ${}^{\mathcal{F}}\boldsymbol{\omega}^{\mathcal{R}}$ from Eq. (5-508), we obtain ${}^{\mathcal{F}}\mathbf{H}_O$ as

$${}^{\mathcal{F}}\mathbf{H}_O = (I_{rr}^O\mathbf{e}_r \otimes \mathbf{e}_r + I_{\theta\theta}^O\mathbf{e}_\theta \otimes \mathbf{e}_\theta + I_{zz}^O\mathbf{e}_z \otimes \mathbf{e}_z) \cdot \dot{\theta}\mathbf{e}_z = I_{zz}^O\dot{\theta}\mathbf{E}_z \tag{5-564}$$

Now for a slender uniform rod we have

$$\bar{I}_{zz} = \frac{ml^2}{12} \tag{5-565}$$

Then, applying the parallel-axis theorem, we obtain

$$I_{zz}^{O} = \bar{I}_{zz} + m(\pi_r^2 + \pi_\theta^2) \tag{5-566}$$

Now since in this case the translation is performed in only the $\mathbf{e}_r$-direction, we have

$$\pi_r = \frac{l}{2} \tag{5-567}$$

$$\pi_\theta = 0 \tag{5-568}$$

Therefore,

$$I_O = \frac{ml^2}{12} + m\left(\frac{l}{2}\right)^2 = \frac{ml^2}{3} \tag{5-569}$$

The angular momentum of the rod relative to the hinge point in reference frame $\mathcal{F}$ is then given as

$${}^{\mathcal{F}}\mathbf{H}_O = \frac{ml^2}{3}\dot{\theta}\mathbf{e}_z \tag{5-570}$$

Computing the rate of change of ${}^{\mathcal{F}}\mathbf{H}_O$ in reference frame $\mathcal{F}$, we obtain which implies that

$$\frac{{}^{\mathcal{F}}d}{dt}\left({}^{\mathcal{F}}\mathbf{H}_O\right) = \frac{ml^2}{3}\ddot{\theta}\mathbf{e}_z \tag{5-571}$$

Setting $\mathbf{M}_O$ from Eq. (5-561) equal to ${}^{\mathcal{F}}d({}^{\mathcal{F}}\mathbf{H}_O)/dt$ from Eq. (5-571), we obtain

$$\frac{mgl}{2}\cos\theta - xN + rK_l(x - x_0) - K_\tau\theta = \frac{ml^2}{3}\ddot{\theta} \tag{5-572}$$

System of Two Differential Equations

We can now use the results of parts (b), (c), and (d) to obtain a system of two differential equations that describes the motion of the rod and the disk. First, substituting the expression for F_R from Eq. (5-552) into Eq. (5-540), we obtain

$$-\frac{mr}{2}\left(\ddot{\theta} + \frac{\ddot{x}}{r}\right) + mg\sin\theta - K_l(x - x_0) = m(\ddot{x} + r\ddot{\theta} - x\dot{\theta}^2) \tag{5-573}$$

Rearranging Eq. (5-573), we have

$$m(\ddot{x} + r\ddot{\theta} - x\dot{\theta}^2) + \frac{mr}{2}\left(\ddot{\theta} + \frac{\ddot{x}}{r}\right) + K_l(x - x_0) - mg\sin\theta = 0 \tag{5-574}$$

Simplifying Eq. (5-574), we obtain the first differential equation as

$$\tfrac{3}{2}m\ddot{x} + \tfrac{3}{2}mr\ddot{\theta} - mx\dot{\theta}^2 + K_l(x - x_0) - mg\sin\theta = 0 \tag{5-575}$$

Next, solving Eq. (5-541) for N, we obtain

$$N = m(2\dot{x}\dot{\theta} + x\ddot{\theta} + r\dot{\theta}^2) - mg\cos\theta \tag{5-576}$$

Substituting the expression for N from Eq. (5-576) into Eq. (5-572), we have

$$\frac{mgl}{2}\cos\theta - x\left[m(2\dot{x}\dot{\theta} + x\ddot{\theta} + r\dot{\theta}^2) + mg\cos\theta\right] + rK_l(x - x_0) - K_T\theta = \frac{ml^2}{3}\ddot{\theta} \tag{5-577}$$

Rearranging Eq. (5-577), we obtain the second differential equation as

$$m\left(\frac{l^2}{3} + x^2\right)\ddot{\theta} + 2mx\dot{x}\dot{\theta} + mrx\dot{\theta}^2 + mg\left(x - \frac{l}{2}\right)\cos\theta - rK_l(x - x_0) + K_T\theta = 0 \tag{5-578}$$

The system of two differential equations for the rod and the disk is then given as

$$\tfrac{3}{2}m\ddot{x} + \tfrac{3}{2}mr\ddot{\theta} - mx\dot{\theta}^2 + K_l(x - x_0) - mg\sin\theta = 0 \tag{5-579}$$

$$m\left(\frac{l^2}{3} + x^2\right)\ddot{\theta} + 2mx\dot{x}\dot{\theta} + mrx\dot{\theta}^2 + mg\left(x - \frac{l}{2}\right)\cos\theta - rK_l(x - x_0) + K_T\theta = 0 \tag{5-580}$$

■

Example 5-10

A system consists of two barrels that roll without slip in the vertical plane as shown in Fig. 5-26. The barrels are modeled as thin cylindrical shells of mass m and radius r. The lower barrel rolls without slip along a fixed horizontal surface while the upper barrel rolls without slip along the lower barrel. Knowing that gravity acts downward, determine a system of two differential equations in terms of the angles θ and ϕ that describes the motion of the barrels.

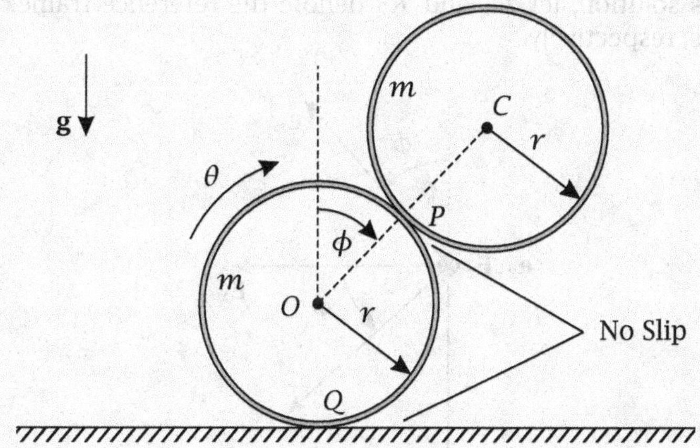

Figure 5-26 Two barrels rolling along one another.

Solution to Example 5–10

Kinematics

First, let $\mathcal{F}$ be a fixed reference frame. Then, we choose the following coordinate system fixed in reference frame $\mathcal{F}$:

$$\begin{array}{ccc} & \text{Origin at } O & \\ & \text{when } t = 0 & \\ \mathbf{E}_x & = & \text{To the right} \\ \mathbf{E}_z & = & \text{Into page} \\ \mathbf{E}_y & = & \mathbf{E}_z \times \mathbf{E}_x \end{array}$$

Next, let $\mathcal{A}$ be a reference frame that is fixed to the direction of OC. Then, we choose the following coordinate system fixed in reference frame $\mathcal{A}$:

$$\begin{array}{ccc} & \text{Origin at } O & \\ & \text{fixed to lower barrel} & \\ \mathbf{e}_r & = & \text{Along } OC \\ \mathbf{e}_z & = & \text{Into page } (= \mathbf{E}_z) \\ \mathbf{e}_\phi & = & \mathbf{e}_z \times \mathbf{e}_r = \mathbf{E}_z \times \mathbf{e}_r \end{array}$$

Using Fig. 5-27, we can solve for $\{\mathbf{e}_r, \mathbf{e}_\phi\}$ in terms of $\{\mathbf{E}_x, \mathbf{E}_y\}$ to give

$$\begin{array}{rcl} \mathbf{e}_r & = & \sin\phi\mathbf{E}_x - \cos\phi\mathbf{E}_y \\ \mathbf{e}_\phi & = & \cos\phi\mathbf{E}_x + \sin\phi\mathbf{E}_y \end{array} \tag{5-581}$$

Similarly, we can solve for $\{\mathbf{E}_x, \mathbf{E}_y\}$ in terms of $\{\mathbf{e}_r, \mathbf{e}_\phi\}$ to give

$$\begin{array}{rcl} \mathbf{E}_x & = & \sin\phi\mathbf{e}_r + \cos\phi\mathbf{e}_\phi \\ \mathbf{E}_y & = & -\cos\phi\mathbf{e}_r + \sin\phi\mathbf{e}_\phi \end{array} \tag{5-582}$$

It is important to understand that, because the direction OC is not fixed in either barrel, reference frame $\mathcal{A}$ is also *not* fixed in either barrel. Finally, for clarity with the remainder of this solution, let $\mathcal{R}_1$ and $\mathcal{R}_2$ denote the reference frames of the lower and upper barrels, respectively.

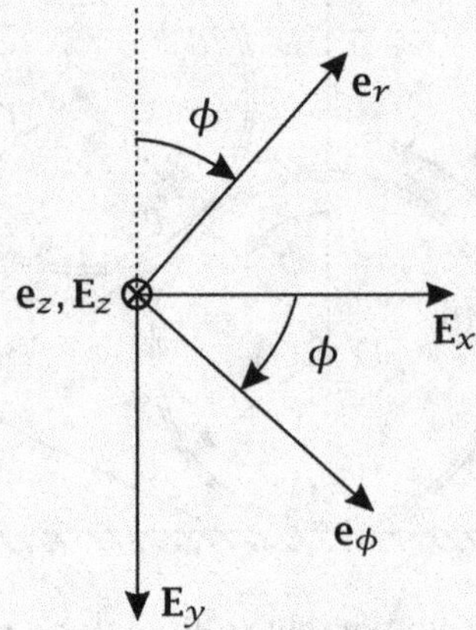

Figure 5-27 Relationship between bases $\{\mathbf{E}_x, \mathbf{E}_y, \mathbf{E}_z\}$ and $\{\mathbf{e}_r, \mathbf{e}_\theta, \mathbf{e}_z\}$ for Example 5-10.

Now, looking ahead to the kinetics of this problem, it is clear that we will need to apply some form of Euler's laws to the system and/or subsystems of the system. Consequently, it is necessary to determine those kinematic quantities that are required to apply Euler's laws. In particular, in this problem the following kinematic quantities are relevant and must be computed in reference frame $\mathcal{F}$: the angular velocity of the lower barrel; the velocity and acceleration of the center of mass of the lower barrel; the angular velocity of the upper barrel; and the velocity and acceleration of the center of mass of the upper barrel. Each of these quantities will now be computed.

Angular Velocity of Lower Barrel

Because the angle θ describes the absolute rotation of the lower barrel relative to reference frame $\mathcal{F}$, the angular velocity of the lower barrel in reference frame $\mathcal{F}$ is given as

$$ {}^{\mathcal{F}}\boldsymbol{\omega}^{\mathcal{R}_1} = \dot{\theta}\mathbf{E}_z \tag{5-583} $$

Velocity and Acceleration of Center of Mass of Lower Barrel

We note that Q is the instantaneous point of contact of the lower barrel with the ground. Because the lower barrel rolls without slip on the ground and the ground is a fixed surface, we have

$$ {}^{\mathcal{F}}\mathbf{v}_Q = \mathbf{0} \tag{5-584} $$

Also, applying Eq. (2-517) on page 106, we have

$$ {}^{\mathcal{F}}\mathbf{v}_O - {}^{\mathcal{F}}\mathbf{v}_Q = {}^{\mathcal{F}}\boldsymbol{\omega}^{\mathcal{R}_1} \times (\mathbf{r}_O - \mathbf{r}_Q) \tag{5-585} $$

In addition, from the geometry we have

$$ \mathbf{r}_O - \mathbf{r}_Q = -r\mathbf{E}_y \tag{5-586} $$

Using ${}^{\mathcal{F}}\boldsymbol{\omega}^{\mathcal{R}_1}$ from Eq. (5-583) and the fact that ${}^{\mathcal{F}}\mathbf{v}_Q = \mathbf{0}$, we obtain

$$ {}^{\mathcal{F}}\mathbf{v}_O = \dot{\theta}\mathbf{E}_z \times (-r\mathbf{E}_y) = r\dot{\theta}\mathbf{E}_x \tag{5-587} $$

Computing the rate of change of ${}^{\mathcal{F}}\mathbf{v}_O$ in reference frame $\mathcal{F}$, we have

$$ {}^{\mathcal{F}}\mathbf{a}_O = \frac{{}^{\mathcal{F}}d}{dt}\left({}^{\mathcal{F}}\mathbf{v}_O\right) = r\ddot{\theta}\mathbf{E}_x \tag{5-588} $$

Angular Velocity of Upper Barrel

First, from the geometry of the problem we have

$$ \mathbf{r}_C - \mathbf{r}_O = 2r\mathbf{e}_r \tag{5-589} $$

Computing the rate of change of $\mathbf{r}_C - \mathbf{r}_O$ in reference frame $\mathcal{F}$, we obtain

$$ {}^{\mathcal{F}}\mathbf{v}_C - {}^{\mathcal{F}}\mathbf{v}_O = \frac{{}^{\mathcal{F}}d}{dt}(\mathbf{r}_C - \mathbf{r}_O) = 2r\dot{\phi}\mathbf{e}_\phi \tag{5-590} $$

Equation (5-590) implies that

$$^{\mathcal{F}}\mathbf{v}_C = {}^{\mathcal{F}}\mathbf{v}_O + 2r\dot{\phi}\mathbf{e}_\phi \tag{5-591}$$

Substituting the result for ${}^{\mathcal{F}}\mathbf{v}_O$ from Eq. (5-587), we obtain

$$^{\mathcal{F}}\mathbf{v}_C = r\dot{\theta}\mathbf{E}_x + 2r\dot{\phi}\mathbf{e}_\phi \tag{5-592}$$

We can obtain a second expression for ${}^{\mathcal{F}}\mathbf{v}_C$ as follows. Noting that P is the point of contact between the lower and upper barrels and that the upper barrel rolls without slip on the lower barrel, we know from Eq. (2-525) that the velocity of point P in reference frame $\mathcal{F}$ is the same on either barrel, i.e.,

$$^{\mathcal{F}}\mathbf{v}_P^{\mathcal{R}_1} = {}^{\mathcal{F}}\mathbf{v}_P^{\mathcal{R}_2} \equiv {}^{\mathcal{F}}\mathbf{v}_P \tag{5-593}$$

Then, because we are considering point P to be fixed to the lower barrel, we have

$$^{\mathcal{F}}\mathbf{v}_P - {}^{\mathcal{F}}\mathbf{v}_O = {}^{\mathcal{F}}\boldsymbol{\omega}^{\mathcal{R}_1} \times (\mathbf{r}_P - \mathbf{r}_O) \tag{5-594}$$

From the geometry we have

$$\mathbf{r}_P - \mathbf{r}_O = r\mathbf{e}_r \tag{5-595}$$

Then, substituting $\mathbf{r}_P - \mathbf{r}_O$ from Eq. (5-595) and ${}^{\mathcal{F}}\boldsymbol{\omega}^{\mathcal{R}_1}$ from Eq. (5-583) into Eq. (5-594), we obtain

$$^{\mathcal{F}}\mathbf{v}_P - {}^{\mathcal{F}}\mathbf{v}_O = \dot{\theta}\mathbf{E}_z \times r\mathbf{e}_r = r\dot{\theta}\mathbf{e}_\phi \tag{5-596}$$

Equation (5-594) implies that

$$^{\mathcal{F}}\mathbf{v}_P = {}^{\mathcal{F}}\mathbf{v}_O + r\dot{\theta}\mathbf{e}_\phi = r\dot{\theta}\mathbf{E}_x + r\dot{\theta}\mathbf{e}_\phi \tag{5-597}$$

Next, because this time we are considering point P to be fixed to the upper barrel, we have

$$^{\mathcal{F}}\mathbf{v}_C - {}^{\mathcal{F}}\mathbf{v}_P = {}^{\mathcal{F}}\boldsymbol{\omega}^{\mathcal{R}_2} \times (\mathbf{r}_C - \mathbf{r}_B) \tag{5-598}$$

where ${}^{\mathcal{F}}\boldsymbol{\omega}^{\mathcal{R}_2}$ is the angular velocity of the upper barrel in reference frame $\mathcal{F}$. Because the motion is planar, we can write ${}^{\mathcal{F}}\boldsymbol{\omega}^{\mathcal{R}_2}$ as

$$^{\mathcal{F}}\boldsymbol{\omega}^{\mathcal{R}_2} = \omega_2\mathbf{E}_z \tag{5-599}$$

Furthermore, from the geometry we have

$$\mathbf{r}_C - \mathbf{r}_B = r\mathbf{e}_r \tag{5-600}$$

Therefore, we obtain

$$^{\mathcal{F}}\mathbf{v}_C - {}^{\mathcal{F}}\mathbf{v}_P = \omega_2\mathbf{E}_z \times r\mathbf{e}_r = r\omega_2\mathbf{e}_\phi \tag{5-601}$$

Solving for ${}^{\mathcal{F}}\mathbf{v}_C$, we obtain

$$^{\mathcal{F}}\mathbf{v}_C = {}^{\mathcal{F}}\mathbf{v}_P + r\omega_2\mathbf{e}_\phi \tag{5-602}$$

Substituting the result for ${}^{\mathcal{F}}\mathbf{v}_P$ from Eq. (5-597), we obtain a second expression for ${}^{\mathcal{F}}\mathbf{v}_C$ as

$$^{\mathcal{F}}\mathbf{v}_C = r\dot{\theta}\mathbf{E}_x + r\dot{\theta}\mathbf{e}_\phi + r\omega_2\mathbf{e}_\phi = r\dot{\theta}\mathbf{E}_x + r(\dot{\theta} + \omega_2)\mathbf{e}_\phi \tag{5-603}$$

Then, setting the expressions for ${}^{\mathcal{F}}\mathbf{v}_C$ from Eqs. (5-592) and (5-603) equal, we obtain

$$r\dot{\theta}\mathbf{E}_x + 2r\dot{\phi}\mathbf{e}_\phi = r\dot{\theta}\mathbf{E}_x + r(\dot{\theta} + \omega_2)\mathbf{e}_\phi \tag{5-604}$$

Simplifying Eq. (5-604), we obtain

$$2r\dot{\phi}\mathbf{e}_\phi = r(\dot{\theta} + \omega_2)\mathbf{e}_\phi \tag{5-605}$$

Then, from Eq. (5-605) we have

$$2r\dot{\phi} = r(\dot{\theta} + \omega_2) \tag{5-606}$$

Solving Eq. (5-606) for ω_2, we obtain

$$\omega_2 = 2\dot{\phi} - \dot{\theta} \tag{5-607}$$

The angular velocity of the upper barrel in reference frame $\mathcal{F}$ is then given as

$${}^{\mathcal{F}}\boldsymbol{\omega}^{\mathcal{R}_2} = (2\dot{\phi} - \dot{\theta})\mathbf{E}_z \tag{5-608}$$

Acceleration of Center of Mass of Upper Barrel

Computing the rate of change of ${}^{\mathcal{F}}\mathbf{v}_C$ as given in Eq. (5-592) in reference frame $\mathcal{F}$, the acceleration of the center of mass of the upper barrel is given as

$${}^{\mathcal{F}}\mathbf{a}_C = \frac{{}^{\mathcal{F}}d}{dt}\left({}^{\mathcal{F}}\mathbf{v}_C\right) = r\ddot{\theta}\mathbf{E}_x - 2r\dot{\phi}^2\mathbf{e}_r + 2r\ddot{\phi}\mathbf{e}_\phi \tag{5-609}$$

Kinetics

For this problem we will apply the following balance laws: (1) Euler's 2nd law relative to the contact point of the lower barrel with the ground, (2) Euler's 1st law to the upper barrel, and (3) Euler's 2nd law relative to the center of mass of the upper barrel.

Application of Euler's 2nd Law to Lower Barrel Relative to Instantaneous Point of Contact with Ground

The free body diagram of the lower barrel is shown in Fig. 5-28. Using Fig. 5-28, we have

$\mathbf{N}_1$	=	Reaction force of ground on lower barrel
$\mathbf{R}_1$	=	Rolling force of ground on lower barrel
$m\mathbf{g}$	=	Force of gravity
$\mathbf{N}_2$	=	Reaction force of upper barrel on lower barrel
$\mathbf{R}_2$	=	Rolling force of upper barrel on lower barrel

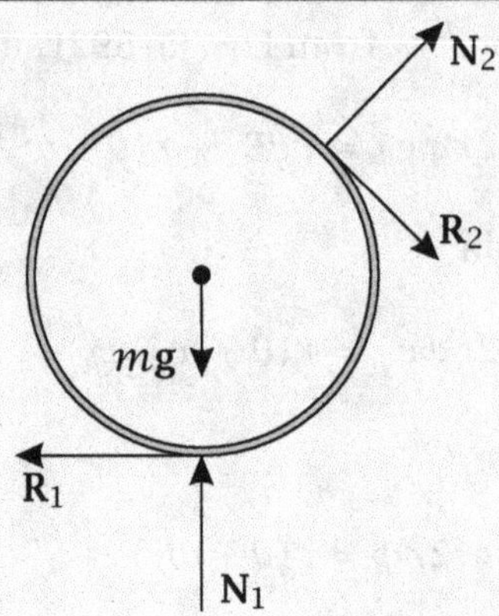

Figure 5–28 Free body diagram of lower barrel for Example 5–10.

Because the lower barrel rolls without slip along the ground and the ground is a fixed surface, the velocity of point Q in reference frame $\mathcal{F}$ is zero, i.e., ${}^{\mathcal{F}}\mathbf{v}_Q = \mathbf{0}$. Applying the general form of Euler's law to the lower barrel about point Q, we have

$$\mathbf{M}_Q - (\mathbf{r}_O - \mathbf{r}_Q) \times m\,{}^{\mathcal{F}}\mathbf{a}_Q = \frac{{}^{\mathcal{F}}d}{dt}\left({}^{\mathcal{F}}\mathbf{H}_Q\right) \tag{5-610}$$

where $\mathbf{r}_O$ is the position of the center of mass of the lower barrel. Using Eq. (2–524) on page 107, the acceleration of points O and Q on the lower barrel in reference frame $\mathcal{F}$ are related as

$${}^{\mathcal{F}}\mathbf{a}_Q = {}^{\mathcal{F}}\mathbf{a}_O + {}^{\mathcal{F}}\boldsymbol{\alpha}^{\mathcal{R}_1} \times (\mathbf{r}_Q - \mathbf{r}_O) + {}^{\mathcal{F}}\boldsymbol{\omega}^{\mathcal{R}_1} \times [\boldsymbol{\omega}_1 \times (\mathbf{r}_Q - \mathbf{r}_O)] \tag{5-611}$$

where ${}^{\mathcal{F}}\boldsymbol{\alpha}^{\mathcal{R}_1}$ is the angular acceleration of the lower barrel in reference frame $\mathcal{F}$. Now ${}^{\mathcal{F}}\boldsymbol{\alpha}^{\mathcal{R}_1}$ is given as

$${}^{\mathcal{F}}\boldsymbol{\alpha}^{\mathcal{R}_1} = \frac{{}^{\mathcal{F}}d}{dt}\left({}^{\mathcal{F}}\boldsymbol{\omega}^{\mathcal{R}_1}\right) = \ddot{\theta}\mathbf{E}_z \tag{5-612}$$

Next, relative position between the center of mass of the lower barrel and the instantaneous point of contact of the lower barrel with the ground is given as

$$\mathbf{r}_Q - \mathbf{r}_O = r\mathbf{E}_y \tag{5-613}$$

Therefore, the acceleration of point Q in reference frame $\mathcal{F}$ is given as

$$\begin{aligned} {}^{\mathcal{F}}\mathbf{a}_Q &= r\ddot{\theta}\mathbf{E}_x + \ddot{\theta}\mathbf{E}_z \times (r\mathbf{E}_y) + \dot{\theta}\mathbf{E}_z \times \left[\dot{\theta}\mathbf{E}_z \times (r\mathbf{E}_y)\right] \\ &= r\ddot{\theta}\mathbf{E}_x - r\ddot{\theta}\mathbf{E}_x - r\dot{\theta}^2\mathbf{E}_y \\ &= -r\dot{\theta}^2\mathbf{E}_y \end{aligned} \tag{5-614}$$

Using the expression for ${}^{\mathcal{F}}\mathbf{a}_Q$ as given in Eq. (5–614), the inertial moment relative to point Q is obtained as

$$-(\mathbf{r}_O - \mathbf{r}_Q) \times m\,{}^{\mathcal{F}}\mathbf{a}_Q = r\mathbf{E}_y \times m(-r\dot{\theta}^2\mathbf{E}_y) = \mathbf{0} \tag{5-615}$$

We see that, *because the lower barrel is modeled as a homogeneous thin cylindrical shell*, the inertial moment of the lower barrel relative to the instantaneous point of contact is *zero*. Using the result from Eq. (5-615), *for this problem* we have

$$\mathbf{M}_Q = \frac{{}^{\mathcal{F}}d}{dt}\left({}^{\mathcal{F}}\mathbf{H}_Q\right) \tag{5-616}$$

Next, examining Fig. 5-28, we see that the forces $\mathbf{N}_1$, $\mathbf{R}_1$, and $m\mathbf{g}$ all pass through point Q. Therefore, the moment about point Q is due entirely to the forces $\mathbf{N}_2$ and $\mathbf{R}_2$ and is given as

$$\mathbf{M}_Q = (\mathbf{r}_P - \mathbf{r}_Q) \times (\mathbf{N}_2 + \mathbf{R}_2) \tag{5-617}$$

where we note that both $\mathbf{N}_2$ and $\mathbf{R}_2$ act at the instantaneous point of contact of the lower barrel with the upper barrel (i.e., point P). Now, because $\mathbf{N}_2$ and $\mathbf{R}_2$ are the reaction force and rolling force, respectively, applied by the upper barrel to the lower barrel, we have

$$\begin{aligned} \mathbf{N}_2 &= N_2\mathbf{e}_r \\ \mathbf{R}_2 &= R_2\mathbf{e}_\phi \end{aligned} \tag{5-618}$$

Furthermore, we have

$$\mathbf{r}_P - \mathbf{r}_Q = r\mathbf{e}_r - r\mathbf{E}_y \tag{5-619}$$

The moment applied to the lower barrel relative to point Q is then given as

$$\mathbf{M}_Q = (r\mathbf{e}_r - r\mathbf{E}_y) \times (N_2\mathbf{e}_r + R_2\mathbf{e}_\phi) \tag{5-620}$$

Expanding Eq. (5-620), we obtain

$$\mathbf{M}_Q = -rN_2\mathbf{E}_y \times \mathbf{e}_r + rR_2\mathbf{e}_r \times \mathbf{e}_\phi - rR_2\mathbf{E}_y \times \mathbf{e}_\phi \tag{5-621}$$

Computing the relevant vector products in Eq. (5-621), we have

$$\begin{aligned} \mathbf{e}_r \times \mathbf{e}_\phi &= \mathbf{E}_z \\ \mathbf{E}_y \times \mathbf{e}_r &= \mathbf{E}_y \times (\sin\phi\mathbf{E}_x - \cos\phi\mathbf{E}_y) = -\sin\phi\mathbf{E}_z \\ \mathbf{E}_y \times \mathbf{e}_\phi &= \mathbf{E}_y \times (\cos\phi\mathbf{E}_x + \sin\phi\mathbf{E}_y) = -\cos\phi\mathbf{E}_z \end{aligned} \tag{5-622}$$

Substituting the expressions from Eq. (5-622) into (5-621), we obtain

$$\mathbf{M}_Q = rN_2\sin\phi\mathbf{E}_z + rR_2\mathbf{E}_z + rR_2\cos\phi\mathbf{E}_z \tag{5-623}$$

Factoring out $\mathbf{E}_z$ from the last expression, we obtain the moment applied to the lower barrel relative to the instantaneous point of contact Q as

$$\mathbf{M}_Q = [rN_2\sin\phi + rR_2(1+\cos\phi)]\,\mathbf{E}_z \tag{5-624}$$

Finally, we need to compute the rate of change of the angular momentum of the lower barrel relative to point Q in reference frame $\mathcal{F}$. First, the angular momentum relative to the center of mass of the lower barrel in reference frame $\mathcal{F}$ is given as

$${}^{\mathcal{F}}\mathbf{H}_O = \mathbf{I}_O^{\mathcal{R}_1} \cdot {}^{\mathcal{F}}\boldsymbol{\omega}^{\mathcal{R}_1} \tag{5-625}$$

where we note that point O is the center of mass of the lower barrel. Now, because $\{\mathbf{E}_x, \mathbf{E}_y, \mathbf{E}_z\}$ is a principal-axis basis, the moment of inertia relative to the center of mass of the lower barrel is given as

$$\bar{\mathbf{I}}^{\mathcal{R}_1} = I_{xx}^O \mathbf{E}_x \otimes \mathbf{E}_x + I_{yy}^O \mathbf{E}_y \otimes \mathbf{E}_y + I_{zz}^O \mathbf{E}_z \otimes \mathbf{E}_z \tag{5-626}$$

Then, using ${}^{\mathcal{F}}\boldsymbol{\omega}^{\mathcal{R}_1}$ from Eq. (5-583), we have

$${}^{\mathcal{F}}\mathbf{H}_O = (I_{xx}^O \mathbf{E}_x \otimes \mathbf{E}_x + I_{yy}^O \mathbf{E}_y \otimes \mathbf{E}_y + I_{zz}^O \mathbf{E}_z \otimes \mathbf{E}_z) \cdot \dot{\theta}\mathbf{E}_z = I_{zz}^O \dot{\theta}\mathbf{E}_z \tag{5-627}$$

Next, using Eq. (5-24), the angular momentum of the lower barrel in reference frame $\mathcal{F}$ relative to point Q is given as

$${}^{\mathcal{F}}\mathbf{H}_Q = {}^{\mathcal{F}}\mathbf{H}_O + (\mathbf{r}_Q - \mathbf{r}_O) \times m({}^{\mathcal{F}}\mathbf{v}_Q - {}^{\mathcal{F}}\mathbf{v}_O) \tag{5-628}$$

Using the fact that ${}^{\mathcal{F}}\mathbf{v}_Q = \mathbf{0}$, we have from Eq. (5-587) that

$${}^{\mathcal{F}}\mathbf{v}_Q - {}^{\mathcal{F}}\mathbf{v}_O = -{}^{\mathcal{F}}\mathbf{v}_O = -r\dot{\theta}\mathbf{E}_x \tag{5-629}$$

Furthermore, substituting Eqs. (5-613), (5-627), and (5-629) into Eq. (5-628), we obtain

$${}^{\mathcal{F}}\mathbf{H}_Q = \bar{I}_{zz}\dot{\theta}\mathbf{E}_z + (r\mathbf{E}_y) \times m(-r\dot{\theta}\mathbf{E}_x) = \bar{I}_{zz}\dot{\theta}\mathbf{E}_z + mr^2\dot{\theta}\mathbf{E}_z = (\bar{I}_{zz} + mr^2)\dot{\theta}\mathbf{E}_z \tag{5-630}$$

Because the barrel is modeled as a thin cylindrical shell, we have

$$I_{zz}^O = mr^2 \tag{5-631}$$

Consequently, ${}^{\mathcal{F}}\mathbf{H}_Q$ simplifies to

$${}^{\mathcal{F}}\mathbf{H}_Q = 2mr^2\dot{\theta}\mathbf{E}_z \tag{5-632}$$

Computing the rate of change of ${}^{\mathcal{F}}\mathbf{H}_Q$ in reference frame $\mathcal{F}$, we obtain

$$\frac{{}^{\mathcal{F}}d}{dt}\left({}^{\mathcal{F}}\mathbf{H}_Q\right) = 2mr^2\ddot{\theta}\mathbf{E}_z \tag{5-633}$$

Then, substituting $\mathbf{M}_Q$ and ${}^{\mathcal{F}}d({}^{\mathcal{F}}\mathbf{H}_Q)/dt$ from Eqs. (5-624) and (5-632), respectively, into (5-616), we have

$$rN_2 \sin\phi + rR_2(1 + \cos\phi) = 2mr^2\ddot{\theta} \tag{5-634}$$

Simplifying this last expression by dividing through by r, we obtain the scalar equation

$$2mr\ddot{\theta} = N_2 \sin\phi + R_2(1 + \cos\phi) \tag{5-635}$$

Application of Euler's 1st Law to Upper Barrel

The free body diagram of the upper barrel is shown in Fig. 5-29, where

$m\mathbf{g}$ = Force of gravity
$-\mathbf{N}_2$ = Reaction force of lower barrel on upper barrel
$-\mathbf{R}_2$ = Rolling force of lower barrel on upper barrel

We note that, from Newton's 3rd law, the forces applied by the lower barrel on the upper barrel are equal and opposite to the forces applied by the upper barrel on the lower barrel (see Fig. 5-28).

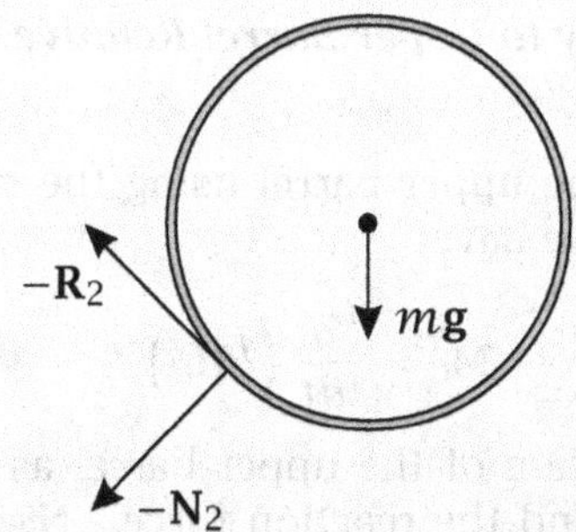

Figure 5–29 Free body diagram of upper barrel for Example 5–10.

Applying Euler's 1st law to the center of mass of the upper barrel, we have

$$\mathbf{F}_2 = m^{\mathcal{F}}\mathbf{a}_C \tag{5-636}$$

where $\mathbf{F}_2$ is the resultant force on the upper barrel. From the free body diagram of Fig. 5–29 we have that

$$\mathbf{F}_2 = m\mathbf{g} - \mathbf{N}_2 - \mathbf{R}_2 \tag{5-637}$$

Substituting the expressions for $\mathbf{N}_2$ and $\mathbf{R}_2$ from Eq. (5–618), we obtain

$$\mathbf{F}_2 = mg\mathbf{E}_y - N_2\mathbf{e}_r - R_2\mathbf{e}_\phi \tag{5-638}$$

Now we can substitute the expression for $\mathbf{E}_y$ in terms of $\mathbf{e}_r$ and $\mathbf{e}_\phi$ from Eq. (5–582) into the last equation. This gives

$$\mathbf{F}_2 = mg(-\cos\phi\mathbf{e}_r + \sin\phi\mathbf{e}_\phi) - N_2\mathbf{e}_r - R_2\mathbf{e}_\phi \tag{5-639}$$

Simplifying this last expression gives

$$\mathbf{F}_2 = -(N_2 + mg\cos\phi)\mathbf{e}_r + (mg\sin\phi - R_2)\mathbf{e}_\phi \tag{5-640}$$

Then, using the result for the acceleration of the center of mass of the upper barrel from Eq. (5–609), we can equate $\mathbf{F}_2$ and $m^{\mathcal{F}}\mathbf{a}_C$ to obtain

$$-(N_2 + mg\cos\phi)\mathbf{e}_r + (mg\sin\phi - R_2)\mathbf{e}_\phi = mr\ddot{\theta}\mathbf{E}_x - 2mr\dot{\phi}^2\mathbf{e}_r + 2mr\ddot{\phi}\mathbf{e}_\phi \tag{5-641}$$

Then, in order to get an equation in terms of a single coordinate system, we can substitute the expression for $\mathbf{E}_x$ in terms of $\mathbf{e}_r$ and $\mathbf{e}_\phi$ from Eq. (5–582) into the last equation. This gives

$$\begin{aligned} -(N_2 + mg\cos\phi)\mathbf{e}_r + (mg\sin\phi - R_2)\mathbf{e}_\phi = {} & mr\ddot{\theta}(\sin\phi\mathbf{e}_r + \cos\phi\mathbf{e}_\phi) \\ & - 2mr\dot{\phi}^2\mathbf{e}_r + 2mr\ddot{\phi}\mathbf{e}_\phi \end{aligned} \tag{5-642}$$

Simplifying Eq. (5–642), we obtain

$$\begin{aligned} -(N_2 + mg\cos\phi)\mathbf{e}_r + (mg\sin\phi - R_2)\mathbf{e}_\phi &= mr(\ddot{\theta}\sin\phi - 2\dot{\phi}^2)\mathbf{e}_r \\ &\quad + mr(\ddot{\theta}\cos\phi + 2\ddot{\phi})\mathbf{e}_\phi \end{aligned} \tag{5-643}$$

Equating components in Eq. (5–643) yields the following two scalar equations:

$$-N_2 - mg\cos\phi = mr\ddot{\theta}\sin\phi - 2mr\dot{\phi}^2 \tag{5-644}$$

$$mg\sin\phi - R_2 = mr\ddot{\theta}\cos\phi + 2mr\ddot{\phi} \tag{5-645}$$

Application of Euler's 2nd Law to Upper Barrel Relative to Center of Mass of Upper Barrel

Applying Euler's 2nd law to the upper barrel using the center of mass of the upper barrel as the reference point, we have

$$\mathbf{M}_C = \frac{^{\mathcal{F}}d}{dt}\left(^{\mathcal{F}}\mathbf{H}_C\right) \tag{5-646}$$

Examining the free body diagram of the upper barrel as shown in Fig. 5-29, we see that the force of gravity $m\mathbf{g}$ and the reaction force $-\mathbf{N}_2$ both pass through point C. Consequently, the moment about point C is due to only the force $-\mathbf{R}_2$ and is given as

$$\mathbf{M}_C = (\mathbf{r}_{R_2} - \mathbf{r}_C) \times (-\mathbf{R}_2) \tag{5-647}$$

Now, from the geometry we have

$$\mathbf{r}_{R_2} - \mathbf{r}_C = -r\mathbf{e}_r \tag{5-648}$$

Therefore,

$$\mathbf{M}_C = -r\mathbf{e}_r \times (-R_2\mathbf{e}_\phi) = rR_2\mathbf{E}_z \tag{5-649}$$

Now we need the rate of change of angular momentum of the upper barrel about point C. First, we have

$$^{\mathcal{F}}\mathbf{H}_C = \mathbf{I}_C^{\mathcal{R}_2} \cdot {}^{\mathcal{F}}\boldsymbol{\omega}^{\mathcal{R}_2} \tag{5-650}$$

Furthermore, the moment of inertia tensor relative to point C for the upper barrel is given in terms of the basis $\{\mathbf{E}_x, \mathbf{E}_y, \mathbf{E}_z\}$ as

$$\mathbf{I}_C^{\mathcal{R}_2} = I_{xx}^C \mathbf{E}_x \otimes \mathbf{E}_x + I_{yy}^C \mathbf{E}_y \otimes \mathbf{E}_y + I_{zz}^C \mathbf{E}_z \otimes \mathbf{E}_z \tag{5-651}$$

Then, using $^{\mathcal{F}}\boldsymbol{\omega}^{\mathcal{R}_2}$ from Eq. (5-608), we obtain the angular momentum of the upper barrel relative to point C as

$$^{\mathcal{F}}\mathbf{H}_C = (I_{xx}^C \mathbf{E}_x \otimes \mathbf{E}_x + I_{yy}^C \mathbf{E}_y \otimes \mathbf{E}_y + I_{zz}^C \mathbf{E}_z \otimes \mathbf{E}_z) \cdot {}^{\mathcal{F}}\boldsymbol{\omega}^{\mathcal{R}_2} = I_{zz}^C(2\dot{\phi} - \dot{\theta})\mathbf{E}_z \tag{5-652}$$

Because the upper barrel is modeled as a thin cylindrical shell and $\{\mathbf{E}_x, \mathbf{E}_y, \mathbf{E}_z\}$ is a principal-axis basis relative to the center of mass of the upper barrel, we have

$$I_{zz}^C = mr^2 \tag{5-653}$$

Consequently, we obtain $^{\mathcal{F}}\mathbf{H}_C$ as

$$^{\mathcal{F}}\mathbf{H}_C = mr^2(2\dot{\phi} - \dot{\theta})\mathbf{E}_z \tag{5-654}$$

Computing the rate of change of $^{\mathcal{F}}\mathbf{H}_C$ in reference frame $\mathcal{F}$ gives

$$\frac{^{\mathcal{F}}d}{dt}\left(^{\mathcal{F}}\mathbf{H}_C\right) = mr^2(2\ddot{\phi} - \ddot{\theta})\mathbf{E}_z \tag{5-655}$$

Setting $\mathbf{M}_C$ from Eq. (5-649) equal to $^{\mathcal{F}}d(^{\mathcal{F}}\mathbf{H}_C)/dt$ from Eq. (5-655) gives

$$rR_2\mathbf{E}_z = mr^2(2\ddot{\phi} - \ddot{\theta})\mathbf{E}_z \tag{5-656}$$

Dropping the dependence on $\mathbf{E}_z$ and simplifying, we obtain the scalar equation

$$R_2 = mr(2\ddot{\phi} - \ddot{\theta}) \tag{5-657}$$

System of Two Differential Equations

We can now use Eqs. (5-635), (5-644), (5-645), and (5-657) to obtain a system of two differential equations. The first differential equation is obtained by substituting R_2 from Eq. (5-657) into (5-645). This gives

$$mg\sin\phi - mr(2\ddot{\phi} - \ddot{\theta}) = mr\ddot{\theta}\cos\phi + 2mr\ddot{\phi} \tag{5-658}$$

Rearranging and simplifying Eq. (5-658), we obtain the first differential equation as

$$r\ddot{\theta}(1+\cos\phi) + 4r\ddot{\phi} - g\sin\phi = 0 \tag{5-659}$$

The second differential equation is obtained as follows. Multiplying Eqs. (5-644) and (5-645) by $\sin\phi$ and $\cos\phi$, respectively, we obtain

$$-N_2\sin\phi - mg\cos\phi\sin\phi = mr\ddot{\theta}\sin^2\phi - 2mr\dot{\phi}^2\sin\phi \tag{5-660}$$

$$mg\sin\phi\cos\phi - R_2\cos\phi = mr\ddot{\theta}\cos^2\phi + 2mr\ddot{\phi}\cos\phi \tag{5-661}$$

Adding Eqs. (5-660) and (5-661) gives

$$-N_2\sin\phi - R_2\cos\phi = mr\ddot{\theta}(\sin^2\phi + \cos^2\phi) - 2mr\dot{\phi}^2\sin\phi + 2mr\ddot{\phi}\cos\phi \tag{5-662}$$

Noting that $\sin^2\phi + \cos^2\phi \equiv 1$, we have

$$-N_2\sin\phi - R_2\cos\phi = mr\ddot{\theta} - 2mr\dot{\phi}^2\sin\phi + 2mr\ddot{\phi}\cos\phi \tag{5-663}$$

Next, augmenting the result of Eq. (5-635) to (5-663), we obtain the following two equations:

$$-N_2\sin\phi - R_2\cos\phi = mr\ddot{\theta} - 2mr\dot{\phi}^2\sin\phi + 2mr\ddot{\phi}\cos\phi \tag{5-664}$$

$$N_2\sin\phi + R_2(1+\cos\phi) = 2mr\ddot{\theta} \tag{5-665}$$

Adding Eqs. (5-664) and (5-665), we have

$$R_2 = 3mr\ddot{\theta} + 2mr\ddot{\phi}\cos\phi - 2mr\dot{\phi}^2\sin\phi \tag{5-666}$$

Substituting the expression for R_2 from Eq. (5-657) into (5-665) gives

$$mr(2\ddot{\phi} - \ddot{\theta}) = 3mr\ddot{\theta} + 2mr\ddot{\phi}\cos\phi - 2mr\dot{\phi}^2\sin\phi \tag{5-667}$$

Rearranging and simplifying Eq. (5-667), we obtain the second differential equation of motion as

$$2\ddot{\theta} - (1-\cos\phi)\ddot{\phi} - \dot{\phi}^2\sin\phi = 0 \tag{5-668}$$

The system of two differential equations without reaction forces is then given as

$$(2\cos\phi - 1)r\ddot{\theta} + 6r\ddot{\phi} - 2g\sin\phi = 0 \tag{5-669}$$

$$2\ddot{\theta} - (1-\cos\phi)\ddot{\phi} - \dot{\phi}^2\sin\phi = 0 \tag{5-670}$$

■

5.9 Rotational Dynamics of a Rigid Body Using Moment of Inertia

Because the angular momentum of a rigid body is conveniently represented in terms of the moment of inertia tensor, it is useful to derive a special set of results for the rotational dynamics of a rigid body in terms of the moment of inertia. It is also useful to understand the stability of motion of a rigid body when it rotates about a direction that is in the neighborhood of a principal axis. In this section we derive the coordinate-free results for the rotational motion of a rigid body. Subsequently, we derive a special set of results in terms of a body-fixed coordinate system. Then, the body-fixed rotational equations are specialized further to the case of a principal-axis body-fixed coordinate system. These latter results are referred to as *Euler's equations*. Finally, using Euler's equations, the stability of the rotational motion in the neighborhood of a principal-axis is analyzed.

5.9.1 Euler's 2nd Law in Terms of Moment of Inertia Tensor

Recall that the angular momentum of a rigid body $\mathcal{R}$ relative to a point Q fixed in $\mathcal{R}$ can be written in terms of the moment of inertia tensor $\mathbf{I}_Q^{\mathcal{R}}$ as

$$^{\mathcal{N}}\mathbf{H}_Q = \mathbf{I}_Q^{\mathcal{R}} \cdot {}^{\mathcal{N}}\boldsymbol{\omega}^{\mathcal{R}} \tag{5-671}$$

Computing the rate of change of $^{\mathcal{N}}\mathbf{H}_Q$ in Eq. (5-671) in an inertial reference frame $\mathcal{N}$, we have

$$\frac{^{\mathcal{N}}d}{dt}\left({}^{\mathcal{N}}\mathbf{H}_Q\right) = \frac{^{\mathcal{N}}d}{dt}\left(\mathbf{I}_Q^{\mathcal{R}} \cdot {}^{\mathcal{N}}\boldsymbol{\omega}^{\mathcal{R}}\right) \tag{5-672}$$

Then, applying the rate of change transport theorem of Eq. (2-128) on page 47, we obtain

$$\frac{^{\mathcal{N}}d}{dt}\left(\mathbf{I}_Q^{\mathcal{R}} \cdot {}^{\mathcal{N}}\boldsymbol{\omega}^{\mathcal{R}}\right) = \frac{^{\mathcal{R}}d}{dt}\left(\mathbf{I}_Q^{\mathcal{R}} \cdot {}^{\mathcal{N}}\boldsymbol{\omega}^{\mathcal{R}}\right) + {}^{\mathcal{N}}\boldsymbol{\omega}^{\mathcal{R}} \times \left(\mathbf{I}_Q^{\mathcal{R}} \cdot {}^{\mathcal{N}}\boldsymbol{\omega}^{\mathcal{R}}\right) \tag{5-673}$$

Furthermore, we have

$$\frac{^{\mathcal{R}}d}{dt}\left(\mathbf{I}_Q^{\mathcal{R}} \cdot {}^{\mathcal{N}}\boldsymbol{\omega}^{\mathcal{R}}\right) = \frac{^{\mathcal{R}}d}{dt}\left(\mathbf{I}_Q^{\mathcal{R}}\right) \cdot {}^{\mathcal{N}}\boldsymbol{\omega}^{\mathcal{R}} + \mathbf{I}_Q^{\mathcal{R}} \cdot \frac{^{\mathcal{R}}d}{dt}\left({}^{\mathcal{N}}\boldsymbol{\omega}^{\mathcal{R}}\right) \tag{5-674}$$

Using the result of Eq. (5-674), we have

$$\frac{^{\mathcal{N}}d}{dt}\left({}^{\mathcal{N}}\mathbf{H}_Q\right) = \frac{^{\mathcal{R}}d}{dt}\left(\mathbf{I}_Q^{\mathcal{R}}\right) \cdot {}^{\mathcal{N}}\boldsymbol{\omega}^{\mathcal{R}} + \mathbf{I}_Q^{\mathcal{R}} \cdot \frac{^{\mathcal{R}}d}{dt}\left({}^{\mathcal{N}}\boldsymbol{\omega}^{\mathcal{R}}\right) + {}^{\mathcal{N}}\boldsymbol{\omega}^{\mathcal{R}} \times \left(\mathbf{I}_Q^{\mathcal{R}} \cdot {}^{\mathcal{N}}\boldsymbol{\omega}^{\mathcal{R}}\right) \tag{5-675}$$

Now we note that

$$\frac{^{\mathcal{N}}d}{dt}\left({}^{\mathcal{N}}\boldsymbol{\omega}^{\mathcal{R}}\right) = {}^{\mathcal{N}}\boldsymbol{\alpha}^{\mathcal{R}}$$

which then becomes

$$\frac{^{\mathcal{N}}d}{dt}\left({}^{\mathcal{N}}\mathbf{H}_Q\right) = \frac{^{\mathcal{R}}d}{dt}\left(\mathbf{I}_Q^{\mathcal{R}}\right) \cdot {}^{\mathcal{N}}\boldsymbol{\omega}^{\mathcal{R}} + \mathbf{I}_Q^{\mathcal{R}} \cdot {}^{\mathcal{N}}\boldsymbol{\alpha}^{\mathcal{R}} + {}^{\mathcal{N}}\boldsymbol{\omega}^{\mathcal{R}} \times \left(\mathbf{I}_Q^{\mathcal{R}} \cdot {}^{\mathcal{N}}\boldsymbol{\omega}^{\mathcal{R}}\right) \tag{5-676}$$

We then have

$$\mathbf{M}_Q - (\bar{\mathbf{r}} - \mathbf{r}_Q) \times m\,{}^{\mathcal{N}}\mathbf{a}_Q = \frac{^{\mathcal{R}}d}{dt}\left(\mathbf{I}_Q^{\mathcal{R}}\right) \cdot {}^{\mathcal{N}}\boldsymbol{\omega}^{\mathcal{R}} + \mathbf{I}_Q^{\mathcal{R}} \cdot {}^{\mathcal{N}}\boldsymbol{\alpha}^{\mathcal{R}} + {}^{\mathcal{N}}\boldsymbol{\omega}^{\mathcal{R}} \times \left(\mathbf{I}_Q^{\mathcal{R}} \cdot {}^{\mathcal{N}}\boldsymbol{\omega}^{\mathcal{R}}\right) \tag{5-677}$$

Now we recall that the moment of inertia tensor $\mathbf{I}_Q^{\mathcal{R}}$ can be expressed in terms of a body-fixed coordinate system as

$$\mathbf{I}_Q^{\mathcal{R}} = \sum_{i=1}^{3}\sum_{j=1}^{3} I_{ij}^{Q}\mathbf{e}_i \otimes \mathbf{e}_j \tag{5-678}$$

Then, because $\{\mathbf{e}_1, \mathbf{e}_2, \mathbf{e}_3\}$ is a body-fixed basis, we have

$$\frac{^{\mathcal{R}}d\mathbf{e}_i}{dt} = \mathbf{0} \quad , \qquad (i = 1, 2, 3) \tag{5-679}$$

Furthermore, recall from Eqs. (5-54) and (5-55) that the moments and products of inertia I_{ij}^{Q} $(i, j = 1, 2, 3)$ in terms of a body-fixed coordinate system are constant. Consequently, the rate of change of $\mathbf{I}_Q^{\mathcal{R}}$ as viewed by an observer in the rigid body is *zero*, i.e.,

$$\frac{^{\mathcal{R}}d}{dt}\left(\mathbf{I}_Q^{\mathcal{R}}\right) = \mathbf{O} \tag{5-680}$$

where $\mathbf{O}$ is the zero tensor. Applying the result of Eq. (5-680) in (5-677), we have

$$\mathbf{M}_Q - (\bar{\mathbf{r}} - \mathbf{r}_Q) \times m\,{}^{\mathcal{N}}\mathbf{a}_Q = \mathbf{I}_Q^{\mathcal{R}} \cdot {}^{\mathcal{N}}\boldsymbol{\alpha}^{\mathcal{R}} + {}^{\mathcal{N}}\boldsymbol{\omega}^{\mathcal{R}} \times \left(\mathbf{I}_Q^{\mathcal{R}} \cdot {}^{\mathcal{N}}\boldsymbol{\omega}^{\mathcal{R}}\right) \tag{5-681}$$

We note in Eq. (5-681) that

$$\frac{^{\mathcal{N}}d}{dt}\left({}^{\mathcal{N}}\mathbf{H}_Q\right) = \mathbf{I}_Q^{\mathcal{R}} \cdot {}^{\mathcal{N}}\boldsymbol{\alpha}^{\mathcal{R}} + {}^{\mathcal{N}}\boldsymbol{\omega}^{\mathcal{R}} \times \left(\mathbf{I}_Q^{\mathcal{R}} \cdot {}^{\mathcal{N}}\boldsymbol{\omega}^{\mathcal{R}}\right) \tag{5-682}$$

Equation (5-677) is a coordinate-free form of Euler's 2nd law in terms of the moment of inertia tensor relative to a point Q fixed in the rigid body.

Suppose now that we choose the reference point to be the center of mass of the rigid body. Then $\bar{\mathbf{r}} - \mathbf{r}_Q = \bar{\mathbf{r}} - \bar{\mathbf{r}} = \mathbf{0}$ and Eq. (5-681) simplifies to

$$\bar{\mathbf{M}} = \bar{\mathbf{I}}^{\mathcal{R}} \cdot {}^{\mathcal{N}}\boldsymbol{\alpha}^{\mathcal{R}} + {}^{\mathcal{N}}\boldsymbol{\omega}^{\mathcal{R}} \times \left(\bar{\mathbf{I}}^{\mathcal{R}} \cdot {}^{\mathcal{N}}\boldsymbol{\omega}^{\mathcal{R}}\right) \tag{5-683}$$

where, in a manner similar to Eq. (5-681), we note that

$$\frac{^{\mathcal{N}}d}{dt}\left({}^{\mathcal{N}}\bar{\mathbf{H}}\right) = \bar{\mathbf{I}}^{\mathcal{R}} \cdot {}^{\mathcal{N}}\boldsymbol{\alpha}^{\mathcal{R}} + {}^{\mathcal{N}}\boldsymbol{\omega}^{\mathcal{R}} \times \left(\bar{\mathbf{I}}^{\mathcal{R}} \cdot {}^{\mathcal{N}}\boldsymbol{\omega}^{\mathcal{R}}\right) \tag{5-684}$$

Equation (5-683) is a coordinate-free form of Euler's 2nd law in terms of the moment of inertia tensor relative to the center of mass of the rigid body.

5.9.2 Euler's Equations for a Rigid Body

While Eq. (5-683) provides a coordinate-free way to describe the rotational motion of a rigid body relative to the center of mass of the body, it is often convenient to write Eq. (5-683) in terms of a specific body-fixed coordinate system. To this end, let $\{\mathbf{e}_1, \mathbf{e}_2, \mathbf{e}_3\}$ be a body-fixed basis. Then in terms of $\{\mathbf{e}_1, \mathbf{e}_2, \mathbf{e}_3\}$ we can write

$${}^{\mathcal{N}}\boldsymbol{\omega}^{\mathcal{R}} = \omega_1\mathbf{e}_1 + \omega_2\mathbf{e}_2 + \omega_3\mathbf{e}_3 \tag{5-685}$$

Computing the rate of change of ${}^{\mathcal{N}}\boldsymbol{\omega}^{\mathcal{R}}$ in reference frame $\mathcal{N}$ (and recalling that the rate of change of ${}^{\mathcal{N}}\boldsymbol{\omega}^{\mathcal{R}}$ is the same in references frames $\mathcal{N}$ and $\mathcal{R}$), we obtain

$$\frac{{}^{\mathcal{N}}d}{dt}\left({}^{\mathcal{N}}\boldsymbol{\omega}^{\mathcal{R}}\right) = \frac{{}^{\mathcal{N}}d}{dt}\left({}^{\mathcal{N}}\boldsymbol{\omega}^{\mathcal{R}}\right) = \dot{\omega}_1\mathbf{e}_1 + \dot{\omega}_2\mathbf{e}_2 + \dot{\omega}_3\mathbf{e}_3 \tag{5-686}$$

Furthermore, the moment of inertia tensor can be written in the body-fixed basis $\{\mathbf{e}_1, \mathbf{e}_2, \mathbf{e}_3\}$ as

$$\begin{aligned} \bar{\mathbf{I}}^{\mathcal{R}} &= \bar{I}_{11}\mathbf{e}_1 \otimes \mathbf{e}_1 + \bar{I}_{12}\mathbf{e}_1 \otimes \mathbf{e}_2 + \bar{I}_{13}\mathbf{e}_1 \otimes \mathbf{e}_3 \\ &+ \bar{I}_{12}\mathbf{e}_2 \otimes \mathbf{e}_1 + \bar{I}_{22}\mathbf{e}_2 \otimes \mathbf{e}_2 + \bar{I}_{13}\mathbf{e}_2 \otimes \mathbf{e}_3 \\ &+ \bar{I}_{13}\mathbf{e}_3 \otimes \mathbf{e}_1 + \bar{I}_{23}\mathbf{e}_3 \otimes \mathbf{e}_2 + \bar{I}_{33}\mathbf{e}_3 \otimes \mathbf{e}_3 \end{aligned} \tag{5-687}$$

Using Eqs. (5-685), (5-686), and (5-687), we obtain

$$\begin{aligned} \bar{\mathbf{I}}^{\mathcal{R}} \cdot {}^{\mathcal{N}}\boldsymbol{\omega}^{\mathcal{R}} &= (\bar{I}_{11}\omega_1 + \bar{I}_{12}\omega_2 + \bar{I}_{13}\omega_3)\mathbf{e}_1 \\ &+ (\bar{I}_{12}\omega_1 + \bar{I}_{22}\omega_2 + \bar{I}_{23}\omega_3)\mathbf{e}_2 \\ &+ (\bar{I}_{13}\omega_1 + \bar{I}_{23}\omega_2 + \bar{I}_{33}\omega_3)\mathbf{e}_3 \end{aligned} \tag{5-688}$$

and

$$\begin{aligned} \bar{\mathbf{I}}^{\mathcal{R}} \cdot {}^{\mathcal{N}}\boldsymbol{\alpha}^{\mathcal{R}} &= (\bar{I}_{11}\dot{\omega}_1 + \bar{I}_{12}\dot{\omega}_2 + \bar{I}_{13}\dot{\omega}_3)\mathbf{e}_1 \\ &+ (\bar{I}_{12}\dot{\omega}_1 + \bar{I}_{22}\dot{\omega}_2 + \bar{I}_{23}\dot{\omega}_3)\mathbf{e}_2 \\ &+ (\bar{I}_{13}\dot{\omega}_1 + \bar{I}_{23}\dot{\omega}_2 + \bar{I}_{33}\dot{\omega}_3)\mathbf{e}_3 \end{aligned} \tag{5-689}$$

Suppose now that we let

$$H_1 = (\bar{I}_{11}\dot{\omega}_1 + \bar{I}_{12}\dot{\omega}_2 + \bar{I}_{13}\dot{\omega}_3) \tag{5-690}$$

$$H_2 = (\bar{I}_{12}\dot{\omega}_1 + \bar{I}_{22}\dot{\omega}_2 + \bar{I}_{23}\dot{\omega}_3) \tag{5-691}$$

$$H_3 = (\bar{I}_{13}\dot{\omega}_1 + \bar{I}_{23}\dot{\omega}_2 + \bar{I}_{33}\dot{\omega}_3) \tag{5-692}$$

Then,

$$\begin{aligned} {}^{\mathcal{N}}\boldsymbol{\omega}^{\mathcal{R}} \times \left(\bar{\mathbf{I}}^{\mathcal{R}} \cdot {}^{\mathcal{N}}\boldsymbol{\omega}^{\mathcal{R}}\right) &= (H_3\omega_2 - H_2\omega_3)\mathbf{e}_1 + (H_1\omega_3 - H_3\omega_1)\mathbf{e}_2 \\ &+ (H_2\omega_1 - H_1\omega_2)\mathbf{e}_3 \end{aligned} \tag{5-693}$$

Also, let

$$\bar{\mathbf{M}} = \bar{M}_1\mathbf{e}_1 + \bar{M}_2\mathbf{e}_2 + \bar{M}_3\mathbf{e}_3 \tag{5-694}$$

Substituting the results of Eqs. (5-689), (5-693), and (5-694) into Eq. (5-683), we obtain the following three scalar equations:

$$\bar{M}_1 = \bar{I}_{11}\dot{\omega}_1 + \bar{I}_{12}\dot{\omega}_2 + \bar{I}_{13}\dot{\omega}_3 + H_3\omega_2 - H_2\omega_3 \tag{5-695}$$

$$\bar{M}_2 = \bar{I}_{12}\dot{\omega}_1 + \bar{I}_{22}\dot{\omega}_2 + \bar{I}_{23}\dot{\omega}_3 + H_1\omega_3 - H_3\omega_1 \tag{5-696}$$

$$\bar{M}_3 = \bar{I}_{13}\dot{\omega}_1 + \bar{I}_{23}\dot{\omega}_2 + \bar{I}_{33}\dot{\omega}_3 + H_2\omega_1 - H_1\omega_2 \tag{5-697}$$

Finally, substituting the expressions in Eqs. (5-690)-(5-692) into Eqs. (5-695)-(5-697), respectively, we have

$$\begin{aligned}\bar{M}_1 &= \bar{I}_{11}\dot{\omega}_1 + \bar{I}_{12}(\dot{\omega}_2 - \omega_1\omega_3) + \bar{I}_{13}(\dot{\omega}_3 + \omega_1\omega_2) \\ &\quad + (\bar{I}_{33} - \bar{I}_{22})\omega_3\omega_2 + \bar{I}_{23}(\omega_2^2 - \omega_3^2) \qquad (5\text{-}698)\\ \bar{M}_2 &= \bar{I}_{12}(\dot{\omega}_1 + \omega_2\omega_3) + \bar{I}_{22}\dot{\omega}_2 + \bar{I}_{23}(\dot{\omega}_3 - \omega_1\omega_2) \\ &\quad + (\bar{I}_{11} - \bar{I}_{33})\omega_1\omega_3 + \bar{I}_{13}(\omega_3^2 - \omega_1^2) \qquad (5\text{-}699)\\ \bar{M}_3 &= \bar{I}_{13}(\dot{\omega}_1 - \omega_2\omega_3) + \bar{I}_{23}(\dot{\omega}_2 + \omega_1\omega_3) + \bar{I}_{33}\dot{\omega}_3 \\ &\quad + (\bar{I}_{22} - \bar{I}_{11})\omega_2\omega_1 + \bar{I}_{12}(\omega_1^2 - \omega_2^2) \qquad (5\text{-}700)\end{aligned}$$

It can be seen that Eqs. (5-698)-(5-700) are quite complicated since they involve all of the moments and products of inertia. Suppose now that $\{\mathbf{e}_1, \mathbf{e}_2, \mathbf{e}_3\}$ is a principal-axis basis. Then Eqs. (5-698)-(5-700) simplify to

$$\bar{M}_1 = \bar{I}_1\dot{\omega}_1 + (\bar{I}_3 - \bar{I}_2)\omega_3\omega_2 \qquad (5\text{-}701)$$
$$\bar{M}_2 = \bar{I}_2\dot{\omega}_2 + (\bar{I}_1 - \bar{I}_3)\omega_1\omega_3 \qquad (5\text{-}702)$$
$$\bar{M}_3 = \bar{I}_3\dot{\omega}_3 + (\bar{I}_2 - \bar{I}_1)\omega_2\omega_1 \qquad (5\text{-}703)$$

where we recall that $\bar{I}_1$, $\bar{I}_2$, and $\bar{I}_3$ are the principal-axis moments of inertia of the rigid body relative to the center of mass of the body. Equations (5-701)-(5-703) are called *Euler's equations* for a rigid body. It is important to remember that Eqs. (5-701)-(5-703) hold only when the reference point is the center of mass and both the moment of inertia tensor and the angular velocity of the rigid body have been expressed in a principal-axis body-fixed basis.

5.9.3 Stability of Rotational Motion About a Principal Axis

Suppose now that no moments are applied to the rigid body (i.e., $\bar{M}_1 = \bar{M}_2 = \bar{M}_3 = 0$). Then, from Eqs. (5-701)-(5-703) we have

$$\bar{I}_1\dot{\omega}_1 + (\bar{I}_3 - \bar{I}_2)\omega_3\omega_2 = 0 \qquad (5\text{-}704)$$
$$\bar{I}_2\dot{\omega}_2 + (\bar{I}_1 - \bar{I}_3)\omega_1\omega_3 = 0 \qquad (5\text{-}705)$$
$$\bar{I}_3\dot{\omega}_3 + (\bar{I}_2 - \bar{I}_1)\omega_2\omega_1 = 0 \qquad (5\text{-}706)$$

Eqs. (5-704)-(5-706) describe the moment-free rotational motion of a rigid body in a principal-axis coordinate system relative to the center of mass of the body. Suppose now that we consider small perturbations in the motion relative to a constant rotation about the first principal axis, i.e.,

$$\omega_1 = \Omega + \delta\omega_1 \qquad (5\text{-}707)$$
$$\omega_2 = \delta\omega_2 \qquad (5\text{-}708)$$
$$\omega_3 = \delta\omega_3 \qquad (5\text{-}709)$$

where Ω is a constant and $\delta\omega_i \ll \Omega$ $(i = 1, 2, 3)$. Computing the rates of change of ω_1, ω_2, and ω_3 in Eqs. (5-707)-(5-709) and substituting the results into Eqs. (5-704)-(5-706), respectively, we have

$$\bar{I}_1\delta\dot{\omega}_1 + (\bar{I}_3 - \bar{I}_2)\delta\omega_2\delta\omega_3 = 0 \qquad (5\text{-}710)$$
$$\bar{I}_2\delta\dot{\omega}_2 + (\bar{I}_1 - \bar{I}_3)(\Omega + \delta\omega_1)\delta\omega_3 = 0 \qquad (5\text{-}711)$$
$$\bar{I}_3\delta\dot{\omega}_3 + (\bar{I}_2 - \bar{I}_1)\delta\omega_2(\Omega + \delta\omega_1) = 0 \qquad (5\text{-}712)$$

Then, dropping all products $\delta\omega_i\delta\omega_j$ $(i, j = 1, 2, 3)$ in Eqs. (5-710)-(5-712), we obtain

$$\begin{aligned} \bar{I}_1\delta\dot{\omega}_1 &= 0 & (5\text{-}713) \\ \bar{I}_2\delta\dot{\omega}_2 + (\bar{I}_1 - \bar{I}_3)\Omega\delta\omega_3 &= 0 & (5\text{-}714) \\ \bar{I}_3\delta\dot{\omega}_3 + (\bar{I}_2 - \bar{I}_1)\Omega\delta\omega_2 &= 0 & (5\text{-}715) \end{aligned}$$

Next, computing the rate of change of Eq. (5-714) and

$$\bar{I}_2\delta\ddot{\omega}_2 + (\bar{I}_1 - \bar{I}_3)\Omega\delta\dot{\omega}_3 = 0 \tag{5-716}$$

Then, solving Eq. (5-715) for $\delta\dot{\omega}_3$ and substituting the result into Eq. (5-716), we obtain

$$\delta\ddot{\omega}_2 + \frac{(\bar{I}_2 - \bar{I}_1)(\bar{I}_3 - \bar{I}_1)}{\bar{I}_2\bar{I}_3}\Omega^2\delta\omega_2 = 0 \tag{5-717}$$

It can be seen that Eq. (5-717) is a second-order linear constant coefficient differential equation in $\delta\omega_2$. In particular, the stability of Eq. (5-717) is governed by the sign of the quantity

$$\omega_n^2 = \Omega^2\frac{(\bar{I}_2 - \bar{I}_1)(\bar{I}_3 - \bar{I}_1)}{\bar{I}_2\bar{I}_3} \tag{5-718}$$

In particular, when $\omega_n^2 < 0$ the general solution of Eq. (5-717) has the form

$$\delta\omega_2(t) = C_1 e^{-\omega_n t} + C_2 e^{\omega_n t} \tag{5-719}$$

whereas when $\omega_n^2 > 0$ the general solution of Eq. (5-717) has the form

$$\delta\omega_2(t) = C_1 \cos\omega_n t + C_2 \sin\omega_n t \tag{5-720}$$

where C_1 and C_2 are constants determined by the initial conditions. It is seen that Eq. (5-719) has a growing exponential term and, thus, is unstable, while Eq. (5-720) is bounded due to the fact that it consists of a sum of two sinusoids. From Eqs. (5-719) and (5-720) it is seen that the linearized motion of the rigid body will be unstable if ω_n^2 is negative and will be stable if ω_n^2 is positive. Now, examining Eq. (5-718) further, it is seen that ω_n^2 will be positive if $\bar{I}_1$ is either the *largest* or *smallest* principal moment of inertia. On the other hand, it is seen that ω_n^2 will be negative if $\bar{I}_1$ is the intermediate moment of inertia (i.e., if either $\bar{I}_2 < \bar{I}_1 < \bar{I}_3$ or $\bar{I}_3 < \bar{I}_1 < \bar{I}_2$). Consequently, the linearized motion will be stable if $\bar{I}_1$ is either the largest or smallest principal moment of inertia.

The stability of the rotational motion of a rigid body about either its largest or smallest principal axis and the instability of motion about the intermediate principal axis can be demonstrated by using commonly found objects. For example, if one spins a tennis racket about either an axis along its handle (corresponding to the smallest principal moment of inertia) or about an axis orthogonal to the plane of the racket (corresponding to its largest principal moment of inertia), the motion will essentially remain about either of these axes. On the other hand, if one spins the racket about the third principal axis (corresponding to its intermediate moment of inertia), the motion will not remain about this axis. Instead, in the latter case, the motion will rapidly depart from its initial rotation about this intermediate principal axis.

Example 5–11

Consider a homogeneous circular disk of radius r rolling without slip along a fixed horizontal surface as shown in Fig. 5–30. Assuming that gravity acts vertically downward, determine a system of three differential equations of motion for the disk in terms of the Type I Euler angles ψ, θ, and ϕ.

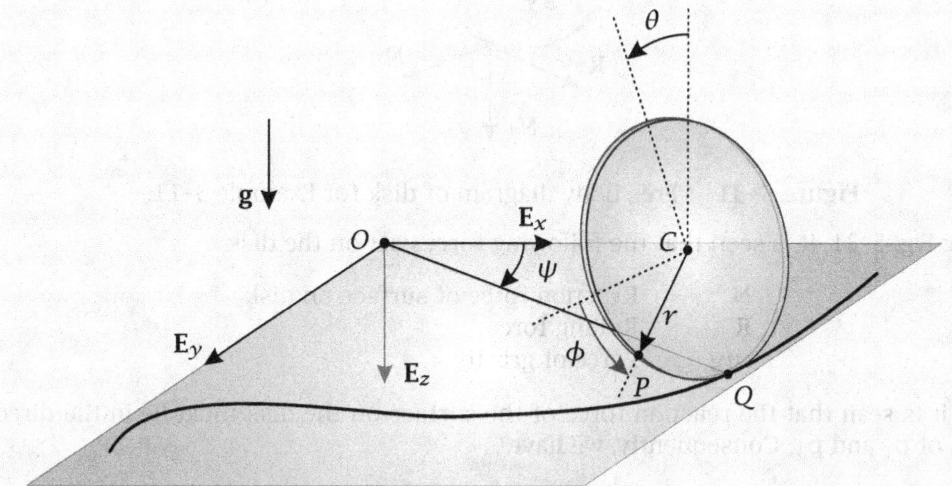

Figure 5–30 Disk rolling on a horizontal surface.

Solution to Example 5–11

Preliminaries

We recall that the kinematics of the center of mass of the disk were determined in Example 2–12. In particular, the acceleration of the center of mass of the disk is given from Eq. (2–634) on page 125 as

$$\begin{aligned} {}^{S}\mathbf{a}_C = &- \left[r\ddot{\theta} + r\dot{\psi}(\dot{\phi} - \dot{\psi}\sin\theta)\cos\theta \right] \mathbf{q}_1 \\ &+ \left[r\ddot{\phi} - r\ddot{\psi}\sin\theta - 2r\dot{\psi}\dot{\theta}\cos\theta \right] \mathbf{q}_2 \\ &+ \left[r\dot{\theta}^2 - r\dot{\psi}(\dot{\phi} - \dot{\psi}\sin\theta)\sin\theta \right] \mathbf{q}_3 \end{aligned} \tag{5-721}$$

where the basis $\{\mathbf{q}_1, \mathbf{q}_2, \mathbf{q}_3\}$ is as defined in Example 2–12. Now, the differential equations of motion will be determined using the center of mass of the disk as the reference point. Consequently, it will be necessary to apply both of Euler's laws.

Kinetics

This problem will be solved by applying both of Euler's laws using the center of mass of the disk as the reference point. The free body diagram of the disk is shown in Fig. 5–31.

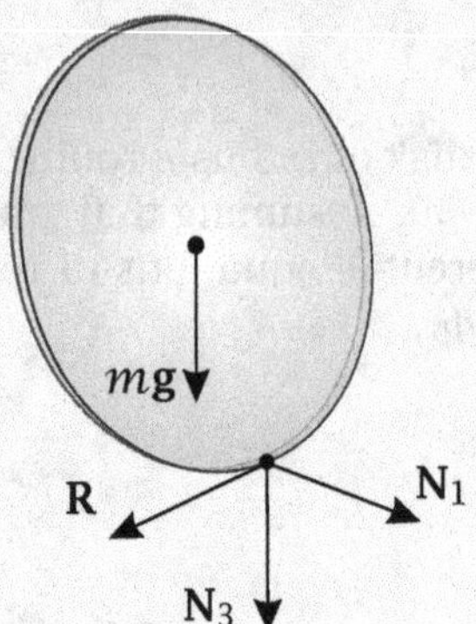

Figure 5–31 Free body diagram of disk for Example 5–11.

Using Fig. 5-31, it is seen that the following forces act on the disk:

$$
\begin{array}{lcl}
\mathbf{N} & = & \text{Reaction force of surface on disk} \\
\mathbf{R} & = & \text{Rolling force} \\
m\mathbf{g} & = & \text{Force of gravity}
\end{array}
$$

Now it is seen that the reaction force of the surface on the disk must lie in the directions of $\mathbf{p}_1$ and $\mathbf{p}_3$. Consequently, we have

$$\mathbf{N} = N_1\mathbf{p}_1 + N_3\mathbf{p}_3 \tag{5–722}$$

Furthermore, the rolling force must act in the direction of $\mathbf{p}_2$. Therefore, we have

$$\mathbf{R} = R\mathbf{p}_2 \tag{5–723}$$

Finally, the force of gravity acts in the direction of $\mathbf{p}_3$, which implies

$$m\mathbf{g} = mg\mathbf{p}_3 \tag{5–724}$$

Adding the forces in Eqs. (5–722), (5–723), and (5–724), the resultant force acting on the disk is given as

$$\mathbf{F} = N_1\mathbf{p}_1 + N_3\mathbf{p}_3 + R\mathbf{p}_2 + mg\mathbf{p}_3 = N_1\mathbf{p}_1 + R\mathbf{p}_2 + (N_3 + mg)\mathbf{p}_3 \tag{5–725}$$

Next, from the geometry of the bases $\{\mathbf{p}_1, \mathbf{p}_2, \mathbf{p}_3\}$ and $\{\mathbf{q}_1, \mathbf{q}_2, \mathbf{q}_3\}$ as given in Fig. 2–45 on page 123, we have

$$
\begin{aligned}
\mathbf{p}_1 &= \cos\theta\,\mathbf{q}_1 + \sin\theta\,\mathbf{q}_3 && (5\text{–}726)\\
\mathbf{p}_2 &= \mathbf{q}_2 && (5\text{–}727)\\
\mathbf{p}_3 &= -\sin\theta\,\mathbf{q}_1 + \cos\theta\,\mathbf{q}_3 && (5\text{–}728)
\end{aligned}
$$

Substituting the results of Eqs. (5–726)-(5–728) into (5–725) gives

$$\mathbf{F} = N_1(\cos\theta\,\mathbf{q}_1 + \sin\theta\,\mathbf{q}_3) + R\mathbf{q}_2 + (N_3 + mg)(-\sin\theta\,\mathbf{q}_1 + \cos\theta\,\mathbf{q}_3) \tag{5–729}$$

Rearranging Eq. (5–729), the resultant force acting on the disk is given as

$$\mathbf{F} = [N_1\cos\theta - (N_3 + mg)\sin\theta]\,\mathbf{q}_1 + R\mathbf{q}_2 + [N_1\sin\theta + (N_3 + mg)\cos\theta]\,\mathbf{q}_3 \tag{5–730}$$

Application of Euler's 1st Law to Disk

Applying Euler's 1st law by setting $\mathbf{F}$ from Eq. (5-730) equal to $m^{S}\mathbf{a}_C$ using ${}^{S}\mathbf{a}_C$ from Eq. (5-721), we obtain the following three scalar equations:

$$N_1\cos\theta-(N_3+mg)\sin\theta = -m\left[r\ddot{\theta}+r\dot{\psi}(\dot{\phi}-\dot{\psi}\sin\theta)\cos\theta\right] \quad (5\text{-}731)$$

$$R = m\left[r\ddot{\phi}-r\ddot{\psi}\sin\theta-2r\dot{\psi}\dot{\theta}\cos\theta\right] \quad (5\text{-}732)$$

$$N_1\sin\theta+(N_3+mg)\cos\theta = m\left[r\dot{\theta}^2-r\dot{\psi}(\dot{\phi}-\dot{\psi}\sin\theta)\sin\theta\right] \quad (5\text{-}733)$$

Application of Euler's 2nd Law to Disk

Euler's 2nd law relative to the center of mass of the disk is applied by using Euler's equations as given in Eqs. (5-701)-(5-703). First, since the force of gravity passes through the center of mass of the disk, the moment applied to the disk relative to the center of mass is given as

$$\bar{\mathbf{M}}=(\mathbf{r}_Q-\bar{\mathbf{r}})\times(\mathbf{N}+\mathbf{R}) \quad (5\text{-}734)$$

Observing that $\mathbf{r}_Q-\bar{\mathbf{r}}=r\mathbf{q}_3$ and using the expressions for $\mathbf{N}$ and $\mathbf{R}$ from Eqs. (5-722) and (5-723), respectively, we obtain

$$\bar{\mathbf{M}}=r\mathbf{q}_3\times(N_1\mathbf{p}_1+N_3\mathbf{p}_3+R\mathbf{p}_2) \quad (5\text{-}735)$$

Furthermore, using the expressions for $\mathbf{p}_1$, $\mathbf{p}_2$, and $\mathbf{p}_3$ in Eqs. (5-726-(5-728), Eq. (5-735) can be written as

$$\bar{\mathbf{M}}=r\mathbf{q}_3\times[N_1(\cos\theta\mathbf{q}_1+\sin\theta\mathbf{q}_3)+N_3(-\sin\theta\mathbf{q}_1+\cos\theta\mathbf{q}_3)+R\mathbf{q}_2] \quad (5\text{-}736)$$

Rearranging Eq. (5-736), we have

$$\bar{\mathbf{M}}=r\mathbf{q}_3\times[(N_1\cos\theta-N_3\sin\theta)\mathbf{q}_1+R\mathbf{q}_2+(N_1\sin\theta+N_3\cos\theta)\mathbf{q}_3] \quad (5\text{-}737)$$

Computing the vector products in Eq. (5-737) gives

$$\bar{\mathbf{M}}=-rR\mathbf{q}_1+r(N_1\cos\theta-N_3\sin\theta)\mathbf{q}_2 \quad (5\text{-}738)$$

Now we note that $\mathbf{q}_1$ and $\mathbf{q}_2$ can be expressed in the body-fixed basis $\{\mathbf{e}_1,\mathbf{e}_2,\mathbf{e}_3\}$ as

$$\mathbf{q}_1 = \mathbf{e}_1 \quad (5\text{-}739)$$

$$\mathbf{q}_2 = \cos\phi\mathbf{e}_2-\sin\phi\mathbf{e}_3 \quad (5\text{-}740)$$

Substituting Eqs. (5-739) and (5-740) into Eq. (5-738), we obtain

$$\bar{\mathbf{M}}=-rR\mathbf{e}_1+r(N_1\cos\theta-N_3\sin\theta)(\cos\phi\mathbf{e}_2-\sin\phi\mathbf{e}_3) \quad (5\text{-}741)$$

Equation (5-741) simplifies to

$$\bar{\mathbf{M}}=-rR\mathbf{e}_1+r(N_1\cos\theta-N_3\sin\theta)\cos\phi\mathbf{e}_2-r(N_1\cos\theta-N_3\sin\theta)\sin\phi\mathbf{e}_3 \quad (5\text{-}742)$$

Therefore, the three components of the moment applied to the disk relative to the center of mass of the disk are given as

$$\bar{M}_1 = -rR \quad (5\text{-}743)$$
$$\bar{M}_2 = r(N_1\cos\theta - N_3\sin\theta)\cos\phi \quad (5\text{-}744)$$
$$\bar{M}_1 = -r(N_1\cos\theta - N_3\sin\theta)\sin\phi \quad (5\text{-}745)$$

Next, since $\{\mathbf{e}_1, \mathbf{e}_2, \mathbf{e}_3\}$ is a principal-axis basis and we have chosen the center of mass as the reference point, the three components of the rates of change of angular momentum relative to the center of mass are obtained by using the right-hand sides of Eqs. (5-701)-(5-703). We then obtain

$$-rR = \bar{I}_1\dot{\omega}_1 + (\bar{I}_3 - \bar{I}_2)\omega_3\omega_2 \quad (5\text{-}746)$$
$$r(N_1\cos\theta - N_3\sin\theta)\cos\phi = \bar{I}_2\dot{\omega}_2 + (\bar{I}_1 - \bar{I}_3)\omega_1\omega_3 \quad (5\text{-}747)$$
$$-r(N_1\cos\theta - N_3\sin\theta)\sin\phi = \bar{I}_3\dot{\omega}_3 + (\bar{I}_2 - \bar{I}_1)\omega_2\omega_1 \quad (5\text{-}748)$$

where, because we are using Type I Euler angles, the quantities ω_1, ω_2, and ω_3 are given in Eqs. (2-585)-(2-587), respectively, on page 119 while the quantities $\dot{\omega}_1$, $\dot{\omega}_2$, and $\dot{\omega}_3$ are given in Eqs. (2-589)-(2-591), respectively, on page 119. Now for a uniform circular disk we have

$$\bar{I}_1 = \tfrac{1}{2}mr^2 \quad (5\text{-}749)$$
$$\bar{I}_2 = \tfrac{1}{4}mr^2 \quad (5\text{-}750)$$
$$\bar{I}_3 = \tfrac{1}{4}mr^2 \quad (5\text{-}751)$$

As a result, Eqs. (5-746)-(5-748) can be written as

$$-rR = \tfrac{1}{2}mr^2\dot{\omega}_1 \quad (5\text{-}752)$$
$$r(N_1\cos\theta - N_3\sin\theta)\cos\phi = \tfrac{1}{4}mr^2(\dot{\omega}_2 + \omega_1\omega_3) \quad (5\text{-}753)$$
$$-r(N_1\cos\theta - N_3\sin\theta)\sin\phi = \tfrac{1}{4}mr^2(\dot{\omega}_3 - \omega_2\omega_1) \quad (5\text{-}754)$$

Determination of System of Three Differential Equations of Motion

The three differential equations that describe the motion of the disk are obtained as follows. First, substituting the expression for R from Eq. (5-732) into Eq. (5-752) gives

$$-rm\left[r\ddot{\phi} - r\ddot{\psi}\sin\theta - 2r\dot{\psi}\dot{\theta}\cos\theta\right] = \tfrac{1}{2}mr^2\dot{\omega}_1 \quad (5\text{-}755)$$

Simplifying Eq. (5-755), we obtain the first differential equation of motion as

$$\dot{\omega}_1 + 2\ddot{\phi} - 2\ddot{\psi}\sin\theta - 4\dot{\psi}\dot{\theta}\cos\theta = 0 \quad (5\text{-}756)$$

Next, multiplying Eqs. (5-753) and (5-754) by $\sin\phi$ and $\cos\phi$, respectively, and adding, we have

$$\tfrac{1}{4}mr^2\left[(\dot{\omega}_2 + \omega_1\omega_3)\sin\phi + (\dot{\omega}_3 - \omega_2\omega_1)\cos\phi\right] = 0 \quad (5\text{-}757)$$

Simplifying Eq. (5-757), we obtain the second differential equation of motion as

$$(\dot{\omega}_2 + \omega_1\omega_3)\sin\phi + (\dot{\omega}_3 - \omega_2\omega_1)\cos\phi = 0 \quad (5\text{-}758)$$

Furthermore, multiplying Eqs. (5-753) and (5-754) by $\cos\phi$ and $\sin\phi$, respectively, and subtracting, we obtain

$$r(N_1\cos\theta - N_3\sin\theta) = \tfrac{1}{4}mr^2\left[(\dot\omega_2 + \omega_1\omega_3)\cos\phi + (\dot\omega_3 - \omega_2\omega_1)\sin\phi\right] \quad (5\text{-}759)$$

Dropping the common factor of r from Eq. (5-759), we have

$$N_1\cos\theta - N_3\sin\theta = \tfrac{1}{4}mr\left[(\dot\omega_2 + \omega_1\omega_3)\cos\phi + (\dot\omega_3 - \omega_2\omega_1)\sin\phi\right] \quad (5\text{-}760)$$

Then, solving Eq. (5-731) for $N_1\cos\theta - N_3\sin\theta$ gives

$$N_1\cos\theta - N_3\sin\theta = mg\sin\theta - m\left[r\ddot\theta + r\dot\psi(\dot\phi - \dot\psi\sin\theta)\cos\theta\right] \quad (5\text{-}761)$$

Substituting the result of Eq. (5-761) into Eq. (5-760) gives

$$\begin{aligned} mg\sin\theta - m\left[r\ddot\theta + r\dot\psi(\dot\phi - \dot\psi\sin\theta)\cos\theta\right] \\ = \tfrac{1}{4}mr\left[(\dot\omega_2 + \omega_1\omega_3)\cos\phi + (\dot\omega_3 - \omega_2\omega_1)\sin\phi\right] \end{aligned} \quad (5\text{-}762)$$

Simplifying Eq. (5-762), we obtain the third differential equation of motion as

$$\begin{aligned} r(\dot\omega_2 + \omega_1\omega_3)\cos\phi + r(\dot\omega_3 - \omega_2\omega_1)\sin\phi \\ + r\ddot\theta + r\dot\psi(\dot\phi - \dot\psi\sin\theta)\cos\theta - 4g\sin\theta = 0 \end{aligned} \quad (5\text{-}763)$$

The system of three differential equations of motion for the disk is then given as

$$\dot\omega_1 + 2\ddot\phi - 2\ddot\psi\sin\theta - 4\dot\psi\dot\theta\cos\theta = 0 \quad (5\text{-}764)$$

$$(\dot\omega_2 + \omega_1\omega_3)\sin\phi + (\dot\omega_3 - \omega_2\omega_1)\cos\phi = 0 \quad (5\text{-}765)$$

$$\begin{aligned} r(\dot\omega_2 + \omega_1\omega_3)\cos\phi + r(\dot\omega_3 - \omega_2\omega_1)\sin\phi \\ + r\ddot\theta + r\dot\psi(\dot\phi - \dot\psi\sin\theta)\cos\theta - 4g\sin\theta = 0 \end{aligned} \quad (5\text{-}766)$$

where we again note that the quantities ω_1, ω_2, and ω_3 are given in Eqs. (2-585)-(2-587), respectively, on page 119 while the quantities $\dot\omega_1$, $\dot\omega_2$, and $\dot\omega_3$ are given in Eqs. (2-589)-(2-591), respectively, on page 119.

■

5.10 Work and Energy for a Rigid Body

5.10.1 Kinetic Energy of a Rigid Body

Let $\mathcal{R}$ be a rigid body and let $\mathcal{N}$ be an inertial reference frame. Furthermore, let $\mathbf{r}$ be the position of a point P on $\mathcal{R}$ and let ${}^{\mathcal{N}}\mathbf{v}$ be the velocity of point P on $\mathcal{R}$ in the inertial reference frame $\mathcal{N}$. Then the kinetic energy of a rigid body $\mathcal{R}$ in the inertial reference $\mathcal{N}$ is defined as

$${}^{\mathcal{N}}T = \frac{1}{2}\int_{\mathcal{R}} {}^{\mathcal{N}}\mathbf{v}\cdot{}^{\mathcal{N}}\mathbf{v}\,dm \quad (5\text{-}767)$$

Now from Eq. (2-517) on page 106, we have

$$ {}^{\mathcal{N}}\mathbf{v} = {}^{\mathcal{N}}\bar{\mathbf{v}} + {}^{\mathcal{N}}\boldsymbol{\omega}^{\mathcal{R}} \times (\mathbf{r} - \bar{\mathbf{r}}) \tag{5-768} $$

Suppose now that we let $\boldsymbol{\rho} = \mathbf{r} - \bar{\mathbf{r}}$. Then Eq. (5-767) can be written as

$$ {}^{\mathcal{N}}T = \frac{1}{2}\int_{\mathcal{R}} \left[{}^{\mathcal{N}}\bar{\mathbf{v}} + {}^{\mathcal{N}}\boldsymbol{\omega}^{\mathcal{R}} \times \boldsymbol{\rho}\right] \cdot \left[{}^{\mathcal{N}}\bar{\mathbf{v}} + {}^{\mathcal{N}}\boldsymbol{\omega}^{\mathcal{R}} \times \boldsymbol{\rho}\right] dm \tag{5-769} $$

Expanding Eq. (5-769), we obtain

$$ \begin{aligned} {}^{\mathcal{N}}T = \frac{1}{2}\int_{\mathcal{R}} \Big\{ & {}^{\mathcal{N}}\bar{\mathbf{v}} \cdot {}^{\mathcal{N}}\bar{\mathbf{v}} + 2\,{}^{\mathcal{N}}\bar{\mathbf{v}} \cdot {}^{\mathcal{N}}\boldsymbol{\omega}^{\mathcal{R}} \times \boldsymbol{\rho} \\ & + \left[{}^{\mathcal{N}}\boldsymbol{\omega}^{\mathcal{R}} \times \boldsymbol{\rho}\right] \cdot \left[{}^{\mathcal{N}}\boldsymbol{\omega}^{\mathcal{R}} \times \boldsymbol{\rho}\right]\Big\} dm \end{aligned} \tag{5-770} $$

which can be rewritten as

$$ \begin{aligned} {}^{\mathcal{N}}T = {} & \frac{1}{2}\int_{\mathcal{R}} {}^{\mathcal{N}}\bar{\mathbf{v}} \cdot {}^{\mathcal{N}}\bar{\mathbf{v}} dm + \int_{\mathcal{R}} {}^{\mathcal{N}}\bar{\mathbf{v}} \cdot \left[{}^{\mathcal{N}}\boldsymbol{\omega}^{\mathcal{R}} \times \boldsymbol{\rho}\right] dm \\ & + \frac{1}{2}\int_{\mathcal{R}} \left[{}^{\mathcal{N}}\boldsymbol{\omega}^{\mathcal{R}} \times \boldsymbol{\rho}\right] \cdot \left[{}^{\mathcal{N}}\boldsymbol{\omega}^{\mathcal{R}} \times \boldsymbol{\rho}\right] dm \end{aligned} \tag{5-771} $$

Since ${}^{\mathcal{N}}\bar{\mathbf{v}}$ is independent of the integral, Eq. (5-771) can be rewritten as

$$ \begin{aligned} {}^{\mathcal{N}}T = {} & \frac{1}{2}{}^{\mathcal{N}}\bar{\mathbf{v}} \cdot {}^{\mathcal{N}}\bar{\mathbf{v}} \int_{\mathcal{R}} dm + {}^{\mathcal{N}}\bar{\mathbf{v}} \cdot \int_{\mathcal{R}} {}^{\mathcal{N}}\boldsymbol{\omega}^{\mathcal{R}} \times \boldsymbol{\rho} dm \\ & + \frac{1}{2}\int_{\mathcal{R}} \left[{}^{\mathcal{N}}\boldsymbol{\omega}^{\mathcal{R}} \times \boldsymbol{\rho}\right] \cdot \left[{}^{\mathcal{N}}\boldsymbol{\omega}^{\mathcal{R}} \times \boldsymbol{\rho}\right] dm \end{aligned} \tag{5-772} $$

Then, since

$$ \int_{\mathcal{R}} dm = m $$

Eq. (5-772) becomes

$$ \begin{aligned} {}^{\mathcal{N}}T = {} & \frac{1}{2}m{}^{\mathcal{N}}\bar{\mathbf{v}} \cdot {}^{\mathcal{N}}\bar{\mathbf{v}} + {}^{\mathcal{N}}\bar{\mathbf{v}} \cdot \int_{\mathcal{R}} {}^{\mathcal{N}}\boldsymbol{\omega}^{\mathcal{R}} \times \boldsymbol{\rho} dm \\ & + \frac{1}{2}\int_{\mathcal{R}} \left[{}^{\mathcal{N}}\boldsymbol{\omega}^{\mathcal{R}} \times \boldsymbol{\rho}\right] \cdot \left[{}^{\mathcal{N}}\boldsymbol{\omega}^{\mathcal{R}} \times \boldsymbol{\rho}\right] dm \end{aligned} \tag{5-773} $$

Next, because ${}^{\mathcal{N}}\boldsymbol{\omega}^{\mathcal{R}}$ is independent of the integral, the second term in Eq. (5-773) can be written as

$$ {}^{\mathcal{N}}\bar{\mathbf{v}} \cdot \int_{\mathcal{R}} {}^{\mathcal{N}}\boldsymbol{\omega}^{\mathcal{R}} \times \boldsymbol{\rho} dm = {}^{\mathcal{N}}\bar{\mathbf{v}} \cdot \left[{}^{\mathcal{N}}\boldsymbol{\omega}^{\mathcal{R}} \times \int_{\mathcal{R}} \boldsymbol{\rho} dm\right] \tag{5-774} $$

Furthermore, since $\boldsymbol{\rho} = \mathbf{r} - \bar{\mathbf{r}}$, we have

$$ \int_{\mathcal{R}} \boldsymbol{\rho} dm = \int_{\mathcal{R}} (\mathbf{r} - \bar{\mathbf{r}}) dm = \mathbf{0} \tag{5-775} $$

which implies that

$$ {}^{\mathcal{N}}\bar{\mathbf{v}} \cdot \int_{\mathcal{R}} {}^{\mathcal{N}}\boldsymbol{\omega}^{\mathcal{R}} \times \boldsymbol{\rho} dm = {}^{\mathcal{N}}\bar{\mathbf{v}} \cdot \left[{}^{\mathcal{N}}\boldsymbol{\omega}^{\mathcal{R}} \times \int_{\mathcal{R}} \boldsymbol{\rho} dm\right] = 0 \tag{5-776} $$

In addition, we have

$$\frac{1}{2}{}^{\mathcal{N}}\bar{\mathbf{v}}\cdot{}^{\mathcal{N}}\bar{\mathbf{v}}\int_{\mathcal{R}}dm=\frac{1}{2}m{}^{\mathcal{N}}\bar{\mathbf{v}}\cdot{}^{\mathcal{N}}\bar{\mathbf{v}} \tag{5-777}$$

The kinetic energy of Eq. (5-773) then simplifies to

$${}^{\mathcal{N}}T=\frac{1}{2}m{}^{\mathcal{N}}\bar{\mathbf{v}}\cdot{}^{\mathcal{N}}\bar{\mathbf{v}}+\frac{1}{2}\int_{\mathcal{R}}\left[{}^{\mathcal{N}}\boldsymbol{\omega}^{\mathcal{R}}\times\boldsymbol{\rho}\right]\cdot\left[{}^{\mathcal{N}}\boldsymbol{\omega}^{\mathcal{R}}\times\boldsymbol{\rho}\right]dm \tag{5-778}$$

Furthermore, using the properties of the scalar triple product on page 9, we have

$$\left[{}^{\mathcal{N}}\boldsymbol{\omega}^{\mathcal{R}}\times\boldsymbol{\rho}\right]\cdot\left[{}^{\mathcal{N}}\boldsymbol{\omega}^{\mathcal{R}}\times\boldsymbol{\rho}\right]=\left[\boldsymbol{\rho}\times({}^{\mathcal{N}}\boldsymbol{\omega}^{\mathcal{R}}\times\boldsymbol{\rho})\right]\cdot{}^{\mathcal{N}}\boldsymbol{\omega}^{\mathcal{R}} \tag{5-779}$$

Next, from the properties of the vector triple product on page 10, we have

$$\boldsymbol{\rho}\times({}^{\mathcal{N}}\boldsymbol{\omega}^{\mathcal{R}}\times\boldsymbol{\rho})=(\boldsymbol{\rho}\cdot\boldsymbol{\rho}){}^{\mathcal{N}}\boldsymbol{\omega}^{\mathcal{R}}-\left(\boldsymbol{\rho}\cdot{}^{\mathcal{N}}\boldsymbol{\omega}^{\mathcal{R}}\right)\boldsymbol{\rho} \tag{5-780}$$

Equation (5-780) can be written more conveniently in tensor form as

$$\boldsymbol{\rho}\times({}^{\mathcal{N}}\boldsymbol{\omega}^{\mathcal{R}}\times\boldsymbol{\rho})=[(\boldsymbol{\rho}\cdot\boldsymbol{\rho})\mathbf{U}-(\boldsymbol{\rho}\otimes\boldsymbol{\rho})]\cdot{}^{\mathcal{N}}\boldsymbol{\omega}^{\mathcal{R}} \tag{5-781}$$

where $\mathbf{U}$ is the identity tensor and $(\boldsymbol{\rho}\cdot{}^{\mathcal{N}}\boldsymbol{\omega}^{\mathcal{R}})\boldsymbol{\rho}=(\boldsymbol{\rho}\otimes\boldsymbol{\rho})\cdot{}^{\mathcal{N}}\boldsymbol{\omega}^{\mathcal{R}}$. Then, substituting the result of Eq. (5-781) into (5-779), we obtain

$$\left[{}^{\mathcal{N}}\boldsymbol{\omega}^{\mathcal{R}}\times\boldsymbol{\rho}\right]\cdot\left[{}^{\mathcal{N}}\boldsymbol{\omega}^{\mathcal{R}}\times\boldsymbol{\rho}\right]=\left\{[(\boldsymbol{\rho}\cdot\boldsymbol{\rho})\mathbf{U}-(\boldsymbol{\rho}\otimes\boldsymbol{\rho})]\cdot{}^{\mathcal{N}}\boldsymbol{\omega}^{\mathcal{R}}\right\}\cdot{}^{\mathcal{N}}\boldsymbol{\omega}^{\mathcal{R}} \tag{5-782}$$

Consequently, the kinetic energy as given in Eq. (5-778) can be written as

$${}^{\mathcal{N}}T=\frac{1}{2}{}^{\mathcal{N}}\bar{\mathbf{v}}\cdot{}^{\mathcal{N}}\bar{\mathbf{v}}\int_{\mathcal{R}}dm+\frac{1}{2}\int_{\mathcal{R}}\left\{[(\boldsymbol{\rho}\cdot\boldsymbol{\rho})\mathbf{U}-(\boldsymbol{\rho}\otimes\boldsymbol{\rho})]\cdot{}^{\mathcal{N}}\boldsymbol{\omega}^{\mathcal{R}}\right\}\cdot{}^{\mathcal{N}}\boldsymbol{\omega}^{\mathcal{R}}dm \tag{5-783}$$

Again using the fact that ${}^{\mathcal{N}}\boldsymbol{\omega}^{\mathcal{R}}$ is independent of the integral, the quantity ${}^{\mathcal{N}}\boldsymbol{\omega}^{\mathcal{R}}$ can be taken out of the integral to give

$${}^{\mathcal{N}}T=\frac{1}{2}m{}^{\mathcal{N}}\bar{\mathbf{v}}\cdot{}^{\mathcal{N}}\bar{\mathbf{v}}+\frac{1}{2}\left[\left\{\int_{\mathcal{R}}[(\boldsymbol{\rho}\cdot\boldsymbol{\rho})\mathbf{U}-(\boldsymbol{\rho}\otimes\boldsymbol{\rho})]\,dm\right\}\cdot{}^{\mathcal{N}}\boldsymbol{\omega}^{\mathcal{R}}\right]\cdot{}^{\mathcal{N}}\boldsymbol{\omega}^{\mathcal{R}} \tag{5-784}$$

Recall from Eqs. (5-43) and (5-44) on page 327 that

$$\left\{\int_{\mathcal{R}}[(\boldsymbol{\rho}\cdot\boldsymbol{\rho})\mathbf{U}-(\boldsymbol{\rho}\otimes\boldsymbol{\rho})]\,dm\right\}\cdot{}^{\mathcal{N}}\boldsymbol{\omega}^{\mathcal{R}}=\bar{\mathbf{I}}^{\mathcal{R}}\cdot{}^{\mathcal{N}}\boldsymbol{\omega}^{\mathcal{R}}={}^{\mathcal{N}}\bar{\mathbf{H}} \tag{5-785}$$

where ${}^{\mathcal{N}}\bar{\mathbf{H}}$ is the angular momentum of the rigid body relative to the center of mass in the inertial reference frame $\mathcal{N}$. Substituting ${}^{\mathcal{N}}\bar{\mathbf{H}}$ from Eq. (5-785) into (5-784), the kinetic energy of a rigid body can be written as

$$\boxed{{}^{\mathcal{N}}T=\tfrac{1}{2}m{}^{\mathcal{N}}\bar{\mathbf{v}}\cdot{}^{\mathcal{N}}\bar{\mathbf{v}}+\tfrac{1}{2}{}^{\mathcal{N}}\bar{\mathbf{H}}\cdot{}^{\mathcal{N}}\boldsymbol{\omega}^{\mathcal{R}}} \tag{5-786}$$

The term

$$\tfrac{1}{2}m{}^{\mathcal{N}}\bar{\mathbf{v}}\cdot{}^{\mathcal{N}}\bar{\mathbf{v}} \tag{5-787}$$

is the *translational kinetic energy of the center of mass* of the rigid body in reference frame $\mathcal{N}$ while the term

$$\tfrac{1}{2}{}^{\mathcal{N}}\bar{\mathbf{H}}\cdot{}^{\mathcal{N}}\boldsymbol{\omega}^{\mathcal{R}} \tag{5-788}$$

is the *rotational kinetic energy* relative to the center of mass of the rigid body in reference frame $\mathcal{N}$. Thus, Eq. (5-786) states that the kinetic energy of a rigid body is the sum of the translational kinetic energy of the center of mass of the rigid body and the rotational kinetic energy about the center of mass of the rigid body. Equation (5-786) is referred to as *Koenig's decomposition* of the kinetic energy of a rigid body.

5.10.2 Work-Energy Theorem for a Rigid Body

Computing the rate of change of kinetic energy using Eq. (5–786), we obtain

$$\begin{aligned}\frac{d}{dt}\left({}^{\mathcal{N}}T\right) = \frac{1}{2}m\left[\frac{{}^{\mathcal{N}}d}{dt}\left({}^{\mathcal{N}}\bar{\mathbf{v}}\right)\cdot\bar{\mathbf{v}}+\bar{\mathbf{v}}\cdot\frac{{}^{\mathcal{N}}d}{dt}\left({}^{\mathcal{N}}\bar{\mathbf{v}}\right)\right]\\ +\frac{1}{2}\left[\frac{{}^{\mathcal{N}}d}{dt}\left({}^{\mathcal{N}}\bar{\mathbf{H}}\right)\cdot{}^{\mathcal{N}}\boldsymbol{\omega}^{\mathcal{R}}+{}^{\mathcal{N}}\bar{\mathbf{H}}\cdot\frac{{}^{\mathcal{N}}d}{dt}\left({}^{\mathcal{N}}\boldsymbol{\omega}^{\mathcal{R}}\right)\right]\end{aligned} \tag{5–789}$$

Recall in Eq. (5–789) that, because ${}^{\mathcal{N}}T$ is a scalar function, its rate of change is independent of the reference frame and, thus, can be computed arbitrarily in the inertial reference frame $\mathcal{N}$. Equation (5–789) simplifies to

$$\frac{d}{dt}\left({}^{\mathcal{N}}T\right) = m\frac{{}^{\mathcal{N}}d}{dt}\left({}^{\mathcal{N}}\bar{\mathbf{v}}\right)\cdot{}^{\mathcal{N}}\bar{\mathbf{v}}+\frac{1}{2}\left[\frac{{}^{\mathcal{N}}d}{dt}\left({}^{\mathcal{N}}\bar{\mathbf{H}}\right)\cdot{}^{\mathcal{N}}\boldsymbol{\omega}^{\mathcal{R}}+{}^{\mathcal{N}}\bar{\mathbf{H}}\cdot\frac{{}^{\mathcal{N}}d}{dt}\left({}^{\mathcal{N}}\boldsymbol{\omega}^{\mathcal{R}}\right)\right] \tag{5–790}$$

Now focus on the two quantities in the second term of Eq. (5–790). In particular, using Eq. (5–684) on page 411, we have

$$\begin{aligned}\frac{{}^{\mathcal{N}}d}{dt}\left({}^{\mathcal{N}}\bar{\mathbf{H}}\right)\cdot{}^{\mathcal{N}}\boldsymbol{\omega}^{\mathcal{R}} &= \left[\bar{\mathbf{I}}^{\mathcal{R}}\cdot{}^{\mathcal{N}}\boldsymbol{\alpha}^{\mathcal{R}}+{}^{\mathcal{N}}\boldsymbol{\omega}^{\mathcal{R}}\times\left(\bar{\mathbf{I}}^{\mathcal{R}}\cdot{}^{\mathcal{N}}\boldsymbol{\omega}^{\mathcal{R}}\right)\right]\cdot{}^{\mathcal{N}}\boldsymbol{\omega}^{\mathcal{R}}\\ &= \left[\bar{\mathbf{I}}^{\mathcal{R}}\cdot{}^{\mathcal{N}}\boldsymbol{\alpha}^{\mathcal{R}}\right]\cdot{}^{\mathcal{N}}\boldsymbol{\omega}^{\mathcal{R}}+\left[{}^{\mathcal{N}}\boldsymbol{\omega}^{\mathcal{R}}\times\left(\bar{\mathbf{I}}^{\mathcal{R}}\cdot{}^{\mathcal{N}}\boldsymbol{\omega}^{\mathcal{R}}\right)\right]\cdot{}^{\mathcal{N}}\boldsymbol{\omega}^{\mathcal{R}}\end{aligned} \tag{5–791}$$

Now it is observed that

$$\left[{}^{\mathcal{N}}\boldsymbol{\omega}^{\mathcal{R}}\times\left(\bar{\mathbf{I}}^{\mathcal{R}}\cdot{}^{\mathcal{N}}\boldsymbol{\omega}^{\mathcal{R}}\right)\right]\cdot{}^{\mathcal{N}}\boldsymbol{\omega}^{\mathcal{R}} = 0 \tag{5–792}$$

Consequently,

$$\frac{{}^{\mathcal{N}}d}{dt}\left({}^{\mathcal{N}}\bar{\mathbf{H}}\right)\cdot{}^{\mathcal{N}}\boldsymbol{\omega}^{\mathcal{R}} = \left[\bar{\mathbf{I}}^{\mathcal{R}}\cdot{}^{\mathcal{N}}\boldsymbol{\alpha}^{\mathcal{R}}\right]\cdot{}^{\mathcal{N}}\boldsymbol{\omega}^{\mathcal{R}} \tag{5–793}$$

Furthermore, since the moment of inertia tensor is symmetric, we have

$$\left[\bar{\mathbf{I}}^{\mathcal{R}}\cdot{}^{\mathcal{N}}\boldsymbol{\alpha}^{\mathcal{R}}\right]\cdot{}^{\mathcal{N}}\boldsymbol{\omega}^{\mathcal{R}} = \left[\bar{\mathbf{I}}^{\mathcal{R}}\cdot{}^{\mathcal{N}}\boldsymbol{\omega}^{\mathcal{R}}\right]\cdot{}^{\mathcal{N}}\boldsymbol{\alpha}^{\mathcal{R}} \tag{5–794}$$

However, we also have

$${}^{\mathcal{N}}\bar{\mathbf{H}}\cdot\frac{{}^{\mathcal{N}}d}{dt}\left({}^{\mathcal{N}}\boldsymbol{\omega}^{\mathcal{R}}\right) = \left[\bar{\mathbf{I}}^{\mathcal{R}}\cdot{}^{\mathcal{N}}\boldsymbol{\omega}^{\mathcal{R}}\right]\cdot{}^{\mathcal{N}}\boldsymbol{\alpha}^{\mathcal{R}} \tag{5–795}$$

As a result, the two quantities in the second term of Eq. (5–790) are equal, i.e., we have

$$\frac{{}^{\mathcal{N}}d}{dt}\left({}^{\mathcal{N}}\bar{\mathbf{H}}\right)\cdot{}^{\mathcal{N}}\boldsymbol{\omega}^{\mathcal{R}} = {}^{\mathcal{N}}\bar{\mathbf{H}}\cdot\frac{{}^{\mathcal{N}}d}{dt}\left({}^{\mathcal{N}}\boldsymbol{\omega}^{\mathcal{R}}\right) \tag{5–796}$$

Using the identity of Eq. (5–796) in (5–790), we obtain

$$\frac{d}{dt}\left({}^{\mathcal{N}}T\right) = m\frac{{}^{\mathcal{N}}d}{dt}\left({}^{\mathcal{N}}\bar{\mathbf{v}}\right)\cdot{}^{\mathcal{N}}\bar{\mathbf{v}}+\frac{{}^{\mathcal{N}}d}{dt}\left({}^{\mathcal{N}}\bar{\mathbf{H}}\right)\cdot{}^{\mathcal{N}}\boldsymbol{\omega}^{\mathcal{R}} \tag{5–797}$$

Next, using ${}^{\mathcal{N}}d({}^{\mathcal{N}}\bar{\mathbf{v}})/dt = {}^{\mathcal{N}}\bar{\mathbf{a}}$, Eq. (5–797) can be written as

$$\frac{d}{dt}\left({}^{\mathcal{N}}T\right) = m{}^{\mathcal{N}}\bar{\mathbf{a}}\cdot{}^{\mathcal{N}}\bar{\mathbf{v}}+\frac{{}^{\mathcal{N}}d}{dt}\left({}^{\mathcal{N}}\bar{\mathbf{H}}\right)\cdot{}^{\mathcal{N}}\boldsymbol{\omega}^{\mathcal{R}} \tag{5–798}$$

Then, substituting Euler's 1st law from Eq. (5–225) on page 355 and Euler's 2nd law from Eq. (5–237) from page 357, Eq. (5–798) becomes

$$\boxed{\frac{d}{dt}\left({}^{\mathcal{N}}T\right) = \mathbf{F}\cdot{}^{\mathcal{N}}\bar{\mathbf{v}} + \bar{\mathbf{M}}\cdot{}^{\mathcal{N}}\boldsymbol{\omega}^{\mathcal{R}}} \tag{5–799}$$

Equation (5–799) is called the *work-energy theorem for a rigid body.*

As it turns out, Eq. (5–799) is not the most useful form of the work-energy theorem for a rigid body. The difficulty in applying Eq. (5–799) arises from the term $\mathbf{F}\cdot{}^{\mathcal{N}}\bar{\mathbf{v}}$. Fortunately, a more useful form of the work-energy theorem can be obtained as follows. From Eq. (5–218) we have

$$\bar{\mathbf{M}} = \sum_{i=1}^{n}(\mathbf{r}_i - \bar{\mathbf{r}})\times\mathbf{F}_i + \boldsymbol{\tau} \tag{5–800}$$

where $\mathbf{F}_1,\ldots,\mathbf{F}_n$ are the forces and the resultant pure torque, respectively, applied to the rigid body. Substituting Eq. (5–800) into (5–799), we obtain

$$\frac{d}{dt}\left({}^{\mathcal{N}}T\right) = \mathbf{F}\cdot{}^{\mathcal{N}}\bar{\mathbf{v}} + \left[\sum_{i=1}^{n}(\mathbf{r}_i - \bar{\mathbf{r}})\times\mathbf{F}_i + \boldsymbol{\tau}\right]\cdot{}^{\mathcal{N}}\boldsymbol{\omega}^{\mathcal{R}} \tag{5–801}$$

Moving the quantity ${}^{\mathcal{N}}\boldsymbol{\omega}^{\mathcal{R}}$ inside the summation and recalling that the scalar product is commutative, Eq. (5–801) can be rearranged to give

$$\frac{d}{dt}\left({}^{\mathcal{N}}T\right) = \mathbf{F}\cdot{}^{\mathcal{N}}\bar{\mathbf{v}} + \sum_{i=1}^{n}{}^{\mathcal{N}}\boldsymbol{\omega}^{\mathcal{R}}\cdot[(\mathbf{r}_i - \bar{\mathbf{r}})\times\mathbf{F}_i] + \boldsymbol{\tau}\cdot{}^{\mathcal{N}}\boldsymbol{\omega}^{\mathcal{R}} \tag{5–802}$$

Next, from the properties of the scalar triple product on page 9, we have

$${}^{\mathcal{N}}\boldsymbol{\omega}^{\mathcal{R}}\cdot[(\mathbf{r}_i - \bar{\mathbf{r}})\times\mathbf{F}_i] = \left[{}^{\mathcal{N}}\boldsymbol{\omega}^{\mathcal{R}}\times(\mathbf{r}_i - \bar{\mathbf{r}})\right]\cdot\mathbf{F}_i \tag{5–803}$$

Furthermore, from Eq. (2–517) on page 106, we have

$${}^{\mathcal{N}}\boldsymbol{\omega}^{\mathcal{R}}\times(\mathbf{r}_i - \bar{\mathbf{r}}) = {}^{\mathcal{N}}\mathbf{v}_i - {}^{\mathcal{N}}\bar{\mathbf{v}} \tag{5–804}$$

Equation (5–803) then becomes

$${}^{\mathcal{N}}\boldsymbol{\omega}^{\mathcal{R}}\cdot[(\mathbf{r}_i - \bar{\mathbf{r}})\times\mathbf{F}_i] = \left[{}^{\mathcal{N}}\mathbf{v}_i - {}^{\mathcal{N}}\bar{\mathbf{v}}\right]\cdot\mathbf{F}_i \tag{5–805}$$

Substituting Eq. (5–805) into (5–802), we obtain

$$\frac{d}{dt}\left({}^{\mathcal{N}}T\right) = \mathbf{F}\cdot{}^{\mathcal{N}}\bar{\mathbf{v}} + \sum_{i=1}^{n}\left[{}^{\mathcal{N}}\mathbf{v}_i - {}^{\mathcal{N}}\bar{\mathbf{v}}\right]\cdot\mathbf{F}_i + \boldsymbol{\tau}\cdot{}^{\mathcal{N}}\boldsymbol{\omega}^{\mathcal{R}} \tag{5–806}$$

Furthermore, we note that

$$\sum_{i=1}^{n}\mathbf{F}_i\cdot{}^{\mathcal{N}}\bar{\mathbf{v}} = \mathbf{F}\cdot{}^{\mathcal{N}}\bar{\mathbf{v}} \tag{5–807}$$

Substituting Eq. (5–807) into (5–806) gives

$$\frac{d}{dt}\left({}^{\mathcal{N}}T\right) = \mathbf{F}\cdot{}^{\mathcal{N}}\bar{\mathbf{v}} + \sum_{i=1}^{n}\mathbf{F}_i{}^{\mathcal{N}}\mathbf{v}_i - \mathbf{F}\cdot{}^{\mathcal{N}}\bar{\mathbf{v}} + \boldsymbol{\tau}\cdot{}^{\mathcal{N}}\boldsymbol{\omega}^{\mathcal{R}} \tag{5–808}$$

Observing that the first and third terms in Eq. (5–808) cancel, we obtain

$$\boxed{\frac{d}{dt}\left({}^{\mathcal{N}}T\right) = \sum_{i=1}^{n} \mathbf{F}_i \cdot {}^{\mathcal{N}}\mathbf{v}_i + \boldsymbol{\tau} \cdot {}^{\mathcal{N}}\boldsymbol{\omega}^{\mathcal{R}}} \tag{5–809}$$

Equation (5–809) is a more useful form of the work-energy theorem for a rigid body. The usefulness of Eq. (5–809) will be seen in examples that follow this section.

5.10.3 Principle of Work and Energy for a Rigid Body

Suppose now that we consider motion of a rigid body $\mathcal{R}$ over a time interval $t \in [t_1, t_2]$. Then, integrating Eq. (5–809) from t_1 to t_2, we have

$$\int_{t_1}^{t_2} \frac{d}{dt}\left({}^{\mathcal{N}}T\right) dt = \int_{t_1}^{t_2} \left\{ \sum_{i=1}^{n} \mathbf{F}_i \cdot {}^{\mathcal{N}}\mathbf{v}_i + \boldsymbol{\tau} \cdot {}^{\mathcal{N}}\boldsymbol{\omega}^{\mathcal{R}} \right\} dt \tag{5–810}$$

Now we know that

$$\int_{t_1}^{t_2} \sum_{i=1}^{n} \mathbf{F}_i \cdot {}^{\mathcal{N}}\mathbf{v}_i dt = {}^{\mathcal{N}}W_{12}^{F} \tag{5–811}$$

where ${}^{\mathcal{N}}WF_{12}$ is the work done by all forces acting on the rigid body in the inertial reference frame $\mathcal{N}$. Furthermore, we have

$$\int_{t_1}^{t_2} \boldsymbol{\tau} \cdot {}^{\mathcal{N}}\boldsymbol{\omega}^{\mathcal{R}} dt = {}^{\mathcal{N}}W_{12}^{\mathcal{T}} \tag{5–812}$$

where ${}^{\mathcal{N}}W_{12}^{\mathcal{T}}$ is the work done by all of the pure torques acting on the rigid body. Finally,

$$\int_{t_1}^{t_2} \frac{d}{dt}\left({}^{\mathcal{N}}T\right) dt = {}^{\mathcal{N}}T(t_2) - {}^{\mathcal{N}}T(t_1) \tag{5–813}$$

Then, substituting the results of Eqs. (5–811), (5–812), and (5–813) into Eq. (5–810), we obtain

$$\boxed{{}^{\mathcal{N}}T(t_2) - {}^{\mathcal{N}}T(t_1) = {}^{\mathcal{N}}W_{12}^{F} + {}^{\mathcal{N}}W_{12}^{\mathcal{T}}} \tag{5–814}$$

which is called the *principle of work and energy* for a rigid body and states that the change in kinetic energy of a rigid body in an inertial reference frame during a time interval $t \in [t_1, t_2]$ is equal to the *sum* of the work done by all forces and all pure torques acting on the rigid body over that time interval.

5.10.4 Alternate Form of Work-Energy Theorem for a Rigid Body

Suppose now that we classify the forces that act on the rigid body as either conservative or nonconservative such that

$$\begin{array}{lcl} \mathbf{F}_1^c, \ldots, \mathbf{F}_p^c & = & \text{Conservative forces} \\ \mathbf{F}_1^{nc}, \ldots, \mathbf{F}_q^{nc} & = & \text{Nonconservative forces} \end{array}$$

where $p + q = n$. Then the resultant force acting on the body can be written as

$$\sum_{i=1}^{n} \mathbf{F}_i = \sum_{i=1}^{p} \mathbf{F}_i^c + \sum_{i=1}^{q} \mathbf{F}_i^{nc} \tag{5-815}$$

Similarly, suppose the resultant pure torque is decomposed into a *conservative* pure torque, $\boldsymbol{\tau}^c$, and a *nonconservative* pure torque, $\boldsymbol{\tau}^{nc}$, i.e.,

$$\boldsymbol{\tau} = \boldsymbol{\tau}^c + \boldsymbol{\tau}^{nc} \tag{5-816}$$

The work-energy theorem for a rigid body as given in Eq. (5-809) can then be written as

$$\frac{d}{dt}\left({}^{\mathcal{N}}T\right) = \sum_{i=1}^{p} \mathbf{F}_i^c \cdot {}^{\mathcal{N}}\mathbf{v}_i^c + \sum_{i=1}^{q} \mathbf{F}_i^{nc} \cdot {}^{\mathcal{N}}\mathbf{v}_i^{nc} + \boldsymbol{\tau}^c \cdot {}^{\mathcal{N}}\boldsymbol{\omega}^{\mathcal{R}} + \boldsymbol{\tau}^{nc} \cdot {}^{\mathcal{N}}\boldsymbol{\omega}^{\mathcal{R}} \tag{5-817}$$

where ${}^{\mathcal{N}}\mathbf{v}_i^c$ $(i = 1, \dots, p)$ and ${}^{\mathcal{N}}\mathbf{v}_i^{nc}$ $(i = 1, \dots, q)$ are the velocities in reference frame $\mathcal{N}$ of the points of application on $\mathcal{R}$ of the conservative forces and nonconservative forces, respectively. Recalling the power of a conservative force from Eq. (3-282), each of the conservative forces $\mathbf{F}_i^c$ $(i = 1, 2, \dots, r)$ satisfies the property

$$\mathbf{F}_i^c \cdot {}^{\mathcal{N}}\mathbf{v}_i^c = -\frac{d}{dt}\left({}^{\mathcal{N}}U_i\right) \quad (i = 1, 2, \dots, p) \tag{5-818}$$

where ${}^{\mathcal{N}}U_1, \dots, {}^{\mathcal{N}}U_p$ are the potential energies of the conservative forces $\mathbf{F}_1^c, \dots, \mathbf{F}_p^c$, respectively, in the inertial reference frame $\mathcal{N}$. Similarly, using Eq. (5-207), the conservative pure torque $\boldsymbol{\tau}^c$ satisfies the property

$$\boldsymbol{\tau}^c \cdot {}^{\mathcal{N}}\boldsymbol{\omega}^{\mathcal{R}} = -\frac{d}{dt}\left({}^{\mathcal{N}}V\right) \tag{5-819}$$

where ${}^{\mathcal{N}}V$ is the potential energy of the conservative pure torque $\boldsymbol{\tau}^c$ in the inertial reference frame $\mathcal{N}$. Substituting Eqs. (5-818) and (5-819) into Eq. (5-817), we obtain

$$\frac{d}{dt}\left({}^{\mathcal{N}}T\right) = \sum_{i=1}^{p}\left[-\frac{d}{dt}\left({}^{\mathcal{N}}U_i\right)\right] + \sum_{i=1}^{q} \mathbf{F}_i^{nc} \cdot {}^{\mathcal{N}}\mathbf{v}_i^{nc} - \frac{d}{dt}\left({}^{\mathcal{N}}V\right) + \boldsymbol{\tau}^{nc} \cdot {}^{\mathcal{N}}\boldsymbol{\omega}^{\mathcal{R}} \tag{5-820}$$

Rearranging Eq. (5-820), we have

$$\frac{d}{dt}\left[{}^{\mathcal{N}}T + \sum_{i=1}^{p} {}^{\mathcal{N}}U_i + {}^{\mathcal{N}}V\right] = \sum_{i=1}^{q} \mathbf{F}_i^{nc} \cdot {}^{\mathcal{N}}\mathbf{v}_i^{nc} + \boldsymbol{\tau}^{nc} \cdot {}^{\mathcal{N}}\boldsymbol{\omega}^{\mathcal{R}} \tag{5-821}$$

The quantity

$${}^{\mathcal{N}}E = {}^{\mathcal{N}}T + {}^{\mathcal{N}}V + \sum_{i=1}^{r} {}^{\mathcal{N}}U_i \tag{5-822}$$

is called the *total energy* of the rigid body in the inertial reference frame $\mathcal{N}$. In terms of the total energy, Eq. (5-821) can be written as

$$\boxed{\frac{d}{dt}\left({}^{\mathcal{N}}E\right) = \sum_{i=1}^{q} \mathbf{F}_i^{nc} \cdot {}^{\mathcal{N}}\mathbf{v}_i^{nc} + \boldsymbol{\tau}^{nc} \cdot {}^{\mathcal{N}}\boldsymbol{\omega}^{\mathcal{R}}} \tag{5-823}$$

Equation (5-823) is called the *alternate form of the work-energy theorem* for a rigid body and states that the rate of change of the total energy of a rigid body is equal to the sum of the power produced by all nonconservative forces *and* the power produced by the resultant nonconservative pure torque acting the body.

5.10.5 Conservation of Energy for a Rigid Body

It is seen that if the power produced by all of the nonconservative forces and resultant nonconservative pure torque is zero, i.e.,

$$\sum_{i=1}^{q} \mathbf{F}_i^{nc} \cdot {}^{\mathcal{N}}\mathbf{v}_i^{nc} + \boldsymbol{\tau}^{nc} \cdot {}^{\mathcal{N}}\boldsymbol{\omega}^{\mathcal{R}} = 0 \qquad (5\text{–}824)$$

then

$$\frac{d}{dt}\left({}^{\mathcal{N}}E\right) = 0 \qquad (5\text{–}825)$$

which implies that

$${}^{\mathcal{N}}E = \text{constant} \qquad (5\text{–}826)$$

In other words, the total energy of a rigid body will be conserved if the *sum* of the power produced by the nonconservative forces and resultant nonconservative pure torque acting on the body is zero.

5.10.6 Alternate Form of Principle of Work and Energy for a Rigid Body

Suppose now that we consider the motion of a rigid body $\mathcal{R}$ over a time interval $t \in [t_1, t_2]$. Then, integrating Eq. (5–823) from t_1 to t_2, we have

$$\int_{t_1}^{t_2} \frac{d}{dt}\left({}^{\mathcal{N}}E\right) dt = \int_{t_1}^{t_2} \left\{ \sum_{i=1}^{q} \mathbf{F}_i^{nc} \cdot {}^{\mathcal{N}}\mathbf{v}_i^{nc} + \boldsymbol{\tau}^{nc} \cdot {}^{\mathcal{N}}\boldsymbol{\omega}^{\mathcal{R}} \right\} dt \qquad (5\text{–}827)$$

Now we know that

$$\int_{t_1}^{t_2} \sum_{i=1}^{q} \mathbf{F}_i^{nc} \cdot {}^{\mathcal{N}}\mathbf{v}_i^{nc} dt = {}^{\mathcal{N}}W_{12}^{F^{nc}} \qquad (5\text{–}828)$$

where ${}^{\mathcal{N}}WF^{nc}{}_{12}$ is the work done by all of the nonconservative forces acting on the rigid body in the inertial reference frame $\mathcal{N}$. Furthermore, we have

$$\int_{t_1}^{t_2} \boldsymbol{\tau}^{nc} \cdot {}^{\mathcal{N}}\boldsymbol{\omega}^{\mathcal{R}} dt = {}^{\mathcal{N}}W_{12}^{\tau^{nc}} \qquad (5\text{–}829)$$

where ${}^{\mathcal{N}}W_{12}^{\tau^{nc}}$ is the work done by the resultant nonconservative pure torque acting on the rigid body. Finally,

$$\int_{t_1}^{t_2} \frac{d}{dt}\left({}^{\mathcal{N}}E\right) dt = {}^{\mathcal{N}}E(t_2) - {}^{\mathcal{N}}E(t_1) \qquad (5\text{–}830)$$

Then, substituting the results of Eqs. (5–828), (5–829), and (5–830) into Eq. (5–827), we obtain

$$\boxed{{}^{\mathcal{N}}E(t_2) - {}^{\mathcal{N}}E(t_1) = {}^{\mathcal{N}}W_{12}^{F^{nc}} + {}^{\mathcal{N}}W_{12}^{\tau^{nc}}} \qquad (5\text{–}831)$$

Equation (5–831) is called the *alternate form of the principle of work and energy* for a rigid body and states that the change in total energy of a rigid body in an inertial reference frame during a time interval $t \in [t_1, t_2]$ is equal to the *sum* of the work done by all nonconservative forces and all nonconservative pure torques acting on the rigid body over that time interval.

Example 5–12

Consider Example 5-5 of a homogeneous disk of mass m and radius r rolling in the vertical plane without slip along an inextensible cord where the differential equation of motion for the disk was obtained using Euler's 2nd law. In this example, we determine the differential equation of motion for the disk using the alternate form of the work-energy theorem for a rigid body.

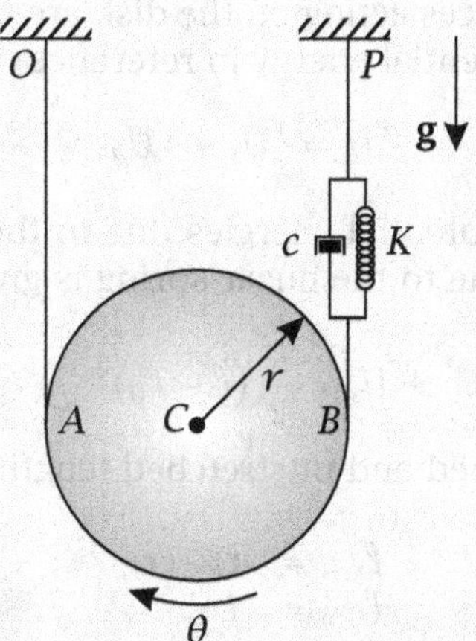

Figure 5-32 Disk rolling on cord attached to linear spring and nonlinear damper.

Solution to Example 5–12

The alternate form of the work-energy theorem for a rigid body is given in Eq. (5-823) as

$$\frac{d}{dt}\left({}^{\mathcal{F}}E\right) = \sum_{i=1}^{q} \mathbf{F}_i^{nc} \cdot {}^{\mathcal{F}}\mathbf{v}_i^{nc} + \boldsymbol{\tau}^{nc} \cdot {}^{\mathcal{F}}\boldsymbol{\omega}^{\mathcal{R}} \tag{5-832}$$

Now the total energy of the system in reference frame $\mathcal{F}$ is given as

$${}^{\mathcal{F}}E = {}^{\mathcal{F}}T + {}^{\mathcal{F}}V + \sum_{i=1}^{p} {}^{\mathcal{F}}U_i \tag{5-833}$$

Applying Eq. (5-786), the kinetic energy of the disk is given as

$${}^{\mathcal{F}}T = \tfrac{1}{2}m\,{}^{\mathcal{F}}\mathbf{v}_C \cdot {}^{\mathcal{F}}\mathbf{v}_C + \tfrac{1}{2}\,{}^{\mathcal{F}}\mathbf{H}_C \cdot {}^{\mathcal{F}}\boldsymbol{\omega}^{\mathcal{R}} \tag{5-834}$$

where ${}^{\mathcal{F}}\mathbf{H}_C$ is the angular momentum relative to the center of mass C of the rigid body. Using the expression for ${}^{\mathcal{F}}\mathbf{v}_C$ from Eq. (5-338) of Example 5-5, we obtain the first term in Eq. (5-834) as

$$\tfrac{1}{2}m\,{}^{\mathcal{F}}\mathbf{v}_C \cdot {}^{\mathcal{F}}\mathbf{v}_C = \tfrac{1}{2}mr^2\dot{\theta}^2 \tag{5-835}$$

Furthermore, the angular momentum of the disk about the center of mass of the disk is given as

$${}^{\mathcal{F}}\mathbf{H}_C = \mathbf{I}_C^{\mathcal{R}} \cdot \boldsymbol{\omega} = \tfrac{1}{2}mr^2\dot{\theta}\mathbf{E}_z \tag{5-836}$$

The second term in Eq. (5-834) is then given as

$$\tfrac{1}{2}{}^{\mathcal{F}}\mathbf{H}_C \cdot {}^{\mathcal{F}}\boldsymbol{\omega}^{\mathcal{R}} = \tfrac{1}{2}\left[\tfrac{1}{2}mr^2\dot{\theta}\mathbf{E}_z \cdot \dot{\theta}\mathbf{E}_z\right] = \tfrac{1}{4}mr^2\dot{\theta}^2 \tag{5-837}$$

Then, adding Eqs. (5-835) and (5-836), the kinetic energy of the disk is given as

$${}^{\mathcal{F}}T = \tfrac{1}{2}mr^2\dot{\theta}^2 + \tfrac{1}{4}mr^2\dot{\theta}^2 = \tfrac{3}{4}mr^2\dot{\theta}^2 \tag{5-838}$$

Now, the only conservative forces acting on the disk are those of the linear spring and gravity. Consequently, the potential energy in reference frame $\mathcal{F}$ is given as

$${}^{\mathcal{F}}U = {}^{\mathcal{F}}U_s + {}^{\mathcal{F}}U_g \tag{5-839}$$

where ${}^{\mathcal{F}}U_s$ and ${}^{\mathcal{F}}U_g$ are the potential energies due to the spring and gravity, respectively. The potential energy due to the linear spring is given from Eq. (3-296) on page 193 as

$${}^{\mathcal{F}}U_s = \frac{K}{2}\left(\ell - \ell_0\right)^2 \tag{5-840}$$

where ℓ and ℓ_0 are the stretched and unstretched lengths of the spring, respectively. For this problem, we have

$$\begin{aligned} \ell &= \mathbf{r}_B - \mathbf{r}_P \\ \ell_0 &= L \end{aligned} \tag{5-841}$$

where $\mathbf{r}_P$ is the position of the attachment point of the spring. Using $\mathbf{r}_B - \mathbf{r}_P$ from Eq. (5-347), the position of point B is given as

$$\mathbf{r}_B - \mathbf{r}_P = (2r\theta + L)\mathbf{E}_x \tag{5-842}$$

which implies that

$$\ell - \ell_0 = \|\mathbf{r}_B - \mathbf{r}_P\| - L = 2r\theta + L - L = 2r\theta \tag{5-843}$$

The potential energy of the linear spring is then given as

$${}^{\mathcal{F}}U_s = \frac{K}{2}(2r\theta)^2 = 2Kr^2\theta^2 \tag{5-844}$$

Next, because the force of gravity acts at the center of mass of the disk, the potential energy due to gravity is given as

$${}^{\mathcal{F}}U_g = -m\mathbf{g} \cdot \mathbf{r}_C \tag{5-845}$$

Substituting the expressions for $m\mathbf{g}$ and $\mathbf{r}_C$ from Eqs. (5-325) and (5-342), respectively, we obtain

$${}^{\mathcal{F}}U_g = -mg\mathbf{E}_x \cdot (r\theta\mathbf{E}_x) = -mgr\theta \tag{5-846}$$

The potential energy is then given as

$${}^{\mathcal{F}}U = 2Kr^2\theta^2 - mgr\theta \tag{5-847}$$

and the total energy is given as

$${}^{\mathcal{F}}E = {}^{\mathcal{F}}T + {}^{\mathcal{F}}U = \tfrac{3}{4}mr^2\dot{\theta}^2 + 2Kr^2\theta^2 - mgr\theta \tag{5-848}$$

Computing the rate of change of ${}^{\mathcal{F}}E$ we obtain

$$\frac{d}{dt}\left({}^{\mathcal{F}}E\right) = \tfrac{3}{2}mr^2\dot{\theta}\ddot{\theta} + 4Kr^2\theta\dot{\theta} - mgr\dot{\theta} \tag{5-849}$$

Next, because no pure torques are applied to the disk, we have

$$\boldsymbol{\tau}^{nc} \cdot {}^{\mathcal{F}}\boldsymbol{\omega}^{\mathcal{R}} = 0 \tag{5-850}$$

Also, because the only nonconservative force that produces any power is that of the damper, we obtain

$$\sum_{i=1}^{q} \mathbf{F}_i^{nc} \cdot {}^{\mathcal{F}}\mathbf{v}_i^{nc} = \mathbf{F}_d \cdot \mathbf{v}_d = \mathbf{F}_d \cdot {}^{\mathcal{F}}\mathbf{v}_B \tag{5-851}$$

Using the expressions for $\mathbf{F}_d$ and ${}^{\mathcal{F}}\mathbf{v}_B$ from Eqs. (5-351) and (5-318), we have

$$\sum_{i=1}^{q} \mathbf{F}_i^{nc} \cdot {}^{\mathcal{F}}\mathbf{v}_i^{nc} = -c(2r\dot{\theta})^3\mathbf{E}_x \cdot (2r\dot{\theta}\mathbf{E}_x) = -16cr^4\dot{\theta}^4 \tag{5-852}$$

Then, setting $d({}^{\mathcal{F}}E)/dt$ in Eq. (5-849) equal to $\sum_{i=1}^{q} \mathbf{F}_i^{nc} \cdot {}^{\mathcal{F}}\mathbf{v}_i^{nc}$ from (5-852), we obtain

$$\tfrac{3}{2}mr^2\dot{\theta}\ddot{\theta} + 4Kr^2\theta\dot{\theta} - mgr\dot{\theta} = -16cr^4\dot{\theta}^4 \tag{5-853}$$

Rearranging this last expression, we obtain

$$\dot{\theta}\left[\tfrac{3}{2}mr^2\ddot{\theta} + 16cr^4\dot{\theta}^3 + 4Kr^2\theta - mgr\right] = 0 \tag{5-854}$$

Observing that $\dot{\theta}$ is not zero as a function of time, the differential equation of motion for the disk is given as

$$\tfrac{3}{2}mr^2\ddot{\theta} + 16cr^4\dot{\theta}^3 + 4Kr^2\theta - mgr = 0 \tag{5-855}$$

It is seen that the result of Eq. (5-855) is identical to that of Eq. (5-366) as obtained on page 367 in Example 5-5.

■

Example 5-13

A uniform slender rod of mass m and length l is hinged at one of its ends to a fixed point O as shown in Fig. 5-33. Also attached to the rod at point O are a torsional spring with spring constant K and a torsional damper with damping coefficient c. The spring and damper both produce resistive pure torques. The spring is uncoiled when the angle θ shown in the figure is zero while the torque of the damper is proportional to $\dot{\theta}$. Knowing that gravity acts downward, determine the differential equation of motion for the rod using the following methods: (a) Euler's 2nd law using the hinge point as the reference, (b) Euler's laws using the center of mass of the rod as the reference, and

(c) the alternate form of the work-energy theorem for a rigid body.

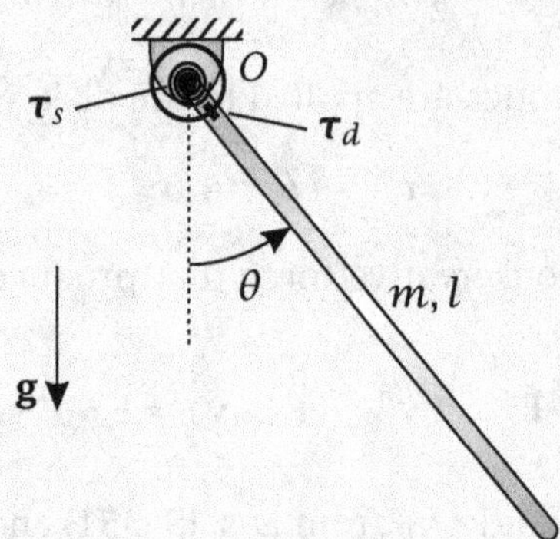

Figure 5-33 Rod of mass m and length l hinged at a fixed point O and connected to a linear torsional spring and linear torsional damper.

Solution to Example 5-13

Kinematics

First, let $\mathcal{F}$ be a fixed reference frame. Then, choose the following coordinate system fixed in reference frame $\mathcal{F}$:

Origin at O		
$\mathbf{E}_x$	=	Along rod when $\theta = 0$
$\mathbf{E}_z$	=	Out of page
$\mathbf{E}_y$	=	$\mathbf{E}_z \times \mathbf{E}_x$

Next, let $\mathcal{R}$ be a reference frame fixed to the rod. Then, choose the following coordinate system fixed in reference frame $\mathcal{R}$:

Origin at O		
$\mathbf{e}_r$	=	Along rod
$\mathbf{e}_z$	=	Out of page
$\mathbf{e}_\theta$	=	$\mathbf{E}_z \times \mathbf{e}_r$

The geometry of the bases $\{\mathbf{E}_x, \mathbf{E}_y, \mathbf{E}_z\}$ and $\{\mathbf{e}_r, \mathbf{e}_\theta, \mathbf{e}_z\}$ is shown in Fig. 5-34, from which we have

$$\mathbf{E}_x = \cos\theta\,\mathbf{e}_r - \sin\theta\,\mathbf{e}_\theta \tag{5-856}$$

$$\mathbf{E}_y = \sin\theta\,\mathbf{e}_r + \cos\theta\,\mathbf{e}_\theta \tag{5-857}$$

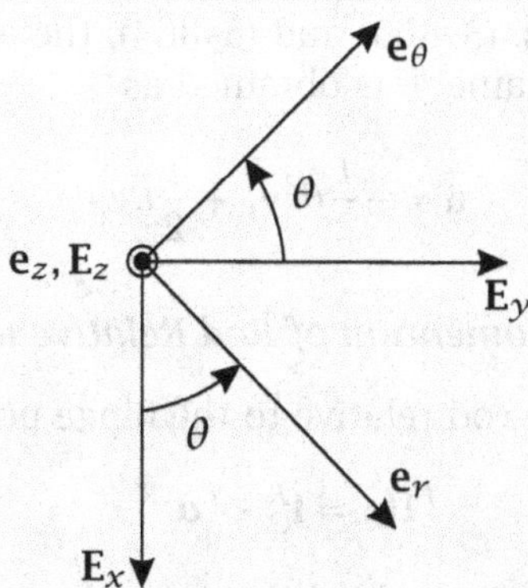

Figure 5-34 Geometry of bases $\{\mathbf{E}_x, \mathbf{E}_y, \mathbf{E}_z\}$ and $\{\mathbf{e}_r, \mathbf{e}_\theta, \mathbf{e}_z\}$ for Example 5-13.

It is seen from the geometry that the angular velocity of reference frame $\mathcal{R}$ in reference frame $\mathcal{F}$ is given as

$$^{\mathcal{F}}\boldsymbol{\omega}^{\mathcal{R}} = \dot{\theta}\mathbf{e}_z \tag{5-858}$$

Now for this problem we will need the following kinematic quantities: (1) the acceleration of the center of mass of the rod; (2) the rate of change of angular momentum relative to the hinge point; and (3) the rate of change of angular momentum relative to the center of mass of the rod. Each of these quantities is now computed.

Acceleration of Center of Mass of Rod

Because the rod has uniform density, the position of the center of mass of the rod is given as

$$\bar{\mathbf{r}} = \frac{l}{2}\mathbf{e}_r \tag{5-859}$$

Then the velocity of the center of mass of the rod in reference frame $\mathcal{F}$ is obtained by applying the rate of change transport theorem between reference frames $\mathcal{R}$ and $\mathcal{F}$ as

$$^{\mathcal{F}}\bar{\mathbf{v}} = \frac{^{\mathcal{F}}d\bar{\mathbf{r}}}{dt} = \frac{^{\mathcal{R}}d\bar{\mathbf{r}}}{dt} + {}^{\mathcal{F}}\boldsymbol{\omega}^{\mathcal{R}} \times \bar{\mathbf{r}} \tag{5-860}$$

Noting that l is constant, we have $^{\mathcal{R}}d\bar{\mathbf{r}}/dt = {}^{\mathcal{R}}\bar{\mathbf{v}} = \mathbf{0}$. Furthermore, using the expression for $^{\mathcal{F}}\boldsymbol{\omega}^{\mathcal{R}}$ from Eq. (5-858), we obtain $^{\mathcal{F}}\bar{\mathbf{v}}$ as

$$^{\mathcal{F}}\bar{\mathbf{v}} = \frac{^{\mathcal{F}}d\bar{\mathbf{r}}}{dt} = \dot{\theta}\mathbf{e}_z \times \frac{l}{2}\mathbf{e}_r = \frac{l}{2}\dot{\theta}\mathbf{e}_\theta \tag{5-861}$$

Then the acceleration of the center of mass of the rod is obtained by applying the rate of change transport theorem to $^{\mathcal{F}}\bar{\mathbf{v}}$ between reference frames $\mathcal{R}$ and $\mathcal{F}$ as

$$^{\mathcal{F}}\bar{\mathbf{a}} = \frac{^{\mathcal{F}}d}{dt}\left(^{\mathcal{F}}\bar{\mathbf{v}}\right) = \frac{^{\mathcal{R}}d}{dt}\left(^{\mathcal{F}}\bar{\mathbf{v}}\right) + {}^{\mathcal{F}}\boldsymbol{\omega}^{\mathcal{R}} \times {}^{\mathcal{F}}\bar{\mathbf{v}} \tag{5-862}$$

Now we have

$$\frac{^{\mathcal{R}}d}{dt}\left(^{\mathcal{F}}\bar{\mathbf{v}}\right) = \frac{l}{2}\ddot{\theta}\mathbf{e}_\theta \tag{5-863}$$

$$^{\mathcal{F}}\boldsymbol{\omega}^{\mathcal{R}} \times {}^{\mathcal{F}}\bar{\mathbf{v}} = \dot{\theta}\mathbf{e}_z \times \frac{l}{2}\dot{\theta}\mathbf{e}_\theta = -\frac{l}{2}\dot{\theta}^2\mathbf{e}_r \tag{5-864}$$

Adding the expressions in Eqs. (5-863) and (5-864), the acceleration of the center of mass of the rod in reference frame $\mathcal{F}$ is obtained as

$$ {}^{\mathcal{F}}\bar{\mathbf{a}} = -\frac{l}{2}\dot{\theta}^2\mathbf{e}_r + \frac{l}{2}\ddot{\theta}\mathbf{e}_\theta \tag{5-865} $$

Rate of Change of Angular Momentum of Rod Relative to Hinge Point

The angular momentum of the rod relative to the hinge point O is given as

$$ {}^{\mathcal{F}}\mathbf{H}_O = \mathbf{I}_O^{\mathcal{R}} \cdot {}^{\mathcal{F}}\boldsymbol{\omega}^{\mathcal{R}} \tag{5-866} $$

Now, because $\{\mathbf{e}_r, \mathbf{e}_\theta, \mathbf{e}_z\}$ is a principal-axis basis, we have

$$ \mathbf{I}_O^{\mathcal{R}} = I_{rr}^O \mathbf{e}_r \otimes \mathbf{e}_r + I_{\theta\theta}^O \mathbf{e}_\theta \otimes \mathbf{e}_\theta + I_{zz}^O \mathbf{e}_z \otimes \mathbf{e}_z \tag{5-867} $$

Substituting $\mathbf{I}_O^{\mathcal{R}}$ and ${}^{\mathcal{F}}\boldsymbol{\omega}^{\mathcal{R}}$ from Eqs. (5-867) and (5-858), respectively, into Eq. (5-866), we obtain ${}^{\mathcal{F}}\mathbf{H}_O$ as

$$ {}^{\mathcal{F}}\mathbf{H}_O = I_{zz}^O \dot{\theta}\mathbf{e}_z \tag{5-868} $$

where I_{zz}^O is the moment of inertia of the rod relative to point O in the $\mathbf{e}_z$-direction. Then, from the parallel-axis theorem we have

$$ I_{zz}^O = \bar{I}_{zz} + m(\pi_r^2 + \pi_\theta^2) \tag{5-869} $$

where $\bar{I}_{zz}$ is the moment of inertia about the center of mass of the rod and π_r and π_θ are the distances by which the axis in the direction of $\mathbf{e}_z$ through the center of mass of the rod is translated in the directions $\mathbf{e}_r$ and $\mathbf{e}_\theta$. Now for a slender rod we have

$$ \bar{I}_{zz} = \frac{ml^2}{12} \tag{5-870} $$

Furthermore, because the axis in the direction of $\mathbf{e}_z$ through the center of mass of the rod is translated to point O purely in the direction $\mathbf{e}_r$, we have $\pi_r = l/2$ and $\pi_\theta = 0$. Consequently,

$$ I_{zz}^O = \frac{ml^2}{12} + m\left(\frac{l}{2}\right)^2 == \frac{ml^2}{12} + \frac{ml^2}{4} = \frac{ml^2}{3} \tag{5-871} $$

Then

$$ {}^{\mathcal{F}}\mathbf{H}_O = \frac{ml^2}{3}\dot{\theta}\mathbf{e}_z \tag{5-872} $$

Computing the rate of change of ${}^{\mathcal{F}}\mathbf{H}_O$ in reference frame $\mathcal{F}$, we have

$$ \frac{{}^{\mathcal{F}}d}{dt}\left({}^{\mathcal{F}}\mathbf{H}_O\right) = \frac{ml^2}{3}\ddot{\theta}\mathbf{e}_z \tag{5-873} $$

Rate of Change of Angular Momentum of Rod Relative to Center of Mass of Rod

The angular momentum relative to the center of mass of the rod is given as

$$ {}^{\mathcal{N}}\bar{\mathbf{H}} = \bar{\mathbf{I}}^{\mathcal{R}} \cdot {}^{\mathcal{F}}\boldsymbol{\omega}^{\mathcal{R}} \tag{5-874} $$

Now, because $\{\mathbf{e}_r, \mathbf{e}_\theta, \mathbf{e}_z\}$ is a principal-axis basis, we have

$$\bar{\mathbf{I}}^{\mathcal{R}} = \bar{I}_{rr}\mathbf{e}_r \otimes \mathbf{e}_r + \bar{I}_{\theta\theta}\mathbf{e}_\theta \otimes \mathbf{e}_\theta + \bar{I}_{zz}\mathbf{e}_z \otimes \mathbf{e}_z \tag{5-875}$$

Substituting $\mathbf{I}_O^{\mathcal{R}}$ and ${}^{\mathcal{F}}\boldsymbol{\omega}^{\mathcal{R}}$ from Eqs. (5-858) and (5-875), respectively, into Eq. (5-874), we obtain ${}^{\mathcal{F}}\bar{\mathbf{H}}$ as

$${}^{\mathcal{F}}\bar{\mathbf{H}} = \bar{I}_{zz}\dot{\theta}\mathbf{e}_z \tag{5-876}$$

Then, using the moment of inertia, $\bar{I}_{zz}$, about the center of mass of a uniform slender rod, we have

$${}^{\mathcal{F}}\bar{\mathbf{H}} = \frac{ml^2}{12}\dot{\theta}\mathbf{e}_z \tag{5-877}$$

The rate of change of ${}^{\mathcal{F}}\bar{\mathbf{H}}$ in reference frame $\mathcal{F}$ is then given as

$$\frac{{}^{\mathcal{F}}d}{dt}\left({}^{\mathcal{F}}\bar{\mathbf{H}}\right) = \frac{ml^2}{12}\ddot{\theta}\mathbf{e}_z \tag{5-878}$$

Kinetics

The free body diagram of the rod is given in Fig. 5-35.

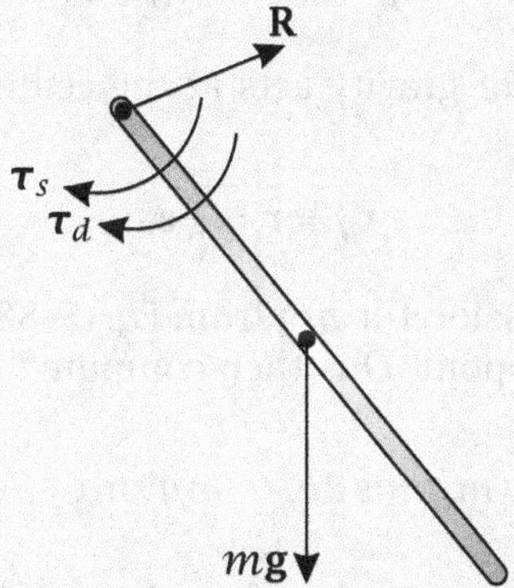

Figure 5-35 Free body diagram for Example 5-13.

Using Fig. 5-35, it is seen that the following forces and pure torques act on the rod:

$\boldsymbol{\tau}_s$	=	Pure torque due to torsional spring
$\boldsymbol{\tau}_d$	=	Pure torque due to torsional damper
$\mathbf{R}$	=	Reaction force on rod at hinge
$m\mathbf{g}$	=	Force of gravity

It is emphasized that the torsional spring and torsional damper apply *pure torques* to the rod. Resolving the forces in the basis $\{\mathbf{e}_r, \mathbf{e}_\theta, \mathbf{e}_z\}$, we have

$$\boldsymbol{\tau}_s = -K\theta\mathbf{e}_z \tag{5-879}$$

$$\boldsymbol{\tau}_d = -c\dot{\theta}\mathbf{e}_z \tag{5-880}$$

$$\mathbf{R} = R_r\mathbf{e}_r + R_\theta\mathbf{e}_\theta \tag{5-881}$$

$$m\mathbf{g} = mg\mathbf{E}_x \tag{5-882}$$

Then, using the expression for $\mathbf{E}_x$ from Eq. (5-856), the force of gravity is given as

$$m\mathbf{g} = mg(\cos\theta\,\mathbf{e}_r - \sin\theta\,\mathbf{e}_\theta) = mg\cos\theta\,\mathbf{e}_r - mg\sin\theta\,\mathbf{e}_\theta \tag{5-883}$$

It is extremely important to understand that the resultant force acting on the rod *does not* include the pure torques due to the torsional spring and the torsional damper because the pure torques do not arise from forces that act on the rod.

(a) Differential Equation Using Hinge as Reference Point

Because O is an inertially fixed point, we can apply Euler's law relative to point O in reference frame $\mathcal{F}$ as

$$\mathbf{M}_O = \frac{^{\mathcal{F}}d}{dt}\left({}^{\mathcal{F}}\mathbf{H}_O\right) \tag{5-884}$$

From the free body diagram of Fig. 5-35 we observe that the reaction $\mathbf{R}$ passes through point O. Consequently, the moment relative to point O is due to gravity, the torsional spring, and the torsional damper and is given as

$$\mathbf{M}_O = (\mathbf{r}_g - \mathbf{r}_O) \times m\mathbf{g} + \boldsymbol{\tau}_s + \boldsymbol{\tau}_d \tag{5-885}$$

Now we know that the force of gravity acts at the center of mass of the rod. Consequently,

$$\mathbf{r}_g = \bar{\mathbf{r}} = \frac{l}{2}\mathbf{e}_r \tag{5-886}$$

Furthermore, using the expression for $m\mathbf{g}$ from Eq. (5-883) and noting that $\mathbf{r}_O = \mathbf{0}$, the moment due to gravity about point O is then computed as

$$(\mathbf{r}_g - \mathbf{r}_O) \times m\mathbf{g} = \frac{l}{2}\mathbf{e}_r \times (mg\cos\theta\,\mathbf{e}_r - mg\sin\theta\,\mathbf{e}_\theta) = -\frac{mgl}{2}\sin\theta\,\mathbf{e}_z \tag{5-887}$$

Then, using the expressions for $\boldsymbol{\tau}_s$ and $\boldsymbol{\tau}_d$ from Eqs. (5-879) and (5-880), respectively, we obtain the moment applied to the rod relative to the hinge point as

$$\begin{aligned}\mathbf{M}_O &= -\frac{mgl}{2}\sin\theta\,\mathbf{e}_z - K\theta\mathbf{e}_z - c\dot{\theta}\mathbf{e}_z \\ &= -\left(\frac{mgl}{2}\sin\theta + K\theta + c\dot{\theta}\right)\mathbf{e}_z\end{aligned} \tag{5-888}$$

Then, setting $\mathbf{M}_O$ in Eq. (5-888) equal to ${}^{\mathcal{F}}d({}^{\mathcal{F}}\mathbf{H}_O)/dt$ in Eq. (5-873), we obtain

$$-\left(\frac{mgl}{2}\sin\theta + K\theta + c\dot{\theta}\right) = \frac{ml^2}{3}\ddot{\theta} \tag{5-889}$$

Rearranging this last expression, we obtain the differential equation of motion as

$$\frac{ml^2}{3}\ddot{\theta} + c\dot{\theta} + K\theta + \frac{mgl}{2}\sin\theta = 0 \tag{5-890}$$

(b) Differential Equation Using Center of Mass of Rod as Reference Point

Using the center of mass of the rod as the reference point, we need to apply both of Euler's laws. Applying Euler's 1st law in reference frame $\mathcal{F}$, we have

$$\mathbf{F} = m^{\mathcal{F}}\bar{\mathbf{a}} \tag{5-891}$$

Using the free body diagram of Fig. 5-35 we have

$$\mathbf{F} = \mathbf{R} + m\mathbf{g} \tag{5-892}$$

Furthermore, using the expressions for $\mathbf{R}$ and $m\mathbf{g}$ from Eqs. (5-881) and (5-883), respectively, the resultant force acting on the rod is given as

$$\mathbf{F} = R_r\mathbf{e}_r + R_\theta\mathbf{e}_\theta + mg\cos\theta\,\mathbf{e}_r - mg\sin\theta\,\mathbf{e}_\theta \tag{5-893}$$

which simplifies to

$$\mathbf{F} = (R_r + mg\cos\theta)\mathbf{e}_r + (R_\theta - mg\sin\theta)\mathbf{e}_\theta \tag{5-894}$$

Setting $\mathbf{F}$ from Eq. (5-894) equal to $m^{\mathcal{F}}\bar{\mathbf{a}}$ from Eq. (5-865), we obtain

$$(R_r + mg\cos\theta)\mathbf{e}_r + (R_\theta - mg\sin\theta)\mathbf{e}_\theta = -\frac{ml}{2}\dot{\theta}^2\mathbf{e}_r + \frac{ml}{2}\ddot{\theta}\mathbf{e}_\theta \tag{5-895}$$

which results in the following two scalar equations:

$$R_r + mg\cos\theta = -\frac{ml}{2}\dot{\theta}^2 \tag{5-896}$$

$$R_\theta - mg\sin\theta = \frac{ml}{2}\ddot{\theta} \tag{5-897}$$

Next, we apply Euler's 2nd law about the center of mass of the rod. First, the moment relative to the center of mass of the rod is given as

$$\bar{\mathbf{M}} = \frac{^{\mathcal{F}}d}{dt}\left(^{\mathcal{F}}\bar{\mathbf{H}}\right) \tag{5-898}$$

Now, because the force of gravity passes through the center of mass, the only moments relative to the center of mass are due to the reaction force, $\mathbf{R}$; the pure torque of the torsional spring $\boldsymbol{\tau}_s$; and the pure torque of the torsional damper, $\boldsymbol{\tau}_d$. Recalling that a pure torque is *independent* of the particular point on the rigid body, it can be translated anywhere on the body without change. Consequently, the moment applied to the rod about the center of mass of the rod is given as

$$\bar{\mathbf{M}} = (\mathbf{r}_R - \bar{\mathbf{r}}) \times \mathbf{R} + \boldsymbol{\tau}_s + \boldsymbol{\tau}_d \tag{5-899}$$

Because $\mathbf{r}_R = \mathbf{r}_O = \mathbf{0}$, we have

$$\mathbf{r}_R - \bar{\mathbf{r}} = -\frac{l}{2}\mathbf{e}_r \tag{5-900}$$

Therefore,

$$(\mathbf{r}_R - \bar{\mathbf{r}}) \times \mathbf{R} = -\frac{l}{2}\mathbf{e}_r \times (R_r\mathbf{e}_r + R_\theta\mathbf{e}_\theta) = -\frac{l}{2}R_\theta\mathbf{e}_z \tag{5-901}$$

We then obtain

$$\bar{\mathbf{M}} = -\frac{l}{2}R_\theta \mathbf{e}_z - K\theta \mathbf{e}_z - c\dot{\theta}\mathbf{e}_z = -\left(\frac{l}{2}R_\theta + K\theta + c\dot{\theta}\right)\mathbf{e}_z \tag{5-902}$$

Then, setting $\bar{\mathbf{M}}$ in Eq. (5-902) with ${}^{\mathcal{F}}d({}^{\mathcal{F}}\bar{\mathbf{H}})/dt$ from Eq. (5-878), we obtain

$$-\left(\frac{l}{2}R_\theta + K\theta + c\dot{\theta}\right) = \frac{ml^2}{12}\ddot{\theta} \tag{5-903}$$

Rearranging this last expression, we obtain

$$\frac{ml^2}{12}\ddot{\theta} + K\theta + c\dot{\theta} = -\frac{l}{2}R_\theta \tag{5-904}$$

Equations (5-896), (5-897), and (5-904) can now be used to determine the differential equation of motion. First, solving Eq. (5-897) for R_θ gives

$$R_\theta = \frac{ml}{2}\ddot{\theta} + mg\sin\theta \tag{5-905}$$

Next, substituting the expression for R_θ into Eq. (5-904), we obtain

$$\frac{ml^2}{12}\ddot{\theta} + K\theta + c\dot{\theta} = -\frac{l}{2}\left[\frac{ml}{2}\ddot{\theta} + mg\sin\theta\right] \tag{5-906}$$

Then, expanding the right-hand side of Eq. (5-906),

$$\frac{ml^2}{12}\ddot{\theta} + K\theta + c\dot{\theta} = -\frac{ml^2}{4}\ddot{\theta} - \frac{mgl}{2}\sin\theta \tag{5-907}$$

Rearranging Eq. (5-906), we obtain the differential equation of motion as

$$\frac{ml^2}{3}\ddot{\theta} + c\dot{\theta} + K\theta + \frac{mgl}{2}\sin\theta = 0 \tag{5-908}$$

We note that Eq. (5-908) is the same as the result as obtained in part (a) above.

(c) Differential Equation Using Alternate Form of Work-Energy Theorem

For this problem we use the alternate form of the work-energy theorem for a rigid body given in Eq. (5-823) on page 425 as

$$\frac{d}{dt}\left({}^{\mathcal{F}}E\right) = \sum_{i=1}^{q} \mathbf{F}_i^{nc} \cdot {}^{\mathcal{F}}\mathbf{v}_i^{nc} + \boldsymbol{\tau}^{nc} \cdot {}^{\mathcal{F}}\boldsymbol{\omega}^{\mathcal{R}} \tag{5-909}$$

Now the total energy of the system is given as

$${}^{\mathcal{F}}E = {}^{\mathcal{F}}T + {}^{\mathcal{F}}U \tag{5-910}$$

where ${}^{\mathcal{F}}T$ and ${}^{\mathcal{F}}U$ are the kinetic energy and potential energy, respectively, in reference frame $\mathcal{F}$. First, the kinetic energy of the rod in reference frame $\mathcal{F}$ is given as

$${}^{\mathcal{F}}T = \tfrac{1}{2}m\,{}^{\mathcal{F}}\bar{\mathbf{v}} \cdot {}^{\mathcal{F}}\bar{\mathbf{v}} + \tfrac{1}{2}\,{}^{\mathcal{F}}\bar{\mathbf{H}} \cdot {}^{\mathcal{F}}\boldsymbol{\omega}^{\mathcal{R}} \tag{5-911}$$

Substituting the expression for ${}^{\mathcal{F}}\bar{\mathbf{v}}$ from Eq. (5-861) and the expression for ${}^{\mathcal{F}}\bar{\mathbf{H}}$ from Eq. (5-877) into (5-911), we obtain

$$ {}^{\mathcal{F}}T = \tfrac{1}{2}m\left(\tfrac{l}{2}\dot{\theta}\mathbf{e}_\theta \cdot \tfrac{l}{2}\dot{\theta}\mathbf{e}_\theta\right) + \tfrac{1}{2}\left(\frac{ml^2}{12}\dot{\theta}\mathbf{e}_z \cdot \dot{\theta}\mathbf{e}_z\right) \tag{5-912} $$

Simplifying Eq. (5-912) gives

$$ {}^{\mathcal{F}}T = \frac{ml^2}{8}\dot{\theta}^2 + \frac{ml^2}{24}\dot{\theta}^2 = \frac{ml^2}{6}\dot{\theta}^2 \tag{5-913} $$

Next, since the only conservative force acting on the rod is that due to gravity, while the only conservative pure torque acting on the rod is that due to the torsional spring, the potential energy of the system in reference frame $\mathcal{F}$ is given as

$$ {}^{\mathcal{F}}U = {}^{\mathcal{F}}U_g + {}^{\mathcal{F}}U_s \tag{5-914} $$

where ${}^{\mathcal{F}}U_g$ is the potential energy of gravity while ${}^{\mathcal{F}}U_s$ is the potential energy of the torsional spring. Now, since gravity is a constant force, we have

$$ {}^{\mathcal{F}}U_g = -m\mathbf{g} \cdot \bar{\mathbf{r}} \tag{5-915} $$

Using the expression for $m\mathbf{g}$ from Eq. (5-883) and the expression for $\bar{\mathbf{r}}$ from Eq. (5-859), we obtain ${}^{\mathcal{F}}U_g$ as

$$ {}^{\mathcal{F}}U_g = -(mg\cos\theta\,\mathbf{e}_r - mg\sin\theta\,\mathbf{e}_\theta)\cdot\left(\tfrac{l}{2}\sin\theta\,\mathbf{e}_r\right) = -\frac{mgl}{2}\cos\theta \tag{5-916} $$

Next, the potential energy due to the torsional spring, ${}^{\mathcal{F}}U_s$, has the general form

$$ {}^{\mathcal{F}}U_s = \tfrac{1}{2}K(\theta - \theta_0)^2 \tag{5-917} $$

Because the torsional spring is uncoiled when $\theta = 0$, we have that $\theta_0 = 0$. Therefore,

$$ {}^{\mathcal{F}}U_s = \tfrac{1}{2}K\theta^2 \tag{5-918} $$

Adding the results of Eqs. (5-916) and (5-918), we obtain the total potential energy of the system as

$$ {}^{\mathcal{F}}U = {}^{\mathcal{F}}U_g + {}^{\mathcal{F}}U_s = -\frac{mgl}{2}\cos\theta + \frac{1}{2}K\theta^2 \tag{5-919} $$

Furthermore, adding Eqs. (5-913) and (5-919), the total energy of the system is given as

$$ {}^{\mathcal{F}}E = {}^{\mathcal{F}}T + {}^{\mathcal{F}}U = \frac{ml^2}{6}\dot{\theta}^2 - \frac{mgl}{2}\cos\theta + \frac{1}{2}K\theta^2 \tag{5-920} $$

Computing the rate of change of ${}^{\mathcal{F}}E$, we obtain

$$ \frac{d}{dt}\left({}^{\mathcal{F}}E\right) = \frac{ml^2}{3}\dot{\theta}\ddot{\theta} + \frac{mgl}{2}\dot{\theta}\sin\theta + K\dot{\theta}\theta \tag{5-921} $$

Factoring out $\dot{\theta}$ from Eq. (5-921) gives

$$ \frac{d}{dt}\left({}^{\mathcal{F}}E\right) = \dot{\theta}\left[\frac{ml^2}{3}\ddot{\theta} + \frac{mgl}{2}\sin\theta + K\theta\right] \tag{5-922} $$

Next, we need to determine the terms on the right-hand side of Eq. (5-909). It can be seen that the only force other than gravity that acts on the rod is the reaction force, **R**. Furthermore, **R** acts at point O where the velocity is zero. Therefore,

$$\sum_{i=1}^{q} \mathbf{F}_i^{nc} \cdot {}^{\mathcal{F}}\mathbf{v}_i^{nc} = \mathbf{R} \cdot \mathbf{0} = 0 \tag{5-923}$$

Also, the only pure torque other than the moment due to the torsional spring that acts on the rod is that of the torsional damper. Therefore,

$$\boldsymbol{\tau}^{nc} \cdot {}^{\mathcal{F}}\boldsymbol{\omega}^{\mathcal{R}} = -c\dot{\theta}\mathbf{e}_z \cdot \dot{\theta}\mathbf{e}_z = -c\dot{\theta}^2 \tag{5-924}$$

Then the power of all nonconservative forces and nonconservative pure torques is given as

$$\sum_{i=1}^{q} \mathbf{F}_i^{nc} \cdot {}^{\mathcal{F}}\mathbf{v}_i^{nc} + \boldsymbol{\tau}^{nc} \cdot {}^{\mathcal{F}}\boldsymbol{\omega}^{\mathcal{R}} = -c\dot{\theta}^2 \tag{5-925}$$

Setting the result of Eq. (5-922) equal to the result of Eq. (5-925), we obtain

$$\dot{\theta}\left[\frac{ml^2}{3}\ddot{\theta} + \frac{mgl}{2}\sin\theta + K\theta\right] = -c\dot{\theta}^2 \tag{5-926}$$

Since $\dot{\theta}$ is not zero as a function of time, we can drop $\dot{\theta}$ from both sides of the last expression. We then obtain

$$\frac{ml^2}{3}\ddot{\theta} + \frac{mgl}{2}\sin\theta + K\theta = -c\dot{\theta} \tag{5-927}$$

Rearranging this last expression gives

$$\frac{ml^2}{3}\ddot{\theta} + c\dot{\theta} + K\theta + \frac{mgl}{2}\sin\theta = 0 \tag{5-928}$$

which is the same result as obtained in parts (a) and (b).

■

Example 5–14

Consider Example 5-8 from page 384 of the imbalanced wheel rolling without slip along an inclined plane of inclination angle β. Derive the differential equation of motion for the wheel using the alternate form of the work-energy theorem for a rigid body.

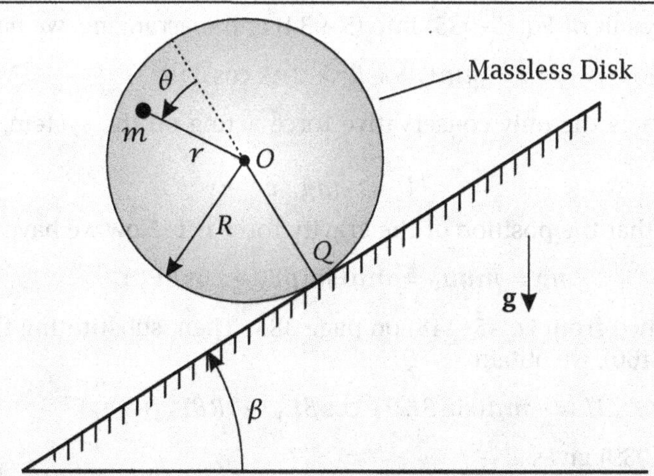

Figure 5-36 Imbalanced wheel rolling along incline with inclination angle β.

Solution to Example 5-14

Using the free body diagram in Fig. 5-21 on page 388, we see that the forces $\mathbf{F}_f$ and $\mathbf{N}$ act at point Q while the force of gravity, $m\mathbf{g}$, acts at the location of the particle. We know that $m\mathbf{g}$ is conservative because it is a constant force. Furthermore, since the wheel rolls without slip along a fixed surface, we have ${}^{\mathcal{F}}\mathbf{v}_Q = \mathbf{0}$. Consequently, the forces $\mathbf{F}_f$ and $\mathbf{N}$ do no work, i.e.,

$$\mathbf{F}_f \cdot {}^{\mathcal{F}}\mathbf{v}_Q = 0 \tag{5-929}$$

$$\mathbf{N} \cdot {}^{\mathcal{F}}\mathbf{v}_Q = 0 \tag{5-930}$$

Then, because the only force acting on the system that produces any power is the conservative force of gravity, energy is conserved, i.e.,

$${}^{\mathcal{F}}E = {}^{\mathcal{F}}T + {}^{\mathcal{F}}U = \text{constant} \tag{5-931}$$

Now, using the results from the kinematics of Example 5-8, the kinetic energy is given as

$${}^{\mathcal{F}}T = \tfrac{1}{2}m\,{}^{\mathcal{F}}\mathbf{v} \cdot {}^{\mathcal{F}}\mathbf{v} \tag{5-932}$$

where we note that the contribution from the term involving the angular momentum is zero because the mass in this problem is concentrated at a single point. Then, substituting ${}^{\mathcal{F}}\mathbf{v}$ from Eq. (5-467) on page 387 into Eq. (5-932), we obtain the kinetic energy as

$${}^{\mathcal{F}}T = \tfrac{1}{2}m(R\dot{\theta}\mathbf{E}_x + r\dot{\theta}\mathbf{e}_\theta) \cdot (R\dot{\theta}\mathbf{E}_x + r\dot{\theta}\mathbf{e}_\theta) \tag{5-933}$$

Expanding Eq. (5-933) gives

$${}^{\mathcal{F}}T = \tfrac{1}{2}m(R^2\dot{\theta}^2 + 2rR\dot{\theta}^2\mathbf{E}_x \cdot \mathbf{e}_\theta + r^2\dot{\theta}^2) \tag{5-934}$$

Then, using Eq. (5-447) on page 385 of Example 5-8), we have

$$\mathbf{E}_x \cdot \mathbf{e}_\theta = \mathbf{E}_x \cdot (\cos\theta\mathbf{E}_x + \sin\theta\mathbf{E}_y) = \cos\theta \tag{5-935}$$

Substituting the result of Eq. (5–935) into (5–934) and rearranging, we have

$$ {}^{\mathcal{F}}T = \tfrac{1}{2}m(r^2 + R^2 + 2rR\cos\theta)\dot\theta^2 \tag{5–936}$$

Next, since gravity is the only conservative force acting on the system, the potential energy is given as

$$ {}^{\mathcal{F}}U = -m\mathbf{g}\cdot\mathbf{r} \tag{5–937}$$

where it is noted that the position of the gravity force is $\mathbf{r}$. Now we have

$$ m\mathbf{g} = mg\mathbf{u}_v = mg(\sin\beta\mathbf{E}_x + \cos\beta\mathbf{E}_y) \tag{5–938}$$

where $\mathbf{u}_v$ is obtained from Eq. (5–448) on page 385. Then, substituting the expression for $\mathbf{r}$ from Eq. (5–460), we obtain

$$ {}^{\mathcal{F}}U = -mg(\sin\beta\mathbf{E}_x + \cos\beta\mathbf{E}_y)\cdot(R\theta\mathbf{E}_x + r\mathbf{e}_r) \tag{5–939}$$

Expanding Eq. (5–939) gives

$$ {}^{\mathcal{F}}U = -mg(R\theta\sin\beta + r\sin\beta\mathbf{E}_x\cdot\mathbf{e}_r + r\cos\beta\mathbf{E}_y\cdot\mathbf{e}_r) \tag{5–940}$$

Now from Eq. (5–446) on page 385 we have

$$\begin{aligned} \mathbf{E}_x\cdot\mathbf{e}_r &= \mathbf{E}_x\cdot(\sin\theta\mathbf{E}_x - \cos\theta\mathbf{E}_y) = \sin\theta \\ \mathbf{E}_y\cdot\mathbf{e}_r &= \mathbf{E}_x\cdot(\sin\theta\mathbf{E}_x - \cos\theta\mathbf{E}_y) = -\cos\theta \end{aligned} \tag{5–941}$$

Substituting the results of Eq. (5–941) into (5–940), we obtain

$$ {}^{\mathcal{F}}U = -mg(R\theta\sin\beta + r\sin\beta\sin\theta - r\cos\beta\cos\theta) \tag{5–942}$$

Then, noting that $\sin\beta\sin\theta - \cos\beta\cos\theta = -\cos(\theta+\beta)$, we obtain the potential energy as

$$ {}^{\mathcal{F}}U = -mg(R\theta\sin\beta - r\cos(\theta+\beta)) = -mgR\theta\sin\beta + mgr\cos(\theta+\beta) \tag{5–943}$$

Adding the kinetic energy in Eq. (5–936) to the potential energy in Eq. (5–943), we obtain the total energy as

$$\begin{aligned} {}^{\mathcal{F}}E = {}^{\mathcal{F}}T + {}^{\mathcal{F}}U &= \tfrac{1}{2}m(r^2 + R^2 + 2rR\cos\theta)\dot\theta^2 \\ &\quad - mgR\theta\sin\beta + mgr\cos(\theta+\beta) \\ &= \text{constant} \end{aligned} \tag{5–944}$$

Computing the rate of change of ${}^{\mathcal{F}}E$, we obtain

$$\begin{aligned} \frac{d}{dt}\left({}^{\mathcal{F}}E\right) &= m(R^2 + r^2 + 2rR\cos\theta)\dot\theta\ddot\theta - mrR\dot\theta\sin\theta\,\dot\theta^2 \\ &\quad - mgR\dot\theta\sin\beta - mgr\dot\theta\sin(\theta+\beta) = 0 \end{aligned} \tag{5–945}$$

Noting that $\dot\theta \neq 0$ as a function of time, we can drop $\dot\theta$ from Eq. (5–945) to give the differential equation of motion as

$$ m(r^2 + R^2 + 2rR\cos\theta)\ddot\theta - mrR\dot\theta^2\sin\theta - mgr\sin(\theta+\beta) - mgR\sin\beta = 0 \tag{5–946}$$

We note that the differential equation of Eq. (5–946) is identical to that of Eq. (5–505) as obtained in Example 5–8.

■

5.11 Impulse and Momentum for a Rigid Body

5.11.1 Linear Impulse and Linear Momentum for a Rigid Body

Recall from Euler's 1st law that the translational motion of the center of mass of a rigid body $\mathcal{R}$ is given from Eq. (5-225) as

$$\mathbf{F} = m^{\mathcal{N}}\bar{\mathbf{a}} \tag{5-947}$$

where $\mathcal{N}$ is an inertial reference frame. Integrating both sides of Eq. (5-947) from t_1 to t_2, we have

$$\int_{t_1}^{t_2} \mathbf{F}dt = \int_{t_1}^{t_2} m^{\mathcal{N}}\bar{\mathbf{a}}dt = m^{\mathcal{N}}\mathbf{v}(t_2) - m^{\mathcal{N}}\mathbf{v}(t_1) = {}^{\mathcal{N}}\mathbf{G}(t_2) - {}^{\mathcal{N}}\mathbf{G}(t_1) \tag{5-948}$$

where ${}^{\mathcal{N}}\mathbf{G}$ is the linear momentum of a rigid body as defined in Eq. (5-6) on page 323. Recalling the definition of a linear impulse from Chapter 3 we have

$$\hat{\mathbf{F}} = {}^{\mathcal{N}}\mathbf{G}_2 - {}^{\mathcal{N}}\mathbf{G}_1 \tag{5-949}$$

where

$$\hat{\mathbf{F}} = \int_{t_1}^{t_2} \mathbf{F}dt \tag{5-950}$$

is the linear impulse of the force $\mathbf{F}$ on the time interval from t_1 to t_2. Equation (5-949) is called the *principle of linear impulse and momentum* for a rigid body and states that the change in linear momentum of the rigid body is equal to the linear impulse applied to the rigid body.

5.11.2 Angular Impulse and Angular Momentum for a Rigid Body

Suppose now that we integrate the result for the balance of angular momentum relative to an arbitrary reference point Q as given in Eq. (5-234) from an initial time t_1 to a final time t_2. We then have

$$\int_{t_1}^{t_2} \mathbf{M}_Q dt - \int_{t_1}^{t_2} (\bar{\mathbf{r}} - \mathbf{r}_Q) \times m^{\mathcal{N}}\mathbf{a}_Q dt = \int_{t_1}^{t_2} \frac{{}^{\mathcal{N}}d}{dt}\left({}^{\mathcal{N}}\mathbf{H}_Q\right) dt \tag{5-951}$$

It is seen that the right-hand side of Eq. (5-951) is given as

$$\int_{t_1}^{t_2} \frac{{}^{\mathcal{N}}d}{dt}\left({}^{\mathcal{N}}\mathbf{H}_Q\right) dt = {}^{\mathcal{N}}\mathbf{H}_Q(t_2) - {}^{\mathcal{N}}\mathbf{H}_Q(t_1) \tag{5-952}$$

Furthermore, the first term on the left-hand side of Eq. (5-951) is the angular impulse on the time interval from t_1 to t_2, i.e.,

$$\hat{\mathbf{M}}_Q = \int_{t_1}^{t_2} \mathbf{M}_Q dt \tag{5-953}$$

We then have

$$\boxed{\hat{\mathbf{M}}_Q - \int_{t_1}^{t_2} (\bar{\mathbf{r}} - \mathbf{r}_Q) \times m^{\mathcal{N}}\mathbf{a}_Q dt = {}^{\mathcal{N}}\mathbf{H}_Q(t_2) - {}^{\mathcal{N}}\mathbf{H}_Q(t_1)} \tag{5-954}$$

Equation (5–954) is called the *principle of angular impulse and angular momentum* for a rigid body $\mathcal{R}$ relative to an arbitrary reference point Q.

Suppose now that we restrict our attention to motion relative to either a point O fixed in an inertial reference frame $\mathcal{N}$ or the center of mass of a rigid body. In either of these two cases, it is seen that the second term in Eq. (5–954) is zero. Consequently, when the reference point is fixed in an inertial reference frame, we have

$$\boxed{\hat{\mathbf{M}}_O = {}^{\mathcal{N}}\mathbf{H}_O(t_2) - {}^{\mathcal{N}}\mathbf{H}_O(t_1)} \tag{5–955}$$

Equation (5–955) is called the *principle of angular impulse and angular momentum* for a rigid body relative to a point fixed in an inertial reference frame $\mathcal{N}$. Similarly, when the reference point is the center of mass, we have

$$\boxed{\hat{\bar{\mathbf{M}}} = {}^{\mathcal{N}}\bar{\mathbf{H}}(t_2) - {}^{\mathcal{N}}\bar{\mathbf{H}}(t_1)} \tag{5–956}$$

Equation (5–956) is called the *principle of angular impulse and angular momentum* for a rigid body relative to the center of mass of the rigid body in an inertial reference frame $\mathcal{N}$.

Example 5–15

A sphere of mass m and radius r is set in motion by placing it in contact with a plane inclined at an angle β with the horizontal and giving its center of mass an initial velocity $\mathbf{v}_0$ in the direction of the incline as shown in Fig. 5–37. Knowing that the coefficient of sliding friction between the sphere and the incline is μ and that gravity acts downward, determine (a) the velocity of the center of mass of the sphere when sliding stops and rolling begins, (b) the angular velocity of the sphere when sliding stops and rolling begins, and (c) the time at which the sphere stops sliding and begins rolling.

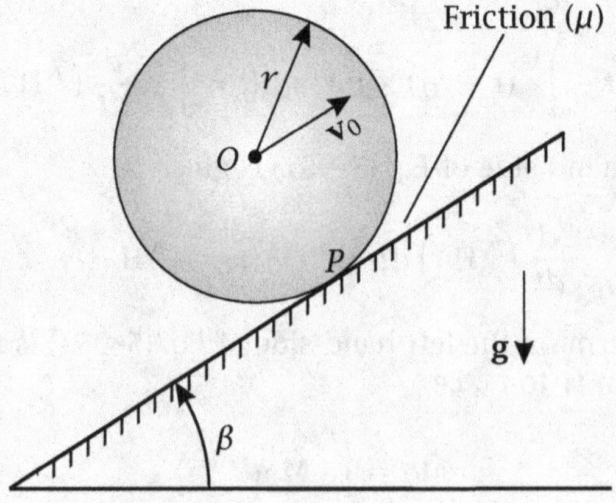

Figure 5–37 Sphere sliding along a surface inclined at an angle β with the horizontal.

Solution to Example 5–15

Kinematics

It is convenient to analyze this problem using a fixed reference frame, $\mathcal{F}$. Corresponding to $\mathcal{F}$, the following coordinate system is chosen to describe the motion:

	Origin at O when $t = 0$	
$\mathbf{E}_x$	=	Up incline
$\mathbf{E}_z$	=	Into page
$\mathbf{E}_y$	=	$\mathbf{E}_z \times \mathbf{E}_x$

It is important to note for this problem that, while the origin of the above chosen coordinate system is a fixed point, the center of mass of the sphere (located at point O) is a *moving* point. In terms of the basis $\{\mathbf{E}_x, \mathbf{E}_y, \mathbf{E}_z\}$, the position of the center of mass of the sphere is given as

$$\bar{\mathbf{r}} = \mathbf{r}_O = x\mathbf{E}_x \tag{5–957}$$

The velocity of the center of mass of the disk is then given as

$${}^{\mathcal{F}}\bar{\mathbf{v}} = \dot{x}\mathbf{E}_x \equiv \bar{v}\mathbf{E}_x \tag{5–958}$$

The acceleration of the center of mass of the disk is then obtained as

$${}^{\mathcal{F}}\bar{\mathbf{a}} = \ddot{x}\mathbf{E}_x \tag{5–959}$$

Furthermore, the position of the point of contact of the sphere with the incline is given as

$$\mathbf{r}_P = x\mathbf{E}_x + r\mathbf{E}_y \tag{5–960}$$

Now, because the sphere rotates about an axis orthogonal to the page, the angular velocity of the sphere in reference frame $\mathcal{F}$ can be written as

$${}^{\mathcal{F}}\boldsymbol{\omega}^{\mathcal{R}} = \omega\mathbf{E}_z \tag{5–961}$$

The velocity of the point of contact on the sphere, denoted ${}^{\mathcal{F}}\mathbf{v}_P$, is then determined using Eq. (2–517) on page 106 as

$${}^{\mathcal{F}}\mathbf{v}_P = {}^{\mathcal{F}}\bar{\mathbf{v}} + {}^{\mathcal{F}}\boldsymbol{\omega}^{\mathcal{R}} \times (\mathbf{r}_P - \bar{\mathbf{r}}) \tag{5–962}$$

Now from Eqs. (5–957) and (5–960) we have

$$\mathbf{r}_P - \bar{\mathbf{r}} = r\mathbf{E}_y \tag{5–963}$$

Substituting ${}^{\mathcal{F}}\boldsymbol{\omega}^{\mathcal{R}}$ and $\mathbf{r}_P - \bar{\mathbf{r}}$ from Eqs. (5–961) and (5–963), respectively, into Eq. (5–962), we obtain

$${}^{\mathcal{F}}\mathbf{v}_P = \bar{v}\mathbf{E}_x + \omega\mathbf{E}_z \times (r\mathbf{E}_y) = (\bar{v} - r\omega)\mathbf{E}_x \tag{5–964}$$

Kinetics

Next, because for this problem we are interested in the velocity and angular velocity of the sphere after a finite time has passed (in this case the time interval from $t = 0$ to $t = t_1$, where t_1 is the time when sliding stops and rolling begins), this problem must be solved using impulse and momentum for a rigid body. In particular, *both* the principle of linear impulse and linear momentum *and* the principle of angular impulse and angular momentum must be applied to the sphere. The free body diagram of the sphere during sliding is shown in Fig. 5-38.

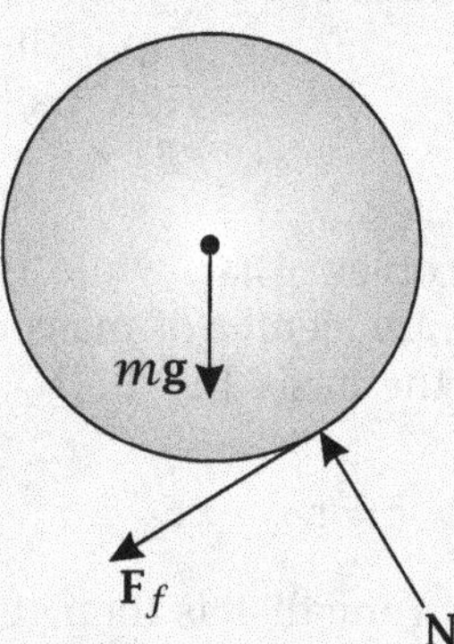

Figure 5-38 Free body diagram of sphere for Example 5-15.

Using Fig. 5-38, the forces acting on the sphere are given as follows:

$$\begin{array}{rcl} m\mathbf{g} & = & \text{Force of gravity} \\ \mathbf{N} & = & \text{Reaction force of surface on sphere} \\ \mathbf{F}_f & = & \text{Force of friction} \end{array}$$

Resolving the forces acting on the sphere in the basis $\{\mathbf{E}_x, \mathbf{E}_y, \mathbf{E}_z\}$, we obtain

$$m\mathbf{g} = mg\mathbf{u}_v \tag{5-965}$$

$$\mathbf{N} = N\mathbf{E}_y \tag{5-966}$$

$$\mathbf{F}_f = -\mu\|\mathbf{N}\|\frac{\mathbf{v}_{\text{rel}}}{\|\mathbf{v}_{\text{rel}}\|} \tag{5-967}$$

where $\mathbf{u}_v$ is the unit vector in the vertically downward direction. It is noted that the force of gravity acts at the center of mass of the sphere while the reaction force and the friction force act at the point of contact of the sphere with the surface of the alley (point P). Therefore, the quantity $\mathbf{v}_{\text{rel}}$ is the velocity of point P relative to the surface, i.e.,

$$\mathbf{v}_{\text{rel}} = {}^{\mathcal{F}}\mathbf{v}_P - {}^{\mathcal{F}}\mathbf{v}_P^S \tag{5-968}$$

Because the surface is stationary, we have ${}^{\mathcal{F}}\mathbf{v}_P^S = \mathbf{0}$. Using the result of Eq. (5-964), we obtain $\mathbf{v}_{\text{rel}}$ as

$$\mathbf{v}_{\text{rel}} = {}^{\mathcal{F}}\mathbf{v}_P = (\bar{v} - r\omega)\mathbf{E}_x \tag{5-969}$$

Next, noting that $N = \|\mathbf{N}\|$, the force of friction is given as

$$\mathbf{F}_f = -\mu\|\mathbf{N}\|\frac{(\bar{v} - r\omega)\mathbf{E}_x}{\|(\bar{v} - r\omega)\mathbf{E}_x\|} = -\mu\|\mathbf{N}\|\frac{\bar{v} - r\omega}{|(\bar{v} - r\omega)|}\mathbf{E}_x \tag{5-970}$$

from which it is seen that the direction of the force of friction is determined by the *sign* of the quantity

$$\frac{\bar{v} - r\omega}{|\bar{v} - r\omega|}$$

For this problem we are given that the initial angular velocity is zero. Therefore, at the initial time $t = 0$, the point P and the center of mass of the sphere are moving with the same velocity. Furthermore, since the initial velocity is such that $v_0 > 0$, *during sliding* it must be the case that $\bar{v} - r\omega > 0$, which implies that

$$\frac{\bar{v} - r\omega}{|\bar{v} - r\omega|}\mathbf{E}_x = \mathbf{E}_x \tag{5-971}$$

Therefore, the force of friction during sliding is given as

$$\mathbf{F}_f = -\mu\|\mathbf{N}\|\mathbf{E}_x \tag{5-972}$$

Now that we have a complete description of the kinematics and a complete description of the forces acting on the sphere, we can proceed to the application of impulse and momentum.

Application of Linear Impulse and Linear Momentum to Sphere During Sliding

Using the principle of linear impulse and linear momentum for a rigid body, we have

$$\hat{\mathbf{F}} = {}^{\mathcal{F}}\mathbf{G}(t_1) - {}^{\mathcal{F}}\mathbf{G}(t_0) = {}^{\mathcal{F}}\mathbf{G}_1 - {}^{\mathcal{F}}\mathbf{G}_0 \tag{5-973}$$

where

$$\hat{\mathbf{F}} = \int_0^{t_1} \mathbf{F}dt \tag{5-974}$$

From the free body diagram of the sphere given in Fig. 5-38 we have

$$\mathbf{F} = m\mathbf{g} + \mathbf{N} + \mathbf{F}_f \tag{5-975}$$

Substituting the expressions for $m\mathbf{g}$, $\mathbf{N}$, and $\mathbf{F}_f$ from Eqs. (5-965)-(5-967), respectively, into Eq. (5-975), we obtain

$$\mathbf{F} = mg\mathbf{u}_v + N\mathbf{E}_y - \mu\|\mathbf{N}\|\mathbf{E}_x \tag{5-976}$$

Now, using the geometry as shown in Eq. (5-39), we have the unit vector $\mathbf{u}_v$ as

$$\mathbf{u}_v = -\sin\beta\mathbf{E}_x + \cos\beta\mathbf{E}_y \tag{5-977}$$

Then the resultant force of Eq. (5-976) can be written as

$$\begin{aligned}\mathbf{F} &= mg(-\sin\beta\mathbf{E}_x + \cos\beta\mathbf{E}_y) + N\mathbf{E}_y - \mu\|\mathbf{N}\|\mathbf{E}_x \\ &= -(\mu\|\mathbf{N}\| + mg\sin\beta)\mathbf{E}_x + (N + mg\cos\beta)\mathbf{E}_y\end{aligned} \tag{5-978}$$

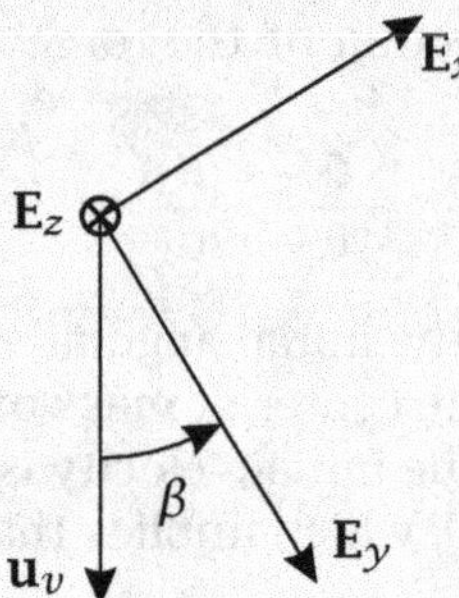

Figure 5–39 Unit vector, $\mathbf{u}_v$, in vertically downward direction for Example 5–15.

Applying Euler's 1st law to the sphere by setting $\mathbf{F}$ from Eq. (5–976) equal to $m^{\mathcal{F}}\bar{\mathbf{a}}$ from Eq. (5–959), we obtain

$$-(\mu\|\mathbf{N}\| + mg\sin\beta)\mathbf{E}_x + (N + mg\cos\beta)\mathbf{E}_y = m\ddot{x}\mathbf{E}_x \tag{5–979}$$

Equating components in Eq. (5–979), we have

$$\begin{aligned} -(\mu\|\mathbf{N}\| + mg\sin\beta) &= m\ddot{x} && (5–980) \\ N + mg\cos\beta &= 0 && (5–981) \end{aligned}$$

which implies that

$$N = -mg\cos\beta \tag{5–982}$$

Furthermore, from Eq. (5–982) we obtain

$$\|\mathbf{N}\| = |N| = mg\cos\beta \tag{5–983}$$

Substituting the result of Eq. (5–983) into Eq. (5–972), the force of friction is

$$\mathbf{F}_f = -\mu\|\mathbf{N}\|\mathbf{E}_x = -\mu mg\cos\beta\mathbf{E}_x \tag{5–984}$$

The resultant force acting on the sphere during sliding is then given as

$$\mathbf{F} = -(\mu\|\mathbf{N}\| + mg\sin\beta)\mathbf{E}_x = -mg(\mu\cos\beta + \sin\beta)\mathbf{E}_x \tag{5–985}$$

Integrating Eq. (5–985) from $t = 0$ to $t = t_1$, the linear impulse $\hat{\mathbf{F}}$ during sliding is obtained as

$$\hat{\mathbf{F}} = \int_0^{t_1} -mg(\mu\cos\beta + \sin\beta)\mathbf{E}_x dt = -mgt_1(\mu\cos\beta + \sin\beta)\mathbf{E}_x \tag{5–986}$$

where we observe that the quantity $mg(\mu\cos\beta - \sin\beta)$ is a constant. Next, the linear momenta at $t = 0$ and $t = t_1$ are given, respectively, as

$$\begin{aligned} {}^{\mathcal{F}}\mathbf{G}(0) &= m^{\mathcal{F}}\bar{\mathbf{v}}(0) = mv_0\mathbf{E}_x && (5–987) \\ {}^{\mathcal{F}}\mathbf{G}(t_1) &= m^{\mathcal{F}}\bar{\mathbf{v}}(t_1) = m\bar{v}_1\mathbf{E}_x && (5–988) \end{aligned}$$

Substituting $\hat{\mathbf{F}}$ from Eq. (5–986) and ${}^{\mathcal{F}}\mathbf{G}(0)$ and ${}^{\mathcal{F}}\mathbf{G}(t_1)$ from Eqs. (5–987) and (5–988), respectively, into Eq. (5–973) gives

$$-mgt_1(\mu\cos\beta + \sin\beta)\mathbf{E}_x = m\bar{v}_1\mathbf{E}_x - m\bar{v}_0\mathbf{E}_x = m(\bar{v}_1 - v_0)\mathbf{E}_x \tag{5–989}$$

Equation (5-989) simplifies to

$$\bar{v}_1 - v_0 = -gt_1(\mu\cos\beta + \sin\beta) \quad (5\text{-}990)$$

Equation (5-990) is the *first* of three equations that will be used to obtain the solution to this problem. It is noted that this equation has two unknowns: $\bar{v}_1$ and t_1.

Application of Angular Impulse and Angular Momentum to Sphere During Sliding

For this problem there are no convenient fixed points about which to apply angular impulse and angular momentum. Consequently, we must apply angular impulse and angular momentum about the center of mass of the sphere. From the principle of angular impulse and angular momentum, we have

$$\hat{\bar{\mathbf{M}}} = {}^{\mathcal{F}}\bar{\mathbf{H}}(t_1) - {}^{\mathcal{F}}\bar{\mathbf{H}}(0) = {}^{\mathcal{F}}\bar{\mathbf{H}}_1 - {}^{\mathcal{F}}\bar{\mathbf{H}}_0 \quad (5\text{-}991)$$

where

$$\hat{\bar{\mathbf{M}}} = \int_0^{t_1} \bar{\mathbf{M}} dt \quad (5\text{-}992)$$

From the free body diagram of Fig. 5-38 we see that the forces $m\mathbf{g}$ and $\mathbf{N}$ pass through the center of mass of the sphere. Consequently, the resultant moment about the center of mass is due entirely to the friction force and is given as

$$\bar{\mathbf{M}} = (\mathbf{r}_P - \bar{\mathbf{r}}) \times \mathbf{F}_f \quad (5\text{-}993)$$

where we note that the friction force acts at point P. Substituting $\mathbf{r}_P - bfrbar$ from Eq. (5-963) and $\mathbf{F}_f$ from Eq. (5-984) into (5-993) gives

$$\bar{\mathbf{M}} = r\mathbf{E}_y \times (-\mu mg\cos\beta\mathbf{E}_x) = r\mu mg\cos\beta\mathbf{E}_z \quad (5\text{-}994)$$

Integrating Eq. (5-994) from $t = 0$ to $t = t_1$, we obtain the angular impulse as

$$\hat{\bar{\mathbf{M}}} = \int_0^{t_1} r\mu mg\cos\beta\mathbf{E}_z dt = r\mu mgt_1\cos\beta\mathbf{E}_z \quad (5\text{-}995)$$

where we observe that the quantity $r\mu mgt_1\cos\beta$ and the unit vector $\mathbf{E}_z$ are both constant. Next, the angular momentum of the sphere relative to the center of mass of the sphere is given as

$$^{\mathcal{F}}\bar{\mathbf{H}} = \bar{\mathbf{I}}^{\mathcal{R}} \cdot {}^{\mathcal{F}}\boldsymbol{\omega}^{\mathcal{R}} \quad (5\text{-}996)$$

Now, in terms of the basis $\{\mathbf{E}_x, \mathbf{E}_y, \mathbf{E}_z\}$, we have

$$\bar{\mathbf{I}}^{\mathcal{R}} = \bar{I}_{xx}\mathbf{E}_x \otimes \mathbf{E}_x + \bar{I}_{yy}\mathbf{E}_y \otimes \mathbf{E}_y + \bar{I}_{zz}\mathbf{E}_z \otimes \mathbf{E}_z \quad (5\text{-}997)$$

Then, substituting the moment of inertia tensor from Eq. (5-997) and the angular velocity from Eq. (5-961) into (5-996), we obtain

$$^{\mathcal{F}}\bar{\mathbf{H}} = \left(\bar{I}_{xx}\mathbf{E}_x \otimes \mathbf{E}_x + \bar{I}_{yy}\mathbf{E}_y \otimes \mathbf{E}_y + \bar{I}_{zz}\mathbf{E}_z \otimes \mathbf{E}_z\right) \cdot \omega\mathbf{E}_z = \bar{I}_{zz}\omega\mathbf{E}_z \quad (5\text{-}998)$$

Using the expression for ${}^{\mathcal{F}}\bar{\mathbf{H}}$ from Eq. (5–998), the angular momenta relative to the center of mass of the sphere at $t = 0$ and $t = t_1$ are given, respectively, as

$$
\begin{aligned}
{}^{\mathcal{F}}\bar{\mathbf{H}}(0) &= \bar{I}_{zz}\omega(0)\mathbf{E}_z = \bar{I}_{zz}\omega_0\mathbf{E}_z && (5\text{–}999)\\
{}^{\mathcal{F}}\bar{\mathbf{H}}(t_1) &= \bar{I}_{zz}\omega(t_1)\mathbf{E}_z = \bar{I}_{zz}\omega_1\mathbf{E}_z && (5\text{–}1000)
\end{aligned}
$$

Because the sphere is homogeneous, we have

$$\bar{I}_{zz} = \tfrac{2}{5}mr^2$$

Consequently,

$$
\begin{aligned}
{}^{\mathcal{F}}\bar{\mathbf{H}}_0 &= \tfrac{2}{5}mr^2\omega_0\mathbf{E}_z && (5\text{–}1001)\\
{}^{\mathcal{F}}\bar{\mathbf{H}}_1 &= \tfrac{2}{5}mr^2\omega_1\mathbf{E}_z && (5\text{–}1002)
\end{aligned}
$$

Substituting ${}^{\mathcal{F}}\bar{\mathbf{H}}_0$, ${}^{\mathcal{F}}\bar{\mathbf{H}}_1$, and $\hat{\mathbf{M}}$ from Eqs. (5–1001), (5–1002), and (5–995), respectively, into Eq. (5–991), we obtain

$$r\mu mgt_1\cos\beta\mathbf{E}_z = \tfrac{2}{5}mr^2\omega_1\mathbf{E}_z - \tfrac{2}{5}mr^2\omega_0\mathbf{E}_z = \tfrac{2}{5}mr^2(\omega_1 - \omega_0)\mathbf{E}_z \qquad (5\text{–}1003)$$

Finally, because the initial angular velocity is zero, Eq. (5–1003) simplifies to

$$r\mu mgt_1\cos\beta\mathbf{E}_z = \tfrac{2}{5}mr^2\omega_1\mathbf{E}_z \qquad (5\text{–}1004)$$

which gives

$$\mu gt_1\cos\beta = \tfrac{2}{5}r\omega_1 \qquad (5\text{–}1005)$$

This is the *second* of three equations that will be used to obtain the solution to this problem. It is noted that this equation has two unknowns: ω_1 and t_1.

Kinematic Constraint at Sliding/Rolling Transition Point

Recall from Eq. (2–531) on page 109 that a rigid body will roll along a fixed surface if the velocity of the point on the body in contact with the fixed surface is zero. Because in this problem we are interested in the point where sliding stops and rolling begins, it is required that the velocity of the point of contact, P, be zero at $t = t_1$, i.e., we require that

$${}^{\mathcal{F}}\mathbf{v}_P(t_1) = \mathbf{0} \qquad (5\text{–}1006)$$

Now, using Eq. (2–517) from page 106 at $t = t_1$, we can express ${}^{\mathcal{F}}\mathbf{v}_P(t_1)$ in terms of the velocity of the center of mass of the sphere as

$${}^{\mathcal{F}}\bar{\mathbf{v}}(t_1) = {}^{\mathcal{F}}\mathbf{v}_P(t_1) + {}^{\mathcal{F}}\boldsymbol{\omega}^{\mathcal{R}}(t_1) \times (\bar{\mathbf{r}}(t_1) - \mathbf{r}_P(t_1)) \qquad (5\text{–}1007)$$

We note from Eqs. (5–961) and (5–963) that

$$
\begin{aligned}
\bar{\mathbf{r}}(t_1) - \mathbf{r}_P(t_1) &= -r\mathbf{E}_y && (5\text{–}1008)\\
{}^{\mathcal{F}}\boldsymbol{\omega}^{\mathcal{R}}(t_1) &= \omega_1\mathbf{E}_z && (5\text{–}1009)
\end{aligned}
$$

Then, substituting ${}^{\mathcal{F}}\mathbf{v}_P(t_1)$ from Eq. (5–1006) and using the expressions in Eqs. (5–1008) and (5–1009), we obtain

$$ {}^{\mathcal{F}}\bar{\mathbf{v}}(t_1) = \omega_1 \mathbf{E}_z \times (-r\mathbf{E}_y) = r\omega_1 \mathbf{E}_x \quad (5\text{–}1010) $$

Furthermore, noting that ${}^{\mathcal{F}}\bar{\mathbf{v}}(t_1) = \bar{v}_1 \mathbf{E}_x$, Eq. (5–1010) simplifies to

$$ \bar{v}_1 \mathbf{E}_x = r\omega_1 \mathbf{E}_x \quad (5\text{–}1011) $$

We then obtain

$$ \bar{v}_1 = r\omega_1 \quad (5\text{–}1012) $$

This is the *third* of three equations that will be used to obtain the solution to this problem. It is noted that this equation has two unknowns: $\bar{v}_1$ and ω_1.

(a) Velocity of Center of Mass of Sphere When Sliding Stops and Rolling Begins

Solving Eq. (5–1005) for t_1, we obtain

$$ t_1 = \frac{2r\omega_1}{5\mu g \cos\beta} \quad (5\text{–}1013) $$

Next, substituting $\bar{v}_1$ from Eq. (5–1012) into (5–1013), we have

$$ t_1 = \frac{2\bar{v}_1}{5\mu g \cos\beta} \quad (5\text{–}1014) $$

Then, substituting t_1 from Eq. (5–1014) into (5–990) gives

$$ \bar{v}_1 - v_0 = -g\frac{2\bar{v}_1}{5\mu g \cos\beta}(\mu \cos\beta + \sin\beta) \quad (5\text{–}1015) $$

Solving Eq. (5–1015) for $\bar{v}_1$, we obtain

$$ \bar{v}_1 = \frac{5\mu v_0 \cos\beta}{7\mu \cos\beta + 2\sin\beta} \quad (5\text{–}1016) $$

Therefore, the velocity of the center of mass of the sphere when sliding stops and rolling begins is given as

$$ {}^{\mathcal{F}}\bar{\mathbf{v}}_1 = \frac{5\mu v_0 \cos\beta}{7\mu \cos\beta + 2\sin\beta}\mathbf{E}_x \quad (5\text{–}1017) $$

(b) Angular Velocity of Sphere When Sliding Stops and Rolling Begins

Substituting $\bar{v}_1$ from Eq. (5–1016) into (5–1012), we obtain

$$ r\omega_1 = \frac{5\mu v_0 \cos\beta}{7\mu \cos\beta + 2\sin\beta} \quad (5\text{–}1018) $$

Solving Eq. (5–1018) for ω_1 gives

$$ \omega_1 = \frac{5\mu v_0 \cos\beta}{r(7\mu \cos\beta + 2\sin\beta)} \quad (5\text{–}1019) $$

Therefore, the angular velocity of the sphere when sliding stops and rolling begins is given as

$$ {}^{\mathcal{F}}\boldsymbol{\omega}^{R}(t_1) = \omega_1 \mathbf{E}_z = \frac{5\mu v_0 \cos\beta}{r(7\mu \cos\beta + 2\sin\beta)}\mathbf{E}_z \quad (5\text{–}1020) $$

(c) Time at Which Sliding Stops and Rolling Begins

Substituting $\bar{v}_1$ from Eq. (5-1016) into (5-1014), we obtain the time at which sliding stops and rolling begins as

$$t_1 = \frac{2v_0}{g(7\mu\cos\beta + 2\sin\beta)} \tag{5-1021}$$

■

5.12 Collision of Rigid Bodies

In Section 4.8, we developed a model for the collision between two objects under the assumption that the objects were particles. It is clear that in a real collision the objects are *not* particles, i.e., the objects have a nonzero size and have some associated shape. In order to account for objects with nonzero dimension, in this section we extend the discussion of Section 4.8 to the collision between two rigid bodies.

5.12.1 Collision Model

Consider a collision between two rigid bodies $\mathcal{R}_1$ and $\mathcal{R}_2$. Assume that, during the collision, the positions and orientations of the two bodies do not change and that the two rigid bodies are in contact at a single point P as shown in Fig. 5-40.
Then, let

$$\begin{aligned}
\bar{\mathbf{r}}_1 &= \text{Position of Center of Mass of } \mathcal{R}_1 \text{ During Collision} \\
\bar{\mathbf{r}}_2 &= \text{Position of Center of Mass of } \mathcal{R}_2 \text{ During Collision} \\
\mathbf{r}_P &= \text{Position of Point of Contact } P \text{ During Collision}
\end{aligned}$$

Also, let $\mathbf{n}$ be the unit normal to the tangent planes of $\mathcal{B}_1$ and $\mathcal{B}_2$ at point P, and let $\mathbf{u}$ and $\mathbf{w}$ be two unit vectors that lie in the plane orthogonal to $\mathbf{n}$ such that $\{\mathbf{u}, \mathbf{w}, \mathbf{n}\}$ forms a right-handed system. Then, denoting t_0 and t_1 as the times at the beginning and the end of the collision, we can define the following quantities in an inertial reference frame $\mathcal{N}$:

$$\begin{aligned}
{}^{\mathcal{N}}\bar{\mathbf{v}}^{\mathcal{R}_1} &= \text{Velocity of center of mass of } \mathcal{R}_1 \text{ at } t_0 \\
{}^{\mathcal{N}}\bar{\mathbf{v}}^{\mathcal{R}_2} &= \text{Velocity of center of mass of } \mathcal{R}_2 \text{ at } t_0 \\
{}^{\mathcal{N}}\mathbf{v}_P^{\mathcal{R}_1} &= \text{Velocity of point } P \text{ on } \mathcal{R}_1 \text{ at } t_0 \\
{}^{\mathcal{N}}\mathbf{v}_P^{\mathcal{R}_2} &= \text{Velocity of point } P \text{ on } \mathcal{R}_2 \text{ at } t_0 \\
\left({}^{\mathcal{N}}\bar{\mathbf{v}}^{\mathcal{R}_1}\right)' &= \text{Velocity of center of mass of } \mathcal{R}_1 \text{ at } t_1 \\
\left({}^{\mathcal{N}}\bar{\mathbf{v}}^{\mathcal{R}_2}\right)' &= \text{Velocity of center of mass of } \mathcal{R}_2 \text{ at } t_1 \\
\left({}^{\mathcal{N}}\mathbf{v}_P^{\mathcal{R}_1}\right)' &= \text{Velocity of point } P \text{ on } \mathcal{R}_1 \text{ at } t_1 \\
\left({}^{\mathcal{N}}\mathbf{v}_P^{\mathcal{R}_2}\right)' &= \text{Velocity of point } P \text{ on } \mathcal{R}_2 \text{ at } t_1 \\
{}^{\mathcal{N}}\bar{\mathbf{H}}^{\mathcal{R}_1} &= \text{Angular momentum relative to center of mass of } \mathcal{R}_1 \text{ at } t_0 \\
{}^{\mathcal{N}}\bar{\mathbf{H}}^{\mathcal{R}_2} &= \text{Angular Momentum relative to center of mass of } \mathcal{R}_2 \text{ at } t_0 \\
\left({}^{\mathcal{N}}\bar{\mathbf{H}}^{\mathcal{R}_1}\right)' &= \text{Angular Momentum relative to center of mass of } \mathcal{R}_1 \text{ at } t_1 \\
\left({}^{\mathcal{N}}\bar{\mathbf{H}}^{\mathcal{R}_2}\right)' &= \text{Angular Momentum relative to center of mass of } \mathcal{R}_2 \text{ at } t_1
\end{aligned}$$

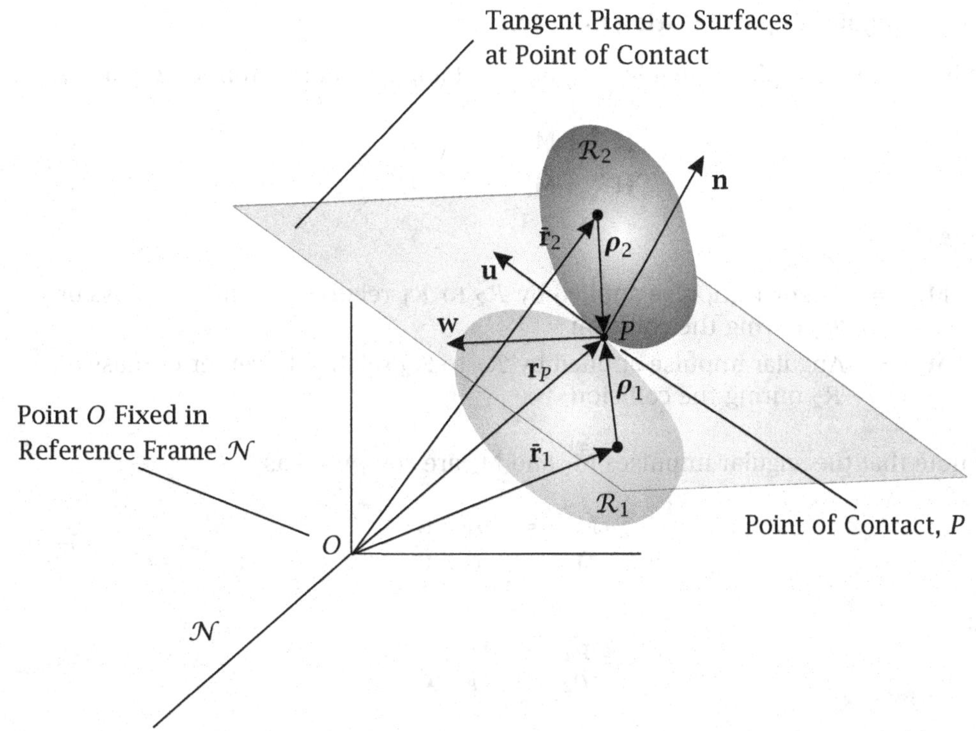

Figure 5-40 Collision between two rigid bodies.

5.12.2 Linear Impulse During Collision

Applying the principle of linear impulse and linear momentum to $\mathcal{R}_1$ and $\mathcal{R}_2$, we have, respectively,

$$\begin{aligned} m_1 {}^{\mathcal{N}}\bar{\mathbf{v}}^{\mathcal{R}_1} + \hat{\mathbf{R}}_1 &= m_1 \left({}^{\mathcal{N}}\bar{\mathbf{v}}^{\mathcal{R}_1}\right)' \\ m_2 {}^{\mathcal{N}}\bar{\mathbf{v}}^{\mathcal{R}_2} + \hat{\mathbf{R}}_2 &= m_2 \left({}^{\mathcal{N}}\bar{\mathbf{v}}^{\mathcal{R}_2}\right)' \end{aligned} \tag{5-1022}$$

where

$\hat{\mathbf{R}}_1$ = Linear impulse applied by $\mathcal{R}_2$ to $\mathcal{R}_1$ during the collision
$\hat{\mathbf{R}}_2$ = Linear impulse applied by $\mathcal{R}_1$ to $\mathcal{R}_2$ During the collision

Then, from Newton's 3rd law we have

$$\hat{\mathbf{R}}_1 = -\hat{\mathbf{R}}_2 \tag{5-1023}$$

Consequently,

$$\begin{aligned} m_1 {}^{\mathcal{N}}\bar{\mathbf{v}}^{\mathcal{R}_1} + \hat{\mathbf{R}}_1 &= m_1 \left({}^{\mathcal{N}}\bar{\mathbf{v}}^{\mathcal{R}_1}\right)' \\ m_2 {}^{\mathcal{N}}\bar{\mathbf{v}}^{\mathcal{R}_2} - \hat{\mathbf{R}}_1 &= m_2 \left({}^{\mathcal{N}}\bar{\mathbf{v}}^{\mathcal{R}_2}\right)' \end{aligned} \tag{5-1024}$$

Adding the two expressions in Eq. (5-1024), we obtain

$$m_1 {}^{\mathcal{N}}\bar{\mathbf{v}}^{\mathcal{R}_1} + m_2 {}^{\mathcal{N}}\bar{\mathbf{v}}^{\mathcal{R}_2} = m_1 \left({}^{\mathcal{N}}\bar{\mathbf{v}}^{\mathcal{R}_1}\right)' + m_2 \left({}^{\mathcal{N}}\bar{\mathbf{v}}^{\mathcal{R}_2}\right)' \tag{5-1025}$$

which states that the linear momentum of the system must be conserved during the collision.

5.12.3 Angular Impulse During Collision

Applying the principle of angular impulse and angular momentum to $\mathcal{R}_1$ and $\mathcal{R}_2$, we have

$$\begin{aligned} {}^{\mathcal{N}}\bar{\mathbf{H}}^{\mathcal{R}_1} + \hat{\mathbf{M}}_1 &= \left({}^{\mathcal{N}}\bar{\mathbf{H}}^{\mathcal{R}_1}\right)' \\ {}^{\mathcal{N}}\bar{\mathbf{H}}^{\mathcal{R}_2} + \hat{\mathbf{M}}_2 &= \left({}^{\mathcal{N}}\bar{\mathbf{H}}^{\mathcal{R}_2}\right)' \end{aligned} \tag{5-1026}$$

where

$\hat{\mathbf{M}}_1$ = Angular impulse applied by $\mathcal{R}_2$ to $\mathcal{R}_1$ relative to center of mass of $\mathcal{R}_1$ during the collision

$\hat{\mathbf{M}}_2$ = Angular impulse applied by $\mathcal{R}_1$ to $\mathcal{R}_2$ relative to center of mass of $\mathcal{R}_2$ during the collision

We note that the angular impulses $\hat{\mathbf{M}}_1$ and $\hat{\mathbf{M}}_2$ are computed as

$$\begin{aligned} \hat{\mathbf{M}}_1 &= \boldsymbol{\rho}_1 \times \hat{\mathbf{R}}_1 \\ \hat{\mathbf{M}}_1 &= \boldsymbol{\rho}_2 \times \hat{\mathbf{R}}_2 \end{aligned} \tag{5-1027}$$

and

$$\begin{aligned} \boldsymbol{\rho}_1 &= \mathbf{r}_P - \bar{\mathbf{r}}_1 \\ \boldsymbol{\rho}_2 &= \mathbf{r}_P - \bar{\mathbf{r}}_2 \end{aligned} \tag{5-1028}$$

5.12.4 Coefficient of Restitution

Similar to the collision between two particles, the collision between two rigid bodies is governed by the material properties of each body at the point of contact. These properties are subsumed into an ad hoc scalar parameter, e, called the *coefficient of restitution.* In particular, the coefficient of restitution condition for the collision between two rigid bodies is given as

$$\boxed{e = \frac{\left({}^{\mathcal{N}}\mathbf{v}_P^{\mathcal{R}_2}\right)' \cdot \mathbf{n} - \left({}^{\mathcal{N}}\mathbf{v}_P^{\mathcal{R}_1}\right)' \cdot \mathbf{n}}{{}^{\mathcal{N}}\mathbf{v}_P^{\mathcal{R}_1} \cdot \mathbf{n} - {}^{\mathcal{N}}\mathbf{v}_P^{\mathcal{R}_2} \cdot \mathbf{n}}} \tag{5-1029}$$

Equation (5-1029) is called *Poisson's hypothesis.*

5.12.5 Solving for Post-Impact Velocities and Angular Velocities

Unlike the collision between two particles, general expressions for the post-impact conditions of two colliding rigid bodies cannot be obtained. In particular, the post-impact angular velocities and velocities of the centers of mass must be obtained for each problem by solving Eqs. (5-1024), (5-1025), (5-1026), and (5-1029). The ease or difficulty in solving these equations is highly problem-dependent. It is noted that, for the case where the motion is planar, Eqs. (5-1024), (5-1025), (5-1026), and (5-1029) simplify drastically and become much more tractable. To demonstrate the applicability of the collision model derived in this section, in the next two examples we consider problems involving collisions with rigid bodies in planar motion.

Example 5–16

A uniform slender rod of mass m and length l is released from rest at an angle θ above a horizontal surface as shown in Fig. 5-41. After release, the rod descends a distance h under the influence of gravity until its lowest point (i.e., point B) strikes the surface. Knowing that the coefficient of restitution between the rod and the surface is e (where $0 \le e \le 1$), determine (a) the velocity of the center of mass of the rod the instant before the rod strikes the surface and (b) the angular velocity and the velocity of the center of mass of the rod the instant after the rod strikes the surface.

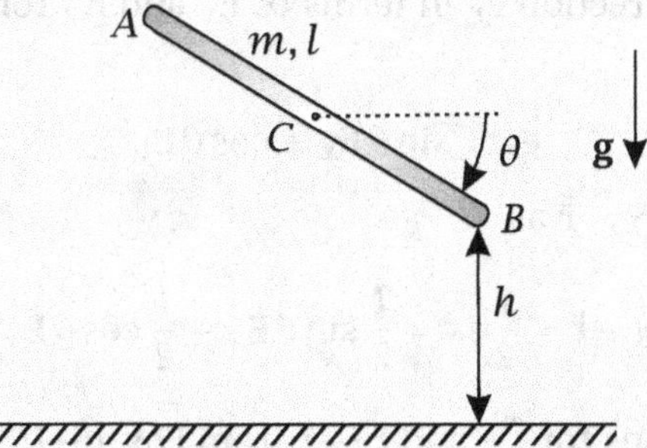

Figure 5–41 Rod striking fixed horizontal surface after descending from a height h above the surface.

Solution to Example 5–16

Kinematics

First, let $\mathcal{F}$ be a fixed reference frame. Then, choose the following coordinate system fixed in reference frame $\mathcal{F}$:

Origin at point C when $t = 0$		
$\mathbf{E}_x$	=	Along $\mathbf{g}$
$\mathbf{E}_z$	=	Out of page
$\mathbf{E}_y$	=	$\mathbf{E}_z \times \mathbf{E}_x$

Then, the velocity of the center of mass of the rod is given as

$$^{\mathcal{F}}\bar{\mathbf{v}} = \bar{v}_x \mathbf{E}_x + \bar{v}_y \mathbf{E}_y \tag{5-1030}$$

Next, denoting the reference frame of the rod by $\mathcal{R}$ and observing that the motion is planar, the angular velocity of the rod in reference frame $\mathcal{F}$ is given as

$$^{\mathcal{F}}\boldsymbol{\omega}^{\mathcal{R}} = \omega \mathbf{E}_z \tag{5-1031}$$

Finally, the velocity of point B located at the lower end of the rod is obtained using Eq. (2-517) on page 106) as

$$^{\mathcal{F}}\mathbf{v}_B = {}^{\mathcal{F}}\bar{\mathbf{v}} + {}^{\mathcal{F}}\boldsymbol{\omega}^{\mathcal{R}} \times (\mathbf{r}_B - \bar{\mathbf{r}}) \tag{5-1032}$$

Now, the direction from the center of mass of the rod to point B on the rod is denoted $\mathbf{e}_r$ and is shown in Fig. 5-42.

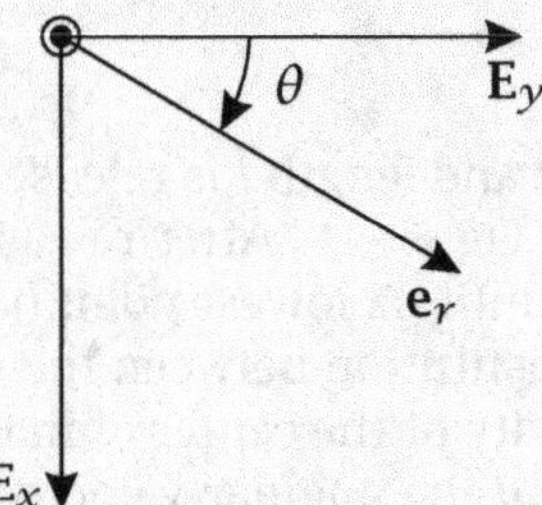

Figure 5-42 Direction $\mathbf{e}_r$ in terms of $\mathbf{E}_x$ and $\mathbf{E}_y$ for Example 5-16.

Using Fig. 5-42, we have

$$\mathbf{e}_r = \sin\theta\,\mathbf{E}_x + \cos\theta\,\mathbf{E}_y \tag{5-1033}$$

Consequently, we obtain $\mathbf{r}_B - \bar{\mathbf{r}}$ as

$$\mathbf{r}_B - \bar{\mathbf{r}} = \frac{l}{2}\mathbf{e}_r = \frac{l}{2}\sin\theta\,\mathbf{E}_x + \frac{l}{2}\cos\theta\,\mathbf{E}_y \tag{5-1034}$$

Substituting the expressions for ${}^{\mathcal{F}}\bar{\mathbf{v}}$ and $\mathbf{r}_B - \bar{\mathbf{r}}$ from Eqs. (5-1030) and (5-1034), respectively, into (5-1032), we obtain

$$\begin{aligned} {}^{\mathcal{F}}\mathbf{v}_B &= \bar{v}_x\mathbf{E}_x + \bar{v}_y\mathbf{E}_y + \omega\mathbf{E}_z \times \left[\frac{l}{2}\sin\theta\,\mathbf{E}_x + \frac{l}{2}\cos\theta\,\mathbf{E}_y\right] \\ &= \left(\bar{v}_x - \frac{l\omega}{2}\cos\theta\right)\mathbf{E}_x + \left(\bar{v}_y + \frac{l\omega}{2}\sin\theta\right)\mathbf{E}_y \end{aligned} \tag{5-1035}$$

Kinetics

The analysis of this problem is most conveniently performed in two distinct phases: (1) the motion during the descent to the surface (before impact) and (2) the motion during the impact. The kinetics associated with each of the two phases of the motion will now be analyzed.

Kinetics During Descent to Surface (Before Impact)

The free body diagram of the rod during the descent of the rod to the surface is shown in Fig. 5-43.

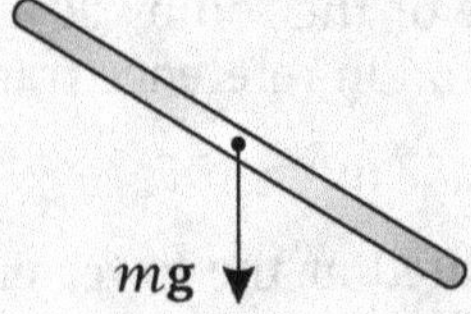

Figure 5-43 Free body diagram of rod during descent for Example 5-16.

Examining Fig. 5-43, it can be seen that the only force acting on the rod during the descent is that due to gravity. Since gravity is conservative, it follows that, during the descent, energy is conserved. Therefore, from the principle of work and energy for a rigid body we have

$$^{\mathcal{F}}T + {}^{\mathcal{F}}U = \text{constant} \tag{5-1036}$$

where $^{\mathcal{F}}T$ is the kinetic energy of the rod and $^{\mathcal{F}}U$ is the potential energy of the rod. Applying Eq. (5-1036) at the beginning and the end of the descent, we have

$$^{\mathcal{F}}T_0 + {}^{\mathcal{F}}U_0 = {}^{\mathcal{F}}T_1 + {}^{\mathcal{F}}U_1 = \text{constant} \tag{5-1037}$$

where

$$\begin{aligned} {}^{\mathcal{F}}T_0 &= \text{Kinetic energy at beginning of descent} \\ {}^{\mathcal{F}}U_0 &= \text{Potential energy at beginning of descent} \\ {}^{\mathcal{F}}T_1 &= \text{Kinetic energy at end of descent} \\ {}^{\mathcal{F}}U_1 &= \text{Potential energy at end of descent} \end{aligned}$$

Because the rod is initially at rest, the initial kinetic energy is zero, i.e.,

$$^{\mathcal{F}}T_0 = {}^{\mathcal{F}}U_0 = 0 \tag{5-1038}$$

Then, from Eq. (5-1037) we have

$$^{\mathcal{F}}T_1 + {}^{\mathcal{F}}U_1 = 0 \tag{5-1039}$$

Now the kinetic energy at the end of the descent, $^{\mathcal{F}}T_1$, is given as

$$^{\mathcal{F}}T_1 = \tfrac{1}{2}m\,{}^{\mathcal{F}}\bar{\mathbf{v}}_1 \cdot {}^{\mathcal{F}}\bar{\mathbf{v}}_1 + \tfrac{1}{2}{}^{\mathcal{F}}\bar{\mathbf{H}}_1 \cdot {}^{\mathcal{F}}\boldsymbol{\omega}_1^{\mathcal{R}} \tag{5-1040}$$

However, because the only force acting on the rod during the descent is that of gravity, and the force of gravity acts at the center of mass, no moment is applied relative to the center of mass of the rod during the descent. Consequently,

$$^{\mathcal{F}}\bar{\mathbf{H}}_1 = \mathbf{0} \tag{5-1041}$$

Therefore, the kinetic energy of Eq. (5-1040) reduces to

$$^{\mathcal{F}}T_1 = \tfrac{1}{2}m\,{}^{\mathcal{F}}\bar{\mathbf{v}}_1 \cdot {}^{\mathcal{F}}\bar{\mathbf{v}}_1 \tag{5-1042}$$

Now, because the rod translates in the $\mathbf{E}_x$-direction during the descent, we have

$$^{\mathcal{F}}\bar{\mathbf{v}}_1 = \bar{v}_1\mathbf{E}_x \tag{5-1043}$$

Substituting $^{\mathcal{F}}\bar{\mathbf{v}}_1$ from Eq. (5-1043) into (5-1042), we obtain

$$^{\mathcal{F}}T_1 = \tfrac{1}{2}m\bar{v}_1^2 \tag{5-1044}$$

Finally, the potential energy due to gravity at the end of the descent is given as

$$^{\mathcal{F}}U_1 = -m\mathbf{g} \cdot \mathbf{r}_1 = -mg\mathbf{E}_x \cdot h\mathbf{E}_x = -mgh \tag{5-1045}$$

(a) Velocity of Center of Mass of Rod the Instant Before Impact

Substituting the results of Eqs. (5-1044) and (5-1045) into Eq. (5-1039), we obtain the following scalar equation:

$$\frac{1}{2}m\bar{v}_1^2 - mgh = 0 \tag{5-1046}$$

Solving Eq. (5-1046) for $\bar{v}_1$, we obtain

$$\bar{v}_1 = \sqrt{2gh} \tag{5-1047}$$

Therefore, the velocity of the center of mass of the rod at the end of the descent is given as

$$^{\mathcal{F}}\bar{\mathbf{v}}_1 = \bar{v}_1\mathbf{E}_x = \sqrt{2gh}\mathbf{E}_x \tag{5-1048}$$

Kinetics During Impact

In order to determine the velocity of the center of mass of the rod and the angular velocity of the rod the instant after impact, we will need to apply the principles of linear impulse and linear momentum along with the coefficient of restitution condition for the collision between two rigid bodies. Each of these steps is now applied.

Application of Linear Impulse and Linear Momentum to Rod During Impact

Applying the principle of linear impulse and momentum to the rod during impact, we have

$$\hat{\mathbf{F}} = m^{\mathcal{F}}\bar{\mathbf{v}}' - m^{\mathcal{F}}\bar{\mathbf{v}} \tag{5-1049}$$

where $^{\mathcal{F}}\bar{\mathbf{v}}$ and $^{\mathcal{F}}\bar{\mathbf{v}}'$ are the velocities of the rod the instant before and after impact, respectively. The free body diagram of the rod during impact with the ground is shown in Fig. 5-44.

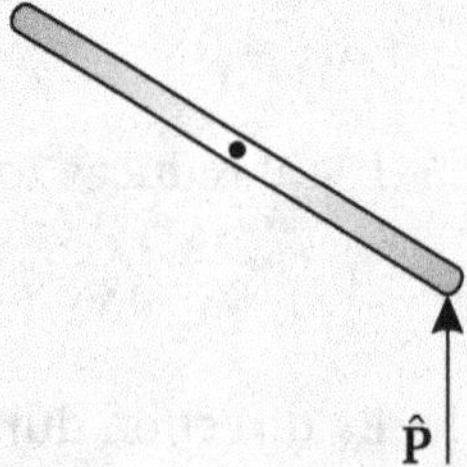

Figure 5-44 Free body diagram of rod during impact for Example 5-16.

It can be seen from Fig. 5-44 that the only impulse applied to the rod during impact is that due to the ground (we note that, because the *position* does not change during impact, i.e., the impact is assumed to occur instantaneously, gravity does not apply an impulse during impact). Denoting the impulse applied by the ground by $\hat{\mathbf{P}}$ and noting that the ground imparts an impulse in the $\mathbf{E}_x$-direction, we have

$$\hat{\mathbf{F}} = \hat{\mathbf{P}} = \hat{P}\mathbf{E}_x \tag{5-1050}$$

Next, because the impact occurs at the same instant of time as the rod terminates its descent, we know that ${}^{\mathcal{F}}\bar{\mathbf{v}}$ is equal to ${}^{\mathcal{F}}\bar{\mathbf{v}}_1$ as given in Eq. (5-1048), i.e.,

$$ {}^{\mathcal{F}}\bar{\mathbf{v}} = {}^{\mathcal{F}}\bar{\mathbf{v}}_1 = \bar{v}_1\mathbf{E}_x = \sqrt{2gh}\mathbf{E}_x \tag{5-1051} $$

Finally, using the general expression for the velocity of the center of mass of the rod as given in Eq. (5-1030), the velocity of the center of mass of the rod the instant *after* impact is given as

$$ {}^{\mathcal{F}}\bar{\mathbf{v}}_1' = \bar{v}_x'\mathbf{E}_x + \bar{v}_y'\mathbf{E}_y \tag{5-1052} $$

Substituting $\hat{\mathbf{F}}$, ${}^{\mathcal{F}}\bar{\mathbf{v}}$, and ${}^{\mathcal{F}}\bar{\mathbf{v}}'$ from Eqs. (5-1050), (5-1051), and (5-1052), respectively, into Eq. (5-1049), we obtain

$$ \hat{P}\mathbf{E}_x = m(\bar{v}_x'\mathbf{E}_x + \bar{v}_y'\mathbf{E}_y) - m\bar{v}_1\mathbf{E}_x \tag{5-1053} $$

Equating components in Eq. (5-1053), we obtain the following two scalar equations:

$$ m\bar{v}_x' - m\bar{v}_1 = \hat{P} \tag{5-1054} $$

$$ m\bar{v}_y' = 0 \tag{5-1055} $$

It is seen from Eq. (5-1055) that

$$ \bar{v}_y' = 0 \tag{5-1056} $$

Application of Angular Impulse and Angular Momentum to Rod During Impact

Applying the principle of angular impulse and angular momentum to the rod during impact relative to the center of mass of the rod, we have

$$ \hat{\bar{\mathbf{M}}} = {}^{\mathcal{F}}\bar{\mathbf{H}}' - {}^{\mathcal{F}}\bar{\mathbf{H}} \tag{5-1057} $$

where

${}^{\mathcal{F}}\bar{\mathbf{H}}$ = Angular momentum of rod relative to center of mass of rod the instant before impact

${}^{\mathcal{F}}\bar{\mathbf{H}}'$ = Angular momentum of rod relative to center of mass of rod the instant after impact

Because before impact the rod is not rotating, we have

$$ {}^{\mathcal{F}}\bar{\mathbf{H}} = \mathbf{0} \tag{5-1058} $$

Next, ${}^{\mathcal{F}}\bar{\mathbf{H}}'$ is given as

$$ {}^{\mathcal{F}}\bar{\mathbf{H}}' = \bar{\mathbf{I}} \cdot \left({}^{\mathcal{F}}\boldsymbol{\omega}_1^{R}\right)' \tag{5-1059} $$

Because $\{\mathbf{E}_x, \mathbf{E}_y, \mathbf{E}_z\}$ is a principal-axis basis, we have

$$ \bar{\mathbf{I}} = \bar{I}_{xx}\mathbf{E}_x \otimes \mathbf{E}_x + \bar{I}_{yy}\mathbf{E}_y \otimes \mathbf{E}_y + \bar{I}_{zz}\mathbf{E}_z \otimes \mathbf{E}_z \tag{5-1060} $$

Therefore, we obtain

$$ {}^{\mathcal{F}}\bar{\mathbf{H}}' = \bar{I}_{zz}\omega'\mathbf{E}_z \tag{5-1061} $$

Noting that $\bar{I}_{zz} = ml^2/12$, we obtain ${}^{\mathcal{F}}\bar{\mathbf{H}}'$ as

$$ {}^{\mathcal{F}}\bar{\mathbf{H}}' = \frac{ml^2}{12}\omega'\mathbf{E}_z \tag{5-1062}$$

Furthermore, because the only impulse applied to the rod during impact is that due to the ground, the angular impulse about the center of mass of the rod during impact is given as

$$\hat{\mathbf{M}} = (\mathbf{r}_B - \bar{\mathbf{r}}) \times \hat{\mathbf{F}} = (\mathbf{r}_B - \bar{\mathbf{r}}) \times \hat{\mathbf{P}} \tag{5-1063}$$

Then, substituting the expression for $\mathbf{r}_B - \bar{\mathbf{r}}$ from Eq. (5-1034) and the expression for $\hat{\mathbf{F}}$ from Eq. (5-1050) into (5-1063), we have

$$\hat{\mathbf{M}} = \left(\frac{l}{2}\sin\theta\mathbf{E}_x + \frac{l}{2}\cos\theta\mathbf{E}_y\right) \times \hat{P}\mathbf{E}_x = -\frac{l}{2}\hat{P}\cos\theta\mathbf{E}_z \tag{5-1064}$$

Finally, substituting ${}^{\mathcal{F}}\bar{\mathbf{H}}'$ and $\hat{\mathbf{M}}$ from Eqs. (5-1059) and (5-1064), respectively, into (5-1057), we obtain

$$-\frac{l}{2}\hat{P}\cos\theta\mathbf{E}_z = \frac{ml^2}{12}\omega_1'\mathbf{E}_z \tag{5-1065}$$

Simplifying Eq. (5-1065 gives

$$\frac{ml}{6}\omega_1' = -\hat{P}\cos\theta \tag{5-1066}$$

Application of Coefficient of Restitution Condition to Rod During Impact

It is seen that Eqs. (5-1054) and (5-1066) together form a system of two equations in *three* unknowns $\bar{v}_x'$, ω', and $\hat{P}$. Consequently, it is necessary that we obtain another independent equation in order to solve the problem. This last equation is obtained by applying the coefficient of restitution condition at the impact point B, i.e., applying

$$e = \frac{\left({}^{\mathcal{F}}\mathbf{v}_B^{\mathcal{R}}\right)' \cdot \mathbf{n} - \left({}^{\mathcal{F}}\mathbf{v}_B^{\mathcal{G}}\right)' \cdot \mathbf{n}}{{}^{\mathcal{F}}\mathbf{v}_B^{\mathcal{G}} \cdot \mathbf{n} - {}^{\mathcal{F}}\mathbf{v}_B^{\mathcal{R}} \cdot \mathbf{n}} \tag{5-1067}$$

where the superscript $\mathcal{G}$ denotes the fixed horizontal surface and $\mathbf{n}$ is the direction of impact. Since the horizontal surface is fixed, we have

$$\left({}^{\mathcal{F}}\mathbf{v}_B^{\mathcal{G}}\right)' = {}^{\mathcal{F}}\mathbf{v}_B^{\mathcal{G}} = \mathbf{0} \tag{5-1068}$$

Furthermore, since the impact occurs in the $\mathbf{E}_x$-direction, the direction of impact, $\mathbf{n}$, is given as

$$\mathbf{n} = \mathbf{E}_x \tag{5-1069}$$

Therefore, Eq. (5-1067) reduces to

$$e = -\frac{\left({}^{\mathcal{F}}\mathbf{v}_B^{\mathcal{R}}\right)' \cdot \mathbf{E}_x}{{}^{\mathcal{F}}\mathbf{v}_B^{\mathcal{R}} \cdot \mathbf{E}_x} \tag{5-1070}$$

which can be rearranged to give

$$\left({}^{\mathcal{F}}\mathbf{v}_B^{\mathcal{R}}\right)' \cdot \mathbf{E}_x = -e\,{}^{\mathcal{F}}\mathbf{v}_B^{\mathcal{R}} \cdot \mathbf{E}_x \tag{5-1071}$$

Now, Eq. (5-1071) can be resolved in terms of components using the expression for ${}^{\mathcal{F}}\mathbf{v}_B$ as given in Eq. (5-1035). First, because the rod is not rotating before impact, we have

$$ {}^{\mathcal{F}}\mathbf{v}_B = {}^{\mathcal{F}}\bar{\mathbf{v}} = \bar{v}_1\mathbf{E}_x = \bar{v}\mathbf{E}_x = \sqrt{2gh}\mathbf{E}_x \tag{5-1072} $$

Next, using Eq. (5-1035), the velocity of point B in reference frame $\mathcal{F}$ the instant after impact is given as

$$ \left({}^{\mathcal{F}}\mathbf{v}_B^{\mathcal{R}}\right)' = \left(\bar{v}_x' - \frac{l\omega'}{2}\cos\theta\right)\mathbf{E}_x + \left(\bar{v}_y' + \frac{l\omega'}{2}\sin\theta\right)\mathbf{E}_y \tag{5-1073} $$

Substituting the expressions for ${}^{\mathcal{F}}\mathbf{v}_B^{\mathcal{R}}$ and $\left({}^{\mathcal{F}}\mathbf{v}_B^{\mathcal{R}}\right)'$ from Eqs. (5-1072) and (5-1073) into Eq. (5-1071), we obtain

$$ \bar{v}_x' - \frac{l}{2}\omega'\cos\theta = -e\bar{v} \tag{5-1074} $$

(b) Angular Velocity and Velocity of Center of Mass of Rod the Instant After Impact

Equations (5-1054), (5-1066), and (5-1074) are a system of three equations in the three unknowns $\bar{v}_{2x}$, ω_2, and $\hat{P}$. Multiplying Eq. (5-1054) by $\cos\theta$, we obtain

$$ m\bar{v}_x'\cos\theta - m\bar{v}\cos\theta = \hat{P}\cos\theta \tag{5-1075} $$

Adding Eq. (5-1075) to (5-1066), we have

$$ m\bar{v}_x'\cos\theta - m\bar{v}\cos\theta + \frac{ml}{6}\omega' = 0 \tag{5-1076} $$

Dropping the dependence on m in Eq. (5-1076) gives

$$ \bar{v}_x'\cos\theta - \bar{v}\cos\theta + \frac{l}{6}\omega' = 0 \tag{5-1077} $$

Next, multiplying Eq. (5-1074) by $\cos\theta$ gives

$$ \bar{v}_x'\cos\theta - \frac{l}{2}\omega'\cos^2\theta = -e\bar{v}\cos\theta \tag{5-1078} $$

Subtracting Eq. (5-1078) from (5-1077) gives

$$ -\bar{v}\cos\theta + \frac{l}{6}\omega' + \frac{l}{2}\omega'\cos^2\theta = e\bar{v}\cos\theta \tag{5-1079} $$

Rearranging Eq. (5-1079) gives

$$ l\omega'\left(\frac{1}{6} + \frac{1}{2}\cos^2\theta\right) = (1+e)\bar{v}\cos\theta \tag{5-1080} $$

Solving Eq. (5-1080) for ω_2 gives

$$ \omega' = \frac{6(1+e)\cos\theta}{l(1+3\cos^2\theta)}\bar{v} \tag{5-1081} $$

Recalling from Eq. (5-1047) that $\bar{v} = \sqrt{2gh}$, we obtain the angular velocity of the rod immediately after impact as

$$\left({}^{\mathcal{F}}\boldsymbol{\omega}^{\mathcal{R}}\right)' = \frac{6(1+e)\cos\theta}{l(1+3\cos^2\theta)}\sqrt{2gh}\mathbf{E}_z \quad (5\text{-}1082)$$

Substituting ω_2 from Eq. (5-1082) into (5-1077) gives

$$\bar{v}'_x\cos\theta - \bar{v}\cos\theta + \frac{l}{6}\frac{6(1+e)\cos\theta}{l(1+3\cos^2\theta)}\bar{v} = 0 \quad (5\text{-}1083)$$

Solving Eq. (5-1083) for $\bar{v}'_x$ gives

$$\bar{v}'_x = \bar{v} - \frac{1+e}{1+3\cos^2\theta}\bar{v} \quad (5\text{-}1084)$$

Equation (5-1084) can be simplified to

$$\bar{v}'_x = \frac{3\cos^2\theta - e}{1+3\cos^2\theta}\bar{v} \quad (5\text{-}1085)$$

Recalling from Eq. (5-1047) that $\bar{v} = \sqrt{2gh}$, we obtain

$$\bar{v}'_x = \frac{3\cos^2\theta - e}{1+3\cos^2\theta}\sqrt{2gh} \quad (5\text{-}1086)$$

We then obtain the post-impact velocity of the center of mass after impact as

$${}^{\mathcal{F}}\bar{\mathbf{v}}' = \frac{3\cos^2\theta - e}{1+3\cos^2\theta}\sqrt{2gh}\mathbf{E}_x \quad (5\text{-}1087)$$

■

Example 5–17

A uniform slender rod of mass M and length L is hinged at point O, located at one of its ends as shown in Fig. 5-45. The rod is initially at rest when it is struck by a particle of mass m traveling with a horizontal velocity of $\mathbf{v}_0$. Assuming a perfectly inelastic impact, determine (a) the distance x (measured from point O) at which the particle must strike the rod in order that the reaction impulse at the pivot is zero and (b) the angular velocity of the rod immediately after impact.

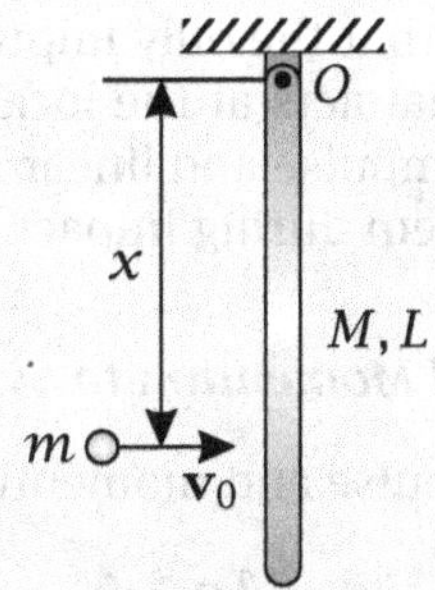

Figure 5-45 Particle of mass m striking rod of mass M and length L.

Solution to Example 5-17

Kinematics

First, let $\mathcal{F}$ be a fixed reference frame. Then, choose the following coordinate system fixed in reference frame $\mathcal{F}$:

	Origin at O	
$\mathbf{E}_x$	=	Along rod at $t = 0^-$
$\mathbf{E}_z$	=	Out of page
$\mathbf{E}_y$	=	$\mathbf{E}_z \times \mathbf{E}_x$

where $t = 0^-$ is the instant before impact. Next, let $\mathcal{R}$ be a reference frame fixed to the rod. Then, it is convenient to use the following coordinate system fixed in reference frame $\mathcal{R}$:

	Origin at O	
$\mathbf{e}_r$	=	Along rod
$\mathbf{e}_z$	=	Out of page $(= \mathbf{E}_z)$
$\mathbf{e}_\theta$	=	$\mathbf{e}_z \times \mathbf{e}_r$

Kinetics

It is noted that this problem can be solved by considering *only* the entire system consisting of the particle and the rod and applying the principles of linear impulse and linear momentum and angular impulse and angular momentum to the system. The free body diagram of the system during impact is shown in Fig. 5-46.

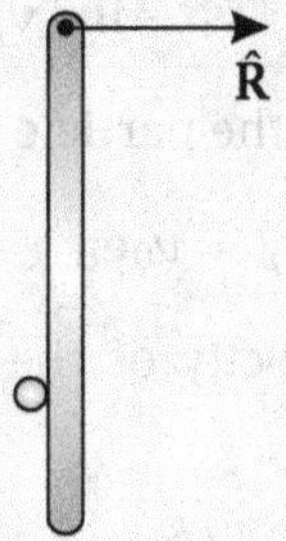

Figure 5-46 Free body diagram of particle-rod system of Example 5-17.

Examining Fig. 5-46, it can be seen that the only impulse applied to the system during impact is the reaction impulse $\hat{\mathbf{R}}$ that acts at the location of the hinge at point O. We now apply the principles of linear impulse and linear momentum and angular impulse and angular momentum to the system during impact.

Application of Linear Impulse and Momentum to System

Applying the principle of linear impulse and momentum to the entire system, we have

$$^{\mathcal{F}}\mathbf{G}' - {}^{\mathcal{F}}\mathbf{G} = \hat{\mathbf{F}} \tag{5-1088}$$

where $^{\mathcal{F}}\mathbf{G}$ and $^{\mathcal{F}}\mathbf{G}'$ are the linear momenta of the system before and after impact, respectively. Now, because the rod is initially motionless, the linear momentum of the system before impact is due entirely to the particle, which implies that

$$^{\mathcal{F}}\mathbf{G} = m{}^{\mathcal{F}}\mathbf{v}_p \tag{5-1089}$$

where $^{\mathcal{F}}\mathbf{v}_p$ is the velocity of the particle in reference frame $\mathcal{F}$ before impact. Furthermore, the linear momentum of the system after impact is due to motion of *both* the rod and the particle, which implies that

$$^{\mathcal{F}}\mathbf{G}' = {}^{\mathcal{F}}\mathbf{G}_p' + {}^{\mathcal{F}}\mathbf{G}_r' \tag{5-1090}$$

where $^{\mathcal{F}}\mathbf{G}_p'$ and $^{\mathcal{F}}\mathbf{G}_r'$ are the linear momenta of the particle and rod, respectively, after impact. The linear momentum of the particle after impact is given as

$$^{\mathcal{F}}\mathbf{G}_p' = m{}^{\mathcal{F}}\mathbf{v}_p' \tag{5-1091}$$

where $^{\mathcal{F}}\mathbf{v}_p'$ is the velocity of the particle after impact. Next, the linear momentum of the rod after impact is given as

$$^{\mathcal{F}}\mathbf{G}_r' = M{}^{\mathcal{F}}\bar{\mathbf{v}}_r' \tag{5-1092}$$

where $^{\mathcal{F}}\bar{\mathbf{v}}_r'$ is the velocity of the center of mass of the rod after impact. Adding the expressions in Eqs. (5-1091) and (5-1092), we obtain the post-impact linear momentum of the system as

$$^{\mathcal{F}}\mathbf{G}' = m{}^{\mathcal{F}}\mathbf{v}_p' + M{}^{\mathcal{F}}\bar{\mathbf{v}}_r' \tag{5-1093}$$

Substituting the results of Eqs. (5-1089) and (5-1093) into Eq. (5-1088), we obtain

$$m{}^{\mathcal{F}}\mathbf{v}_p' + M{}^{\mathcal{F}}\bar{\mathbf{v}}_r' - m{}^{\mathcal{F}}\mathbf{v}_p = \hat{\mathbf{F}} \tag{5-1094}$$

Now, we are given the initial velocity of the particle as

$$^{\mathcal{F}}\mathbf{v}_p = v_0\mathbf{e}_\theta \tag{5-1095}$$

Furthermore, denoting the angular velocity of the rod immediately after impact by $({}^{\mathcal{F}}\boldsymbol{\omega}^{\mathcal{R}})'$, from kinematics we have

$$^{\mathcal{F}}\bar{\mathbf{v}}_r' = \left(\frac{^{\mathcal{F}}d\bar{\mathbf{r}}}{dt}\right)' = \left(\frac{^{\mathcal{R}}d\bar{\mathbf{r}}}{dt} + {}^{\mathcal{F}}\boldsymbol{\omega}^{\mathcal{R}} \times \bar{\mathbf{r}}\right)' \tag{5-1096}$$

where $\bar{\mathbf{r}}$ is the position of the center of mass of the rod. Now we know that

$$\frac{{}^{\mathcal{R}}d\bar{\mathbf{r}}}{dt} = \mathbf{0} \tag{5-1097}$$

Consequently,

$${}^{\mathcal{F}}\bar{\mathbf{v}}_r' = \left({}^{\mathcal{F}}\boldsymbol{\omega}^{\mathcal{R}}\right)' \times \bar{\mathbf{r}} \tag{5-1098}$$

Then, since the motion is planar, we have

$${}^{\mathcal{F}}\boldsymbol{\omega}^{\mathcal{R}} = \omega\mathbf{e}_z \tag{5-1099}$$

which implies that

$$\left({}^{\mathcal{F}}\boldsymbol{\omega}^{\mathcal{R}}\right)' = \omega'\mathbf{e}_z \tag{5-1100}$$

Also, because the rod is uniform, the position of the center of mass of the rod is given as

$$\bar{\mathbf{r}} = \frac{L}{2}\mathbf{e}_r \tag{5-1101}$$

Substituting the expressions for $({}^{\mathcal{F}}\boldsymbol{\omega}^{\mathcal{R}})'$ and $\bar{\mathbf{r}}$ from Eqs. (5-1100) and (5-1101), respectively, into Eq. (5-1098), we obtain ${}^{\mathcal{F}}\bar{\mathbf{v}}_r'$ as

$${}^{\mathcal{F}}\bar{\mathbf{v}}_r' = \omega'\mathbf{e}_z \times \frac{L}{2}\mathbf{e}_r = \omega'\frac{L}{2}\mathbf{e}_\theta \tag{5-1102}$$

Next, because the particle sticks to the rod on impact, the particle and rod must move as a single rigid body after impact. Therefore, the angular velocity of the particle and rod are the same after impact. Using this last fact, we have

$${}^{\mathcal{F}}\mathbf{v}_p' = \left({}^{\mathcal{F}}\boldsymbol{\omega}^{\mathcal{R}}\right)' \times \mathbf{r} \tag{5-1103}$$

where $\mathbf{r}$ is the position of the particle. We note that the position of the particle is given as

$$\mathbf{r} = x\mathbf{e}_r \tag{5-1104}$$

Consequently,

$${}^{\mathcal{F}}\mathbf{v}_p' = \omega'\mathbf{e}_z \times x\mathbf{e}_r = \omega' x\mathbf{e}_\theta \tag{5-1105}$$

Finally, we have

$$\hat{\mathbf{F}} = \hat{\mathbf{R}} = \hat{R}\mathbf{e}_\theta \tag{5-1106}$$

where $\hat{\mathbf{R}}$ is the reaction impulse of the hinge. Substituting the results of Eqs. (5-1095), (5-1102), (5-1105), and (5-1106) into Eq. (5-1094), we obtain

$$m\omega' x\mathbf{e}_\theta + M\omega'\frac{L}{2}\mathbf{e}_\theta = mv_0\mathbf{e}_\theta + \hat{R}\mathbf{e}_\theta \tag{5-1107}$$

which yields the scalar equation

$$mv_0 + \hat{R} = \left(mx + \frac{ML}{2}\right)\omega' \tag{5-1108}$$

Application of Angular Impulse and Momentum to System Relative to Point O

Observing that point O is inertially fixed, we can apply the principle of angular impulse and angular momentum relative to point O as

$$\hat{\mathbf{M}}_O = {}^{\mathcal{F}}\mathbf{H}'_O - {}^{\mathcal{F}}\mathbf{H}_O \tag{5-1109}$$

Now from the free body diagram we see that the angular impulse applied to the system about point O is zero, i.e.,

$$\hat{\mathbf{M}}_O = \mathbf{0} \tag{5-1110}$$

Equation (5-1109) then simplifies to

$${}^{\mathcal{F}}\mathbf{H}'_O - {}^{\mathcal{F}}\mathbf{H}_O = \mathbf{0} \tag{5-1111}$$

which implies that

$${}^{\mathcal{F}}\mathbf{H}_O = {}^{\mathcal{F}}\mathbf{H}'_O \tag{5-1112}$$

Before impact the rod is motionless. Hence the angular momentum of the system before impact is due to only the particle. Therefore,

$${}^{\mathcal{F}}\mathbf{H}_O = \mathbf{r} \times m {}^{\mathcal{F}}\mathbf{v}_0 \tag{5-1113}$$

Substituting the earlier expressions for $\mathbf{r}$ and ${}^{\mathcal{F}}\mathbf{v}_O$, we obtain

$${}^{\mathcal{F}}\mathbf{H}_O = x\mathbf{e}_r \times m v_0 \mathbf{e}_\theta = m v_0 x \mathbf{e}_z \tag{5-1114}$$

After impact, both the rod and the particle move. Consequently,

$${}^{\mathcal{F}}\mathbf{H}'_O = \left({}^{\mathcal{F}}\mathbf{H}'_O\right)_p + \left({}^{\mathcal{F}}\mathbf{H}'_O\right)_r \tag{5-1115}$$

where $\left({}^{\mathcal{F}}\mathbf{H}'_O\right)_p$ and $\left({}^{\mathcal{F}}\mathbf{H}'_O\right)_r$ are the angular momenta of the particle and rod, respectively, immediately after impact. The angular momentum of the particle after impact is given as

$$\left({}^{\mathcal{F}}\mathbf{H}'_O\right)_p = m\mathbf{r} \times {}^{\mathcal{F}}\mathbf{v}'_p \tag{5-1116}$$

Then, using the expression for ${}^{\mathcal{F}}\mathbf{v}'_p = \omega' x \mathbf{e}_\theta$ from Eq. (5-1105), we obtain

$$\left({}^{\mathcal{F}}\mathbf{H}'_O\right)_p = m x \mathbf{e}_r \times \omega' x \mathbf{e}_\theta = m x^2 \omega' \mathbf{e}_z \tag{5-1117}$$

The angular momentum of the rod after impact is given as

$$\left({}^{\mathcal{F}}\mathbf{H}'_O\right)_r = \mathbf{I}_O^{\mathcal{R}} \cdot \left({}^{\mathcal{F}}\boldsymbol{\omega}^{\mathcal{R}}\right)' \tag{5-1118}$$

Because $\{\mathbf{e}_r, \mathbf{e}_\theta, \mathbf{e}_z\}$ is a principal-axis basis, we have

$$\mathbf{I}_O^{\mathcal{R}} = I_{rr}^O \mathbf{e}_r \otimes \mathbf{e}_r + I_{\theta\theta}^O \mathbf{e}_\theta \otimes \mathbf{e}_\theta + I_{zz}^O \mathbf{e}_z \otimes \mathbf{e}_z \tag{5-1119}$$

Furthermore, using the post-impact angular velocity of the rod from Eq. (5-1100), we obtain

$$\left({}^{\mathcal{F}}\mathbf{H}'_O\right)_r = (I_{rr}^O \mathbf{e}_r \otimes \mathbf{e}_r + I_{\theta\theta}^O \mathbf{e}_\theta \otimes \mathbf{e}_\theta + I_{zz}^O \mathbf{e}_z \otimes \mathbf{e}_z) \cdot \omega' \mathbf{e}_z = I_{zz}^O \omega' \mathbf{e}_z \tag{5-1120}$$

Now for a uniform slender rod we have

$$\bar{I}_{zz} = \frac{ML^2}{12} \tag{5-1121}$$

Then, applying the parallel-axis theorem, we obtain

$$I_{zz}^O = \bar{I}_{zz} + M(\pi_r^2 + \pi_\theta^2) \tag{5-1122}$$

We see in this case that the only translation occurs in the $\mathbf{e}_r$-direction. Therefore,

$$\begin{aligned} \pi_r &= \frac{L}{2} \\ \pi_\theta &= 0 \end{aligned} \tag{5-1123}$$

We then obtain the moment of inertia about the $\mathbf{e}_z$-direction through the point O as

$$I_{zz}^O = \frac{ML^2}{12} + M\left(\frac{L}{2}\right)^2 = \frac{ML^2}{3} \tag{5-1124}$$

The post-impact angular momentum of the rod is then given as

$$\left({}^{\mathcal{F}}\mathbf{H}'_O\right)_r = \frac{ML^2}{3}\omega'\mathbf{e}_z \tag{5-1125}$$

Substituting the results of Eqs. (5-1117) and (5-1125) into Eq. (5-1115), we obtain the angular momentum of the system after impact as

$${}^{\mathcal{F}}\mathbf{H}'_O = mx^2\omega'\mathbf{e}_z + \frac{ML^2}{3}\omega'\mathbf{e}_z = \left(mx^2 + \frac{ML^2}{3}\right)\omega'\mathbf{e}_z \tag{5-1126}$$

Furthermore, substituting ${}^{\mathcal{F}}\mathbf{H}_O$ and ${}^{\mathcal{F}}\mathbf{H}'_O$ from Eqs. (5-1114) and (5-1126), respectively, into Eq. (5-1112), we obtain

$$mv_0x = \left(mx^2 + \frac{ML^2}{3}\right)\omega' \tag{5-1127}$$

(a) Value of x Such That $\hat{\mathbf{R}} = \mathbf{0}$

We can now use the results from Eqs. (5-1108) and (5-1127) to solve for the distance x where $\hat{\mathbf{R}} = \mathbf{0}$. Setting $\hat{R} = 0$ in Eq. (5-1108), we have

$$mv_0 = \left(mx + \frac{ML}{2}\right)\omega' \tag{5-1128}$$

Next, solving Eq. (5-1127) for ω', we obtain

$$\omega' = \frac{mv_0x}{mx^2 + \dfrac{ML^2}{3}} \tag{5-1129}$$

Substituting ω' from Eq. (5-1129) into (5-1128), we have

$$mv_0 = \left(mx + \frac{ML}{2}\right)\frac{mv_0x}{mx^2 + \dfrac{ML^2}{3}} \tag{5-1130}$$

Rearranging and simplifying Eq. (5-1130), we obtain

$$\frac{ML^2}{3} + mx^2 = \left(\frac{ML}{2} + mx\right)x \tag{5-1131}$$

Simplifying Eq. (5-1131) gives

$$\frac{ML^2}{3} = x\frac{ML}{2} \tag{5-1132}$$

Solving Eq. (5-1132) for x, we obtain

$$x = \tfrac{2}{3}L \tag{5-1133}$$

(b) Post-Impact Angular Velocity of Rod

Substituting x from Eq. (5-1133) into Eq. (5-1129), we obtain the post-impact angular velocity of the rod as

$$\omega' = \frac{mv_0\left(\frac{2}{3}L\right)}{m\left(\frac{2}{3}L\right)^2 + \frac{ML^2}{3}} \tag{5-1134}$$

Simplifying Eq. (5-1134), we obtain

$$\omega' = \frac{6m}{4m + 3M}\frac{v_0}{L} \tag{5-1135}$$

Substituting the result of Eq. (5-1135) into (5-1100), we obtain the post-impact angular velocity of the system as

$$\left({}^{\mathcal{F}}\boldsymbol{\omega}^{\mathcal{R}}\right)' = \frac{6m}{4m + 3M}\frac{v_0}{L}\mathbf{e}_z \tag{5-1136}$$

■

Summary of Chapter 5

This chapter was devoted to developing the framework for solving and analyzing rigid body kinetics problems. The first topics covered were the center of mass and linear momentum of a rigid body. The position of the center of mass of a rigid body was defined as

$$\bar{\mathbf{r}} = \frac{\int_{\mathcal{R}} \mathbf{r} dm}{m} \tag{5-1}$$

where $\mathbf{r}$ is measured relative point O fixed in an inertial reference frame $\mathcal{N}$. In terms of the center of mass, the linear momentum of a rigid body was derived as

$$^{\mathcal{N}}\mathbf{G} = m^{\mathcal{N}}\bar{\mathbf{v}} \tag{5-9}$$

where $^{\mathcal{N}}\bar{\mathbf{v}}$

$$^{\mathcal{N}}\bar{\mathbf{v}} = \frac{^{\mathcal{N}}d\bar{\mathbf{r}}}{dt} \tag{5-10}$$

is the velocity of the center of mass of the rigid body in the inertial reference frame $\mathcal{N}$. Finally, the acceleration of the center of mass of a rigid body as viewed by an observer in an inertial reference frame $\mathcal{N}$ was given as

$$^{\mathcal{N}}\bar{\mathbf{a}} = \frac{^{\mathcal{N}}d}{dt}\left(^{\mathcal{N}}\bar{\mathbf{v}}\right)$$

The next topic covered in this chapter was the angular momentum of a rigid body. The angular momentum of a rigid body in an inertial reference frame $\mathcal{N}$ relative to an arbitrary reference point Q was defined as

$$^{\mathcal{N}}\mathbf{H}_Q = \int_{\mathcal{R}} (\mathbf{r} - \mathbf{r}_Q) \times (^{\mathcal{N}}\mathbf{v} - {^{\mathcal{N}}\mathbf{v}_Q}) dm \tag{5-13}$$

The angular momentum of a rigid body in the inertial reference frame $\mathcal{N}$ relative to a point O fixed in $\mathcal{N}$ was defined as

$$^{\mathcal{N}}\mathbf{H}_O = \int_{\mathcal{R}} (\mathbf{r} - \mathbf{r}_O) \times {^{\mathcal{N}}\mathbf{v}} dm \tag{5-14}$$

The angular momentum of a rigid body in the inertial reference frame $\mathcal{N}$ relative to the center of mass of the rigid body was defined as

$$^{\mathcal{N}}\bar{\mathbf{H}} = \int_{\mathcal{R}} (\mathbf{r} - \bar{\mathbf{r}}) \times (^{\mathcal{N}}\mathbf{v} - {^{\mathcal{N}}\bar{\mathbf{v}}}) dm \tag{5-15}$$

Using the three definitions of angular momentum, it was shown that $^{\mathcal{N}}\mathbf{H}_Q$ and $^{\mathcal{N}}\mathbf{H}_O$ are related as

$$^{\mathcal{N}}\mathbf{H}_Q = {^{\mathcal{N}}\mathbf{H}_O} - (\bar{\mathbf{r}} - \mathbf{r}_Q) \times m^{\mathcal{N}}\mathbf{v}_Q - (\mathbf{r}_Q - \mathbf{r}_O) \times m^{\mathcal{N}}\bar{\mathbf{v}} \tag{5-21}$$

Next, it was shown that $^{\mathcal{N}}\bar{\mathbf{H}}$ and $^{\mathcal{N}}\mathbf{H}_O$ are related as

$$^{\mathcal{N}}\bar{\mathbf{H}} = {^{\mathcal{N}}\mathbf{H}_O} - (\bar{\mathbf{r}} - \mathbf{r}_O) \times m^{\mathcal{N}}\bar{\mathbf{v}} \tag{5-22}$$

Finally, it was shown that ${}^{\mathcal{N}}\mathbf{H}_Q$ and ${}^{\mathcal{N}}\bar{\mathbf{H}}$ are related as

$$ {}^{\mathcal{N}}\mathbf{H}_Q = {}^{\mathcal{N}}\bar{\mathbf{H}} + (\mathbf{r}_Q - \bar{\mathbf{r}}) \times m({}^{\mathcal{N}}\mathbf{v}_Q - {}^{\mathcal{N}}\bar{\mathbf{v}}) \tag{5-24} $$

The next topic that was covered in this chapter was the derivation of the angular momentum of a rigid body in terms of the moment of inertia tensor. The moment of inertia tensor relative to a body-fixed point Q was derived as

$$ \mathbf{I}_Q^{\mathcal{R}} = \int_{\mathcal{R}} \left[\left(\boldsymbol{\rho}_Q \cdot \boldsymbol{\rho}_Q \right) \mathbf{U} - \left(\boldsymbol{\rho}_Q \otimes \boldsymbol{\rho}_Q \right) \right] dm \tag{5-36} $$

It was then shown that the angular momentum of a rigid body $\mathcal{R}$ relative to a body-fixed point Q can be written as

$$ {}^{\mathcal{N}}\mathbf{H}_Q = \mathbf{I}_Q^{\mathcal{R}} \cdot {}^{\mathcal{N}}\boldsymbol{\omega}^{\mathcal{R}} \tag{5-39} $$

where $\boldsymbol{\rho}_Q = \mathbf{r} - \mathbf{r}_Q$. Next, it was shown that the moment of inertia tensor relative to the center of mass of a rigid body can be written as

$$ \bar{\mathbf{I}}^{\mathcal{R}} = \int_{\mathcal{R}} [(\boldsymbol{\rho} \cdot \boldsymbol{\rho}) \mathbf{U} - (\boldsymbol{\rho} \otimes \boldsymbol{\rho})] \, dm \tag{5-43} $$

where $\boldsymbol{\rho} = \mathbf{r} - \bar{\mathbf{r}}$. It was then shown that the angular momentum of a rigid body $\mathcal{R}$ relative to the center of mass of a rigid body is given as

$$ {}^{\mathcal{N}}\bar{\mathbf{H}} = \bar{\mathbf{I}}^{\mathcal{R}} \cdot {}^{\mathcal{N}}\boldsymbol{\omega}^{\mathcal{R}} \tag{5-44} $$

Then, for a body-fixed coordinate system with origin at the body-fixed point Q, the moments of inertia of the rigid body were derived as

$$ \begin{aligned} I_{11}^Q &= \int_{\mathcal{R}} (\rho_2^2 + \rho_3^2) dm \\ I_{22}^Q &= \int_{\mathcal{R}} (\rho_1^2 + \rho_3^2) dm \\ I_{33}^Q &= \int_{\mathcal{R}} (\rho_1^2 + \rho_2^2) dm \end{aligned} \tag{5-54} $$

Similarly, the products of inertia relative to the body-fixed point Q were derived as

$$ \begin{aligned} I_{12}^Q &= -\int_{\mathcal{R}} \rho_1 \rho_2 dm \\ I_{13}^Q &= -\int_{\mathcal{R}} \rho_1 \rho_3 dm \\ I_{23}^Q &= -\int_{\mathcal{R}} \rho_2 \rho_3 dm \end{aligned} \tag{5-55} $$

For the case where the reference point is the center of mass, the moments of inertia were derived as

$$ \begin{aligned} \bar{I}_{11} &= \int_{\mathcal{R}} (\rho_2^2 + \rho_3^2) dm \\ \bar{I}_{22} &= \int_{\mathcal{R}} (\rho_1^2 + \rho_3^2) dm \\ \bar{I}_{33} &= \int_{\mathcal{R}} (\rho_1^2 + \rho_2^2) dm \end{aligned} \tag{5-56} $$

Furthermore, the products of inertia relative to the center of mass of the rigid body were derived as

$$\begin{aligned}\bar{I}_{12} &= -\int_{\mathcal{R}} \rho_1\rho_2 dm \\ \bar{I}_{13} &= -\int_{\mathcal{R}} \rho_1\rho_3 dm \\ \bar{I}_{23} &= -\int_{\mathcal{R}} \rho_2\rho_3 dm\end{aligned} \tag{5-57}$$

After defining the moments and products of inertia for a rigid body, it was shown that the moments and products of inertia relative to a body-fixed point Q are related to the moments and products of inertia relative to the center of mass via the *parallel-axis theorem* as

$$\begin{aligned}I_{11}^Q &= \bar{I}_{11} + m(\pi_2^2 + \pi_3^2) \\ I_{22}^Q &= \bar{I}_{22} + m(\pi_1^2 + \pi_3^2) \\ I_{33}^Q &= \bar{I}_{33} + m(\pi_1^2 + \pi_2^2) \\ I_{12}^Q = I_{21}^Q &= \bar{I}_{12} - m\pi_1\pi_2 \\ I_{13}^Q = I_{31}^Q &= \bar{I}_{13} - m\pi_1\pi_3 \\ I_{23}^Q = I_{32}^Q &= \bar{I}_{23} - m\pi_2\pi_3\end{aligned} \tag{5-78}$$

where $\boldsymbol{\pi} = \mathbf{r}_Q - \bar{\mathbf{r}} = \pi_1\mathbf{e}_1 + \pi_2\mathbf{e}_2 + \pi_3\mathbf{e}_3$. Finally, it was discussed that, for the case of a *principal-axis coordinate system*, the products of inertia are zero, i.e., $I_{ij}^Q = 0\ (i \neq j,\ i, j = 1, 2, 3)$

The next topics covered were Euler's laws for a rigid body. Euler's laws were stated as follows:

$$\mathbf{F} = m\,{}^{\mathcal{N}}\bar{\mathbf{a}} \tag{5-225}$$

and

$$\mathbf{M}_O = \frac{{}^{\mathcal{N}}d}{dt}\left({}^{\mathcal{N}}\mathbf{H}_O\right) \tag{5-226}$$

where O is a point fixed point in an inertial reference frame $\mathcal{N}$. It was then shown that, for an arbitrary reference point Q, Euler's 2nd law is given as

$$\mathbf{M}_Q - (\bar{\mathbf{r}} - \mathbf{r}_Q) \times m\,{}^{\mathcal{N}}\mathbf{a}_Q = \frac{{}^{\mathcal{N}}d}{dt}\left({}^{\mathcal{N}}\mathbf{H}_Q\right) \tag{5-235}$$

Finally, for the case where the reference point is the center of mass of the rigid body, it was shown that

$$\bar{\mathbf{M}} = \frac{{}^{\mathcal{N}}d}{dt}\left({}^{\mathcal{N}}\bar{\mathbf{H}}\right) \tag{5-237}$$

In terms of the moment of inertia tensor relative to a point Q fixed in the rigid body, it was shown that Euler's 2nd law can be written as

$$\mathbf{M}_Q - (\bar{\mathbf{r}} - \mathbf{r}_Q) \times m\,{}^{\mathcal{N}}\mathbf{a}_Q = \mathbf{I}_Q^{\mathcal{R}} \cdot {}^{\mathcal{N}}\boldsymbol{\alpha}^{\mathcal{R}} + {}^{\mathcal{N}}\boldsymbol{\omega}^{\mathcal{R}} \times \left(\mathbf{I}_Q^{\mathcal{R}} \cdot {}^{\mathcal{N}}\boldsymbol{\omega}^{\mathcal{R}}\right) \tag{5-681}$$

Similarly, in terms of the moment of inertia tensor relative to the center of mass of the rigid body, it was shown that Euler's 2nd law can be written as

$$\bar{\mathbf{M}} = \bar{\mathbf{I}}^{\mathcal{R}} \cdot {}^{\mathcal{N}}\boldsymbol{\alpha}^{\mathcal{R}} + {}^{\mathcal{N}}\boldsymbol{\omega}^{\mathcal{R}} \times \left(\bar{\mathbf{I}}^{\mathcal{R}} \cdot {}^{\mathcal{N}}\boldsymbol{\omega}^{\mathcal{R}}\right) \tag{5-683}$$

Finally, in terms of a body-fixed principal-axis coordinate system located at the center of mass of the rigid body, it was shown that Eq. (5-683) leads to the following three scalar equations:

$$\bar{M}_1 = \bar{I}_1\dot{\omega}_1 + (\bar{I}_3 - \bar{I}_2)\omega_3\omega_2 \tag{5-701}$$

$$\bar{M}_2 = \bar{I}_2\dot{\omega}_2 + (\bar{I}_1 - \bar{I}_3)\omega_1\omega_3 \tag{5-702}$$

$$\bar{M}_3 = \bar{I}_3\dot{\omega}_3 + (\bar{I}_2 - \bar{I}_1)\omega_2\omega_1 \tag{5-703}$$

It was stated that Eqs. (5-701)–(5-703) are called *Euler's equations* for a rigid body.

The next topics covered were power, work, and energy of a rigid body. The kinetic energy of a rigid body as viewed by an observer in an inertial reference frame $\mathcal{N}$ was defined as

$$^{\mathcal{N}}T = \frac{1}{2}\int_{\mathcal{R}} {^{\mathcal{N}}\mathbf{v}} \cdot {^{\mathcal{N}}\mathbf{v}}dm \tag{5-767}$$

A more useful expression for the kinetic energy of a rigid body, called *Koenig's decomposition*, was derived as

$$^{\mathcal{N}}T = \tfrac{1}{2}m{^{\mathcal{N}}\bar{\mathbf{v}}} \cdot {^{\mathcal{N}}\bar{\mathbf{v}}} + \tfrac{1}{2}{^{\mathcal{N}}\bar{\mathbf{H}}} \cdot {^{\mathcal{N}}\boldsymbol{\omega}^{\mathcal{R}}} \tag{5-786}$$

The work-energy theorem for a rigid body was then derived as

$$\frac{d}{dt}\left({^{\mathcal{N}}T}\right) = \mathbf{F} \cdot {^{\mathcal{N}}\bar{\mathbf{v}}} + \bar{\mathbf{M}} \cdot {^{\mathcal{N}}\boldsymbol{\omega}^{\mathcal{R}}} \tag{5-799}$$

A more useful form of the work-energy theorem for a rigid body was then derived as

$$\frac{d}{dt}\left({^{\mathcal{N}}T}\right) = \sum_{i=1}^{n} \mathbf{F}_i \cdot {^{\mathcal{N}}\mathbf{v}_i} + \boldsymbol{\tau} \cdot {^{\mathcal{N}}\boldsymbol{\omega}^{\mathcal{R}}} \tag{5-809}$$

Using the work-energy theorem for a rigid body, the principle of work and energy for a rigid body was derived as

$$^{\mathcal{N}}T_2 - {^{\mathcal{N}}T_1} = {^{\mathcal{N}}W_{12}^{F}} + {^{\mathcal{N}}W_{12}^{\tau}} \tag{5-814}$$

where ${^{\mathcal{N}}W_{12}^{F}}$ is the work done by all of the forces acting on the body while ${^{\mathcal{N}}W_{12}^{\tau}}$ is the work done by all of the pure torques acting on the body over a time interval $t \in [t_1, t_2]$. Then, using the work-energy theorem and the concept of a conservative force, an alternate form of the work-energy theorem was derived as

$$\frac{d}{dt}\left({^{\mathcal{N}}E}\right) = \sum_{i=1}^{q} \mathbf{F}_i^{nc} \cdot {^{\mathcal{N}}\mathbf{v}_i^{nc}} + \boldsymbol{\tau}^{nc} \cdot {^{\mathcal{N}}\boldsymbol{\omega}^{\mathcal{R}}} \tag{5-823}$$

where q is the number of nonconservative forces acting on the rigid body. Finally, using the alternate form of the work-energy theorem for a rigid body, the alternate form of the principle of work and energy for a rigid body was derived as

$$^{\mathcal{N}}E_2 - {^{\mathcal{N}}E_1} = {^{\mathcal{N}}W_{12}^{F^{nc}}} + {^{\mathcal{N}}W_{12}^{\tau^{nc}}} \tag{5-831}$$

where ${^{\mathcal{N}}W_{12}^{F^{nc}}}$ is the work done by all of the nonconservative forces acting on the body while ${^{\mathcal{N}}W_{12}^{\tau^{nc}}}$ is the work done by all of the nonconservative pure torques acting on the body over a time interval $t \in [t_1, t_2]$.

The next topics covered in this chapter were the principles of linear impulse and linear momentum and angular impulse and angular momentum for a rigid body. The principle of linear impulse and linear momentum for a rigid body was stated as

$$\hat{\mathbf{F}} = {}^{\mathcal{N}}\mathbf{G}_2 - {}^{\mathcal{N}}\mathbf{G}_1 \tag{5-949}$$

where ${}^{\mathcal{N}}\mathbf{G} = m{}^{\mathcal{N}}\bar{\mathbf{v}}$ is the linear momentum of the rigid body in the inertial reference frame $\mathcal{N}$ and $\hat{\mathbf{F}}$ is the resultant external linear impulse applied to the rigid body. Then, the principle of angular impulse and momentum was derived as

$$\hat{\mathbf{M}}_Q - \int_{t_1}^{t_2} (\bar{\mathbf{r}} - \mathbf{r}_Q) \times m{}^{\mathcal{N}}\mathbf{a}_Q dt = {}^{\mathcal{N}}\mathbf{H}_Q(t_2) - {}^{\mathcal{N}}\mathbf{H}_Q(t_1) \tag{5-954}$$

It was then discussed that, for the case where the reference point is a point O fixed in an inertial reference frame $\mathcal{N}$, the principle of angular impulse and angular momentum simplifies to

$$\hat{\mathbf{M}}_O = {}^{\mathcal{N}}\mathbf{H}_O(t_2) - {}^{\mathcal{N}}\mathbf{H}_O(t_1) \tag{5-955}$$

where $\hat{\mathbf{M}}_O$ is the resultant external angular impulse applied to the rigid body relative to the point O fixed in the inertial reference frame $\mathcal{N}$. Similarly, for the case where the reference point is the center of mass of the rigid body, we have

$$\hat{\mathbf{M}} = {}^{\mathcal{N}}\bar{\mathbf{H}}(t_2) - {}^{\mathcal{N}}\bar{\mathbf{H}}(t_1) \tag{5-956}$$

where $\hat{\mathbf{M}}$ is the resultant external angular impulse applied to the rigid body relative to the center of mass of the rigid body.

The last topic that was covered in this chapter was the collision between two rigid bodies. In particular, it was assumed that the coefficient of restitution condition is applied at the point of contact P between the rigid bodies during the collision, i.e.,

$$e = \frac{\left({}^{\mathcal{N}}\mathbf{v}_P^{\mathcal{R}_2}\right)' \cdot \mathbf{n} - \left({}^{\mathcal{N}}\mathbf{v}_P^{\mathcal{R}_1}\right)' \cdot \mathbf{n}}{{}^{\mathcal{N}}\mathbf{v}_P^{\mathcal{R}_1} \cdot \mathbf{n} - {}^{\mathcal{N}}\mathbf{v}_P^{\mathcal{R}_2} \cdot \mathbf{n}} \tag{5-1029}$$

In addition, it was shown that it is necessary to apply linear impulse and momentum and angular impulse and momentum to each rigid body in order to obtain the velocity of the center of mass and angular velocity of each body immediately after impact.

Problems for Chapter 5

5–1 A homogeneous circular cylinder of mass m and radius r is at rest atop a massless slab as shown in Fig. P5-1. The slab initially lies stationary on a flat horizontal surface when it is suddenly pulled with a very large velocity to the right and removed from under the cylinder. Assuming that the removal of the slab takes place in a time t, that the coefficient of dynamic friction between all surfaces is μ, that gravity acts downward, and that sliding occurs throughout the time when the slab is in contact with the cylinder, determine the following quantities at the instant that the slab is removed: (a) the velocity of the center of mass of the cylinder and (b) the angular velocity of the cylinder.

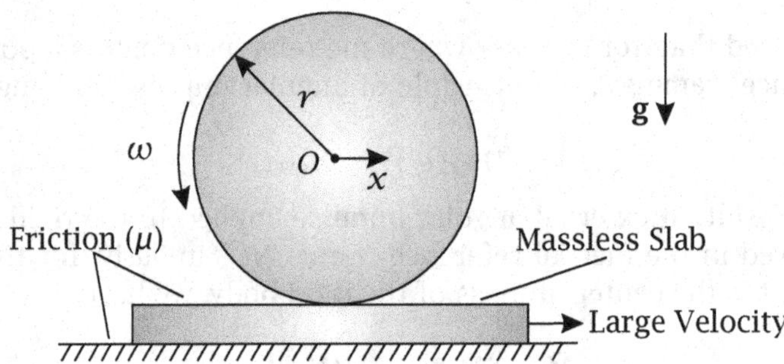

Figure P5-1

5–2 A collar of mass m_1 is attached to a rod of mass m_2 and length l as shown in Fig. P5-2. The collar slides without friction along a horizontal track while the rod is free to rotate about the pivot point Q located at the collar. Knowing that the angle θ describes the orientation of the rod with the vertical, that x is the horizontal position of the cart, and that gravity acts downward, determine a system of two differential equations for the collar and the rod in terms of x and θ.

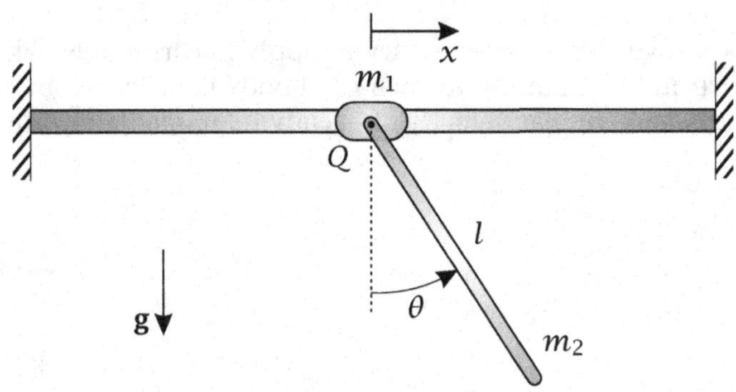

Figure P5-2

5–3 A bulldozer pushes a boulder of mass m with a known force **P** up a hill inclined at a constant inclination angle β as shown in Fig. P5-3. For simplicity, the boulder is modeled as a uniform sphere of mass m and radius r. Assuming that the boulder rolls without slip along the surface of the hill, that the coefficient of dynamic Coulomb friction between the bulldozer and the boulder is μ, that the force **P** is along the direction of the incline and passes through the center of mass of the boulder, and that gravity acts downward, determine the differential equation of motion of the boulder in terms of the variable x.

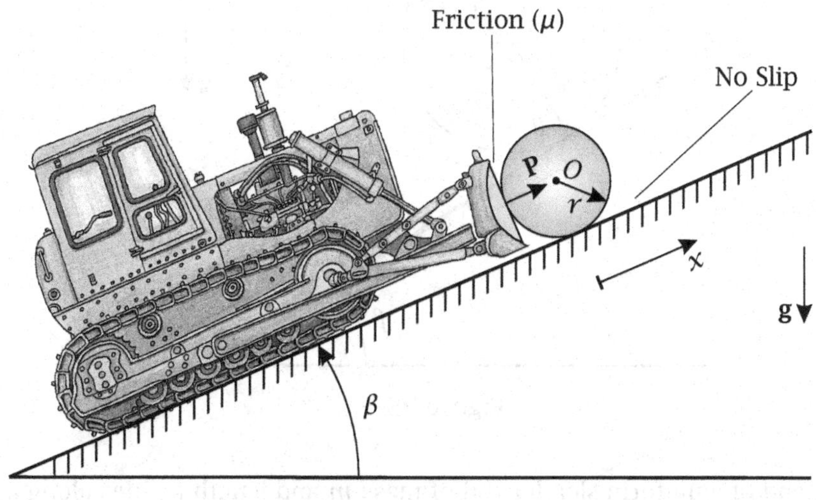

Figure P5-3

5–4 A uniform slender rod of mass m and length l pivots about its center at the fixed point O as shown in Fig. P5-4. A torsional spring with spring constant K is attached to the rod at the pivot point. The rod is initially at rest and the spring is uncoiled when a linear impulse $\hat{\mathbf{P}}$ is applied transversely at the lower end of the rod. Determine (a) the angular velocity of the rod immediately after the impulse $\hat{\mathbf{P}}$ is applied and (b) the maximum angle $\theta_{\max}$ attained by the rod after the impulse is applied.

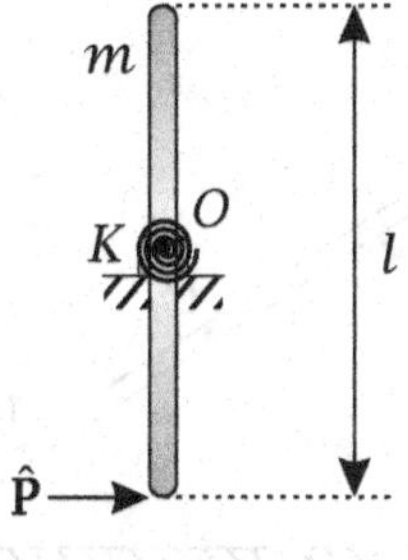

Figure P5-4

5–5 A homogeneous cylinder of mass m and radius r moves along a surface inclined at a constant inclination angle β as shown in Fig. P5-5. The surface of the incline is composed of a frictionless segment of known length x between points A and B and a segment with a coefficient of friction μ from point B onwards. Knowing that the cylinder is released from rest at point A and that gravity acts vertically downward, determine (a) the velocity of the center of mass and the angular velocity when the disk reaches point B, (b) the time (measured from point B) when sliding stops and rolling begins, and (c) the velocity of the center of mass and the angular velocity of the disk when sliding stops and rolling begins.

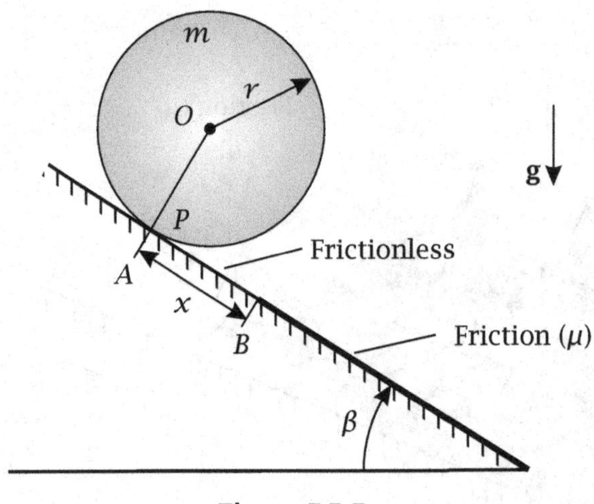

Figure P5-5

5–6 One end of a uniform slender rod of mass m and length l slides along a frictionless vertical surface while the other end of the rod slides along a frictionless horizontal surface as shown in Fig. P5-6. The angle θ formed by the rod is measured from the vertical. Knowing that gravity acts vertically downward, determine (a) the differential equation of motion for the rod while it maintains contact with both surfaces and (b) the value of the angle θ at which the rod loses contact with the vertical surface. In obtaining your answers, you may assume that the initial conditions are $\theta(t = 0) = 0$ and $\dot{\theta}(t = 0) = 0$.

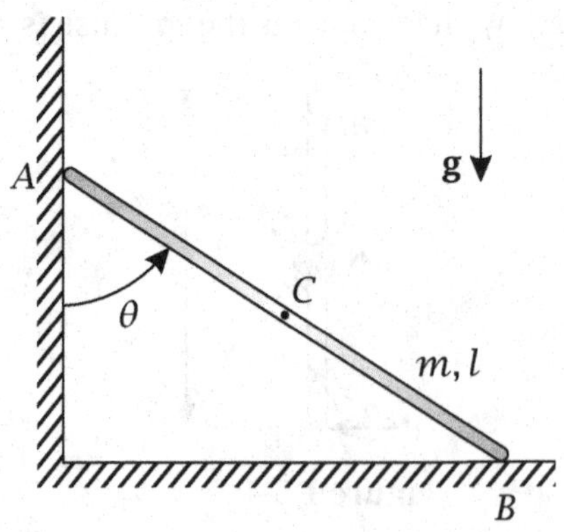

Figure P5-6

5–7 A homogeneous semi-circular cylinder of mass m and radius r rolls without slip along a horizontal surface as shown in Fig. P5-7. The center of mass of the cylinder is located at point C while point O is located at the center of the main diameter of the cylinder. Knowing that the angle θ is measured from the vertical and that gravity acts downward, determine the differential equation of motion for the cylinder. In obtaining your answers, you may assume that $4r/(3\pi) \approx 0.42r$.

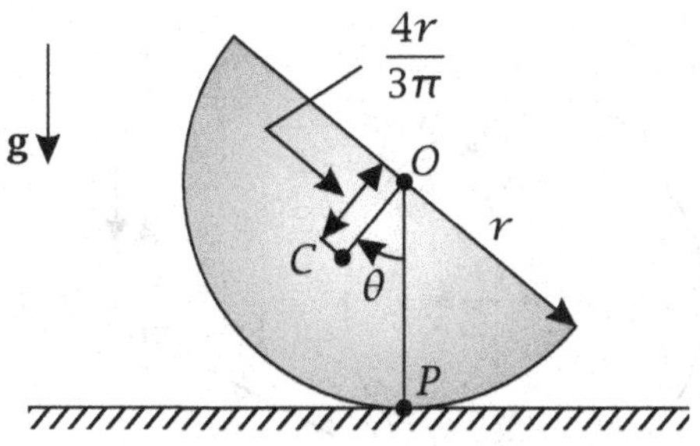

Figure P5-7

5–8 A homogeneous sphere of radius r rolls without slip along a fixed spherical surface of radius R as shown in Fig. P5-8. The angle θ measures the amount by which the sphere has rotated from the vertical direction. Knowing that gravity acts downward and assuming the initial conditions $\theta(0) = 0$ and $\dot{\theta}(0) = 0$, determine the differential equation of motion while the sphere maintains contact with the spherical surface.

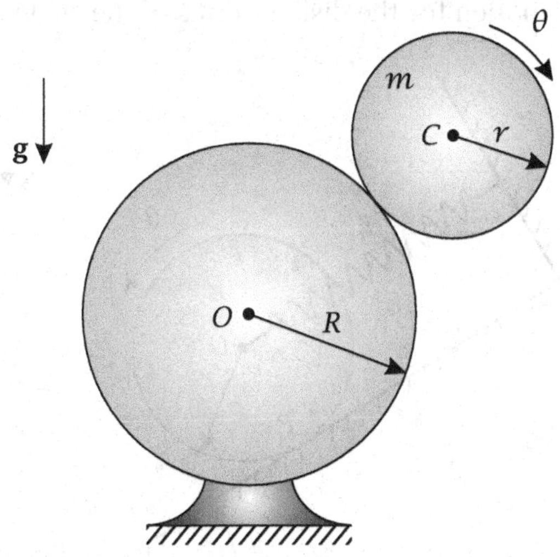

Figure P5-8

5–9 A uniform cylinder of mass m and radius r is set in motion by placing it in contact with a horizontal surface as shown in Fig. P5-9. The coefficient of sliding friction between the cylinder and the surface is μ. Knowing that x is the horizontal position of the center of mass of the cylinder and that $\boldsymbol{\omega}$ is the angular velocity of the cylinder (with corresponding rotation rate ω), determine the time at which sliding stops and rolling begins for the following four sets of initial conditions:

(a) $\dot{x}(t=0) = \dot{x}_0 > 0,\ \omega(t=0) = \omega_0 = 0$

(b) $\dot{x}(t=0) = 0,\ \omega(t=0) = \omega_0 > 0$

(c) $\dot{x}(t=0) = \dot{x}_0$, $\omega(t=0) = \omega_0 > 0$, $\dot{x}_0 - r\omega_0 > 0$

(d) $\dot{x}(t=0) = \dot{x}_0$, $\omega(t=0) = \omega_0 > 0$, $\dot{x}_0 - r\omega_0 < 0$

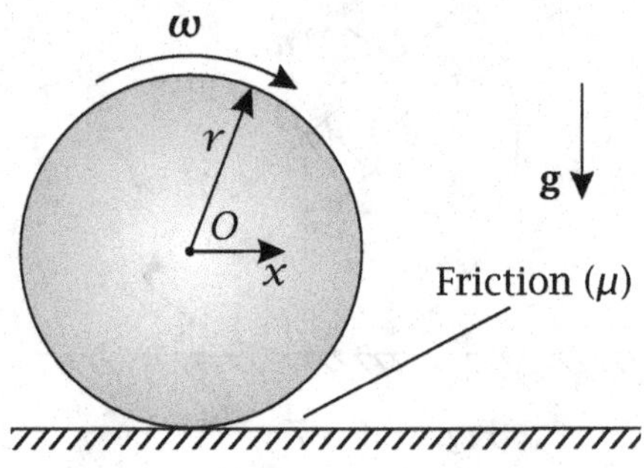

Figure P5-9

5–10 A uniform circular disk of mass m and radius r rolls without slip along a plane inclined at a constant angle β with horizontal as shown in Fig. P5-10. Attached at the center of the disk is a linear spring with spring constant K. Knowing that the spring is unstretched when the angle θ is zero and that gravity acts downward, determine the differential equation of motion for the disk in terms of the angle θ.

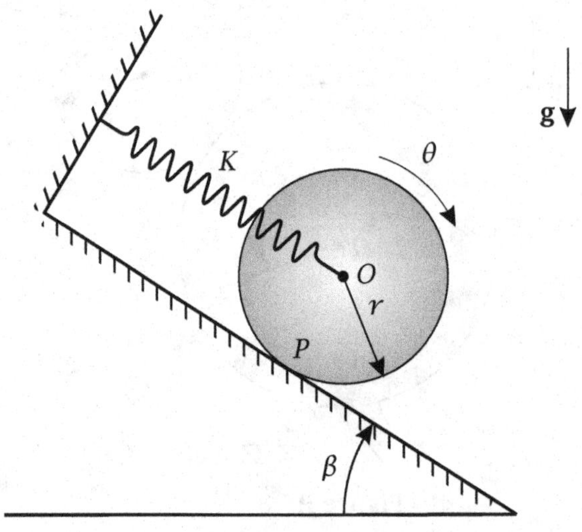

Figure P5-10

5–11 A slender rod of mass m and length l is suspended from a massless collar at point O as shown in Fig. P5-11. The collar in turn slides without friction along a horizontal track. The position of the collar is denoted as x while the angle formed by the rod with the vertical is denoted θ. Given that a *known* horizontal force **P** is applied to the rod at the point O and that gravity acts downward, determine a system of two differential equations in terms of x and θ that describes the motion of the rod.

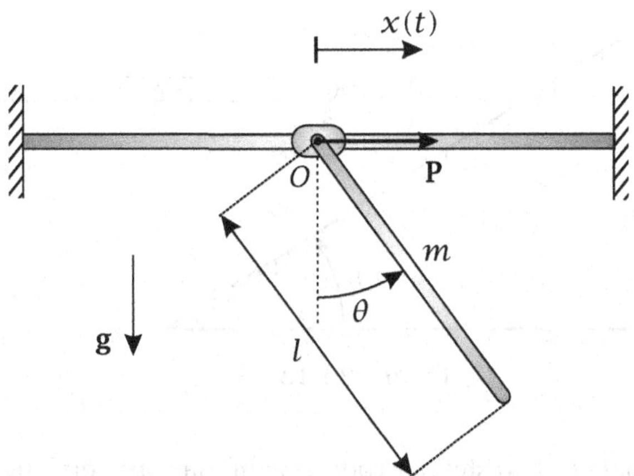

Figure P5-11

5–12 A uniform slender rod of mass m and length $2l$ slides without friction along a fixed circular track of radius R as shown in Fig. P5-12. Knowing that θ is the angle from the vertical to the center of the rod and that gravity acts downward, determine the differential equation of motion for the rod.

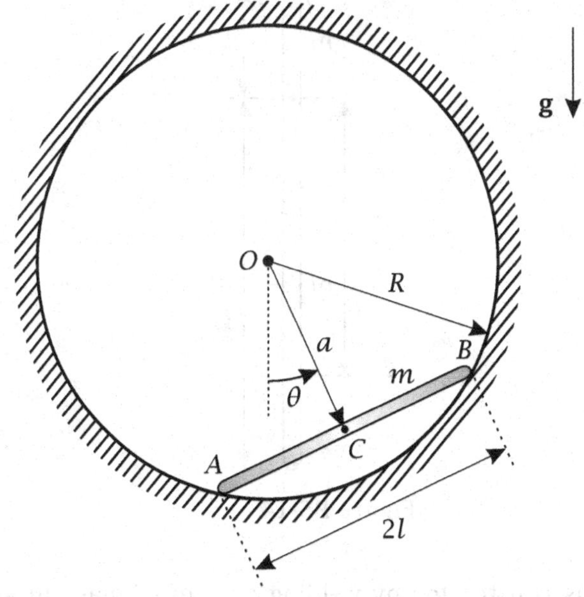

Figure P5-12

5–13 A uniform slender rod of mass m and length l is translating horizontally with a velocity $\mathbf{v}_0$ and is not rotating when it strikes a fixed surface inclined at an angle β with the horizontal as shown in Fig. P5-13. Assuming that the coefficient of restitution between the rod and the incline is e, determine the following quantities relative to the incline at the instant after the impact has occurred: (a) the velocity of the center of mass of the rod and (b) the angular velocity of the rod.

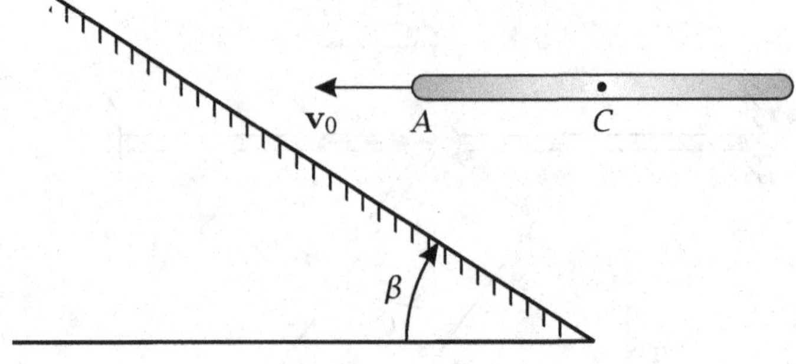

Figure P5-13

5–14 A system consists of two slender rods, one of mass m and length l and the other of mass m and length $3l$, attached to form a double pendulum as shown in Fig. P5-14. The rods are initially motionless when the lower rod is struck by a horizontal impulse $\hat{\mathbf{P}}$ at a distance x from the hinge point connecting the two rods. Assuming no gravity, determine the value of x for which the two rods will rotate with the same angular velocity (relative to the ground) the instant after the impulse is applied.

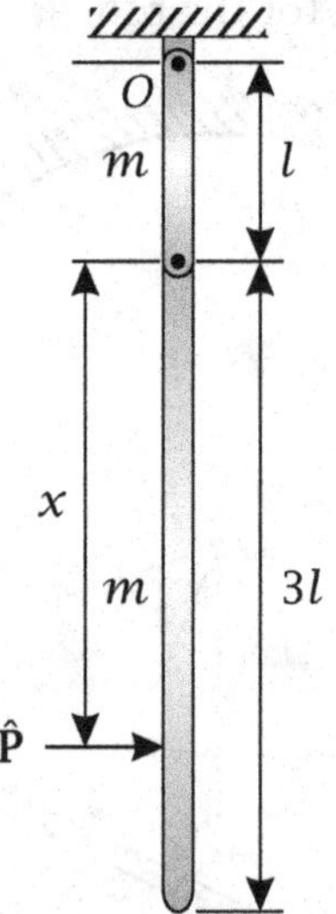

Figure P5-14

5–15 A rigid body is constructed by welding an arm of mass m and length l to a circular disk of mass M and radius r as shown in Fig. P5-15. Knowing that the disk rolls without slip along a horizontal surface, that θ describes the orientation of the arm relative to the vertically downward direction, and that gravity acts downward, determine the differential equation of motion for the rigid body.

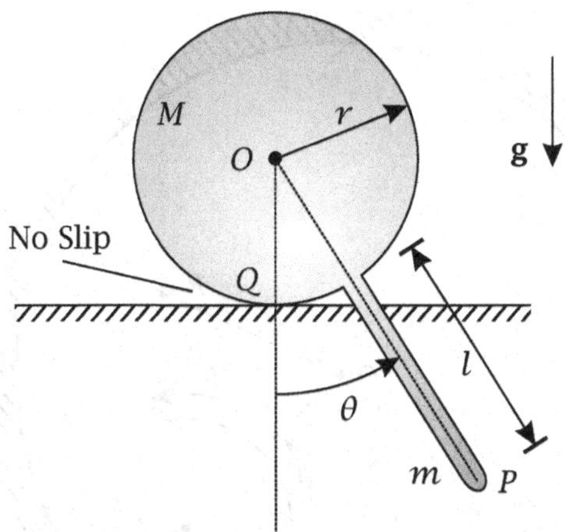

Figure P5-15

5–16 A homogeneous sphere of mass m and radius r rolls without slip along a horizontal surface as shown in Fig. P5-16. The variable x describes the position of the center of mass of the sphere. A horizontal force **P** is then applied at a height h from the surface such that **P** lies in the vertical plane that contains the center of mass of the sphere. Knowing that gravity acts downward, determine the differential equation of motion for the sphere in terms of the variable x.

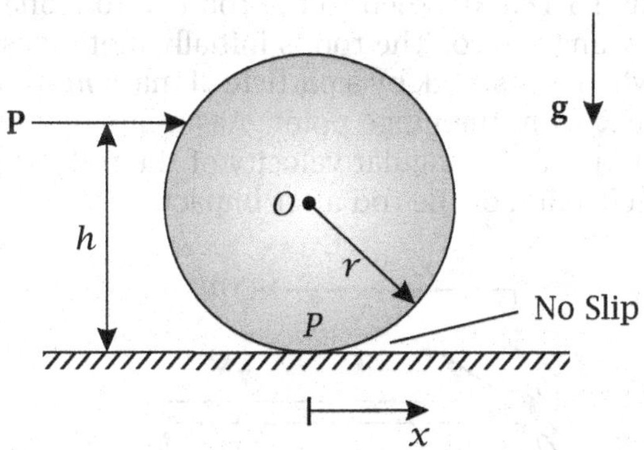

Figure P5-16

5–17 A uniform disk of mass m and radius r rolls without slip along the inside of a fixed circular track of radius R as shown in Fig. P5-17. The angles θ and ϕ measure the position of the center of the disk and the angle of rotation of the disk, respectively, relative to the vertically downward direction. Knowing that the angles θ and ϕ are simultaneously zero and that gravity acts downward, determine the differential equation of motion for the disk *in terms of the angle* θ.

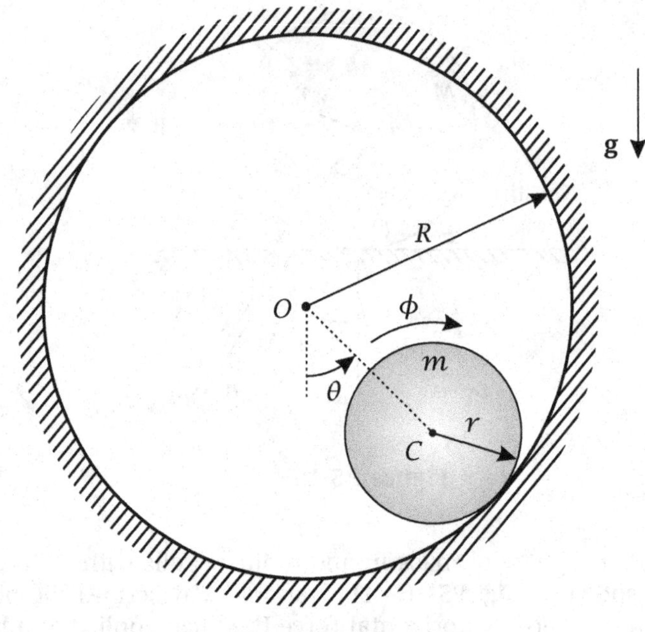

Figure P5-17

5–18 A slender uniform rod of mass M and length L is hinged at one of its ends at the point O as shown in Fig. P5-18. Attached to the rod is a torsional spring with spring constant K and uncoiled angle zero. The rod is initially motionless at the equilibrium position of the spring when it is struck by a particle of mass m moving with velocity $\mathbf{v}_0$ at a horizontal distance x from the hinge point. Assuming no gravity and a perfectly inelastic impact, determine (a) the angular velocity of the rod the instant after impact and (b) the maximum deflection of the rod after impact.

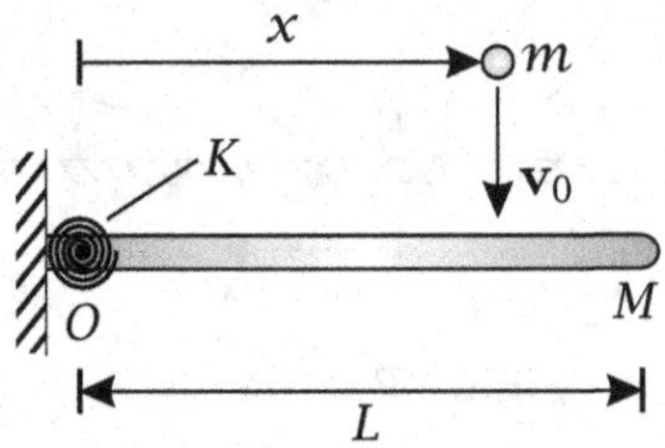

Figure P5-18

5–19 A triangular wedge of mass m and wedge angle β slides without friction along a horizontal surface. A circular disk of mass M and radius R rolls without slip along the surface of the wedge as shown in Fig. P5-19. Knowing that x describes the displacement of the wedge, that θ describes the orientation of the disk relative to the wedge, and that gravity acts vertically downward, determine a system of two differential equations for the wedge and disk in terms of the variables x and θ.

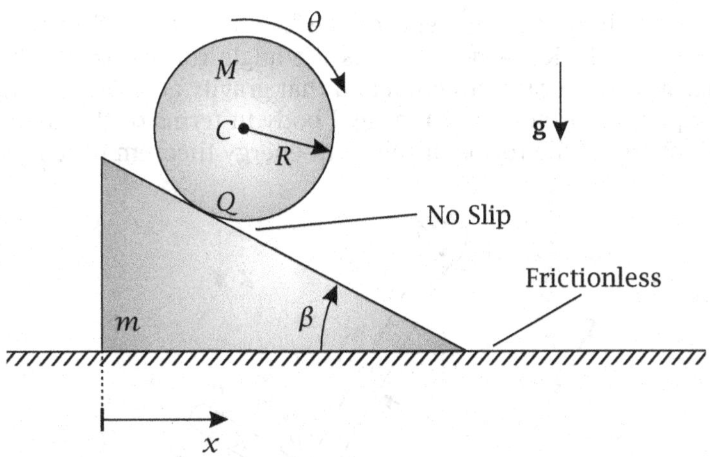

Figure P5-19

5–20 A uniform circular disk of mass m and radius r rolls without slip along a plane inclined at a constant angle β with horizontal as shown in Fig. P5-20. Attached to the disk at the point A (where A lies in the direction of PO) is a linear spring with spring constant K and a nonlinear damper with damping constant c. The damper exerts a force of the form

$$\mathbf{F}_d = -c\|\mathbf{v}_A\|^3 \frac{\mathbf{v}_A}{\|\mathbf{v}_A\|}$$

where $\mathbf{v}_A$ is the velocity of point A as viewed by an observer fixed to the ground. Knowing that the spring is unstretched when the angle θ is zero and that gravity acts downward, determine the differential equation of motion for the disk.

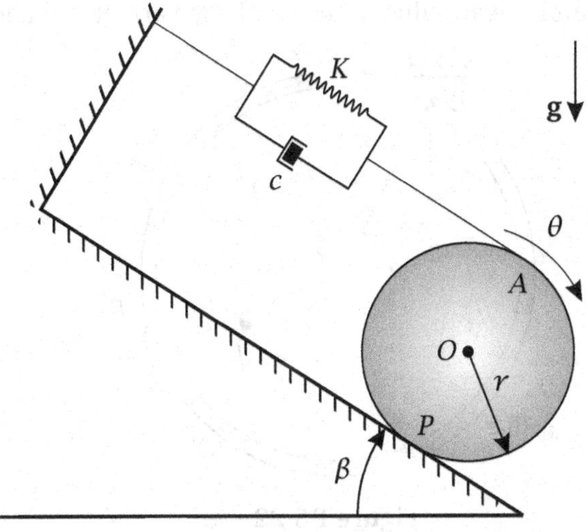

Figure P5-20

5–21 A rigid body of mass $2m$ is constructed by welding together a rod of mass m and length l to a rod of mass m and length $2a$ to form the shape of the letter

"T". The rigid body is hinged at one end of the height of the "T" to the fixed point O as shown in Fig. P5-21. Knowing that θ is the angle formed by the height of the "T" with the vertically downward direction and that gravity acts downward, determine the differential equation of motion of the rigid body in terms of the angle θ using (a) Euler's laws and (b) one of the forms of the work-energy theorem for a rigid body.

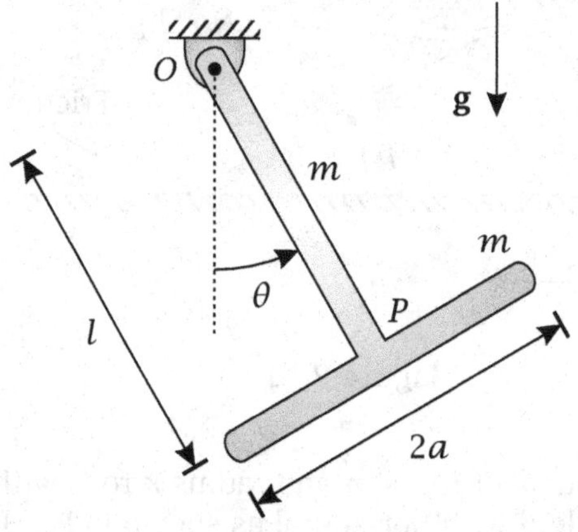

Figure P5-21

5–22 A collar of mass m_1 slides without friction along a circular annulus of mass m_2 and radius R as shown in Fig. P5-22. The annulus rotates without friction about a point O located on its diameter. Knowing that θ is the angle between the vertically downward direction and the direction from O to C (where C is the center of the annulus), that ϕ is the angle that describes the location of the collar relative to the direction OC, and that gravity acts downward, determine a system of two differential equations of motion for the collar and the annulus in terms of the variables θ and ϕ.

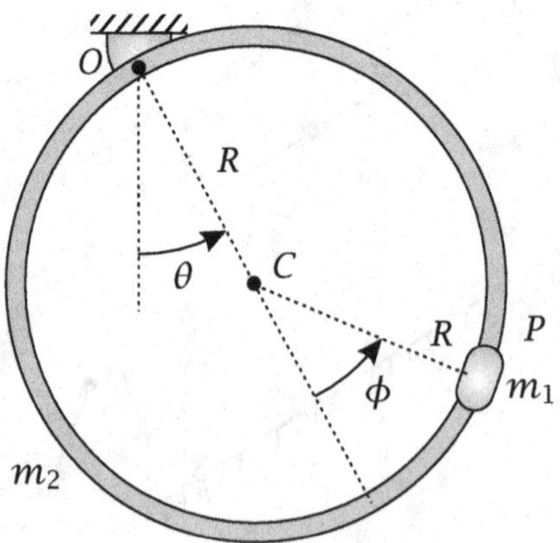

Figure P5-22

5–23 A slender uniform rod of mass m and length l is hinged at one of its ends at point Q to a massless collar. The collar simultaneously slides with known displacement $y(t)$ along a vertical track and rotates with constant angular velocity Ω as shown in Fig. P5-23. Knowing that θ is the angle between the track and the rod and that gravity acts downward, determine the differential equation of motion for the rod in terms of the angle θ.

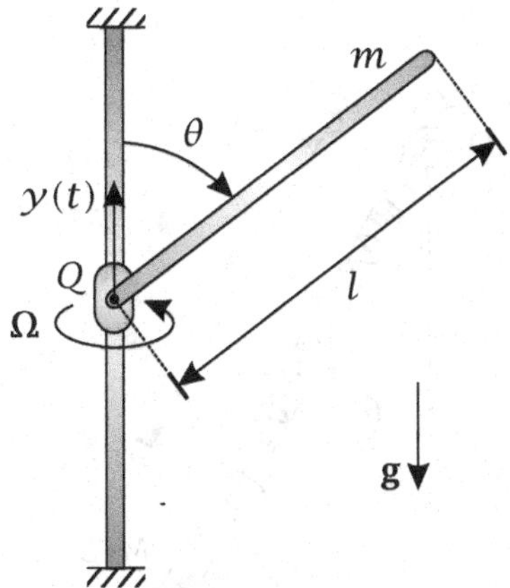

Figure P5-23

5–24 A block of mass M, with a semicircular section of radius R cut out of its center, slides without friction along a horizontal surface. A homogeneous circular disk of radius r rolls without slip along the circular surface cut out of the block as shown in Fig. P5-24. Knowing that x describes the displacement of the block, that θ describes the location of the center of the disk (point P) relative to the center of the block (point O), that ϕ describes the rotation of the disk relative to the direction OP, and that gravity acts downward, determine a system of two differential equations of motion for the block and disk in terms of the variables x and θ.

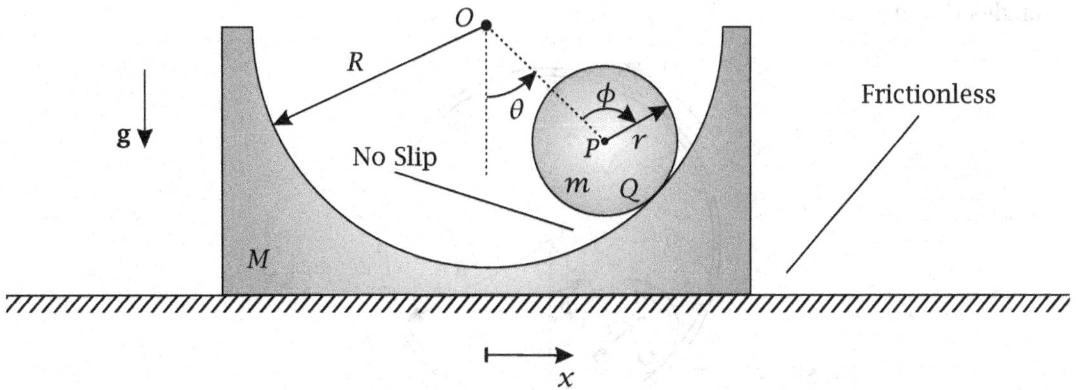

Figure P5-24

5–25 A particle of mass m slides without friction along a rod of mass M and length L as shown in Fig. P5-25. The particle is attached to a linear spring with spring constant K and unstretched length r_0 while the rod is free to pivot about one of its ends at the fixed hinge at point O. Knowing that the distance r describes the position of the particle relative to point O, that the angle θ describes the orientation of the rod relative to the downward direction, and that gravity acts downward, determine a system of two differential equations of motion for the particle and the rod in terms of the variables r and θ.

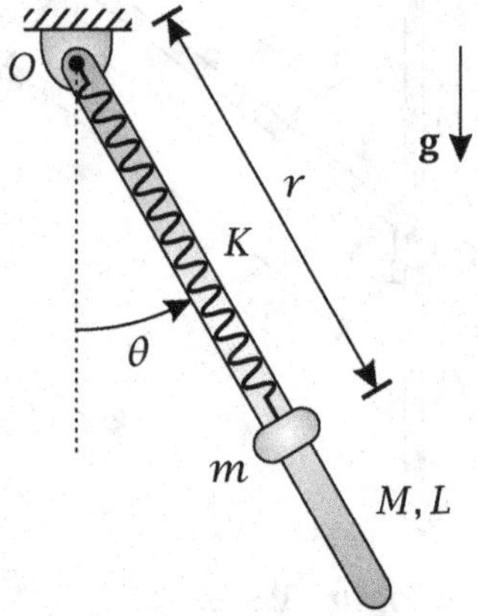

Figure P5-25

5–26 A system consists of an annulus of mass m and radius R and a uniform circular disk of mass m and radius r as shown in Fig. P5-26. The annulus rolls without slip along a fixed horizontal surface while the disk rolls without slip along the inner surface of the annulus. Knowing that the angle θ describes the rotation of the annulus relative to the ground, that the angle ϕ describes the location of the center of mass of the disk relative to the center of the annulus, and that gravity acts downward, determine a system of two differential equations for the annulus and the disk in terms of the angles θ and ϕ.

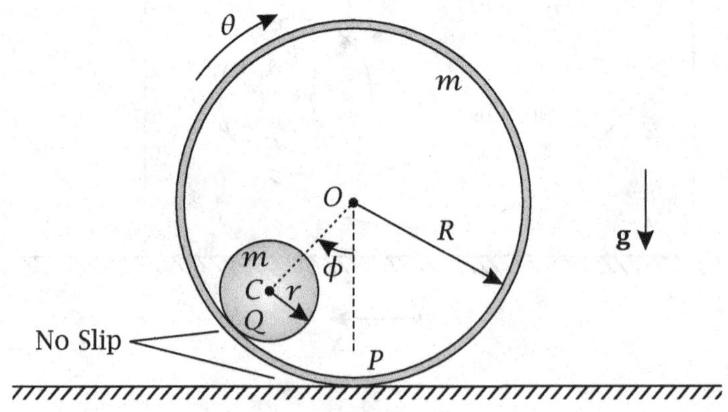

Figure P5-26

5–27 A rigid body consists of a massless disk of radius r with a particle embedded in it a distance l from the center of the disk as shown in Fig. P5-27. The disk rolls without slip along a fixed circular track of radius R and center at point O. The angle θ

describes the orientation of the direction OC (where C is the geometric center of the disk) with the vertically downward direction while the angle ϕ describes location of the particle relative to the direction OC. Knowing that gravity acts downward, determine the differential equation of motion for the disk in terms of the angle θ.

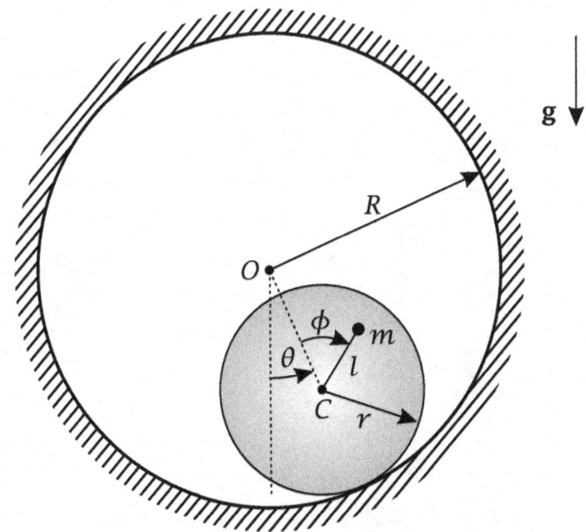

Figure P5-27

Appendix A

Principal-Axis Moments of Inertia of Homogeneous Bodies

Table A–1 Circular cylinder of radius r and height h.

$$I_{xx} = I_{yy} = \frac{m(3r^2 + h^2)}{12}$$

$$I_{zz} = \frac{mr^2}{2}$$

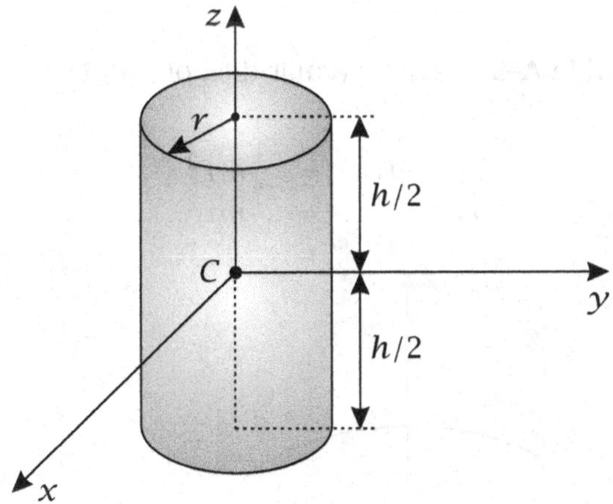

Table A–2 Sphere of radius r.

$$I_{xx} = I_{yy} = I_{zz} = \frac{2}{5}mr^2$$

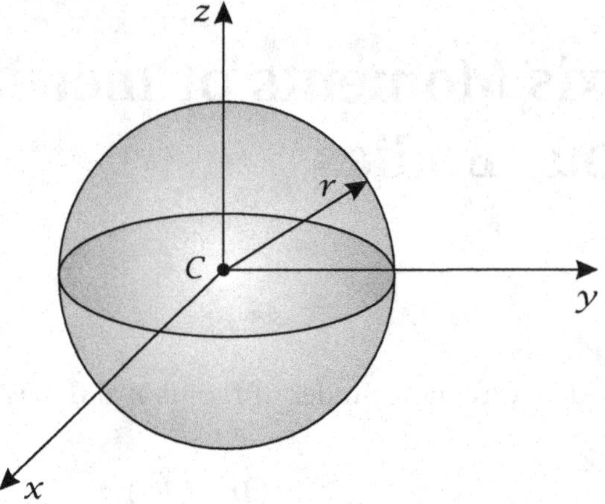

Table A–3 Thin circular disk of radius r.

$$\begin{aligned} I_{xx} = I_{yy} &= \tfrac{1}{4}mr^2 \\ I_{zz} &= \tfrac{1}{2}mr^2 \end{aligned}$$

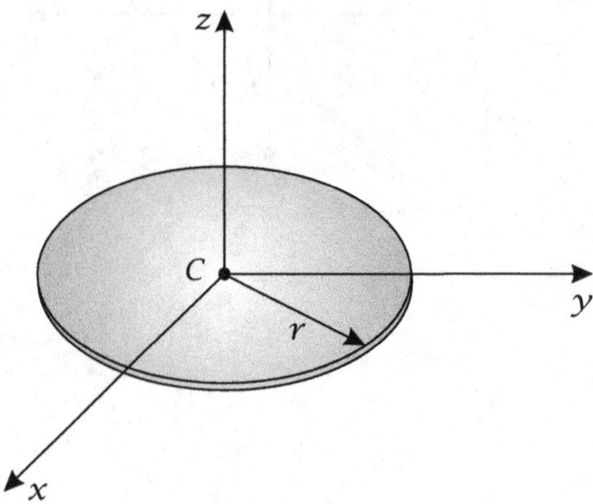

Table A–4 Slender rod of length l.

$$
\begin{aligned}
I_{xx} &= 0 \\
I_{yy} = I_{zz} &= \frac{1}{12} m l^2
\end{aligned}
$$

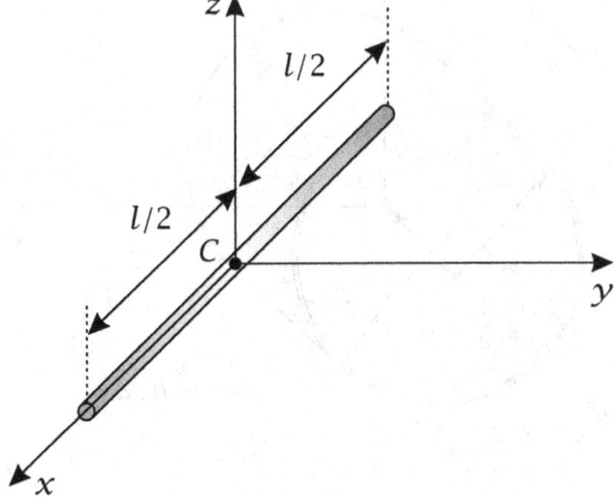

Table A–5 Hemisphere of radius r.

$$
\begin{aligned}
I_{xx} = I_{yy} &= \tfrac{83}{320} m r^2 \\
I_{zz} &= \tfrac{2}{5} m r^2
\end{aligned}
$$

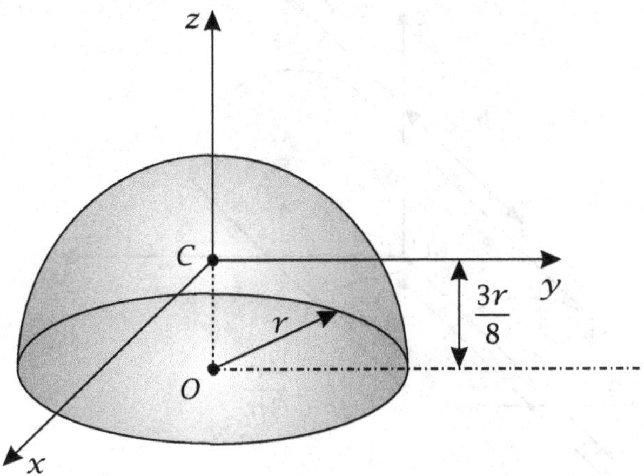

Table A–6 Thin cylindrical shell of radius r and Length h.

$$\begin{aligned} I_{xx} &= mr^2 \\ I_{yy} = I_{zz} &= \tfrac{1}{12}m(6r^2 + h^2) \end{aligned}$$

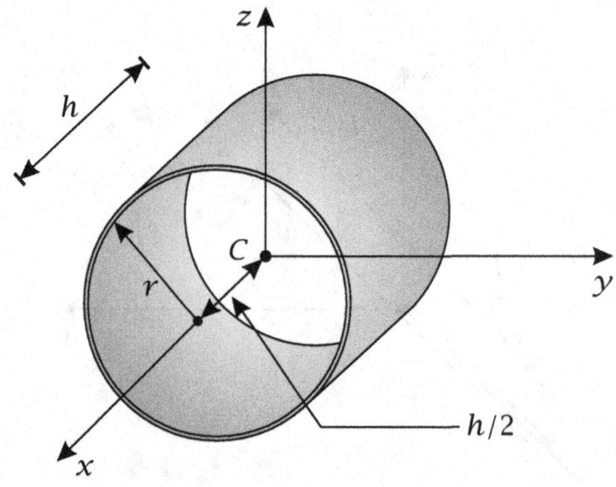

Table A–7 Semicircular cylinder of radius r and length h.

$$\begin{aligned} I_{xx} &= 0.320mr^2\,;\ I_{x'x'} = \tfrac{1}{2}mr^2 \\ I_{yy} &= 0.0699mr^2 + \tfrac{1}{12}mh^2 \\ I_{zz} &= \tfrac{1}{12}m(3r^2 + h^2) \end{aligned}$$

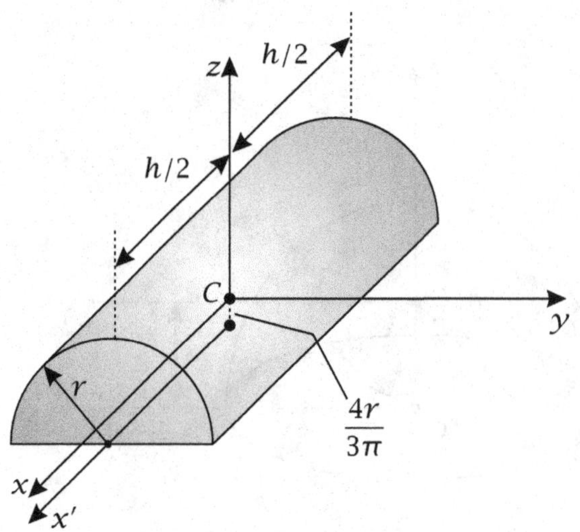

Table A–8 Ellipsoid with axes of length a, b, and c.

$$I_{xx} = \frac{m}{5}(b^2 + c^2)$$

$$I_{yy} = \frac{m}{5}(a^2 + c^2)$$

$$I_{zz} = \frac{m}{5}(a^2 + b^2)$$

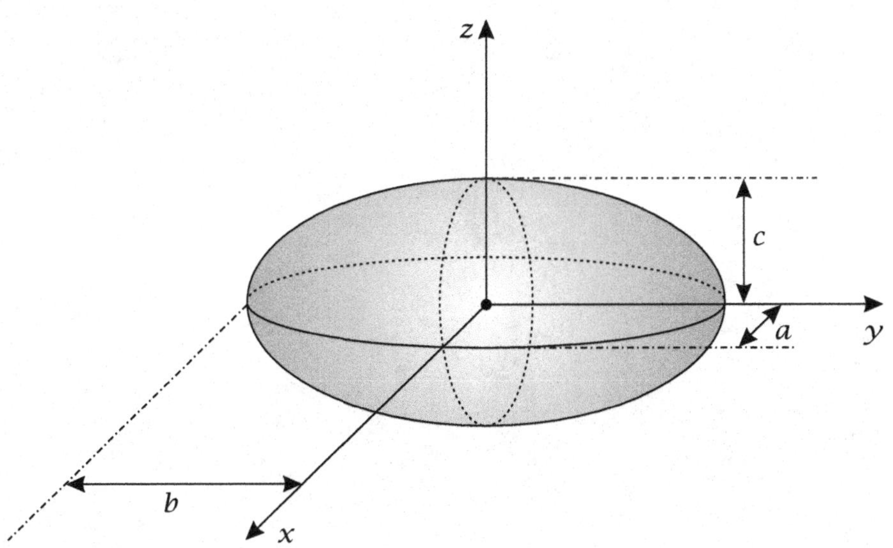

Appendix B

Identities, Derivatives, Integrals, and Gradient

B.1 Identities

$$\sin^2 u + \cos^2 u = 1$$
$$\sin(u \pm v) = \sin u \cos v \pm \sin v \cos u$$
$$\cos(u \pm v) = \cos u \cos v \mp \sin v \sin u$$
$$\sin 2u = 2 \sin u \cos u$$
$$\cos 2u = \cos^2 u - \sin^2 u$$
$$\cos^2 u = \frac{1}{2} + \frac{1}{2}\cos 2u$$
$$\sin^2 u = \frac{1}{2} - \frac{1}{2}\cos 2u$$
$$1 + \tan^2 u = \sec^2 u$$
$$1 + \cot^2 u = \csc^2 u$$

B.2 Derivatives and Integrals

B.2.1 Derivatives

$$\frac{d}{dt}(uv) = \frac{du}{dt}v + u\frac{dv}{dt}$$
$$\frac{d}{dt}\left(\frac{u}{v}\right) = \frac{\frac{du}{dt}v - u\frac{dv}{dt}}{v^2}$$
$$\frac{d}{dt}(u(v)) = \frac{du(v)}{dv}\frac{du}{dt}$$
$$\frac{d}{dt}(u\dot{u}) = \dot{u}^2 + u\ddot{u}$$

$$\frac{d}{dt}\left(\frac{1}{2}\dot{u}^2\right) = \dot{u}\ddot{u}$$

$$\frac{d}{dt}(u^n) = nu^{n-1}\dot{u}$$

$$\frac{d}{dt}(e^{-au}) = -a\dot{u}e^{-au}$$

$$\frac{d}{dt}(\sin au) = a\dot{u}\cos au$$

$$\frac{d}{dt}(\cos au) = -a\dot{u}\sin au$$

$$\frac{d}{dt}(\sec au) = a\dot{u}\sec au\tan au$$

$$\frac{d}{dt}(\csc au) = -a\dot{u}\csc au\cot au$$

B.2.2 Integrals

$$\int u dv = uv - \int v du$$

$$\int u^n du = \frac{u^{n+1}}{n+1} + C$$

$$\int \cos au du = \frac{\sin au}{a} + C$$

$$\int \sin au du = -\frac{\cos au}{a} + C$$

$$\int a^u du = \frac{a^u}{\ln a} + C$$

$$\int \ln au du = u\ln au - u + C$$

$$\int \frac{u}{u^2 + a^2} du = \frac{1}{2}\ln|u^2 + a^2| + C$$

B.3 Gradient of a Scalar Function

B.3.1 Gradient in Cartesian Coordinates

$$f = f(x, y, z) = 0$$

$$\nabla f = \frac{\partial f}{\partial x}\mathbf{e}_x + \frac{\partial f}{\partial y}\mathbf{e}_y + \frac{\partial f}{\partial z}\mathbf{e}_z$$

B.3.2 Gradient in Cylindrical Coordinates

$$f = f(r, \theta, z) = 0$$

$$\nabla f = \frac{\partial f}{\partial r}\mathbf{e}_r + \frac{1}{r}\frac{\partial f}{\partial \theta}\mathbf{e}_\theta + \frac{\partial f}{\partial z}\mathbf{e}_z$$

B.3.3 Gradient in Spherical Coordinates

$$f = f(r, \phi, \theta) = 0$$

$$\nabla f = \frac{\partial f}{\partial r}\mathbf{e}_r + \frac{1}{r}\frac{\partial f}{\partial \phi}\mathbf{e}_\phi + \frac{1}{r \sin\phi}\frac{\partial f}{\partial \theta}\mathbf{e}_\theta$$

Appendix C

Answers to Selected Problems

Chapter 2

2–1

$$^{\mathcal{F}}\mathbf{v} = v_0\mathbf{e}_x + \Omega r\mathbf{e}_y$$

$$^{\mathcal{F}}\mathbf{a} = -\Omega^2 r\mathbf{e}_x + 2\Omega v_0\mathbf{e}_y$$

2–2

$$^{\mathcal{F}}\mathbf{v} = \dot{r}\mathbf{e}_r + r(\Omega + \dot{\theta})\mathbf{e}_\theta$$

$$^{\mathcal{F}}\mathbf{a} = \left[\ddot{r} - r(\Omega + \dot{\theta})^2\right]\mathbf{e}_r + \left[r\ddot{\theta} + 2\dot{r}(\Omega + \dot{\theta})\right]\mathbf{e}_\theta$$

2–3

$$^{\mathcal{F}}\mathbf{v} = \dot{r}\mathbf{e}_r + r\dot{\theta}\mathbf{e}_\theta + r\Omega\sin\theta\,\mathbf{e}_z$$

$$^{\mathcal{F}}\mathbf{a} = (\ddot{r} - r\dot{\theta}^2 - r\Omega^2\sin^2\theta)\mathbf{e}_r + (2\dot{r}\dot{\theta} + r\ddot{\theta} - r\Omega^2\cos\theta\,\sin\theta)\mathbf{e}_\theta$$
$$+2\Omega(\dot{r}\sin\theta + r\dot{\theta}\cos\theta)\mathbf{e}_z$$

2–4

$$^{\mathcal{F}}\mathbf{v} = \dot{x}\mathbf{e}_x + (2x\dot{x}/a)\mathbf{e}_y - \Omega x\mathbf{e}_z$$

$$^{\mathcal{F}}\mathbf{a} = (\ddot{x} - \Omega^2 x)\mathbf{e}_x + [2(\dot{x}^2 + x\ddot{x})/a]\,\mathbf{e}_y - 2\Omega\dot{x}\mathbf{e}_z$$

2–5

$$^{\mathcal{F}}\mathbf{v} = (\dot{x} - \Omega y)\mathbf{e}_x + (\dot{y} + \Omega x)\mathbf{e}_y + \dot{z}\mathbf{e}_z$$

$$^{\mathcal{F}}\mathbf{a} = (\ddot{x} - 2\Omega\dot{y} - \Omega^2 x)\mathbf{e}_x + (\ddot{y} + 2\Omega\dot{x} - \Omega^2 y)\mathbf{e}_y + \ddot{z}\mathbf{e}_z$$

2–8

(a) $^{\mathcal{F}}s = {}^{\mathcal{F}}s_0 + R\sec\phi(\theta - \theta_0)$

(b) $\mathbf{e}_t = \cos\phi\mathbf{e}_\theta + \sin\phi\mathbf{e}_z$; $\mathbf{e}_n = -\mathbf{e}_r$; $\mathbf{e}_b = -\sin\phi\mathbf{e}_\theta + \cos\phi\mathbf{e}_z$

(c) $^{\mathcal{F}}\mathbf{v} = R\dot{\theta}\sec\phi\mathbf{e}_t$; $^{\mathcal{F}}\mathbf{a} = R\ddot{\theta}\sec\phi\mathbf{e}_t + R\dot{\theta}^2\mathbf{e}_n$

2–9

$$^{\mathcal{F}}\mathbf{v}_P = \frac{l}{2}\dot{\theta}\cos\theta\,\mathbf{e}_x - \frac{l}{2}\dot{\theta}\sin\theta\,\mathbf{e}_y - \frac{l}{2}\Omega\sin\theta\,\mathbf{e}_z$$

$$^{\mathcal{F}}\mathbf{a}_P = -\frac{l\Omega^2}{2}\sin\theta\,\mathbf{e}_x - \frac{v_0^2}{2l\cos^3\theta}\mathbf{e}_y - \left[v_0\Omega + \frac{l\dot{\Omega}\sin\theta}{2}\right]\mathbf{e}_z$$

2–10 $\,^{\mathcal{F}}\mathbf{v} = R\dot{\theta}\mathbf{u}_2 - R\Omega\cos\theta\,\mathbf{u}_3$

$$^{\mathcal{F}}\mathbf{a} = -(R\Omega^2\cos^2\theta + R\dot{\theta}^2)\mathbf{u}_1 + (R\ddot{\theta} + R\Omega^2\cos\theta\,\sin\theta)\mathbf{u}_2 + 2R\Omega\dot{\theta}\sin\theta\,\mathbf{u}_3$$

2–13 (a) $\mathbf{e}_t = \dfrac{-a\mathbf{e}_r + \mathbf{e}_\theta}{\sqrt{1+a^2}};\ \mathbf{e}_n = -\dfrac{\mathbf{e}_r + a\mathbf{e}_\theta}{\sqrt{1+a^2}};\ \mathbf{e}_b = \mathbf{e}_z$

(b) $\kappa = \dfrac{1}{r_0 e^{-a\theta}\sqrt{1+a^2}}$

(c) $^{\mathcal{F}}\mathbf{v} = r_0\dot{\theta}e^{-a\theta}\sqrt{1+a^2}\mathbf{e}_t$

$$^{\mathcal{F}}\mathbf{a} = r_0 e^{-a\theta}\sqrt{1+a^2}(\ddot{\theta} - a\dot{\theta}^2)\mathbf{e}_t + r_0\dot{\theta}^2 e^{-a\theta}\sqrt{1+a^2}\mathbf{e}_n$$

2–17 (a) $\kappa = \dfrac{2+\theta^2}{a(1+\theta^2)^{3/2}}$

(b) $\mathbf{e}_t = \dfrac{\mathbf{e}_r + \theta\mathbf{e}_\theta}{\sqrt{1+\theta^2}};\ \mathbf{e}_n = \dfrac{-\theta\mathbf{e}_r + \mathbf{e}_\theta}{\sqrt{1+\theta^2}};\ \mathbf{e}_b = \mathbf{e}_z$

(c) $^{\mathcal{F}}\mathbf{v} = a\dot{\theta}\sqrt{1+\theta^2}\mathbf{e}_t$

$$^{\mathcal{F}}\mathbf{a} = \frac{a}{\sqrt{1+\theta^2}}\left[\left(\ddot{\theta}(1+\theta^2) + \dot{\theta}^2\theta\right)\mathbf{e}_t + (2+\theta^2)\dot{\theta}^2\mathbf{e}_n\right]$$

2–19 $\,^{\mathcal{F}}\mathbf{v} = R\dot{\phi}\mathbf{e}_\phi + R\dot{\theta}\sin\phi\,\mathbf{e}_\theta$

$$^{\mathcal{F}}\mathbf{a} = -(R\dot{\phi}^2 + R\dot{\theta}^2\sin^2\phi)\mathbf{e}_r + (R\ddot{\phi} - R\dot{\theta}^2\cos\phi\sin\phi)\mathbf{e}_\phi + (R\ddot{\theta}\sin\phi + 2R\dot{\theta}\dot{\phi}\cos\phi)\mathbf{e}_\theta$$

2–21 $\,^{\mathcal{F}}\mathbf{v}_P = \dot{x}\mathbf{E}_x + l\dot{\theta}\mathbf{e}_\theta$

$$^{\mathcal{F}}\mathbf{a}_P = \ddot{x}\mathbf{E}_x - l\dot{\theta}^2\mathbf{e}_r + l\ddot{\theta}\mathbf{e}_\theta$$

Chapter 3

3–1 (a) $\ddot{\theta} + \dfrac{g}{r_0}\cos 2\theta = 0$

(b) $\mathbf{N} = \left[2mr_0\dot{\theta}_0^2 - 3mg\sin 2\theta\right]\mathbf{e}_n$

3–3 (a) $mR(1+\tan^2\phi)\ddot{\theta} + mg\tan\phi = 0$

(b) $^{\mathcal{F}}s = \dfrac{R^2\dot{\theta}_0^2}{2g\alpha}\left(1+\tan^2\phi\right)^{3/2}$

3–5 $\ddot{\theta} - a\dot{\theta}^2 = \dfrac{g}{r_0(1+a^2)}e^{a\theta}(a\sin\theta - \cos\theta)$

3–7 $r\ddot{\theta} + 2\dot{r}\dot{\theta} = 0;\ \ddot{r} - r\dot{\theta}^2\sin^2\beta + g\cos\beta\sin\beta = 0$

3–9 (a) $\dot{\theta}^2 + \theta\ddot{\theta} - \Omega^2 = 0$

(b) $\mathbf{T} = [4mR\Omega^3 t]\,\mathbf{e}_y$

3–10 $m \sec x\,[\ddot{x} + \dot{x}^2 \tan x] = -mg \sin x$

3–11 $\ddot{r} - r\dot{\theta}^2 + \frac{K}{m}(r - L) = 0;\ r\ddot{\theta} + 2\dot{r}\dot{\theta} = 0$

3–12 $m[\ddot{r} - r\dot{\theta}^2 - l\omega^2 \cos(\theta - \omega t)] + K(r - r_0) = 0$

$m[2\dot{r}\dot{\theta} + r\ddot{\theta} + l\omega^2 \sin(\theta - \omega t)] = 0$

3–13 $r\ddot{\theta} + 2\dot{r}\dot{\theta} = 0;\ \left[1 + \left(\frac{r}{R}\right)^2\right]\ddot{r} + \frac{r}{R^2}\dot{r}^2 - r\dot{\theta}^2 + \frac{gr}{R} = 0$

3–17 $(l - R\theta)\ddot{\theta} - R\dot{\theta}^2 - g\cos\theta = 0$

3–19 (a) $\ddot{\theta} - \dfrac{2K}{m}\sin\theta = 0$

(b) $\mathbf{N} = \left[-mR\dot{\theta}_0^2 - 4KR\cos\theta_0 + 6KR\cos\theta + KR\right]\mathbf{e}_r$

3–20 $m(\ddot{r} - r\dot{\theta}^2) = -F;\ 2\dot{r}\dot{\theta} + r\ddot{\theta} = 0$

3–23 $\ddot{r} - r\Omega^2 \sin^2\beta + c\dot{r} + g\cos\beta = 0$

Chapter 4

4–2 (a) $\hat{\mathbf{F}}_m = mv_0 \cos\theta \dfrac{M}{m+M}\mathbf{E}_x + mv_0 \sin\theta\,\mathbf{E}_y$

(b) ${}^{\mathcal{F}}\mathbf{v}'_m = {}^{\mathcal{F}}\mathbf{v}'_M = \dfrac{m}{m+M} v_0 \cos\theta\,\mathbf{E}_x$

4–3 (a) ${}^{\mathcal{F}}\mathbf{v}_m = {}^{\mathcal{F}}\mathbf{v}_m = \dfrac{m}{m+M}\sqrt{2gh}\,\mathbf{E}_x$

(b) $x_{\max} = \dfrac{(m+M)g}{3K}\left[1 + \sqrt{1 + \dfrac{6Kh}{(m+M)g}}\right]$

4–6 (a) ${}^{\mathcal{F}}\mathbf{v}'_A = \dfrac{m_A - em_B}{m_A + m_B} v_0 \mathbf{E}_x;\ {}^{\mathcal{F}}\mathbf{v}'_B = \dfrac{m_A(1+e)}{m_A + m_B} v_0 \mathbf{E}_x$

(b) $v_0 = \dfrac{2(m_A + m_B)}{m_A(1+e)}\sqrt{gr}$

4–8 $(m_A \cos^2\theta + m_B \sin^2\theta)l\ddot{\theta} + (m_A - m_B)l\dot{\theta}^2 \cos\theta \sin\theta - m_A g \cos\theta = 0$

4–12 $\ddot{\theta} + \dfrac{gl}{l^2 + a^2}\sin\theta = 0$

4–17 $\ddot{\theta} + \dfrac{ga}{a^2 + l^2}\sin\theta = 0$

4–19 $x = \dfrac{2}{3}l$

4–22
$$m_1\ddot{x}_1 + Kx_1\left(1 + \frac{\ell_0}{\sqrt{x_1^2 + x_2^2}}\right) - m_1 g = 0$$
$$m_2\ddot{x}_2 + Kx_2\left(1 + \frac{\ell_0}{\sqrt{x_1^2 + x_2^2}}\right) = 0$$

Chapter 5

5–1 (a) ${}^{\mathcal{F}}\mathbf{v}_O(t) = \mu g t \mathbf{E}_x$

(b) ${}^{\mathcal{F}}\boldsymbol{\omega}^{\mathcal{R}}(t) = -\dfrac{2\mu g}{r}t\mathbf{E}_z$

5–2
$$l\ddot{\theta} + \frac{3}{2}\ddot{x}\cos\theta + \frac{3g}{2}\sin\theta = 0$$
$$(m_1 + m_2)\ddot{x} - \frac{m_2 l}{2}\dot{\theta}^2\sin\theta + \frac{m_2 l}{2}\ddot{\theta}\cos\theta = 0$$

5–3 $\frac{7}{5}m\ddot{x} + mg\sin\beta = P(1 - \mu)$

5–4 (a) $({}^{\mathcal{F}}\boldsymbol{\omega}^{\mathcal{R}})' = \dfrac{6\hat{F}}{ml}\mathbf{e}_z$

(b) $\theta_{\max} = \dfrac{6\hat{F}}{ml}\sqrt{\dfrac{ml^2}{12K}}$

5–5 (a) ${}^{\mathcal{F}}\bar{\mathbf{v}}(t_B) = \sqrt{2gx\sin\beta}\mathbf{E}_x$; ${}^{\mathcal{F}}\boldsymbol{\omega}^{\mathcal{R}}(t_B) = \mathbf{0}$

(b) $t - t_B = \dfrac{\bar{v}(t_1)}{g(3\mu\cos\beta - \sin\beta)}$

(c) ${}^{\mathcal{F}}\bar{\mathbf{v}}(t - t_B) = \dfrac{2\mu\cos\beta\sqrt{2gx\sin\beta}}{(3\mu\cos\beta - \sin\beta)}\mathbf{E}_x$

${}^{\mathcal{F}}\boldsymbol{\omega}^{\mathcal{R}}(t - t_B) = \dfrac{2\mu\cos\beta\sqrt{2gx\sin\beta}}{r(3\mu\cos\beta - \sin\beta)}\mathbf{E}_z$

5–6 (a) $\ddot{\theta} - \dfrac{3g}{2l}\sin\theta = 0$

(b) $\theta = \cos^{-1}(2/3)$

5-7 $$mr^2\ddot{\theta}\,[1.5 - 0.84\cos\theta\,] + 0.42mr^2\dot{\theta}^2\sin\theta + 0.42mgr\sin\theta = 0$$

5-8 $$\ddot{\theta} - \frac{5g}{7r}\sin\left(\frac{r\theta}{R+r}\right) = 0$$

5-10 $$\tfrac{3}{2}mr^2\ddot{\theta} + Kr^2\theta - mgr\sin\beta = 0$$

5-11 $$m\ddot{x} + \frac{ml\ddot{\theta}}{2}\cos\theta - \frac{ml\dot{\theta}^2}{2}\sin\theta = P$$

$$\frac{ml}{6}\left(1 + 3\sin^2\theta\right)\ddot{\theta} + \frac{ml\dot{\theta}^2}{2}\cos\theta\,\sin\theta + mg\sin\theta = -P\cos\theta$$

5-12 $$(3ma^2 + ml^2)\ddot{\theta} + 3mga\sin\theta = 0$$

5-16 $$\ddot{x} = \frac{5hP}{7mr}$$

5-17 $$\frac{3}{2}m(R-r)\ddot{\theta} + mg\sin\theta = 0$$

5-20 $$\frac{3}{2}mr^2\ddot{\theta} + 16cr^4\dot{\theta}^3 + 4Kr\theta - mgr\sin\beta = 0$$

5-21 $$\ddot{\theta} + \frac{6gl}{4l^2 + a^2}\sin\theta = 0$$

Bibliography

Baruh, H. (1999), *Analytical Dynamics*, McGraw-Hill, New York.

Bedford, A. and Fowler, W. (2005), *Engineering Dynamics: Mechanics Principles*, Prentice-Hall, Upper Saddle River, NJ.

Beer, F. P. and Johnston, E. R. (1997), *Vector Mechanics for Engineers: Dynamics*, Sixth Edition, McGraw-Hill, New York.

Coulomb, C. A. (1785), "Théorie Des Machines Simples En Ayant Égard Au Frottement et à La Roideur Des Cordages," *Mémoires de Mathématique et de Physique Présentés à L'Académie Royale des Sciences Par Divers Savans, et Lus Dan Ses Assemblés*, Vol. 10, pp. 161–332.

Frenet, J-F. (1852), "Sur Quelques Propriétés Des Courbes à La Double Courbure," *Journal De Mathématiques Pures et Appliquées*, Vol. 17, pp. 437–447.

Greenwood, D. T. (1977), *Classical Dynamics*, Prentice-Hall, Upper Saddle River, NJ.

Greenwood, D. T. (1988), *Principles of Dynamics*, Prentice-Hall, Upper Saddle River, NJ.

Greenwood, D. T. (2002), *Advanced Dynamics*, Cambridge University Press, Upper Saddle River, NJ.

Hibbeler, R. C. (2001), *Engineering Mechanics: Dynamics*, Ninth Edition, Prentice-Hall, Upper Saddle River, NJ.

Huygens, C. (1673), "Horologium Oscillatorium, Sive de Motu Pendulorum ad Horologia Aptato," Paris. English translation in by Blackwell, R. J. (1986) "The Pendulum Clock, or, Geometrical Demonstrations Concerning the Motion of Pendula as Applied to Clocks, Series in the History of Technology and Science," Iowa State University Press, Ames, IA.

Kane,T. R. and Levinson, D. A. (1985), *Dynamics: Theory and Applications*, McGraw-Hill, New York.

Kreyszig, E. (1988), *Advanced Engineering Mathematics*, John Wiley & Sons, New York.

Merriam, J. L. and Kraige, L. G. (1997), *Engineering Mechanics: Dynamics*, Fourth Edition, John Wiley & Sons, New York.

Newton, I. (1687), *Philosophiae Naturalis Principia Mathematica*, English Translation by Motte, A. (1729), Revised Translation by Cajori, F. (1934), published by The University of California Press, Berkeley, California.

O'Reilly, O. M. (2001), *Engineering Dynamics: A Primer*, Springer-Verlag, New York.

O'Reilly, O. M. and Srinivasa, A. R. (2002), *On Potential Energies and Constraints of Rigid Bodies and Particles, Mathematical Problems in Engineering*, Vol. 8, No. 1, pp. 169–180.

O'Reilly, O. M. (2004), *Lecture Notes on the Dynamics of Particles and Rigid Bodies: Class Notes for Engineering Mechanics III and Intermediate Dynamics ME 175*, Department of Mechanical Engineering, University of California, Berkeley.

Papastavridis, J. G. (2002), *Analytical Mechanics: A Comprehensive Treatise on the Dynamics of Constrained Systems: For Engineers, Physicists, and Mathematicians*, Oxford University Press, New York.

Serret, J. A. (1851), "Sur Quelque Formule Reletives à La Théorie Des Courbes à Double Courbure," *Journal De Mathematiques Pures et Appliquees*, Vol. 16, pp. 193–207.

Synge, J. L. and Griffith, B. L. (1959), *Principles of Mechanics*, McGraw-Hill, New York.

Tenenbaum, R. A. (2004), *Fundamentals of Applied Dynamics*, Springer-Verlag, New York.

Thomas, G. B. and Finney, R. L. (2002), *Calculus and Analytic Geometry*, Addison-Wesley Publishing Company, New York.

Thornton, S. T. and Marion, J. B. (2004), *Classical Dynamics of Particles and Systems*, Brooks/Cole Publishing, Belmont, California.

Index

Printed in the United States
By Bookmasters